Sustainable Management of Potato Pests and Diseases

Swarup Kumar Chakrabarti •
Sanjeev Sharma • Mohd Abas Shah
Editors

Sustainable Management of Potato Pests and Diseases

Springer

Editors
Swarup Kumar Chakrabarti
ICAR-Central Potato Research Institute
Shimla, Himachal Pradesh, India

Sanjeev Sharma
Division of Plant Protection
ICAR-Central Potato Research Institute
Shimla, Himachal Pradesh, India

Mohd Abas Shah
ICAR-Central Potato Research Institute-Regional Station
Jalandhar, Punjab, India

ISBN 978-981-16-7694-9 ISBN 978-981-16-7695-6 (eBook)
https://doi.org/10.1007/978-981-16-7695-6

This Springer imprint is published by the registered company Springer Nature Singapore Pte Ltd.
The registered company address is: 152 Beach Road, #21-01/04 Gateway East, Singapore 189721, Singapore

Preface

Potato (*Solanum tuberosum* L.) is the third most important crop in terms of human consumption after wheat and rice. Cultivated in all continents except Antarctica, potato is an essential part of the diet of billions of people. Though developed countries dominated potato production globally till 2005, there was a remarkable shift afterwards towards developing countries with a strong growth in production in Asia and Africa. Both potato production and consumption are accelerating in most of the developing countries with Africa showing the maximum growth now. The two emerging Asian economies, viz. China and India together contribute nearly one third of the global potato production at present. Potato is preferred in these densely populated countries largely because of its high productivity, flexibility in terms of fitting into many prevailing cropping systems, and stable yields under conditions in which other crops may fail. Potato consumption in this region is increasing due to increasing industrialization and participation of women in the job market that created demand for processed, ready-to-eat convenience food, particularly in urban areas.

Though potato emerged as an important food crop globally within about 400 years of its introduction from South America, it has also faced serious biotic threats in the process. In fact, a large number of pests and pathogens cause appreciable economic losses now to this crop globally. A large number of fungal diseases including the infamous late blight and a plethora of insect pests are particularly important causing significant direct and indirect losses to this crop thereby threatening global food sustainability. Besides, vector-virus complexes constitute the most important challenge worldwide for both seed and ware potato production. The already challenging pest and disease scenario in potato is getting exacerbated due to intensive crop cultivation, climate change and its consequences on host resistance, and pathogen virulence as well as globalization of commodity trade.

On the other side, considerable progress has now been made in the understanding of the molecular basis of plant–pathogen interactions, epidemiology of diseases and their causal agents, and the deployment of this knowledge to design suitable control and management methods. There has been an exponential growth in the number of studies unravelling the genetic and genomic basis of species interaction and pest biology. Newer tools like RNAi and CRISPR-Cas9 systems hold much potential to widen the gene pool for utilization in breeding pest and disease-resistant varieties now. As of now, the management of pests and diseases in potato is heavily reliant on the use of plant protection products, which is not environmentally sustainable. Moreover, management schedules based on synthetic pesticides are rendered ineffective due to continuous evolution of pathogen variants, resistance development, and emergence of new pests and diseases which usually were characterized as of minor importance. The global phenomenon of climate change is complicating the situation further.

In this book, an attempt has been made to bring together information on such aspects of pest and disease management which are believed to be more sustainable in the long run. The 21 chapters of the book cover the essential aspects of pest and disease management. The book is organized into two main themes—an overview of the major diseases like late bight of potato, other fungal, bacterial, nematode, and viral diseases with latest updates, and pests like aphids which are of global significance and recently emerged whiteflies as major pests. The other theme includes chapters on areas of research and knowledge which have potential for sustainable potato production, e.g. biological control, use of microbial secondary metabolites, green chemicals, and role of nutrition. Issues relating to pesticide residues, international phytosanitary measures, new chemistry pesticides which are mostly safer than earlier generation chemicals, and management of pesticide resistance has also been dealt with in separate chapters. This book would be an excellent source of up-to-date information for the researchers working in the disciplines of plant pathology, agricultural entomology, and nematology in general, and potato researchers in particular. It would also be a good source book for industries dealing with plant protection, disease diagnostics, plant breeding, and agronomy of potato. It will serve as an important source of information for students, academicians, and policy makers on the aspects of sustainable pest and disease management in potato.

It will be our privilege to have critical views and constructive criticism from our readers for further improvement of this book.

Shimla, Himachal Pradesh, India — Swarup Kumar Chakrabarti
Shimla, Himachal Pradesh, India — Sanjeev Sharma
Jalandhar, Punjab, India — Mohd Abas Shah

Acknowledgements

First of all, the Editors would like to unreservedly thank all the authors who generously devoted part of their busy time to contribute several versions and revisions of the draft chapters that became the final chapters assembled in this volume. We are equally grateful to the organizations that they are affiliated with for allowing them to take up this important task.

We wish to acknowledge the various resources in public domain that have been used for preparing various chapters e.g., the IPPC (FAO), CODEX (FAO), to mention a few.

We would like to express our gratitude to many former and current mentors and colleagues for sharing their insight and expertise in compiling this book. We would also like to convey our gratitude to the top management of our respective organizations for their support and encouragement.

We are thankful to the reviewers for their generous efforts to improve the book further. We also thank our editorial contact and production editors at Springer for their support.

Finally, we express a deep sense of gratitude to our parents, families, and friends, living and passed on, whose love and friendship are priceless. It is to them that we dedicate this book.

Contents

Editors and Contributors

About the Editors

Swarup Kumar Chakrabarti is the *Vice Chancellor*, Uttar Banga Krishi Viswavidyalaya Pundibari, Cooch Behar, West Bengal. Earlier, he served as Director of ICAR-Central Potato Research Institute, Shimla (Himachal Pradesh) and ICAR-Central Tuber Crops Research Institute, Thiruvananthapuram (Kerala). He obtained his doctorate from ICAR-Indian Agricultural Research Institute, New Delhi, in Plant Pathology; and post-docs from Waksman Institute, Rutgers, USA, and CIRAD-AMIS, IGEPAM, France. He is a world renowned expert in molecular pathology of late blight of potato, potato viruses, and bacterial wilt of potato. He was the Country Leader of the Potato Genome Sequencing Consortium comprising of 26 international institutes belonging to 14 countries. The consortium deciphered the complex genome of potato that has been published in the high impact journal "Nature". He has more than 200 peer-reviewed publications of national and international repute and has delivered numerous oral and poster presentations in numerous national/international meetings. He is a fellow of the prestigious National Academy of Agricultural Sciences (NAAS), New Delhi, the Indian Potato Association, Shimla, the Indian Phytopathological Society, New Delhi, and the Confederation of Horticulture Associations of India, New Delhi. He has been awarded with Hari Om Ashram Trust Award of ICAR for the biennium 2016–2017, Shri L.C. Sikka Endowment Award (2013–2014) from the National Academy of Agricultural Sciences, New Delhi; S. Ramanujan Award (2008–2011) from ICAR-Central Potato Research Institute, Shimla; Dr. J.P. Verma Memorial Award (2010) from the Indian Phytopathological Society, New Delhi, besides numerous other recognitions from national and international organizations.

Sanjeev Sharma is Principal Scientist and Head of Plant Protection Division at ICAR-Central Potato Research Institute, Shimla (Himachal Pradesh). He did his doctorate from Dr. YS Parmar University of Horticulture and Forestry, Solan (HP) in Plant Pathology. He specializes in population biology, epidemiology, and management of potato pathogens; *Phytophthora infestans* (the late blight pathogen) in particular. He was part of the team that developed the pan-India late blight forecasting model "IndoBlightCast". Dr. Sharma was honoured with Young

Scientist Award, Hill Society of Agriculture, Uttarakhand; ICAR Research Award for Outstanding Interdisciplinary Research in Agricultural and Allied Sciences 2015–2016 (Crop and Horticultural Sciences), New Delhi; IPA-Kaushalya Sikka Memorial Award (2013–2017), Indian Potato Association, Shimla; IPA Gold Medal, Indian Potato Association, Shimla, Best Worker Award 2020 (Scientific Category), ICAR-Central Potato Research Institute, Shimla. As a pathologist, he has contributed to the development of late blight-resistant cultivars and in identification and registration of elite germplasm possessing late blight resistance. He is a fellow of the Indian Society of Hill Agriculture, Uttarakhand, and the Indian Potato Association, Shimla. Dr. Sharma has over 150 research papers, book chapters, review papers, and technical bulletins to his credit. He represented the country at various international fora pertaining to late blight of potato like Bangladesh (2012), Indonesia (2013), Nepal (2014), and China (2019), to mention a few.

Mohd Abas Shah is a Scientist at ICAR-Central Institute of Temperate Horticulture, Srinagar, Jammu, and Kashmir. Earlier he served in the capacity of Scientist at ICAR-Central Potato Research Institute-Regional Station, Jalandhar (Punjab). He did his doctorate in Agricultural Entomology from Sher-e-Kashmir University of Agricultural Sciences and Technology of Kashmir (SKUAST), Shalimar, Srinagar. He is working on the population biology, ecology, and management of sweetpotato whitefly (*Bemisia tabaci*), the potato aphids, and the viruses they transmit in potato crops. He has more than 30 peer-reviewed publications of national and international repute and has delivered numerous oral and poster presentations in various national/ international fora. He is the member of various professional societies including the Indian Potato Association, Shimla, and the Entomological Society of India, New Delhi. He has been awarded the Best Paper Award from the Indian Potato Association, Shimla, besides various Best Poster awards and gold medal for outstanding doctoral research from SKUAST-Kashmir.

Contributors

R. Asokan Division of Biotechnology, Indian Institute of Horticultural Research, Bangalore, Karnataka, India

Aarti Bairwa Division of Plant Protection, ICAR-Central Potato Research Institute, Shimla, Himachal Pradesh, India

Rekha Balodi ICAR-National Research Centre for Integrated Pest Management, Pusa Campus, New Delhi, India

Ashutosh Bhaik Corteva Agriscience, Hyderabad, Telangana, India

Vinay Bhardwaj Division of Crop Improvement, ICAR-Central Potato Research Institute, Shimla, Himachal Pradesh, India

Chaitra Bhat ICAR-Indian Agricultural Research Institute, Pusa, New Delhi, India

Anuj Bhatnagar ICAR-Central Potato Research Institute-Regional Station, Modipuram, Meerut, Uttar Pradesh, India

S. K. Chakrabarti ICAR-Central Potato Research Institute, Shimla, Himachal Pradesh, India
Uttar Banga Krishi Viswavidyalaya Pundibari, Cooch Behar, West Bengal, India

Sorabh Chaudhry ICAR-Central Potato Research Institute-Regional Station, Modipuram, Meerut, Uttar Pradesh, India

Anil K. Choudhary Division of Plant Protection, ICAR-Central Potato Research Institute, Shimla, Himachal Pradesh, India

Dalamu Division of Crop Improvement, ICAR-Central Potato Research Institute, Shimla, Himachal Pradesh, India

Deepika Laboratory of Herbal Pesticides, Department of Botany, Institute of Science, Banaras Hindu University, Varanasi, India

Bhawna Dipta Division of Plant Protection, , ICAR-Central Potato Research Institute, Shimla, Himachal Pradesh, India

V. K. Dua Division of Plant Protection, ICAR-Central Potato Research Institute, Shimla, Himachal Pradesh, India

J. K. Dubey Department of Entomology, Dr Y S Parmar University of Horticulture and Forestry, Nauni, Solan, Himachal Pradesh, India

Nawal Kishore Dubey Laboratory of Herbal Pesticides, Department of Botany, Institute of Science, Banaras Hindu University, Varanasi, India

S. C. Dubey Indian Council of Agricultural Research, New Delhi, India

R. Ellango Division of Biotechnology, Indian Institute of Horticultural Research, Bangalore, Karnataka, India

Jean-Luc Genet Corteva Agriscience, Colmar, France

Kavita Gupta ICAR-National Bureau of Plant Genetic Resources, New Delhi, India

Aravinda B. Hubballi Department of Entomology, GKVK, University of Agriculture Sciences, Bengaluru, India

A. Jeevalatha ICAR-Central Potato Research Institute, Shimla, Himachal Pradesh, India

Priyanka Kaundal ICAR-Central Potato Research Institute, Shimla, Himachal Pradesh, India

Thomas P. Kuhar Department of Entomology, Virginia Tech, Blacksburg, VA, USA

Arun Kumar CSIR-Institute of Himalayan Bioresource Technology, Palampur, Himachal Pradesh, India
Academy of Scientific and Innovative Research, Ghaziabad, Uttar Pradesh, India

Manoj Kumar ICAR-Central Potato Research Institute-Regional Station, Modipuram, Meerut, Uttar Pradesh, India

Prince Kumar ICAR-Central Potato Research Institute-Regional Station, Jalandhar, Punjab, India

Ravinder Kumar Division of Plant Protection, ICAR-Central Potato Research Institute, Shimla, Himachal Pradesh, India

Vinod Kumar Division of Crop Improvement, ICAR-Central Potato Research Institute, Shimla, Himachal Pradesh, India

Mehi Lal ICAR-Central Potato Research Institute-Regional Station, Modipuram, Meerut, Uttar Pradesh, India

Chandan Maharana Crop Protection Section, Vivekananda Institute of Hill Agriculture (VPKAS), Almora, Uttarakhand, India

Vikas Mangal Division of Crop Improvement, ICAR-Central Potato Research Institute, Shimla, Himachal Pradesh, India

Mamadou Kane Mboup Corteva Agriscience, Colmar, France

Chris McCullough Department of Entomology, Virginia Tech, Blacksburg, VA, USA

Priyank H. Mhatre ICAR-Central Potato Research Institute-Regional Station, Muthorai, Udhagamandalam, The Nilgiris, Tamil Nadu, India

Kailash C. Naga Division of Plant Protection, ICAR-Central Potato Research Institute, Shimla, Himachal Pradesh, India

M. Nagesh ICAR-National Bureau of Agriculturally Important Insect Resources, Bengaluru, Karnataka, India

M. Nikhil Raj ICAR-Indian Agricultural Research Institute, Pusa, New Delhi, India

Vinod Kumar Padala ICAR-Research Complex for Eastern Region, Research Centre for Makhana, Darbhanga, Bihar, India

Amit Paschapur Crop Protection Section, Vivekananda Institute of Hill Agriculture (VPKAS), Almora, Uttarakhand, India

K. V. Raghavendra ICAR-National Research Centre for Integrated Pest Management, Pusa Campus, New Delhi, India

Baswaraj Raigond Division of Plant Protection, ICAR-Central Potato Research Institute, Shimla, Himachal Pradesh, India

Vinay Sagar Division of Plant Protection, , ICAR-Central Potato Research Institute, Shimla, Himachal Pradesh, India

Mohd Abas Shah ICAR-Central Potato Research Institute-Regional Station, Jalandhar, Punjab, India
ICAR-Central Institute of Temperate horticulture, Rangreth, Srinagar, Jammu and Kashmir, India

Ajay Sharma Department of Entomology, Dr Y S Parmar University of Horticulture and Forestry, Nauni, Solan, Himachal Pradesh, India

Anil Sharma ICAR-Central Potato Research Institute-Regional Station, Jalandhar, Punjab, India

Ashwani Kumar Sharma ICAR-Central Potato Research Institute, Shimla, Himachal Pradesh, India

Jagdev Sharma Division of Plant Protection, ICAR-Central Potato Research Institute, Shimla, Himachal Pradesh, India

Sanjeev Sharma Division of Plant Protection, ICAR-Central Potato Research Institute, Shimla, Himachal Pradesh, India

S. Sundaresha Division of Plant Protection, ICAR-Central Potato Research Institute, Shimla, Himachal Pradesh, India

Ashish Kumar Singh Crop Protection Section, Vivekananda Institute of Hill Agriculture (VPKAS), Almora, Uttarakhand, India

Birpal Singh ICAR-Central Potato Research Institute, Shimla, Himachal Pradesh, India

Salej Sood Division of Crop Improvement, ICAR-Central Potato Research Institute, Shimla, Himachal Pradesh, India

J. Sridhar ICAR-Central Potato Research Institute, Shimla, Himachal Pradesh, India

A. R. N. S. Subbanna Crop Protection Section, Vivekananda Institute of Hill Agriculture (VPKAS), Almora, Uttarakhand, India

S. Subhash ICAR-Central Potato Research Institute-Regional Station, Modipuram, Meerut, Uttar Pradesh, India

R. S. Tegg Tasmania Institute of Agriculture, University of Tasmania, 13 St. Johns Avenue, New Town, Hobart, TAS, Australia

Rahul Kumar Tiwari Division of Plant Protection, ICAR-Central Potato Research Institute, Shimla, Himachal Pradesh, India

E. P. Venkatasalam ICAR-Central Potato Research Institute-Regional Station, Muthorai, Udhagamandalam, The Nilgiris, Tamil Nadu, India

V. Venkatesverulu ICAR-Central Potato Research Institute, Shimla, Himachal Pradesh, India

Gaurav Verma Division of Plant Protection, ICAR-Central Potato Research Institute, Shimla, Himachal Pradesh, India

C. R. Wilson Tasmania Institute of Agriculture, University of Tasmania, 13 St. Johns Avenue, New Town, Hobart, TAS, Australia

Potato Pests and Diseases: A Global Perspective

1

S. K. Chakrabarti, Sanjeev Sharma, and Mohd Abas Shah

Abstract

Potato is among the most important food crops round the world. Potato yield and quality is constrained by a myriad of insect pests and pathogens. Beyond doubt, the late blight of potato continues to be the most important disease of potato which can cause loses worth billions of dollars annually. Apart from the changing composition and intensity of the pests and diseases, the emergence of newer species/strains continues at an alarming rate. The impact of climate change is going to be profound on the crop as well as the pests and pathogens, of course the response is species specific. The risk of the spread of invasive organisms of quarantine significance has increased many folds due to international trade and exchange of germplasm. On the other hand, the potato pest and disease management has come a long way, and several innovative approaches are being adopted for diagnostics and detection, monitoring and forecasting and management of various pest and diseases. In this chapter, an outline of the current issues of the potato pest and disease management is given along with a discussion on sustainability of the management practices.

S. K. Chakrabarti
ICAR-Central Potato Research Institute, Shimla, Himachal Pradesh, India

S. Sharma
Division of Plant Protection, ICAR-Central Potato Research Institute, Shimla, Himachal Pradesh, India

M. A. Shah (✉)
ICAR-Central Potato Research Institute-Regional Station, Jalandhar, Punjab, India
e-mail: mabas.shah@icar.gov.in

S. Kumar Chakrabarti et al. (eds.), *Sustainable Management of Potato Pests and Diseases*, https://doi.org/10.1007/978-981-16-7695-6_1

Keywords

Pest management · Quarantine · Pesticide resistance · Economic losses · Emerging diseases · Climate change · Disease diagnosis · Innovation · Sustainability

1.1 Introduction

Potato (*Solanum tuberosum*) is regarded as one of the most important crops in addressing the challenge of food security, especially in developing countries. Presently, the crop is raised in 19 million hectares with 378 million tonnes global production. The world has witnessed notable increase in potato production in many countries during the last two decades because of area expansion and improvements in yield. Global statistics indicate that potato production is shifting towards developing countries, especially in Asia and Africa, and potato production in the developing countries has surpassed the developed world (FAOSTAT 2020). Worldwide, in 2018, China was the largest producer of potatoes, with India at second number (Fig. 1.1) (FAOSTAT 2020). More than a billion people consume potato on a regular basis, and it is a vital source of income for millions of farmers (Devaux et al. 2014). Though potato production has increased enormously since the 1960s in the developing world, but the sustainability of potato production globally is threatened by adverse abiotic conditions, pests and pathogens. Potato is prone to more than 100 pests and diseases including insects, nematodes, viruses, bacteria,

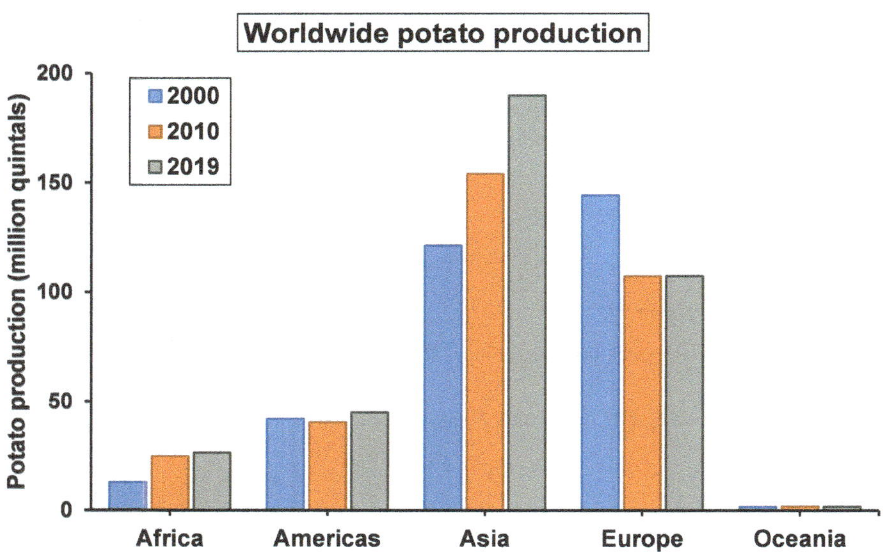

Fig. 1.1 Potato production from 2000 to 2019 (Source: FAOSTAT 2020)

oomycete and fungi which cause direct yield losses and decrease of farmer's incomes by downgrading the quality of affected tubers. Due to climate change, global trade and increasing cropping intensity, pathogens are evolving at a faster rate and adapting to the new climate and hosts. As a consequence, health management is becoming more and more complex. Therefore, knowledge about the pathogens as well as factors influencing disease severity is needed to setup efficient control strategies.

The pest and disease management in potato is heavily reliant on the use of synthetic pesticides. Continued use of such plant protection products has led to development of resistance, secondary pest outbreaks and pest resurgence on frequent basis. Besides, the use of pesticides is generally harmful to mankind and the environment. Safer alternatives of pest and disease management are continuously being explored. Besides, the pest and disease management in potato is constrained due to the effects of climate change and the risk of invasive pests. In this chapter, current issues of pest and disease management in potato are discussed to lay open the platform for thorough discussions on the sustainable management of potato pets and diseases which are described throughout the book.

1.2 Pests and Diseases of Potato: Economic Importance

The potato is susceptible to many diseases. The oomycetes and the fungi ranging from moulds to the smuts and rusts perhaps constitute the most important biotic threats to potato. Potato is susceptible to numerous bacterial diseases, phytoplasmas, more than three dozen viruses of the yellow and mosaic groups and several parasitic nematodes. Diseases of potatoes include arguably the most historically significant crop disease, late blight, which is still the most important potato disease. Haverkort et al. (2009) estimated that the global costs and losses due to late blight may take 16% of all global potato production. At 100 €/t, the world potato production represents a value of €38 billion today. The 16% loss then represents an annual financial loss of €6.1 billion per annum today. After potato late blight, early blight represents one of the most important fungal diseases of potato today. Enormous yield losses are reported worldwide: 18–39% in the United States (Harrison and Venette 1970), 2–40% in parts of Europe (Leiminger and Hausladen 2014), 20–50% in South Africa (Van der Waals et al. 2001) and >20% in Australia (Horsfield et al. 2010). In India, yield loss to the tune of 79% is recorded under severe condition. An increasing emphasis on the cosmetic appearance of potatoes has recently brought hitherto non-significant diseases into prominence. Unless effective methods of control are practiced, some of the diseases have the potential to cause the total loss of a crop. The economic importance and losses caused by various fungal, bacterial, viral and nematode diseases in potato crops are described in detail in respective chapters.

Insect pests are among the major constraints affecting the yield and quality of produce in potato. Variable losses are estimated across locations, with global losses estimated on average at 16% (Oerke et al. 1994). Losses as high as 30% and 70% are recorded for various pests under different crop management regimes; complete crop losses or complete loss of quality seed crops is not a surprise if proper management practices are not adopted (Kroschel and Schaub 2013; Mujica and Kroschel 2013; Kroschel et al. 2020).

A large number of insect pests infest potato crops. The insect may either damage the tubers directly or feed on leaves or stems and transmit debilitating pathogens (Radcliffe 1994). The latter result in loss of seed potato quality with severe consequences for the seed chain at multiple levels. Insect pests are also assigned to classes as per convenience, namely, above-ground (indirect) and below-ground (direct) pests. Kroschel et al. (2020) described a total of 49 species of insect pests infesting potato crops in different parts of the world. Out of these, 6 major and 32 minor species are prevalent throughout the temperate, tropical and subtropical regions; 9 major species are prevalent in tropical and subtropical regions and 2 major species affect potato crops in the temperate regions. Although the composition of major pests varies with many other factors, the Colorado potato beetle, aphids, leafhoppers, the potato psyllid, potato tuber moth, whitefly and wireworms are considered the deadliest insect pests of potato (Kroschel et al. 2012).

Some of these pest species specialize on Solanaceae and have moved around the world along with the spread of potato cultivation. The Colorado potato beetle and potato tuber moth are the classic examples of this phenomenon. Other species are generalists with near cosmopolitan distribution like the aphids and mites. Most regions in the world are also affected by regionally important pests such as the beet leafhopper (*Circulifer tenellus* (Baker)) in the north-west USA and the 28 spotted ladybird (*Henosepilachna vigintioctopunctata* (Fabr.)) in China. On the other hand, aphids are the major concern in prominent potato production areas in Europe, and the Colorado potato beetle is of secondary importance. The reverse is true in most states of the United States and Canada. In India, aphids, whitefly (*Bemisia tabaci*), numerous species of leaf hoppers, white grubs and cutworms are major pests (Shah et al. 2020). Among the global pests of potato, aphids are among the most important. Aphids are sap feeding insects, but the major damage inflicted by aphids in potato crops is by transmission of numerous potato viruses. The resulting viral diseases lead to considerable yield reductions, limit the production of disease-free seed potatoes and cause a progressive degeneration of seed stocks. Worldwide, the most important concern for the production of quality seed potatoes is the aphid-transmitted *Potyvirus, potato virus Y* (PVY). Being a non-persistent virus that has been demonstrated of being transmitted by more than 60 species of aphids, PVY infection is the most difficult to contain and is the most actively researched problem in potato pest management (Gray et al. 2010; Karasev and Gray 2013).

Waters and Jensen (2014) proposed that the insect pest scenario for potatoes at a given location are related to the end-use markets (e.g. subsistence agriculture, table stock, processing, long-term storage etc.) and the level of industrialization and access to insecticides and other crop protection technology. For example, potato tuber worm can be a pest of life-threatening proportions for subsistence farmers using rustic storage for food preservation, whereas it is a pest controlled easily by one or two well-timed insecticide treatments for technologically advanced farms with refrigerated storage. Conversely, an aphid-related disease such as potato virus Y (PVY) may be a minor yield-reducing concern for subsistence farming, but represents a nearly intractable problem and the difference between profit and loss for technologically intensive farms.

1.2.1 Emerging Pests and Diseases

Many of the important pests and diseases affecting potato can be regarded as emerging due to the attributes: (1) increase in incidence, geographical distribution or host range; (2) change in pathogenesis; (3) have newly evolved or (4) have been discovered or newly described (Anderson et al. 2004). The primary reasons for the occurrence of emerging pests and diseases are related to increased trade and travel, intensified and expanded land use, changes in agricultural practices, planting of new varieties and extreme weather events linked to climate change (Fry et al. 2015). Though late blight (*Phytophthora infestans*) has historically been an important disease of potatoes, it is regarded as re-emerging disease due to regular emergence of novel strains of the pathogen with increasing virulence and appearance in new locations with surprising intensity (Fry et al. 2015). Multiple clonal lineages have been found across the globe that indicates the history of the displacement of lineages over time. Migrations of exotic clonal lineages contribute substantially to change in the population composition in most locations worldwide (Fry 2020). A complex population structure is observed in Europe with ~70% population dominated by a few widely disseminated clonal lineages (www.euroblight.net). Recently, declining trend in the combined frequency of the clones EU_13_A2, EU_6_A1 and EU_1_A1 and increasing trend in the frequency of EU_36_A2, EU_37_A2, EU_41_A2 and EU_43_A2 have been observed in Europe (Cooke et al. 2019). Migration of the pathogen followed by selection has been the major causes of population change. The epidemiological consequences of sexually reproducing populations have made late blight management more difficult due to early set of epidemics thereby demanding more fungicide applications (Hannukkala et al. 2007). The occurrence of fungicide resistance to fluazinam (EU_37_A2) has further aggravated its management. Since the pathogen population is continually evolving and novel clonal lineages with new traits are emerging, it is necessary to tailor management schedule to the local pathogen population.

In recent years with warm and dry summers, early blight has become widespread and turned into one of most important diseases after late blight. In the past, potato diseases caused by *Alternaria* species were described as early blight (caused by *A. solani*) and brown spot (caused by *A. alternata*) disease. Recent studies, however, showed that the *Alternaria* population on potato is much more elaborate than a two-species disease complex. Ten species of *Alternaria* have been implicated to cause foliar diseases of potato worldwide. These include *A. solani*, *A. alternata*, *A. tenuissima*, *A. dumosa*, *A. arborescens*, *A. infectoria*, *A. grandis*, *A. interrupta*, *A. longipes* and *A. arbusti* (Taheri et al. 2009; Ardestani et al. 2010; Rodrigues et al. 2010; Shoaib et al. 2014; Leiminger et al. 2015; Tymon et al. 2016).

Bacterial diseases are other important biotic constraints, especially in tropical and subtropical regions and in some warm temperate regions of the world. About seven bacterial diseases affect potato worldwide and cause severe damages especially on tubers. Bacterial wilt and black leg are considered the most important diseases, whereas potato ring rot, pink eye and common scab are the minor ones. Recently, *Ralstonia* species complex has been reclassified on the basis of whole genome comparisons into three distinct species: *R. solanacearum* (Phylotype II), *R. pseudosolanacearum* (Phylotypes I and III) and *R. syzygii* (Phylotype IV) (Safni et al. 2014; Prior et al. 2016). Phylotype I strains are regarded to be of Asian origin, Phylotype II strains are thought to be of South American origin, whereas Phylotype III appears to have evolved in Africa and Phylotype IV in Indonesia. The *R. solanacearum* species complex is widely designated as a quarantine organism in many countries in an effort to prevent its movement across geographical borders. Nevertheless, the PIIB1 strain has spread from its origin to many potato-growing areas worldwide, presumably with movement in trade of infected seed tubers (Elphinstone 2005).

Pectobacterium spp. are important bacterial potato pathogens and can be aggressive on tubers and stems causing wilting and eventually plant death. The *Pectobacterium* species most commonly found on potato include *P. atrosepticum*, *P. brasiliense*, *P. carotovorum*, *P. odoriferum*, *P. parmentieri*, *P. peruviense*, *P. polaris* and *P. punjabense*. Though there are some regional differences in the species distribution, some appear to be ubiquitous. For example, *P. atrosepticum*, *P. brasiliense*, *P. parmentieri* and *P. carotovorum* are found on multiple continents (Pérombelon and Kelman 1987; Duarte et al. 2004; Kim et al. 2009; Pitman et al. 2008, 2010; van der Merwe et al. 2010; De Boer et al. 2012; Ngadze et al. 2012; She et al. 2017; van der Wolf et al. 2017; Wang et al. 2017a, b, c). In Europe, *P. atrosepticum* has been the predominant species responsible for blackleg disease on potato, while *P. carotovorum* is often associated with soft rot in storage. *P. brasiliense* was originally identified as causing disease on potato in Brazil (Duarte et al. 2004) and has been common in the United States since 2001, as also *P. parmentieri* (Yap et al. 2004; Kim et al. 2009). This species was not known to cause disease on potato in Europe prior to 2012–2013 but has since increased greatly in its incidence in many European countries (de Werra et al. 2015) and is now recognized as an important pathogen in Africa as well (van der Merwe et al. 2010).

Dickeya species (formerly *Erwinia chrysanthemi*) have emerged as a new threat to potato production in Europe. Like *Pectobacterium*, *Dickeya* species have also a wide host range with global distribution (Samson et al. 2005). Out of eight *Dickeya* species, only *D. dianthicola* and *D. solani* are of concern to potato (Toth et al. 2011). In some cases, *D. dianthicola* replaced *P. atrosepticum* as the dominant blackleg pathogen (Parkinson et al. 2009; Toth et al. 2011). *D. solani* has been recognized as a new *Dickeya* pathogen on potato by several groups from 2004 through 2010 (Laurila et al. 2008; Parkinson et al. 2009; Slawiak et al. 2009) and spreading across Europe on seed tubers.

Zebra chip (ZC), a new and economically important disease of potato in the United States, Mexico, Central America and New Zealand, is caused by the bacterium "*Candidatus* Liberibacter solanacearum (Lso)" and transmitted to potato by the potato psyllid, *Bactericera cockerelli*. Although Lso is only spread in potato by *B. cockerelli*, it can also be found in other *Bactericera* species, suggesting that vector feeding preferences limit the species of vectors important for zebra chip and not Lso-vector interactions (Borges et al. 2017). These observations suggest that the bacterium can easily be introduced in many parts of the world along with its insect vectors.

Insect pests such as the potato tuber moth, *Phthorimaea operculella* (Zeller), and the leafminer fly, *Liriomyza huidobrensis* (Blanchard), have become invasive and occur today as serious pests in many tropical and subtropical regions. Around 2006, the tomato leafminer, *Tuta absoluta* Meyrick, although a minor pest in potato, was unintentionally introduced to Spain, from where it continued its devastating journey across Africa and into Asia and reached India within less than 10 years (Rahman et al. 2012; Caparros Megido et al. 2013; Sridhar et al. 2015; Kanle Satishchandra et al. 2019). As farmers had not been prepared and no control measures had been in place, the pest caused large production losses in tomato (*Lycopersicon esculentum* Mill.); under certain conditions, potato was more heavily infested as known from South America. The bud midge, *Prodiplosis longifolia* Gagne, currently with a restricted distribution in Florida and Virginia and South America (Colombia, Peru, and Ecuador) could become an invasive species supported by its very polyphagous feeding habit. Lately, the fall armyworm (*Spodoptera frugiperda* J. E. Smith) (Lepidoptera: Noctuidae) is threatening crops across globe with very fast rate of spread. Potato could be a potential host for the fall armyworm in tropical and subtropical potato crops, e.g. in India (Yu 1982; Sidana et al. 2018; Nagoshi et al. 2019).

Potato is a temperate crop; however, efforts at breeding day neutral and high temperature tolerance have enabled tropicalisation of the crops. This phenomenon exposed the potato crops to some of the insect pests which are mainly prevalent in the warmer regions located in tropics or the subtropics. The potato cultivation in India is a major example of this type of pest shift. More than 85% of potato production in India is realized from the tropical and subtropical regions where the sweetpotato whitefly, *Bemisia tabaci*, is a major pest of the potato now (Shah et al. 2021a). To worsen the situation, we saw the evolution of the potato-specific *Begomovirus*, *tomato leaf curl New Delhi virus* [potato] around two decades back (Usharani et al. 2004; Kumar et al. 2021). The whitefly-virus complex is the major concern for quality seed potato production in India (Shah et al. 2020, 2021a).

1.3 Effect of Climate Change on Occurrence and Distribution of Potato Pests and Diseases

Climate change is a reality, and it has impact on infection, reproduction, dispersal and survival between seasons and other critical stages in the life cycle of a pathogen. Studies have shown that, assuming global temperature rise by 2 °C, there will be lower risk of late blight in warmer areas (<22 °C) and higher risk in cooler areas

(>13 °C) at global level. However, earlier onset of warm temperatures could result in an early appearance of late blight disease in temperate regions with the potential for more severe epidemics and increased number of fungicide applications needed for its control. Similar predictions have been made for Finland where for each 1 °C warming, late blight would occur 4–7 days earlier, and with the susceptibility period extended by 10–20 days (Kaukoranta 1996) resulting in 1–4 additional fungicide applications. In United Kingdom, hotter and drier summers are likely to reduce the importance of late blight, although earlier disease onset may act in opposite direction. An empirical climate-disease model has suggested that under the climate change scenario of 1 °C temperature increase with 30% reduction in precipitation in Germany, incidence of potato late blight will decrease to a mere 16% of its current level. Increase in both temperature and RH has added new dimension to late blight across the world. Under such a situation, *P. infestans* is likely to attack potato stems more often than foliage. In fact, in recent years it is more of "stem blight" than the foliar blight. This phase of the disease is more serious than the foliar stage as it affects the very crop plant. In India, Lahaul valley of HP, which was earlier free from late blight because of lack of precipitation, has now experienced attack of late blight due to occurrence of rainfall (Singh et al. 2013). Studies on the potential impact of climate change on late blight outbreak in Punjab and western Uttar Pradesh revealed that a higher number of sprays would be required in Punjab, whereas there would be no change or it is likely to be reduced in western Uttar Pradesh under future scenario (Dua et al. 2015). Shortest incubation period at higher temperature (28 °C) is on records for *Phytophthora infestans* (Becktell et al. 2005).

The effect of climate change on soilborne pathogens would vary from pathogen to pathogen. Black scurf and common scab diseases are favoured by moderate temperature and are likely to remain insulated from global warming in the near future. However, as the ambient temperatures are likely to increase by 1.4–5.8 °C by the end of the century, the severity of these two diseases may decrease substantially. Charcoal rot, which is favoured by high temperature and moisture, is currently endemic in eastern Uttar Pradesh, Bihar, Chhattisgarh and Madhya Pradesh, and the severity of this disease is likely to increase in these regions under global warming. Moreover, it is also expected that it may expand to other parts of north-central plains as well. *Sclerotium* wilt is restricted to plateau regions and is favoured by high temperature and moisture. With the increase in temperature due to global warming, the disease may also become prevalent in eastern Indo-Gangetic plains and may also enter into other areas like mid hills. Similarly, bacterial wilt may also advance to higher altitudes in hilly regions due to global warming, making them unfit for seed production.

Diseases caused by *Synchytrium endobioticum* (wart) and *Spongospora subterranea* (powdery scab) are favoured by low temperature and high soil moisture. Although wart spores can cause infection in the range of 10–28 °C with an optimum of 21 °C, there is hardly any infection beyond 23 °C. Therefore, warmer climates are likely to reduce wart infestation. Similarly, powdery scab infestation is also likely to be reduced with increase in temperature and reduction in rainfall under climate change scenario. The optimum temperature for powdery scab is 12 °C, and moisture requirement is 100%; the global warming may either lead to elimination of this disease or be pushed to higher altitudes making high hills (2500 m asl) free of powdery scab (Singh et al. 2013). The *Pectobacterium* species also differ in optimal and upper limits of growth temperatures. For example, *P. atrosepticum* and *P. parmentieri* die above 33 °C, while *P. carotovorum* and *P. brasiliense* can grow at temperatures up to 39 °C; hence the former two species may get eliminated, whereas the later may expand to new horizons (Charkowski 2015) under climate change scenario.

Among other phenomena, the impact of climate change on the crop and the associated insect pest complex is likely to be substantial (Hijmans 2003; Minhas et al. 2018; Raymundo et al. 2018; Rana et al. 2020; Shah et al. 2021b). It is being realized that insect species can respond to climate change in a multitude of ways, mainly in anticipation of a generally warmer and drier environment rich in CO_2 predicted through climate change general circulation models (IPCC 2013). Insect species may shift their geographic distributions or phenology in an attempt to track changes in their optimal conditions. Within ectotherms and endotherms alike, there is substantial evidence of range shifts already occurring, particularly towards the poles where temperatures are increasing; with recent evidence also suggesting, this is the case for pest species (Bebber et al. 2013). There are several reported cases of phenological change with the environment in insect pests. Crop damage due to insect pests may increase due to higher number of generations in a season, i.e. increased voltinism. Based on a day-degree model, an increased number of suitable days for development could allow faster generation time and, therefore, an additional generation (or possibly even two) to develop within a growing season (Barton and Terblanche 2014).

Climatic response phenology models have been used to assess the effect of temperature increase under projected changes in global temperature for the year 2050 and beyond for a wide range of potato pests (as summarized by Kroschel et al. 2020): *P. operculella* (Kroschel et al. 2013, 2016); *L. huidobrensis* (Mujica et al. 2016); Guatemalan potato tuber moth, *Tecia solanivora* (Povolny) (Schaub et al. 2016); Andean potato tuber moth, *Symmetrischema tangolias* (Gyen) (Sporleder et al. 2016); the whiteflies, *Bemisia tabaci* (Gennadius); and *Trialeurodes vaporariorum* (Westwood) (Gamarra et al. 2016a, b). As mentioned above, the geographical range of most of these is likely to expand with increased damage potential. However, the response is likely to be species specific. The deadly leafminer fly, *L. huidobrensis*, will expand in range to temperate regions of Asia, North and South America and Europe, as well as into subtropical and tropical mountainous regions with a moderate increase of its establishment and damage potential (Mujica et al. 2016). Even smaller changes in temperature predicted for tropical regions compared to temperate regions will have stronger consequences on pest development due to already higher existing metabolism rates of organisms such as insects (Dillon et al. 2010).

1.4 Pests and Pathogens of Quarantine Significance

All the pests and pathogens are not present everywhere. However, chances of their introduction into new geographical areas are very high due to globalisation. Challenges like population growth, globalization, climate change, bioterrorism and changing agribusiness infrastructure hamper plant bio-security at the local, regional and global levels. It is important for each nation to develop a plant bio-security infrastructure that ensures a safe and constant supply of food, feed and fibre. It is equally important to develop an international framework for cooperation that maintains plant bio-security without compromising trade. The devastating effects resulting from diseases and pests introduced along with international movement of planting material, agricultural produce and products are well documented. The historical Irish famine of 1845, caused by late blight of potato introduced from Central America, coffee rust introduced in Sri Lanka in 1875, fluted scale on citrus, San Jose scale in apples, banana bunchy top, the dreaded Golden nematode infesting potatoes, wart pathogen and the noxious weed *Lantana camara* introduced in 1809 from Central America are glaring examples that clearly demonstrate that introduction and establishment of quarantine pests including weeds into new areas can severely damage the crop production and economy of a region/country. Movement of plants and plant products between bio-geographical zones by human activities is now generally accepted to be the primary mode of introduction of exotic pathogens and pests.

Potato is usually propagated through tubers, and this vegetative mode of propagation is beset with many problems; hence, there is continuous threat of their introduction into new areas with planting material. In most of the countries, the following pests and pathogens are regarded as quarantined viz., potato tuber nematode (*Ditylenchus destructor*), stem and bulb nematode (*Ditylenchus dipsaci)*, potato cyst nematodes (*Globodera rostochiensis*, *G. pallida*), gangrene (*Phoma exigua* var. *foeta*), potato wart (*Synchytrium endobioticum*), potato smut (*Thecaphora solani*), bacterial ring rot (*Clavibacter michiganensis* ssp. *sepedonicus*), Colorado potato beetle (*Leptinotarsa decemlineata*), Andean potato weevil (*Premnotrypes* spp.) and viruses (Andean potato latent, Andean potato mottle, arracacha B virus, potato deforming mosaic, potato T, potato yellow dwarf, potato yellow vein, potato calico strain of tobacco ring spot virus, potato strains of tobacco streak virus and potato purple-top wilt and stolbur *Phytoplasmas*), which might establish in the country and hamper potato production programme. The Colorado potato beetle, *Leptinotarsa decemlineata* (Say), native to Mexico, has spread across most of the United States and was introduced into France in the 1920s from where it spread further reaching also parts of China (CABI 2017). Insect pests such as the potato tuber moth, *Phthorimaea operculella* (Zeller), and the leafminer fly, *Liriomyza huidobrensis* (Blanchard), with their centre of origin in the Andes have spread to and established in many tropical and subtropical regions.

1.5 Innovations in Disease Diagnostics

The conventional pathogen detection is totally dependent on morphological, serological and molecular based technologies, which are complicated, skill intensive and time consuming and require sophisticated lab facilities. However, with acceleration of computer technology and proliferation of smartphones coupled with software-based information and widespread access, the plant disease diagnosis and management can be deployed in an effective manner (Mohanty et al. 2016). An innovative artificial intelligence (AI)-based mobile phone app may provide an effective, easy-to-use diagnostic and low-cost solution. The recent advances in artificial intelligence coupled with increasing global smartphone penetration have paved the way for smartphone-assisted disease diagnosis (Johnson et al. 2021). There are android-based mobile apps available for download on smartphones which can diagnose a disease based on the pictures of the disease symptoms captured on the device. These apps have an expert-based algorithm of expert system which not only diagnoses the disease but gives advisories for the suitable management. One such plant disease diagnosis app "plantix" can detect more than 400 pathogens of 30 crops and has about ten million users worldwide (https://plantix.net).Interestingly, the app may further provide the advisories to minimize the disease outbreak in the next cropping season.

The first ever deep learning neural network platform in detection of potato diseases has been developed recently. Oppenheim et al. (2019) reported the detection of four economically important potato diseases viz., black scurf, common scab, silver scurf and black dot with more than 90% accuracy. The two devastating diseases of potato viz., early blight and late blight, were also being detected using this neural network. Recently, a cost-effective smartphone-based volatile organic compound (VOC) fingerprinting platform has been developed which allows non-invasive diagnosis of late blight caused by *Phytophthora infestans* (Li et al. 2019). This VOC-sensing platform is portable and easy to handle and can perform multiplex detection by classification of ten plant volatiles at a time. In fact, the symptomless plants having fungal infection may also be detected using this sensor-based platform. The approach will supplement the lab-based routine molecular detection and will become a viable additional method to help prevent major yield losses in potato.

E-nose technology has gained popularity for its applications in a field of human diagnostics, food quality and environmental safety. The technology is a quick and easily operated, rapid responding flexible tool to recognize signature gas samples (Cellini et al. 2017). The main operating principle relies in chemical interaction and electrical conductivity of sensors and resultant variations in the signature molecules. The e-nose basically comprises of an array of sensors with different sensibilities against diverse chemicals with diverse functional groups. The sensor generates electrical signals which correspond to the gas composition of the sample, and hence unique e-nose profiles are created which may be matched with a diverse pool of reference samples. Several plant materials infected with pathogenic bacteria or fungi have been subjected to e-nose-based detection, and potato is no exception.

The earliest detection of potato pathogen using e-nose tool was utilized in 2004 where researchers developed generic system for detection of statuary potato pathogen causing ring rot and bacterial wilt in potato (Stinson 2007). An advancement in e-nose-based detection system of the same disease came in 2014, when a commercial electronic nose (e-nose) having metal oxide sensor array was developed that recognizes volatile compounds emitted by potatoes experimentally infected with *Ralstonia solanacearum* or *Clavibacter michiganensis* subsp. *sepedonicus*, which are bacterial agents of potato brown and ring rot, respectively (Cellini et al. 2017). An e-nose gas chamber and sampling device was designed to detect the volatile profiles in soft rot infected potato tubers. The developed device utilized RBF NN algorithm and SVM algorithm and detected soft rot in tubers with accuracy as high as 89%. Likewise, there is a report of detection of soft rot in potatoes caused by the bacterium *Pectobacterium carotovorum* through the use of an array of low-cost gas sensors. These researchers utilized the strong odour emitted by rot infected potato in the cold stores and devised a low-cost gas sensor to detect the disease based on signature molecules. Under lab conditions, they demonstrated that out of 11 sensors used, 3 were able to detect the soft rot infected tubers with 100% accuracy (Rutolo et al. 2018). The identified sensors therefore offer promise for an automated in-store monitoring system. E-nose technology has significantly progressed since the first attempts of application to plant diagnosis and may become mature in a near future.

It is worthwhile to point to some valuable publications on the subject of potato entomology, e.g. Giordanengo et al. (2013), the insect-related portions of Potato Health Management (Johnson 2007) and the basic biology of several insect-transmitted diseases as discussed in Stevenson et al. (2001). Numerous field manuals for identifying potato pests have been published, e.g. Zehnder et al. (1994), Strand and Rude (2006) and Johnson (2007). Other suggested readings include Navarre and Pavek (2014) and Kroschel et al. (2020).

1.6 Development in the Area of Pest/Disease Forecasting

Though the chemical-based management of pests and diseases are still dominating, their use can be rationalized. This can be achieved through the use of forecasting models and decision support systems (DSSs). Forecasting allows a better control of a disease and a more efficient use of fungicides by making informed disease management decisions; whereas, the DSS-based strategy can deliver general or site-specific information to the stakeholders enabling them to take firm decisions on the management thereby resulting in economic gains and environment protection (Cooke et al. 2011; Sharma 2019). Various late blight forecasting models and DSSs have been developed across the globe for the management of late blight in different agro-ecologies and are discussed in Chap. 7.

Degree-day models are routinely used for predicting the emergence or build-up of insect population for various insect pests like potato tuber moth, the Andean tuber worms (*Symmetrischema tangolias* and *Tecia solanivora*), Colorado potato beetle, other potato beetles, potato psyllids, leaf hoppers etc. in different parts of the world

(Keller 2003; Sporleder et al. 2004; Giordanengo et al. 2013). The most impactful is perhaps monitoring of the flight activity of aphids and forecasting the extent of the risk of PVY spread in seed potato crops. Extensive aphid monitoring programmes using suction traps have been running successfully in European countries, the United States and New Zealand, for example. The oldest network is in the United Kingdom, which has been running for more than 50 years. Each week results of trap catch (species composition and abundance) with a cumulative vector pressure index are published and made available to the farmers and others involved with this sector. This index is designed to give the user an assessment of the risk to their crop of PVY spread and helps in decision-making processes when considering the need for insecticide treatments and to decide the best time to burn down/cutting of haulms of potato crops (details in Chap. 9).

1.7 Innovations in Pest and Disease Management

Dynamics of pathogens and insect pests keep on changing over time and space. This warrants the development of novel and future-oriented management strategies which are rapid, easy to operate and implicate, cost effective, widely applicable and well automated in the era of computer-based modernization. There is certainly good development in this direction, and novel futuristic techniques are emerging in potato farming, some are being adopted, while others are expected to be followed in the near future (Sharma and Tiwari 2021). The development and manufacture of effective, safe to humans and environmentally friendly pesticides have been a challenge and an important target. The general trends and strategies for novel pesticides include development of pesticides that are effective at an extremely low dosage, development of pesticides that are readily degradable and less residual in the environment and development of selective toxic agrochemicals. These development strategies have become increasingly prominent. The most developed fungicides are succinate dehydrogenase inhibitors (SDHI), demethylation inhibitors (DMI) and inhibitors of the mitochondrial electron transport chain complex III, i.e. quinone outside inhibitors (QoI) and quinone inside inhibitors (QiI). Due to the development of resistance to fungicides with existing modes of action, the second general trend is the development of fungicides with a novel mode of action and a unique chemical structure, and many fungicides possessing these traits have been launched or are under development. Other trends are the development of novel plant defence activators and novel natural product origin fungicides (Umetsu and Shirai 2020). "Oxathiapiprolin" (Zorvec™, Orondis™) is a new class of piperidinyl thiazole isoxazoline fungicide effective against several fungal diseases, including downy mildew and late blight on crops such as vegetables, ornamentals and turf. Its mode of action involves binding to the oxysterol-binding protein in oomycetes (Hagiwara et al. 2019).

Plant-associated bacteria contribute to their host's health in diverse ways, among which the emission of disease inhibiting volatile organic compounds (VOCs) is one option. Volatile organic compounds (VOCs) produced by the plant microbiota have been demonstrated to elicit plant defences and inhibit the growth and development of numerous plant pathogens. The inhibitory impact of volatiles emitted by *Pseudomonas* species against late blight has been shown by impeding mycelial growth and sporangia germination of *P. infestans* (Bailly and Weisskopf 2017). The VOCs containing sulphur compound S-methyl methane thiosulfonate (MMTS) had shown high *in planta* protective potential against late blight without phytotoxic effects. This protective activity of MMTS is not mediated by the plant immune system but is due to its anti-oomycete activity (Chinchilla et al. 2019). This provides new perspectives for plant protection by opening new research avenues on the role of VOCs in the interaction between plants and their microbiome and thus could help select for efficient biocontrol strategies and lead to a greener chemical disease management in the field.

Genome editing or gene editing has opened new opportunities to introduce sequence-specific modifications into the genomes of organisms and in the identification, characterization and validation of resistant genes coupled with their deployment in suitable cultivars for the development of disease resistance (Schenke et al. 2020). Genome editing technologies such as transcription activator-like effector nucleases (TALENs) and clustered regularly interspaced short palindromic repeats/CRISPR-associated protein 9 (Cas9) have become powerful genetic tools for increasing pathogen resistance in plants (Zhan et al. 2019). The novel opportunities provided by genome editing, synthetic biology and gene drive may give immense support in managing diseases caused by viral, bacterial and fungal pathogens in potato. A recent study revealed that mutation of a single gene in Arabidopsis, *DMR6* (downy mildew resistance 6), led to increased salicyclic acid levels and resistance to several plant pathogens, including bacteria and oomycetes (Zeilmaker et al. 2014). Interestingly, the tomato orthologous SlDMR6–1 is also upregulated in response to infection by *Pseudomonas syringae* pv *tomato* and *Phytophthora capsici*. Null mutants of SlDMR6–1 generated via CRISPR/Cas9 system showed resistance to *P. syringae*, *P. capsici* and *Xanthomonas* spp. without detrimental effects on tomato growth and development (Thomazella et al. 2016). Together, these results suggest that knocking out *DMR6* may be a promising strategy to confer broad spectrum disease resistance to plants. A related but advanced system CRISPR/Cas13a is effective to cleave single-stranded RNA, thus providing protection against RNA viruses in plant. Recently, a study reported the reframing of CRISPR/Cas13a to protect the potato plants from *Potato virus Y* (PVY) (Zhan et al. 2019). Four of the PVY target region P3 protein (potyviral membrane protein), CI (cytoplasmic inclusion bodies), NIb (RNA-dependent RNA polymerase) and CP (coat protein) were selected and used for sgRNA design. There was a suppressed PVY accumulation, and symptoms were attenuated in transgenic potato lines expressing small guide RNA (Cas13a/sgRNA). Another innovative approach is to target susceptibility genes (S-genes) which are beneficial for pathogens' growth and development. Silencing of S genes attenuates the disease symptoms as observed in several economically important crops such as rice, wheat, tomatoes, citrus etc. A recent study by Kieu et al. (2021) reported that CRISPR/Cas9 genome editing-mediated induced mutations in susceptibility genes confer increased late blight resistance in potatoes. A tetra-allelic deletion mutant (conferring co-expression of two guide RNAs) was generated using CRISPR/Cas9 system targeting seven putative S genes including two DMR6 potato homologues. The mutant plants assayed for late blight resistance have shown significantly high level of resistance as compared to control.

Presently, the exogenous application of double-stranded RNA, small interfering RNA (siRNA) and hpRNA-mediated post-transcriptional gene silencing (i.e. RNA interference or gene silencing) have emerged as sustainable strategies in activating plant defence against phytopathogenic diseases (Machado et al. 2018). The management of late blight disease in potato was also made possible using dsRNA-based spray formulations. A latest study reported that spraying of *Phytophthora infestans* derived dsRNA molecules as an effective plant protection strategy for the management of potato late blight (Sundaresha et al. 2021). Overall dsRNA-based spray formulation was highly effective in mitigating late blight disease symptoms, and this can be an effective strategy alternative to chemical pesticides. In fact, many studies have been found effective in managing vector populations using dsRNA-based spray, and this can be utilized in managing whiteflies and aphids in potato production system (Arif et al. 2012; Thakur et al. 2014). It is well reported that whitefly mortality has been induced via oral delivery of dsRNA, targeting five important genes namely actin ortholog, ADP/ATP translocase, α-tubulin, ribosomal protein L9 (rpl9) and v-ATPase A. There was enhanced whitefly resistance in transgenic tobacco plants expressing dsRNA of v-ATPase gene (Thakur et al. 2014). Similarly, dsRNA-based spray formulations targeting potential RNAi target genes (TREH, ATPD, ATPE and CHS1) were selected and cloned and further tested through the transdermal dsRNA delivery system against soybean aphids. The delivered dsRNA silenced the gene expressions of target genes with mortality rate up to 81.67%. Such kind of novel, ecofriendly and highly target-specific formulations are the need of the hour for simultaneous management of virus and vector populations.

Unmanned aerial vehicles (UAVs) and drones are becoming increasingly popular in the era of precision agriculture. The drones are excellent in terms of speed of its coverage of large area much faster than humans so they can be a valuable scouting tool especially if other sensors are also equipped in it. The advancement in image processing and machine learning tools has evolved to such an extent that it can generate useful information for the management of the farm (Sugiura et al. 2018). There are several successful examples of drone-based monitoring and redressal of various abiotic and biotic stresses in commercial crop plants. Multispectral, thermal and hyperspectral images are the preferred methods of acquiring information in disease monitoring. The drone-based disease diagnostics has been used for potato diseases. Sugiura et al. (2018) devised RGB sensor-based UAV for the detection of PVY-infected plants. Likewise, the late blight phenotyping, disease progression was also assessed in potato using similar RGB-based UAVs (Sugiura et al. 2016). The spectral imaging based system also captured disease progression and provided area under disease progress curve (AUDPC) as phenotypic data. Identification of the onset of potato black leg disease (*Pectobacterium atrosepticum*) within using off-the-shelf digital cameras equipped on UAV was also made possible in United

Kingdom (Gibson-Poole et al. 2017). Besides disease monitoring, drones are being used for application of pesticides. The advantages of drone-based application of pesticides are on account of savings on pesticides and spraying volume per unit area, less application time and more penetration and coverage of pesticides.

Among the pest control options, potato crops are highly reliant on the use of synthetic pesticides. Potato has been infamously nicknamed "one of the most chemically-dependent crops in the world". Since the mid-twentieth century, intense use of insecticides has led to the selection of resistant insect pest populations. Potato pests include some of the species that are most prone to evolving resistance to a wide variety of chemicals. The Arthropod Pesticide Resistance Database (2018) lists 469 cases of green peach aphid (*Myzus persicae* (Sulzer)) resistance to a total of 80 active ingredients, 300 cases of Colorado potato beetle (*L. decemlineata*) resistance to a total of 56 active ingredients, 111 cases of greenhouse whitefly (*T. vaporariorum*) resistance to 27 active ingredients and 501 cases of two-spotted spider mite (*Tetranychus urticae* Koch) resistance to rather impressive 95 active ingredients (as reported in Kroschel et al. 2020). The extent of resistance is likely to be underestimated because not every case of its development is entered into the database. It is possible that the ability to deal with toxic glycoalkaloids contained in potato foliage serves as a preadaptation to resisting synthetic pesticides (Alyokhin and Chen 2017). Not every population of a given pest species is resistant to all compounds that have been recorded to fail against that species. However, these statistics vividly illustrate the seriousness of the problem. On several occasions, potato growers already experienced the situation when virtually all commercially available chemicals failed to control their targets (Alyokhin et al. 2013). Today, as with other major crops, potato culture has to deal with increasing environmental and public concerns that lead to the reduction of new chemical discoveries and development, while also supporting a rapidly rising world demand. In recent decades, there has been a shift in insecticide development to safer and more targeted (or narrow spectrum) insecticides that are often less toxic to non-target species. Today, there is a wide diversity of insecticide mode of actions and products that can aid in a more sustainable approach to pest management for potatoes (details in Chap. 16).

Despite the fact that chemical control is the most popular form of insect pest management in the potato industry, yet, this does not necessarily have to be the case. Very basic understanding of the pest population dynamics revealed the periods of maximum spread of virus like PVY. This information is routinely used to decide the time of haulms-cutting or destruction. This simple cultural practice is perhaps the most important virus disease control tactic that is widely adopted round the world in seed potato crops. Similarly, in subtropical conditions in India, studies on population dynamics of the most efficient vector of potato viruses, the peach-potato aphid, *M. persicae*, were used to decide the dates of planting and haulms-cutting during a period of lowest aphid activity for quality seed potato production (Pushkarnath 1967). Thorough understanding of the potato ecosystem using the systems approach is the way forward. Similarly, high hopes are associated with novel genomics techniques to provide breakthroughs for major problems in potato pest management, along with other aspects.

The management tactics are rendered ineffective due to contiguous evaluation of pathogen variants, resistance development and emergence of new pests and diseases which usually were characterized as of minor importance or were under the radar. Climate change is/will complicate the situation further. And needless to say, the synthetic plant protection chemicals have always been under criticism due to their ill effects to human beings and the environment. In this book, an attempt has been made to bring together information on such aspects of pest management which are believed to be more sustainable in the long run.

1.8 Conclusion

Potato is among the most important food crops which has a pivotal role to play in food and nutritional security of the populace in the foreseeable times. The yield and quality of potato is contained by a myriad of insect pests, fungi, bacterial, viruses and nematode pathogens. The pest management regime is currently almost entirely dependent on the use of plant protection products, the use of which is not sustainable in the long run. Newer issue like that of climate change, invasive species and newer species/strains of pest and pathogens are straining the pest management practices further. As time witnessed, the potato pets and disease management has come a long way, its high time to think on the lines to make the practices sustainable for the future generations.

References

Alyokhin A, Chen YH (2017) Adaptation to toxic hosts as a factor in the evolution of insecticide resistance. Curr Opin Insect Sci 21(1):33–38

Alyokhin A, Chen YH, Udalov M, Benkovskaya G, Lindström L (2013) Evolutionary considerations in potato pest management. In: Giordanengo P, Vincent C, Alyokhin A (eds) Insect pests of potato: global perspectives on biology and management. Academic Press, Oxford, pp 543–571

Anderson PK, Cunningham AA, Patel NG et al (2004) Emerging infectious diseases of plants: pathogen pollution, climate change and agrotechnology drivers. Trends Ecol Evol 19:535–544. https://doi.org/10.1016/j.tree.2004.07.021

Ardestani ST, Sharifnabi B, Zare R, Moghadam AA (2010) New *Alternaria* species associated with potato leaf spot in various potato growing regions of Iran. Iranian J Plant Pathol 45:83–86

Arif M, Azhar U, Arshad M et al (2012) Engineering broad-spectrum resistance against RNA viruses in potato. Transgenic Res 21:303–311

Arthropod Pesticide Resistance Database (2018). https://www.pesticideresistance.org. Accessed 3 Jul 2018

Bailly A, Weisskopf L (2017) Mining the volatilomes of plant-associated microbiota for new biocontrol solutions. Front Microbiol 8:1638. https://doi.org/10.3389/fmicb.2017.01638

Barton MG, Terblanche JS (2014) Predicting performance and survival across topographically heterogeneous landscapes: the global pest insect *Helicoverpa armigera* (Hübner, 1808) (Lepidoptera: Noctuidae). Austral Entomol 53(3):249–258

Bebber DP, Ramotowski MA, Gurr SJ (2013) Crop pests and pathogens move pole wards in a warming world. Nat Clim Change 3(11):985

Becktell M, Daughtrey M, Fry W (2005) Temperature and leaf wetness requirements for pathogen establishment, incubation period, and sporulation of *Phytophthora infestans* on *Petunia hybrida*. Plant Dis 89(9):975–979

Borges KM, Cooper WR, Garczynski SF, Thinakaran J, Jensen AS, Horton DR et al (2017) "Candidatus Liberibacter solanacearum" associated with the psyllid, *Bactericera maculipennis* (Hemiptera: Triozidae). Environ Entomol 46(2):210–216. https://doi.org/10.1093/ee/nvw174

CABI (2017) *Leptinotarsa decemlineata* (Colorado potato beetle) data sheet. In: Invasive species compendium. CAB International, Wallingford

Caparros Megido R, Brostaux Y, Haubruge E et al (2013) Propensity of the tomato leafminer, *Tuta absoluta* (Lepidoptera: Gelechiidae), to develop on four potato plant varieties. Am J Potato Res 90:255–260. https://doi.org/10.1007/s12230-013-9300-9

Cellini A, Blasioli S, Biondi E et al (2017) Potential applications and limitations of electronic nose devices for plant disease diagnosis. Sensors 17:2596

Charkowski AO (2015) Biology and control of Pectobacterium in potato. Am J Potato Res 92(2): 223–229

Chinchilla D, Bruisson S, Meyer S et al (2019) A sulfur-containing volatile emitted by potato-associated bacteria confers protection against late blight through direct anti-oomycete activity. Sci Rep 9(1):18778

Cooke DEL, Kessel GJT, Lassen P, Hansen JG (2019) The European population of Phytophthora infestans in a global context. Seventeenth Euroblight workshop, York, 12-15 May, 2019. WUR-Special Report No 19, pp 35–36

Cooke LR, Schepers HTAM, Hermansen A et al (2011) Epidemiology and integrated control of potato late blight in Europe. Potato Res 54:183–222

De Boer SH, Li X, Ward LJ (2012) Pectobacterium spp. associated with bacterial stem rot syndrome of potato in Canada. Phytopathology 102(10):937–947

de Werra P, Bussereau F, Keiser A, Ziegler D (2015) First report of potato blackleg caused by Pectobacterium carotovorum subsp Brasiliense in Switzerland. Plant Dis 99(4):551

Devaux A, Kromann P, Ortiz O (2014) Potatoes for sustainable global food security. Potato Res 57: 185–199. https://doi.org/10.1007/s11540-014-9265-1

Dillon ME, Wang G, Huey RB (2010) Global metabolic impacts of recent climate warming. Nature 467:704–706

Dua VK, Singh BP, Islam A, Sanjeev S (2015) Potential impact of climate change on late blight outbreak in western Uttar Pradesh and Punjab using Jhulsacast model. Potato J 42(1):58–71

Duarte V, De Boer SH, Ward LJ, De Oliveira AMR (2004) Characterization of atypical *Erwinia carotovora* strains causing blackleg of potato in Brazil. J Appl Microbiol 96(3):535–545

Elphinstone JG (2005) The current bacterial wilt situation: a global overview. In: Allen C, Piror P, Hayward AC (eds) Bacterial wilt disease and the *Ralstonia solanacearum* species complex. APS Press, St Paul, pp 9–28

FAOSTAT (2020) Food and Agriculture Organization of the United Nations. http://www.fao.org/faostat/en/#data/QC

Fry WE (2020) Phytophthora infestans: the itinerant invader; "late blight": the persistent disease. Phytoparasitica 48:87–94. https://doi.org/10.1007/s12600-019-00778-3

Fry WE, Birch PRJ, Judelson HS et al (2015) Five reasons to consider *Phytophthora infestans* a re-emerging pathogen. Phytopathology 105:966–981

Gamarra H, Carhuapoma P, Mujica N, Kreuze J, Kroschel J (2016b) Greenhouse whitefly, *Trialeurodes vaporariorum* (Westwood 1956). In: Kroschel J, Mujica N, Carhuapoma P, Sporleder M (eds) Pest distribution and risk atlas for Africa—potential global and regional distribution and abundance of agricultural and horticultural pests and associated biocontrol agents under current and future climates. International Potato Center (CIP), Lima, pp 154–168. https://doi.org/10.4160/9789290604761

Gamarra H, Mujica N, Carhuapoma P, Kreuze J, Kroschel J (2016a) Sweetpotato whitefly, *Bemisia tabaci* (Gennadius 1989) (biotype B). In: Kroschel J, Mujica N, Carhuapoma P, Sporleder M (eds) Pest distribution and risk atlas for Africa—potential global and regional distribution and abundance of agricultural and horticultural pests and associated biocontrol agents under current and future climates. International Potato Center (CIP), Lima, pp 85–99. https://doi.org/10.4160/9789290604761

Gibson-Poole S, Humphris S, Toth I, Hamilton A (2017) Identification of the onset of disease within a potato crop using a UAV equipped with un-modified and modified commercial off-the-shelf digital cameras. Adv Anim Biosci 8:812–816

Giordanengo P, Vincent C, Alyokhin A (eds) (2013) Insect pests of potato. Elsevier, Amsterdam

Gray S, De Boer S, Lorenzen J, Karasev A, Whitworth J, Nolte P, Singh R, Boucher A, Xu H (2010) Potato virus Y: an evolving concern for potato crops in the United States and Canada. Plant Dis 94(12):1384–1397

Hagiwara H, Ezaki R, Hamada T et al (2019) Development of a novel fungicide, tolprocarb. J Pestic Sci 44:1–6

Hannukkala AO, Kaukoranta T, Lehtinen A, Rahkonen A (2007) Late-blight epidemics on potato in Finland, 1933-2002; increased and earlier occurrence of epidemics associated with climate change and lack of rotation. Plant Pathol 56(1):167–176

Harrison MD, Venette JR (1970) Chemical control of potato early blight and its effect on potato yield. Am J Potato Res 47(3):81–86

Haverkort AJ, Struik PC, Visser RGF, Jacobsen E (2009) Applied biotechnology to combat late blight in potato caused by *Phytophthora infestans*. Potato Res 52(3):249–264

Hijmans RJ (2003) The effect of climate change on global potato production. Am J Potato Res 80: 271–279. https://doi.org/10.1007/BF02855363

Horsfield A, Wicks T, Davies K, Wilson D, Paton S (2010) Effect of fungicide use strategies on the control of early blight (*Alternaria solani*) and potato yield. Australas Plant Pathol 39(4): 368–375

IPCC (2013) Climate change 2013: mitigation of climate change, vol 3. Cambridge University Press, Cambridge

Johnson DA (ed) (2007) Potato health management. The American Phytopathological Society Press, St Paul

Johnson J, Sharma G, Srinivasan S et al (2021) Enhanced field-based detection of potato blight in complex backgrounds using deep learning. Plant Phenomics 2021:1–13. https://doi.org/10.34133/2021/9835724

Kanle Satishchandra N, Chakravarthy AK, Özgökçe MS, Atlihan R (2019) Population growth potential of *Tuta absoluta* (Meyrick) (Lepidoptera: Gelechiidae) on tomato, potato, and eggplant. J Appl Entomol 143(5):518–526

Karasev AV, Gray SM (2013) Continuous and emerging challenges of potato virus Y in potato. Annu Rev Phytopathol 51:571–586

Kaukoranta T (1996) Impact of global warming on potato late blight: risk, yield loss, and control. Agric Food Sci Finl 5:311–327

Keller S (2003) Integrated pest management of the potato tuber moth in cropping systems of different agro-ecological zones. In: Kroschel J (ed) Advances in crop research, vol 1. Margraf, Weikersheim

Kieu NP, Lenman M, Wang ES et al (2021) Mutations introduced in susceptibility genes through CRISPR/Cas9 genome editing confer increased late blight resistance in potatoes. Sci Rep 11: 4487

Kim H-S, Ma B, Perna NT, Charkowski AO (2009) Phylogeny and virulence of naturally occurring type III secretion system-deficient *Pectobacterium* strains. Appl Environ Microbiol 75(13): 4539–4549

Kroschel J, Mujica N, Alcazar J, Canedo V, Zegarra O (2012) Developing integrated pest management for potato: experiences and lessons from two distinct potato production systems of Peru. In: He Z, Larkin RP, Honeycutt CW (eds) Sustainable potato production: global case studies. Springer, London, pp 419–450

Kroschel J, Mujica N, Carhuapoma P, Sporleder M (2016) Pest distribution and risk atlas for Africa—potential global and regional distribution and abundance of agricultural and

horticultural pests and associated biocontrol agents under current and future climates. International Potato Center (CIP), Lima, p 650. https://doi.org/10.4160/9789290604761
Kroschel J, Mujica N, Okonya J, Alyokhin A (2020) Insect pests affecting potatoes in tropical, subtropical, and temperate regions. In: Campos H, Ortiz O (eds) The potato crop. Springer, Cham, pp 251–306
Kroschel J, Schaub B (2013) Biology and ecology of potato tuber moths as major pests of potato. In: Giordanengo P, Vincent C, Alyokhin A (eds) Insect pests of potato: global perspectives on biology and management. Academic Press, Oxford, pp 165–192
Kroschel J, Sporleder M, Tonnang HEZ, Juarez H, Carhuapoma P, Gonzales JC, Simon R (2013) Predicting climate change caused changes in global temperature on potato tuber moth *Phthorimaea operculella* (Zeller) distribution and abundance using phenology modeling and GIS mapping. Agric For Meteorol 170:228–241
Kumar R, Tiwari RK, Jeevalatha A et al (2021) Potato apical leaf curl disease: current status and perspectives on a disease caused by tomato leaf curl New Delhi virus. J Plant Dis Prot 128:897–911. https://doi.org/10.1007/s41348-021-00463-w
Laurila J, Ahola V, Lehtinen A et al (2008) Characterization of *Dickeya* strains isolated from potato and river water samples in Finland. Eur J Plant Pathol 122(2):213–225. https://doi.org/10.1007/s10658-008-9274-5
Leiminger J, Bassler E, Knappe C (2015) Quantification of disease progression of *Alternaria* spp. on potato using real-time PCR. Eur J Plant Pathol 141:295–309
Leiminger JH, Hausladen H (2014) Untersuchungen zur Befallsentwicklung und Ertragswirkung der Dürrfleckenkrankheit (Alternaria spp.) in Kartoffelsorten unterschiedlicher Reifegruppe. Gesunde Pflanzen 66:29–36
Li Z, Paul R, Ba Tis T et al (2019) Non-invasive plant disease diagnostics enabled by smartphone-based fingerprinting of leaf volatiles. Nature Plants 5:856–866
Machado AK, Brown NA, Urban M et al (2018) RNAi as an emerging approach to control fusarium head blight disease and mycotoxin contamination in cereals. Pest Manag Sci 74:790–799
Minhas JS, Kumar P, Kumar D, Dua VK, Gupta YK (2018) Response of potato to elevated CO_2 under short days: growth, physiological parameters and tuber yield. Indian J Hortic 75(1):82–86
Mohanty SP, Hughes DP, Salathé M (2016) Using deep learning for image-based plant disease detection. Front Plant Sci 7:1419
Mujica N, Carhuapoma P, Kroschel J (2016) Serpentine leafminer fly, *Liriomyza huidobrensis* (Blanchard 1926). In: Kroschel J, Mujica N, Carhuapoma P, Sporleder M (eds) Pest distribution and risk atlas for Africa—potential global and regional distribution and abundance of agricultural and horticultural pests and associated biocontrol agents under current and future climates. International Potato Center (CIP), Lima, pp 114–125. https://doi.org/10.4160/9789290604761
Mujica N, Kroschel J (2013) Pest intensity-crop loss relationships for the leafminer fly *Liriomyza huidobrensis* (Blanchard) in different potato (*Solanum tuberosum* L.) varieties. Crop Prot 47:6–16
Nagoshi RN, Dhanani I, Asokan R, Mahadevaswamy HM, Kalleshwaraswamy CM, Sharanabasappa et al (2019) Genetic characterization of fall armyworm infesting South Africa and India indicate recent introduction from a common source population. PLoS One 14(5): e0217755. https://doi.org/10.1371/journal.pone.0217755
Navarre R, Pavek MJ (eds) (2014) The potato: botany, production and uses. CAB International, Wallingford
Ngadze E, Brady CL, Coutinho TA et al (2012) Pectinolytic bacteria associated with potato soft rot and blackleg in South Africa and Zimbabwe. Eur J Plant Pathol 134:33–549. https://doi.org/10.1007/s10658-012-0036-z
Oerke EC, Dehne HW, Schönbeck F, Weber A (1994) Crop production and crop protection. Estimated losses in major food and cash crops. Elsevier, Amsterdam, p 808
Oppenheim D, Shani G, Erlich O, Tsror L (2019) Using deep learning for image-based potato tuber disease detection. Phytopathology 109:1083–1087

Parkinson N, Stead D, Bew J et al (2009) Dickeya species relatedness and clade structure determined by comparison of RecA sequences. Int J Syst Evol Microbiol 59(10):2388–2393. https://doi.org/10.1099/ijs.0.009258-0

Pérombelon MCM, Kelman A (1987) Blackleg and other potato diseases caused by soft rot erwinias: proposal for revision of terminology. Plant Dis 71(3):283–285

Pitman AR, Harrow SA, Visnovsky SB (2010) Genetic characterisation of *Pectobacterium wasabiae* causing soft rot disease of potato in New Zealand. Eur J Plant Pathol 126(3):423–435

Pitman AR, Wright PJ, Galbraith MD, Harrow SA (2008) Biochemical and genetic diversity of pectolytic enterobacteria causing soft rot disease of potatoes in New Zealand. Australas Plant Pathol 37(6):559–568

Prior P, Ailloud F, Dalsing BL et al (2016) Genomic and proteomic evidence supporting the division of the plant pathogen *Ralstonia solanacearum* into three species. BMC Genomics 17(1):90

Pushkarnath (1967) Seed potato production in the sub-tropical plains of India. Am Potato J 44:429–441

Radcliffe EB, Johnson KB (1994) Biology and management of leafhoppers on potato. In: Zehnder GW, Powelson ML, Jansson RK, Raman KV (eds) Advances in potato pest biology and management. APS Press, Saint Paul, pp 71–82

Rana A, Dua VK, Chauhan S et al (2020) Climate change and potato productivity in Punjab—impacts and adaptation. Potato Res 63:597–613. https://doi.org/10.1007/s11540-020-09460-2

Raymundo R, Asseng S, Robertson R, Petsakos A, Hoogenboom G, Quiroz R, Hareau G, Wolf J (2018) Climate change impact on global potato production. Eur J Agron 100:87–98

Rehman A, Saeed AM, Khan AM, Rafique A, Ashraf M (2012) Potato: a new host plant of *Tuta absoluta* Povolny (Lepidoptera: Gelechiidae) in Turkey. Pak J Zool 44(4):1183–1184

Rodrigues TTMS, Berbee ML, Siommons EG et al (2010) First report of *Alternaria tomatophila* and *A. grandis* causing early blight on tomato and potato in Brazil. New Dis Rep 22:28

Rutolo MF, Clarkson JP, Covington JA (2018) The use of an electronic nose to detect early signs of soft-rot infection in potatoes. Biosyst Eng 167:137–143

Safni I, Cleenwerck I, De Vos P et al (2014) Polyphasic taxonomic revision of the *Ralstonia solanacearum* species complex: proposal to emend the descriptions of *Ralstonia solanacearum* and *Ralstonia syzygii* and reclassify current *R. syzygii* strains as *Ralstonia syzygii* subsp. *syzygii* subsp. nov., *R. solanacearum* phylotype IV strains as *Ralstonia syzygii* subsp. *indonesiensis* subsp. nov., banana blood disease bacterium strains as *Ralstonia syzygii* subsp. *celebesensis* subsp. nov. and *R. solanacearum* phylotype I and III strains as *Ralstonia pseudosolanacearum* sp. nov. Int J Syst Evol Microbiol 64:3087–3103. https://doi.org/10.1099/ijs.0.066712-0

Samson R, Legendre JB, Christen R et al (2005) Transfer of *Pectobacterium chrysanthemi* (Burkholder et Al. 1953) Brenner et Al. 1973 and *Brenneria paradisiaca* to the genus *Dickeya* gen. nov. as *Dickeya chrysanthemi* comb. nov. and *Dickeya paradisiaca* comb. nov. and delineation of four novel species, *Dickeya dadantii* sp. nov., *Dickeya dianthicola* sp. nov., *Dickeya dieffenbachiae* sp. nov. and *Dickeya* zeae sp. nov. Int J Syst Evol Microbiol 55(Pt 4): 1415–1427. https://doi.org/10.1099/ijs.0.02791-0

Schaub B, Carhuapoma P, Kroschel J (2016) Guatemalan potato tuber moth, *Tecia solanivora* (Povolny 1973). In: Kroschel J, Mujica N, Carhuapoma P, Sporleder M (eds) Pest distribution and risk atlas for Africa—potential global and regional distribution and abundance of agricultural and horticultural pests and associated biocontrol agents under current and future climates. International Potato Center (CIP), Lima, pp 24–38. https://doi.org/10.4160/9789290604761

Schenke D, Cai D (2020) Applications of CRISPR/Cas to improve crop disease resistance: beyond inactivation of susceptibility factors. iScience 23:101478

Shah MA, Aarti B, Naga KC, Subhash S et al (2020) Important potato pests and their management. In: Singh AK, Chakrabarti SK, Singh B, Sharma J, Dua VK (eds) Potato science and technology for sub-tropics. NIPA, New Delhi, pp 295–326

Shah MA, Al-Tawaha AR, Naga KC, Sharma S (2021b) Insect pest management in the era of climate change. In: Fahad S, Sonmez O, Saud S, Wang D, Wu C, Adnan M, Tarun V (eds)

Climate change and plants: biodiversity, growth and interactions. CRC Press, New York, pp 133–146

Shah MA, Sharma S, Kumar R (2021a) Seasonal population dynamics and flight activity of sweetpotato whitefly, Bemisia tabaci on potato in the sub-tropical plains of India. Arthropod-Plant Interact 15:605–613. https://doi.org/10.1007/s11829-021-09835-3

Sharma S (2019) Scheduling fungicide application for economic management of late blight using indo-Blightcast model. Potato J 46(2):124–131

Sharma S, Tiwari RK (2021) Futuristic strategies for management of potato diseases. In: Bharat NK, Gautam HR (eds) Technology strides in plant health management. Neoti Book Agency, New Delhi, pp 63–90

She XM, Lan GB, Tang YF, He ZF (2017) Pectobacterium carotovorum subsp. *brasiliense* causing pepper black spot disease in China. J Plant Pathol 99(3):769–772

Shoaib A, Akhtar N, Akhtar S, Hafeez R (2014) First report of *Alternaria longipes* causing leaf spot of potato cv. Sante in Pakistan. Plant Dis 98:1742

Sidana JA, Singh BI, Sharma O (2018) Occurrence of the new invasive pest, fall armyworm, *Spodoptera frugiperda* (JE smith) (Lepidoptera: Noctuidae), in the maize fields of Karnataka, India. Curr Sci 115:621–623

Singh BP, Dua VK, Govindakrishnan PM, Sharma S (2013) Impact of climate change on potato. In: Singh HP (ed) Climate resilient horticulture: adaptation and mitigation strategies. Springer, New Delhi, pp 125–135

Sławiak M, van Beckhoven JRCM, Speksnijder AGCL et al (2009) Biochemical and genetical analysis reveal a new clade of biovar 3 *Dickeya* spp. strains isolated from potato in Europe. Eur J Plant Pathol 125(2):245–261. https://doi.org/10.1007/s10658-009-9479-2

Sporleder M, Carhuapoma P, Kroschel J (2016) Andean potato tuber moth, *Symmetrischema tangolias* (Gyen 1913). In: Kroschel J, Mujica N, Carhuapoma P, Sporleder M (eds) Pest distribution and risk atlas for Africa—potential global and regional distribution and abundance of agricultural and horticultural pests and associated biocontrol agents under current and future climates. International Potato Center (CIP), Lima, pp 39–53. https://doi.org/10.4160/9789290604761

Sporleder M, Kroschel J, Gutierrez Quispe MR, Lagnaoui A (2004) A temperature-based simulation model for the potato tuberworm, *Phthorimaea operculella* Zeller (Lepidoptera: Gelechiidae). Environ Entomol 33:477–486

Sridhar V, Nitin KS, Nagaraja T (2015) Comparative biology of south American tomato moth, *Tuta absoluta* (Meyrick) (Lepidoptera: Gelechiidae) on three solanaceous host plants. Pest Manag Hortic Ecosys 21(2):159–161

Stevenson WR, Loria R, Franc GD, Weingartner DP (2001) Compendium of potato diseases, 2nd edn. The American Phytopathological Society, St Paul

Stinson JA (2007) Sensing potato pathogens: an automated approach. The University of Manchester, Manchester

Strand LL, Rude PA (2006) Integrated pest management for potatoes in the western United States, 2nd edn. University of California Agriculture and Natural Resources Publications, Davis

Sugiura R, Tsuda S, Tamiya S et al (2016) Field phenotyping system for the assessment of potato late blight resistance using RGB imagery from an unmanned aerial vehicle. Biosyst Eng 148:1–10

Sugiura R, Tsuda S, Tsuji H, Murakami N (2018) Virus-infected plant detection in potato seed production field by UAV imagery. American Society of Agricultural and Biological Engineers, Washington, p 1

Sundaresha S, Sharma S, Bairwa A et al (2021) Spraying of dsRNA molecules derived from *Phytophthora infestans*, as a plant protection strategies for the management of potato late blight

Taheri AS, Sharif NB, Zare R, Abbasi MA (2009) Alternaria interrupta, a new pathogen causing potato early blight in Iran. Rostaniha 10:72–73

Thakur N, Upadhyay SK, Verma PC et al (2014) Enhanced whitefly resistance in transgenic tobacco plants expressing double stranded RNA of v-ATPase a gene. PLoS One 9:e87235

Thomazella DPT, Brail Q, Dahlbeck D, Staskawicz B (2016) CRISPR-Cas9 mediated mutagenesis of a *DMR6* ortholog in tomato confers broad-spectrum disease resistance bioRxiv https://doi.org/10.1101/064824

Toth IK, van der Wolf JM, Saddler et al (2011) Dickeya species: an emerging problem for potato production in Europe. Plant Pathol 60(3):385–399. https://doi.org/10.1111/j.1365-3059.2011.02427.x

Tymon LS, Peever TL, Johnson DA (2016) Identification and enumeration of small-spored *Alternaria* species associated with potato in the US northwest. Plant Dis 100:465–472

Umetsu N, Shirai Y (2020) Development of novel pesticides in the 21st century. J Pestic Sci 45(2): 54–74

Usharani KS, Surendranath B, Paul-Khurana SM, Garg ID, Malathi VG (2004) Potato leaf curl—a new disease of potato in northern India caused by a strain of tomato leaf curl New Delhi virus. Plant Pathol 253:235

van der Merwe JJ, Coutinho TA, Korsten L, van der Waals JE (2010) Pectobacterium *carotovorum* subsp. *brasiliensis* causing blackleg on potatoes in South Africa. Eur J Plant Pathol 126(2): 175–185

Van der Waals JE, Korsten L, Aveling TAS (2001) A review of early blight of potato. Afr Plant Protect 7(2):91–102

Van der Wolf JM, de Haan EG, Kastelein P et al (2017) Virulence of *Pectobacterium carotovorum* subsp. *brasiliense* on potato compared with that of other *Pectobacterium* and *Dickeya* species under climatic conditions prevailing in the Netherlands. Plant Pathol 66(4):571–583

Wang J, Haapalainen M, Schott T et al (2017a) Genomic sequence of 'Candidatus Liberibacter solanacearum' haplotype C and its comparison with haplotype a and B genomes. PLoS One 12(2):e0171531

Wang J, Wang YH, Dai PG et al (2017c) First report of tobacco bacterial leaf blight caused by *Pectobacterium carotovorum* subsp. *brasiliense* in China. Plant Dis 101(5):830–830

Wang N, Pierson EA, Setubal JC et al (2017b) The Candidatus Liberibacter–host interface: insights into pathogenesis mechanisms and disease control. Annu Rev Phytopathol 55:451–482

Wates TD, Jensen AS (2014) Insect pests of potato. In: Navarre R, Pavek MJ (eds) The potato, botany, production and uses. CABI, Wallingford, pp 133–147

Yap M-N, Barak JD, Charkowski AO (2004) Genomic diversity of Erwinia carotovora subsp. carotovora and its correlation with virulence. Appl Environ Microbiol 70(5):3013–3023

Yu SJ (1982) Host plant induction of glutathione S-transferase in the fall armyworm. Pestic Biochem Physiol 18(1):101–106

Zehnder GW, Powelson ML, Jansson RK, Raman KV (1994) Advances in potato pest biology and management. The American Phytopathological Society Press, St Paul

Zeilmaker T, Ludwig NR, Elberse J et al (2014) Downy mildew resistant 6 and DMR6-like oxygenase 1 are partially redundant but distinct suppressors of immunity in Arabidopsis. Wiley Online Libr 81:210–222

Zhan X, Zhang F, Zhong Z et al (2019) Generation of virus-resistant potato plants by RNA genome targeting. Plant Biotechnol J 17:1814–1822

Phytosanitary Standards and International Exchange of Potato

2

Kavita Gupta and S. C. Dubey

Abstract

International Plant Protection Convention (IPPC) is recognized by Agreement on Application of Sanitary and Phytosanitary (SPS) measures of WTO for developing phytosanitary standards—International standards for phytosanitary measures (ISPMs). These aim to facilitate trade and avoid use of unjustified measures during trade. Till date, 42 ISPMs have been developed, viz., pest risk analysis (PRA), establishment of pest-free areas (PFA), export certification, pest eradication, pest reporting, etc., which are periodically reviewed and amended. *Article* 3 of SPS Agreement dealing with harmonization encourages members to establish, recognize, and apply common SPS measures whereby countries can adopt ISPMs or have a higher standard provided they scientifically justify such measures. For effective compliance, India needs to harmonize/develop national standards on phytosanitary measures (NSPM). The Directorate of Plant Protection Quarantine and Storage has developed 24 standards on PRA, operation manuals on import and export inspection, post-entry quarantine, import and export of biocontrol agents, certification of facilities for forced hot air treatment of packaging material requirements, establishment of pest-free areas, etc. The NSPM 19 on Survey and Monitoring Protocols for Establishment of PFA for brown rot (*Ralstonia solanacearum*) of potato provides guidelines for undertaking survey and monitoring of brown rot to identify PFA to facilitate export of table potato to European Union. Also, new NSPMs are required for phosphine fumigation, surveillance for regulated pests, consignments in transit, pest

K. Gupta
ICAR-National Bureau of Plant Genetic Resources, New Delhi, India

S. C. Dubey (✉)
Indian Council of Agricultural Research, New Delhi, India
e-mail: sunil.dubey@icar.gov.in

S. Kumar Chakrabarti et al. (eds.), *Sustainable Management of Potato Pests and Diseases*, https://doi.org/10.1007/978-981-16-7695-6_2

reporting and cold disinfestation treatments, etc. Besides, outdated laboratory manuals also need review and updating. The development of national standards for PRA is a vital breakthrough in terms of compliance to SPS Agreement in order to facilitate trade.

Keywords

Phytosanitary · Standards · ISPM · NSPM · Regulation · Pest risk analysis · Pest-free area

2.1 Background

Phytosanitary measures are adopted for both domestically produced and imported goods to protect human or animal life or health from food-borne risks, humans from animal and plant-carried diseases, plants from pests or diseases, and the territory of a country from the spread of a pests or diseases. Phytosanitary measures are not new to global trade. Because of the global concern that they might be used for trade protection, the Agreement on Application of Sanitary and Phytosanitary Measures (SPS Agreement) was negotiated during the Uruguay Round (https://www.wto.org/). The Agreement recognizes that countries have the right to maintain SPS measures for the protection of the population and the agricultural sector. However, it requires them to base their SPS measures on scientific principles and not to use them as disguised restrictions to trade. All countries maintain measures to ensure that food is safe for consumers and to prevent the spread of pests or diseases among animals and plants. These SPS measures can be in the forms of requiring products to come from a pest-free area, inspection of products, specific treatment or processing of products, setting of allowable maximum levels of pesticide residues, or permitted use of only certain additives in food. Sanitary (human and animal health) and phytosanitary (plant health) measures apply to both domestically produced food or local animal and plant diseases and products coming from other countries (Khetarpal and Gupta 2002).

The present situation, where consumers are increasingly requesting governments to be vigilant and make efforts to minimize the risks of marketing and importing products which could jeopardize the health of people or animals or harm agriculture, is the result of several episodes such as the case of contamination by dioxin of a large number of agricultural products (and of the spreading of contamination through international trade) – where consumers felt that health and safety were at risk. The increasing use of genetically modified seeds and the perception that GM crops might negatively affect human and animal health and the environment contribute to a strong request for strict measures in the SPS field. For developing countries, the best option therefore is to become capable of responding to the exigencies, which implies building up technical and infrastructural capabilities.

2.2 Role of Standards and Regulations

Countries require that domestically produced and imported goods both conform to regulations and adhere to standards. The number of standards and regulations is constantly increasing in most countries because of the expansion in volume, variety, and technical sophistication of products manufactured and traded. Nowadays, standards and regulations target at complying with a variety of aims and tasks (Gupta and Sathyapala 2020). Some of them are traditional—such as minimizing risks, providing information to consumers about the characteristics of products, providing information to producers about market needs and expectations, facilitating market transactions, raising efficiency, and contributing to economies of scale. Others are less—such as serving as benchmarks for technological capability and networking and increasing technology diffusion. Standards and regulations also respond to growing public demand, often raised by consumer associations and environmental groups, to have products with minimum detrimental effect on the environment and on health while conforming to quality requirements (Gupta and Khetarpal 2005).

Although standards and regulations can promote economic development and trade, they may also be used as tools to impede international trade and protect domestic producers, mainly through the following:

- Unjustified different requirements in different markets
- Unnecessary costly or time-consuming tests
- Duplicative conformity assessment procedures

The risk that countries resort to standards and regulations for domestic protection is increasing, since the trade barriers, such as tariffs, were reduced during the several rounds of multilateral negotiations. This risk is particularly high in the agricultural sector where reducing the level of protection by tariffs and non-tariff barriers would increase the importance of SPS measures as trade protection instruments. Probably, the major difficulty in dealing with standards and regulations is to distinguish those measures, which are justified by a legitimate goal from those, which are applied for protectionist purposes.

Compliance with regulations is mandatory; therefore, products, which do not comply with regulations, cannot be sold in a given market. On the other hand, standards are voluntary; therefore, no product can be stopped at the border or refused access to the domestic market because of noncompliance with standards. However, in practical terms, the distinction between standards and regulations is fading away, since conformity to standards is often a precondition for the acceptability of products by consumers and/or distributors.

The divergence of standards and regulations creates costs for international trade. In some cases, these costs are justified, since they arise from legitimate differences in societal preferences, technological development, and environmental and health conditions. In these cases, instead of harmonization of standards, mutual recognition of standards would provide a better solution. On the other hand, where divergences are not justified, international harmonization of standards seems to be a better solution. However, it is the efficiency and fairness of the international standard development process that is crucial for minimizing distortions to international trade. The benefits of harmonization may be impeded if the process is captured by special interests in order to exclude other market participants or if it is not adequately transparent.

Article 3 of the SPS Agreement encourages countries to use international standards as a basis for their regulations. It recognizes for food safety the standards, guidelines, and recommendations established by the Codex Alimentarius Commission (CAC), for animal health those developed by the World Organization for Animal Health (OIE), and for plant protection those developed under the auspices of the Secretariat of the International Plant Protection Convention (IPPC). For matters not covered by these organizations, standards developed by "other relevant international organizations open for membership to all Members," as identified by the SPS Committee, are recognized (IPPC 2021a). However, the Agreement does not specify the procedures that the relevant international organizations should adhere to in order to produce genuine international standards. As of March 2021, there are 44 adopted ISPMs (ISPM 30 being revoked), 29 Diagnostic Protocols, and 39 Phytosanitary Treatments (IPPC 2021b).

These international standards:

- Protect sustainable agriculture and enhance global food security
- Protect the environment, forests, and biodiversity
- Facilitate economic and trade development

2.3 The International Plant Protection Convention (IPPC)

The Secretariat of the IPPC was formed in 1993 and the standard-setting activity started the same year. The IPPC is responsible for phytosanitary standard setting and the harmonization of phytosanitary measures affecting trade. To date, 43 standards have been completed, and several others are at different stages of development. The Commission on Phytosanitary Measures has the responsibility for identifying the topics and priorities for the standard-setting activity. The IPPC is an international treaty for plant protection to which 184 countries currently adhere (https://www.ippc.int/en/countries/all/list-countries/). The Convention came into force in 1952 and has been amended once in 1979 and again in 1997.

2.3.1 International Standards on Phytosanitary Measures

International Standards for Phytosanitary Measures (ISPMs) are prepared by the Secretariat of the IPPC as part of the United Nations Food and Agriculture Organization's Global program of policy and technical assistance in plant quarantine. This program makes available to interested countries the standards, guidelines, and recommendations to achieve international harmonization of phytosanitary measures with the aim to facilitate trade and avoid the use of unjustified measures as barriers to trade. The standards once developed are periodically harmonized/ brought in line with the other more recently developed ISPMs and phytosanitary concepts within the framework of IPPC (Table 2.1).

The ISPMs are intended to harmonize phytosanitary measures used in international trade. They provide guidance to member countries to assist them in implementing national phytosanitary programs that fulfill requirements of the IPPC and contribute to harmonization between contracting parties. Although WTO member countries are required to base their phytosanitary measures on international standards where they exist, national phytosanitary measures do not necessarily violate the WTO-SPS Agreement if they differ from international standards. The application of phytosanitary measures that result in higher standards must be technically justified.

2.3.2 Process for Developing International Standards for Phytosanitary Measures (ISPMs)

The suggestion to develop an ISPM can originate in several places, including individuals, industry, National Plant Protection Organizations (NPPOs), and Regional Plant Protection Organizations (RPPOs). Priorities are established by the Commission on Phytosanitary Measures in consultation with the Secretariat.

Draft Stage NPPOs or RPPOs submit draft standards to the Secretariat of the IPPC. These drafts are reviewed, edited, and referred by the Secretariat to the Committee of Experts on Phytosanitary Measures (CEPM). Alternatively, the IPPC Secretariat might form an international working group or enlist experts to help draft a standard.

The CEPM considers the proposals and recommends action. The CEPM may also suggest that the secretariat develops new standards and guidelines. The IPPC secretariat may, after consultation with the CEPM, arrange for a technical working group or a consultant to modify draft standards if necessary. The CEPM continues to review progress on the document and recommends to the IPPC Secretariat the timing of submissions to governments for technical comments.

Consultation Stage Individual member countries and RPPOs review and comment on the draft. Input is provided to the CEPM through the IPPC Secretariat which determines the nature and extent of changes to be made in drafts based on comments from consultation. Acceptance of the redrafted standard by the CEPM results in submission of the standard to the Commission. If the CEPM recommends that is not relevant, the final text may be published (Fig. 2.1).

Approval Stage Endorsement by the commission results in final adoption of an ISPM. The standard is published and distributed by FAO. The member countries are expected to adhere to the standard.

Table 2.1 International standards for phytosanitary measures

ISPM	Title	Year of adoption/revision
ISPM 1	Phytosanitary principles for the protection of plants and the application of phytosanitary measures in international trade	Adopted in 1993, revised in 2006
ISPM 2	Framework for pest risk analysis	Adopted in 1995, revised in 2007
ISPM 3	Guidelines for the export, shipment, import and release of biological control agents and other beneficial organisms	Adopted in 1995, revised in 2005
ISPM 4	Requirements for the establishment of pest free areas	Adopted in 1995
ISPM 5	Glossary of phytosanitary terms	Updated as needed
	Supplement 1: Guidelines on the interpretation and application of the concept of "official control" and "not widely distributed" (2012)	
	Supplement 2: Guidelines on the understanding of "potential economic importance" and related terms including reference to environmental considerations (2003)	
	Appendix 1: Terminology of the convention on biological diversity in relation to the glossary of phytosanitary terms	2009
ISPM 6	Surveillance	Adopted in 1997, revised in 2018
ISPM 7	Phytosanitary certification system	Adopted in 1997, revised in 2011
ISPM 8	Determination of pest status in an area	Adopted in 1998
ISPM 9	Guidelines for pest eradication programmes	Adopted in 1998
ISPM 10	Requirements for the establishment of pest free places of production and pest free production sites	Adopted in 1999
ISPM 11	Pest risk analysis for quarantine pests	Adopted in 2001, revised in 2004 and 2013
ISPM 12	Phytosanitary certificates	Adopted in 2001, revised in 2011
	Appendix 1: Electronic phytosanitary certificates, information on standard XML schemas and exchange mechanisms	2014
ISPM 13	Guidelines for the notification of noncompliance and emergency action	Adopted in 2001
ISPM 14	The use of integrated measures in a systems approach for pest risk management	Adopted in 2002
ISPM 15	Regulation of wood packaging material in international trade	Adopted in 2002, revised in 2009, Annexes 1 and 2 revised in 2013 and in 2018
ISPM 16	Regulated non-quarantine pests: Concept and application	Adopted in 2002
ISPM 17	Pest reporting	Adopted in 2002

(continued)

Table 2.1 (continued)

ISPM	Title	Year of adoption/revision
ISPM 18	Guidelines for the use of irradiation as a phytosanitary measure	Adopted in 2003
ISPM 19	Guidelines on lists of regulated pests	Adopted in 2003
ISPM 20	Guidelines for a phytosanitary import regulatory system	Adopted in 2004, revised in 2017
	Annex 1: Arrangements for verification of compliance of consignments by the importing country in the exporting country	2017
ISPM 21	Pest risk analysis for regulated non-quarantine pests	Adopted in 2004
ISPM 22	Requirements for the establishment of areas of low pest prevalence	Adopted in 2005
ISPM 23	Guidelines for inspection	Adopted in 2005
ISPM 24	Guidelines for the determination and recognition of equivalence of phytosanitary measures	Adopted in 2005
ISPM 25	Consignments in transit	Adopted in 2006
ISPM 26	Establishment of pest-free areas for fruit flies (Tephritidae)	Adopted in 2006, revised in 2014 and 2015
	Appendix 1: Fruit fly trapping	2011
	Annex 2: Control measures for an outbreak within a fruit fly-pest free area	2014
	Annex 3: Phytosanitary procedures for fruit fly (Tephritidae) management	2015
ISPM 27	Diagnostic protocols for regulated pests	Adopted in 2006
	DP 1: Diagnostic protocol for Thrips palmi Karny	2010
	DP 2: Diagnostic protocol for plum pox virus	2012, revised in 2018
	DP 3: Diagnostic protocol for Trogoderma granarium everts	2012
	DP 4: Diagnostic protocol for Tilletia indica Mitra	2014
	DP 5: Diagnostic protocol for Phyllosticta citricarpa (McAlpine) aa on fruit	2014
	DP 6: Diagnostic protocol for Xanthomonas citri subsp. citri	2014
	DP 7: Diagnostic protocol for potato spindle tuber viroid	2015
	DP 8: Diagnostic protocol for Ditylenchus dipsaci and Ditylenchus destructor	2015
	DP 9: Diagnostic protocol for genus Anastrepha Schiner	2015
	DP 10: Diagnostic protocol for Bursaphelenchus xylophilus	2016
	DP 11: Diagnostic protocol for Xiphinema americanum sensu lato	2016
	DP 12: Diagnostic protocol for Phytoplasmas	2016
	DP 13: Diagnostic protocol for Erwinia amylovora	2016
	DP 14: Diagnostic protocol for Xanthomonas fragariae	2016
	DP 15: Diagnostic protocol for citrus tristeza virus	2016

(continued)

Table 2.1 (continued)

ISPM	Title	Year of adoption/revision
	DP 16: Diagnostic protocol for genus Liriomyza	2016
	DP 17: Diagnostic protocol for Aphelenchoides besseyi, A. fragariae and A. ritzemabosi	2016
	DP 18: Diagnostic protocol for Anguina spp.	2017
	DP 19: Diagnostic protocol for Sorghum halepense	2017
	DP 20: Diagnostic protocol for Dendroctonus ponderosae	2017
	DP 21: Diagnostic protocol for 'Candidatus Liberibacter solanacearum'	2017
	DP 22: Diagnostic protocol for fusarium circinatum	2017
	DP 23: Diagnostic protocol for Phytophthora ramorum	2017
	DP 24: Diagnostic protocol for tomato spotted wilt virus, impatiens necrotic spot virus and watermelon silver mottle virus	2017
	DP 25: Diagnostic protocol for Xylella fastidiosa	2018
	DP 26: Diagnostic protocol for Austropuccinia psidii	2018
	DP 27: Diagnostic protocol for Ips spp.	2018
	DP 28: Diagnostic protocol for Conotrachelus nenuphar	2018
	DP 29: Diagnostic protocol for Bactrocera dorsalis	2019
ISPM 28	Phytosanitary treatments for regulated pests	Adopted in 2007
	PT 1: Irradiation treatment for Anastrepha ludens	2009
	PT 2: Irradiation treatment for Anastrepha obliqua	2009
	PT 3: Irradiation treatment for Anastrepha serpentina	2009
	PT 4: Irradiation treatment for Bactrocera jarvisi	2009
	PT 5: Irradiation treatment for Bactrocera tryoni	2009
	PT 6: Irradiation treatment for Cydia pomonella	2009
	PT 7: Irradiation treatment for fruit flies of the family Tephritidae (generic)	2009
	PT 8: Irradiation treatment for Rhagoletis pomonella	2009
	PT 9: Irradiation treatment for Conotrachelus nenuphar	2010
	PT 10: Irradiation treatment for Grapholita molesta	2010
	PT 11: Irradiation treatment for Grapholita molesta under hypoxia	2010
	PT 12: Irradiation treatment for Cylas formicarius elegantulus	2011
	PT 13: Irradiation treatment for Euscepes postfasciatus	2011
	PT 14: Irradiation treatment for Ceratitis capitata	2011
	PT 15: Vapour heat treatment for Bactrocera cucurbitae on Cucumis melo var. reticulatus	2014
	PT 16: Cold treatment for Bactrocera tryoni on Citrus sinensis	2015
	PT 17: Cold treatment for Bactrocera tryoni on Citrus reticulata x C. sinensis	2015

(continued)

Table 2.1 (continued)

ISPM	Title	Year of adoption/revision
	PT 18: Cold treatment for Bactrocera tryoni on Citrus Limon	2015
	PT 19: Irradiation treatment for Dysmicoccus neobrevipes, Planococcus lilacinus and Planococcus minor	2015
	PT 20: Irradiation treatment for Ostrinia nubilalis	2016
	PT 21: Vapour heat treatment for Bactrocera melanotus and Bactrocera xanthodes on Carica papaya	2016
	PT 22: Sulfuryl fluoride fumigation treatment for insects in debarked wood	2017
	PT 23: Sulfuryl fluoride fumigation treatment for nematodes and insects in debarked wood	2017
	PT 24: Cold treatment for Ceratitis capitata on Citrus sinensis	2017
	PT 25: Cold treatment for Ceratitis capitata on Citrus reticulata x C. sinensis	2017
	PT 26: Cold treatment for Ceratitis capitata on Citrus Limon	2017
	PT 27: Cold treatment for Ceratitis capitata on Citrus paradisi	2017
	PT 28: Cold treatment for Ceratitis capitata on Citrus reticulata	2017
	PT 29: Cold treatment for Ceratitis capitata on Citrus Clementina	2017
	PT 30: Vapour heat treatment for Ceratitis capitata on Mangifera indica	2017
	PT 31: Vapour heat treatment for Bactrocera tryoni on Mangifera indica	2017
	PT 32: Vapour heat treatment for Bactrocera dorsalis on Carica papaya	2018
ISPM 29	Recognition of pest-free areas and areas of low pest prevalence	Adopted in 2007
ISPM 30:	Revoked. Establishment of areas of low pest prevalence for fruit flies (Tephritidae	Adopted in 2008. Incorporated as an annex to ISPM 35 in 2018
ISPM 31	Methodologies for sampling of consignments	Adopted in 2008
ISPM 32	Categorization of commodities according to their pest risk	Adopted in 2009
ISPM 33	Pest-free potato (solanum spp.) micro-propagative material and minitubers for international trade	Adopted in 2010
ISPM 34	Design and operation of post-entry quarantine stations for plants	Adopted in 2010
ISPM 35	Systems approach for pest risk management of fruit flies (Tephritidae)	Adopted in 2012
ISPM 36	Integrated measures for plants for planting	Adopted in 2012

(continued)

Table 2.1 (continued)

ISPM	Title	Year of adoption/revision
ISPM 37	Determination of host status of fruit to fruit flies (Tephritidae)	Adopted in 2016
ISPM 38	International movement of seeds	Adopted in 2017
ISPM 39	International movement of wood	Adopted in 2017
ISPM 40	International movement of growing media in association with plants for planting	Adopted in 2017
ISPM 41	International movement of used vehicles, machinery and equipment	Adopted in 2017
ISPM 42	Requirements for the use of temperature treatments as a phytosanitary measures	Adopted in 2018
ISPM 43	Requirements for the use of fumigation as a phytosanitary measure	Adopted in 2019

The CAC and OIE adopt standards, guidelines, and recommendations by a simple majority of votes cast, when adoption by consensus proves to be impossible to achieve. Because of the simple majority rule, some codex standards were adopted or rejected by a relatively small majority with a large number of countries not voting in favor, for example, the standard on maximum residue limits for growth hormones was approved by 33 votes in favor, 29 against, and 7 abstentions. This has led to questions regarding genuine international nature of the standards. However, in case of IPPC, a two-thirds majority for the establishment of a standard and a procedure for adoption of a standard as given above is being followed (IPPC 2011).

The Standards Committee of the IPPC also recognizes the relationships between the ISPMs. For instance, for detailed risk identification and communication components, ISPM-2 is to be referred to and ISPM 11 or 21 for risk assessment and management components. Also, the ISPM-3 on the code of conduct for import and release of exotic biocontrol agents would use the hazard identification component of ISPM-2 and the risk assessment component of ISPM-11. These standards act as broad guidelines based on which countries need to either develop their own national standards or adopt the international standard for trade.

2.4 Phytosanitary Standards Relevant to Exchange of Potato

2.4.1 ISPM 1: Phytosanitary Principles for the Protection of Plants and the Application of Phytosanitary Measures in International Trade (Adopted in 1993, Revised in 2006)

This standard was earlier named "Principles of plant quarantine as related to international trade" and described the general and specific principles of plant quarantine as related to international trade to facilitate the process of developing international standards for plant quarantine. The objective is to reduce or eliminate the use of unjustifiable phytosanitary measures as barriers to trade. The FAO Conference

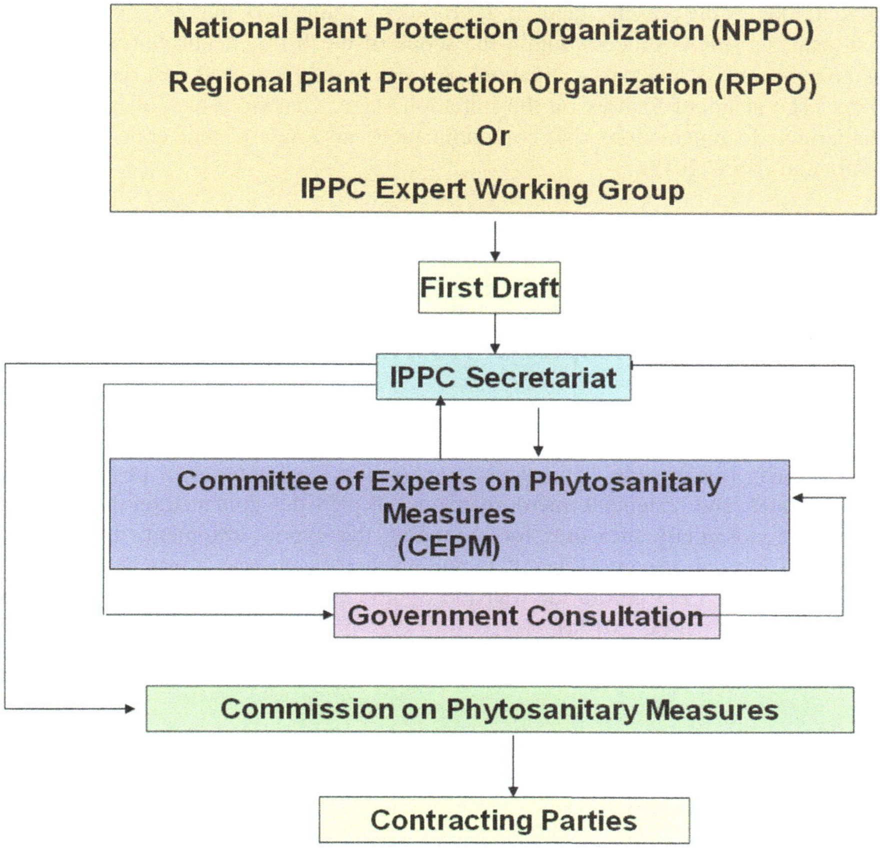

Fig. 2.1 Flowchart of the process for developing ISPMs

adopted this ISPM in 1993 before the completion of the GATT Uruguay Round negotiations that resulted in the SPS Agreement and the establishment of WTO. The adoption and coming into force of the SPS Agreement in 1995 and the adoption of IPPC in 1997 represent further development of the original concepts that formed the basis for ISPM No.1. Hence, it has subsequently been revised and updated (IPPC 2016a).

2.4.2 ISPM 2: Framework for Pest Risk Analysis (Adopted in 1995, Revised in 2007)

This standard was first adopted by the Twenty-Eighth Session of the FAO Conference in November 1995 as Guidelines for pest risk analysis. This first revision was adopted by the Second Session of the Commission on Phytosanitary Measures in

March 2007 as the present standard. It provides a framework that describes the pest risk analysis (PRA) process within the scope of the IPPC. It introduces the three stages of pest risk analysis – initiation, pest risk assessment, and pest risk management. The standard focuses on the initiation stage. Generic issues of information gathering, documentation, risk communication, uncertainty, and consistency are addressed (IPPC 2019a).

2.4.3 ISPM 3: Guidelines for the Export, Shipment, Import, and Release of Biological Control Agents and Other Beneficial Organisms (Adopted in 1995, Revised in 2005)

The standard describes responsibilities of authorities of governments, importers, and exporters for the import and release of biological control agents capable of self-replication (parasitoids, predators, parasites, nematodes, phytophagous invertebrates, and beneficial microorganisms). It provides guidance on the application of phytosanitary measures for regulating the export, shipment, import, and release of biological control agents including their packaging, formulation, etc. It was adopted in 1995 before the revision of IPPC in 1997 and was reviewed and updated in 2005 (IPPC 2017a).

2.4.4 ISPM 4: Requirements for the Establishment of Pest-Free Areas (Adopted in 1995)

The standard outlines the requirements for the establishment and use of pest-free areas (PFAs) as a risk management option for phytosanitary certification of plants and plant products and other regulated articles exported from the PFA or to support the scientific justification for phytosanitary measures taken by an importing country for protection of an endangered PFA. A "pest-free area" is "an area in which a specific pest does not occur as demonstrated by scientific evidence and which, where appropriate, this condition is being officially maintained." Broadly, the PFAs are defined as three types:

- An entire country
- A part of a country in which a limited infested area is present
- An uninfected part of a country situated within a generally infested area

The main components or stages considered in the establishment and subsequent maintenance of a PFA are suitable systems to establish freedom, phytosanitary measures to maintain that freedom, and subsequent checks to verify that freedom has been maintained. All this needs to be properly documented (IPPC 2017b).

2.4.5 ISPM 5: Glossary of Phytosanitary Terms (Updated as Needed)

- Supplement 1: Guidelines on the interpretation and application of the concept of "official control" and "not widely distributed" (2012).
- Supplement 2: Guidelines on the understanding of "potential economic importance" and related terms including reference to environmental considerations (2003).
- Appendix 1: Terminology of the Convention on Biological Diversity in Relation to the Glossary of Phytosanitary Terms (2009)

The Glossary of phytosanitary terms is a reference standard listing harmonized terms, definitions, and abbreviations in each of the FAO languages. It also provides cross-references and includes supplements where necessary to explain the interpretations and applications of certain terms. The first edition of the standard was formulated in 1990, and subsequently, it has been revised six times to update and add new definitions. A small working group of five experts along with the secretariat meet annually to review changes proposed and recommend modifications so that the understanding of phytosanitary systems in different geographical regions and languages is improved. The purpose of this standard is to increase clarity and consistency in the use and understanding of terms and definitions which are used by NPPOs for official phytosanitary purposes, in phytosanitary legislation and regulations, as well as for official information exchange. Presently, it defines >100 terms related to plant protection with a multilingual index.

Supplement 1 provides guidance on the official control of regulated pests, and on determination of when a pest is considered to be present but not widely distributed, for the decision on whether a pest qualifies as a quarantine pest. The purpose of this supplement is to describe more precisely the interpretation of the concept of official control and its application in practice for quarantine pests that are present in an area as well as for regulated non-quarantine pests and the concept of "present but not widely distributed and under official control" for quarantine pests (IPPC 2019b).

Supplement 2 provides the background and other relevant information to clarify potential economic importance and related terms, so that such terms are clearly understood and their application is consistent with the International Plant Protection Convention (IPPC) and the International Standards for Phytosanitary Measures (ISPMs). These also show the application of certain economic principles as they relate to the IPPC's objectives, in particular, in protecting uncultivated/unmanaged plants, wild flora, habitats, and ecosystems with respect to invasive alien species that are pests. These also clarify, with respect to pests, that the scope of the IPPC covers the protection of cultivated plants in agriculture, horticulture and forestry, uncultivated/unmanaged plants, wild flora, habitats, and ecosystems.

Appendix 1 gives explanations on terms and definitions available from the CBD which have been noted to be based on concepts different from those of the IPPC, so similar terms are given distinctly different meanings. The CBD terms and definitions could not accordingly be used directly in the Glossary. It was decided by IPPC to present these terms and definitions in the separate Appendix to the Glossary, providing explanations of how they differ from IPPC terminology. In relation to each term considered, the CBD definition is first provided. This is placed alongside an "Explanation in IPPC context," in which, as usual, Glossary terms (or derived forms of Glossary terms) are shown in bold. The explanations constitute the main body of this Appendix.

2.4.6 ISPM 6: Surveillance (Adopted in 1997, Revised in 2018)

This standard describes the components of survey and monitoring systems for the purpose of pest detection and the supply of information for use in PRA, the establishment of PFAs, and appropriate preparation of pest lists. The collection and recording of pest information is fundamental to the countries to appropriately justify the SPS measures taken on the basis of PRA and also for establishing PFAs. The implication is that the NPPO should be in a position to validate declarations of the absence or limited distribution of quarantine pests. There are two major types of surveillance systems described: general surveillance whereby information on particular pests which are of concern is gathered from many sources and specific surveys by which NPPOs obtain information on pests of concern on specific sites in an area over a defined period of time. Emphasis is also laid on the training of the personnel for surveys, sampling methods, preservation, transportation, and record keeping. Transparency should be adopted by reporting/publication of reports of pest presence, distribution, or absence as derived from the surveys conducted (IPPC 2018a).

This standard may contribute to the protection of biodiversity and the environment by helping countries develop systems to provide reliable and well-structured information on the presence, absence, or distribution of pests in an area and information about hosts or commodities as pathways. These pests could include organisms relevant to biodiversity (e.g., invasive alien species).

2.4.7 ISPM 7: Phytosanitary Certification System (Adopted in 1997, Revised in 2011)

The national system for issuance of valid and credible phytosanitary certificates is described under this standard. The basic elements of the phytosanitary certification include ascertaining the relevant phytosanitary requirements of the importing country including the import permit requirements, verification of the consignment conformity to the requirements at the time of certification, and finally issuing the phytosanitary certificate. To fulfil such requirements as those for a certification system are legal authority and management responsibility, including resources, documentation, communication, and review mechanism. Also, the NPPO of the exporting country should maintain a system for documenting the relevant certification procedures. Guidance and instruction material for all procedures should be available. Records of all activities leading to issuance of phytosanitary certificates need to be maintained (IPPC 2016b).

2.4.8 ISPM 8: Determination of Pest Status in an Area (Adopted in 1998)

This standard describes the content of a pest record and the use of such pest records and other information in the determination of pest status in an area. All importing and exporting countries need the information concerning the status of pests for risk analyses, the establishment of and the compliance with import regulations, and establishment and maintenance of PFAs. As per the standard, pest record information should include the presence or absence of a pest, the time and location of observation, hosts, damage caused, and any other relevant information. The determination of pest status also requires expert judgement concerning the information available on the present-day occurrence of a pest from pest records and other relevant data from surveillance, publications, databases, etc. The pest status as described by the standard is in three categories, the presence either in all or few parts, the absence that could be due to eradication also, and transience of pest as present in isolated pockets which may survive and require application of phytosanitary measures (IPPC 2017c).

2.4.9 ISPM 9: Guidelines for Pest Eradication Programs (Adopted in 1998)

This standard describes the components of a pest eradication program, which can lead to the establishment, or re-establishment of the absence of a pest in an area. It provides guidance on the development of a pest eradication program and for reviewing the procedures of an existing eradication program. Here the pests are considered as newly entered into the area where eradication is undertaken and emergency eradication measures may be needed. However, eradication program may also be directed towards established exotic pests or indigenous pests in defined areas. The eradication program development is broadly outlined to include firstly gathering general information and planning the program. This would include evaluation of pest reports, contingency plans to address specific pest or pest groups, reporting requirements, and information sharing with broader audiences such as growers, residents, and local governments. Secondly, the decision to undertake an eradication program should be based on evaluation of the circumstances of detection of pests, their identification, the risk identified by a pest-initiated PRA, estimated present, potential distribution, and feasibility of the eradication program. The third or the main part would include the actual process of establishment of a management team followed by the conduct of the eradication program. The three main activities to be included in the program are as follows:

- Surveillance, to fully investigate the distribution of the pest
- Containment, to prevent the spread of the pest
- Treatment, to eradicate the pest when it is found

The direction and coordination would be under the management authority usually the NPPO. Ensuring the criteria to determine when eradication has been achieved and that appropriate documentation, which is the fourth requirement, has been done to ensure that records have been maintained supporting all stages of the eradication process. The entire program should also be subject to periodic review to analyze and assess information gathered, to check that objectives are being achieved (IPPC 2016c).

2.4.10 ISPM 10: Requirements for the Establishment of Pest-Free Places of Production and Pest-Free Production Sites (Adopted in 1999)

The standard describes the requirements for the establishment and use of pest-free places of production and pest-free production sites as risk management options for meeting phytosanitary requirements for the export of plants, plant products, and other regulated articles. The standard is meant to facilitate export by using the concept of pest freedom to provide the assurance to importing countries that plants and plant products are free from a specific pest or pests and meet the phytosanitary requirements of the importing country. In cases where a defined portion of a place of production is managed as a separate unit and can be managed as pest free, it may be regarded as a pest-free production site. The use of such sites depends on and the use of criteria concerning the biology of the pest, the characteristics of the place of production, and the operational capabilities of the producer and is the responsibility of the NPPO. As per the standard, the requirements for the establishment and maintenance of the pest-free place of production or site as a phytosanitary measure by the NPPO include the following:

- Systems to establish and maintain pest freedom
- Verification that pest freedom has been attained or maintained, product identity, consignment integrity, and phytosanitary security
- Where necessary, establishment and maintenance of appropriate buffer zone on the basis of distance over which the pest is likely to spread naturally during the course of the growing season

Documentation and maintenance of records concerning the measures taken along with periodic review and audit procedures undertaken by the NPPO are essential to support the assurance of pest freedom and system appraisal (IPPC 2016d).

2.4.11 ISPM 11: Pest Risk Analysis for Quarantine Pests (Adopted in 2001, Revised in 2004 and 2013)

The objectives of a PRA are, for a specific area, to identify pests and/or pathways of quarantine concern and evaluate their risk, to identify endangered areas, and if appropriate to identify risk management options. This came up in 2001 when the IPPC realized the inadequacy of the previous standard ISPM 2 on pest risk analysis to tackle all the issues related to a PRA. PRA for quarantine pest also follows the process defined by the three stages:

- Stage I (initiating the process) involves identifying the pest(s) and pathways that are of quarantine concern and should be considered for risk analysis in relation to the identified PRA area.
- Stage II (risk assessment) begins with the categorization of individual pests to determine whether the criteria for a quarantine pest are satisfied. Risk assessment continues with an evaluation of the probability of pest entry, establishment, and spread and of their potential economic consequences.
- Stage III (risk management) involves identifying management options for reducing the risks identified at the second stage. These are evaluated for efficacy, feasibility, and impact in order to select those that are appropriate.

The requirements for all the three stages have been described in detail in the standard, and the fourth stage of documentation of the whole process from initiation to pest risk management should be sufficiently recorded, so that whenever a review or a dispute arises, the source of information and the rationale used in reaching the management decision can be clearly demonstrated. In April 2004, the standard was revised to include analysis of environmental risks by weeds/invasive plants and living modified organisms which have been integrated into the various stages described above. The various annexes give the details of the scope of IPPC with regard to the environmental risks, PRA for living modified organisms (LMOs), and determination of its potential as a pest (IPPC 2019c).

2.4.12 ISPM 12: Phytosanitary Certificates (Adopted in 2001, Revised in 2011)—Appendix 1: Electronic Phytosanitary Certificates, Information on Standard XML Schemas and Exchange Mechanisms (2014)

This standard describes principles and guidelines for the preparation and issue of phytosanitary certificates for export and re-export. Model certificates are provided in the annex of the revised text of IPPC adopted in 1997 and are appended to this standard for reference. It also contains detailed explanations on the various components of the model certificates indicating the information needed for their appropriate completion.

Details are also given on how phytosanitary certificates could accompany the consignment or may be transmitted by mail or other means, or where agreed between countries, NPPOs may use electronic phytosanitary certificates, using standardized language, structure of the message, and exchange protocols.

Special consideration is given to situations of re-export, particularly when the issuance of a phytosanitary certificate for export is not required by the country of re-export and when specific phytosanitary measures need to be conducted in the country of origin (IPPC 2017d).

2.4.13 ISPM 13: Guidelines for the Notification of Noncompliance and Emergency Action (Adopted in 2001)

This standard describes the actions to be taken by countries regarding the notification of the following:

- A significant/specific instance of failure of an imported consignment to comply with specified phytosanitary/documentary requirements by detection of specified regulated pests/phytosanitary certification.
- An emergency action taken on the detection in an imported consignment of a regulated pest not listed as being associated with the commodity from the exporting country or posing a potential phytosanitary threat.

There is a provision under the revised IPPC for countries to report significant instances of noncompliance of imported consignments with phytosanitary requirements, including those related to documentation or to report emergency action. The required information includes the reference number, date of notification, identity of the NPPOs of the importing and exporting countries, identity of the consignment and date of first action, reasons for action taken, etc. this should also include supporting information such as diagnostic results, pest association, etc. This allows the importing country to investigate any new phytosanitary situation where emergency action is taken to ascertain justification of phytosanitary action taken and if any changes are required for them. Also, for consignments in transit with instances of noncompliance, the action to be taken by the transit country would be notifying to the country of final destination (IPPC 2016e).

2.4.14 ISPM 14: The Use of Integrated Measures in a Systems Approach for Pest Risk Management (Adopted in 2002)

This standard provides for the development and evaluation of integrated measures in a systems approach as an option for pest risk management under the relevant international standards for PRA designed to meet phytosanitary requirements for the import of plants, plant products, and other regulated articles. The ISPM on PRA provides general guidelines on measures for pest risk management. In systems approach, integration of various measures of risk management is integrated in a defined manner to meet the appropriate level of protection of an importing country which acts independently with a cumulative effect. They can also be developed to provide phytosanitary protection in situations where no single measure is available. It provides an alternative to procedures such as disinfestation treatments or replaces more restrictive measures like prohibition. The application of critical control points in a systems approach may be useful to identify and evaluate points in a pathway where specified pest risks can be reduced and monitored. The development and evaluation of such a systems approach may use quantitative or qualitative methods. The decisions regarding the acceptability of a systems approach lies with the importing country, subject to consideration of technical justification, minimal impact, transparency, non-discrimination, equivalence, and operational feasibility. The many options that can be taken include those at pre-planting, pre-harvest, harvest, post-harvest treatments and handling, transportation, and distribution.

Various types of systems approaches are described in the standard with details on its development, evaluation, etc. It also outlines the responsibilities of both importing as well as exporting countries. The procedure for critical control including safety and hazard evaluation is to effectively monitor the system used (IPPC 2019d).

2.4.15 ISPM 16: Regulated Non-quarantine Pests: Concept and Application (Adopted in 2002)

This standard describes the concept of regulated non-quarantine pests (RNQPs) and identifies their characteristics. It describes the application of the concept in practice and the relevant elements for regulatory systems. As per the standard pests that are not quarantine pests may be subjected to phytosanitary measures because their presence in plants for planting results in economically unacceptable impacts. They have been defined under the IPPC (1997) ArticleVI.2, "contracting parties shall not require phytosanitary measures for non-regulated pests," which also distinguishes it from the quarantine pests (Table 2.2).

The standard allows the members to apply the concept of RNPQs if it is technically justified, based on risk analysis, managed risk, minimal impact,

Table 2.2 Comparison of quarantine pests and RNQPs

Defining criteria	Quarantine pest	RNQP
Pest status	Absent or of limited distribution	Present and may be widely distributed
Pathway	Phytosanitary measures for any pathway	Phytosanitary measures only on plants for planting
Economic impact	Impact is predicted	Impact is known
Official control	Under official control if present with the aim of eradication or containment	Under official control with respect to the specified plants for planting with the aim of suppression

equivalence, non-discrimination, and transparency. The NPPO must take into consideration the above points prior to designating any pest as RNQPs. In addition, some specific issues such as host-pest interactions and the existence of certification programs (e.g., seed certification) for plants for planting may be considered. Various options on phytosanitary action that can be taken for noncompliance with phytosanitary requirements for RNQPs are also listed (IPPC 2016f).

2.4.16 ISPM 17: Pest Reporting (Adopted in 2002)

This standard described the responsibilities of and requirements for contracting parties in reporting the occurrence, outbreak, and spread of pests in areas, an obligation under the IPPC (1997, Article VIII, Ia). It also provides guidance on reporting successful eradication of pests and establishment of PFAs. The IPPC requires countries to communicate immediate and potential dangers on the occurrence, outbreak, and spread of pests, and the NPPOs have the responsibility to collect pest information by surveillance and to verify pest records. Pest records should contain information on identification of pests, location, pest status, and immediate or potential danger. Reports of successful eradication, the establishment of PFAs, and other information may also be provided utilizing the same reporting procedure. Details on the content of reports have also been elaborated in the standard. The countries are advised to follow good reporting practices set out in ISPM-8-(determination of pest status in an area), and the reports should not be confidential. Countries may have in place requirements regarding confidentiality of certain information, e.g., identity of growers. The NPPOs should undertake periodic review of their pest surveillance and reporting systems to ensure that they are meeting their reporting obligations and to identify possibilities for improving reliability and timeliness. The national pest surveillance and reporting systems should be adequately described and documented so that this information is available upon request (IPPC 2017e).

2.4.17 ISPM 18: Guidelines for the Use of Irradiation as a Phytosanitary Measure (Adopted in 2003)

This standard provides technical guidance on application of ionizing radiation as a phytosanitary treatment for regulated pests or articles. This does not include treatments used for production of sterile organisms for pest control/sanitary treatments (food safety and animal health)/preservation or improvement of commodity quality (e.g., shelf-life extension) or inducing mutagenesis. The NPPOs should be assured that the efficacy of the treatment is scientifically demonstrated for the regulated pest(s) of concern and the required response. Application of the treatment requires dosimetry and dose mapping to ensure that the treatment is effective in particular facilities and with specific commodity configurations. The NPPO is responsible for ensuring that facilities are appropriately designed for phytosanitary treatments. List of irradiation treatments approved for specified applications includes radioactive isotopes (gamma rays from cobalt-60 or cesium-137), electrons generated from machine source (up to 10 Mev), and X-rays (up to 5 Mev). Record keeping by the treatment facility and documentation are a requirement. A checklist for facility approval requires the premises with appropriate building, equipment, safety requirements, trained and competent personnel, product handling, storage and segregation, conformity and proper dosage of irradiation treatment, identification by proper packaging and labeling, and documentation (IPPC 2019e).

2.4.18 ISPM 19: Guidelines on Lists of Regulated Pests (Adopted in 2003)

This standard describes the procedures to prepare, maintain, and make available lists of regulated pests. The IPPC requires the NPPOs to establish, update, and make available lists of regulated pests to specify all currently regulated pests for which phytosanitary measures may be taken. Specific lists are also provided by the NPPOs of exporting countries as the means to specify the regulated pests for the certification of particular commodities. All quarantine pests including those subjects to provisional or emergency measures and regulated non-quarantine pests should be listed. Information required on listed pests includes name of pest, categories of regulated pests to which it belongs, and their association with regulated article(s). Updating of the lists is required when pests are added or deleted or when required information or supplementary information changes (IPPC 2016g).

2.4.19 ISPM 20: Guidelines for a Phytosanitary Import Regulatory System (Adopted in 2004, Revised in 2017)

2.4.19.1 Annex 1 Arrangements for verification of compliance of consignments by the importing country in the exporting country (2017)

The standard describes the structure and operation of a phytosanitary import regulatory system and the rights, obligations, and responsibilities which should be considered in establishing, operating, and revising the system. Its objective is to prevent introduction of quarantine pests or limit the entry of regulated non-quarantine pests with imported commodities and other regulated articles. The NPPO has the responsibilities identified under Article IV.2 of IPPC (1997) relating to import including surveillance, inspection, disinfestation or disinfection, the conduct of PRA, and training and development of staff. These responsibilities involve related functions in areas such as administration, audit and compliance checking, action taken on noncompliance, emergency action, authorization of personnel, and settlement of disputes (IPPC 2019f).

2.4.20 ISPM 21: Pest Risk Analysis for Regulated Non-quarantine Pests (Adopted in 2004)

This standard provides guidelines for conducting PRA for RNQPs to identify pests associated with plants for planting, to evaluate their risk, and if appropriate to identify risk management options to achieve a tolerance level. PRA for RNQPs follows the same three stages as described in the other ISPMs (No. 2 and11) on PRA. The difference being the risk assessment to determine if the plants for planting are the main source for pest infestation and if economic impacts of the pest on the intended use of those plants for planting are unacceptable. The risk management may identify tolerance levels to avoid the unacceptable economic impacts at stage 2 and management options to achieve that tolerance (IPPC 2019g).

2.4.21 ISPM 22: Requirements for the Establishment of Areas of Low Pest Prevalence (Adopted in 2005)

This standard describes the requirements and procedures for the establishment of areas of low pest prevalence (ALPP) for regulated pests in an area and, to facilitate export, for pests regulated by an importing country only. This includes the identification, verification, maintenance, and use of those ALPPs. The establishment of an area of low pest prevalence is a pest management option used to maintain or reduce a pest population below a specified level in an area which may be used to facilitate exports or to limit pest impact in the area.

A specified low pest level should be determined taking into consideration the overall operational and economic feasibility of establishing a program to meet or maintain this level and the objective for which an ALPP is to be established by the NPPO. ALPPs may be established and maintained for regulated pests or for pests regulated by an importing country only.

Surveillance of the relevant pest should be conducted according to appropriate protocols (ISPM 6 (Guidelines for surveillance)). Additional phytosanitary procedures may be required to establish and maintain an ALPP.

Once established, the ALPP should be maintained by the continuation of the measures used for its establishment and the necessary documentation and verification procedures. In most cases, an official operational plan which specifies the required phytosanitary procedures is needed. If there is a change in the status of the ALPP, a corrective action plan should be initiated (IPPC 2016h).

2.4.22 ISPM 23: Guidelines for Inspection (Adopted in 2005)

This standard describes procedures for the inspection of consignments of plants, plant products, and other regulated articles at import and export. It is focused on the determination of compliance with phytosanitary regulations, based on visual examination, documentary checks, and identity and integrity checks.

National plant protection organizations (NPPOs) have the responsibility for "the inspection of consignments of plants and plant products moving in international traffic and, where appropriate, the inspection of other regulated articles, particularly with the object of preventing the introduction and/or spread of pests" (Article IV.2 (c) of the IPPC).

Inspectors determine compliance of consignments with phytosanitary regulations, based on visual examination for detection of pests and regulated articles, and documentary checks and identity and integrity checks. The result of inspection should allow an inspector to decide whether to accept, detain, or reject the consignment, or whether further analysis is required.

NPPOs may determine that consignments should be sampled during inspection. The sampling methodology used should depend on the specific inspection objectives (IPPC 2019h).

2.4.23 ISPM 24: Guidelines for the Determination and Recognition of Equivalence of Phytosanitary Measures (Adopted in 2005)

This standard describes the principles and requirements that apply for the determination and recognition of equivalence of phytosanitary measures. It also describes a procedure for equivalence determinations in international trade.

Equivalence is one of the IPPC basic principles (ISPM 1 (Phytosanitary principles for the protection of plants and the application of phytosanitary measures in international trade)).

Equivalence generally applies to cases where phytosanitary measures already exist for a specific pest associated with trade in a commodity or commodity class. Equivalence determinations are based on the specified pest risk, and equivalence may apply to individual measures, a combination of measures, or integrated measures in a systems approach.

A determination of equivalence requires an assessment of phytosanitary measures to determine their effectiveness in mitigating a specified pest risk. The determination of equivalence of measures may also include an evaluation of the exporting contracting party's phytosanitary systems or programs that support implementation of those measures. Normally, the determination involves a sequential process of information exchange and evaluation and is generally an agreed procedure between importing and exporting contracting parties. Information is provided in a form that allows the evaluation of existing and proposed measures for their ability to meet the importing contracting party's appropriate level of protection.

The exporting contracting party may request information from the importing contracting party on the contribution that its existing measures make to meet its appropriate level of protection. The exporting contracting party may propose an alternative measure, indicating how this measure achieves the required level of protection, and this is evaluated by the importing contracting party. In some cases, such as where technical assistance is provided, importing contracting parties may make proposals for alternative phytosanitary measures. Contracting parties should endeavor to undertake equivalence determinations and to resolve any differences without undue delays (IPPC 2017f).

2.4.24 ISPM 25: Consignments in Transit (Adopted in 2006)

This standard describes procedures to identify, assess, and manage pest risks associated with consignments of regulated articles which pass through a country without being imported, in such a manner that any phytosanitary measures applied in the country of transit are technically justified and necessary to prevent the introduction into and/or spread of pests within that country.

International trade may involve the movement of consignments of regulated articles which pass through a country without being imported, under Customs1 control. Such movements may present a pest risk to the country of transit. Contracting parties to the IPPC may apply measures to consignments in transit through their territories (Article VII.1(c) and VII.2(g) of the IPPC), provided that the measures are technically justified and necessary to prevent the introduction and/or spread of pests (Article VII.4 of the IPPC).

This standard provides guidelines by which the national plant protection organization (NPPO) of the country of transit may decide which movements require intervention of the NPPO and are subject to the application of phytosanitary measures, and if so, the type of phytosanitary measures to be applied.

In such cases, the responsibilities and elements of the transit system are described, together with the need for cooperation and communication, non-discrimination, review, and documentation (IPPC 2016i).

2.4.25 ISPM 27: Diagnostic Protocols for Regulated Pests (Adopted in 2006)

This standard provides guidance on the structure and content of the International Plant Protection Convention (IPPC) diagnostic protocols for regulated pests. The protocols describe procedures and methods for the official diagnosis of regulated pests that are relevant for international trade. They provide at least the minimum requirements for reliable diagnosis of regulated pests.

This standard sets the framework for the content of diagnostic protocols, their purpose and use, their publication and development. Diagnostic protocols for 29 regulated pests are included as annexes to this standard. Information relevant for diagnosis is provided in the diagnostic protocol on the specified regulated pest, its taxonomic position, and the methods to detect and identify it. Diagnostic protocols contain the minimum requirements for reliable diagnosis of the specified regulated pests and provide flexibility to ensure that methods are appropriate for use in the full range of circumstances. The methods included in diagnostic protocols are selected on the basis of their sensitivity, specificity, and reproducibility, and information related to these factors is provided for each of these methods. Detailed information and guidance for the detection of pests is provided on, for example, signs and/or symptoms associated with the pest, illustrations (where appropriate), developmental stages of the pest, and methods for detecting the pest in a commodity, as well as methods for extracting, recovering, and collecting the pests from plants. Information and guidance for the identification of pests includes detailed information on morphological and morphometric methods, methods based on biological properties, and methods based on biochemical and molecular properties of the pest. Furthermore, detailed guidance is provided on the records that should be kept. Diagnostic protocols are intended to be used by laboratories performing pest diagnosis as part of phytosanitary measures. They are subject to review and amendment to take into account new developments in pest diagnosis. The standard also provides guidance on how these protocols are initiated, developed, reviewed, and published. Under this ISPM, diagnostic protocols for 29 different pests have been notified (IPPC 2016j).

2.4.26 ISPM 28: Phytosanitary Treatments for Regulated Pests (Adopted in 2007)

Harmonized phytosanitary treatments support efficient phytosanitary measures in a wide range of circumstances and enhance the mutual recognition of treatment efficacy.

This standard presents as annexes phytosanitary treatments evaluated and adopted by the Commission on Phytosanitary Measures (CPM). It also describes the requirements for submission and evaluation of the efficacy data and other relevant information on a phytosanitary treatment that can be used as a phytosanitary measure and that will be annexed to this standard after its adoption. The treatments are for the control of regulated pests on regulated articles, primarily those moving in international trade. The adopted treatments provide the minimum requirements necessary to control a regulated pest at a stated efficacy. The scope of this standard does not include issues related to pesticide registration or other domestic requirements for approval of treatments (e.g., irradiation).

Annexes to this standard contain those phytosanitary treatments which have been adopted by the CPM. National plant protection organizations (NPPOs) and regional plant protection organizations (RPPOs) may submit data and other information for the evaluation of efficacy, feasibility, and applicability of treatments. The information should include a detailed description of the treatment, including efficacy data, the name of a contact person, and the reason for the submission. Treatments that are eligible for evaluation include mechanical, chemical, irradiation, physical, and controlled atmosphere treatments. The efficacy data should be clear and should preferably include data on the treatment under laboratory or controlled conditions as well as under operational conditions. Information on feasibility and applicability of the proposed treatment(s) should include items on cost, commercial relevance, level of expertise required to apply the treatment, and versatility (IPPC 2016k). Submissions with complete information will be considered by the Technical Panel on Phytosanitary Treatments (TPPT), and if the treatment is deemed acceptable, it will be recommended to the CPM for adoption. Under this standard, there are 32 different phytosanitary treatments notified against various pests on different commodities.

2.4.27 ISPM 29: Recognition of Pest-Free Areas and Areas of Low Pest Prevalence (Adopted in 2007)

This standard provides guidance and describes a procedure for the bilateral recognition of pest-free areas and areas of low pest prevalence. This standard does not include specified timelines for the recognition procedure. This standard also provides some considerations regarding pest-free places of production and pest-free production sites.

Recognition of pest-free areas (PFAs) and areas of low pest prevalence (ALPPs) is a technical and administrative process to achieve acceptance of the status of the relevant pest in a delimited area. Technical requirements for establishment of PFAs and ALPPs, as well as certain elements relating to recognition, are addressed in other ISPMs. In addition, many principles of the IPPC are relevant. Contracting parties to the IPPC should proceed with a recognition process without undue delay. The process should be applied without discrimination between contracting parties.

Contracting parties should endeavor to maintain transparency in all aspects of the recognition process. The procedure described in this standard deals with those cases where detailed information and verification may be required, such as in areas in which eradication or suppression of a pest has recently been achieved. This procedure includes the various steps for the contracting parties, viz., request for recognition, acknowledgement of receipt of the request and the accompanying information package, description of the process, assessment of the information provided, communication of the results of assessment, and provision of official recognition.

Both exporting and importing contracting parties have specific responsibilities relating to the recognition of PFAs and ALPPs. The recognition process should be sufficiently documented by contracting parties (IPPC 2017g).

2.4.28 ISPM 31: Methodologies for Sampling of Consignments (Adopted in 2008)

This standard provides guidance to NPPOs in selecting appropriate sampling methodologies for inspection or testing of consignments to verify compliance with phytosanitary requirements.

This standard provides the statistical basis for, and complements, ISPM 20 (Guidelines for a phytosanitary import regulatory system) and ISPM 23 (Guidelines for inspection). Inspection of consignments of regulated articles moving in trade is an essential tool for the management of pest risks and is the most frequently used phytosanitary procedure worldwide to determine if pests are present and/or the compliance with phytosanitary import requirements. This standard does not give guidance on field sampling as required for surveys, etc. These are usually developed at the national level by the NPPOs or other agencies for internal use.

The sampling methodologies used by NPPOs in selecting samples for the inspection of consignments of commodities moving in international trade are based on a number of sampling concepts. These include parameters such as acceptance level, level of detection, confidence level, efficacy of detection, and sample size.

The application of statistically based methods, such as simple random sampling, systematic sampling, stratified sampling, sequential sampling, or cluster sampling, provides results with a statistical confidence level. Other sampling methods that are not statistically based, such as convenience sampling, haphazard sampling, or selective sampling, may provide valid results in determining the presence or absence of a regulated pest(s), but no statistical inference can be made on their basis. Operational limitations would have an effect on the practicality of sampling under one or another method.

In using sampling methodologies, NPPOs accept some degree of risk that non-conforming lots may not be detected. Inspection using statistically based methods can provide results with a certain level of confidence only and cannot prove the absence of a pest from a consignment (IPPC 2016l).

2.4.29 ISPM 32: Categorization of Commodities According to their Pest Risk (Adopted in 2009)

This standard provides criteria for NPPOs of importing countries on how to categorize commodities according to their pest risk when considering import requirements. This categorization would help in identifying whether further PRA is required and if phytosanitary certification is needed.

The first stage of categorization is based on whether the commodity has been processed and, if so, the method and degree of processing to which the commodity has been subjected before export. The second stage of categorization of commodities is based on their intended use after import. Contaminating pests or storage pests that could become associated with the commodity after processing are not considered in this standard.

The concept of categorization of commodities according to their pest risk takes into account whether the product has been processed and, if so, the method and degree of processing to which it has been subjected and the commodity's intended use and the consequent potential for the introduction and spread of regulated pests.

This allows pest risks associated with specific commodities to be assigned to categories. The objective of such categorization is to provide importing countries with criteria to better identify the need for a pathway-initiated PRA and to facilitate the decision-making process regarding the possible establishment of import requirements.

Four categories are identified, which group commodities according to their level of pest risk (two for processed commodities, two for unprocessed commodities).

Lists of the methods of processing and the associated resultant commodities are also provided (IPPC 2016m).

2.4.30 ISPM-33: Pest-Free Potato (Solanum Spp.) Micropropagative Material and Minitubers for International Trade

This standard which provides guidance on the production, maintenance, and phytosanitary certification of pest-free potato (*Solanum tuberosum* and related tuber-forming species) micropropagative material and minitubers intended for international trade.

The standard requires that facilities used for the production of potato micropropagative material and minitubers for export to be authorized or operated directly by the national plant protection organization (NPPO) of the exporting country.

Pest risk analysis (PRA) is carried out by the NPPO of the importing country, to provide the justification for establishing phytosanitary import requirements for regulated pests in trade of potato micropropagative material and minitubers.

The phytosanitary measures opted for managing risks related to potato micropropagative material should include testing for the pests regulated by the importing country, and management systems for the maintenance and propagation of potato micropropagative material derived from candidate plants that have been determined to be pest-free in closed, aseptic conditions (likely in vitro).

For the production of minitubers, measures need to include derivation from pest-free potato micropropagative material and production in a pest-free production site/conditions.

To establish pest-free potato micropropagative material, candidate plants to be tested in a testing laboratory are authorized or operated directly by the NPPO. This laboratory should meet general requirements for ensuring that all material moved into a maintenance and propagation facility is free from pests regulated by the importing country.

Facilities for the establishment of pest-free potato micropropagative material and testing for pest freedom are subject to strict requirements to prevent contamination or infestation of material.

Facilities for maintenance and propagation of pest-free potato micropropagative material and minituber production are also subject to stringent requirements to maintain pest freedom. Staff should be trained and competent in techniques for the establishment and maintenance of pest-free potato micropropagative material, the production of pest-free minitubers, diagnostic testing as required, and in following administrative, management, and record-keeping procedures.

Throughout all production and testing processes, the identity of all propagative material should be preserved and traceability to be maintained through adequate documentation.

All facilities should be officially audited to ensure that they continue to meet requirements.

In addition, inspections should ensure that the potato micropropagative material and minitubers meet the importing country's phytosanitary import requirements.

Pest-free potato micropropagative material and minitubers moving in international trade should be accompanied by a phytosanitary certificate (IPPC 2019i).

2.4.31 ISPM 42: Requirements for the Use of Temperature Treatments as a Phytosanitary Measures (Adopted in 2018)

This standard provides technical guidance on the application of various temperature treatments as phytosanitary measures for regulated pests on regulated articles. This standard does not provide details on specific treatments.

This standard provides guidance on how temperature treatments may be used for pest management to comply with phytosanitary import requirements.

This standard provides guidance on the main operational requirements for the application of each type of temperature treatment to achieve pest mortality at a specified efficacy. This standard also provides guidance on monitoring and recording systems and temperature mapping of facilities to ensure that the specific facility–commodity configuration will enable the treatment to be effective.

The NPPO to be responsible for approving the treatment facilities and procedures should be in place to ensure the accurate measuring, recording, and documentation of treatments applied. Phytosanitary treatments based on temperature are considered to be effective when the specific temperature–time combination required for the stated efficacy to be achieved is attained.

The purpose of this standard is to provide generic requirements for the application of phytosanitary temperature treatments, specifically those adopted under ISPM 28 (phytosanitary treatments for regulated pests).

The Standards Committee of the IPPC recognizes the relationships between the various ISPMs. For instance, for detailed hazard identification and risk communication components, ISPM-2 is to be referred and either ISPM 11 or 21 for risk assessment and management components. Also, the ISPM-3 on the code of conduct for import and release of exotic biocontrol agents would use the hazard identification component of ISPM-2 and the risk assessment component of ISPM-11 (IPPC 2018b).

2.5 National Standards: The Case of India

Presently, with varied agricultural produce and food production, India is in a position to expand its agricultural trade, but requires credible National Standards for all critical phytosanitary activities. It is the responsibility of Directorate of Plant Protection, Quarantine, and Storage (DPPQS), Department of Agriculture and Cooperation being the NPPO to develop the National Standards for Phytosanitary Measures (NSPMs). These are aimed to serve as guidelines for application of phytosanitary measures as harmonized with international standards so as to facilitate trade and avoid rejection due to noncompliance (PQIS 2021).

Standards developed/in process of development are listed in Table 2.3.

Few of the standards listed above such as Import Export and Post–entry Inspection manuals were developed several years ago and need to be revised to be in tune with the Plant Quarantine (Regulation of Import into India) Order 2003, while NSPM 6 and 16 need to be adopted. When comparing the ISPMs with the NSPMs, it is clear that many of the national standards such as the one on "Pest Risk Analysis" and the "Guidelines on Certification of Forced Hot-air Treatment Facilities for Wood Packaging Material" have been developed on the lines of the ISPMs and the appropriate level of protection perceived/required at the national level. The latter was developed to ensure that approved measures are applied consistently to the wood packaging material with the marking as per the ISPM 15 and would facilitate proper assessment and certification of heat treatment facilities. However, standard operating procedures (SOPs) are under development for only a few NSPMs and need to be developed urgently for "Import Inspection and Quarantine Clearance" and for various treatment facilities such as hot air for wood packaging and hot water immersion and for vapor heat. Also, new NSPMs are required for phosphine fumigation, surveillance for regulated pests, consignments in transit, pest reporting, and cold disinfestation treatments. Besides, the outdated laboratory manuals need to be reviewed and updated for all the disciplines of plant protection and for general seed health testing in the light of advanced techniques developed in the area (Gupta and Dubey 2017).

The NSPM 19 is on Survey and Monitoring Protocols for Establishment of Pest-Free Areas for Brown Rot (*Ralstonia solanacearum*) of Potato. It provides necessary guidelines for undertaking survey and monitoring of brown rot and to meet the requirements of establishment, maintenance, and verification of pest-free areas for brown rot of potato and use as a risk management option for undertaking phytosanitary certification of export of table potato from pest-free areas to European Union or provide scientific justification for phytosanitary measures for protection of endangered pest-free area. This standard would enable the recognition of pest-free areas in line with provisions of international agreements and thus facilitate the trade (PQIS 2005).

2.6 Perspectives

Although India is reasonably well placed in terms of facilities and technical expertise, we need to reprioritize and re-orient our line of action. The aim is that the desired data is generated and ultimately amalgamated with the existing knowledge to scientifically justify trade decisions through the development of appropriate standards. The changing scenario is indicative of the fact that the science of plant protection is no more limited to augmenting yield but also in boosting the economy of the country by effective compliance with the phytosanitary standards during agricultural trade. Those engaged in research activities related to SPS Agreement

Table 2.3 National standards for phytosanitary measures (NSPMs) developed in India

NSPM No.	Title of standard	Adopted
1	Plant quarantine operation systems manual	July, 1999
2	Import inspection manual	July, 1999
3	Export inspection manual	July, 1999
4	Post-entry quarantine inspection manual	July, 1999
5	Pest risk analysis: administrative process manual	January, 2004
6	Pest risk analysis-technical methodology	November, 1999
7	Guidelines for reporting plant quarantine activities	March, 2003
8	Guidelines for auditing of plant quarantine activities	November, 2002
9	Guidelines for Certification of Forced Hot-Air Treatment Facilities for Wood Packaging Material Revision 1 on: May, 2011	August, 2004
10	Guidelines for export inspection and phytosanitary certification of fresh mango (*Mangifera indica*) fruits to P. R. China	January, 2005
11	Quarantine treatments and application Procedures-1. Methyl bromide fumigation	February, 2005
12	Guidelines for Assessment, Accreditation, and Auditing of Fumigation Agencies Revision 1 on: May, 2011	February, 2005
13	Requirements for establishment of pest-free areas for mango nut weevil (*Sternochetus mangiferae*) and pulp weevil (*Sternochetus frigidus*)	May, 2005
14	Requirements for establishment of pest-free areas for tephritid fruit flies	May, 2005
15	Guidelines for certification of hot-water immersion treatment facilities	May, 2005
16	Guidelines for development of National Standards for phytosanitary measures	February, 2005
17	Guidelines for regulating export, import, and release of biological control agents and other beneficial organisms	December, 2004
18	Guidelines for certification of heat treatment facilities for Niger seed	December, 2004
19	Requirements for establishment of pest-free areas for brown rot (*Ralstonia solanacearum*) of potato	February, 2005
20	Guidelines for certification of vapor heat treatment facilities for fresh fruits	December, 2005
21	Guidelines for certification of irradiation treatment facilities for fresh fruits	January, 2006
22	Guidelines for assessment, audit and accreditation of fumigation agencies for undertaking aluminum phosphide fumigation	August, 2011
23	Guidelines for phytosanitary service agency and phytosanitary service provider for inspection of plants/plant products and other regulated articles in export	June, 2020
24	Guidelines for establishment of Pest-free area	April, 2020

and their implementation need to further identify the focused areas of research to compliment the national efforts. An enormous amount of work, within a fixed time frame, needs to be carried out (Dubey and Gupta 2019).

Although the task is to be accomplished by the NPPO, it would be extremely difficult without the involvement and support of Research Institutes and Universities. The Ministry of Agriculture needs to envisage a program and policy to benefit from the synergy of the available expertise in different research organizations and Ministries in the country.

References

Dubey SC, Kavita Gupta (2019) current practice of pest risk analysis on seeds and ISPM 38 implementation. 5th APSA expert consultation on phytosanitary issues held at Bangkok, Thailand by Asia Pacific seed association from august 28 to 29, 2019 held at W Bangkok hotel, Bangkok

Gupta K, Dubey SC (2017) Biosecurity policies influencing international exchange of PGR. Indian J Plant Genet Res 30(3):258–266

Gupta K, Khetarpal RK (2005) International standards of phytosanitary measures. In: Gadewar AV, Singh BP (eds) Plant protection in the new millennium. Satish, Delhi, pp 109–139

Gupta K, Sathyapala S (2020) Training manual for phytosanitary standards. Under FAOTCP/NEP/3702 building capacities to improve and sustain forest health to enhance the resilience of forests and livelihoods of forest dependent communities in Nepal, p 74

IPPC (2011) Procedure manual 2011. https://www.ippc.int/ar/publications/46/. Accessed 31 Mar 2021

IPPC (2016a) ISPM 1: phytosanitary principles for the protection of plants and the application of phytosanitary measures in international trade. https://assets.ippc.int/static/media/files/publication/en/2016/01/ISPM_01_2006_En_2015-12-22_PostCPM10_InkAmReformatted.pdf. Accessed 31 Mar 2021

IPPC (2016b) ISPM 7: phytosanitary certification system. https://assets.ippc.int/static/media/files/publication/en/2016/01/ISPM_07_2011_En_2015-12-22_PostCPM10_InkAmReformatted.pdf. Accessed 31 Mar 2021

IPPC (2016c) ISPM 9: guidelines for pest eradication programmes. https://assets.ippc.int/static/media/files/publication/en/2016/01/ISPM_09_1998_En_2015-12-22_PostCPM10_InkAmReformatted.pdf. Accessed 31 Mar 2021

IPPC (2016d) ISPM 10: requirements for the establishment of pest free places of production and pest free production sites. https://assets.ippc.int/static/media/files/publication/en/2017/03/ISPM_10_1999_En_2015-12-22_PostCPM10_InkAmReformatted.pdf. Accessed 31 Mar 2021

IPPC (2016e) ISPM 13: guidelines for the notification of non-compliance and emergency action. https://assets.ippc.int/static/media/files/publication/en/2016/01/ISPM_13_2001_En_2015-12-22_PostCPM10_InkAmReformatted.pdf. Accessed 31 Mar 2021

IPPC (2016f) ISPM 16: regulated non-quarantine pests: concept and application. https://assets.ippc.int/static/media/files/publication/en/2016/01/ISPM_16_2002_En_2015-12-22_PostCPM10_InkAmReformatted.pdf. Accessed 31 Mar 2021

IPPC (2016g) ISPM 19: guidelines on lists of regulated pests. https://assets.ippc.int/static/media/files/publication/en/2016/01/ISPM_19_2003_En_2015-12-22_PostCPM10_InkAmReformatted.pdf. Accessed 31 Mar 2021

IPPC (2016h) ISPM 22: requirements for the establishment of areas of low pest prevalence. https://assets.ippc.int/static/media/files/publication/en/2016/01/ISPM_22_2005_En_2015-12-22_PostCPM10_InkAmReformatted.pdf. Accessed 31 Mar 2021

IPPC (2016i) ISPM 25: consignments in transit. https://assets.ippc.int/static/media/files/publication/en/2016/01/ISPM_25_2006_En_2015-12-22_PostCPM10_InkAmReformatted.pdf. Accessed 31 Mar 2021
IPPC (2016j) ISPM 27: diagnostic protocols for regulated pests. https://assets.ippc.int/static/media/files/publication/en/2016/01/ISPM_27_2006_WithoutApp2_En_2015-12-22_PostCPM10_InkAmReformatted.pdf. Accessed 31 Mar 2021
IPPC (2016k) ISPM 28: phytosanitary treatments for regulated pests. https://assets.ippc.int/static/media/files/publication/en/2016/01/ISPM_28_2007_WithoutApp1_En_2015-12-22_PostCPM10_InkAmReformatted.pdf. Accessed 31 Mar 2021
IPPC (2016l) ISPM 31: methodologies for sampling of consignments. https://assets.ippc.int/static/media/files/publication/en/2016/01/ISPM_31_2008_En_2015-12-22_PostCPM10_InkAmReformatted.pdf. Accessed 31 Mar 2021
IPPC (2016m) ISPM 32: categorization of commodities according to their pest risk. https://assets.ippc.int/static/media/files/publication/en/2016/01/ISPM_32_2009_En_2015-12-22_PostCPM10_InkAmReformatted.pdf. Accessed 31 Mar 2021
IPPC (2017a) ISPM 3: guidelines for the export, shipment, import and release of biological control agents and other beneficial organisms. https://assets.ippc.int/static/media/files/publication/en/2017/05/ISPM_03_2005_En_2017-05-23_PostCPM12_InkAm.pdf. Accessed 31 Mar 2021
IPPC (2017b) ISPM 4: requirements for the establishment of pest free areas. https://assets.ippc.int/static/media/files/publication/en/2017/05/ISPM_04_1995_En_2017-05-23_PostCPM12_InkAm.pdf. Accessed 31 Mar 2021
IPPC (2017c) ISPM 8: determination of pest status in an area. https://assets.ippc.int/static/media/files/publication/en/2017/06/ISPM_08_1998_En_2017-05-23_PostCPM12_InkAm.pdf. Accessed 31 Mar 2021
IPPC (2017d) ISPM 12: phytosanitary certificates. https://assets.ippc.int/static/media/files/publication/en/2017/06/ISPM_17_2002_En_2017-05-25_PostCPM12_InkAm.pdf. Accessed 31 Mar 2021
IPPC (2017e) ISPM 17: pest reporting. https://assets.ippc.int/static/media/files/publication/en/2017/06/ISPM_17_2002_En_2017-05-25_PostCPM12_InkAm.pdf. Accessed 31 Mar 2021
IPPC (2017f) ISPM 24: guidelines for the determination and recognition of equivalence of phytosanitary measures. https://assets.ippc.int/static/media/files/publication/en/2017/05/ISPM_24_2005_En_2017-05-25_PostCPM12_InkAm.pdf. Accessed 31 Mar 2021
IPPC (2017g) ISPM 29: recognition of pest free areas and areas of low pest prevalence https://assets.ippc.int/static/media/files/publication/en/2017/05/ISPM_29_2007_En_2017-05-25_PostCPM12_InkAm.pdf. Accessed 31 Mar 2021
IPPC (2018a) ISPM 6: surveillance. https://assets.ippc.int/static/media/files/publication/en/2019/02/ISPM_06_2018_En_Surveillance_2018-05-20_PostCPM13.pdf. Accessed 31 Mar 2021
IPPC (2018b) ISPM 42: requirements for the use of temperature treatments as phytosanitary measures. https://assets.ippc.int/static/media/files/publication/en/2019/02/ISPM_42_2018_En_Temperature_2018-04-21_WithCover.pdf. Accessed 31 Mar 2021
IPPC (2019a) ISPM 2: framework for pest risk analysis. https://assets.ippc.int/static/media/files/publication/en/2019/05/ISPM_02_2007_En_Framework_PRA_2019-04-30_PostCPM14_InkAm.pdf. Accessed 31 Mar 2021
IPPC (2019b) ISPM 5: glossary of phytosanitary terms. https://assets.ippc.int/static/media/files/publication/en/2019/06/ISPM_05_2019_En_Glossary_2019-06-26_PostCPM-14-Fixed.pdf. Accessed 31 Mar 2021
IPPC (2019c) ISPM 11: pest risk analysis for quarantine pests. https://assets.ippc.int/static/media/files/publication/en/2019/05/ISPM_11_2013_En_PRA_QPs_2019-04-30_PostCPM14_InkAm.pdf. Accessed 31 Mar 2021
IPPC (2019d) ISPM 14: the use of integrated measures in a systems approach for pest risk management. https://assets.ippc.int/static/media/files/publication/en/2019/05/ISPM_14_2002_En_SystemsApproach_2019-04-30_PostCPM14_InkAm.pdf. Accessed 31 Mar 2021

IPPC (2019e) ISPM 18: guidelines for the use of irradiation as a phytosanitary measure. https://assets.ippc.int/static/media/files/publication/en/2019/05/ISPM_18_2003_En_Irradiation_2019-04-30_PostCPM14_InkAm.pdf. Accessed 31 Mar 2021

IPPC (2019f) ISPM 20: Guidelines for a phytosanitary import regulatory system. https://assets.ippc.int/static/media/files/publication/en/2019/06/ISPM_20_2017_En_Import_2019-06-10_PostCPM14_InkAm.pdf. Accessed 31 Mar 2021

IPPC (2019g) ISPM 21: pest risk analysis for regulated non-quarantine pests. https://assets.ippc.int/static/media/files/publication/en/2019/05/ISPM_21_2004_En_PRA_RNQPs_2019-04-30_PostCPM14_InkAm.pdf. Accessed 31 Mar 2021

IPPC (2019h) ISPM 23: guidelines for inspection. https://assets.ippc.int/static/media/files/publication/en/2019/05/ISPM_23_2005_En_Inspection_2019-04-30_PostCPM14_InkAm.pdf. Accessed 31 Mar 2021

IPPC (2019i) ISPM 33: pest free potato (solanum spp.) micropropagative material and minitubers for international trade. https://assets.ippc.int/static/media/files/publication/en/2019/05/ISPM_33_2010_En_Potato_Minitubers_2019-04-30_PostCPM14_InkAm.pdf. Accessed 31 Mar 2021

IPPC (2021a) International Plant Protection Convention (IPPC). https://www.ippc.int/en/. Accessed 31 Mar 2021

IPPC (2021b) Adopted Standards (ISPMs). https://www.ippc.int/en/core-activities/standards-setting/ispms/. Accessed 31 Mar 2021

Khetarpal RK, Gupta K (2002) Implications of sanitary and phytosanitary agreement of WTO on plant protection in India. Annu Rev Plant Pathol 1:1–26

PQIS (2005) NSPM 19: Survey and monitoring protocols for establishment of pest-free areas for brown rot (*Ralstonia solanacearum*) of potato. https://phytosanitarysolutions.com/wp-content/uploads/2015/06/NSPM-19-Survey-Monitoring.pdf. Accessed 31 Mar 2021

PQIS (2021) Standards: plant quarantine information system. http://164.100.79.121/PQIS/PQISPub/html/standards.htm. Accessed 31 Mar 2021

3 Fungicide Resistance: Threats and Management Approaches

Ashutosh Bhaik, Mamadou Kane Mboup, and Jean-Luc Genet

Abstract

Fungal pathogens are responsible for significant reduction of crop yield globally when left untreated. The spread of fungal diseases can be controlled by crop management, rotation systems, the use of resistant cultivars, and the application of fungicides. Fungicides are often employed by farmers due to their simplicity and efficiency, thereby making it as an integral part of disease management strategies. However, resistance against these chemicals due to indiscriminate use leads to loss of efficacy. This loss of efficacy of a fungicide against a plant pathogen poses a serious problem, because of limited option of chemistries available for managing a particular set of pathogens. Also, discovery of new actives has become a key challenge due to the cost of development and stringent regulations over the past few decades. In this chapter, an attempt has been made to understand the approaches that are employed to delay or prevent the resistance development against fungicides in respect to potato crop.

Keywords

Resistance management · FRAC · Mode of action · Rotation · Fungicide mixture · Multisite fungicide · Resistance risk

A. Bhaik (✉) · M. K. Mboup · J.-L. Genet
Corteva Agriscience, Hyderabad, Telangana, India
e-mail: Ashutosh.Bhaik@corteva.com

S. Kumar Chakrabarti et al. (eds.), *Sustainable Management of Potato Pests and Diseases*, https://doi.org/10.1007/978-981-16-7695-6_3

3.1 Introduction

Ever since humans started practicing crop cultivation about 10,000 years ago, pests and diseases have been a constant threat for sustainable food supply. Fungicides are invariably being used to control plant diseases which still stand at 20% (Jørgensen et al. 2017). Historic records suggest sulfur compounds being used as early as 4500 years ago by Sumarians. Agriculture sector witnessed rapid use of pesticides after World War II with the advent of advanced chemistry. Many companies ventured into the business of pesticide manufacturing and synthesis during late twentieth century which helped farmers to produce sufficient food not only to themselves but for ever burgeoning human population.

The rapid advances in chemical disease control in agriculture lead to the following concerns:-

a) Chemical products are being phased out or restricted due to safety and environmental concerns.
b) Resistance is building up in target population due to repeated or indiscriminate use of same mode of action of chemicals.

The loss of a fungicide to agriculture through resistance is a problem that affects us all. It may lead to unexpected and costly crop losses to farmers causing local shortages and increased food prices. Manufacturers lose revenue vital to funding the enormous development costs of new products. Without reinvestment there would be no new compounds. This would cause serious disease management problems and endanger the world food supply.

3.2 History of Chemical Control and Resistance

Humans have battled plant diseases since ages by using different chemicals, with oldest report being the use of lime sulfur sprays (Forsyth 1802). Use of Bordeaux mixture by Millardet (1885) marked the beginning of use of copper compounds as fungicide for control of more dreaded diseases like grape downy mildew (*Plasmopara viticola*) and potato late blight (*Phytophthora infestans*). Later research led to the development of organic compounds especially dithiocarbamates (mancozeb), phthalimides (captafol), and chlorothalonil, which are now very widely used as non-systemic protectant fungicides. Owing to their multisite modes of action, these chemicals are still prevalent. Despite extensive use over many years, resistance has not been an issue with these fungicides.

Last few decades of twentieth century saw rapid development of new-generation fungicides which are site-specific, selective, and more systemic and had more potency against targeted pathogens. Following new set of criteria for fungicide development has been followed: (1) development of pesticides that are effective at an extremely low dosage, (2) development of pesticides that are readily degradable and less residual in the environment, and (3) development of selectively toxic

Table 3.1 Historic references of fungicide resistance in crop diseases (Adapted from Hewitt (1998) and Brent (2012))

Date	Fungicide class	Time to resistance (approx. yrs)	Disease example	References
1960	Aromatic hydrocarbons	20	Citrus storage rots, *Penicillium* spp.	Eckert (1982)
1964	Organomercury	40	Cereal leaf spot, *Pyrenophora* spp.	Noble et al. (1966)
1969	Dodine (guanidine)	10	Apple scab, *Venturia inaequalis*	Szkolnik and Gilpatrick (1969)
1970	Benzimidazoles (MBCs)	2	Many pathogens	Dekker (1986)
1971	2-Aminopyrimidines	2	Powdery mildews	Brent (1982)
1980	Phenylamides	2	Potato late blight, grape downy mildew	Staub (1994)
1982	Demethylation inhibitors (DMIs)	7	Cereal powdery mildew and other diseases	De Waard (1994)
1998	Quinone outside inhibitors (QoIs)	2	Cereal powdery mildew	Chin et al. (2001)
2007	Succinate dehydrogenase inhibitors	4–5	*Alternaria alternata* (nuts), early blight of potato (*Alternaria solani*)	Avenot and Michailides (2007) and Miles et al. (2014)

agrochemicals. Even with these development strategies, the risk of resistance to these chemistries has not come down rather the cost of product development has increased many times thereby limiting the number of new chemistries in the market. To date 50 different modes of action (FRAC 2021) are identified for fungicides.

During the first half of twentieth century, resistance against the fungicides was unknown. It was only during the 1960s when reports against performance failure started being reported from across the world. Usually, the first indications of a resistance problem came from reports from growers of failure of disease control following fungicide treatment.

During late 1980s and early 1990s, the concept of fungicide resistance was well documented and perceived by the farmers as well as the academia (Table 3.1). It was during the 1990s when Fungicide Resistance Action Committee (FRAC), an inter-company committee, was formed to manage the situation after the onset of phenylamide resistance. Phenylamide working group was established which soon after issued a set of guidelines for resistance management. The recommendations included using mixtures for foliar application, avoiding curative use, and limiting the number of sprays per season. These guidelines were implemented by all the companies involved, and this fungicide class continued in use against all target diseases (Staub 1994).

Nowadays, manufacturing companies as well as regulators are concerned about the resistance risk of new actives even before registration of the compound against a particular disease. For companies, it is about extending the product life cycle as the cost of development has increased many folds when compared to that was during last century. European regulations make it mandatory to conduct resistance risk assessment, and if appropriate, systems for risk management can be proposed, in the context of official registration of plant protection products. FRAC has also established fungicide resistance management practices, which serve as a guideline for the use and labeling of the new crop protection products. Even novel mode of action chemistries like group 49 has its own OSBPI working group which is accessing the resistance situations annually.

3.3 Resistance Defined

The loss of performance of a plant protection product because of the development of resistance in the target pest can be costly to the grower, the crop protection company, and the environment. The first reports from growers' field during the early 1960s–1970s were of failure of disease control following fungicide applications. Many other reasons than resistance could be attributed such as incorrect application (doze, time of application), use of expired or deteriorated or wrong products, wrongly identified pathogen, or application at the time of exceptionally heavy disease pressure, and at times growers and advisers attribute difficulties of disease control to fungicide resistance in the absence of evidence that resistance was the main cause. However, in many cases there was no other obvious explanation, and loss of control was soon shown to be associated with greatly decreased sensitivity of the pathogen, as revealed by laboratory or glasshouse tests on samples taken from the problem sites.

The term fungicide resistance, as used by the Fungicide Resistance Action Committee (FRAC), refers to an acquired, heritable reduction in sensitivity of a fungus to a specific anti-fungal agent (or fungicide). To manage resistance effectively, scientists study fungicide resistance on many different levels including the cellular, organismal, or population/field level. Reports of "resistance" from the field (i.e., where growers observed reduced efficacy of a product that has been effective against that particular pathogen) must be confirmed by studies at the organismal level showing a reduction in sensitivity of the fungal isolate(s) to the specific fungicide. Some scientists use the terms reduced sensitivity or tolerance when referring to smaller reductions in sensitivity which may have little or no impact on fungicide usage in the field and save the term "resistance" for large reductions in sensitivity of individual isolates which are likely to

(continued)

affect the efficacy of a specific fungicide under field conditions if the resistant isolates become widespread in the pathogen population. The term field resistance may also be used to indicate this loss of control under field conditions.

FRAC states that the term "resistance" should only be used in situations where:

- Development of resistance leads to failure of disease control under practical field conditions following proper labeled use of a fungicide.
- Demonstration that loss of control is due to the presence of pathogenic strain(s) with reduced fungicide sensitivity.

The development of fungicide resistance is a population evolutionary process. Fungi, like other organisms, are constantly changing. Occasionally, under certain conditions, these changes provide an advantage or disadvantage in terms of the progeny's ability to survive and reproduce. Advantageous changes allow the individual containing the change to survive and reproduce resulting in their progeny constituting a greater percentage of the population over subsequent generations. This can happen relatively rapidly in fungi as their *reproductive frequency* (i.e., the number of progenies produced from a single individual and the speed with which they complete their life cycle) is high. For example, a single *Phytophthora infestans* lesion can produce thousands of spores, and a spore can produce a new sporulating lesion in 3–5 days. The change may be evolutionarily neutral, or even slightly disadvantageous, under most conditions and only be advantageous when certain factors are present. This is the case with fungicide resistance. In most cases of fungicide resistance, the change leading to reduced sensitivity is evolutionarily neutral except when the specific fungicide is applied. The fungicide is exerting *selection pressure* on the pathogen population since it is killing the initial (or *wild type*) population but does not kill the changed (or *mutant*) population (Fig. 3.1). When changes are slightly disadvantageous under normal conditions (i.e., in the absence of the fungicide), the frequency of the changed population may decrease when the selection pressure is removed. This disadvantage is termed as *fitness penalty*.

Some Definitions

Qualitative resistance describes when fungicide resistance results from modification of a single major gene, and pathogen subpopulations are either sensitive or fully resistant to the pesticide. Resistance in this case is seen as complete loss of disease control that cannot be regained by using higher rates or more frequent fungicide applications, e.g., is resistance developed by several pathogens to the strobilurin (FRAC code 11) fungicides.

Quantitative resistance describes when fungicide resistance results from modification of several interacting genes, and pathogen isolates exhibit a

(continued)

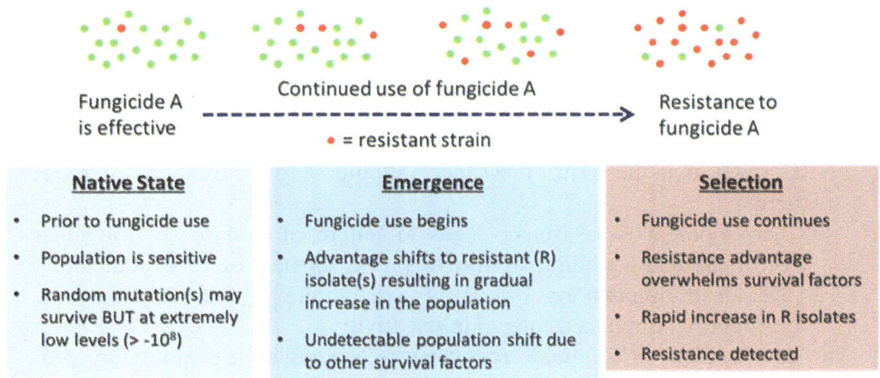

Fig. 3.1 Different stages of fungicide resistance when fungicide A is used repeatedly over several generations

range in sensitivity to the fungicide depending on the number of gene changes. Variation in sensitivity within the population is continuous. Resistance in this case is seen as an erosion of disease control that can be regained by using higher rates, more frequent applications, or a fungicide in the chemical class that has inherently higher activity. Long-term selection for resistance in the pathogen by repeated applications may eventually result in the highest labeled rates and/or shortest application intervals not being able to adequately control the disease. An example is resistance in the cucurbit powdery mildew pathogen to the DMI (FRAC code 3) fungicides.

Cross resistance describes fungal isolates that are resistant to one fungicide and also resistant to other closely related fungicides, even when they have not been exposed to these other fungicides, because these fungicides all have similar mode of action. Sometimes fungicides in the same chemical class act at a slightly different point in the biosynthetic pathway in the pathogen which sufficiently limits cross resistance.

Fungicides with **single-site mode of action** (targeted) are generally at medium to high risk for resistance development. Single-site means the fungicide acts at a specific point in a biosynthetic pathway in the pathogen. These fungicides are at risk for resistance development because a change in the pathogen at this point can render the fungicide less effective or ineffective. A simple change of just one base pair in the DNA molecule can be sufficient to lead to full resistance (fungicide completely ineffective), as occurs with the strobilurin (FRAC code 11) fungicides.

Most fungicides being developed today have a single-site mode of action because this is associated with lower potential for negative impact on the environment, including non-target organisms. Their targeted activity means

(continued)

that they can safely move into the plant (fungicide is not toxic to the plant), resulting in better rain fastness than contact fungicides and better activity as they can move across lower surface where pathogens often develop best. A few modern fungicides have systemic activity and can move more widely in plants, especially when applied to roots. Older fungicides, such as copper and chlorothalonil, have low potential for resistance to develop because they have **multi-site mode of action.**

3.4 Mechanism of Resistance

Fungi have been described as "a mutable and treacherous tribe," but that even this is something of an understatement is abundantly evident.. .

E. W. Buxton, Heterokaryosis, Saltation and Adaptation (1960).

There are four main mechanisms by which fungi can become resistant to fungicides.

3.4.1 Alteration of the Target Site so that Sensitivity to the Fungicide Is Reduced

By far the most common way that fungi can become resistant to a specific fungicide is via a change at the target site. As fungi grow, their DNA is replicated when new cells are created. This process of replication is imperfect, and errors can occur. These errors are known as mutations. Since DNA is the code used to produce enzymes in the cell, some mutations result in changes to the amino acid sequence of the target site which in turn alters the shape of the lock/target site. The fungicide/key may not fit as well anymore or may not fit at all in the target site/lock. This results in a reduction in sensitivity that may range from small to very large.

3.4.2 Detoxification or Metabolism of the Fungicide

The fungal cell contains a vast array of metabolic machinery for normal cellular processes. This metabolic machinery may be able to modify the fungicide to a non-toxic form that is no longer harmful to the cell. Some fungicides are applied as inactive pro-fungicides which require further metabolism by the fungal cell to become the active form. If fungal metabolism is altered such that the activation step does not occur, the active form of the fungicide is not produced.

3.4.3 Overexpression of the Target

As discussed above, the fungicide is "competing" with the natural substrate for the target site. As more and more fungicide enter the cell, it out-competes the natural substrate for the target and as a result shuts down critical cellular processes. The production of additional target site enzyme (i.e., overexpression of the target) may increase the likelihood that enough of the fungal substrate will be able to bind with the target site enzyme such that cellular processes such as respiration can occur to some degree. Higher doses of the fungicide in in vitro experiments may restore the balance in favor of the fungicide, but higher doses may not always be practical under field conditions.

3.4.4 Exclusion or Expulsion from the Target Site

Efflux pumps exist naturally within the cell to exclude or expel foreign substances or to export endogenous substances. In fungi, the most common efflux pumps are ABC and MFS transporters. Despite these efflux pumps, most fungicides can reach effective concentrations inside the cell and inhibit cellular processes. Occasionally, these transporters are successful in expelling enough of the fungicide such that an isolate has reduced sensitivity. The fungicides expelled from the cell by a specific transporter may or may not be active at the same target site; i.e., there is not a direct relationship between the transporter that expels a specific fungicide and the target site of the fungicide. Multidrug resistance (MDR) develops when a specific transporter is able to exclude multiple fungicides from different target site groups. Application of the fungicides in question may exert enough selection pressure that isolates containing these fungicide-exporting transporters become more prevalent in the population as is the case in *Botrytis cinerea* (Kretschmer et al. 2009).

3.5 Evaluating Resistance Risk

Risk of resistance buildup and loss of efficacy to fungicide is of greater concern to all the new actives that are commercialized by crop protection companies. Resistance risk assessment therefore is a regulatory requirement in European Union. Risk is assessed by combining risk values for the fungicide, pathogen, and agronomic system under consideration, using a risk matrix (triangle) defined by Kuck and Russel (2006). The risk of resistance developing to a fungicide is driven by three components (Fig. 3.2). The first is the inherent pathogen risk. Pathogens are classified as high risk if they have a history of developing fungicide resistance, or if they would be expected to develop resistance rapidly, based on their biological characteristics. These are generally pathogens with short life cycles that have multiple generations in a single crop,and that produce spores in huge numbers. The second consideration is the risk associated with the mode of action (MoA) of the active ingredient. For example, multisite fungicides, which interfere with

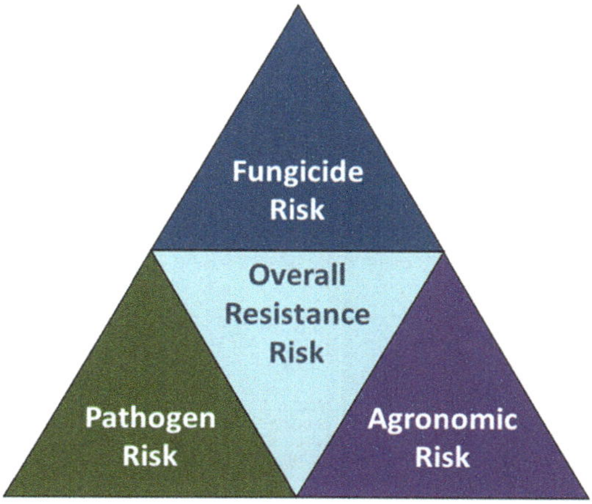

Fungicide risk

- Efficacy
- Site of action (single vs multisite)
- Novel mode of action
- Resistance mechanisms

Pathogen risk

- Epidemiology (cycles per season)
- Opportunity for recombination
- History with other fungicides

Agronomic factors

- Environment influence resistance development
- Cultural effect specific to crop and geography

Fig. 3.2 Fungicide resistance risk triangle

multiple biochemical pathways, are considered as low risk, and resistance to these active ingredients is less likely to evolve. The final consideration is the agronomic risk, which encompasses the environment in which the fungicide is used, including farmer practices for managing disease.

The scheme has been refined lately by including numerical values for "fungicide risk" and "disease risk." These risk values are multiplied together to get the combined risk value. The resistance risk scheme was further refined by adding agronomic risk as the third factor in the matrix (Fig. 3.3). These ratings are based on the experience over the year of resistance development against the actives and published/shared data during the meetings. The agronomic risk is disproportionately high in many Asian countries. In tropical and subtropical climates, the warm and humid environment is conducive to pathogen development, and farmers may exacerbate this favorable environment through practices that allow the pathogen to survive continuously, such as planting the same crop in consecutive seasons, with limited crop rotation. Up-to-date assessments of fungicide-associated risk are presented on the FRAC website for all fungicides in current use.

3.6 Detecting and Monitoring Resistance

The first and most obvious detection of resistance is from the farmer field where he observes decline in fungicide performance, response for which is increasing dose rate and/or frequency applications of fungicides. Poor performance though can be attributed to several causes like poor application and timing, wrong dose rate, or very exceptional disease pressure. If resistance is a problem, it needs confirming trials of

Fungicide Risk	Risk	Combined Risk			Agronomic Risk
Benzimidazoles Dicarboximides Phenylamides QoI fungicides	High = 3	3 1.5 0.75	6 3 1.5	9 4.5 2.25	High = 1 Medium = 0.5 Low = 0.25
Carboxanilides SBI fungicides Anilinopyrimidines Phenylpyrroles	Medium = 2	2 1 0.5	4 2 1	6 3 2.25	High = 1 Medium = 0.5 Low = 0.25
Multi-site fungicide MBI-R inhibitors SAR inducers	Low = 1	0.5 0.25 0.125	1 0.5 0.25	1.5 0.75 0.3	High = 1 Medium = 0.5 Low = 0.25
		Low = 1	Medium = 2	High = 3	Pathogen Risk
		Seed borne and soil borne pathogens, Rust fungi	Eyespot, Rhynchos-porium, Septoria	Blumeria, Botrytis, Plasmopara, Magoporthe, Venturia	

Fig. 3.3 A "risk matrix" fungicide resistance risk assessment scheme (from Kuck and Russell 2006)

more than just a single season's work. So anecdotal evidence from growers must be backed up by a program of field work supported by glasshouse and laboratory assays.

3.6.1 Bioassays

Designed to reveal difference in phenotypical response to pesticide molecule. Live pest colonies, lines, populations, or isolates of interest is exposed to pesticide and then compared to those of sensitive references. A most common approach to confirm resistance is by comparing sensitivity of isolates obtained from sites where performance has eroded with the sensitivity of isolates never exposed to the at-risk fungicide. Ideally the existence of a sensitivity distribution of the target fungal population established prior to widespread use of a new fungicide will allow a meaningful confirmation of resistance. The benefit of a "baseline" sensitivity distribution in various aspects of resistance management was described in detail by Russell (2005), and its importance is recognized in many countries where a baseline sensitivity distribution is a requirement for registration of a new fungicide. The ability to confirm resistance through comparison with a baseline sensitivity will depend on the sample size from the suspected resistant population and inclusion of at least one reference isolate to check for variation between assay tests. In practice where baseline sensitivity data do not exist, comparisons can be made between isolates obtained from at-risk sites with those collected from untreated areas. Often researchers obtain baseline data using "historic" isolates which have been maintained in culture collections, sometimes for many years, and which were isolated before the at-risk fungicide was used.

3.6.2 Molecular Assays

Molecular, nucleic acid-based assays detecting genes or mutations involved in resistance. The starting material can be living or dead population. Sufficient DNA or RNA of suitable quality should be extracted for further analysis. Literature is full of different molecular techniques used to monitor resistance, and certainly the most well documented is perhaps detection of the mutation generating the G143A amino acid change in the target b-type cytochrome of complex III of respiration, causing resistance to QoI fungicides.

3.6.3 Biochemical Assays

As most of the pesticides bind to and inactivate a vital protein (enzyme) which is responsible for some essential biochemical reaction in the pathogen, the efficiency is dependent on the pesticide reaching the binding site. Biochemical assays are usually used to situation where the mechanism of resistance has been elucidated.

3.7 Fungicide Resistance Management Approaches

Fungicide resistance management strategies aim to delay the evolution and spread of resistance in a sensitive pathogen population, while ensuring effective disease control. If fungicide resistance is confirmed or highly suspected, diverse approaches to managing resistance need to be incorporated into disease management strategies.

3.7.1 Managing the Application Dose

- FRAC recommends to strictly follow the country label for the crop-disease for effectively managing the disease and prevent the resistance development.
- The majority of the evidence suggests that an increased dose of fungicide increases selection for fungicide resistance (but note here that the primary aim of effective disease control may make it impossible to reduce fungicide dosages).
- A number of possible mechanisms by which an increased dose may reduce selection have, however, not been studied. Partial resistance and multi-gene/multi-mutation cases are the key examples of this.

3.7.2 Managing the Number of Sprays

- All current evidence suggests that increasing the number of fungicide applications increases selection for fungicide resistance.
- Most evidence suggests that splitting fungicide dosage between two or more applications increases selection.

3.7.3 The Use of Fungicide Mixtures

- The vast majority of the evidence shows that adding a mixing partner effective on the target diseases to a high resistance- risk fungicide reduces selection for fungicide resistance, even when the dose of the high-risk fungicide stays the same in the mixture.
- Adding a mixing component to a high-risk fungicide and reducing the dose of the high-risk fungicide further reduces selection for fungicide resistance.
- There is too little evidence on the use of mixtures of two at-risk fungicides, and work in this area is needed. The evidence that does exist suggest that mixing two at-risk fungicides is a valid anti-resistance strategy.

3.7.4 The Use of Fungicide Alternations

- Limited evidence suggests that alternating with a fungicide that has a different mode of action does not alter selection for the high-risk fungicide, if the number of applications of the high-risk fungicide remains constant with and without alternation.
- The evidence suggests that replacing part of the fungicide program with a fungicide with a different MOA reduces selection.

3.7.5 Alternations Versus Mixtures

- It depends on the balance between increased selection due to dose splitting and decreased selection due to mixing whether mixing reduces selection to a greater or lesser extent than alternation. The experimental and modeling evidence shows that in many cases, mixing is the better strategy, but for any single case, this needs to be established before conclusions can be reached.

3.7.6 Protective Versus Curative Use

- There is no evidence that protective or curative use consistently results in a lower rate of selection for fungicide resistance (but note that protective fungicide applications may be needed for effective disease control).

- The existing evidence suggests that the specific circumstances will determine whether a shift in spray timing will increase or decrease selection for fungicide resistance.

Resistance management should be based on evidence interpreted within a sound experimental, theoretical, and practical framework. In discussions on resistance management, it is often not explicit what the evidence is. With this chapter, we hope to contribute to evidence-based resistance management (Adopted from van den Bosch et al. 2015).

3.8 Fungicide Resentence Action Committee (FRAC)

Industry recognizes its responsibility in safeguarding new chemistries that are brought to market. Through FRAC and its Working Groups, companies are striving to establish more effective communications to alert all those involved in the research, production, marketing, registration, and use of fungicides to the problems of resistance.

With enlightened stakeholders, effective strategies can be conceived and adopted. Cooperative action is essential if we are to preserve the option of chemical disease control for our crops.

FRAC is a Specialist Technical Group of CropLife International (CLI; Formerly Global Crop Protection Federation, GCPF). As such, we work within the legal framework defined by CLI and take care to ensure that strict anti-trust guidelines are observed. In few regions Fungicide Resistance Action Group (FRAG) a are mixed groups with industry and non-industry members is also functional.

3.8.1 Purpose

The purpose of FRAC is to provide fungicide resistance management guidelines to prolong the effectiveness of "at-risk" fungicides and to limit crop losses should resistance occur.

The main aims of FRAC are to as follows:

1. Identify existing and potential fungicide resistance problems.
2. Collate information and distribute it to those involved with fungicide research, distribution, registration, and use.
3. Provide guidelines and advice on the use of fungicides to reduce the risk of resistance developing and to manage it should it occur.
4. Recommend procedures for use in fungicide resistance studies.
5. Stimulate open links and collaborations with universities, government agencies, advisors, extension workers, distributors, and farmers.

3.8.2 FRAC Guidelines

FRAC Guidelines for resistance management are produced by the individual FRAC Working Groups and Expert Fora. These Guidelines provide information on how to use specific areas of fungicide chemistry for control of plant diseases on various crops while maintaining a good anti-resistance strategy.

The Guidelines should be regarded as the minimum resistance management strategy required, and it is possible that a more stringent strategy should be used in individual cases. FRAC recommends that you seek advice from your local resistance management organization (e.g., local country FRAC or FRAG), your local crop advisor or extension agent, or the manufacturer or distributor of the product to see if a more restrictive strategy is recommended.

The FRAC Guidelines deal only with areas of fungicide chemistry. FRAC is not allowed to make recommendations for the use of individual products. If you require advice on which active ingredients to use in your disease control program, please consult your local crop advisor or extension agent, or the distributor or manufacturer of potential products.

3.9 FRAC Guidelines: Potato Diseases

FRAC guidelines with respect to use of different fungicides in potato crops and resistance management are outlined in this section. Table 3.2 enlists common fungicide groups with their mode of action and FRAC classification.

3.9.1 Oxysterol Binding Protein Homologue Inhibitor (OSBPI) Fungicides (FRAC Code 49) (Updated April 2021)

- Fungicide programs must deliver effective disease management. Apply OSBPIs at effective rates and intervals according to manufacturers' recommendations. Effective disease management throughout the season is a critical component to delay the build-up and spread of resistant pathogen populations.
- Apply OSBPIs only preventatively and in mixtures with effective fungicides from different cross-resistance groups.
- The mixture partner should give effective control of the target disease(s) at the rate and interval selected.
- Foliar exposure to OSBPI products should not exceed thirty-three percent (33%) of the total period of protection needed per crop.
- Fungicide programs must deliver effective disease management. Apply OSBPIs at effective rates and intervals according to manufacturers' recommendations. Effective disease management throughout the season is a critical component to delay the build-up and spread of resistant pathogen populations.
- Apply OSBPIs only preventatively and in mixtures with effective fungicides from different cross-resistance groups.

Table 3.2 Classification of fungicides used in potato crops based on mode of action groups, target site, and risk of resistance

FRAC code	Target site and code	Group name	Chemical group	Common name	Comments
49	F9 lipid homeostasis and transfer/storage	OSBPI oxysterol binding protein homologue inhibition	Piperidinyl thiazole isoxazolines	Oxathiapiprolin fluoxapiprolin	Resistance risk assumed to be medium to high (single site inhibitor). Resistance management required
11	C3 complex III: Cytochrome bc1 (ubiquinol oxidase) at Qo site (cyt b gene)	Complex III; cytochrome bc1 (ubiquinol oxidase) at Qo site (cyt b gene)	Methoxy-acrylates	Azoxystrobin coumoxystrobin enoxastrobin flufenoxystrobin picoxystrobin pyraoxystrobin	Resistance known in various fungal species. Target site mutations in cyt b gene (G143A, F129L) and additional mechanisms cross resistance shown between all members of the QoI group. High risk. (see FRAC QoI guidelines for resistance management)
			Methoxy-acetamide	Mandestrobin	
			Methoxy-carbamates	Pyraclostrobinpyrametostrobintriclopyricarb	
			Oximino-acetates	Kresoxim-methyl	
			Oximino-acetamides	Dimoxystrobinfenaminostrobin metominostrobin	
			Oxazolidine-dones	Famoxadone	
			Dihydro-dioxazines	Fluoxastrobin	
			Imidazolinones	Fenamidone	
			Benzyl-carbamates	Pyribencarb	

(continued)

Table 3.2 (continued)

FRAC code	Target site and code	Group name	Chemical group	Common name	Comments
40	H5 cellulose synthase	CAA-fungicides (carboxylic acid amides)	Cinnamic acid amides	Dimethomorphflumorph	Cross resistance between all members of the CAA group. Low to medium risk. See FRAC CAA guidelines for resistance management
			Valinamide carbamates	Benthiavalicarbiprovalicarb	
			Mandelic acid amides	Mandipropamid	
4	RNA polymerase I	PA-fungicides (Phenylamides)	Acyl- alanines	Benalaxyl benalaxyl-M (=kiralaxyl) furalaxyl metalaxyl metalaxyl-M (=mefenoxam)	Resistance and cross-resistance well known in various oomycetes but mechanism unknown. High risk. (see FRAC Phenylamide guidelines for resistance management)
		Oxazolidinones	Oxadixyl		
		Butyrolactones	Ofurac		
21	C4 complex III: Cytochrome bc1 (ubiquinone reductase) at qi site	QiI - fungicides (Quinone inside inhibitors)	Cyano-imidazole	Cyazofamid	Resistance risk unknown but assumed to be medium to high (mutations at target site known in model organisms). Resistance management required. No spectrum overlaps with the oomycete-fungicides cyazofamid and amisulbrom
			Sulfamoyl-triazole	Amisulbrom	

29	C5 uncouplers of oxidative phosphorylation		2,6-dinitro-anilines	Fluazinam	Low risk
27	Unknown	Cyanoacetamide-oxime	Cyanoacetamide-oxime	Cymoxanil	Resistance claims described. Low to medium risk. Resistance management required
45	C8 complex III: Cytochrome bc1 (ubiquinone reductase) at Qo site, stigmatellin binding sub-site	QoSI fungicides (Quinone outside inhibitor, stigmatellin binding type)	Triazolo-pyrimidylamine	Ametoctradin	Not cross resistant to QoI fungicides. Resistance risk assumed to be medium to high (single site inhibitor). Resistance management required

Adapted from FRAC (2021).

- The mixture partner should give effective control of the target disease(s) at the rate and interval selected.
- Foliar exposure to OSBPI products should not exceed thirty-three percent (33%) of the total period of protection needed per crop.
- In case of non-cucurbit multiple crops, do not make more than six (6) foliar applications of OSBPI product per year on the same acreage or greenhouse, targeting the same pathogen.

3.9.2 QoI Fungicides (FRAC Code 11) (Updated June 2020)

Fundamental principles that must be adhered to when applying resistance management strategies for QoI fungicides is that:

- The QoI fungicides (azoxystrobin, coumoxystrobin, dimoxystrobin, enoxastrobin, famoxadone, fenamidone, fenaminostrobin, fluoxastrobin, flufenoxystrobin, kresoxim-methyl, mandestrobin, metominostrobin, orysastrobin, pyraoxystrobin picoxystrobin, pyraclostrobin, pyrametastrobin, pyribencarb, triclopyricarb trifloxystrobin) are in the same cross-resistance group, FRAC Code 11.
- The QoI fungicide in subgroup A (metyltetraprole), Code 11A fungicide, is not cross resistant with Code 11 fungicides on the pathogens with G143A mutation.
- Fungicide programs must deliver effective disease management. Apply QoI fungicide-based products at effective rates and intervals according to manufacturers' recommendations. Effective disease management is a critical component to delay the build-up of resistant pathogen populations.
- The number of applications of QoI fungicide-based products within a total disease management program must be limited whether applied solo or in mixtures with other fungicides. This limitation is inclusive to all QoI fungicides. Limitation of QoI fungicides within a spray program provides time and space when the pathogen population is not influenced by QoI fungicide selection pressure.
- Limitation of the total number of QoI applications is detailed in the specific crop recommendations. In consideration of the cross-resistance profile of subgroups 11 and 11A, the maximum allowed number of QoI-containing sprays is increased by one, where both QoI fungicides (code 11) and QoI fungicides in subgroup A (code 11A) are included in a spray program in a given cropping season. All crop-specific recommendations will be regularly reviewed based on sensitivity monitoring.
- A consequence of limitation of QoI fungicide-based products is the need to alternate them with effective fungicides from different cross-resistance groups (refer to the specific crop recommendations).
- QoI fungicides, containing only the solo product, should be used in single or block applications in alternation with fungicides from a different cross-resistance group. Specific recommendation on size of blocks is given for specific crops.

- QoI fungicides, applied as tank mix or as a co-formulated mixture with an effective mixture partner, should be used in single or block applications in alternation with fungicides from a different cross-resistance group. Specific recommendations on size of blocks are given for specific crops.
- Mixture partners for QoI fungicides should be chosen carefully to contribute to effective control of the targeted pathogen(s). The mixture partner must have a different mode of action, and in addition it may increase spectrum of activity or provide needed curative activity. Use of mixtures containing only QoI fungicides (including two-way mixtures of code 11 fungicide and code 11A fungicide) must not be considered as an anti-resistance measure.
- Where local regulations do not allow mixtures, then strict alternations with non-cross resistant fungicides (no block applications) are necessary.
- An effective partner for a QoI fungicide is one that provides satisfactory disease control when used alone on the target disease.
- QoI fungicides are very effective at preventing spore germination and should therefore be used at the early stages of disease development (preventive treatment).

3.9.3 Late Blight (Phytophthora Infestans)

- Apply QoI fungicides according to manufacturer's recommendations for the target disease (or complex) at the specific crop growth stage indicated. Effective disease management is a critical parameter in delaying the build-up of resistant pathogen populations.
- Where QoI fungicide products are applied alone do not exceed 1 spray out of 3 with a maximum of 3 sprays per crop. Do not use more than 2 consecutive applications.
- Where QoI fungicide products are applied in mixtures (co-formulations or tank mixes) do not exceed 50% of the total number of sprays or a maximum of 6 QoI fungicide applications whichever is the lower. Do not use more than 3 consecutive QoI fungicide containing sprays.

3.9.4 Early Blight (Alternaria Solani, Alternaria Alternata)

- Where QoI fungicide products are applied solo do not exceed 33% of the total number of sprays or a maximum of 4. Where mixtures (co-formulations or tank mixes) are used do not exceed 50% of the total number of sprays or a maximum of 6 QoI fungicide applications, whichever is the lower.
- Where resistance has been confirmed, QoI fungicides must be applied only in mixture with partners contributing to the effective control of the target pathogens.

3.9.5 CAA Fungicides (FRAC Code 40) (Updated April 2020)

- Apply CAA fungicides preferably in a preventive manner.
- Alternation with fungicides having other modes of action is recommended in spray programs.
- No resistant isolates from field populations have been found since the introduction of CAA fungicides in 1993.
- Phytophthora infestans is classified by FRAC as a medium risk pathogen. Long-term experience with CAA fungicides demonstrates that the resistance risk of Phytophthora infestans to this fungicide group is low to moderate. For effective resistance management, a precautionary strategy has to be implemented.
- Apply a maximum of 50% of the total number of intended applications for late blight control.
- For more detailed product recommendations, refer to the use guidelines published by the respective CAA manufacturers.

3.9.6 Phenylamide Fungicides (FRAC Code 4) (Updated March 2020)

- The phenylamides should be used on a preventive and not curative or eradicative basis.
- For foliar applications, the phenylamides should be used in a pre-packed mixture containing an unrelated effective partner and used in a sound management program. Where residual partners are used, it is recommended to use between three quarters and full recommended rates. The phenylamide dosage in the mixture depends on its intrinsic activity and is defined by the respective company.
- The number of phenylamide applications should be limited (two to four applications per crop and year, with a maximum of two consecutive applications). The application intervals should not exceed 14 days and may be shorter in cases of high disease pressure. If rates and application intervals are reduced, the total amount of the phenylamide fungicide used per season should not exceed that of the full rate, and the total exposure time should remain the same. The rate of the mixing partners should remain the same for both intervals.
- Phenylamide sprays are recommended early season or during the period of active vegetative growth of the crop. The farmer should switch to non-phenylamide products not later than the normal standard application interval of the non-phenylamide product.

3.9.7 QiI Fungicides (FRAC Code 21) (Updated February 2021)

- Apply QiI fungicides preferably in a preventive manner.
- Apply a maximum of 50% of the total number of intended applications for late blight control during one crop cycle.

- Alternation with fungicides having other modes of action are recommended in spray programs.
- Apply QiI fungicides according to manufacturers' instructions.

3.9.8 Fluzinam Fungicides (FRAC Code 29) (Updated June 2020)

- Apply fluazinam preventatively.
- Maximum of six applications.
- In regions with reported resistance, it is recommended to limit the number of fluazinam applications to max. 50% of all applications, and use mixtures with fungicides belonging to other modes of action that provide satisfactory efficacy against Phytophthora infestans.
- No more than three sequential applications of fluazinam. In regions with resistance or reduced sensitivity, apply a maximum of two sequential applications if product is used solo.
- Refer to manufacturer's recommendations for rates and intervals.

3.9.9 Cymoxanil Fungicides (FRAC Code 27) (Updated February 2021)

- Use always in mixture with another fungicide active on the target diseases.
- Apply preventatively.
- The number of applications of cymoxanil-containing products should be restricted: Potato and Tomato: 6.
- Always follow product specific label recommendations for resistance management.

3.9.10 Ametoctradin Fungicides (FRAC Code 45) (Updated June 2021)

- Apply ametoctradin containing products in a preventative manner.
- Always follow product-specific recommendations for resistance management.

3.10 Importance of Multisite Fungicides in Managing Pathogen Resistance

One of the key recommendations of FRAC is to make use of multisite fungicides (FRAC Group M) in spray programs, especially in crops with multiple sprays such as fruits and vegetables, or certain arable crops. Due to their mode of action, multisite fungicides are considered as a low resistance risk group. Therefore, they offer the possibility for use as mixing partners or alternating with single site and other medium- to high resistance-risk fungicides. Over the past decades, no cases of field resistance against multisite have been reported. There are clear benefits to recommending multi-site fungicides in spray programs:

- Multisite fungicides display a low risk to develop resistance and are effective mixing/alternating partners for medium- to high-risk fungicides.
- Beyond protecting and prolonging the lifespan of highly effective medium- to high resistance-risk fungicides, multisite fungicides provide added levels and spectrum of disease control. With this they can also support the single sites to be even more efficient.
- Multisite fungicides are considered a valuable tool to manage resistance by preventing or delaying its development to many pathogens in many crops.
- In some crops, multisites play an increasing role in spray programs to sustain effective disease control and resistance management, e.g., for *Zymoseptoria tritici* in wheat, *Ramularia collo-cygni* in barley, and *Phakopsora pachyrhizi* in soybeans.

Restricting the use of multisite fungicides from use in important crops could result in faster development of resistance to single site mode of action fungicides. This in turn could lead to epidemic disease development, serious crop losses, and finally the loss of highly effective fungicides for a sustainable disease management.

3.11 Conclusion and Future Outlook

To sustain ever burgeoning human population and with eminent climate change risk of invasive or lesser-known pathogens, disease management practices will play a key role in coming few decades for maintain sustainable food supply. Disease control chemicals (fungicides) will be the effective and sure shot tool in the arsenal for integrated disease management practices for minimizing crop losses. With recent advancements in advanced digital tools and predictive tools, the applications will be much more precisely targeted in a field at disease spots/patches thereby reducing the environmental loading of the active making it more sustainable. Even with newer modes of action, safer chemistries are being discovered; resistance management program still plays a pivotal role for enhancing the product life cycle. Generating baselines before launching actives, having monitoring program, and regularly reviewing the resistance management strategies will continue to play key role. The advent of new technologies and real-time detections will help organizations like FRAC not only to communicate and manage things faster but the collaboration among the stakeholders be it private corporates or public sector researchers by sharing of the data or publications will be faster and will surely help the farmers and the end consumers.

References

Avenot HF, Michailides TJ (2007) Resistance to boscalid fungicide in *Alternaria alternata* isolates from pistachio in California. Plant Dis 91:1345–1350

Brent KJ (1982) Case study 4: powdery mildews of barley and cucumber. In: Dekker J, Georgopoulos SG (eds) Fungicide resistance in crop protection. Pudoc, Wageningen, pp 219–230

Brent KJ (2012) Historical perspectives of fungicide resistance. In: Thind TS (ed) Fungicide resistance in crop protection: risk and management. CAB International, Wallingford, pp 3–18

Chin KM, Chavaillaz D, Kasbohrer M, Staub T, Felsenstein FG (2001) Characterising resistance risk of *Erysiphe graminis* f. sp. *tritici* to strobilurins. Crop Prot 20:87–96

De Waard MA (1994) Resistance to fungicides which inhibit sterol 14a-demethylation, an historical perspective. In: Heaney S, Slawson D, Hollomon DW, Smith M, Russell PE, Parry DW (eds) Fungicide resistance. British Crop Protection Council, Farnham, Surrey, pp 3–10

Dekker J (1986) Preventing and managing fungicide resistance. In: committee on strategies management of pesticide resistant pest population, national research council (eds) Pesticide resistance: strategies and tactics for management. National Academic Press, Washington, DC, pp 347–354

Eckert JW (1982) Fungicide resistance in crop protection. In: Dekker J, Georgopoulos SG (eds) Case study 5: penicillium decay of citrus fruits. Pudoc, Wageningen, The Netherlands, pp 231–250

Forsyth W (1802) A treatise on the culture and management of fruit trees. Nichols, London

FRAC (Fungicide Resistance Action Committee) (2021) FRAC list of plant pathogenic organisms resistant to disease control agents. www.frac.info. Accessed 30 Aug 2021

Hewitt HG (1998) Fungicides in crop protection. CAB International, Cambridge

Jørgensen LN, van den Bosch F, Oliver RP, Heick TM, Paveley N (2017) Targeting fungicide inputs according to need. Annu Rev Phytopathol 55:181–203

Kretschmer M, Leroch M, Mosbach A, Walker A-S, Fillinger S, Mernke D et al (2009) Fungicide-driven evolution and molecular basis of multidrug resistance in field populations of the grey mould fungus *Botrytis cinerea*. PLoS Pathog 5(12):e1000696. https://doi.org/10.1371/journal.ppat.1000696

Kuck KH, Russell EP (2006) FRAC: combined resistance risk assessment. Asp Appl Biol 78:3–10

Miles TD, Miles LA, Fairchild KL, Wharton PS (2014) Screening and characterization of resistance to succinate dehydrogenase inhibitors in *Alternaria solani*. Plant Pathol 63:155–164

Millardet PMA (1885) Traitement du mildou par le melange de sulphate de cuivre et de chaux. J d'Agric Practique 49:513–516. (In French)

Noble M, Maggarvie QD, Hams AF, Leaf LL (1966) Resistance to mercury of Pyrenophora avenae in Scottish seed oats. Plant Pathol 15:23–28

Russell PE (2005) Sensitivity Baselines in Fungicide Resistance Research and Management. FRAC Monograph No. 3. Aimprint, UK

Staub T (1994) Early experiences with phenylamide resistance and lessons for continued successful use. In: Heaney S, Slawson D, Hollomon DW, Smith M, Russell PE, Parry DW (eds) Fungicide resistance. British Crop Protection Council, Farnham, Surrey, pp p131–p138

Szkolnik M, Gilpatrick JD (1969) Apparent resistance of *Venturia inaequalis* in New York apple orchards. Plant Dis Rep 53:861–884

van den Bosch F, Paveley N, Fraaije B, van den Berg F, Oliver R (2015) Evidence-based resistance management: a review of existing evidence. In: Ishii H, Hollomon DW (eds) Fungicide resistance in plant pathogens: principles and a guide to practical management. Springer, Tokyo, pp 63–76

Role of Plant Nutrition in Disease Development and Management

4

Jagdev Sharma, V. K. Dua, Sanjeev Sharma, Anil K. Choudhary, Prince Kumar, and Anil Sharma

Abstract

Potato is one of the major contributors in combating malnutrition across the globe besides ushering in global food security. However, potato as well as production systems involving potato face many challenges from numerous biotic and abiotic stresses. Plant diseases are the major biotic stress for this crop for centuries across the agro-ecologies and farming situations. Relationship between resistance and susceptibility to plant diseases and plant nutrition is receiving more attention than ever to devise advanced and safe precision disease management prescription across the globe. Many studies have pointed out the definite role of plant nutrients and beneficial elements in promotion or suppression of disease resistance/tolerance, susceptibility, and severity and thereby their use in disease management options. Among essential and beneficial elements, N, P, K, Ca, S, B, Zn, Cu, Cl, and Si have a proven role in plant diseases. Phenolic compounds and lignin content constitute the defense system of plants against infection, and plant nutrients affect their synthesis by affecting growth and chemical composition of the tissues. Besides induction of resistance reactions in plants against pathogens, there is an enduring, non-specific resistance against pathogens' systemic resistance acquired after application of synthetic compounds including plant nutrients. This resistance is related to the formation of structural barriers such as lignification, induction of pathogenesis-related proteins, and conditioning of the plants. Thus, systemic induced resistance

J. Sharma (✉) · V. K. Dua · S. Sharma · A. K. Choudhary
Division of Plant Protection, ICAR–Central Potato Research Institute, Shimla, Himachal Pradesh, India
e-mail: jagdev.sharma@icar.gov.in

P. Kumar · A. Sharma
ICAR–Central Potato Research Institute-Regional Station, Jalandhar, Punjab, India

S. Kumar Chakrabarti et al. (eds.), *Sustainable Management of Potato Pests and Diseases*, https://doi.org/10.1007/978-981-16-7695-6_4

(SIR) caused by application of plant nutrients in balanced amounts could be an alternative strategy to manage the diseases and reduce pesticide residues.

Keywords

Disease resistance · Nutrients · Plant health · Potato · Systemic induced resistance (SIR) · Soils

4.1 Introduction

Potato is the third most important food crop after two major cereals, wheat and rice, in many parts of the world and, thus, plays a vital role in global food security and economy as well. In general, the plant diseases continue to play a major limiting role in the agricultural production in intensively managed crops like potato. The potato crop is reported to be affected by numerous insect pests and diseases which are responsible for the significant yield losses in potato production systems in different parts of the world. Conventional disease management approaches are known to cause degradation of the environment and contamination of the food with synthetic chemicals and toxins. Indiscriminate use of fungicides may also escalate the production costs and farm expenditures besides inducing resistance development against fungicides. Increasing global concerns about food safety, environmental quality and pesticide resistance are also pressing upon devising the alternative pest management techniques (Choudhary and Rahi 2018). Research reports indicate that the mechanisms for disease tolerance are multicomponent (Reuveni et al. 1998), and plant disease resistance many a times is induced in response to the pathogen. Apart from new virulent phytopathogenic races, the abiotic stresses and cultural factors such as intensive use of fertilizers, irrigation water, and pesticides may also modify the rate of development of diseases and more importantly the disease resistance reactions in the crops (Lambert et al. 2005).

Relationship between resistance and susceptibility to plant diseases and the fertilizer types and their application doses has been extensively investigated world over, but currently it is receiving more attention than ever to devise advanced precision disease management prescription across the globe. Over 2440 global studies, which included $\geq$400 diseases and pests, have reported some relationships between potassium (K) alone or combined with other elements and the plant health status (Perrenoud 1990). Thus, there exists the role of plant nutrients and beneficial elements in promotion and suppression of disease resistance/tolerance, susceptibility and severity, and thereby their use in disease management options. Resistance to diseases and insect pests can be systemically induced in plants by non-pathogens or with chemicals such as phosphate salts (Doubrava et al. 1988; Descalzo et al. 1990; Kuc 1995; Ye et al. 1995). There are numerous factors that may explain variations including rates and types of fertilizer applied, soil NPK status, and crop susceptibility to pests and diseases as well as the congenial environment (Fig. 4.1). Thus, the balance between various nutrients is as important as their absolute applied rates for better crop health status. For example, effect of increased N rates on reduction in crop resistance is less when K

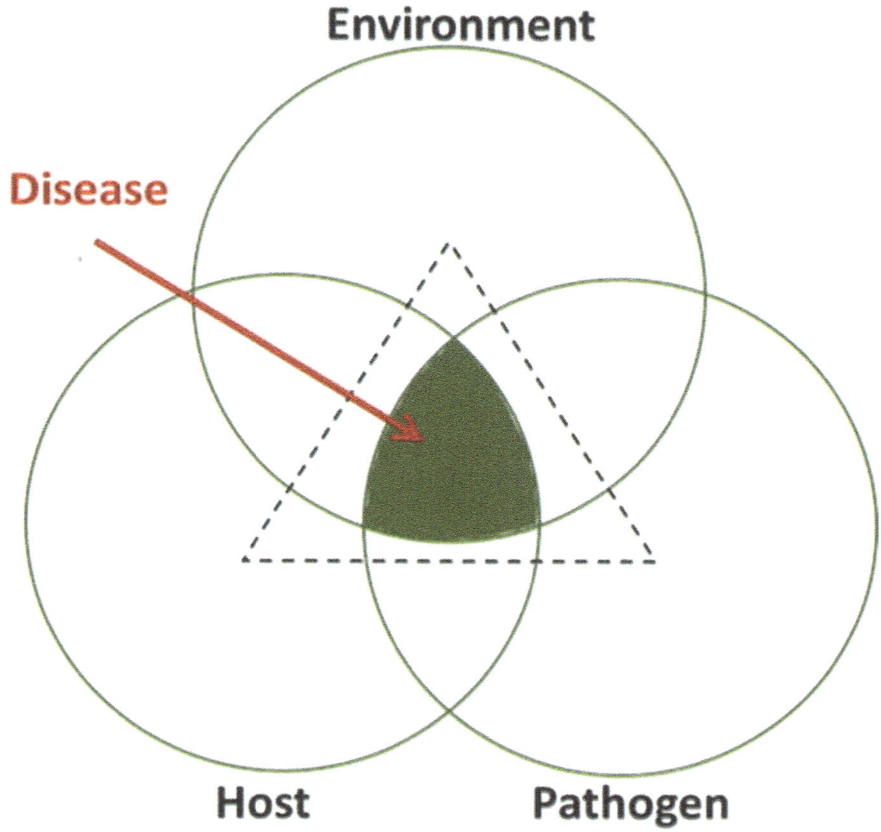

Fig. 4.1 Factors responsible for disease development (Source: Huber et al. 2012)

nutrition is applied in adequate amounts (Lee 1966; Webb and George 1968; Singh 1978; Jensen and Munk 1997). In general, P and K tend to improve the plant health, and their balanced application may reduce the plant diseases in ~65% of cases and may increase the diseases or insect pests in ~28% of cases (Perrenoud 1990) by direct and indirect effects. The P and K may affect the reaction of a plant to pests through (1) direct effects on the pathogen multiplication, development, and survival, (2) direct effects on the internal metabolism of the plant affecting food supply to the pathogen, and (3) the effects on the establishment of the pathogen and its spread within the plant, through the influence of the elements on plant defense responses and cell wall ultrastructures and function of stomata. In case of host resistance to diseases, it is generally observed that the plant nutrition and health is being overlooked. However, disease development rate can be reduced by balanced mineral nutrition in many crops. Thus, optimal fertilization can be an integral component of an integrated program for disease management. Strengthening the natural plant resistance is an important aspect of fertilization practice, and the precision nutrient management like foliar fertilization is the new plant nutrition innovations which may regulate the host-pathogen interactions and the plant resistance to diseases. In this chapter, we have currently reviewed and discussed the role of plant nutrition in disease occurrence, severity, and disease management *vis-à-vis* resistance.

4.2 Plant Nutrition, Disease Occurrence, and Host-Pathogen Interactions in Relation to Nutrient Acquisition

Plants require 17 essential nutrient elements for normal growth and development usually grouped as primary nutrients, secondary nutrients, and micronutrients. Out of 17, 14 nutrients are taken up from the soil. These primary macronutrients [nitrogen (N), phosphorus (P), and potassium (K)], secondary macronutrients [calcium (Ca), sulfur (S), and magnesium (Mg)], and micronutrients [manganese (Mn), iron (Fe), zinc (Zn), copper (Cu), boron (B), molybdenum (Mo), chlorine (Cl), nickel (Ni)] besides being essential for plant growth are also important for disease resistance in plants (Datnoff et al. 2007). It has been observed that fertilization reduces the disease severity when plants have deficiency of the particular nutrient being supplied through fertilizer; hence, optimal fertilization may reduce the disease development to have optimized plant growth (Huber and McCay-Buis 1993). For example, optimal-N in cereal crops reduced the incidence of take-all (Huber and McCay-Buis 1993). The P reduced both take-all and *Pythium* root rot infection in cereals (Kiraly 1976; Huber 1980). Similarly, when a plant is infected by a pathogen, its physiology gets impaired *vis-à-vis* nutrient uptake, assimilation, translocation from the root to the top of the plant shoot, and finally the utilization (Marschner 1995).

Some plant pathogens immobilize the nutrients in the rhizosphere and in infected tissues such as roots, while others interfere with translocation causing nutrient deficiency in some plant parts or the excess accumulation causing nutrient toxicity (Huber and Graham 1999). Some soil-borne pathogens cause root infection and reduce the root water and nutrient acquisition ability (Huber and Graham 1999). Some plant diseases also attack the vascular system, thus impairing nutrient translocation and utilization. Pathogens can also affect membrane permeability or mobilization towards infected sites, which can induce nutrient deficiency or toxicity. The fungus *Fusarium oxysporum* fsp. *Vasinfectum* can increase the concentration of P in plant leaves and may decrease the concentration of N, K, Ca, and Mg as well (Huber and Graham 1999). On the other hand, a reverse trend was observed for cereal rusts and powdery mildew with increasing N application causing increased incidence. Thus, there exists a critical role of balanced crop nutrition in preventing diseases.

Fertilizer application affects the development of plant disease under field conditions directly by improving nutritional status of the plant and indirectly by affecting the conditions which can influence the development of the disease such as dense stands, changes in light interception, and humidity within the crop stand. It is important to provide a balanced nutrition and at the time when the nutrient can be most effective for disease control and also for higher yield. Not only the application

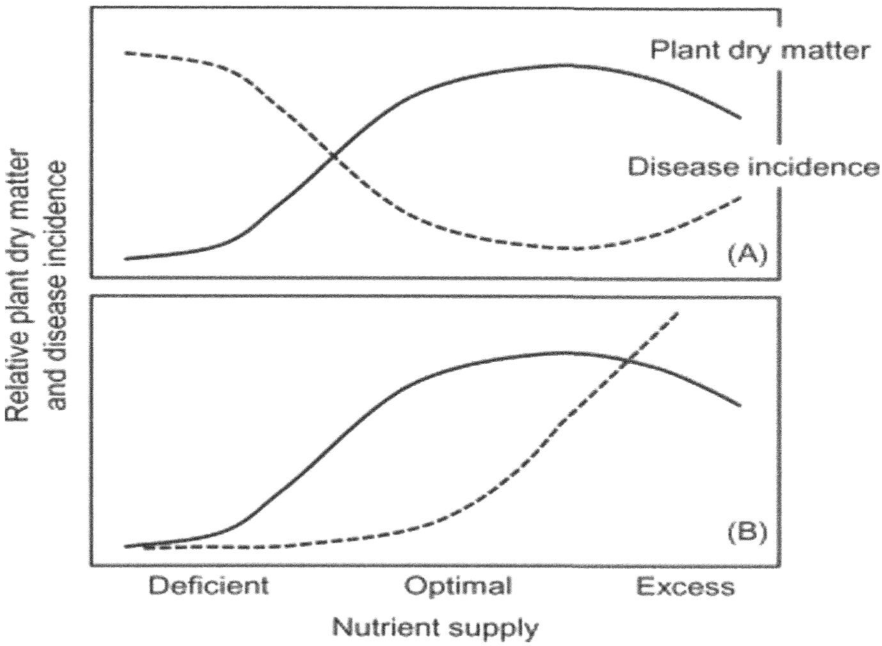

Fig. 4.2 Relationship in plant nutrition and disease occurrence (Source: Huber et al. 2012)

of the fertilizer can affect the disease development but also anything that affects the soil environment such as pH modification through liming, tillage, seedbed firmness, moisture control (irrigation or drainage), crop rotation, cover crops, green manures, manures, and intercropping. There are several examples of disease control through nutrient manipulation which can be achieved by either modifying nutrient availability or modifying nutrient uptake (Huber and Graham 1999). However, the application of fertilizers to the soil is not always effective, such as in the case of Mn, Zn, and Fe in high pH soils with high concentrations of free $CaCO_3$, or where rapid oxidation by microorganisms makes Mn unavailable in the soil. Many times, it is recommended to use foliar applications which relieve aboveground deficiency symptoms, but Mn is not well translocated in the phloem so the root tissues which are attacked by the pathogens remain Mn-deficient (Huber and McCay-Buis 1993). Apart from mineral fertilizers, addition of microorganisms such as bacteria, fungi which form mycorrhizae, and any plant growth-promoting organisms can increase nutrient uptake (P, Zn, Mn) by influencing minor element availability through their oxidation-reduction reactions or siderophore release (Huber and McCay-Buis 1993; Kumar et al. 2016, 2017, 2018). There exists a critical role of balanced crop nutrition in preventing diseases. The nutritional plant status may substantially influence the degree of disease infection by affecting growth and chemical composition of the tissues with respect to plant nutrients (Fig. 4.2). The risk of infection is minimal under optimal nutrient supply although the interactions of different nutrients and disease pathogens are complex and depend on many factors. Hence, here we have described the effect of each nutrient on certain diseases along with their possible *modus operandi* for the tolerance or resistance to the particular pathogen.

4.3 Plant Nutrition and Systemic Induced Resistance or Systemic Acquired Resistance

In general, resistance to diseases in plants can be increased by increased degree of lignification and/or silification, production of inhibitory substances in higher amounts through physiological and biochemical changes, and restricting nutrient transfer to the pathogen which it requires for growth or development. The induction of resistance reactions in plants against pathogens is a well-known phenomenon in plant pathology. It was first described as a resistance to an attack from a non-virulent pathogen. Thus, it is an enduring, non-specific resistance against pathogens, induced by pathogens that cause a necrotic reaction on the infected leaves, and it is called systemic acquired resistance (SAR) if the resistance is systemically distributed within the plant. SAR can be induced by avirulent pathogens and also by chemical compounds such as salicylic acid (SA), which is involved in the signal transduction pathway leading to SAR. Structural analogues of SA can also induce SAR. Wiese et al. (2003) introduced the term chemically induced resistance (CIR), which is used to describe the systemic resistance after application of synthetic compounds. This resistance is related to the formation of structural barriers such as lignification, induction of pathogenesis related proteins, and conditioning of the plants (Graham and Webb 1991). Systemic induced resistance (SIR) has been found to be induced by foliar sprays of nutrients such as phosphates, K, and N. It has been hypothesized that during SIR an immunity signal released or synthesized at the induction site of the inducer leaf is systemically translocated to the challenged leaves, where it activates the mechanisms for defense (Reuveni and Reuveni 1998). Salicylic acid (SA) has been hypothesized as a possible signal, and its exogenous application induces resistance and pathogenesis related (PR) proteins, which typically accompany SIR (Reuveni and Reuveni 1998). However, SA was found in the phloem sap of non-infected upper leaves when it could not be detected in the phloem sap collected from petioles of the lower leaves infected with *Pseudomonas syringae*. This indicates that SA may not be the primary systemic signal for SIR. A single phosphate foliar application can induce high levels of systemic protection against powdery mildew caused by *Sphaerotheca fuliginea* in cucumbers (Reuveni et al. 1997a, 1997b). A similar response was found in maize, where foliar spray with phosphates induced a systemic protection against common rust (caused by *Puccinia sorghi*) and northern leaf blight (caused by *Exserohilum turcicum*).

Trace elements may also play an important role in plants, affecting their susceptibility to fungal or bacterial phytopathogens (Graham 1983). Foliar spray with H_3BO_3, $CuSO_4$, $MnCl_2$, or $KMnO_4$ separately induced systemic protection against powdery mildew in cucumber plants. Similar results were found in wheat, where application of B, Mn, and Zn separately increased the resistance of plants to tan spot

(Simoglou and Dordas 2006). The mechanism of SIR development is still unknown, and it was proposed that the chemicals trigger a release and rapid movement of the *immunity signal* from the infected leaves to the unchallenged ones (Reuveni and Reuveni 1998). The mechanism might involve an increase in both solute and ionically bound components of peroxidase activity and β-1,3-glucanase in protected leaves above those sprayed with $MnCl_2$. Mn and Cu might act as cofactors of metalloprotein enzymes such as peroxidase, for which Mn ions serve as an inducing agent (Marschner 1995; Mengel and Kirkby 2001). Peroxidase and β-1,3-glucanase are involved in the cross-linking of the cell wall components, polymerization of lignin and suberin monomers, and subsequent resistance to pathogens. SA is proposed to be a translocatable signal compound in SIR and interacts with intercellular Ca^{2+} in the induction of chitinase in carrot suspension culture. Application of cations such as Mn, Cu, and B can increase the Ca^{2+} cations and interact with SA and activate SIR (Reuveni and Reuveni 1998). These findings indicate that the mechanism for resistance is present in susceptible plants and it can be induced by simple inorganic chemicals and that this induced resistance is not pest-specific. There is no doubt that plant nutrition affects the disease incidence; however, the effect is disease and element specific. The role of important elements including essential and beneficial ones is described in this chapter.

4.4 Role of Nutrient Elements in Disease Incidence

4.4.1 Nitrogen

Nitrogen (N) is the most important plant nutrient for growth and productivity. The two forms of nitrogen (i.e., NO_3^- and NH_4^+) absorbed by the plant are assimilated differently (Dhillon et al. 2018). These two forms have a profound effect on plant diseases. There is an extensive literature about the effect of N on plant disease avoidance and disease development (Marschner 1995). There are several reports on effect of N on disease development contradicting among them where real causes of this inconsistency are poorly understood (Marschner 1995; Hoffland et al. 2000). These differences may be due to the form of N applied to host (Harrison and Shew 2001), the type of pathogen (Marschner 1995), or the N application timings (Carballo et al. 1994). The variable N effects on disease development in the literature may be due to differential responses depending on pathogen type and their pathogenicity *modus operandi*. In case of obligate parasites like *Puccinia graminis* and *Erysiphe graminis*, high N supply may increase the disease severity. In case of facultative parasites like *Alternaria*, *Fusarium*, and *Xanthomonas* spp., high N supply may decrease the infection severity. In case of soil-borne pathogens, the situation is more complex as the microorganism diversity on root surface is more than the bulk soil. Likewise, there is competition among different microorganisms, and there are chemical barriers such as high concentration of polyphenols in the rhizodermis and physical barriers such as silicon depositions on the endodermis (Huber 1980). The difference between the obligate and facultative parasites is due to their different nutritional requirements. Obligate parasites require assimilates supplied directly from living cells, while facultative parasites are semi-saprophytes which prefer senescing tissue or which release toxins in order to damage or kill the host plant cells. Therefore, all factors which support the metabolic activities of the host cells and which delay the senescence of the host plant can increase resistance or tolerance to facultative parasites (Agrios 2005; Vidhyasekaran 2004). In case of obligate fungal parasites, the nutritional requirements of the parasites cause changes in the anatomy and physiology of the host plant in response to N. At high N rates, there is a higher growth rate during the vegetative stage making plants more disease susceptible. Also, there is an increase in amino acid concentration in the apoplast and on the leaf surface, which promotes the germination and growth of conidia (Robinson and Hodges 1981).

In potatoes, nitrogen deficiency is related to early blight (*Alternaria solani*) incidence. Decrease in early blight infection with the higher rates of nitrogen has been reported by Horsfall and Heuberger (1942) and Walker et al. (1944). According to Horsfall and Dimond (1957), early blight is a low sugar disease and that any factors reducing the sugar content of leaves would increase early blight infection. High nitrogen delays plant maturity especially when other elements are not adequately supplied, while high phosphorus hastens plant maturity. *Phytophthora infestans* (Mont.) de Bary is the oomycete, which was responsible for infamous Irish potato famine, and it still continues to cause worldwide devastation of the potato. Avoiding use of excess nitrogen and use of moderate nitrogen fertilization is often recommended as cultural practices to delay the development of late blight. The possibility of controlling *Alternaria* diseases by surplus application of nitrogen has been studied, and various trials were also successful (Barclay et al. 1973; Soltanpour and Harrison 1974; MacKenzie 1981). However, in practice, control of *Alternaria* by adding more nitrogen to the soil is economically not advantageous as there are large differences between the fertilizer rates for optimal disease suppression and the rate for optimal yield. Other drawbacks associated with high levels of fertilizer are a reduction in tuber quality of potato and the hazard to groundwater. Thus, the crop should be fertilized for optimum yield, and *Alternaria* should be managed by properly timed applications of fungicides during the growing season (MacKenzie 1981).

In general, the susceptibility of potato, tomato, and cotton to *Alternaria* changes with host plant age and coincident with the changes in the contents of nutrients like nitrogen and potassium in the foliage. Plants in the vegetative phase are relatively resistant to the pathogen, and mature plants are highly susceptible to *Alternaria*. After the initiation of tuberization in potatoes, susceptibility to diseases increases gradually, and the shift in host response to *Alternaria* with age is not solely governed by the nutritional content of the foliage. Nevertheless, it can be hypothesized that higher contents of the foliage with nitrogen and potassium might enhance host resistance to *Alternaria* and thus reduce disease severity. There are reports which indicate that disease susceptibility depends on N-supply, and the effect of N-supply on susceptibility is pathogen-specific. At high N rates, some key enzymes of phenol metabolism have lower activity, the content of the phenolics decreases, and the lignin content may be lower – all these constitute defense system of plants against infection. In addition, at high N rates Si content decreases. Therefore, the main reason for the increased susceptibility to obligate parasites at high N rates is the various anatomical and biochemical changes together with the increase in low-molecular-weight organic nitrogen compounds which are used as substrates for parasites. It is believed that plants grown under conditions of low N-availability are better defended against pathogens because there is an increase in the synthesis of defense-related compounds (Hoffland et al. 1999; Wilkens et al. 1996; Hoffland et al. 2000). In case of obligate pathogens, viz., *Pseudomonas syringae* pv. *tomato*, *Ustilago maydis*, and *Oidium lycopersicum*, the increased susceptibility was observed when plants were grown with high N supply (Hoffland et al. 2000).

The form of N is also important in plant diseases (Harrison and Shew 2001). At high NO_3, *Fusarium oxysporum*, *Botrytis cinerea*, *Rhizoctonia solani*, and *Pythium* spp. get decreased. While at high NH_4, disease is decreased with respect to *Pyricularia*, *Thielaviopsis basicola*, *Sclerotium rolfsii*, and *Gibberella zeae*, since the form of N affects soil pH and so the availability of other nutrients such as Mn. The N level can affect the phenolics content of plants, which are precursors of lignin (Bana et al. 2018). At high N levels, there is a decrease in Si content which may affect the disease tolerance. That is why some studies show that adding N-rich soy-meal, meat-meal, and bone-meal to soil led to an increase in ammonia, nitrite, nitrate, pH, and bacterial quantity and suppressed the common scab in potato (Kopecky et al. 2021). The interaction between disease and host depends on several factors, viz., host response, previous crop, N rate, residual-N, time of N application, soil microflora, ratio of ammoniacal to nitrate N and disease complex presence, etc. In nutshell, the N nutrition and host-pathogen interactions are quite complex, and thus, to find a specific mechanism, more research is needed that may explain these observations.

4.4.2 Phosphorus

Phosphorus (P) is the second most important nutrient required for many organic molecules of the cell (phospholipids, DNA, RNA, ATP, etc.) and is involved in many plant metabolic processes (Kumar et al. 2017) and equally required in the pathogens too (Gupta et al. 2016).

One of the best defenses against root diseases is a vigorous and well-developed plant root system, and phosphorus is most beneficial when applied to control seedlings and fungal diseases where vigorous root development permits plants to escape disease (Huber and Graham 1999). Phosphate fertilization of wheat can have a significant effect and almost eliminate economic losses from *Pythium* root rot (Huber 1980). In maize, P application can reduce root rot, especially when it is grown on P-deficient soils (Huber and Graham 1999). A number of studies have shown that P application can reduce bacterial leaf blight in rice, downy mildew, and blue mold, leaf curl virus disease in tobacco, pod and stem blight in soybean, yellow dwarf virus disease in barley, and blast disease in rice (Kirkegaard et al. 1999;

Reuveni et al. 1998, 2000). However, some studies show that P application may increase the disease severity, *Sclerotinia* in many garden plants, *Bremia* in lettuce, and flag smut in wheat (Huber 1980). Foliar P application can induce local and systemic protection against powdery mildew in cucumber, roses, wine grapes, mango, and nectarines (Reuveni and Reuveni 1998). Plants under nutrient stress are more susceptible to disease attack, therefore, balancing P with other nutrient is essential in reducing the risk of disease occurrence. High levels of nitrogen (N) relative to P and other nutrients have been found resulting in severe outbreaks of Pythium, Rhizoctonia, and other diseases in turfgrass. Thomas (1948) found that low nitrogen reduced early blight and high phosphorus increased the disease in two out of three trials under greenhouse conditions. The work of Barclay et al. (1973) suggests that increased nitrogen and decreased phosphorus may be associated with disease resistance by extending the period of meristematic activity during which the plant can wall of the invading fungus. Higher dose of phosphorus and potassium has been found to produce higher potato yield in a late blight year (Roy et al. 2001). Although the role of phosphorus in resistance is variable and seemingly inconsistent (Kiraly 1976) under adverse or stressful conditions, early root development is especially important. Since phosphorus plays an important role in promoting rapid root development in young plants, its application can reduce the negative impact of root diseases in crops.

4.4.3 Potassium

Potassium (K) is the third important primary macronutrients required by the plants. Potassium is well known to induce the resistance against many biotic and abiotic stresses in the plants (Hamim and Choudhary 2019). Potassium decreases the susceptibility of host plants up to the optimal level for growth: beyond this point, there is no further increase in resistance which can be achieved by increasing the K supply (Huber and Graham 1999). The high susceptibility in K-deficient plant to parasitic diseases is due to the metabolic functions of K in plant physiology. Under K-deficiency, synthesis of high molecular-weight compounds (proteins, starch and cellulose) is impaired, while the accumulation of low-molecular-weight organic compounds increases. Potassium promotes the development of thicker outer walls in epidermal cells and thus prevents disease attack. K-deficient plants have impaired protein synthesis and accumulate simple N compounds such as amides which are used by invading plant pathogens. Tissue hardening and stomatal opening patterns are closely related to infestation intensity (Marschner 1995). In addition, the balance between N and K affects disease susceptibility in plants. K-fertilization can reduce the intensity of several infectious obligate and facultative parasite diseases. Potassium reduces the incidence of diseases like bacterial leaf blight, sheath blight, stem rot, sesamum leaf spot in rice, black rust in wheat, sugary disease in sorghum, bacterial leaf blight in cotton, *Cercospora* leaf spot in cassava, tikka leaf spot in peanut, red rust in tea, *Cercospora* leaf spot in mungbean, and seedling rot caused by *Rhizoctonia solani* (Sharma and Duveiller 2004; Sharma et al. 2005) and powdery mildew in grapes (Sharma et al. 2009; Sharma et al. 2012). Application of K can decrease *Helminthosporium* leaf blight severity and increase grain yields in wheat (Sharma et al. 2005). It has been shown that K dissolves the cell walls of the conducting vessels leading to wilting symptoms. Kowalska and Drożdżyński (2018) found that the percentage of potato late blight symptoms with foliar-K was higher than soil applied and foliar-K combined treatments. Since the K has direct synergistic relationships with Fe and Mn. Mn is an important component of photosynthesis, N metabolism, and N assimilation; it activates decarboxylase, dehydrogenase, and oxidase enzymes. Thus, higher K uptake under combined treatment observed less *P. infestans* in potato. High K has been reported to reduce *Alternaria solani* incidence in potato (Blachinski et al. 1996). Foliar application of urea and potassium nitrate, however, did not affect *Alternaria* severity as compared with the untreated control in any of the experiments. It is obvious that correlations between plant K nutritional status and disease incidence exist, but more information on physiological, metabolic, and hormonal processes crucial for plant susceptibility and sensitivity to pathogens parameters is required to establish these correlations. Once a clear correlation between specific K-dependent processes and responses and disease is established, then these features can be used as a diagnostic tool and form a basis for fertilizer and fungicide recommendations.

4.4.4 Calcium

Calcium (Ca) is another important secondary macronutrient that affects the susceptibility to diseases in two ways: (1) Ca is important for the stability and function of plant membranes, and when there is Ca deficiency, there is membrane leakage of low-molecular weight compounds, e.g., sugars and amino acids, from the cytoplasm to the apoplast, which stimulate the pathogen infection (Marschner 1995) and (2) Ca is an important component of cell wall structure as Ca-polygalacturonates are required in the middle lamella for cell wall stability. When Ca concentration drops, there is an increased susceptibility to fungi which preferentially invade the xylem. In addition, plant tissues low in Ca are more susceptible to parasitic diseases during storage. Adequate soil Ca is needed to protect peanut pods from infections by *Rhizoctonia* and *Pythium*, and application of Ca to the soil eliminates the occurrence of these diseases (Huber 1980). Ca induces resistance against *Pythium*, *Sclerotinia*, *Botrytis*, and *Fusarium* (Graham 1983). Ca can be mobilized in lesions of alfalfa caused by *Colletotrichum trifolii* and supports the growth of the pathogen by stimulating the macerating action of pectolytic enzyme polygalacturonic acid transeliminase (Kiraly 1976). Good calcium levels in potato tubers can reduce multiple quality problems including internal rust spot (IRS), internal browning, and hollow heart. Calcium also plays a role in reducing susceptibility to bruising and post-harvest diseases. A putative mechanism by which Ca is believed to provide protection against *Sclerotinia sclerotiorum* is by binding of oxalic acid or by strengthening the cell wall.

4.4.5 Sulfur and Magnesium

Sulphur is used as a fungicide in controlling many plant diseases.. The concept of sulfur-induced-resistance (SIR) was developed after a relationship between the S status and the disease incidence with *Pyrenopeziza brassicae* was uncovered and since then a lot of research was carried out to to identify metabolites, enzymes and reactions, which are potentially activated by the S metabolism to combat fungal pathogens. Sulphur deficient oilseed rape showed higher susceptibility for different pathogens and term 'sulfur enhanced defence' was introduced in the year 2005 . Elemental sulphur may enter directly through the fungal cell wall and diturb redox reactions in the metabolism of the pathogen. It is suggested that the fungicidal action of S is mainly related to the oxidation of important sulfhydryl groups (Beffa 1993). Sulphur can reduce the severity of potato scab (Klikocka et al. 2005). Recently a sulfur-containing volatile emitted by potato-associated bacteria confers protection against late blight through direct anti-oomycete activity (Chinchilla et al. 2019). Identification of the sulfur induced resistance and sulfur enhanced defense mechanisms can minimize input of fungicides by crop specific S fertilization since a higher resistance due to S will not be rapidly broken by new pathotypes.Magnesium (Mg) decreases the Ca content of peanut pods and may predispose them to pod breakdown by *Rhizoctonia* and *Pythium* (Huber 1980). Magnesium plays a major role in photosynthesis being a central atom of chlorophyll that captures the light energy (Marschner 1995). Magnesium is vital for transporting the phloem export of photosynthates, however, in the deficient conditions, the products like sucrose and amino acids get deposited in the leaves which create conducive environment for various disease-causing pathogens to attack (Huber and Jones 2013). The factors governing Mg availability in soils and its uptake may influence the Mg-induced resistance and/or susceptibility in host plants (Marschner 1995; Kumar et al. 2016). The effect of Mg has been investigated in some studies in reducing the disease severity in crops like rice, wheat, citrus, potato, poppy, and groundnut (Moreira et al. 2015).

4.4.6 Micronutrients

The effect of micronutrients on reducing the severity of diseases can be attributed to their involvement in physiology and biochemistry of the plant, as many of the essential micronutrients participate in many processes that can affect the response of plants to pathogens (Marschner 1995; Paul et al. 2016a; Heba et al. 2021). Micronutrients can affect disease resistance indirectly, as nutrient-deficient plants not only exhibit an impaired defense response but often may also become more suitable for feeding as many metabolites such as reducing sugars and amino acids leak outside the plant cell. Systemic acquired resistance (SAR) may be involved in the suppression of plant diseases by micronutrients. Reduction in disease severity has been reported in other crops after a single foliar application of H_3BO_3, $CuSO_4$, $MnCl_2$, or $KMnO_4$, which provided systemic protection against powdery mildew in cucumber plants (Reuveni et al. 1997a, 1997b; Reuveni and Reuveni 1998). These authors also suggested that application of nutrients such as Mn, Cu, and B can exchange and therefore release Ca^{2+} cations from cell walls, which interact with salicylic acid and activate systemic acquired resistance mechanisms. Micronutrients play an important role in plant metabolism by affecting the phenolics and lignin content and also membrane stability (Graham and Webb 1991; Heba et al. 2021). Micronutrients can affect resistance indirectly, as in deficient plants they become more suitable feeding substrate.

4.4.6.1 Zinc

Zinc (Zn) plays a vital role in plant metabolic and enzymatic processes affecting growth and development besides tolerance to many biotic and abiotic stresses (Paul et al. 2016a; Heba et al. 2016, 2021; Pooniya et al. 2019). Zn was found to have a number of different effects as in some cases it decreased, in others increased, and in others had no effect on plant susceptibility to disease (Graham and Webb 1991; Grewal et al. 1996). In most cases, the Zn application reduced disease severity, which could be because of the toxic effect of Zn on the pathogen directly and not through the plant's metabolism (Graham and Webb 1991). For example, plants suffering from a Zn-deficiency showed increased disease severity after infection by *Oidium* spp. (Bolle Jones and Hilton 1956). Zn plays an important role in protein and starch synthesis, and therefore a low Zn concentration induces accumulation of amino acids and reduces sugars in plant tissue (Römheld and Marschner 1991; Marschner 1995). As an activator of Cu/Zn-SOD, Zn is involved in membrane protection against oxidative damage through the detoxification of superoxide radicals (Cakmak 2000). Impairments in membrane structure caused by free radicals lead to increased membrane leakage of low-molecular-weight compounds, the presence of which favors pathogenesis (Graham and Webb 1991; Marschner 1995; Mengel and Kirkby 2001). Application of Zn to the soil reduced infections by *Fusarium graminearum* and root rot diseases, e.g., caused by *G. graminis* in wheat (Graham and Webb 1991; Grewal et al. 1996). Zn has a protective role against the damaging attack of highly toxic oxygen free radicals (Marschner 2011).

4.4.6.2 Boron

Boron (B) is the least understood essential micronutrient for plant growth and development, and at the same time, B deficiency is the most widespread micronutrient deficiency in the world (Brown et al. 2002; Blevins and Lukaszewski 1998; Römheld and Marschner 1991). Boron has a direct function in cell wall structure and stability and has a beneficial effect in reducing disease severity. In several diseases, however, the function of B in disease resistance or tolerance is the least understood of all the essential micronutrients for plants. The function that B in reducing disease susceptibility could be because of (1) the function in cell wall structure; (2) the function in cell membrane permeability, stability, or function; or (3) its role in metabolism of phenolics or lignin (Blevins and Lukaszewski 1998; Brown et al. 2002). Boron promotes stability and rigidity of the cell wall structure and therefore supports the shape and strength of the plant cell (Marschner 1995; Brown et al. 2002) thus possibly involved in the integrity of the plasma membrane (Marschner 1995; Brown et al. 2002; Dordas and Brown 2005).

Boron has been shown to reduce diseases caused by *Plasmodiophora brassicae* in crucifers, *Fusarium solani* and tobacco mosaic virus in bean, *Verticillium albo-atrum* in tomato and cotton, tomato yellow leaf curl virus in tomato, *G. graminis* (Graham and Webb 1991), and *Blumeria graminis* in wheat (Marschner 1995). Patrícia et al. (2018) reported that B and Zn could reduce the *Alternaria grandis* incidence significantly. Seed treatment with 3% boric acid as dip treatment before cold storage has been effective as a safe and effective chemical treatment for the control of black scurf (*Rhizoctonia solani*) and common scab (*Streptomyces scabies*) of potato (*Solanum tuberosum*). Frenkel et al. (2010) reported that when B was applied alone to field-grown potato plants, it did not reduce the severity of late blight, but together with a reduced rate of the fungicide propineb + iprovalicarb, B improved late blight suppression compared with plants treated with the fungicide alone. Under both B and Zn deficiencies, structural integrity of cell membranes is substantially impaired causing membranes to become leaky and unstable (Marschner 2011). Any impairment in membrane stability can cause a massive release of organic compounds from cells to the outside (Huber and Haneklaus 2007), representing a very suitable feeding medium for *Alternaria grandis*. Boron has a protective role also against the damaging attack of highly toxic oxygen free radicals (Marschner 2011). It is clear from the above findings that B and Zn may play a role in determining the intensity of early blight. However, research into the effects of B and Zn application either alone or in combination on early blight disease severity in potato plants is still incipient, and the results are inconclusive.

4.4.6.3 Manganese

Manganese (Mn) is the most studied micronutrient about its effects on diseases and is important in the development of resistance in plants to both root and foliar diseases (Graham and Webb 1991; Huber and Graham 1999; Heckman et al. 2003). Mn-availability in the soil varies and depends on many environmental and soil biotic factors. Mn is required in much higher concentration by higher plants than by fungi and bacteria, and there is opportunity for the pathogen to exploit this difference in requirement (Marschner 1995). Mn-fertilization can control a number of pathogenic diseases such as powdery mildew, downy mildew, take-all, tan spot, and several others (Brennan 1992; Huber and Graham 1999; Heckman et al. 2003; Simoglou and Dordas 2006). Despite the fact that Mn application can affect disease resistance because of the complex soil biochemistry of Mn, the use of Mn is limited due to the ineffectiveness and poor residual effect of Mn fertilizers on most soils that need Mn supplements. In most soils that require addition of Mn such as calcareous soils, 90–95% of added Mn is immobilized within a week. Mn has an important role in lignin biosynthesis, phenol biosynthesis, photosynthesis, and several other functions (Marschner 1995; Graham and Webb 1991). Mn inhibits the induction of amino-peptidase, an enzyme which supplies essential amino acids for fungal growth and pectin methylesterase, a fungal enzyme that degrades host cell walls. Mn controls lignin and suberin biosynthesis (Römheld and Marschner 1991;

Vidhyasekaran 1997) through activation of several enzymes of the shikimic acid and phenylpropanoid pathways (Marschner 1995). Both lignin and suberin are important biochemical barriers to fungal pathogen invasion (Kolattukudy et al. 1994; Rioux and Biggs 1994; Hammerschmidt and Nicholson 2000; Vidhyasekaran 1997, 2004), since they are phenolic polymers resistant to enzymatic degradation (Agrios 2005). Lignin and suberin are believed to contribute to wheat resistance against powdery mildew and to all diseases caused by *Gaeumannomyces graminis* (Rovira et al. 1983; Graham and Webb 1991; Huber 1996; Krauss 1999). It has also been shown that Mn soil applications reduce common scab of potato (Keinath and Loria 1996), *Fusarium* spp. infections in cotton, and *Sclerotinia sclerotiorum* (Lib. de Bary) in squash (Graham and Webb 1991; Agrios 2005).

4.4.6.4 Copper

Copper (Cu) is another important micronutrient which is a component of many enzymes (polyphenol oxidase, diamine oxidase, etc.) important for the synthesis of lignin that impart strength and rigidity to the cell wall, thus affecting disease tolerance (Marschner 1995; Broadley et al. 2012). Cu deficiency also dilates lipid structure in cell membranes, hence influencing resistance to biotic stress (Broadley et al. 2012). Reduced lignification in plants due to low Cu causes higher disease incidence like stem melanosis, take-all root rot, and ergot infection that can occur in Cu-deficient small grains (Marschner 1995). Cu nutrition may decrease many fungal and bacterial diseases associated by cell wall stability and lignification (Marschner 1995; Broadley et al. 2012). The Cu application to soil reduces leaf infections, like powdery mildew in wheat and ergot (*Claviceps* sp.) in wheat (Evans et al. 2007).

4.4.6.5 Iron

Iron (Fe) is an essential micronutrient for plants and their associated microbes; however, the role of Fe in disease resistance is not well studied in plants. Iron plays a very important role in chlorophyll formation; and peroxidase and catalase both of which are plant defense enzymes. Iron cofactors such as heme and Fe-sulfur clusters function in all primary metabolic processes, including respiration, DNA synthesis and repair, and cell proliferation and differentiation. In plants, iron is also essential for chlorophyll and hormone synthesis and photosynthesis. Plant genes encoding iron-binding ferritins (*FER*) are upregulated in many plants following infection, including potato tubers during infection by the oomycete *Phytophthora infestans* and in *Arabidopsis* during infection by the bacteria *D. dadantii* (Mata et al. 2001; Dellagi et al. 2005). Several plant pathogens, e.g., *Fusarium*, have higher requirements for Fe or higher utilization efficiency compared with higher plants. Therefore, compared to other micronutrients such as Mn, Cu, and B, microbes have requirement for Fe. Iron can control or reduce the disease severity of several diseases such as rust in wheat leaves, smut in wheat, and *Colletotrichum musae* in banana (Graham and Webb 1991; Graham 1983). Application of Fe to disease-suppressive soils increased take-all of barley, and in soils with a high disease score, Fe had no effect (Christos 2008). Iron can promote antimycosis or interfere with it. Iron does not seem to affect lignin synthesis, even though Fe is a component of peroxidase and stimulates other enzymes involved in the biosynthetic pathway. Iron can activate enzymes that are involved in the infection of the host by the pathogen or the defense and can promote synthesis of fungal antibiotics by soil bacteria (Graham and Webb 1991). Rhizosphere microorganisms can synthesize siderophores which can lower Fe level in the soils. These siderophores can suppress germination of chlamydospores of *Fusarium oxysporum* sp. *cucumerinum* in vitro. However, the production of siderophores and the antagonisms for Fe are not only mechanisms to limit the growth of parasitic fungus. The effect of Fe application is not as straightforward like Mn, Cu, and B as it can have both positive and negative effect on the host.

4.4.6.6 Chlorine

Chlorine (Cl) is required in very small amounts for plant growth, and Cl deficiency has rarely been reported as a problem in agriculture. However, there are reports showing that Cl application can enhance host plants' resistance to disease in which fairly large amounts of Cl are required, which are much higher than those required to fulfill its role as a micronutrient but far less than those required to induce toxicity (Mann et al. 2004). It has also been suggested that Cl might interact with other nutrients such as Mn. Cl has been shown to control a number of diseases such as stalk rot in corn, stripe rust in wheat, take all in wheat, northern corn leaf blight and downy mildew of millet, and septoria in wheat (Graham and Webb 1991; Mann et al. 2004). The mechanism of Cl's effect on resistance is not well understood. It appears to be non-toxic in vitro and does not stimulate lignin synthesis in wounded wheat leaves. It was suggested that Cl can compete with NO_{3-} absorption and influences the rhizosphere pH: it can suppress nitrification and increase the availability of Mn. Furthermore, Cl ions can mediate reduction of Mn *III*, *IV* oxides and increase Mn for the plant, increasing the tolerance to pathogens.

4.4.7 Beneficial Elements

Although Si is the second most abundant element in the earth's soil and is a component of plants, it is not considered to be an essential element as defined by Arnon and Stout, except for members of the Equisetaceae family (Marschner 1995). However, when Si is added to the soil, plants low in soluble Si show an improved growth, higher yield, reduced mineral toxicities, and better disease and insect resistance (Alvarez and Datnoff 2001; Seebold et al. 2004). Also, in many countries, crops such as rice and sugarcane which accumulate high levels of Si in plant tissue are fertilized routinely with calcium silicate slag to produce higher yields and higher disease resistance. Si has been shown to control a number of diseases such as blast (*Magnaporthe grisea*) in St. Augustine grass, brown spot (*Cochliobolus miyabeanus*) in rice, and sheath blight (*Thanatephorus cucumeris*) in rice and increase the tolerance of various turfgrasses to *Rhizoctonia solani*, *Pythium* spp., *Pyricularia grisea*, and *Blumeria graminis* (Carver et al. 1998; Savant et al. 1997; Alvarez and Datnoff 2001; Seebold et al. 2004; Zhang et al. 2006). The mechanism by which Si confers disease suppression is not well understood. It is believed that Si creates a physical barrier which can restrict fungal hyphae penetration, or it may induce accumulation of antifungal compounds such as flavonoid and diterpenoid phytoalexins which can degrade fungal and bacterial cell walls (Alvarez and Datnoff 2001; Brecht et al. 2004). In a field study, Shah et al. (2019) reported a significant reduction in virus incidence (mild and severe mosaic, leaf roll, and apical leaf curl) in potato plants with Si (potassium silicate). Both white fly and aphids play an important role in spreading virus in potato crop. The reduction in virus incidence as a result of silicon application was attributed to significant reduction in white fly (61.42%) and aphid population (48.98%) after 7 days of foliar application. Dry rot, caused by *Fusarium* spp., is one of the most important disease in storage and seed tubers after planting. *Fusarium* spp. cannot penetrate the sound periderm of tubers; consequently, infection can only occur through wounds or breaks in the periderm. Si has been used to enhance plant resistance against a broad range of bacterial and fungal pathogens; however, the enhanced late blight resistance and the molecular mechanisms involving the plant hormone pathways remain unclear. The mechanism of Si action in plant resistance is still unclear. Its deposition in plant cell walls raised the hypothesis of a possible physical barrier to pathogen penetration. However, the increased activity of phenolic compounds, polyphenol oxidases, and peroxidases in plants treated with Si demonstrates the involvement of this element in the induction of plant defense responses. As per reports, the mechanisms by which Si protects plants against pathogens mainly comprise of physical (Sun et al. 2010), biochemical, and molecular aspects increasing the activity of defense-related enzymes (Datnoff et al. 2007), stimulating the production of antimicrobial compounds and activating the expression of defense-related genes, and regulating the hormone signaling pathways, such as salicylic acid, jasmonic acid, and ethylene. Application of sodium silicate at 100 and 200 mM effectively controlled dry rot of tubers that were challenged by inoculation with a *F. sulphureum* spore suspension thereby suggesting that sodium silicate has direct fungitoxic activity against the pathogen (Li et al. 2009). Xue et al. (2021) reported that treatment of potato plants with Si was found to enhance late blight resistance in both detached leaves and living plants accompanied by induction of reactive oxygen species (ROS) production and pathogenesis-related genes expression.

Apart from the essential nutrients, there are a number of other elements like Li, Na, Be, Al, Ge, F, Br, I, Co, Cr, Cd, Pd, and Hg that are found in plant tissue in trace amounts and have occasionally been linked with host-pathogen relationships. Li and Cd through their marked suppressive effects on powdery mildews are the most noteworthy. Cd was found to inhibit spore germination and development at a concentration of 3 mg kg^{-1}, which is not toxic but elicits a response to infection in the host. Cd and Hg can also promote synthesis of lignin in wheat (Graham and Webb 1991). The mechanism of Li is not known, and it is quite possible that it catalyzes a metabolic pathway which can function in defense.

4.5 Cultural Methods for Improved Plant Nutrition and Disease Resistance

Not only the application of nutrients as fertilizers can increase the tolerance to the disease, but any measure that can increase the availability and limit the imbalance of certain elements can affect growth and the tolerance of diseases. Most of the approaches that are used in sustainable agriculture have been found to provide a balanced plant nutrition and at the same time to increase the availability of certain elements and improve the tolerance of plants to disease (Oborn et al. 2003). Approaches such as soil test-based nutrition, balanced fertilization, biofertilizers, foliar fertilization, crop rotations, green manuring, manures, intercropping and tillage, etc. can affect the plant nutrition (Fig. 4.3) and hence may further induce the disease resistance. Most of these approaches can significantly increase soil organic matter, which is very important in sustainable agriculture.

Fig. 4.3 Nitrogen deficiency increases early blight incidence in potato

4.5.1 Soil Organic Matter

Soil organic matter (SOM) content and quality affects many soil functions which are related to soil health such as soil microbial diversity, moisture retention, infiltration, release, and also plant health (Paul et al. 2016b; Choudhary and Rahi 2018; Singh et al. 2021). Field applied organic residues (crop residues, cover crops, and organic wastes) can affect soil-borne pathogens and diseases, and it is a cultural practice that can affect the availability of nutrients (Stone et al. 2004). Practices such as addition of sphagnum peat, green manures, and animal manures have been shown to produce suppressive soils on which pathogens do not establish or persist and do not affect the crop plants. Addition of sphagnum peat to soil has been shown to suppress disease caused by *Pythium* spp. (Hu et al. 1997). Also, addition of different organic amendments has been shown to reduce *Phytophthora* root rot in a number of species (Hoitink et al. 1977; Spencer and Benson 1982; Szczech et al. 1993; Dixon et al. 1990; Hu et al. 1997). Organic manure can suppress a number of pathogens in sweet corn (causal agents *Drechslera* spp., *Phoma* spp., and *Pythium arrhenomanes*) and snap bean (causal agents *Fusarium solani* and *Pythium* spp.). There are several mechanisms that are proposed to be involved in biologically and organic material-mediated disease suppression such as microbiostasis, microbial colonization of pathogen propagules, and destruction of pathogen propagules, antibiosis, and competition for substrate colonization, competition for root infection sites, and induced systemic resistance (or systemic acquired resistance SAR). SOM can impact not only the total soil nutrient content but also nutrient availability through the activity of soil microorganisms (Kumar et al. 2016; Singh et al. 2020, 2021). Therefore, nutrients can affect disease incidence by increasing plant resistance, improving plant growth (allowing the plant to escape the disease), and influencing the pathogen's environment. Although quantity and quality can have dramatic impacts on soil and plant nutrient content; there are only a few studies which focus on soil properties and disease incidence which investigate the contribution of soil or tissue nutrient contents to disease-suppressive effects. Fields with a history of annual organic amendments had higher microbial activity and K contents. Lower NO_3 content and corky root incidence were positively correlated with soil NO_3 and plant tissue N and negatively correlated with soil N mineralization potential, microbial activity, total soil N, and soil pH. In another study, composed bio-solids improved ryegrass establishment, growth, and tolerance to leaf rust (caused by *Puccinia* spp.) by improving N nutrition in the amended soil (Loschinkohl and Boehm 2001).

4.5.2 Crop Rotation and Cover Crops

Long-term experiments (>100 years) showed that crop rotation together with other fertility management practices is fundamental to long-term agricultural productivity and sustainability (Reid et al. 2001; Stone et al. 2004). The most straightforward principle underlying rotation as a disease control strategy is that plant pathogen propagules have a lifetime in soils, and rotation with non-host crops starves them out

(Reid et al. 2001). In bean crops, rotation is the most powerful and effective practice to control bean diseases (Choudhary et al. 2020). Crop rotation can increase N levels and can also affect the availability of other nutrients which can affect the disease severity (Reid et al. 2001; Huber and Graham 1999). Crop rotations affect the survival of pathogens and have been used extensively to reduce the severity of many diseases. A nutrient that is affected by crop rotation is Mn. It was found that crop rotation with lupins increases the availability of Mn (Graham and Webb 1991). Not only crop rotation but also cover crops can change soil chemical, physical, and biological properties, including the composition of the soil microbial community (Singh et al. 2020, 2021), and can therefore reduce or increase the severity of plant diseases. Cover crops can increase the content of active SOM in the soil, microbial biomass, and microbial activity and contribute to suppression. Cover crops affect the rhizosphere and also the soil microbial community composition and in that indirect way can affect plant health (Biswakarma et al. 2021; Kumar et al. 2021). Crop rotation can influence the severity of soil-borne diseases by increasing the buffering capacity of the soil, denying the pathogen with a host during the interim of unsuitable species, and affecting nitrification, which influences the form of N predominant in the soil (Huber and Graham 1999; Graham and Webb 1991). Green manure can affect the availability of N and also other nutrients such as P and K. Most of the green manure species that are used can fix N with N-fixing bacteria and can increase soil N levels by 459 kg N/ha (Cherr et al. 2006), having significant effect on disease development. Also, green manures can affect the availability of other nutrients such as P, Mn, and Zn, which can affect the tolerance of disease (Huber and Graham 1999; Graham and Webb 1991).

4.5.3 Intercropping

Intercropping systems have the potential to reduce the incidence of diseases (Anil et al. 1998). There are four mechanisms involved in an intercropping system that can reduce disease incidence, all of which lower the population growth rate of the attacking organisms: (1) the associate crop causes plants of the attacked component to be poorer hosts, (2) the associate crop interferes directly with the attacking organism, (3) the associate crop changes the environment of the host such that natural enemies of the attacking organism are favored, and (4) the presence of non-host or resistant plants growing in between susceptible plants can physically block inoculum from reaching the susceptible hosts (i.e., the non-host serves as a physical barrier to the pathogen inoculum). Francis (1989) found that intercropping reduced pests and diseases in ~53% of experiments and increased them in ~18% of experiments. The reasons for this increase in pests include reduced cultivation and increased shading, favoring some pests and pathogens, associate species serving as alternative hosts, and crop residues serving as a source of pathogen inoculums. In addition, intercropping was found to improve nutrients by increasing N from legumes, or increasing the uptake of phosphorus and potassium (Anil et al. 1998).

4.5.4 Soil Tillage

Reduced tillage systems or zero-tillage can increase SOM content in many agricultural systems. Reduced tillage has the advantage that it conserves SOM and reduces erosion, energy consumption, and production costs (Carter 1994; Fernandez et al. 1999; Choudhary et al. 2020). However, reduced tillage can alter the soil environment, and these changes can result in an increase, decrease, or no change in disease incidence or severity, depending on the cropping system and disease. Minimum tillage concentrates residues on the soil surface and therefore concentrates the pathogen propagule number on the soil surface, which might or might not impact disease incidence. Minimum and zero-tillage do not disrupt the plant residues in the soil as much as conventional tillage (i.e., since they tend not to bury them), thereby leaving more stubble on the soil surface (Kumar et al. 2021). The adoption of conservation tillage by farmers has led to an increase in the incidence and severity of many stubble-borne diseases. Standing residues or residues lying on the soil surface are colonized by soil organisms much more slowly, and pathogen survival and growth in the undisturbed residues are favored in these systems. Residue-colonizing pathogens are therefore favored over the reduced tillage system and can generate significant yield reduction (Bockus and Schroyer 1998). Conservation tillage systems concentrate plant residues in the surface soil layer, and microbial biomass and activity are higher in that layer (Singh et al. 2020, 2021).

4.6 Future Perspectives

- More research is needed in order to find the nutrients or nutrient combinations which can help to reduce disease severity. It is also necessary to find the best integrated pest management approaches with disease-resistant varieties which can be combined with specific cultural management techniques that can efficiently control plant diseases.
- In addition, more research is required to find how the nutrients increase or decrease disease tolerance or resistance, what are the changes in plant metabolism, and how this can be used to control plant disease. It is also important to understand the biochemical pathways by which the nutrients can affect disease.
- Despite the fact that each nutrient has several functions, mild deficiency can usually be linked to one or more processes that are most sensitive, and these processes are linked to the secondary metabolism, which is not immediately necessary for the survival of the organism.
- The secondary metabolism is involved in the defense against pathogens, and some of the roles are well understood and others remain to be elucidated. Also, the evidence that an element has a role in the defense mechanisms not yet regarded as essential in higher plants could lead to recognition of their essentiality. This may require a slight modification of the criteria of essentiality to cover the situation in which yield increases, and indeed survival is due to the element in question which is manifested only in the presence of a pathogen.

- Systemic induced resistance (SIR) caused by application of nutrients could be an alternative strategy to reduce disease severity. In addition, there is a commercially available product containing acibenzolar-S-methyl (with the commercial name Actigard) that activates the same defense response of SAR. The best option for SIR will be a chemical which can minimize adverse effects on the host and has high levels of efficacy. NPK fertilizers together with disease-resistant cultivars can be used in this way; however, other nutrients can be used together with NPK in order to reduce disease.
- In addition, any measure such as crop rotations, application of manures, green manures, and cover crops can be used to increase nutrient availability and reduce disease incidence and can be used in the IPM system in sustainable agriculture. Also, the reduction in the crop production cost, the conservation of beneficial biological enemies of pests, preservation of environmental quality, and slowing the rate of development of pesticide-resistant strains are some of the benefits that the use of fertilizer can have on IPM and on sustainable agriculture.

4.7 Conclusion

In most of the studies conducted so far, the addition of nutrients or application of fertilizers has decreased the incidence of diseases in crop plants. This probably is due to the involvement of these nutrients in the tolerance or resistance mechanisms of the host plants. Nutrient application had a much greater effect on reducing disease when the plants were near deficiency levels. Supra-optimal rates of nutrients can also decrease the disease incidence in some cases. In cases where the addition of a nutrient has exacerbated the disease, it is possibly because of toxicity rather than deficiency or nutrient imbalance leading to deficiency of key nutrients. In sustainable agriculture, balanced nutrition is an essential component of any integrative crop protection program; in most cases, it is more cost-effective and also environmentally friendly to control plant disease with the adequate amount of nutrients and reduced usage of pesticides. Crops should be optimally fertilized for targeted yield and not from disease suppression point of view only. Systemic induced resistance (SIR) caused by application of nutrients could be an alternative strategy to reduce disease severity, and any measure such as crop rotation, application of manures, green manures, and cover crops can be used to increase nutrient availability and reduce disease incidence and can be used in the IPM system in sustainable agriculture. In summary, nutrients can reduce disease incidence to an acceptable level, or at least to a level at which further control by other cultural practices or conventional organic biocides is successfully possible and less expensive.

References

Agrios NG (2005) Plant pathology, 5th edn. Elsevier-Academic Press, Cambridge

Alvarez J, Datnoff LE (2001) The economics of silicon for integrated management and sustainable production of rice and sugarcane. Stud Plant Sci 8:221–239. https://doi.org/10.1016/S0928-3420(01)80018-5

Anil L, Park J, Phipps RH, Miller FA (1998) Temperate intercropping of cereals for forage: a review of the potential for growth and utilization with particular reference to the UK. Grass Forage Sci 53:301–317

Bana RS, Sepat S, Rana KS et al (2018) Moisture-stress management under limited and assured irrigation regimes in wheat (*Triticum aestivum*): effects on crop and water productivity, grain quality, nutrient acquisition and soil fertility. Indian J Agric Sci 88(10):606–1612

Barclay GM, Murphy HJ, Manzer FE et al (1973) Effects of different rates of nitrogen and phosphorus on early blight in potatoes. Am Potato J 50:42–48

Beffa T (1993) Inhibitory action of elemental Sulphur (S^0) on fungal spores. Can J Microbiol 39: 731–735

Biswakarma N, Pooniya V, Zhiipao RR et al (2021) Five years integrated crop management in direct seeded rice–zero till wheat rotation of NW India: effects on soil carbon dynamics, crop yields, water productivity and economic profitability. Agric Ecosyst Environ 318:107492. https://doi.org/10.1016/j.agee.2021.107492

Blachinski D, Shtienberg D, Dinoor A et al (1996) Influence of foliar application of nitrogen and potassium on Alternaria diseases in potato, tomato and cotton. Phytoparasitica 24:281–292

Blevins DG, Lukaszewski KM (1998) Boron in plant structure and function. Annu Rev Plant Physiol 49:481–500

Bockus WW, Schroyer JP (1998) The impact of reduced tillage on soilborne plant pathogens. Annu Rev Phytopathol 36:485–500

Bolle Jones EW, Hilton RN (1956) Zinc-deficiency of *Hevea brasiliensis* as a predisposing factor to *Oidium* infection. Nature 177:619–620

Brecht MO, Lawrence E, Datnoff TA et al (2004) Influence of silicon and Chlorothalonil on the suppression of gray leaf spot and increase plant growth in St Augustine grass. Plant Dis 88(4): 338–344. https://doi.org/10.1094/PDIS.2004.88.4.338

Brennan RF (1992) The role of manganese and nitrogen nutrition in the susceptibility of wheat plants to take-all in western Australia. Fertilizer Res 31:35–41

Broadley M, Brown P, Cakmak I, Rengel Z, Zhao F (2012) Function of nutrient: micronutrients. In: Marschner P (ed) Marschner's mineral nutrition of higher plants. Academic Press, Sydney, pp 191–248

Brown PH, Bellaloui N, Wimmer MA et al (2002) Boron in plant biology. Plant Biol 4:205–223

Cakmak IM (2000) Possible roles of zinc in protecting plant cells from damage by reactive oxygen species. New Phytol 146:185–205

Carballo SJ, Blankenship SM, Sanders DC (1994) Drip fertigation with nitrogen and potassium and postharvest susceptibility to bacterial soft rot of bell peppers. J Plant Nutr 17:1175–1191

Carter MR (1994) Strategies to overcome impediments to adoption of conservation tillage. In: Carter MR (ed) Conservation tillage in temperate agroecosystems. Lewis Publishers, Boca Raton, pp 1–22

Carver TLW, Thomas BJ, Robbins MP, Zeyen RJ (1998) Phenylalanine ammonia-lyase inhibition, autofluorescence, and localized accumulation of silicon, calcium and manganese in oat epidermis attacked by the powdery mildew fungus Blumeria graminis (DC) speer. Agri 52:223–243

Cherr CM, Scholberg JMS, McSorley R (2006) Green manure approaches to crop production: a synthesis. Agron J 98:302–319

Chinchilla D, Bruisson S, Meyer S et al (2019) A sulfur-containing volatile emitted by potato-associated bacteria confers protection against late blight through direct anti-oomycete activity. Sci Rep 9:18778. https://doi.org/10.1038/s41598-019-55218-3

Choudhary AK, Rahi S (2018) Organic cultivation of high yielding turmeric (*Curcuma longa* L.) cultivars: a viable alternative to enhance rhizome productivity, profitability, quality and resource-use efficiency in monkey–menace areas of North-Western Himalayas. Ind Crop Prod 124:495–504

Choudhary AK, Varatharajan T, Rohullah BRS (2020) Integrated crop management technology for enhanced productivity, resource-use efficiency and soil health in legumes. Indian J Agric Sci 90(10):1839–1849

Christos D (2008). Role of nutrients in controlling plant diseases in sustainable agriculture. A review. Agronomy for sustainable development, Springer New York, 28 (1): 33–46

Datnoff LE, Elmer WH, Huber DM (2007) Mineral nutrition and plant disease. American Phytopathological Society, Paul

Dellagi M, Rigault D, Segond C et al (2005) Siderophore-mediated upregulation of *Arabidopsis* ferritin expression in response to *Erwinia chrysanthemi* infection. Plant J 43:262–272

Descalzo RC, Rahe JE, Mauza B (1990) Comparative efficacy of induced resistance for selected diseases of greenhouse cucumber. Can J Plant Path 12:16–24

Dhillon BS, Sharma PK, Choudhary AK (2018) Influence of staggered sown spring sunflower (*Helianthus annuus* L.) at varying intra–row spacing and applied–N on pre– and post–anthesis N dynamics and dry matter partitioning in Indo–Gangetic plains region. Commun Soil Sci Plant Anal 49(16):2002–2015

Dixon KW, Frost K, Sivasithamparan K (1990) The effect of amendment of soil with organic matter, a herbicide, and a fungicide on the mortality of seedlings of two species of Bankia inoculated with *Phytophthora cinnamomic*. Acta Hort 264:123–131

Dordas C, Brown PH (2005) Boron deficiency affects cell viability, phenolic leakage and oxidative burst in rose cell cultures. Plant and Soil 268:293–301

Doubrava N, Dean R, Kuc J (1988) Induction of systemic resistance to anthracnose caused by *Colletotrichum lagenarium* in cucumber by oxalates and extracts from spinach and rhubarb leaves. Physiol Mol Plant Path 33:69–79

Evans I, Solberg E, Huber DM (2007) Copper and plant disease. In: Datnoff LE, Elmer WH, Huber DM (eds) Mineral nutrition and plant disease. APS Press, St Paul, pp 177–188

Fernandez MR, McConkey BG, Zentner RP (1999) Effects of tillage method and fallow frequency on leaf spotting diseases of spring wheat in the semiarid Canadian prairies. Soil Tillage Res 50: 259–269

Frenkel O, Yermiyahu U, Forbes GA, Fry WE, Shtienberg D (2010) Restriction of potato and tomato late blight development by sub-phytotoxic concentrations of boron. Plant Pathol 59(4): 626–633

Francis CA (1989) Biological efficiency in multiple cropping systems. Adv Agron 42:1–42

Graham DR (1983) Effects of nutrients stress on susceptibility of plants to disease with particular reference to the trace elements. Adv Bot Res 10:221–276

Graham DR, Webb MJ (1991) Micronutrients and disease resistance and tolerance in plants. In: Mortvedt JJ, Cox FR, Shuman LM, Welch RM (eds) Micronutrients in agriculture, 2nd edn. Soil Science Society of America, Madison, pp 329–370

Grewal HS, Graham RD, Rengel Z (1996) Genotypic variation in zinc efficiency and resistance to crown rot disease (*fusarium graminearum* Schw. Group 1) in wheat. Plant and Soil 186:219–226

Gupta N, Debnath S, Sharma S, Sharma P et al (2016) Role of nutrients in controlling the plant diseases in sustainable agriculture. In: Meena VS et al (eds) Agriculturally Important Microbes for Sustainable Agriculture. Springer, Cham

Hamim H, Choudhary AK (2019) Influence of varying potassium levels on yield, water productivity, profitability and resource-use efficiency in *Kharif* mungbean (*Vigna radiata*) under semi-arid conditions of Afghanistan. Indian J Agric Sci 89(3):551–555

Hammerschmidt R, Nicholson RL (2000) A survey of plant defense responses to pathogens. In: Agrawal AA, Tuzun S, Bent E (eds) Induced plant defenses against pathogens and herbivores. APS Press, Minneapolis, p 390

Harrison UJ, Shew HD (2001) Effects of soil pH and nitrogen fertility on the population dynamics of *Thielaviopsis basicola*. Plant and Soil 228:147–155

Heba MN, Rana DS, Choudhary AK et al (2016) Sulphur and Zn management in groundnut *(Arachis hypogaea)*–wheat (*Triticum aestivum*) cropping system: direct effects on system productivity and residual effects on yield, energetics and Zn biofortification in wheat. Indian J Agric Sci 86(4):441–447

Heba MN, Rana DS, Choudhary AK et al (2021) Influence of Sulphur and zinc nutrition on productivity, quality and biofortification in groundnut (*Arachis hypogaea* L.) in south-Asian alluvial soil. J Plant Nutr 44(8):1151–1170

Heckman JR, Clarke BB, Murphy JA (2003) Optimizing manganese fertilization for the suppression of take-all patch disease on creeping bentgrass. Crop Sci 43:1395–1398

Hoffland E, Jegger MJ, van Beusichem ML (2000) Effect of nitrogen supply rate on disease resistance in tomato depends on the pathogen. Plant and Soil 218:239–247

Hoffland E, van Beusichem ML, Jegger MJ (1999) Nitrogen availability and susceptibility of tomato leaves to *botrytis cinereal*. Plant and Soil 210:263–272

Hoitink HAJ, Van Doren DM, Schmitthenner AF (1977) Suppression of *Phytophthora cinnamomi* in a composted hardwood bark potting medium. Phytopathology 67:561–565

Horsfall JG, Dimond AE (1957) Interaction of tissue sugars, growth substances and disease susceptibility. Z Pflanzenk Pflanzen Suchutz 64:416–421

Horsfall JG, Heuberger JW (1942) Causes, effects, and control of defoliation on tomatoes. Conn Agr Exp Sta Bull 456:222

Hu S, van Bruggen AHC, Wakeman RJ, Grunwald NJ (1997) Microbial suppression of in vitro growth of *Pythium ultimum* and disease incidence in relation to soil C and N availability. Plant and Soil 195:43–52

Huber DM (1980) In: Horsfall JG, Cowling EB (eds) The role of mineral nutrition in defense. In plant disease, an advanced treatise. Academic Press, New York, pp 381–406

Huber DM, Graham RD (1999) The role of nutrition in crop resistance and tolerance to disease. In: Rengel Z (ed) Mineral nutrition of crops fundamental mechanisms and implications. Food Product Press, New York, pp 205–226

Huber DM, Haneklaus S (2007) Managing nutrition to control plant disease. Landbaufors Volken 57:313–322

Huber DM, Jones JB (2013) The role of magnesium in plant disease. Plant and Soil 368:73–85

Huber DM, McCay-Buis TS (1993) A multiple component analysis of the take-all disease of cereals. Plant Dis 77:437–447

Huber DM, Römheld V, Weinmann M (2012) Relationship between nutrition, plant, diseases and pests. In: Marschner P (ed) Marschner's mineral nutrition of higher plants. Academic Press, Sydney, pp 2836–2299

Huber MD (1996) Role of nutrition in the take-all disease of wheat and other small grains. In: Engelhard WA (ed) Management of diseases with macro- and microelements. APS Press, Minneapolis, pp 46–74

Jensen B, Munk L (1997) Nitrogen-induced changes in colony density and spore production of *Erisiphe graminis f.sp. hordei on* seedlings of six spring barley cultivars. Plant Path 46:101–202

Keinath PA, Loria R (1996) Management of Common Scab of potato with plant nutrients. In: Engelhard WA (ed) Management of diseases with macro- and microelements. APS Press, Minneapolis, pp 152–166

Kiraly Z (1976) Plant disease resistance as influenced by biochemical effects of nutrients and fertilizers. In: Fertilizer use and plant health, proc colloquium. Atlanta, International Potash Institute, pp 33–46

Kirkegaard JA, Munns R, James RA, Neate SM (1999) Does water and phosphorus uptake limit leaf growth of Rhizoctonia-infected wheat seedlings. Plant and Soil 209:157–166

Klikocka H, Haneklaus S, Bloem E et al (2005) Influence of sulfur fertilization on infection of potato tubers with *Rhizoctonia solani* and *Streptomyces scabies*. J Plant Nutr 28:819–833

Kolattukudy EP, Kämper J, Kämper U et al (1994) Fungus-induced degradation and reinforcement of defensive barriers of plants. In: Petrini O, Guellete GB (eds) Host Wall alterations by parasitic fungi. APS Press, Minneapolis, pp 67–79

Kopecky J, Rapoport D, Sarikhani E et al (2021) Micronutrients and soil microorganisms in the suppression of potato common scab. Agronomy 11:383

Kowalska J, Drożdżyński D (2018) Effects of potassium fertilisation on late potato blight and yield. Plant Protect Sci 54:87–91

Krauss A (1999) Balanced nutrition and biotic stress. In: IFA agricultural conference on managing plant nutrition, Barcelona, June–July 1999, Spain

Kuc J (1995) Systemic induced resistance as a novel strategy for plant disease control. In: Biotechnology to feed industry. Nottingham University Press, Nottingham, pp 87–95

Kumar A, Choudhary AK, Pooniya V et al (2016) Soil factors associated with micronutrient acquisition in crops- biofortification perspective. In: Singh et al (eds) Biofortification of food crops. Springer, Cham, pp 159–176

Kumar A, Choudhary AK, Suri VK (2017) Agronomic bio–fortification and quality enhancement in okra–pea cropping system through arbuscular mycorrhizal fungi at varying phosphorus and irrigation regimes in Himalayan acid Alfisol. J Plant Nutr 40(8):1213–1229

Kumar A, Rana KS, Choudhary AK et al (2021) Energy budgeting and carbon footprints of zero-tilled pigeon pea–wheat cropping system under sole or dual crop basis residue mulching and Zn-fertilization in a semi-arid agro-ecology. Energy 231:120862. https://doi.org/10.1016/j.energy.2021.120862

Kumar S, Choudhary AK, Rana KS et al (2018) Bio-fortification potential of global wild annual lentil core collection. PLoS One 13(1):e0191122

Lambert DH, Powelson ML, Stevenson WR (2005) Nutritional interactions influencing diseases of potato. Am J Potato Res 82:309–319

Lee SC (1966) The influence of potash on the resistance of the rice plant against diseases. In: Potassium symposium, The Korean Agricultural Chemical Society, pp 77–83

Li YC, Bi Y, Ge YH et al (2009) Antifungal activity of sodium silicate on *fusarium sulphureum* and its effect on dry rot of potato tubers. J Food Sci 74:213–218

Loschinkohl C, Boehm MJ (2001) Composted biosolids incorporation improves turfgrass establishment on disturbed urban soil and reduces leaf rust severity. Hort Sci 36:790–794

MacKenzie DR (1981) Association of potato early blight, nitrogen fertilizer rate and potato yield. Plant Dis 65:575–577

Mann RL, Kettlewell PS, Jenkinson P (2004) Effect of foliar-applied potassium chloride on septoria leaf blotch of winter wheat. Plant Pathol 53:653–659

Marschner H (1995) Mineral nutrition of higher plants, 2nd edn. Academic Press, London, p 889

Marschner P (2011) Marschner's mineral nutrition of higher plants, 3rd edn. Academic Press, London

Mata CG, Lamattina L, Cassia RO (2001) Involvement of iron and ferritin in the potato–*Phytophthora infestans* interaction. Eur J Plant Pathol 107:557–562

Mengel K, Kirkby EA (2001) Principles of plant nutrition, 5th edn. Kluwer Academic Publishers, Amsterdam, p 847

Moreira WR, Bispo WMS, Rios JA et al (2015) Magnesium induced alterations in the photosynthetic performance and resistance of the plants infected with *Bipolaris oryzae*. Sci Agric 72:328. https://doi.org/10.1590/0103-9016-2014-0312

Oborn I, Edwards AC, Witter E et al (2003) Element balances as a toll for sustainable nutrient management: a critical appraisal of their merits and limitations within an agronomic and environmental context. Eur J Agron 20:211–225

Patrícia PM, Fábio S, Alan M et al (2018) Could the supply of boron and zinc improve resistance of potato to early blight? Potato Res 61:169–182

Paul J, Choudhary AK, Sharma S et al (2016b) Potato production through bio–resources: long–term effects on tuber productivity, quality, carbon sequestration and soil health in temperate Himalayas. Scientia Hort 213:152–163

Paul T, Rana DS, Choudhary AK et al (2016a) Crop establishment methods and Zn nutrition in *Bt*-cotton: direct effects on system productivity, economic–efficiency and water–productivity in *Bt*-cotton–wheat cropping system and their residual effects on yield and Zn biofortification in wheat. Indian J Agric Sci 86(11):1406–1412

Perrenoud S (1990) Potassium and plant health. International Potash Institute, Bern, p p364

Pooniya V, Choudhary AK, Bana RS et al (2019) Zinc bio-fortification and kernel quality enhancement in elite *basmati* rice cultivars of South-Asia through legume residue-recycling and Zn fertilization. Indian J Agric Sci 89(2):279–287

Reid LM, Zhu X, Ma BL (2001) Crop rotation and nitrogen effects on maize susceptibility to gibberella (*fusarium graminearum*) ear rot. Plant and Soil 237:1–14

Reuveni M, Agapov V, Reuveni R (1997a) A foliar spray of micronutrient solutions induces local and systemic protection against powdery mildew (*Sphaerotheca fuliginea*) in cucumber plants. Eur J Plant Pathol 103:581–588

Reuveni M, Agapov V, Reuveni R (1997b) Controlling powdery mildew caused by *Sphaerotheca fuliginea* in cucumber by foliar sprays of phosphate and potassium salts. Crop Prot 15:49–53

Reuveni M, Oppernheim D, Reuveni R (1998) Integrated control of powdery mildew on apple trees by foliar sprays of mono-potassium phosphate fertilizer and sterol inhibiting fungicides. Crop Prot 17:563–568

Reuveni R, Dor G, Raviv M et al (2000) Systemic resistance against *Sphaerotheca fuliginea* in cucumber plants exposed to phosphate in hydroponics system, and its control by foliar spray of mono-potassium phosphate. Crop Prot 19:355–361

Reuveni R, Reuveni M (1998) Foliar-fertilizer therapy – a concept in integrated pest management. Crop Prot 17:111–118

Rioux D, Biggs AR (1994) Cell wall changes and nonhost systems: microscopic aspects. In: Pertini O, Quellette B (eds) Host Wall alterations by parasitic fungi. APS Press, Minneapolis, pp 31–44

Robinson PW, Hodges CF (1981) Nitrogen-induced changes in the sugars and amino acids of sequentially senescing leaves of *Poa pratensis* and pathogenesis by *Drechslera sorokiniana*. Phytopathol Z 101:348–361

Römheld V, Marschner H (1991) Function of micronutrients in plants. In: Mortvedt JJ, Cox FR, Shuman LM, Welch RM (eds) Micronutrients in agriculture. Soil Science Society of America, Madison, pp 297–328

Rovira AD, Graham RD, Ascher JS (1983) Reduction in infection of wheat roots by *Gaeumanomyces graminis* var. *tritici* with application of manganese to soil. In: Parker CA, Rovira AD, Moore KJ, Wong PTW, Kollmorgen JF (eds) Ecology and management of soilborne plant pathogens. APS Press, Minneapolis

Roy SK, Sharma RC, Trehan SP (2001) Integrated nutrient management by using farmyard manure and fertilizers in potato-sunlower-paddy rice rotation in the Punjab. J Agric Sci 137:271–278

Savant NK, Snyder GH, Datnoff LE (1997) Silicon management and sustainable rice production. Adv Agron 58:151–199

Seebold K, Lawrence E, Datnoff F, Kucharek C-VTA, Snyder GH (2004) Effects of silicon and fungicides on the control of leaf and neck blast in upland Rice. Plant Dis 88(3):253–258

Shah MA, Sharma S, Sharma J (2019) Bio-efficacy of potassium silicate against aphids and whitefly in potato. Potato J 46(2):132–137

Sharma J, Upadhyay AK, Sawant IS, Sawant SD (2009) Association of mineral nutrients with vein reddening and necrosis in Thompson seedless grape. Indian J Hortic 66:154–162

Sharma J, Upadhyay AK, Sawant IS, Sawant SD (2012) Relationship of nutritional status of field grown Thompson seedless grapevines with powdery mildew incidence. J Appl Hortic 14(2): 114–117

Sharma RC, Duveiller E (2004) Effect of helminthosporium leaf blight on performance of timely and late-seeded wheat under optimal and stressed levels of soil fertility and moisture. Field Crop Res 89:205–218

Sharma S, Duveiller E, Basnet R et al (2005) Effect of potash fertilization on helminthosporium leaf blight severity in wheat, and associated increases in grain yield and kernel weight. Field Crop Res 93:142–150

Simoglou K, Dordas C (2006) Effect of foliar applied boron, manganese and zinc on tan spot in winter durum wheat. Crop Prot 25:657–663

Singh A (1978) Protection against fungal attack by fertilizers. Fert Abstr 11(2):462

Singh U, Choudhary AK, Sharma S (2020) Comparative performance of conservation agriculture Vis-a-Vis organic and conventional farming in enhancing plant attributes and rhizospheric bacterial diversity in *Cajanus cajan*: a field study. Eur J Soil Biol 99:103197

Singh U, Choudhary AK, Sharma S (2021) Agricultural practices modulate the bacterial communities, and nitrogen cycling bacterial guild in rhizosphere: field experiment with soybean. J Sci Food Agric 101:2687–2695

Soltanpour PN, Harrison MD (1974) Interrelations between nitrogen and phosphorus fertilization and early blight control of potatoes. Am Potato J 51:1–7

Spencer S, Benson DM (1982) Pine bark, hardwood bark compost, and peat amendment effects on development of *Phytophthora* spp and lupine root rot. Phytopathology 72:346–351

Stone AG, Scheuerell SJ, Darby HD (2004) In: Magdoff F, Weil RR (eds) Suppression of soilborne diseases in field agricultural systems: organic matter management, cover cropping and other cultural practices, soil organic matter in sustainable agriculture. CRC Press, London

Sun W, Zhang J, Fan Q et al (2010) Silicon-enhanced resistance to rice blast is attributed to silicon-mediated defence resistance and its role as physical barrier. Eur J Pl Path 128:39–49

Szczech M, Rondomański W, Brzeski MW, Smolińska U, Kotowski JF (1993) Suppressive effect of a commercial earthworm compost on some root infecting pathogens of cabbage and tomato. Biol Agric Hortic 10(1):47–52

Thomas HR (1948) Effect of nitrogen, phosphorus and potassium on susceptibility of tomatoes to *Alternaria solani*. J Agric Res 76:289–306

Vidhyasekaran P (1997) Fungal pathogenesis in plants and crops. In: Molecular biology and host defense mechanisms. Marcel Dekker, New York, p 568

Vidhyasekaran P (2004) Concise encyclopaedia of plant pathology. Food Products Press, The Haworth Reference Press, New York, p 619

Walker JC, Whipple OC, Jolivette JP, Hooker WJ (1944) The limitations of spraying tomatoes in Wisconsin. Wis Agr Exp Sta Res Bull 1994:152

Webb J, George R (1968) Potassium reduces corn leaf disease. Better Crops 2:2–4

Wiese J, Bagy MMK, Shubert S (2003) Soil properties, but not plant nutrients (N, P, K) interact with chemically induced resistance against powdery mildew in barley. J Plant Nutr Soil Sci 166: 379–384

Wilkens RT, Spoerke JM, Stamp NE (1996) Differential response of growth and two soluble phenolics of tomato to resource availability. Ecology 77:247–258

Xue X, Lui H, Yang W et al (2021) Foliar application of silicon enhances resistance against *Phytophthora infestans* through the ET/JA- and NPR1-dependent signaling pathways in potato. Front Plant Sci. https://doi.org/10.3389/fpls.2021.609870

Ye XS, Strobel N, Kuc J (1995) Induced systemic resistance (ISR): activation of natural defense mechanisms for plant disease control as part of integrated pest management (IPM). In: Reuveni R (ed) Novel approaches to integrated pest management. CRC Press, Boca Raton, pp 95–13

Zhang Q, Fry J, Lowe K, Tisserat N (2006) Evaluation of calcium silicate for brown patch and dollar spot suppression on turf grasses. Crop Sci 46:1635–1643

Secondary Metabolites of Microbials as Potential Pesticides

5

Chandan Maharana, Vinod Kumar Padala, Aravinda B. Hubballi, M. Nikhil Raj, Amit Paschapur, Chaitra Bhat, Ashish Kumar Singh, and A. R. N. S. Subbanna

Abstract

Crop protection has become an integral part of production system with substantial price tag. Overreliance on chemical pesticides masked the effects of natural pest controlling factors like microbial pathogens. The importance and safety associated with these efficient pest suppression options led to their increased use in recent past. However, being living organisms their formulations, shelf life, persistence and potential in different agroecological regions, etc. are a bottleneck. At this juncture, deep insights into the modes of action lead to the discovery of metabolites that are actually and actively involved in the pathogenicity and killing of the host species. Further advances in organic and synthetic chemistry escorted the commercial facets of these pesticidal secondary metabolites. Recent past has seen the discovery of a variety of novel microbial origin pesticidal compounds and has become an evergrowing science in view of the existing diversity of microbial pathogens and strains. Some of them also saw the status of commercial pesticides with huge success. The target specificity, structural distinctiveness, novel modes of action, and environmental safety are the chief contributing factors

C. Maharana · A. Paschapur · A. K. Singh · A. R. N. S. Subbanna (✉)
Crop Protection Section, Vivekananda Institute of Hill Agriculture (VPKAS), Almora, Uttarakhand, India

V. K. Padala
ICAR-Research Complex for Eastern Region, Research Center for Makhana, Darbhanga, Bihar, India

A. B. Hubballi
Department of Entomology, GKVK, University of Agriculture Sciences, Bangalore, Karnataka, India

M. Nikhil Raj · C. Bhat
ICAR-Indian Agricultural Research Institute, Pusa, New Delhi, India

S. Kumar Chakrabarti et al. (eds.), *Sustainable Management of Potato Pests and Diseases*, https://doi.org/10.1007/978-981-16-7695-6_5

for their success as potential pest suppression options. In this context, this chapter discusses the pesticidal (insecticidal, antifungal, antibacterial, and nematicidal) activities (target pests, modes of action, chemical structures, etc.) of different metabolites produced by diverse pathogenic microorganisms of agricultural importance.

Keywords

Secondary metabolites · Microbes · Biopesticides · Insecticidal · Antifungal · Nematicides · Formulations · Pest control

5.1 Introduction

Microbes are considered as smallest life forms on earth that exits everywhere. They are involved in all the fundamental life processes of the earth's biosphere either directly or indirectly and make it conducive to other life forms. The exploration studies on these miniature life forms lead to identification of some excellent microbes and development of diverse products and processes with huge implications to humans. Thus, the searching for novel strains or species of microbes has become an interesting arena of research among the naturalists more specifically the microbiologists. Moreover, the diversity in microbial community established till date is only a minute share of the whole existing diversity (Gibbons and Gilbert 2015). Unraveling this huge natural diversity offers viable implications in medical, agricultural, and industrial sectors.

Application and use of microbials in management of insect pests is known since time immemorial. With the larger understanding of negative impacts associated with chemical pesticides, use of microbial biocontrol agents against different pest problems is increasing due to their environmental safety. Infact the adoptations of these invisible pests control options by farming community is nominal due to their slow modes of action in comparison with chemical pesticides. Moreover, availability of commercial products, shelf life, consistency in performance over locations, etc. are also major issues which hinder their regular use (Arthurs and Dara 2019). The advents of different molecular tools in recent past lead to identification of active compounds produced by different microbial pathogens and their modes of action against target pests (Subbanna et al. 2020). The host specificity and environmental competency exhibited by these microbial origin metabolites opened up a new arena of research to develop them as biological origin pesticides.

The market of microbial pesticides expected to grow about $4.5 billion by 2023 (Olson 2015) among which microbial-based metabolites are the fastest growing segment (Dunham 2015). Since ages, the microbial biopesticide market is dominated by products of *Bacillus thuringiensis* (Mnif and Ghribi 2015) followed by entomopathogenic fungi (*Beauveria bassiana*, *Metarhizium anisopliae*, *B. brongniartii*, *Lecanicillium lecanii*, etc.) and recently by antagonistic fungi like *Trichoderma* (Berg 2009). The biological efficiency of these microbial pathogens comes from the production of compounds that negative impacts the growth and development of the host. The advancements in molecular techniques facilitated the identification of bioactive compounds in majority of these biopesticides and other pathogenic microbes. Further, the chemical biology studies enabled the production of analogues or parachemical compounds with similar bioactivity and greater environmental stability thus enabling the application of these chemically synthesized compounds or natural products as chemical pesticides. The success of avermectins and spinosyns proved the potential of microbial origin metabolites as environmentally safe, biodegradable, target-specific, and competent pesticides (Tanaka and Omura 1993).

5.2 Secondary Metabolites in Pest Management

Secondary metabolites are also considered as natural products produced by living organisms and are structurally carbon compounds. They are in prodigious numbers and hard to characterize due to their vast structural differences and metabolic activities (Bennett and Bentley 1989). However, there is an evolutionary association with the containing organisms to facilitate its role as living organisms with survival values (Demain and Fang 2000). During the last decades, interest in organic chemistry led to discovery of a huge array of compounds associated with microbes at an exponential manner (Bennett and Bentley 1989). As a whole, the secondary metabolites are categorized as (1) competitive weapons used against other bacteria, fungi, amoebae, plants, insects, and large animals; (2) metal transporting agents; (3) agents of symbiosis between microbes and plants, nematodes, insects, and higher animals; (4) sexual hormones; and (5) differentiation effectors (Demain and Fang 2000). The tendency to produce these bioactive compounds is independent of species or organisms but can be correlated with the existing competitive environment. Besides, horizontal transfer of genes among the microbes also played greater role in predominance and diversity of secondary metabolites (Vining 1992). So, the production and possible viabilities associated with the secondary metabolites is a complex selection process in competence with the existing biotic, abiotic, physiological, and ecological environment.

The advantages associated with secondary metabolites as pesticides are versatility in structure, unique modes of action, target specificity, and biodegradability (Tanaka and Omura 1993). A variety of secondary metabolites with pesticidal activity in field applications has been documented. They include kasugamycin, blasticidin S, mildiomycin, validamycin, and polyoxin as fungicides, tetranactin as a miticide, and avermectin and spinosyn as insecticides. In addition, control of weeds (herbicides) is also possible using the secondary metabolites (Duke et al. 2002). These successful compounds exemplified the interest on pesticidal properties of secondary metabolites from other pathogenic organisms (Saxena and Pandey 2001).

In recent past, many studies reported novel secondary metabolites from a variety of microbial agents. In most of these studies, conventional activity monitoring and further characterization using basic molecular tools were adopted to study the metabolites from existing and proven microbial pathogens. However, the uniqueness comes from the screening of novel microbial groups like myxobacteria, basidiomycetes, and blue-green algae (Tanaka and Omura 1993). Moreover, adoption of different media and fermentation techniques and screening techniques may also yield substantial bioactive secondary metabolites. Advancements in organic chemistry and synthetic chemistry are also an adoptive advantage associated with product purification and commercialization of potent metabolites. Moreover, classical genetic methods like mutation, recombination, etc. can also be adopted in improvising the efficacy and bioactivity of metabolites (Subbanna et al. 2020).

5.3 Fungal Secondary Metabolites Against Pests of Agricultural Importance

The fungal antagonists restrict the growth of plant pathogens by the three suggested mechanisms: antibiosis, competition, and parasitism. Besides, they also induce the defense responses in host plants, termed "induced systemic resistance" (van Loon et al. 1998). Among the above-mentioned mechanisms, antibiosis is considered the most important, in which the antagonists produce an array of secondary metabolites such as antibiotics and toxins, which contribute to the antagonistic activity of fungal biocontrol agents against plant pathogens. Antagonistic strains belonging to the *Trichoderma* and *Fusarium* genera were able to produce various secondary metabolites which can play a role in the mechanism of their biological activity. Production of antimicrobial secondary metabolites has also been reported in many different other fungal genera. The details were given in Table 5.1.

Insects constitute the largest and most diverse group of animals on earth. More than 700 known fungal species from 100 genera have adapted to entomopathogenic fungi (EPF) lifestyle. The largest numbers of fungal species that are pathogenic to insects belong to the order Hypocreales (Dikarya, Ascomycota, Pezizomycotina, Sordariomycetes, Hypocreomycetidae). Typical EPF strains belong to the families of Cordycipitaceae, Clavicipitaceae, and Ophiocordycipitaceae. The Cordycipitaceae incorporate geographically widespread species from the genera *Beauveria*, *Cordyceps*, and *Isaria*. The fungi chiefly rely on a battery of entomotoxins viz., secondary metabolites falling in the categories of non-ribosomal peptides, polyketides, lysine derived alkaloids, and terpenoids. Species from *Beauveria* along with *Metarhizium* (from the family Clavicipitaceae) provide the most important commercial strains of bio-insecticides and are known to infect more than 200 species of different insects that are important agricultural pests or vectors for human and animal diseases. A list of newly identified secondary metabolites and their commercially available products are detailed in Table 5.2.

Table 5.1 Fungal secondary metabolites and their biological activities against plant pathogens

Secondary metabolite	Class of natural compounds	Producing fungus	Activity/antagonistic against/ application	Reference
Trichodecenins, trichorovins, trichocellins	Peptaibols	*Trichoderma sp.*	Antifungal, antibacterial, antiviral	Szekeres et al. (2005)
Trichocaranes, demethylsorbicillin, oxysorbicillinol, trichodenones, harzialactone A and B, trichoviridin, 6-n-pentyl pyrone	Polypeptide, pyroness	*Trichoderma sp.*	Antifungal, antibacterial, antiviral	Macias et al. (2000)
Fusapyrone, deoxyfusapyrone	Alpha pyrones	*F. semitectum*	*Alternaria alternata*, *Ascochyta rabiei*, *aspergillus flavus*, *Botrytis cinerea*, etc.	Altomare et al. (2000), (2004)
Mycophenolic acid, patulin, 3-O-methylfunicone		Penicillium sp.	*R. solani* in vitro	Nicoletti et al. (2004)
Sphaeropsidins A	Pimarane diterpene	*Diplodia* spp.	Antimycotic; anticancer; antibacterial Insecticidal (*Botrytis cinerea*, *fusarium*, *Verticillium dahlia*, *Seiridium*, *Phytophthora*)	Lallemand et al. (2012), Mathieu et al. (2015), Ingels et al. (2017), Evidente et al. (2011), Cimmino et al. (2013)
Oxysporone	Pyranone	*D. africana*	Phytotoxic; Anti-oomycetes; antifungal	Andolfi et al. (2014)
12β-Hydroxy-13α-methoxyverruculogen TR-2; Fumitremorgin C	Indole diketopiperazine alkaloid	*Aspergillus fumigatus* LN-4	Antifungal, antibacterial	Li et al. (2012)
3-ydroxyfumiquinazoline A	Fumiquinazoline	*A. fumigatus* LN-4	Antifungal, antibacterial	Li et al. (2012)
Verruculogen TR-2	Indole diketopiperazine alkaloid	*A. fumigatus* LN-4	Antifungal, antibacterial	Pan et al. (2014)
Neosartins A-C	Dihydropyrazino[1,2-a] indole	*N. Pseudofischeri*	Antifungal, antibacterial	Liang et al. (2014)

(continued)

Table 5.1 (continued)

Secondary metabolite	Class of natural compounds	Producing fungus	Activity/antagonistic against/ application	Reference
Boydines A	Epipolythiodioxopiperazine	*Pseudallescheria boydii*	Antifungal, antibacterial	
4a-*epi*-9α Methoxydihydrodeoxybostrycin	Hydroanthraquinone	*Nigrospora* sp.	Antifungal, antibacterial	Yang et al. (2012)
Nigrosporin B	Anthracene compound	*Nigrospora* sp.	Antibacterial	Yang et al. (2012)
Spiromastixones A–E	Depsidone analogue	*Spiromastix* sp.	Antibacterial	Niu et al. (2014)
Spiromastixones F–J	Depsidone analogue	*Spiromastix* sp.	Antibacterial	Niu et al. (2014)

Table 5.2 The major entomopathogenic fungus, entomotoxic metabolites, and host insect range (Source: Singh et al. 2016)

Entomopathogenic fungi	Entomotoxic metabolites	Host	Commercial formulation	Reference
Cordyceps	Cordycepins	Lepidopteran larvae	–	Kim et al. (2002), Kryukov et al. (2014)
Hypocrella/Aschersonia	Ergosterol, Dustanin, Hypocrellins, 3-hopane-triterpenes	Aleyrodidae, Coccidae families of Hemiptera, and nematodes	–	Isaka et al. (2003), Jin-Ming (2006), Buttachon et al. (2013)
Beauveria	Beauvericin, Bassianin, Oosporein, and bassianolide	Lepidoptera, Coleoptera, Hemiptera, Homoptera, and hymenoptera	Naturalis, Botanigard, Mycontrol O, Boverol, Brocaril, Ostrinil, betel, Engerlingspilz, Beevicide	Elsworth and Grove (1977), Uma Devi et al. (2008), Sandhu et al. (2012)
Metarhizium	Swainsonine and Destruxins	Coleoptera, Hemiptera, Isoptera, Homoptera, Heteroptera, Diptera (mosquitoes), hymenoptera, Siphonaptera and Lepidoptera	MET52, bioblast, BioPath, biogreen, green guard ULV, and green muscle	Goettel et al. (2001), Quarles (2013), Singh, Sandhu et al. (2012)
Paecilomyces (Isaria)	Beauvericin, Beauverolides, and dipicolinic acid (DPA)	Hemiptera	Pfr-97, Pfr 21, Pelicide, PreFeRal, and Pae-sin	Vey et al. (2001), Sandhu et al. (2012)
Verticillium	Hydroxycarboxylic acid, cyclosporine, and dipicolinic acid, Bassianolide	Hemiptera and Thysanoptera (thrips)	Biomite, Bioter, Mycotal, Verelac, and bio-catch	Vey et al. (2001), Sandhu et al. (2012)
Tolypocladium	Efrapeptins, Tolypin, Diketopiperazines	Diptera (mosquitoes), Ephemeroptera (mayflies)	–	Bandani (2004), Bandani (2008)
Hirsutella	Hirsutellin A and B	Mites (citrus rust mites-*Phyllocoptruta oleivora*), Lepidotera	Mycar	McCoy et al. (1992), Aghajanzadeh et al. (2006)
Nomuraea rileyi (Cordycep)	Ergosterol peroxide	Lepidoptera, Coleoptera, Hemiptera	AGO biocontrol nomuraea 50, PreFeRal	Prompiboon et al. (2008), Onofre et al. (2002)

(continued)

Table 5.2 (continued)

Entomopathogenic fungi	Entomotoxic metabolites	Host	Commercial formulation	Reference
Torrubiella	Torrubiellin B (2)	Hemiptera (Coccoidea)	–	Isaka et al. (2012)
Entomophaga	–	Orthoptera, Coleoptera	–	Milner (1997)
Erynia	–	Hemiptera (aphids)	–	Milner (1997), Pell et al. (2001)
Entomophthora	–	Thysanoptera (thrips), Diptera (houseflies)	–	Pell et al. (2001)
Zoophthora	–	Coleoptera, Diptera, Hemiptera, hymenoptera, Lepidoptera, Orthoptera, Trichoptera	–	Glare and Milner (1991), Pell et al. (2001)
Coelomycidium	–	Dipterans (specially black flies)	–	
Myiophagus	–	Dipterans	–	Araújo and Hughes (2016)
Lagendium	–	Dipterans (mosquito)	Laginex AS, LAGINEX 25, LAGINEX	Kerwin et al. (1994), Hallmon et al. (2000), Vyas et al. (2007)
Leptogenia	–	Dipterans (mosquito larvicidal)	–	Lastra et al. (2004), Pelizza et al. (2007, 2013)
Pythium	–	Dipterans (mosquito larvicidal)	–	Su et al. (2001)
Penicillium sp.	Preaustinoid A,A2,B; Dehydroaustin, Acetoxydehydroaustin, Neoaustin, Austin, Okaramine	Larvicidal activities	–	Wang et al. (2017); Kato et al. (2018)

5.4 Bacterial Secondary Metabolites Against Pests of Agricultural Importance

Bacteria produce an array of secondary metabolites, many of which are antagonistic and inhibitory in nature towards insect pests and plant pathogens. They are low molecular weight compounds which are below 2.5 KDa and act in multiple ways to inhibit pest and disease occurrence on plants. Entomopathogenic bacteria like *Bacillus* spp., *Xenorhabdus*, *Photorhabdus*, etc. and soil bacteria like *Pseudomonas* spp. are the most widely researched bacteria for their diverse secondary metabolites and their pest and disease control ability. These bacteria are known to produce secondary metabolites which interact with their external environment; in case of entomopathogenic bacteria like *Bacillus* spp., several important secondary metabolites viz. Cry toxins, Cytolysins, Vegetative Insecticidal Proteins (VIPs), Thurungiensin (β-exotoxin), Phospholipase C, etc. (Crickmore et al. 2011; Chakroun et al. 2016; Liu et al. 2014) are well known for their insecticidal effects and currently used for insect pest management operations (Table 5.3). These secondary metabolites act in multiple ways; some of them may be disrupting midgut cell wall and causing osmotic imbalance and leading to cessation of feeding and ultimately death of insect pest (e.g., cry toxins). In addition, some will synergize activity of other insecticidal proteins and increase the efficacy of other toxic metabolites (e.g., Cytolysins), while others like thuringiensin will interfere the RNA polymerase activity by competing with ATP binding sites and thus hampering natural growth and development of these insect pests. In case of entomopathogenic bacteria like *Xenorhabdus* and *Photorhabdus* which are closely associated with nematodes, after successful invasion of insect hemolymph, it starts producing an array of secondary metabolites which suppresses immune system of the insect host. For example, benzylideneacetone (BZA) and p-hydroxyphenylpropionic acid (PHPP) inhibit phospholipase A2 (PLA2) and thus shutting down eicosanoid biosynthesis which leads to immunity suppression (Mullah et al. 2020; Vatanparast et al. 2019). Another mode of action is inhibition of phenol-oxidase (PO) activity seen in rhabducin, rhabdopeptide, and xenortide peptide which also in-turn suppress immunity of insect host and lets the invading bacteria-nematode complex flourish in a less restrictive environment (Crawford et al. 2012). Then there are phurealipids which are reported to prevent expression of antimicrobial peptide genes and also enhance JH level or reduce JH degradation (Nollmann et al. 2015).

In case of plant pathogen management using secondary metabolites, fluorescent *Pseudomonas* spp. is one important soil bacterium that has recorded an array of secondary metabolites that can act against multiple plant pathogenic microorganisms. Phenazines, phloroglunicoles, dialkylresorcinols, pyrolnitrin, mupirocin, rhizoxins, pyoluteorin and hydrogen cyanide, etc. are some of the known plant pathogen inhibitory secondary metabolites from different strains of *Pseudomonas* spp. (Table 5.4). Most of these compounds show antibacterial and antifungal action against a variety of plant pathogens (Shahid et al. 2017). In case of pyrrolnitrins, they have an inhibitory effect on fungal respiratory chains and are lethal for wide range of Deuteromycete, Ascomycete, and Basidiomycete fungi.

Table 5.3 Bacterial secondary metabolites against insect pest

Secondary metabolite	Source	Biological activity	Reference
Cry toxin	*Bacillus thuringiensis*, *B. papillae*, *Clostridium bifermentans*	Disruption of midgut cell and causing osmotic imbalance	Schnepf et al. (1998), Crickmore et al. (2011)
Cytolysins	*Bacillus thuringiensis*	Causes colloidal osmotic lysis and synergize activity other insecticidal proteins	Sayyed et al. (2001), Knowles and Ellar (1987)
Vegetative insecticidal proteins (VIPs)	*Bacillus thuringiensis*, *B. cereus*	Prevent microfilament formation (VIP 2), apoptotic cell death of midgut epithelial cells (VIP 3)	Chakroun et al. (2016)
Thuringiensin/β-exotoxin	*Bacillus thuringiensis*	Interfering the RNA polymerase activity by competing with ATP binding sites	Liu et al. (2014)
Phospholipase C	*Bacillus cereus*	Hydrolysis of glycerophospholipids which directly influence membrane dynamics and cellular signaling	Binnington and Baule (1993)
Benzylideneacetone (BZA), p-hydroxyphenylpropionic acid (PHPP), 2-oxindole, 4-hydroxyphenylacetic acid (HPA), acetylated phenylalanine-glycine-valine (Ac-FGV), proline-tyrosine (PY), cyclo-proline-tyrosine (cPY)	*Photorhabdus luminescens*, *Xenorhabdus nematophila*	Inhibit phospholipase A2 (PLA2) and shut down eicosanoid biosynthesis	Mullah et al. (2020), Vatanparast et al. (2019)
Rhabducin	*Photorhabdus luminescens*, *Xenorhabdus nematophila*	Inhibit activity of phenoloxidase (PO)	Crawford et al. (2012)
Rhabdopeptide Xenortide peptide	*Photorhabdus luminescens*, *Xenorhabdus nematophila*	Inhibit activity of phenoloxidase (PO)	Cai et al. (2016), Sussmuth and Mainz (2017)
Phurealipids	*Photorhabdus luminescens*, *Xenorhabdus nematophila*	Inhibit JH degradation Enhance JH level Prevent expression of antimicrobial peptide genes	Nollmann et al. (2015)

Similarly, phenazines and dialkylresorcinols have broad antibacterial and antifungal activity (Shahid et al. 2017). Soil bacterium *Pseudomonas* spp. colonize rhizosphere region of crop plants and inhibit root pathogens from attacking the host plant and

Table 5.4 Bacterial secondary metabolites against plant pathogens

Secondary metabolite	Source	Biological activity	Reference
Acylhomoserine lactone	*Bacillus thuringiensis*	Quench pathogenicity of plant pathogenic bacteria	Dong et al. (2001)
Zwittermicin A	*Bacillus thuringiensis*	High activity against the oomycetes and their relatives and gram negative bacteria	Silo-Suh et al. (1998)
Thuricin-17 and bacthuricin-F4 (class IId bacteriosins)	*Bacillus thuringiensis*	Highly effective elicitors for activity of defense-related enzymes leading to improved plant resistance against soil-borne plant diseases	Jung et al. (2011)
Polyketides	Bacteria and fungi	Antibacterials, antifungals, antivirals, and antiparasitics	
Benzaldehyde, non-anal, benzothiazole, and acetophenone	*Bacillus* spp.	Antagonism against *Ralstonia solanacearum*	
Bacillomycin D	*Bacillus* spp.	Cause severe injury to both cell wall and cell membrane of fungal spores and hypha as observed in the killing of *aspergillus flavus*	Gong et al. (2014)
1-Undecene	*Pseudomonas* spp.	Antibacterial against *Phytophthora infestans*	Bailly and Weisskopf (2017)
Dimethyl disulfide and 2-methylpentanoate	*Pseudomonas* spp.	Inducing systemic resistance (ISR) in plants	
Carvacrol and trans-2-hexenal		Hampering conidia germination of *Monilinia laxa*, the agent of brown rot of stone fruit	
Phenazine-1-carboxylate	*P. fluorescens*, *P. Chlororaphis*, and *P. aeruginosa*	Broad antibacterial and antifungal activities	Shahid et al. (2017)
Phenazine-1-carboxamide	*Pseudomonas* spp., such as *P. aeruginosa* and *P. Chlororaphis*	Broad antibacterial and antifungal activities	Mavrodi et al. (2001)
Pyocyanin	*P. aeruginosa*	Broad antibacterial and antifungal activities	Guttenberger et al. (2017)
2,4-diacetyl Phloroglucinol	*P. fluorescens*	Toxic to a wide range of plant pathogenic fungi, exhibits antibacterial and anthelmintic activities	Rezzonico et al. (2007)
Dialkylresorcinols	*Pseudomonas* spp.	Antifungal and antibacterial activities	
2-Hexyl-5-propylalkylresorcinol (DB-2703)	*Pseudomonas* spp.	Showed antibiotic activity against gram-positive bacteria, mycobacteria, yeasts, and fungi.	

(continued)

Table 5.4 (continued)

Secondary metabolite	Source	Biological activity	Reference
Pyrrolnitrin	*P. fluorescens* and *P. chlororaphis (Van Pée and Ligon* 2000	Inhibitor of fungal respiratory chains and is active against a wide range of Deuteromycete, ascomycete, and basidiomycete fungi	Van Pée and Ligon (2000)
Pyoluteorin	*P. fluorescens*Pf-5	Responsible for the control of many soil-borne diseases	Brodhagen et al. (2005)
Mupirocin (pseudomonic acid)	*P. fluorescens*	Antibacterial activity towards gram-positive pathogens	El-Sayed et al. (2001)
Hydrogen cyanide	*Gram-negative bacteria,* including *P. fluorescens*, *P. aeruginosa*, and *Chromobacterium violaceum*	Protect many plants from fungal root diseases, inhibition of cytochrome C oxidase and other metalloproteins	Ramette et al. (2003)
Rhizoxins	*Rhizopus microsporus* *P. fluorescens*Pf-5	Binding to β-tubulin, thereby interfering with microtubule dynamics during mitosis	
Carbapenem	*Serratia plymuthica*	a ß-lactam antibiotic that inhibits bacterial cell wall biosynthesis	Levenfors et al. (2004)
Oocydin A	Some strains of *S. plymuthica*	Antifungal and anti-oomycete haterumalide	Levenfors et al. (2004)
Prodigiosin	*Serratia plymuthica*, *S. rubidaea*, and some strains of *S. marcescens*	Shows antifungal, antibacterial, and antiprotozoal activities	Berg (2000), Williamson et al. (2006)
Zeamine	Produced by plant-associated *S. plymuthica* and other *Serratia* species	Broad-spectrum antibacterial and antifungal compound	Masschelein et al. (2013), Hellberg et al. (2015)
Difficidin	*Bacillus* spp.	Efficient in suppressing plant pathogenic bacterium *E. amylovara*, which causes fire blight disease at orchard trees	Chen et al. (2009)
Polymyxin	*Paenibacillus polymyxa*	Efficient against phytopathogenic *Erwinia amylovara*, the causative agent of fire blight, and *E. carotovora*, the causative agent of soft rot	Choi et al. (2009), Niu et al. (2013)
Mersacidin	*Bacillus sp. HIL Y-85*	Antibacterial activity by inhibition of cell wall synthesis.	Chatterjee et al. (1992)

also induce resistance in crop plants against these soil-borne pathogens. These pseudomonades are also reported to improve nutrient availability by solubilization of mineral nutrient thus improving overall health of plant. *Bacillus* spp. is also as important as the *Pseudomonas* spp. in case of pathogen suppression with secondary metabolites like Aceyl homoserine lactone lactonas, Zwittermicin A, Benzaldehyde, nonanal, benzothiazole, and acetophenone acting against various plant pathogens (Silo-Suh et al. 1998). Similar beneficial soil bacteria which colonize the rhizosphere may involve in suppression of plant pathogenic microorganisms and subsequently develop disease suppressive soils.

5.5 Secondary Metabolites Against Plant Parasitic Nematodes

Biocontrol agents for nematode suppression have shown ability to release a wide range of secondary metabolites as their chief constituents for nematode mortality either directly or indirectly. The fungal bioagents kill nematode directly through the action of competition for space, using lytic enzymes, or antibiosis using secondary metabolites. Fungal metabolites released in culture filtrates are very effective in inhibition of egg hatching and juvenile mortality of *M. incognita*. Fungal bioagents like *Chaetomium globosum* produce various nematicidal secondary metabolites such as 4,5,6-trihydroxy-7-methylphthalide, flavipin, 3-methoxyepicoccone, chaetoglobosin A, and chaetoglobosin B, which is very effective in killing *M. incognita*. Similarly, *Alternaria* sp. produces alternariol 9-mehyl ether which causes mortality of *Bursaphelenchus xylophilus* effectively. *Pochonia chlamydosporia* produces 34 different kinds of secondary metabolites belonging to phenolics, alkaloids, and pheromones effective in management of nematodes. Metabolites secreted by *Fusarium oxysporum* have shown nematicidal potential against *Radopholus similis*, *M. incognita*, and *Pratylenchus zeae* (Hallman et al. 1994).

Bacterial bioagents directly affect nematode through the release of lytic enzymes (chitinases, proteases, and glucanases), gases, cry toxins, and organic compounds. Lytic enzymes produced by plant growth-promoting rhizobacteria (PGPRs) directly breakdown the outer covering of nematode that mainly consists of collagen/keratin like protein compounds. Similarly, the egg shell (chitin layer) is breakable by the lytic enzymes and enabling the premature juveniles to premature hatching and starvation to death. *Bacillus megaterium* strain PSB2, *Lysobacter capsici*, and *Streptomyces cacaoi* GY525 produce chitinases and β-1,3-glucanase which break down the egg shell of RKN species (*Meloidogyne* spp.) and mortality of juveniles (El-Hadad et al. 2010, Jung et al. 2014, Yoon et al. 2012). Proteases (serine and cysteine) act as a vital virulence factor against nematodes. The bacteria *Pseudomonas fluorescens* CHA0 and *Brevibacillus laterosporus* release alkaline serine proteases which degrade the cuticle of nematode and kill nematode effectively (Tian et al. 2006). Cry protein like endotoxin (Cry5, Cry6, Cry12, Cry13, Cry14, Cry21, and Cry55) produced during sporulation by *Bacillus* spp. has been found toxic to the nematodes. These toxins are reported to kill nematode by creating pores in epithelial cell of intestine and further causing intestinal degradation of nematodes

Table 5.5 Bacterial secondary metabolites against plant pathogens

S. No.	Secondary metabolite	Origin species	Targeted nematodes	References
1	Uracil, Dihydrouracil and 9H-purine	*Bacillus subtilis* and *B. cereus*	*Meloidogyne exigua*	Oliveira et al. (2014)
2	Plantazolicin	*B. Amyloliquefaciens* strain FZB42		Liu et al. (2013)
3	C16 sphingosine and phytosphingosine	*B. cereus* strain S2	*C. elegans*	Gao et al. (2016)
4	Cry toxins	*B. Thuringiensis*	*M. incognita*	Liu et al. (2010)
5	2,4-diacetylphloroglucinol	*Pseudomonas fluorescens*	*M. incognita*	Siddiqui and Shaukat (2003)
6	Lactic acid (2-hydroxypropanoic acid)	*L. Capsici* YS1215	*M. incognita*	Lee et al. (2014)
8	Alkaline serine protease BGL4	*Brevibacillus Laterosporus*	*Panagrellus redivivus*	Huang et al. (2005)
9	Collagenase, chitinases, lipases	*P. fluorescens FP805PU*	*Xiphinema index* and *M. ethiopica*	Aballay et al. (2017)

(Bravo et al. 1998; Marroquin et al. 2000; Frankenhuyzen 2009). Similarly, uracil, dihydrouracil, and 9H-purine compound isolated from *Bacillus subtilis* and *B. cereus* showed in vitro mortality of *Meloidogyne exigua* (Oliveira et al. 2014). 2,4-diacetylphloroglucinol produced by *Pseudomonas fluorescens* has been reported to reduce the gall farming ability of *M. incognita* on different crops (Siddiqui and Shaukat 2003). PGPR-mediated metabolism produces large number of gaseous compounds such as H_2S by sulfate-reducing bacteria *Tsukamurella paurometabola* C-924 that directly have nematicidal ability in soil (Marin et al. 2010). There are various bacterial metabolites which possess ability to induce the induced systemic resistance in plants against nematodes. *B. subtilis* have been reported to trigger ISR in eggplants against *M. javanica* by enhancing the level of peroxidases, superoxide dismutase, and phenylalanine ammonia lyase (Abbasi et al. 2014). Some recent secondary metabolites reported against PPNs are detailed in Table 5.5.

5.6 Mode of Action of Secondary Metabolites Against Insect Pest and Diseases

5.6.1 Mode of Action of Secondary Metabolites of Entomopathogenic Bacteria

Bacillus thuringiensis (Bt) are gram-positive spore-forming bacteria, which produce insecticidal proteins called δ-endotoxins such as Cry and Cyt toxins during its sporulation phase. Structurally, Cry proteins are three domain components, based on amino acid sequence. Cry toxins are classified into 67 families (Cry1 to Cry67) with more than 500 genes (Crickmore et al. 2011). These proteins are toxic to specific insect groups and safer to humans, other vertebrates, and plant species. Cry proteins are toxic to insect orders such as Lepidoptera, Coleoptera, Hymenoptera, and Diptera. Cyt proteins are mainly toxic against Diptera with cytolytic activity.

5.6.1.1 Mode of Action of Cry Toxins in Lepidoptera

It all starts with ingestion of Cry protoxin (130-kDa) by susceptible larvae; when Cry protoxin reached midgut along with food, it gets solubilized and activated by gut proteases, finally forming three domain toxic fragments of approximately 60 kDa. The activated 3d-Cry toxin binds with different receptors such as cadherin-like proteins (CAD), aminopeptidase N (APN), and alkaline phosphatase (Soberon et al. 2009; Pigott and Ellar 2007). After successful binding with receptors, Cry toxins form the pores in apical microvilli of the midgut cells (Soberon et al. 2009). This leads to formation of nonselective channels, which is permeable to cations, anions, and neutral solutes, and inflow of excess water results in cell swelling and eventual lysis (Knowles and Ellar 1987).

5.6.1.2 Mode of Action of Cyt Toxins in Diptera

In contrast to Cry proteins, Cyt proteins are mainly toxic against Diptera. Cyt toxins are also generated as protoxins, while during activation process, a small portion of the N-terminus and C-terminus is removed (Li et al. 1996). In contrast to Cry toxins, Cyt toxins directly interact with membrane lipids and form the pores without binding into specific protein receptors (Thomas and Ellar 1983; Li et al., 1996; Promdonkoy and Ellar 2003) or destroying the midgut cells by detergent-like action (Butko 2003). In the case of Cyt 2Aa, proteinase K removes the 32 amino acid residues from the N-terminal end, and 15 amino acid residues from C-terminal end lead to generation of monomeric protein with hemolytic activity (Koni and Ellar 1994).

5.6.1.3 Mode of Action of Vegetative Insecticidal Proteins (VIPs)

The VIPs are produced during vegetative growth phase of bacteria such as *B. thuringiensis* and *B. cereus*. They are divided into four families based on their amino acid sequence. The Vip1 and Vip2 proteins act as binary toxins, and they show toxicity towards members of Coleoptera and Hemiptera, Vip 3 specific to lepidopteran pests, whereas Vip 4 family proteins with unknown toxicity (Chakroun et al. 2016). In the complex of Vip1 and Vip 2, each protein has different functions, where Vip1 component bind to receptors present in midgut cell membrane and Vip 2 component actually enter the cell and prevent microfilament formation with the help of ADP-ribosyltransferase. Even though, Vip 3 proteins structurally differ from Cry proteins, but they show mode of action similar to Cry toxins, in terms of activation by proteolytic enzymes, binding to midgut membrane and pore formation. The binding receptors of Vip 3 proteins being completely different from Cry proteins enhance the success of transgenic crops with pyramiding of gens for delaying resistance and increasing the target pests (Chakroun et al. 2016).

5.6.1.4 Mode of Action of Thuringiensin

Thuringiensin, also called Thu, is a thermostable β-exotoxin, which can retain its bioactivity at 121 °C for 15 min, and is produced by *B. thuringiensis* during vegetative growth (Sharma et al. 1976; Liu et al. 2010). The target insect orders for thuringiensin include Lepidoptera, Coleoptera, Hymenoptera, Diptera, and orthoptera and several nematode species. The insect killing mechanism of thuringiensin is still not fully understood. However, it mainly affects the insect pupation and molting by inhibiting synthesis of RNA by interfering with RNA polymerase, where it acts as ATP anolog and competes with ATP binding sites (Farkas et al. 1969; Beebee et al. 1972; Sebesta et al. 1970; Burgerjon et al. 1969; Espinasse et al. 2002).

5.6.2 Mode of Action of Secondary Metabolites of Entomopathogenic Fungi

5.6.2.1 Mechanism of Insecticidal Activity of Destruxins

Destruxins are the only mycotoxins identified in substantial quantity at advanced stage of infection in insects, which cause mortality (Dumas et al. 1996). These are produced by entomopathogenic fungi such as *Metarhizium anisopliae*, *Aschersonia aleyrodis*, *Alternaria brassicae*, *Beauveria feline*, and *Nigrosabulum globosum* (Rao et al. 2006; Hu et al. 2006; Krasnoff and Gibson 1996; Parada et al. 2007; Che et al. 2001). Destruxins are divided into five families such as destruxins A, B, C, D, and E with five amino acids and an α-hydroxyl acid as structural backbone. Destruxins are mostly toxic against Lepidopteran insects and cause instant, tetanic muscular paralysis upon injection and other effects such as inhibition of fluid secretion in Malpighian tubule in *Schistocerca gregaria* (James et al. 1993) and secretion of ecdysteroid hormone by the prothoracic glands of *Manduca sexta*. They are also known to be inhibiting nucleic acid and protein synthesis (Binnington and Baule 1993). Numerous studies proved that immunomodulatory effects and inhibition of the cellular immune reaction by destruxins are the main reasons for pathogenesis in insects (Vilcinskas et al. 1997).

5.6.2.2 Mode of Action of Beauvericin

Beauvericin is produced by fungus species such as *Beauveria bassiana* and *Fusarium* sp. It is considered as regular toxin of cereals and cereal-based products. The mode of action is similar to enniatin, an antibiotic which increases ion permeability in membrane that leads to oxidative stress at molecular level. Apart from this, Beauvericin induces DNA fragmentation, chromosomal aberrations, and apoptosis (Steinrauf 1985; Ojcious et al. 1991).

The modes of activity against target pests and pathogen are summarized in Tables 5.6 and 5.7.

Table 5.6 Mode of action of secondary metabolites of microbial origin

Pesticidal metabolite	Origin	Target organism	Mode of action	References
Spinosyns	*Saccharopolyspora spinosa*	Lepidoptera, Diptera, Thysanoptera, and some species of Coleoptera and Orthoptera	Activation of nicotinic acetylcholine receptor and GABA receptors	
Avermectins	*Streptomyces avermitilis*	Hemiptera, Thysanoptera, Diptera, Lepidoptera, Coleoptera	Disruption of GABA-gated chloride channel's receptors	Strong and Brown (1987)
Polyoxins and Nikkomycins	*Streptomyces* Sp.	*Mamestra brassicae*, *Mythimna separata*, and *Spodoptera litura*	Inhibit chitin synthetase enzyme thereby inhibiting chitin formation in fungi and insects	Binnington and Baule (1993), Arakawa et al. (2008)
Chitinases	*Streptomyces* sp., bacillus, and *pseudomonas* sp.	*Orgyia pseudotsugata*, *Spodoptera littoralis*	Damaging peritrophic membrane of mid gut; affect feeding rate, insect growth; antifeeding effects, and developmental deformities	Horn et al. (2006), Avupati et al. (2017)
Toxin complex (Tc) proteins	*Photorhabdus luminescens*, *Xenorhabdus nematophila*, *Serratia entomophila*, *pseudomonas* spp.	Coleopteran and Lepidoptera	Destroying mid gut epithelial cells, apoptic cell death	Morgan et al. (2001), Marshall et al. (2012), Waterfield et al. (2001), Vodovar et al. (2006)
Photorhabdus insect-related (Pir) binary toxins	*P. Luminescens* and *P. asymbiotica*	*Plutella xylostella* and mosquito larvae	Pore formation like cry protein	Duchaud et al. (2003)
Makes caterpillars floppy (Mcf) toxins	*P. fluorescens*, *Providencia* sp., and *vibrio* spp	*Manduca sexta*	Apoptosis in insect midgut epithelial cells and emocytes which may cause disturbance in	Hinchliffe et al. (2010), Daborn et al. (2002)

(continued)

Table 5.6 (continued)

Pesticidal metabolite	Origin	Target organism	Mode of action	References
			osmoregulation leading to typical floppy phenology	

Table 5.7 secondary metabolites of microbial origin with antimicrobial properties

Pesticidal metobolite	Origin	Target pathogen	Mode of action	References
Kasugamycin	*Streptomyces kasugaensis*	*Pyricularia oryzae*, *Cercospora* sp., and *Venturia* sp.	Inhibition of translation by blocking initiator t-RNA binding to the 30S subunit	Okuyama et al. (1971), Poldermans et al. (1979)
Streptomycin	*Streptomyces griseus*	*Xanthomonas oryzae*, *X. citri*, *Pseudomonas tabaci*	Streptomycin induces structural changes in small subunit of ribosome causing misread the sequence leads to synthesis of randomn proteins	Demirci et al. (2013)
Macrolactin A	*Bacillus* sp. *sunhua*	*Streptomyces scabies*	Inhibition of sporulation and disruption of mycelium	Han et al. (2005)
Syringomycin	*Pseudomonas syringae* pv. *Syringae*	*Penicillium Digitatum*	Causes pores in plasma membranes leading to electrolyte leakage	Bull et al. (1998)

5.7 Mass Production of Secondary Metabolites of Microbes

A characteristic of secondary metabolism is that they are not produced during the rapid growth phase (trophophase), but are synthesized during the subsequent production stage (idiophase). Production of SM starts when growth is limited by the exhaustion of one key nutrient source, i.e., carbon, nitrogen, or phosphate. For example, penicillin biosynthesis by *Penicillium chrysogenum* starts when glucose is exhausted from the culture medium and the fungus starts consuming lactose, a less readily utilized sugar. Besides, these metabolites have distinctive molecular skeleton due to which about 40% of the microbial metabolites cannot be synthesized.

The pathways for production of SM involve either single enzyme or multienzyme complex. Intermediates or end-products of primary metabolic pathways are channeled through systematic metabolic pathways for synthesis of secondary metabolites. The genes encoding these synthetic pathways are generally present in chromosomal DNA and often are arranged in clusters. For example, *Streptomyces griseus* and *Streptomyces glaucescens* chromosomal DNA contain 30 or more *str/sts* and *blu* genes that participate in streptomycin biosynthesis.

In general, the secondary metabolites are produced in industry by submerged fermentation by batch or fed-batch culture techniques. An improved strain of the producing microorganism is inoculated into a growth medium in flasks and then transferred to a relatively small fermenter or "seed culture." This culture, when in rapid growth phase, is used to inoculate a fermenter tank, in the range of 30,000 to 200,000 liters, with production medium. Several parameters, like medium composition, pH, temperature, agitation and aeration rate, etc., are controlled. Different regulatory mechanisms of SM production by a given microbe are bypassed by manipulation or adjusting these important parameters. In some instances, a precursor is used to increase one specific desirable metabolite. Similarly, exit gases are also analyzed to monitor O_2 and CO_2 concentrations to monitor the metabolic information happening during fermentation.

Solid-state fermentation (SSF) also holds an important potential for the production of secondary metabolites (Barrios-Gonzalez et al. 1988). This system has been used in several oriental countries since antiquity, to prepare diverse fermented foods from grains like soybeans or rice (Hesseltine 1977a, b). Two types of SSF can be distinguished, depending on the nature of solid phase used (Barrios-Gonzalez and Mejia 1996).

1. *Solid Culture of One Support-Substrate Phase*: Solid phase is constituted by a material that assumes, simultaneously, the functions of support and of nutrient source. Agricultural or even animal goods or wastes are used as support substrate.
2. *Solid Culture of Two Substrate-Support Phase*: Solid phase is constituted by an inert support impregnated with a liquid medium. Inert support serves as a reservoir for the nutrients and water. Materials as sugarcane bagasse pith or polyurethane can be used as inert support.

In general, fungi and actinomycetes grow well is SSF, because of the technical similarity between SSF and their natural habitats (soil and organic waste materials). Moreover, the energy requirements of the SSF process are relatively low, since oxygen is transferred directly to the microorganism. Most importantly, SSF system allows much higher yields of targeted SMs in shorter times and often does not require sterile conditions (Rosenblitt et al. 2000).

The microbial production strain is regarded as the heart of the fermentation industry, so improvements in the production strain(s) offer the greatest opportunities for cost reduction without significant capital outlay. The improvement usually resides in increased yields of the desired metabolite. However, other strain characteristics can also be improved. Typical examples include removal of unwanted cometabolites, improved utilization of inexpensive carbon and nitrogen sources, or alteration of cellular morphology to a form better suited for separation of the microbe from the product and/or for improved oxygen transfer in the fermenter. Nowadays, strain improvement can be performed either by classical genetic methods (including genetic recombination) and molecular genetics methods.

5.8 Pros and Cons to the Use of Secondary Metabolites in Pest Management

5.8.1 Pros of Secondary Metabolites in Pest Management

- Eco-friendly in nature: Synthetic insecticides pose some hazards, whereas microbial secondary metabolites offer adequate levels of pest control and pose fewer hazards to environment.
- Microbial secondary metabolites utilized as insecticides are highly valuable because their toxicity to non-target organisms and humans is extremely low.
- Compared with synthetic organic and inorganic insecticides, they are safe for both the pesticide applicator and consumers of treated crops.
- The mode of action of microbial secondary metabolites is often specific to a single group or species of insects, and this specificity indicates that most microbial insecticides do not naturally affect beneficial insects (including predators or parasitoids of pests) in treated areas.
- The issues of residue, resistance, and resurgence are near to zero with respect to microbial secondary metabolites.
- The unique and novel mode of action (molecular target sites) of these metabolites on the insects prevents or delays resistance or cross resistance development.
- Deciphering the chemical nature of microbial secondary metabolites paves path for large-scale industrial production and supply for pest management.
- They form good combination products with synthetic insecticides in integrated pest management programs.
- They are suitable alternative for pesticides in organic pest management programs (e.g., Spinosad).

5.8.2 Cons of Secondary Metabolites in Pest Management

- Speed of kill: Although the microbial secondary metabolites are efficient insecticides, their speed of kill is a major drawback, which is being addressed through several biotechnological tools like gene editing and CRISPR-Cas systems.
- Difficulty in large-scale production: The secondary metabolites are mixture of large arena of chemicals which are difficult to identify and mass multiply on industrial scale.
- Cost issues: The costs of conventional insecticides are much lower and fit the budget of a small-scale farmer in developing countries, when compared to novel molecules of microbial secondary metabolites which are much more costly.

- Availability: The microbial secondary metabolites are not easily available in the local markets due to lesser production and difficulty in meeting the growing demands of farming community.
- Resistance development: Few plant pathogens have already developed resistance to synthetic derivatives of strobularins (Kim and Hwang 2007). However, development of cross resistance can be avoided by following pesticide rotation strategies.

5.9 Success Stories on the Use of Secondary Metabolites for Pest Management

5.9.1 Secondary Metabolites in Insect Pest Management

5.9.1.1 Spinosads

These are the fermentation products of the soil dwelling actinomycetes *Saccharopolyspora spinosa*. The spinosad consists of two major components, Spinosyn A and Spinosyn D, wherein, Spinosyn A forms 80% of the constituent (Orr et al. 2009). The spinosad acts as both contact and ingestion-based toxicant, and it shows the unique mode of action by excitation of the insect nervous system. The insecticide activates the nicotinic acetylcholine receptors on the post synaptic junction thus leading to involuntary muscle contractions, prostration with tremors, and finally paralysis (Sparks et al. 2001; Snyder et al. 2007). The insecticide is sold under green label and thus accepted by both conventional and organic farmers in pest management programs (Stephen et al. 2012; Reddy and Paschapur et al. 2020).

5.9.1.2 Avermectins

These are the active metabolites obtained from the broth cultures of soil bacterium *Streptomyces* sp. The avermectins form a large group of compounds with two major insecticidal compounds like abamectin and milbemycin. The abamectin is a fermentation product of *Streptomyces avermitilis*, constituting of 80% avermectin B1a and 20% avermectin B1b. The abamectin compounds have had a huge impact in animal health as insecticides against worms, ticks, and flies (Campbell et al. 1989). Moreover, the milbemectins are derived from the broth cultures of *Streptomyces hygroscopicus* subsp. *Aureolacrimosus*, and they are constituted by 70% milbemycin A4 and 30% milbemycin A3. They are very potential insecticides against a vast number of insect pests in agriculture as well as animal health. Both abamectin and milbemectin have similar mode of action, wherein they target glutamate and GABA-gated chloride channels in the nervous system of insects (Kornis et al. 1995).

5.9.2 Secondary Metabolites in Disease Management

5.9.2.1 Strobularins

Few of the well-known fungicides commercially available in the markets like trifloxystrobiin, azoxystrobin, and fluoxastrobin are the synthetic derivatives of the strobilurins isolated from a basidiomycetes wood decaying fungi (*Strobilurus tenacellus*) (Wedge and Duke 2006). The strobilurin A and strobilurin B were the first parent compounds isolated from the fungi. They are known for their unique mode of action of targeting and inhibiting the respiration at the complex III of cytochrome bc1 site in fungi (Sauter et al. 1996). Strobilurin-based fungicides are the only group of fungicides that are derived from microbes, and they almost make up almost 23–25% of the global fungicide sales (Juliet et al. 2017). However, a drawback about these fungicides is that certain plant pathogens have already developed resistance to these synthetic fungicides and need immediate research to avoid development of cross resistance and formation of super strains (Kim and Hwang 2007).

5.9.3 Secondary Metabolites in Weed Management

5.9.3.1 Glufosinate

The commercial herbicide glufosinate is a racemic mixture of L-phosphinothricin and D-phosphinothricin isolated by a fermentation product of soil bacterium *Streptomyces hygroscopicus* and *S. viridochromogenes*. The actual mode of action involves irreversible inhibition of glutamine synthetase (GS) leading to excessive accumulation of ammonia and rapid inhibition of photorespiration in plants (Duke et al. 2002). Although several other microbial metabolites like cyclic tetrapeptide tentoxin, tripeptide bialaphos, and AAL-toxin are identified and patented as herbicides, their toxicity to cultivated plants, humans, and animals has restricted their use as commercial products (Abbas et al. 1995; Lydon and Duke 1999).

5.10 Conclusion and Future Prospects

As seen in the previous sections, secondary metabolites form a new group of agrochemicals with novel modes of action targeting new molecular target sites in insects, pathogens, and weeds in modern agriculture. With the advancement in biotechnological tools like gene editing, the development of genetically modified microbes, and CRISPR-cas technologies, the scope to improve toxicity and effectivity of these metabolites forms a long optimistic way ahead. However, the complex biophysical and biochemical mechanisms associated with these secondary metabolites have hindered the studies on these compounds. As mentioned earlier, these compounds are a mixture of large number of chemical components which are difficult to decipher, and they require an interdisciplinary approach to identify, characterize, and further improve them for exploitation as commercial pesticides.

The market surveys conducted by Markets and Markets (2016) and Thakore (2006) showed the increasing demand for microbial pesticides in the US markets, wherein the demand for biorational pesticides is expected to increase from 2% to 17% in the year 2022, which is higher than the demand for conventional pesticides (3%). However, the major drawbacks associated with commercialization of secondary metabolites are their patenting and maintaining secrecy of the pesticidal toxins by both researchers and the industries.

Considering the lacunas associated with the conventional insecticides like residues in the environment, ill effects on non-target organisms, resistance development by insects and pathogens, and public awareness about the environmental effects are enhancing interest of farmers towards use of novel chemistry insecticides like secondary metabolites produced by microbes. Additionally, the increasing interest about these microbial pesticides among scientific community and genome-wide analysis and genome sequencing of microbial pesticides are bound to answer the long awaiting questions about pathogenesis besides revealing metabolic complexes involved with the secondary metabolites. Thus, the concrete interdisciplinary knowledge involving organic chemistry, biotechnology, plant protection, and environmental sciences would help to decipher the importance of these microbial secondary metabolites in pest management and further introduce novel pesticides for commercial production and use in modern agriculture.

Acknowledgments The authors are thankful to Directors of IACR-VPKAS, ICAR-RCER, and ICAR-IARI and Head of Department, Department of Entomology, UAS, Bangalore for their constant guidance and support throughout the manuscript preparation. We also acknowledge the authors of different scientific papers whose results made this book chapter possible.

References

Aballay E, Prodan S, Zamorano A, Castaneda-Alvarez C (2017) Nematicidal effect of rhizobacteria on plant-parasitic nematodes associated with vineyards. World J Microbiol Biotechnol 33(7): 131

Abbas HK, Tanaka T, Duke SO, Boyette CD (1995) Susceptibility of various crop and weed species to AAL-toxin, a natural herbicide. Weed Technol 9:125–130

Abbasi M, Ahmed N, Zaki M, Shuakat S, Khan D (2014) Potential of *Bacillus* species against *Meloidogyne javanica* parasitizing eggplant (*Solanum melongena* L.) and induced biochemical changes. Plant and Soil 375(1/2):159–173

Aghajanzadeh S, Mallik B, Chandrashekar SC (2006) Toxicity of culture filtrate of hirsutella thompsoniifisher against citrus rust mite, *Phyllocoptruta oleivora* Ashmead (Acari: Eriophyidae) and two spotted spider mite, *Tetranychus urticae* Koch (Acari: Tetranychidae). Int J Agric Biol 8:276–279

Altomare C, Pengue R, Favilla M, Evidente A, Visconti A (2004) Structure-activity relationships of derivatives of fusapyrone, an antifungal metabolite of *fusarium semitectum*. J Agric Food Chem 52:2997–3001

Altomare C, Perrone G, Zonno MC, Evidente A, Pengue R, Fanti F, Polonelli L (2000) Biological characterization of fusapyrone and deoxyfusapyrone, two bioactive secondary metabolites of *fusarium semitectum*. J Nat Prod 63:1131–1135

Andolfi A, Maddau L, Linaldeddu BT, Scanu B, Cimmino A, Basso S, Evidente A (2014) Bioactivity studies of oxysporone and several derivatives. Phytochem Lett 10:40–45

Arakawa T, Yukuhiro F, Noda H (2008) Insecticidal effect of a fungicide containing polyoxin B on the larvae of *Bombyx mori* (Lepidoptera: Bombycidae), *Mamestra brassicae*, *Mythimna separata*, and *Spodoptera litura* (Lepidoptera: Noctuidae). Appl Entomol Zool 43(2):173–181

Araújo JP, Hughes DP (2016) Chapterone-diversity of entomopathogenic fungi: which groups conquered the insect body? Adv Genet 94:1–39

Arthurs S, Dara SK (2019) Microbial biopesticides for invertebrate pests and their markets in the United States. J Invertebr Pathol 165:13–21

Avupati RS, Khan MS, Johnson S, Yogi MK (2017) Diversity and functional annotation of chitinolytic *bacillus* and associated chitinases from north western Indian Himalayas. Appl Soil Ecol 119:46–55

Bailly A, Weisskopf L (2017) Mining the volatilomes of plant-associated microbiota for new biocontrol solutions. Front Microbiol 8:1638–1638

Bandani AR (2004) Effect of entomopathogenic fungus *Tolypocladium* species metaboliteefrapeptinon *Galleriamellonella* agglutinin. Commun Agric Appl Biol Sci 69:165–169

Bandani AR (2008) The effects of entomopathogenic fungus, *Tolypocladium cylindrosporum* on cellular defence system of *galleria mellonella*. J Agric Sci Technol 10:135–146

Barrios-González J, Castillo TE, Mejía A (1988) Penicillin production by solid state fermentation. Biotechnoi Leu 10:793–798

Barrios-Gonzalez J, Mejia A (1996) Production of secondary metabolites by solid-state fermentation. Biotechnol Annu Rev 2:85–121

Beebee T, Korner A, Bond RP (1972) Differential inhibition of mammalian ribonucleic acid polymerases by an exotoxin from *Bacillus thuringiensis*. The direct observation of nucleoplasmic ribonucleic acid polymerase activity in intact nuclei. Biochem J 127:619–634

Bennett JW, Bentley R (1989) What's in a name? Microbial secondary metabolism. Adv Appl Microbiol 34:1–28

Berg G (2000) Diversity of antifungal and plant-associated *Serratia plymuthica* strains. J Antimicrob Chemother 88:952–960

Berg G (2009) Plant microbe interactions promoting plant growth and health: perspectives for controlled use of microorganisms in agriculture. Appl Microbiol Biotechnol 84:11–18

Binnington KC, Baule VJ (1993) Naturally occurring insecticidal molecules as candidates for genetic engineering. In: Molecular approaches to fundamental and applied entomology. Springer, New York, pp 38–89

Bravo A, Sarabia S, Lopez L, Ontiveros H, Abarca C, Ortiz A (1998) Characterization of cry genes in a Mexican *Bacillus thuringiensis* strain collection. Appl Environ Microbiol 64:4965–4972

Brodhagen M, Paulsen I, Loper JE (2005) Reciprocal regulation of pyoluteorin production with membrane transporter gene expression in *Pseudomonas fluorescens* Pf-5. Appl Environ Microbiol 71:6900–6909

Bull C, Wadsworth M, Sorensen K, Takemoto J, Austin R, Smilanick J (1998) Syringomycin E produced by the biological control agents control green mold on lemons. Biol Control 12:89–95

Burgerjon A, Biache G, Cals P (1969) Teratology of the Colorado potato beetle, *Leptinotarsa decemlineata*, as provoked by larval administration of the thermostable toxin of Bacillus thuringiensis. J Invertebr Pathol 14:274–278

Butko P (2003) Cytolytic toxin Cyt1A and its mechanism of membrane damage: data and hypotheses. Appl Environ Microbiol 69:2415–2422

Buttachon S, Chandrapatya A, Himaman W, Kijjoa A (2013) Acaricidal activityof *Hypocrella raciborskii* Zimm. (Hypocreales: Clavicipitaceae) crude extract and some pure compounds on *Tetranychusurticae* Koch (Acari: Tetranychidae). Afr J Microbiol Res 7:557–585

Cai X, Nowak S, Wesche F, Bischoff I, Kaiser M, Furst R, Bode HB (2016) Entomopathogenic bacteria use multiple mechanisms for bioactive peptide library design. Nat Chem 9:379–386

Campbell WC (1989) Ivermectin and abamectin. Springer, New York, p 361

Chakroun M, Banyuls N, Bel Y, Escriche B, Ferré J (2016) Bacterial vegetative insecticidal proteins (Vip) from entomopathogenic bacteria. Microbiol Mol Biol Rev 80(2):329–350

Chatterjee S, Chatterjee DK, Lad SJ, Phansalkar MS, Rupp RH, Ganguli BN, Fehlhaber HW, Kogler H (1992) Mersacidin, a new antibiotic from bacillus: fermentation, isolation, purification and chemical characterization. J Antibiot 45:832–838

Che Y, Swenson DC, Gloer JB, Koster B, Malloch D (2001) Pseudodestruxins a and B: new cyclic depsipeptides from the coprophilous fungus *Nigrosabulum globosum*. J Nat Prod 64:555–558

Chen XH, Scholz R, Borriss M, Junge H, Mögel G, Kunz S, Borriss R (2009) Difficidin and bacilysin produced by plant-associated *bacillus amyloliquefaciens* are efficient in controlling fire blight disease. J Biotechnol 140:38–44

Choi SK, Park SY, Kim R, Kim SB, Lee CH, Kim JF, Park SH (2009) Identification of a polymyxinsynthetase gene cluster of *Paenibacillus polymyxa* and heterologous expression of the gene in Bacillus subtilis. J Bacteriol 191:3350–3358

Cimmino A, Andolfi A, Avolio F, Ali A, Tabanca N, Khan IA, Evidente A (2013) Cyclopaldic acid, seiridin, and sphaeropsidin a as fungal phytotoxins, and larvicidal and biting deterrents against *Aedes aegypti* (Diptera: Culicidae): structure-activity relationships. Chem Biodivers 10:1239–1251

Crawford JM, Portmann C, Zhang X, Roeffaers MB, Clardy J (2012) Small molecule perimeter defense in entomopathogenic bacteria. Proc Natl Acad Sci U S A 109:10821–10826

Crickmore N, Zeigler DR, Schnepf E, Van Rie J, Lereclus D, Baum J, Bravo A, Dean DH (2011) Bacillus thuringiensis. Toxin nomenclature. www.lifesci.sussex.ac.uk/home/Neil_Crickmore/Bt/data

Daborn PJ, Waterfield N, Silva CP, Au CP, Sharma S, Ffrench-Constant RH (2002) A single Photorhabdus gene, makes caterpillars floppy (mcf), allows *Escherichia coli* to persist within and kill insects. Proc Natl Acad Sci U S A 99(16):10742–10747

Demain AL, Fang A (2000) The natural functions of secondary metabolites. History of modern biotechnology I:1–39

Demirci H, Murphy F, Murphy E, Gregory ST, Dahlberg AE, Jog G (2013) A structural basis for streptomycin-induced misreading of the genetic code. Nat Commun 4:1355

Dong YH, Wang LH, Xu JL, Zhang HB, Zhang XF, Zhang LH (2001) Quenching quorum-sensing-dependent bacterial infection by an N-acyl homoserine lactonase. Nature 411(6839):813–817

Duchaud E, Rusniok C, Frangeul L, Buchrieser C, Givaudan A, Taourit S, Bocs S, Boursaux-Eude C, Chandler M, Charles JF, Dassa E (2003) The genome sequence of the entomopathogenic bacterium *Photorhabdus luminescens*. Nat Biotechnol 21(11):1307

Duke SO, Dayan FE, Rimando AM, Schrader KK, Aliotta G, Oliva A, Romagni JG (2002) Chemicals from nature for weed management. Weed Sci 50:138–151

Dumas C, Matha V, Quiot JM, Vey A (1996) Effects of destruxins, cyclic depsipeptide mycotoxins, on calcium balance and phosphorylation of intracellular proteins in lepidopteran cell lines. Comp Biochem Physiol 114C:213–219

Dunham B (2015) Microbial biopesticides: a key role in the multinational portfolio. http://dunhamtrimmer.com/wp-content/uploads/2015/01/Products-and-Trends.pdf. Accessed 22 Aug 2017

El-Hadad ME, Mustafa MI, Selim SM, Mahgoob AEA, El-Tayeb TS, Abdel Aziz NH (2010) In vitro evaluation of some bacterial isolates as biofertilizers and biocontrol agents against the second stage juveniles of *Meloidogyne incognita*. World J Microbiol Biotechnol 26:2249–2256

El-Sayed AK, Hothersall J, Thomas CM (2001) Quorum-sensingdependent regulation of biosynthesis of the polyketide antibiotic mupirocin in *Pseudomonas fluorescens* NCIMB 10586. Microbiology 147:2127–2139

Elsworth JF, Grove JF (1977) Cyclodepsipeptides from *Beauveria bassiana* Bals. Part1. Beauverolides HandI. J Chem Soc Perkin Trans 1:207–273

Espinasse S, Gohar M, Chaufaux J, Buisson C, Perchat S, Sanchis V (2002) Correspondence of high levels of beta-exotoxin I and the presence of cry1B in *Bacillus thuringiensis*. Appl Environ Microbiol 68:4182–4186

Evidente A, Maddau L, Scanu B, Andolfi A, Masi M, Motta A, Tuzi A (2011) Sphaeropsidones, phytotoxic dimedone methyl ethers produced by Diplodia cupressi: a structure—activity relationship study. J Nat Prod 74:757–763
Farkas J, Sebesta K, Horska K, Samek Z, Dollijs J, Storm F (1969) The structure of exotoxin of *bacillus thuringiensis* var. gelechiae. Collect Czechslov Chem Commun 34:1118–1120
Frankenhuyzen K (2009) Insecticidal activity of *bacillus thuringiensis* crystal proteins. J Invertebr Pathol 101(1):1–16
Gao H, Qi G, Yin R, Zhang H, Li C, Zhao X (2016) Bacillus cereus strain S2 shows high nematicidal activity against *Meloidogyne incognita* by producing sphingosine. Sci Rep 6:28756
Gibbons SM, Gilbert JA (2015) Microbial diversity—exploration of natural ecosystems and microbiomes. Curr Opin Genet Dev 35:66–72
Glare TR, Milner RJ (1991) Ecology of entomopathogenic fungi. In: Arora DK, Mukeriji KG, Pugh JEF (eds) Handbook of applied mycology. Dekker, NewYork, pp 547–612
Goettel MS, Hajek AE, Siegel JP, Evans HC (2001) Safety of fungal biocontrol agents. In: Butt TM, Jackson CW, Magan N (eds) Fungi as biocontrol agents, progress, problems, and potential. CAB International, Wallingford, pp 347–375
Gong Q, Zhang C, Lu F, Zhao H, Bie X, Lu Z (2014) Identification of bacillomycin D from *Bacillus subtilis* and its inhibition effects against Aspergillusflavus. Food Control 36:8–14
Guttenberger N, Blankenfeldt W, Breinbauer R (2017) Recent developments in the isolation, biological function, biosynthesis, and synthesis of phenazine natural products. Bioorg Med Chem 25:6149–6166
Hallmann J, Sikora RA (1994) Occurrence of plant parasitic nematodes and non-pathogenic species of *fusarium* in tomato plants in Kenya and their role as mutualistic synergists for biological control of root-knot nematodes. Int J Pest Manag 40:321–325
Hallmon CF, Schreiber ET, Vo T, Bloomquist A (2000) Field trials of three concentrations of Laginex as biological larvicide compared to Vectobac- 12AS AS a biocontrol agent for *Culex quinquefasciatus*. J Amer Mosq Control Assoc 16:5–8
Han JS, Cheng JH, Yoon TM, Song J, Rajkarnikar A, Kim WG, Yoo ID, Yang YY, Suh JW (2005) Biological control of common scab diseases by antagonistic strain Bacillus sp. sunhua. J Appl Microbiol 99:213–221
Hellberg J, Matilla MA, Salmond GPC (2015) The broad-spectrum antibiotic, zeamine, kills the nematode worm Caenorhabditiselegans. Front Microbiol 6:137
Hesseltine CW (1977a) Solid-state fermentation part 1. Process Biochem 12:24–27
Hesseltine CW (1977b) Solid-state fermentation part 2. Process Biochem 12:29–32
Hinchliffe SJ, Hares MC, Dowling AJ (2010) Insecticidal toxins from the Photorhabdus and Xenorhabdus bacteria. Open Toxicol J 3(1):101–118
Horn SJ, Sørbotten A, Synstad B et al (2006) Endo/exo mechanism and processivity of family 18 chitinases produced by Serratia marcescens. FEBS J 273:491–503
Hu QB, Ren SX, Wu JH, Chang JM, Musa PD (2006) Investigation of destruxin A and B from 80 Metarhizium strains in China, and the optimization of cultural conditions for the strain MaQ10. Toxicon 48:491–498
Huang X, Tian B, Niu Q, Yang J, Zhang L, Zhang K (2005) An extracellular protease from Brevibacillus laterosporus G4 without parasporal crystals can serve as a pathogenic factor in infection of nematodes. Res Microbiol 156:719–727
Ingels A, Dinhof C, Garg AD, Maddau L, Masi M, Evidente A, Berger W, Dejaegher B, Mathieu V (2017) Computed determination of the in vitro optimal chemocombinations of sphaeropsidin A with chemotherapeutic agents to combat melanomas. Cancer Chemother Pharmacol 79:971–983
Isaka M, Kittakoop P, Thebtaranonth Y (2003) Secondary metabolites of Clavicipitalean fungi. In: White JF, Bacon CW, Hywel-Jones NL, Spata JW (eds) Clavicipitalean fungi: evolutionary biology, chemistry, biocontrol, and culture impacts (mycology series). CRC Press, Boca Raton, pp 355–398

Isaka M, Palasarn S, Tobwor P, Boonruangprapa T, Tasanathai K (2012) Bioactive anthrax quinine dimmers from the leafhopper pathogenic fungus *Torrubiella* sp. BCC28517. J Antibiot 65:571–574. https://doi.org/10.1038/ja.2012.76
James PJ, Kershaw MJ, Reynolds SE, Charnley AK (1993) Inhibition of desert locust (Schistocerca gregaria) Malpighian tubule fluid secretion by destruxins, cyclic peptide toxins from the insect pathogenic fungus *Metarhizium anisopliae*. J Insect Physiol 39:797–804
Jin-Ming G (2006) New biologically active metabolites from Chinese higher fungi. Curr Org Chem 10:849–871. https://doi.org/10.2174/138527206776894393
Juliet DT, Tina C, Maria TP, Susan VD (2017) Activity of two Strobilurin fungicides against three species of decay fungi in agar plate tests. In Proc IRG annual meeting: IRG/WP 17-30704
Jung WJ, Kim KY, Park YS et al (2014) Purification and properties of a Meloidogyne antagonistic chitinase from Lysobacter capsici YS1215. Nematolog 16:63–72
Jung WJ, Mabood F, Souleimanov A, Smith DL (2011) Induction of defense-related enzymes in soybean leaves by class IIdbacteriocins (thuricin 17 and bacthuricin F4) purified from Bacillus strains. Microbiol Res 167(1):14–19
Kato N, Furutani S, Otaka J, Noguchi A, Kinugasa K, Kai K, Hayashi H, Ihara M, Takahashi S, Matsuda K (2018) Biosynthesis and structure–activity relationship studies of okaramines that target insect glutamate-gated chloride channels. ACS Chem Biol 13:561–566
Kerwin JL, Dritz DA, Washino RK (1994) Pilot scale production and application in wildlife ponds of *Lagenidium giganteum* (oomycetes: Lagenidiales). J Amer Mosq Control Assoc 10:451–455
Kim BS, Hwang BK (2007) Microbial fungicides in the control of plant diseases. J Phytopathol 155:641–653
Kim JR, Yeon SH, Kim HS, Ahn YJ (2002) Larvicidal activity against *Plutella xylostella* of cordycepin from the fruiting body of *Cordycepsmilitaris*. Pest Manag Sci 58:713–717
Knowles BH, Ellar DJ (1987) Colloid-osmotic lysis is a general feature of the mechanism of action of bacillus thuringiensis δ-endotoxins with different insect specificity. Biochem BiophysActa 924(3):509–518
Koni PA, Ellar DJ (1994) Biochemical characterization of *Bacillus thuringiensis* cytolytic δ-endotoxins. Microbiology 140:1869–1880
Kornis GI (1995) Avermectins and milbemycins. In: Godfrey CRA (ed) Agrochemicals from natural products. Marcel Dekker, New York, pp 215–255
Krasnoff SB, Gibson DM (1996) New destruxins from the entomopathogenic fungus Aschersoniasp. J Nat Prod 59:485–489
Kryukov VIU, Yaroslavtseva ON, Dubovskiy IM, Tyurin MV, Kryukova NA, Glupov VV (2014) Insecticidal and immunosuppressive effect of ascomycete *Cordycepsmilitaris* on the larvae of the Colorado potato beetle *Leptinotarsa decemlineata*. Biol Bull 41:3
Lallemand B, Masi M, Maddau L, De Lorenzi M, Dam R, Cimmino A, Moreno Y, Banuls L, Andolfi A, Kiss R (2012) Evaluation of in vitro anticancer activity of sphaeropsidins A–C, fungal rearranged pimarane diterpenes, and semisynthetic derivatives. Phytochem Lett 5:770–775
Lastra CL, Scorsetti AC, Marti GA, García JJ (2004) Host range and specificity of an argentine an isolate of the aquatic fungus *Leptolegnia chapmanii* (oomycetes: Saprolegniales), a pathogen of mosquito larvae (Diptera: Culicidae). Mycopathologia 158:311–315
Lee YS, Nguyen XH, Naing KW et al (2014) Role of lytic enzymes secreted by *Lysobacter capsici* YS1215 in the control of root-knot nematode of tomato plants. Indian J Microbiol 55:74–80
Levenfors JJ, Hedman R, Thaning C, Gerhardson B, Welch CJ (2004) Broad-spectrum antifungal metabolites produced by the soil bacterium *Serratia plymuthica* A 153. Soil BiolBiochem 36: 677–685
Li J, Pandelakis AK, Ellar DJ (1996) Structure of the mosquitocidal δ-endotoxin CytB from *Bacillus thuringiensis* sp kyushuensis and implications for membrane pore formation. J Mol Biol 257:129–152

Li XJ, Zhang Q, Zhang AL, Gao JM (2012) Metabolites from *Aspergillus fumigatus*, an endophytic fungus associated with *Melia azedarach*, and their antifungal, antifeedant, and toxic activities. J Agric Food Chem 60:3424–3431

Liang WL, Le X, Li HJ, Yang XL, Chen JX, Xu J, Liu HL, Wang LY, Wang KT, Hu KC (2014) Exploring the chemodiversity and biological activities of the secondary metabolites from the marine fungus Neosartorya pseudofischeri. Mar Drugs 12:5657–5676

Liu X, Ruan L, Peng D, Li L, Sun M, Yu Z (2014) Thuringiensin: a thermostable secondary metabolite from *bacillus thuringiensis* with insecticidal activity against a wide range of insects. Toxins 6(8):2229–2238

Liu XY, Ruan LF, Hu ZF, Peng DH, Cao SY, Yu ZN, Liu Y, Zheng JS, Sun M (2010) Genome-wide screening reveals the genetic determinants of an antibiotic insecticide in *Bacillus thuringiensis*. J Biol Chem 285(50):39191–39200

Liu Z, Budiharjo A, Wang P, Shi H, Fang J, Borriss R et al (2013) The highly modified microcin peptide plantazolicin is associated with nematicidal activity of *Bacillus amyloliquefaciens* FZB42. Appl Microbiol Biotechnol 97:10081–10090

Lydon J, Duke SO (1999) Inhibitors of glutamine biosynthesis. In: Singh BK (ed) Plant amino acids: biochemistry and biotechnology. Marcel Dekker, New York, pp 445–464

Macias FA, Varela RM, Simonet AM, Cutler HG, Cutler SJ, Eden MA, Hill RA (2000) Bioactive carotanes from *Trichoderma virens*. J Nat Prod 63:1197–2000

Marin M, Mena J, Franco R, Pimentel E, Sánchez I (2010) Effects of the bacterial-fungal interaction between *Tsukamurella paurometabola* C 924 and Glomus fasciculatum and Glomus clarum fungi on lettuce mycorrhizal colonization and foliar weight. Biotecnol Apl 27(1):48–51

Markets, Markets (2016) Biopesticides market – global forecast to 2022. By type (bioinsecticides, biofungicides, bioherbicides, and bionematicides), origin (beneficial insects, microbials, plant-incorporated protectants, and biochemicals), mode of application, formulation, crop type and region. http://www.marketsandmarkets.com/. Accessed 24 Nov 2018

Marroquin LD, Elyassnia D, Griffitts JS, Feitelson JS, Aroian RV (2000) *Bacillus thuringiensis* (Bt) toxin susceptibility and isolation of resistance mutants in the nematode *Caenorhabditis elegans*. Genetics 155:1693–1699

Marshall SD, Hares MC, Jones SA, Harper LA, James VR, Harland DP, Jackson TA, Hurst MR (2012) Histopathological effects of the yen-Tc toxin complex from *Yersina entomophaga* MH96 (Enterobacteriaceae) on the midgut of *Costelytra zealandica* (Coleoptera: Scarabaeidae) larvae. Appl Environ Microbiol 78:4835–4847

Masschelein J, Mattheus W, Gao LJ, Moons P, Van Houdt R, Uytterhoeven B, Lamberigts C, Lescrinier E, Rozenski J, Herdewijn P, Aertsen A, Michiels C, Lavigne R (2013) A PKS/ NRPS/ FAS hybrid gene cluster from Serratiaplymuthica RVH1 encoding the biosynthesis of three broad spectrum, zeaminerelated antibiotics. PLoS One 8(1):e54143

Mathieu V, Chantôme A, Lefranc F, Cimmino A, Miklos W, Paulitschke V, Mohr T, Maddau L, Kornienko A, Berger W (2015) Sphaeropsidin a shows promising activity against drug-resistant cancer cells by targeting regulatory volume increase. Cell Mol Life Sci 72:3731–3746

Mavrodi OV, McSpadden Gardener BB, Mavrodi DV, Bonsall RF, Weller DM, Thomashow LS (2001) Genetic diversity of phlD from 2,4-diacetylphloroglucinol-producing fluorescent pseudomonas species. Phytopathology 91:35–43

McCoy CW, Omoto C, Mazet I, Vey A (1992) Biological activity of crude filtrates and hirsutellin A to mosquitoes and phytophagous arthropods. In: Proceedings of the abstracts 120 XXV annual meeting. Society of Invertebrate Pathology, Heidelberg

Milner RJ (1997) Prospects for biopesticides for aphid control. Entomophaga 42:227

Mnif I, Ghribi D (2015) Potential of bacterial derived biopesticides in pest management. Crop Prot 77:52–64

Mollah MMI, Yeasmin F, Kim Y (2020) Benzylideneacetone and other phenylethylamide bacterial metabolites induce apoptosis to kill insects. J of Asia-Pacific Ento 23(2):449–457

Morgan JAW, Sergeant M, Ellis D, Ousley M, Jarrett P (2001) Sequence analysis of insecticidal genes from *Xenorhabdus nematophilus* PMFI296. Appl Environ Microbiol 67(5):2062–2069

Nicoletti R, De Stefano M, De Stefano S, Trincone A, Marziano F (2004) Antagonism against *Rhizoctonia solani* and fungitoxic metabolite production by some *Penicillium* isolates. Mycopathologia 158:465–474

Niu B, Vater J, Rueckert C, Blom J, LehmannM RJJ, Chen XH, Wang Q, Borriss R (2013) Polymyxin P is the active principle in suppressing phytopathogenic *Erwinia* spp. by the biocontrol rhizobacterium *Paenibacillus polymyxa* M-1. BMC Microbiol 13(137)

Niu S, Liu D, Hu X, Proksch P, Shao Z, Lin W (2014) Spiromastixones A–O, antibacterial chlorodepsidones from a deep-sea-derived *Spiromastix* sp. fungus. J Nat Prod 77:1021–1030

Nollmann FI, Heinrich AK, Brachmann AO, Morrisseau C, Mukherjee K, Casanova-Torres AM, Kleinhans D, Kinski S, Schultz K, Beeton M, Kaiser M, Chu YY, Ke LP, Thanwisai A, Bozhuyuk KAJ, Chantratita N, Gotz F, Waterfield NR, Vilcinskas A, Stelzer EHK, Goodrich-Blair H, Hammock BD, Bode HB (2015) A photorhabdus natural product inhibits insect juvenile hormone epoxide hydrolase. Chembiochem 16:766–771

Ojcious DM, Zychlinsky A, Zheng LM, Young JDE (1991) Ionophore-induced apoptosis: role of DNA fragmentation and calcium fluxes. Exp Cell Res 197:43–49

Okuyama A, Machiyama N, Kinoshita T, Tanaka N (1971) Inhibition by kasugamycin of initiation complex formation on 30S ribosomes. Biochem Biophys Res Commun 43:196–199

Oliveira DF, Santos Junior HM, Dos Nunes AS et al (2014) Purification and identification of metabolites produced by *Bacillus cereus* and *B. subtilis* active against *Meloidogyne exigua*, and their in-silico interaction with a putative phosphoribosyltransferase from M. incognita. An Acad Bras Cienc 86:525–538

Olson S (2015) An analysis of the biopesticide market now and where it is going. Outlooks Pest Manage 26:203–206. https://doi.org/10.1564/v26_oct_04

Onofre SB, Riveros Gonzalez R, Messias CL, Azevedo JL, Barros NMD (2002) LC50 of the peptide produced by the entomopathogenic fungus *Nomuraea rileyi* (Farlow) Samson active against third instar larvae of *Anticarsia gemmatalis* (Lep.: Noctuidae). Braz Arch Biol Technol 45:156–163

Orr N, Shaffner AJ, Richey K, Crouse GD (2009) Novel mode of action of spinosad: receptor binding studies demonstrating lack of interaction with known insecticidal target sites. Pestic Biochem Physiol 95:1–5

Pan X, Yang Y, Zhang JR (2014) Molecular basis of host specificity in human pathogenic bacteria. Emerg Microbes Infect 3:23

Parada RY, Oka K, Yamagishi D, Kodama M, Otani H (2007) Destruxin B produced by Alternaria brassicae does not induce accessibility of host plants to fungal invasion. Physiol Mol Plant Pathol 71:48–54

Pelizza SA, LopezLastra CC, Becnel JJ, Bisaro V, Garcia JJ (2007) Biotic and abiotic factors affecting *Leptolegnia chapmanii* infectionin *Aedesaegypti*. J Amer Mosq Control Assoc 23: 177–181

Pelizza SA, Scorsetti AC, Tranchida MC (2013) The sublethal effects of the entomopathicfungus *Leptolegnia chapmanii* on some biological parameters of the dengue vector *Aedes aegypti*. J Insect Sci 13:22

Pell JK, Eilenberg J, Hajek AE, Steinkraus DC (2001) Biology, ecology and pest management potential of Entomophthorales. In: Butt TM (ed) Fungi as biocontrol agents: progress, problems and potential. CAB International, Wallingford, pp 71–153

Pigott CR, Ellar DJ (2007) Role of receptors in *Bacillus thuringiensis* crystal toxin activity. MicrobiolMol Biol Rev 71:255–281

Poldermans B, Goosen N, Van Knippenberg PH (1979) Studies on the function of two adjacent N6, N6- dimethyladenosines near the 3′ end of 16 S ribosomal RNA of Escherichia coli. I. the effect of kasugamycin on initiation of protein synthesis. J Biol Chem 254:9085–9089

Promdonkoy B, Ellar DJ (2003) Investigation of the pore-forming mechanism of a cytolytic δ-endotoxin from *Bacillus thuringiensis*. Biochem J 374:255–259

Prompiboon P, Bhumiratana A, Ruchirawat S, Boucias DG, Wiwat C (2008) Isolation of ergosterol peroxide from *Nomura eariley* iinfected larvae of tobacco cutworm. World J Microbiol Biotechnol 24:2909

Quarles W (2013) New bio pesticides for IPM and organic production. IPM Pract 33:1–20

Ramette A, Frapolli M, Défago G, Moënne-Loccoz Y (2003) Phylogeny of HCN synthase-encoding hcnBC genes in biocontrol fluorescent pseudomonads and its relationship with host plant species and HCN synthesis ability. Mol Plant Microbe Interact 16:525–535

Rao YK, Tsou CH, Tzeng YM (2006) Antioxidants enhanced production of destruxin E from cultivation of *Metarhizium anisopliae*. Appl Microbiol Biotechnol 73:519–524

Reddy KVN, Paschapur A (2020) Employing environmentally safer and novel synthetic insecticides in organic farming for eco-friendly pest management. Indian Farmer 7(04):267–272

Rezzonico F, Zala M, Keel C, Duffy B, Moënne-Loccoz Y, Défago G (2007) Is the ability of biocontrol fluorescent pseudomonads to produce the antifungal metabolite 2,4-diacetylphloroglucinol really synonymous with higher plant protection? New Phytol 173: 861–872

Rosenblitt A, Agosin E, Delgado J, Pérez-Correa R (2000) Solid substrate fermentation of Monascus purpureus; growth, carbon balance and consistency analysis. Biotechnol Prog 16: 152–162

Sandhu SS, Sharma AK, Beniwal V, Goel G, Batra P, Kumar A, Malhotra S (2012) Myco-biocontrol of insect pests: factors involved, mechanism, and regulation. J Pathol 2012:126819

Sauter H, Ammermann E, Roehl F (1996) Strobilurins—from natural products to a new class of fungicides. In: Copping LG (ed) Crop protection agents from nature: natural products and analogues. Royal Soc. Chem, Cambridge, pp 50–81

Saxena S, Pandey AK (2001) Microbial metabolites as eco-friendly agrochemicals for the next millennium. Appl Microbiol Biotechnol 55(4):395–403

Sayyed AH, Crickmore N, Wright DJ (2001) Cyt1Aa from *Bacillus thuringiensis* subspisraelensis is toxic to the diamondback moth, *Plutella xylostella*, and synergizes the activity of Cry1Ac towards a resistant strain. Appl Environ Microbiol 67:5859–5861

Schnepf E, Crickmore NV, Van Rie J, Lereclus D, Baum J, Feitelson J, Zeigler DR, Dean DH (1998) *Bacillus thuringiensis* and its pesticidal crystal proteins. Microbiol Mol Biol Rev 62(3): 775–806

Sebesta K, Horska K (1970) Mechanism of inhibition of DNA-dependent RNA polymerase by exotoxin of *Bacillus thuringiensis*. Biochem Biophys Acta 209:357–376

Shahid I, Rizwan M, Baig DN, Salem RS, Mali KA, Mehnaz S (2017) Secondary metabolites production and plant growth promotion by *Pseudomonas chlororaphis* and *P. aurantiaca* strains isolated from cactus, cotton, and para grass. J Microbiol Biotechnol 27:480–491

Sharma CB, Prasad SS, Pai SB, Sharma S (1976) The exotoxin of *bacillus thuringiensis* A new C-mitotic agent. Experientia 32:1465–1466

Siddiqui IA, Shaukat SS (2003) Plant species, host age and host genotype effects on Meloidogyne incognita biocontrol by *Pseudomonas fluorescens* strain CHA0 and its genetically-modified derivatives. J Phytopathol 151:231–238

Silo-Suh LA, Stabb EV, Raffel SJ, Handelsman J (1998) Target range of zwittermicin A, an aminopolyol antibiotic from *Bacillus cereus*. Curr Microbiol 37(1):6–11

Singh D, Son SY, Lee CH (2016) Perplexing metabolomes in fungal-insect trophic interactions: a Terra incognita of mycobiocontrol mechanisms. Front Microbiol 7:1678

Snyder DE, Meyer J, Zimmermann AG, Qiao M, Gissendanner SJ, Cruthers LR, Slone RL, Young DR (2007) Preliminary studies on the effectiveness of the novel pulicide, spinosad, for the treatment and control of fleas on dogs. Vet Parasitol 150:345–351

Soberon M, Gill SS, Bravo A (2009) Signaling versus punching hole: how do *Bacillus thuringiensis* toxins kill insect midgut cells? Cell Mol Life Sci 66:1337–1349

Sparks TC, Crouse GD, Durst G (2001) Natural products as insecticides: the biology, biochemistry and quantitative structure–activity relationships of spinosyns and spinosoids. Pest Manag Sci 57:896–905

Steinrauf LK (1985) Beauvericin and other enniatins. In: Sigel H (ed) Metal ions in biological systems. Dekker, New York, pp 140–171
Stephens EJ, Losey JE, Allee LL, DiTommaso A, Bodner C, Breyre A (2012) The impact of Cry3Bb Bt-maize on two guilds of beneficial beetles. Agric Ecosyst Environ 156:72–81
Strong L, Brown TA (1987) Avermectins in insect control and biology: a review. Bull Entomol Res 77:357–389
Su X, Zou F, Guo Q, Huang J, Chen TX (2001) A report on a mosquito-killing fungus *Pythium carolinianum*. Fungal Divers 7:129–133
Subbanna ARNS, Stanley J, Rajasekhara H, Mishra KK, Pattanayak A, Bhowmick R (2020) Perspectives of microbial metabolites as pesticides in agricultural pest management. In: Mérillon J-M, Ramawat KG (eds) Co-evolution of secondary metabolites, reference series in phytochemistry. Springer, Cham
Sussmuth RD, Mainz A (2017) Nonribosomal peptide synthesis-principles and prospects. Angew Chem Int Ed Engl 56:3770–3821
Szekeres A, Leitgeb B, Kredics L, Antal Z, Hatvani L, Manczinger L, Vagvolgyi C (2005) Peptaibols and related peptaibiotics of *Trichoderma*—a review. Acta Microbiol Immunol Hung 52:137–168
Tanaka Y, Omura S (1993) Agroactive compounds of microbial origin. Annu Rev Microbiol 47(1): 57–87
Thakore Y (2006) The biopesticide market for global agricultural use. Ind Biotechnol 23:192–208
Thomas WE, Ellar DJ (1983) Mechanism of action of bacillus thuringiensis var israelensis insecticidal deltaendotoxin. FEBS Lett 154:362–368
Tian BY, Li N, Lian LH, Liu JW, Yang JK, Zhang KQ (2006) Cloning, expression and deletion of the cuticle-degrading protease BLG4 from nematophagous bacterium Brevibacillus laterosporus G4. Arch Microbial 186:297–305
Uma Devi K, Padmavathi J, UmaMaheswara Rao C, Khan AAP, Mohan MC (2008) A study of host specificity in the entomopathogenic fungus *Beauveria bassiana* (Hypocreales, Clavicipitaceae). Biocontrol Sci Tech 18:975
Van Loon LC, Bakker PAHM, Pieterse CMJ (1998) Systemic resistance induced by rhizosphere bacteria. Annu Rev Phytopathol 36:453–483
Van Pée KH, Ligon JM (2000) Biosynthesis of pyrrolnitrin and other phenylpyrrole derivatives by bacteria. Nat Prod Rep 17:157–164
Vatanparast M, Ahmed S, Sajjadian SM, Kim Y (2019) A prophylactic role of a secretory PLA2 of Spodoptera exigua against entomopathogens. Dev Comp Immunol 95:108–117
Vey A, Hoagland R, Butt TM (2001) Toxic metabolites of fungal biocontrol agents. In: Butt TM et al (eds) Fungi as biocontrol agents: progress, problems, and potential. CAB International, Wallingford, pp 311–346
Vilcinskas A, Matha V, Goetz P (1997) Inhibition of phagocytic activity of plasmatocytes isolated from galleria melonella by entomogenous fungi and their secondary metabolites. J Insect Physiol 43:475–483
Vining LC (1992) Role of secondary metabolites from microbes. In: Secondary metabolites: their function and evolution. Wiley, Chichester, pp 184–198
Vodovar N, Vallenet D, Cruveiller S, Rouy Z, Barbe V, Acosta C, Cattolico L, Jubin C, Lajus A, Segurens B, Vacherie B (2006) Complete genome sequence of the entomopathogenic and metabolically versatile soil bacterium Pseudomonas entomophila. Nat Biotechnol 24(6):673
Vyas N, Dua KK, Prakash S (2007) Efficacy of *Lagenidium giganteum* metabolites on mosquito larvae with reference to non target organisms. Parasitol Res 101:385–390
Wang X, Li Y, Zhang X, Lai D, Zhou L (2017) Structural diversity and biological activities of the cyclodipeptides from fungi. Molecules 22:20–26
Waterfield NR, Bowen DJ, Fetherston JD, Perry RD (2001) The tc genes of Photorhabdus: a growing family. Trends Microbiol 9(4):185–191

Wedge DE, Duke SO (2006) Finding new fungicides from natural sources. Am Chem Soc Symp Ser 927:152–167

Williamson NR, Fineran PC, Leeper FJ, Salmond GPC (2006) The biosynthesis and regulation of bacterial prodiginines. Nat Rev Microbiol 4:887–899

Yang KL, Wei MY, Shao CL, Fu XM, Guo ZY, Xu RF, Zheng CJ, She ZG, Lin YC, Wang CY (2012) Antibacterial anthraquinone derivatives from a sea anemone-derived fungus *Nigrospora* sp. J Nat Prod 75:935–941

Yoon GY, Lee YS, Lee SY, Park RD, Hyun HN, Nam Y, Kim KY (2012) Effects on of chitinase, glucanase and a secondary metabolite from GY525. Nematology 14:175–184

Advances and Approaches in Mitigating Bacterial Diseases of Potato

6

Vinay Sagar, Sanjeev Sharma, and S. K. Chakrabarti

Abstract

Potatoes are affected by relatively a few bacterial diseases, viz., bacterial wilt or brown rot, soft rot of stem and tubers, common scab, ring rot, pink eye, and leaf spot; of which ring rot (*Clavibacter michiganensis* sub sp. *sepedonicus*) and pink eye (*Pseudomonas* species) do not occur in India, whereas leaf spot (*Xanthomonas vesicatoria*) is a disease of minor importance. Among other bacterial diseases, bacterial wilt or brown rot (*Ralstonia solanacearum*) is the most destructive disease followed by common scab (*Streptomyces* sp.) and soft rot (*Pectobacterium* sp., *Dickeya* sp.). These diseases are prevalent throughout the world and in most potato growing areas in India, inflicting heavy losses to the crop. The control of these diseases has proven to be very difficult because of both the seed and soil-borne nature of these pathogens. Chemical control is nearly impossible. Soil fumigants have shown either slight or no effects on these diseases. Biological control has been investigated, but is still in its infancy. Presently, these diseases are managed through an integrated approach.

Keywords

Bacterial wilt · Common scab · *Pectobacterium* · Potato · *Ralstonia solanacearum* · Soft rot · *Streptomyces* sp.

V. Sagar (✉) · S. Sharma · S. K. Chakrabarti
Division of Plant Protection, ICAR-Central Potato Research Institute, Shimla, Himachal Pradesh, India

S. Kumar Chakrabarti et al. (eds.), *Sustainable Management of Potato Pests and Diseases*, https://doi.org/10.1007/978-981-16-7695-6_6

6.1 Introduction

Potato is the world's most important non-grain food commodity that ranks fourth as main food crop in the world after rice, wheat, and maize. The crop is grown in more than 100 countries, mainly in Asia (195.67 million tons) and Europe (121.76 million tons) (FAOSTAT 2019). Because of its efficiency in producing high quantity of dry matter, energy, and edible protein per unit area per unit time, it holds promise for food security in the scenario of ever-growing world population. The full potential of this crop, however, can only be realized if diseases and pests are kept under control, especially in a subtropical country like India, where the weather is highly conducive for a number of pathogens.

Potato is affected by relatively a few bacterial diseases, viz., bacterial wilt or brown rot, soft rot of stem and tubers, ring rot, common scab, pink eye, and leaf spot. In India, ring rot (*Clavibacter michiganensis* sub sp. *sepedonicus*) and pink eye (*Pseudomonas* species) do not occur, whereas leaf spot (*Xanthomonas vesicatoria*) is a disease of minor importance. Among the other bacterial diseases, bacterial wilt/brown rot is the most destructive disease followed by common scab and soft rot which are discussed in the following pages with special reference to India.

6.2 Bacterial Wilt

Bacterial wilt or brown rot is caused by *Ralstonia solanacearum* (Smith 1896; Yabuuchi et al. 1995). It is one of the most damaging pathogens of potato and has been estimated to affect potato crop in 3.75 million acres in approximately 80 countries (Floyd 2007) with global damage estimates exceeding $950 million per year (Elphinstone 2005). Strains of this pathogen affect more than 450 plant species in over 54 botanical families throughout the world, including a wide range of crop plants, ornamentals, and weeds (Wicker et al. 2007). In India, losses up to 75 per cent have been recorded under extreme conditions (Gadewar et al. 1991). With increase in global temperature, the disease is likely to spread to new areas and affect potato cultivation there.

The disease causes wilting of plants in standing crop and also causes rot of infected tubers in field, storage, and transit. Another indirect loss results from the spread of the disease through latently infected tubers (infected tubers without exhibiting visible symptoms) when used as seed. Potato breeder seed production cannot be undertaken in fields having even slightest bacterial wilt incidence. There is zero tolerance to this disease in most international seed certification systems. Seed produced in bacterial wilt infested areas cannot be used domestically or exported, and therefore, spreading of the disease to seed production areas can provide a great set back to the seed industry.

6.2.1 Symptoms

The earliest symptom of the disease is slight wilting in leaves of top branches during hot sunny days. The leaves show drooping due to loss of turgidity followed by total unrecoverable wilt (Fig. 6.1). In well-established infections, cross sections of stems reveal brown discoloration of infected tissues. In advanced stages of wilt, cut end of base of the stem may show dull white ooze on squeezing. Bacterial wilt in field can be distinguished from other fungal wilts by placing the stem cut sections in clear water. Within a few minutes, a whitish thread like streaming can be observed coming out from cut end into water. This streaming represents the bacterial ooze exuding from the cut ends of colonized vascular bundles. The same test can also be carried out to see infection in tuber (Shekhawat et al. 2000).

Symptoms on potato tubers appear as vascular rot and pitted lesions formed on tuber surface (Fig. 6.2). In vascular rot, the vascular tissues of a transversely cut tuber show dirty white glistening sticky drops of bacterial ooze appearing from the

Fig. 6.1 Symptoms of bacterial wilt (**a**) wilting of plants infected with *R. solanacearum*, (**b**) brown discoloration of stem tissues, and (**c**) bacterial streaming in clear water from cut section of potato stem

Fig. 6.2 Symptoms of *R. solanacearum* infection on potato tubers (**a**) external symptoms on tubers, (**b**) vascular browning of tubers, and (**c**) oozing of bacterial mass through eye

brownish vascular ring within about 2–3 min. Another type of symptom is observed as lesions formed on tuber. Such lesions are produced due to infection of the bacterium through lenticels (skin pore) (Smith and Ramsey 1947). Initially, water soaked spots develop which enlarge in the form of pitted lesion. In advanced stages of wilt, bacterial mass may ooze out from eyes. Such sprouts may carry soil glued with the bacterial ooze. The tubers may not rot in storage and also may not show vascular browning but still carry the pathogen. Such latently infected seed potato tubers may serve as a fresh source of inoculum (Shekhawat et al. 2000).

6.2.2 Causal Organism

Ralstonia solanacearum is a Gram-negative, rod-shaped, strictly aerobic bacterium that measures 0.5–0.7 × 1.5–2.5 μm in size (Denny and Hayward 2001). This is a non-spore-forming, non-encapsulated, nitrate-reducing, and ammonia-forming bacterium. It is sensitive to desiccation and has low tolerance to sodium chloride (up to 2%) as compared to other species of *Ralstonia*. The pathogen under oxygen stress conditions in culture media shift to avirulent form. Lipopolysaccharides of the pathogen play an important role in determination of virulence (Hendrick and Sequira 1984). Virulent isolates are mainly non-flagellate and thus non-motile where as avirulent forms bear 1 to 4 polar flagella and are motile (Kelmen and Hruschka 1973). Virulent isolates on tetrazolium chloride medium develop fluidal irregular-shaped colonies with white to pinkish centre, whereas avirulent types produce small round, dark red dry colonies (Kelman 1954).

For most strains, the optimal growth temperature is between 28 and 32 °C; however, some strains have a lower optimal growth temperature of 27 °C (EPPO 2004). Strains of *R. solanacearum* have conventionally been classified into five races (related to the ability to wilt members of the family *Solanaceae* (r1), banana (r2), potato, and tomato in temperate conditions (r3), ginger (r4), and mulberry (r5) (Buddenhagen et al. 1962; He et al. 1983; Pegg and Moffett 1971), and six biovars (metabolic profiles related to the ability to metabolize a panel of three sugar alcohols and three disaccharides) (Hayward 1964, 1991; He et al. 1983). Based on this classification, potatoes are known to be affected by either r1 (bv 1, 3 and 4), frequent at warmer areas and lower elevations in the tropics, or r3 (bv 2), more common in higher elevations or latitudes (Martin and French 1985).

A new classification scheme was described for strains of *R. solanacearum*, based on variation of DNA sequences (Fegan and Prior 2005). Four phylotypes were identified within the species that broadly reflect the ancestral relationships and geographical origin of the strains. Phylotype I contains strains of Asiatic origin which belong to bv 3, 4, and 5. Phylotype II (American origin) contains r1bv1, r2bv1 (Moko disease causing strains), r3bv2, and bv2T strains. Phylotype III contains strains from Africa and Indian Ocean, which belong to bv1 and bv2T. Phylotype IV contains strains from Indonesia and Japan and a single strain from Australia. Each phylotype can further be subdivided into sequevars based on differences in the sequence of a portion of the *endoglucanase (egl)* gene. In India, the bacterial wilt of potato is known to be caused by strains of phylotype I, IIB, and IV of *R. solanacearum* (Sagar et al. 2014).

Recently using a polyphasic taxonomic approach on an extensive set of strains of *Ralstonia solanacearum* species complex (RSSC) representing all four phylotypes, Safni et al. (2014) divided the RSSC into three genospecies. According to this study, the *R. solanacearum* is restricted to strains of *R. solanacearum* phylotype II only. The second genospecies includes the type strain of *R. syzygii* and contains only phylotype IV strains. This genospecies is subdivided into three distinct groups, namely, *R. syzygii* subsp. *syzygii* (the causal agent of Sumatra disease on clove trees in Indonesia), *R. syzygii* subsp. *celebesensis* (the causal agent of the banana blood disease), and *R. syzygii* subsp. *indonesiensis* (phylotype IV strains isolated from potato, tomato, chili pepper, clove). The third genospecies is designated as *R. pseudosolanacearum* and includes *R. solanacearum* strains belonging to phylotypes I and III. This division has been in the meantime supported by the outcome of proteomic and genomic data (Prior et al. 2016).

6.2.3 Disease Occurrence and Distribution

Bacterial wilt or brown rot has a worldwide distribution (Elphinstone 2005). It is a destructive disease of potato especially in tropical and subtropical parts of Asia, Africa, and South and Central America and in some soils and waterways in Europe and Australia. In India, the disease is endemic to Karnataka, Western Maharashtra, Madhya Pradesh, eastern plains of Assam, Orissa and West Bengal, Chhota Nagpur plateau, north-western Kumaon hills, eastern hills of West Bengal, Meghalaya, Manipur, Tripura, Mizoram, Arunachal Pradesh, and Nilgiris, Annamalai, and Palani hills of Tamil Nadu (Shekhawat et al. 2000). Bacterial wilt is a serious problem in Malwa region and adjoining areas in Madhya Pradesh where potato is grown for processing industry (Sagar et al. 2013). However, it has not been noticed in the north-western high hills (excluding Kumaon hills) and in the north-western and north-central plains which are major seed producing zones of the country and need to be protected from the introduction of the disease.

6.2.4 Disease Cycle

Infected tubers and plant debris in infested soil are two major sources of inoculum. The pathogen infects roots of healthy plants through wounds. Nematodes such as *Meloidogyne incognita* which affect potato roots and tubers increase wilt incidence. Inoculum potential of about 10^7 cfu/g soil favours infection which however is dependent on other predisposing factors. Race 1 has greater ability to survive in soil than race 3 because of the better competitiveness, wide host range, and higher aggressiveness of race 1. Mean soil temperature below 15 °C and above 35 °C does not favour the disease development (Keshwal 1980).

Soil moisture influences the disease in at least four ways: (1) increasing survival of the bacterium in the soil, (2) increasing infection, (3) increasing disease development after infection, and (4) increasing exit of the bacterium from host and spread through the soil. *Ralstonia solanacearum* is capable of causing brown rot in a wide range of soil types and levels of acidity. In majority of the cases, the disease has been reported in acidic soils (pH 4.3–6.8) and only in a few cases in alkaline soils (Shekhawat et al. 1992).

Several other factors that affect pathogen survival in soil and water also affect disease development. The soil type and physicochemical properties have significant influence on survival of the pathogen. Soils having high clay and silt content with higher water holding capacity are favourable for long survival, while high sand contents disfavour its survival. Also, soil moisture and temperature exert a combined effect on survival of the pathogen. The congenial conditions for slow decline of population and virulence for race 1 and 3 are temperature between 10 and 30 °C, soil moisture between 20 and 60 WHC, heavy soils, and aerobic conditions (Shekhawat et al. 1992).

6.2.5 Survival

The pathogen survives through infected seed tubers and in plant debris in soil. Symptomless plants may harbour the bacterium and transmit it to progeny tubers as latent infection. This could lead to severe disease outbreaks when the tubers are grown at disease-free sites. High soil moisture, temperature, oxygen stress, and soil type affect the survival of the pathogen. The pathogen population declines gradually in soil devoid of host plants and their debris (Shekhawat et al. 1992).

6.2.6 Spread

Transmission of *R. solanacearum* from one area to another occurs through infected seed, irrigation water, and farm implements. Under favourable conditions, potato plants infected with *R. solanacearum* may not show any disease symptoms. In this case, latently infected tubers used for potato seed production may play a major role in spread of the bacterium from infected potato seed production sites to healthy potato-growing sites (Elphinstone 2005).

6.2.7 Management

The control of bacterial wilt has proven to be very difficult because of both the seed and soil-borne nature of the pathogen and especially in the case of race 1 due to its broad host range. Chemical control is nearly impossible. Soil fumigants have shown either slight or no effects. Antibiotics such as streptomycin, ampicillin, tetracycline, and penicillin hardly have any effect; in fact, streptomycin application increased the incidence of bacterial wilt in Egypt (Farag et al. 1986). Biological control has been investigated, but is still in its infancy. Potato cultivars developed in Colombia with a *Solanum phureja* and *S. demissum* background showed resistance to *R. solanacearum* (French 1985; Hartman and Elphistone 1994), but the race and strain diversity of the pathogen made it difficult to utilize these in other countries. The absolute control of bacterial wilt, at present, is difficult to achieve; however, economic losses can be brought down considerably using the following eco-friendly package of practices.

6.2.7.1 Healthy Seed

Use of healthy planting material can take care of almost 80% of bacterial wilt problem. Fortunately, bacterial wilt free areas in western and central Indo-Gangetic plains can be the source of disease-free seed in India. Tubers should not be cut since the cutting knife spreads the disease and also cut tubers can contact disease from soil easily.

6.2.7.2 Field Sanitation and Cultural Management

Crop Rotation Following a 2–3-year crop rotation using crops like maize, cereals, garlic, onion, cabbage, and sanai (sun hemp) can help in reduction of the disease. Do not rotate vegetables like brinjal, ginger, chillies, and other solanaceous crops which may act as alternate hosts. Paddy and sugarcane although are not host, still they can carry pathogen and thus contribute to the disease perpetuation (Shekhawat et al. 1992).

Avoid Tillage Operations The pathogen can enter in plant through root or stolon injuries (Nirula and Paharia 1970). Such injuries cannot be avoided during intercultural operations. Therefore, by restricting tillage to the minimum together with full soil cover at planting can help in restricting the disease.

Off-Season Management of Field The pathogen perpetuates in the root system of many weeds and crops. Therefore, it is recommended to clean the field from weeds and root/foliage remnants and burn them. The pathogen in remnants of plants can also be exposed to high temperature above 40 °C in summer in plains. Similarly in hills, at low temperature below 5 °C, and deep ploughing in winter can help in reduction of the pathogen from the field. Soil solarization and deep ploughing of fields together during summer season in subtropical plains can help in reduction of field inoculum.

6.2.7.3 Chemical Control

Soil application of stable bleaching powder @ 12 kg/ ha mixed with fertilizers at planting gives good control of bacterial wilt. Soil fumigants have also shown slight effect. Antibiotics such as streptomycin, ampicillin, tetracycline, and penicillin have been tried but not found practical at field level (Farag et al. 1986). Recently, Biswal and Dhal (2018) have reported that tuber treatments with streptocycline together with basal application of the same antibiotics at 10 days intervals for four times after planting have reduced wilting and increased tuber yield under Odisha conditions.

6.2.7.4 Biological Control

Biological control of bacterial wilt has been investigated both in India and elsewhere. Biocontrol of bacterial wilt by use of antagonists such as *Pseudomonas fluorescens*, *Bacillus* spp., avirulent *P. solanacearum*, and actinomycetes has been found to be effective against the pathogen under controlled conditions. *Bacillus subtilis* strain B5 has been reported to be effective against bacterial wilt pathogen (Sunaina et al. 2006).

6.2.7.5 Breeding for Resistance

Potato cultivars developed in Colombia with a *Solanum phureja* and *S. demissum* background showed resistance to *R. solanacearum* (French 1985; Hartman and Elphistone 1994). However, the race and strain diversity of the pathogen made it difficult to utilize these parents in other countries. Breeding for resistance has not been very successful especially under subtropical and tropical highlands. Cultivars derived from *S. phureja* which exhibit resistance under cool highland subtropics usually succumb to disease under high temperature in the tropics.

Based on intensive ecological and epidemiological studies at ICAR-Central Potato Research Institute, Shimla, the following practices are recommended for checking the bacterial wilt in different agro-climatic zones of the country:

- *Zone I*: This zone comprises of non-endemic areas like Gujarat, Maharashtra, and north-western and north-central plains. This zone is characterized by hot and dry summer with scanty vegetation (April–June); temperature may go up to 40–43 °C. The bacterial wilt is no more a major problem. Therefore, deep ploughing in summer and use of disease-free seed are adequate for the disease control.
- *Zone II*: It includes north-western mid hills (up to 2200 masl), north-eastern hills, and the Nilgiris. The zone is characterized by mild summer, profuse vegetation with a maximum temperature range of 26–30 °C. Winter temperature may go as low as 3–6 °C. Many weed hosts can provide perpetual niche for colonization and survival of the bacteria. The use of disease-free seed and application of stable bleaching powder @ 12 kg/ha mixed with fertilizer at planting, ploughing the field in September–October, and exposing the soil to winter temperature are adequate for disease control. The application of bleaching powder can be substituted by a 2-year crop rotation with crops like wheat, barley, finger millet, cabbage, cauliflower, knol-khol, carrot, onion, garlic, etc. Early planting preferably in February and early harvesting are recommended to minimize the exposure of the crop to high temperature which favours the disease.
- *Zone III*: This zone includes eastern plains and Deccan plateau. The area is relatively rich in vegetation. Day temperature sometimes reaches 38 °C. Heavy precipitation occurs due to western disturbances. Eastern plains and Deccan plateau have many symptomless carriers of the pathogen. Therefore, management

of the disease is most difficult. However, the disease can be kept under check with practices like use of disease-free seed, application of bleaching powder, blind earthing-up, and ploughing in March and leaving the soil exposed to summer temperatures during April–May and crop rotations along with clean cultivation.

- *Zone IV*: This zone includes north western high hills (above 2200 masl excluding Kumaon hills). This zone has a temperate climate with severe winters; daily temperature ranges from −10 to 5 °C during December–January. Snow is common during these months. Bacterial wilt is not endemic, and the use of disease-free seed alone is adequate.

Overall, an integrated approach involving use of pathogen-free seed potato obtained from disease-free areas, reduction of field inoculum through soil solarization and crop rotation, growing crop under right environmental conditions, and application of stable bleaching powder in soil can help in effective management of the bacterial wilt of potato.

6.3 Common Scab of Potato

Common scab caused by *Streptomyces* species causes superficial lesions on the surface of potato tubers and affects the quality of the produce. The affected tubers fetch low price in the market due to poor appearance and also because deeper peeling is required before consumption. Seed lots exceeding 5% incidence is rejected by seed certification agencies (in India) causing huge loss to seed industry. This disease was first recorded in Patna during 1958. Since then, it has become endemic in various potato growing states (Nagaich and Dutt 1972).

6.3.1 Symptoms

Scab begins as small reddish or brownish spot on the surface of the potato tubers and its initial infection takes place during juvenile period of tubers (Paharia and Pushkarnath 1963). Infection takes place mainly through lenticels, and surrounding periderm turns brown and rough. Lesion becomes corky due to elongation and division of invaded cells. Under Indian conditions, multiple kinds of symptoms have been recorded, and they are grouped as (1) a mere brownish roughening or abrasion of tuber skin, (2) proliferated lenticels with hard corky deposition, might lead to star shaped lesion, (3) raised rough and corky pustules, (4) 3–4 mm deep pits surrounded by hard corky tissue, and (5) concentric series of wrinkled layers of cork around central black core (Fig. 6.3) (Nagaich and Dutt 1972).

Fig. 6.3 Various types of scab symptoms caused by *Streptomyces* species on potato tubers

6.3.2 Causal Organism

At least 13 different *Streptomyces* spp. have been found to cause common scab on potato worldwide (Hao et al. 2009). The prominent among them are *Streptomyces scabies* (Thaxter) Lambert and Loria, *S. acidiscabies* Lambert and Loria, *S. turgidiscabies* Takeuchi, and *S. collinus* Lindenbein (Dey et al. 1981), *S. griseus* (Krainsky) Waksman and Henrici (Jeswani et al. 1987), and *S. longisporoflavus*, *S. cinereus*, *S. violaceoruber*, *S. albogriseolus*, *S. griseoflavus*, *S. catenulae*, and others (CPRI 1983). Plant pathogenesis by *Streptomyces* has been reviewed by Loria et al. (1997). *Streptomyces* are bacteria which resemble fungi due to formation of vegetative substrate mycelium that develop aerial filaments. However, the filaments are of smaller dimensions than the true fungi. These filaments produce spores through fragmentation. *Streptomyces* spp. may be pathogenic or non-pathogenic. The pathogenic species produce thaxtomins which are phytotoxins and cause hypertrophy and cell death (Loria et al. 1995).

Considerable variation exists within the pathogen with respect to their pigment production in media, colour, and shape of sporulating filaments and use of specific sugars (Afanasiev 1937; Leach et al. 1939; Schall 1940). *S. scabies* form grey, spiral spore chains on several media and produce brown pigment, whereas *S. acidiscabies* produce peach-coloured wavy chains of spores and brown pigment in medium. The identification and taxonomy of *Streptomyces* spp. have been based on morphological and physiological characteristics combined with thaxtomin production and pathogenicity tests in vitro and in vivo (Wanner 2004). The ability to produce thaxtomin toxin is strongly correlated with the pathogen's pathogenicity. Different species of *Streptomyces* have been found associated with various types of scab lesions (Faucher et al. 1992, 1993).

6.3.3 Disease Occurrence and Distribution

Common scab occurs in most potato-producing areas in Africa, Asia, Europe, and North and South America. In India, it was known to occur in Lahaul Valley (Himachal Pradesh) in severe form since 1969; its frequent occurrence in plains was reported in 1979–1980 (Sharma 1984). Afterwards, it became a major problem in almost all agro-climatic zones of India (Nagaich 1983). Now, the disease has covered almost all the potato-growing areas of the country and is posing a serious threat to successful potato cultivation. In Eastern Uttar Pradesh, common scab on potato has been reported every year in moderate to severe form (Mishra and Srivastava 1999, 2001, 2005). Its real impact is felt in states like Punjab, Uttar Pradesh, and Lahaul valley of Himachal Pradesh where potato production is for seed industry (Paharia and Pushkarnath 1963; Nagaich and Dutt 1972; Jeswani et al. 1987). The disease is spreading fast in some areas in Indo Gangetic plains due to cultivation of potato year after year in the same land (Chakraborty 2012).

6.3.4 Disease Cycle

Potato is physiologically most susceptible to *Streptomyces* spp. in the period following tuber initiation. *Streptomyces* spp. infect the newly formed tubers through stomata and immature lenticels. Once the periderm has differentiated, tubers are no longer susceptible to the pathogen (Loria et al. 1997). The pathogen is both seed and soil-borne. It can survive in soil for several years in plant debris and infested soil (Lutman 1945). Soil conditions greatly influence the pathogen. Favourable conditions include pH between 5.2 to 8.0 or more (Butler and Jones 1961), temperature in the range of 20–30 °C (Gaumann and Hafliger 1945), and low soil moisture (Sanford 1962; Singh and Singh 1981). The pathogen is aerobic in nature, and maintaining high soil moisture for 10–20 days after tuber initiation can help in reducing the common scab (Lapwood et al. 1973).

However, scab outbreaks have been reported in irrigated or wet soil conditions in northern Europe, Israel, and Canada (Doering-Saad et al. 1992; Goyer et al. 1996; Lindholm et al. 1997). The organism is a tuber-borne and is well-adapted saprophyte that persists in soil on decaying organic matter and manure for several years. Infected tubers serve as source of inoculum in the field, giving rise to infected progeny tubers. The pathogenic *Streptomyces* species are both soil- and tuber-borne. Tuber-borne inoculum is likely to be involved in the distribution of new strains or species (Stevenson et al. 2001).

6.3.5 Management

The pathogen is difficult to eradicate because of long survival both on seed tubers and in soils. As common scab pathogen is a bacterium, it is not controlled by seed-applied fungicides. Earlier, formaldehyde, urea formaldehyde, and manganese sulphate were used for control of common scab, but are no longer applied in fields (Locci 1994). In a 2-year Canadian study, seed-applied fludioxonil resulted in a 57.8% reduction of common scab severity, and use of a seed-applied biopesticide containing *Bacillus subtilis* resulted in a 56.1% reduction (Al-Mughrabi et al. 2016).

A perusal through literature provides information on effectiveness of various soil amendments and foliar-applied treatments from many locations, over many years. Unfortunately, few to no treatments provide highly effective and reliable control of the disease across locations (Powelson and Rowe 2008; Lerat et al. 2009). In general, common scab can be managed by use of disease-free seed tubers; tuber treatment with boric acid (3% for 30 min.) before or after cold storage (before sprouting); keeping the moisture near to field capacity right from tuber initiation until the tubers measure 1 cm in diameter; following 3–4 year crop rotation with wheat, pea, oats, barley, soybean, sorghum, and bajra; green manuring and deep ploughing the potato fields in April; and leaving the soil exposed to high temperatures during summer (May to June) in the North Indian plains (Lapwood et al. 1973; Singh and Singh 1981; Kagawa and Hosaka 1991; Arora et al. 2006a, b; Shekhawat et al. 1993).

6.4 Bacterial Soft Rot

Bacterial soft rot can cause significant loss of potato tubers at harvest, transit, and storage. Losses due to poor handling of the produce, poorly ventilated storage, or transit may go up to 100 per cent (Somani and Shekhawat 1990). Soft rot bacteria usually infect potato tubers which have been damaged by mechanical injury or in the presence of other tuber-borne pathogens. Bacterial soft rot develops much faster under warm and humid conditions. The disease also results in blackleg of foliage during the crop growing season.

6.4.1 Symptoms

Initially, a small area of tuber tissue around lenticels or stolon attachment point becomes water soaked and develops soft lesions. Under low humidity, the initial soft rot lesions may become dry and sunken. Under high humidity, the lesions may enlarge and spread to larger area. Tubers in advanced stages of decay are usually invaded by other organisms, and the decaying tissue becomes slimy with foul smell and brown liquid ooze. The tuber skin remains intact, and sometimes the rotten tubers are swollen due to gas formation. At harvest, many small rotten tubers with intact skin can be seen. The infected seed tubers rot before emergence resulting in poor stand of the crop. In cooler regions, another kind of symptoms called blackleg phase develops from soft rot infected seed tubers. The affected haulms become black at collar region just above the ground. Infected plants develop yellowing, start wilting, and die early without producing any tubers. Water-soaked lesions develop on succulent stems, petioles, and leaves. On stem and petioles, the lesions first enlarge into stripes, turn black, and then invade the affected parts causing soft rot and toppling of the stem and leaves (Perombelon and Kelman 1980; Somani and Shekhawat 1990).

6.4.2 Causal Organism

Pectobacterium atrosepticum (van Hall) Gardan et al. 2003 (syn. *Erwinia carotovora* subsp. *atroseptica*), *Pectobacterium carotovorum* sub sp. *carotovorum* (Jones) Hauben et al. 1998 (syn. *Erwinia carotovora* subsp. *carotovora*), *Pectobacterium carotovorum* subsp. *brasiliense* (*Erwinia carotovora* subsp. *brasiliense*) (Pcb) (Duarte et al. 2004), *Pectobacterium wasabiae* (*Erwinia carotovora* subsp. *wasabiae*) (Pwa) (Pitman et al. 2008), and several *Dickeya* spp. (*Erwinia chrysanthemi*), including *D. dianthicola* (*Erwinia chrysanthemi* pv. *dianthicola*), *Dickeya dadantii*, *Dickeya zeae* (*Erwinia chrysanthemi* pv. *zeae*), and the new species *Dickeya solani* (Toth et al. 2011; van der Wolf et al. 2013), are known to cause potato blackleg disease in field and tuber soft rots in storage and in transit (Czajkowski et al. 2015).

P. atrosepticum, the primary enterobacteria causing soft rots, produce pectolytic enzymes and degrade pectin in the middle lamella of host cells, breakdown tissues, and cause soft rot and the decay. The decaying tissue becomes slimy and foul smelling, and brown liquid oozes out from the soft rot affected tubers. About 1500 strains of pectinolytic *Erwinia* have been isolated from infected plants and tubers (Sledz et al. 2000). The pathogen produce certain volatile compounds such as ammonia, trimethylamine, and several volatile sulphides (Lacy-Costello et al. 2001), and early detection of such volatile compounds in storage could be used as a method to detect the disease at initial stage (Lyew et al. 2001).

6.4.3 Disease Occurrence and Distribution

Bacterial soft rot of potato is found wherever potatoes are grown. The disease affect the crop at all stages of growth, but it is more serious on potato tubers under poor storage conditions especially in warm and wet climate. Blackleg (*Pectobacterium atrosepticum*) phase of the disease is not common in India. It occurs only rarely in the Shimla hills in HP, the Kumaon hills in Uttarakhand, Ootacamund in Nilgiris, and also Bihar plains (Shekhawat et al. 2000).

6.4.4 Disease Cycle

Soft rot bacteria may be carried latently in lenticels and wounds and on surface of tubers without any visible symptoms and spread to healthy tubers in stores and during seed cutting, handling, and planting (Perombelon and Kelman 1980; Weber 1990). Water film on surface of tuber which causes proliferation of lenticels and creates anaerobic conditions and injury on surface of tuber predisposes potatoes to soft rot. From soft rot infected seed tubers, bacteria may enter vascular tissues of developing stems and can develop blackleg under favourable conditions. From blackleg infected plants, the pathogen can reach daughter tubers through stolons and initiate tuber decay at the site of tuber attachment (Shekhawat et al. 1984). Decaying tubers in soil could serve as source of contamination for healthy tubers (Perombelon 2000). The threshold level for disease development is about 103 cells of *E. carotovora* sub sp. *atroseptica* per tuber. Tubers harvested in wet soil and with poor ventilation in transit and storage promote the rot (Hingorni and Andy 1953).

In warm climates, where one potato crop follows another or where only short rotation cycles are applied, the bacteria can pass easily from one crop to the next, especially in poorly drained soil. The bacteria can be disseminated in the potato fields by irrigation water, insects, rain, or bacterial aerosols. The pathogen may also spread through water during washing of the produce with contaminated water. Soft rot causing bacteria spread easily from diseased to healthy tubers during storage, handling, and grading (Elphinstone 1987). Insects especially maggots of *Hylemyia* species may also transmit the bacteria from one tuber to another (Agrios 1969).

There are controversial reports on survival of Erwinias in soil. In temperate zones, the bacteria can survive the winter on plant residues; however, no survival has been observed in fields rotated with non-hosts crops. The bacteria can survive at places where rotten potatoes and vegetables are dumped (Elphinstone 1987).

6.4.5 Management

Soft rot bacteria are carried deep inside the tuber and in lenticels and surface wounds making it difficult to eradicate. These quiescent bacteria proliferate in high moisture condition and require water film that cause anaerobic conditions leading to disease development. Surface injury predisposes the tubers to soft rot infection.

An integrated approach involving practices like planting of whole seed potato or well-suberized seed pieces in well-drained soil with temperature around 10–13 °C at less planting depth, tuber treatment with 3% boric acid (Somani and Shekhawat 1985) or 0.05% copper sulphate (Zhang et al. 1993), restricting nitrogen dose to minimum (150 kg/ha), application of stable bleaching powder before planting (Karwasra and Prashar 1998) and during last irrigation (Parashar et al. 1986.), crop rotation following green manure–potato–wheat (Shekhawat et al. 1984), avoiding bruises and cuts to potato tubers during harvest, handling, and proper aeration during storage and transit can minimize soft rots. Adjusting planting time to avoid hot weather during plant emergence and harvesting the crop before soil temperature rises above 28 °C are also recommended to minimize the losses due to soft rot.

6.5 Conclusion and Future Outlook

Bacterial diseases inflict heavy losses in potato crop throughout the world. The diseases like bacterial wilt and soft rot are big constraint in seed potato production. There is zero tolerance to bacterial wilt disease in most international seed certification systems. Similarly, soft rot has also got international attention as emerging threat to potato crop. Seed potato infected with these diseases cannot be used domestically as spread of these diseases to seed production areas can provide a great set back to the seed industry nor be exported and, therefore, pose a big constraint in trade worldwide. These diseases are difficult to manage in want of effective chemicals, bio-control agent, and resistant varieties. The present focus to manage these diseases is on using healthy seed tubers, planting in clean soils, sanitation, cultural practices, crop rotation with non-host plants, and the use of tolerant or resistant varieties.

References

Afanasiev MM (1937) Comparative physiology of Actinomycetes in relation to potato scab. Neb Agr Exp Sta Res Bull 92:1–63

Agrios GN (1969) Plant pathology. Academic Press, New York

Al-Mughrabi KIA, Vikram R, Poirier K, Jayasuriya MG (2016) Management of common scab of potato in the field using biopesticides, fungicides, soil additives, or soil fumigants. Biocont Sci Technol 26:125–135

Arora RK, Sharma J, Garg ID, Singh RK, Somani AK (2006a) Boric acid for control of tuber borne diseases. In: Technologies in aid of healthy seed production. Central Potato Research Institute, Shimla

Arora RK, Sharma J, Singh RK (2006b) Soil solarization. In: Technologies in aid of healthy seed production. Central Potato Research Institute, Shimla

Biswal G, Dhal NK (2018) Management of bacterial wilt disease of potato in coastal plains of Odisha. Afr J Microbiol Res 12(12):284–289. https://doi.org/10.5897/AJMR2017.8714

Buddenhagen IW, Sequeira L, Kelman A (1962) Designation of races in *pseudomonas solanacearum*. Phytopathology 52:726

Butler EJ, Jones SG (1961) Plant pathology. Mac Milan, London and New York, p 979

Chakraborty A (2012) Prevalence of diseases of potato in indo- Gangetic plains of West Bengal. J Mycopathol Res 50:117–119

Czajkowski R, MCM P, Jafra S, Lojkowska E, Potrykus M, van der Wolf JM, Sledz W (2015) Detection, identification and differentiation of *Pectobacterium* and *Dickeya* species causing potato blackleg and tuber soft rot: a review. Ann Appl Biol 166:18–38

Denny TP, Hayward AC (2001) Ralstonia. In: Schaad NW, Jones JB, Chun W (eds) Laboratory guide for identification of plant pathogenic bacteria, 3rd edn. American Phytopathological Society, St. Paul, pp 151–174

Dey SK, Singh N, Singh H (1981) Streptomyces collinus Lin.: a new species causing common scab of potato. Curr Sci 50:830

Doering-Saad C, Kampfer P, Manulis S, Kritzman G, Schneider J, Zakrzewska-Czerwinska J, Schrempf H, Barash I (1992) Diversity among Streptomyces strains causing potato scab. Appl Environ Microbiol 58:3932–3940

Duarte V, De Boer SH, Ward LJ, De Oliveira AMR (2004) Characterization of atypical *Erwinia carotovora* strains causing blackleg of potato in Brazil. J Appl Microbiol 96:535–545

Elphinstone JG (1987) Soft rot and black leg of potato; *Erwinia* spp. technical information bulletin 21. International Potato Centre, Lima, p 18

Elphinstone JG (2005) The current bacterial wilt situation: a global overview. In: Allen C, Prior P, Hayward AC (eds) Ln: Bacterial wilt disease and the Ralstonia solanacearum species complex. American Phytopathological Society, Saint Paul, pp 9–28

EPPO (2004) EPPO standards PM 7/21. Diagnostic protocols for regulated pests: *Ralstonia solanacearum*. EPPO Bull 34:155–157

FAO (2019). http://www.fao.org/faostat/en/#data/QC. Accessed 10 June 2021

Farag NS, Fawzi FG, El-Said SIA, Mikhail MS (1986) Streptomycin in relation to potato brown rot control. Acta Phytopathol Entomol Hung 21:115–122

Faucher E, Otrysko B, Paradis E, Hodge NC, Stall RE, Beaulien C (1993) Characterization of *Streptomyces* causing russet scab in Quebec. Plant Dis 77:1217–1220

Faucher E, Savard T, Beaulieu C (1992) Characterization of actinomycetes isolated from common scab lesions of potato tubers. Can J Plant Pathol 14:197–202

Fegan M, Prior P (2005) How complex is the *Ralstonia solanacearum* species complex? In: Allen C, Prior P, Hayward AC (eds) Bacterial wilt: the disease and the *Ralstonia solanacearum* species complex. APS Press, St Paul, pp 449–461

Floyd J (2007) New pest response guidelines: ralstonia solanacearum race 3 biovar 2. USDAAPHIS – PPQ, emergency and domestic programmes, Riverdale. http://www.aphis.usda.gov/import_export/plants/manuals/index.html. Accessed 10 Aug 2009

French ER (1985) Multiple disease resistance in potato cultivars with Solanum phureja and S. demissum background. Phytopathology 75:1288

Gadewar AV, Trivedi TP, Shekhawat GS (1991) Potato in Karnataka. In: Technical bulletine-17. Central Potato Research Institute, Shimla, p 33

Gardan L, Gouy C, Christen R, Samson R (2003) Elevation of three subspecies of *Pectobacterium carotovorum* to species level: *Pectobacterium atrosepticum* sp. nov., *Pectobacterium betavasculorum* sp. nov. and *Pectobacterium wasabiae* sp. nov. Int J Syst Evol Microbiol 53: 381–391

Gaumann E, Haflinger E (1945) Der Einfluss der Bodentemperatue auf die Entwicklung und den Schorfbelfall der kartoffelknollen. Phytopathol Z 15:23

Goyer C, Faucher E, Beaulieu C (1996) Streptomyces caviscabies sp. nov. from deep-pitted lesions in potatoes in Québec, Canada. Int J Syst Bacteriol 46:635–639

Hao JJ, Meng QX, Yin JF, Kirk WW (2009) Characterization of a new *Streptomyces* strain, DS3024, that causes potato common scab. Plant Dis 93:1329–1334

Hartman GL, Elphinstone JG (1994) Advances in the control of *pseudomonas solanacearum* race 1 in major food crops. In: Hayward AC, Hartman GL (eds) Bacterial wilt: the disease and its causative agent, *Pseudomonas solanacearum*. CAB International, Wallingford, pp 157–177

Hauben L, Moore ERB, Vauterin L, Steenackers M, Mergaert J, Verdonck L, Swings J (1998) Phylogenetic position of phytopathogens within the Enterobacteriaceae. Syst Appl Microbiol 21:384–397

Hayward AC (1964) Characteristics of *pseudomonas solanacearum*. J Appl Microbiol 27:265–277

Hayward AC (1991) Biology and epidemiology of bacterial wilt caused by *Pseudomonas solanacearum*. Annu Rev Phytopathol 29:65–89

He LY, Sequiera L, Kelman A (1983) Characteristics of strains of *Pseudomonas solanacearum* from China. Plant Dis 67:1357–1361

Hendrick C, Sequeria L (1984) Lipopolysaccharide defective mutants of *Pseudomonas solanacearum*. Appl Environ Microbiol 48:94–101

Hingorani MK, Andy SK (1953) Factors influencing bacterial soft rot of potatoes. Indian Phytopath 6:110–115
Jeswani MD, Gadewar AV, Singh BP (1987) Studies on mixed infection of *Streptomyces griseus* and *fusarium oxysporum* on potato tubers. J Soil Ecol 7:73–78
Kaqawa H, Hosaka K (1991) Irrigation and mulch removal treatment as practical means to control russet scab of potato caused by *Streptomyces cheloniumii* (Japanese). Proc Kanto Tosan Plant Protec Soc 38:53–54
Karwasra SS, Prashar RD (1998) Chemical control of incipient infection, weight loss and sprouting of potato tubers during storage. Plant Dis Res 13:49–51
Kelman A (1954) The relationship of pathogenicity of *Pseudomonas solnacearum*, to colony appearance in a tetrazolium medium. Phytopathology 44:693–695
Kelman A, Hruschka J (1973) The role of motility and aerotaxis in the selective increase of avirulent bacteria in still broth culture *pseudomonas solanacearum*. J Gen Microbiol 76:177–188
Keshwal RL (1980) Studies on bacterial wilt of solanaceous crops caused by pseudomonas solanacearum with special reference to its control. Ph.D. thesis, University of Jabalpur, Madhya Pradesh, p 201
Lacy-Costello BPJ, de Even RJ, Gunson HE, Ratcliffe NM, Spencer-Philips PTN (2001) Identification of volatiles generated by potato tubers (*Solanum tuberosum* cv. Marris piper) infection by *Erwinia carotovora*, *Bacillus polymyxa* and *Arthrobacter* sp. Plant Pathol 18:315–351
Lapwood DH, Welling LW, Hawking JH (1973) Irrigation as a practical means to control potato common scab (*Streptomyces scabies*): final experiment and conclusion. Plant Pathol 23:35–41
Leach JG, Decker P, Becker H (1939) Pathogenic races of *Actinomyces scabies* in relation to scab. Phytopathology 29:204–209
Lerat SA, Simao-Beaulieu, Beaulieu C (2009) Genetic and physiological determinants of *Streptomyces scabies* pathogenicity. Mol Plant Pathol 10:579–585
Lindholm P, Kortemaa H, Haahtela K, Kokkola M, Salkinoja-Salonen M, Valkonen JPT (1997) Streptomyces spp. isolated from potato scab lesions under Nordic conditions in Finland. Plant Dis 81:1317–1322
Locci R (1994) Actinomycetes as a plant pathogen. Eur J Plant Pathol 100:179–200
Loria R, Bukhalid RA, Creath RA, Leiner RH, Olivier M, Steffens JC (1995) Differential production of thaxtomins by pathogenic *Streptomyces* species *in vitro*. Phytopathology 85:537–541
Loria R, Bukhalid RA, Fry BA, King RR (1997) Plant pathogenicity in the genus *Streptomyces*. Plant Dis 81:836–846
Lutman BF (1945) The spread of potato scab in soil by potato plant humus. Vt Agricultural Experimental Station Bulletin No 528, p 40
Lyew D, Gariepy Y, Raghavan GSV, Kushalappa AC (2001) Changes in volatile production during an infection of potatoes by *Erwinia carotovora*. Food Res Int 34:807–813
Martin C, French ER (1985) Bacterial wilt of potato, Pseudomonas solanacearum. In: Technical Bulletin-13. International Potato Centre, Lima, p 16
Mishra KK, Srivastava JS (1999) Severity of prevalence of common scab of potato in eastern U P. Indian Potato J 26:23–24
Mishra KK, Srivastava JS (2001) Screening potato cultivars for common scab of potato in naturally infested field. Potato Res 44:19–24
Mishra KK, Srivastava JS (2005) Soil amendments to control common scab of potato. Potato Res 47:101–109
Nagaich BB (1983) Disease resistance in potato in India. Indian Phytopath 36(1):1–10
Nagaich BB, Dutt BL (1972) Common potato diseases in India. Indian Farming 21:20–23
Nirula KK, Paharia KD (1970) Role of root - knot nematodes in spread of brown rot in potatoes. Indian Phytopath 39:600–601
Paharia KD, Pushkarnath (1963) Occurrence of potato scab in Bihar. Indian Potato J 5:104–105
Parashar RD, Sindhan GS, Singh JP (1986) Field application of Klorocin for the control of soft rot of potato in field and storage. J Indian Potato Assoc 13:81–82

Pegg KG, Moffett M (1971) Host range of the ginger strain of *Pseudomonas solanacearum* in Queensland. Aust J Exp Agric Anim Husb 11:696–698

Perombelon MCM (2000) Blackleg risk potential of seed potatoes determined by quantification of tuber contamination by the causal agent and *Erwinia carotovora* spp. *atroseptica*: a critical review. Bulletin OEPP 30:413–420

Perombelon MCM, Kelman A (1980) Ecology of soft rot *Erwinias*. Annu Rev Pathol 18:361–387

Pitman AR, Wright PJ, Galbraith MD, Harrow SA (2008) Biochemical and genetic diversity of pectolytic enterobacteria causing soft rot disease of potatoes in New Zealand. Australas Plant Pathol 37:559–568

Powelson ML, Rowe RC (2008) Managing diseases caused by seed-borne and soil-borne fungi and fungus-like pathogens. In: Johnson D (ed) Potato health management, 2nd edn. APS Press, St. Paul, pp 183–195

Prior P, Ailloud F, Dalsing BL, Remenant B, Sanchez B, Allen C (2016) Genomic and proteomic evidence support the division of the plant pathogen *Ralstonia solanacearum* into three species. BMC Genomics 17:90

Safni I, Cleenwerck I, De Vos P, Fegan M, Sly L, Kappler U (2014) Polyphasic taxonomic revision of the *Ralstonia solanacearum* species complex: proposal to emend the descriptions of *Ralstonia solanacearum* and *Ralstonia syzygii* and reclassify current *R. syzygii* strains as *Ralstonia syzygii* subsp. *syzygii* subsp. nov., *R. solanacearum* phylotype IV strains as *Ralstonia syzygii* subsp. *indonesiensis* subsp. nov., banana blood disease bacterium strains as *Ralstonia syzygii* subsp. *celebesensis* subsp. nov. and *R. solanacearum* phylotype I and III strains as *Ralstonia pseudosolanacearum* sp. nov. Int J Syst Evol Microbiol 64:3087–3103

Sagar V, Jeevalatha A, Mian S, Chakrabarti SK, Gurjar MS, Arora RK, Sharma S, Bakade RR, Singh BP (2014) Potato bacterial wilt in India caused by strains of phylotype I, II and IV of *Ralstonia solanacearum*. Eur J Plant Pathol 138(1):51–65

Sagar V, Somani AK, Arora RK, Sharma S, Chakrabarti SK, Tiwari SK, Chaturvedi R, Singh BP (2013) Status of bacterial wilt of potato in Malwa region of Madhya Pradesh in India. J Plant Pathol 95:321–328

Sanford GB (1962) Some factors affecting the pathogencity of *Actinomyces scabies*. Phytopathology 16:525–547

Schaal LA (1940) Variations in the tolerance of certain physiological races of *Actinomyces scabies* to hydrogen ion concentration. Phytopathology 30:699–700

Sharma KD (1984) Studies on common scab of potato. PhD thesis, GB Pant University of Agriculture and Technology, Pantnagar, Nainital

Shekhawat GS, Chakrabarti SK, Gadevar AV (1992) Potato bacterial wilt in India. Central Potato Research Institute, Shimla, p 52

Shekhawat GS, Gadewar AV, Chakrabarti SK (2000) Potato bacterial wilt in India. Central Potato Research Institute, Shimla, India, p 45. Technical Bulletin-38 (revised)

Shekhawat GS, Piplani S, Ansari MM (1984) Endophytic bacterial flora of potato plant in relation to soft rot disease. Indian Phytopathol 37:501–505

Shekhawat GS, Singh BP, Jeswani MD (1993) Soil and tuber borne diseases of potato. C.P.R.I, Shimla, p 47

Singh N, Singh H (1981) Irrigation as a cultural practice to control common scab of potato. J Indian Potato Assoc 8:35–36

Sledz W, Jafia S, Waleron M, Loz Kowka E (2000) Genetic diversity of *Erwinia carotovora* strain isolated from infected plants grown in Poland. Bulletin- OEPP 30:403–407

Smith EF (1896) A bacterial disease of the tomato, eggplant and Irish potato. US Dept Agric Div Veg Phys Path Bull 12:1–26

Smith MA, Ramsey GB (1947) Bacterial lenticel infection of early potatoes. Phytopathology 37: 225–242

Somani AK, Shekhawat GS (1985) Evaluation of chemicals to prevent soft rot of potatoes. J Indian Potato Assoc 12:214–216

Somani AK, Shekhawat GS (1990) Bacterial soft rot of potato. Central Potato Research Institute, Shimla, p 323

Stevenson WR, Loria R, Franc GD, Weingartner DP (2001) Compendium of potato diseases, 2nd edn. The American Phytopathological Society, St. Paul

Sunaina V, Pandy SK, Singh BP (2006) Bio-B5, an ecofriendly biopesticide-cum-biofertilizer for management of soil and tuber borne diseases of potato and other crops. Central Potato Research Institute Campus, Modipuram, Meerut, p 10

Toth IK, van der Wolf JM, Saddler G, Lojkowska E, Hélias V, Pirhonen M, Tsror (Lahkim) L, Elphinstone JG (2011) Dickeya species: an emerging problem for potato production in Europe. Plant Pathol 60:385–399

van der Wolf JM, Nijhuis EH, Kowalewska MJ, Saddler GS, Parkinson N, Elphinstone JG, Pritchard L, Toth IK, Lojkowska E, Potrykus M, Waleron M, de Vos P, Cleenwerck I, Pirhonen M, Garlant L, Hélias V, Pothier JF, Pflüger V, Duffy B, Tsror L, Manulis S (2013) Dickeya solani sp. nov., a pectinolytic plant pathogenic bacterium isolated from potato (*Solanum tuberosum*). Int J Syst Evol Microbiol 64:768–774

Wanne LA (2004) Field isolates of *Streptomyces* differ in pathogenicity and virulence on radish. Plant Dis 88:785–796

Weber J (1990) Erwinia—a review of recent research. In: EAPR proceedings of the 11th triennial conference of the European Association for Potato Research, Edinburg, 9–13th July 1990. pp 112–121

Wicker E, Grassart L, Coranson-Beaudu R, Mian D, Guilbaud C, Fegan M et al (2007) Ralstonia solanacearum strains from Martinique (French West Indies) exhibiting a new pathogenic potential. Appl Environ Microbiol 71:6790–6801

Yabbuuchi E, Kosako Y, Yano I, Hotta H, Nishiunchi Y (1995) Transfer of two *Burkholderia* and an *Alcaligenes* species to *Ralstonia* gen. Nov. *Ralstonia solnacearum* (Smith, 1986) comb. Nov. and *Ralstonia eutropha* (Davis, 1969) comb. Nov Microbiol Immunol 39:897–904

Zhang XJ, Wang JS, Fang ZD (1993) Control of potato soft rot (*Erwinia carotovora* (Jones) dye) by copper sulphate and its effect on some enzymes in tubers. Acta Phytopathol Sin 23:75–79

Advances in Management of Late Blight of Potato

7

Sanjeev Sharma and Mehi Lal

Abstract

This chapter discusses about late blight, caused by the oomycete *Phytophthora infestans*, the main biotic threat to potato production. The pathogen evolves continuously, mainly through recombination and migration; hence, monitoring of *P. infestans* populations is critical for the development of effective management strategies. The population structure and its monitoring, symptomatology, and pathogenesis are discussed in the present chapter. No single approach is effective; hence, combination of approaches in an integrated manner is essential to combat this disease and is discussed here.

Keywords

Phytophthora infestans · Population structure · Symptoms · Host resistance · Disease forecasting · Alternative approaches · Decision support system

7.1 Introduction

Late blight, caused by the oomycete *Phytophthora infestans* (Mont.) de Bary, has historically been an important disease of potatoes and tomatoes worldwide. It continues to be the main biotic constraint of potato production and has been considered a threat to global food security (Cooke et al. 2012). Losses due to *P. infestans* have been estimated to € 12 billion per annum of which the losses in developing countries have been estimated around € 10 billion per annum (Haverkort et al. 2009). Studies conducted in the United States to estimate the impact of late

S. Sharma (✉) · M. Lal
Division of Plant Protection, ICAR-Central Potato Research Institute, Shimla, Himachal Pradesh, India

S. Kumar Chakrabarti et al. (eds.), *Sustainable Management of Potato Pests and Diseases*, https://doi.org/10.1007/978-981-16-7695-6_7

blight on potato yield and fungicide use revealed that use of the fungicides alone cost $ 77.1 million at an average cost of around $507 per ha which do not include non-fungicide control practices (Guenthner et al. 2001). Region-wise, economic importance of late blight shows that the disease takes highest toll of potato in sub-Saharan Africa (44% crop losses) followed by Latin America (36%), Caribbean (36%), South-East Asia (35%), South-West Asia (19%), and Middle East and North Africa (9%) (CIP 1997). *Phytophthora infestans* is considered as re-emerging pathogen due to regular emergence of its novel strains with increased virulence and its appearance in new locations with surprising intensity (Fry et al. 2015). Management of this devastating pathogen is challenged by its remarkable speed of adaptation to control strategies such as genetically resistant cultivars and fungicides. In the present communication, efforts have been made to discuss about the pathogen, its population structure, symptoms, pathogenesis, and recent advances in the management of the pathogen/disease.

7.2 The Causal Organism

Oomycetes are a diverse group of organisms that morphologically resemble fungi, yet are members of the *Straminipile* (= Stramenopile), and are more closely related to organisms in aquatic environments such as brown algae and diatoms. These are the members of the Kingdom Chromista (Dick 2001; Cavalier-Smith and Chao 2006; Beakes et al. 2012) under Super Kingdom Chromalveolata (Baldauf et al. 2000; Yoon et al. 2002). *P. infestans* is a heterothallic oomycete with both sexual and asexual reproductive cycles. With few exceptions, for example, Toluca Valley, Mexico, Scandinavia, and the Netherlands (Bruberg et al. 2011; Drenth et al. 1993a, b; Fry et al. 2015; Yuen and Andersson 2013), the asexual reproductive cycle dominates resulting in the development of distinct clonal lineages. The vegetative stage of the mycelium in *P. infestans* is diploid, while in true fungi, it is haploid. However, recent studies have shown that progenies from sexual *P. infestans* populations in the modern-day lineages are diploid, but the most important pandemic clonal lineages are triploid (Li et al. 2017). The size of the *P. infestans* genome is considerably larger (240 Mb) and by far the largest and most complex genome sequenced so far in the chromalveolates and even in true fungi. A total of 17,797 protein-coding genes have been detected within the *P. infestans* genome. Overall, the genome is having an extremely high repeat content (~74%) and to have an unusual gene distribution, which is thought to contribute to *P. infestans* evolutionary potential by promoting genome plasticity, thus enhancing genetic variation of effector genes leading to host adaptation (Haas et al. 2009).

Virulence of oomycetes depends on rapidly evolving protein families including extracellular toxins, hydrolytic enzymes, and cell entering effectors that help the pathogen suppress the host plant defenses and gain nutrition from the host (Jiang and Tyler 2012). *P. infestans* secretes large numbers of effectors: apoplastic effectors that accumulate in the plant intercellular space (apoplast) and cytoplasmic effectors that are translocated directly into the plant cell by a specialized infection structure called the haustorium (Whisson et al. 2007). Apoplastic effectors include secreted hydrolytic enzymes such as proteases, lipases, and glycosylases that probably degrade plant tissue, enzyme inhibitors to protect against host defense enzymes, and necrotizing toxins such as the Nep1-like proteins (NLPs) and PcF-like small cysteine-rich proteins (SCRs). At least 563 RxLR genes have been predicted in the *P. infestans* genome. RxLR effectors act as activators of plant immunity, resulting in effector triggered immunity (ETI) (Oh et al. 2009; Wang et al. 2017), while the apoplastic effectors act as activators of the PAMP-triggered immunity (PTI) (Domazakis et al. 2017). All oomycete avirulence genes (encoding products recognized by plant hosts and resulting in host immunity) discovered so far encode RxLR effectors that define a domain required for delivery inside plant cells, followed by diverse, rapidly evolving carboxy-terminal effector domains (Jiang et al. 2008). CRN cytoplasmic effectors were originally identified from *P. infestans* transcripts encoding putative secreted peptides that elicit necrosis *in planta*, a characteristic of plant innate immunity (Torto et al. 2003). Analysis of the *P. infestans* genome sequence revealed an enormous family of 196 CRN genes of unexpected complexity and diversity. Like RXLRs, CRNs are modular proteins and are defined by a highly conserved N-terminal ~50-amino-acid LFLAK domain and an adjacent diversified DWL domain. The effector genes locate mostly in the gene sparse regions of the genome that are rich in repetitive sequences and are rapidly evolving, probably enabling the evolutionary arms race between *P. infestans* and the host plant (Haas et al. 2009; Dong et al. 2015).

7.3 Population Structure of *P. Infestans*

It is imperative to understand the diversity of the pathogen to devise efficient management strategies. Knowledge on the pathogen population structure and its relation to phenotypic characteristics, such as fungicide sensitivity or aggressiveness, is important to develop effective management strategies for the disease (Saville et al. 2015). *Phytophthora infestans* is highly variable and has undergone a drastic change in structure over the period of time. Pathological specializations (races) within potato isolates were reported by Schick (1932) after almost 7 years of introduction of resistant hybrids/cultivars having R genes. However, universal appearance of races did not occur until resistance genes from *Solanum demissum* were transferred to commercial potato, *S. tuberosum*. Since then, the racial complexity has reached its zenith in different countries/regions (Guo et al. 2009; Li et al. 2009; Runno-Paurson et al. 2009; Arora et al. 2014). Up to 1984, only one mating type (A1) was known to occur throughout the world, except Mexico (Tooley et al. 1985). However, there had been worldwide migration as a result of which A2 mating type was introduced other parts of the world. First report of A2 mating type outside Mexico was from Switzerland (Hohl and Iselin 1984). Subsequently, A2 mating type was detected in USSR during the 1990s (Vorobev et al. 1991); the United States (Deahl et al. 1991); Belarus (Ivanyuk and Konstantinovich 1992); the Netherlands (Drenth et al. 1993a, b); India (Singh et al. 1994); Pakistan (Ahmed and Mirza 1995); Northern Ireland (Cooke et al. 1995);

Canada (Chycoski and Punja 1996); France (Gilet 1996); China (Zhiming et al. 1996); Hungary (Bakonyi and Ersek 1997); Italy (Cristinzio and Testa 1997); Ecuador (Oyarzun et al. 1997); Indonesia (Nishimura et al. 1999); Myanmar (Myint 2002); Colombia (Vargas et al. 2009); Sri Lanka (Kelaniyangoda 2011); Tunisia (Harbaoui et al. 2014); Scandinavia and Estonia in 1987 (Vorobyeva et al. 1991); Bolivia, Argentina, Uruguay, and Brazil (Plata 1998; Deahl et al. 2003; Forbes et al. 1998; Casa-Coila et al. 2017); and Algeria (Rekad et al. 2017). Though existence of both mating types has opened up the possibility of sexual reproduction, no evidence of frequent sexual reproduction has been found, suggesting that the sexual populations are ephemeral (Fry et al. 2015). Nevertheless, there are reports (e.g., the Nordic countries) which indicated the frequent occurrence of sexual reproduction in the field and survival of oospores that led to earlier onset of epidemics (Widmark et al. 2007; Schepers 2019). *P. infestans* is generally heterothallic requiring two different mating types for sexual reproduction. The presence of both mating types in central Mexico and in the Nordic countries of Europe and the Netherlands has led to sexual reproduction and high genetic diversity (Drenth et al. 1993a, b; Sjoholm et al. 2013; Wang et al. 2017). However, there are reports of occurrence of homothallic isolates which are self-fertile and constitute a new threat to potato and tomato crops because of their increased genotypic variability, better fitness, and greater aggressiveness (Zhu et al. 2016; Tian et al. 2016; Casa-Coila et al. 2017).

There are platforms. Viz., EuroBlight (http://euroblight.net/), USABlight (http://www.usablight.org/), Tizon Latino (https://tizonlatino.github.io/), AsiaBlight (https://www.asiablight.org), and AfricaBlight, which are carrying out monitoring of *P. infestans* populations across the globe. The findings have revealed that *P. infestans* populations are constantly evolving, and novel, usually more aggressive, genotypes appear periodically replacing the previously dominating genotypes (Schepers 2017). New genotypes can emerge through divergence from other genotypes, through recombination, or migration from other areas (Knaus et al. 2016). The main mode of reproduction of *P. infestans* is asexual, and variable numbers of clonal lineages exist in different countries and regions. Several studies have confirmed that appearance of new genotypes can often be attributed to migration (Fry et al. 2015; Knaus et al. 2016; Saville et al. 2016).

Multiple clonal lineages have been found in the United States since the 1990s, revealing the history of the displacement of lineages over time (Fry and Goodwin 1997; Hu et al. 2012a, b). Genetic analysis using simple sequence repeats (SSRs) of *P. infestans* from herbarium samples from the nineteenth century historic outbreaks revealed the presence of a single dominant clonal lineages FAM-1 that caused disease in both the continents, i.e., the United States and Europe (Saville et al. 2016), suggesting the migration of the pathogen from a similar point of origin (Yoshida et al. 2013). With the emergence of the US-1 lineage during the 1930s in the United States, the historic FAM-1 lineage subsequently declined (Saville et al. 2016). The dominance of US-1 clonal lineage lasted in the United States until the 1980s, when new lineages of the pathogen emerged that were insensitive to mefenoxam (Goodwin et al. 1996). The new genotypes were US-6 (A1 mating type), US-7 (A2 mating type), US-8 (A2 mating type), and US-11, emerged out of Mexico, majority as a result of sexual recombination and some as clonal derivatives of earlier lineages (Goodwin et al. 1998). US-11, which is thought to be the progeny of US-6 and US-7 lineages (Gavino et al. 2000), still occurs in the fields of the Pacific Northwest and Florida.

Although US-1 was a dominant lineage in the United States for 60 years, it declined in the mid-1990s, probably because of its sensitivity to the fungicide mefenoxam. Majority of US lineages, with the exception of US-6, were detected in the 1990s in the United States, and many were resistant to mefenoxam (Saville and Ristaino 2019). The US-8 clonal lineage was responsible for the first pandemic during the 1990s in the United States (Fry and Goodwin 1997; Johnson et al. 1997). The second pandemic was in 2009 due to wide spread of US-22 clonal lineage with infected tomato seedlings throughout northeastern USA (Fry et al. 2013). The population of *P. infestans* in the United States continues to be dominated by relatively few clonal lineages (Hu et al. 2012a, b; Fry et al. 2013). The most recent dominant strains are US-8, US-11, US-22, US-23, and US-24 (Fry et al. 2015). Generally, lineages differ in terms of their response to mefenoxam, and pathogenicity and common lineages in the United States during 1990 to 2009 were largely resistant to mefenoxam (Fry et al. 2015), and growers were not using this molecule to manage late blight. However, the dominant lineage (US-22) in 2009 was sensitive to mefenoxam, and some dominant lineages since 2009 have also been sensitive to mefenoxam (Hu et al. 2012a, b; Saville et al. 2015). Further, lineages in the United States differ in terms of their pathogenicity. US-11 and US-23 are very good pathogens of both tomatoes and potatoes, whereas US-8 and US-24 are not good pathogens of tomatoes. The US-23 lineage has dominated the *P. infestans* population in the United States since 2012 by replacing the previously dominant lineages, including US-8 and US-22. The possible reasons for dominance could be its aggressiveness on both foliage and tubers (Danies et al. 2013) and its pathogenicity on both potatoes and tomatoes (Danies et al. 2013). Studies on genetic structure and sub-clonal variation of extant and recent US lineages revealed that many clonal lineages in the United States have come from Mexico via introduction, but US-23 (from Bolivia and Brazil) and US-1 (from Peru) lineages were introduced from other sources (Saville and Ristaino 2019). However, a survey for the presence of RXLR effector *PiAVR2* revealed the presence of lineages that carried either *PiAVR2*, its resistance-breaking variant *PiAVR2-like*, or both, suggesting lineages have experienced different levels of selection to the *R2* gene in potato, thereby indicating that populations of *P. infestans* in the US are the result of introductions from both South America and Mexico (Saville and Ristaino 2019).

The South and Central America can harbor divergent genotypes as these regions are rich in solanaceous species biodiversity and are centers of origin of the economically important crops that are potential alternative hosts of *P. infestans*. No sexual reproduction of *P. infestans* has been reported in South America; hence populations maintain strictly clonal structures, and A1 mating type is mostly dominant (Acuna et al. 2012; Cardenas et al. 2011). In Mexico, recombination is frequent and the population is extremely divergent (Wang et al. 2017), and it is also considered the origin of the newly emerged genotypes in the United States (Goss et al. 2014; Saville et al. 2016).

EuroBlight is continuously investigating the evolution of potato late blight pathogen in the Europe. A complex population structure is observed in Europe with population dominance (70%) by a few widely disseminated clonal lineages. The clonal lineage 13_A2 was first detected in 2004 in the Netherlands and Germany which has now emerged in regions beyond Europe (Cooke et al. 2012). Some clones are widespread and have been present for more than a decade, but recently the frequency occurrence of three clones (EU_37_A2, EU_36_A2, and EU_41_A2) have increased from 10% (2016) to 40% (2019) by displacing the established clones (EU_13_A2, EU_6_A1, and EU_1_A1) from 60 to 40% of the population. Besides, 20–30% of the sampled European population is genetically diverse and consistent with local, ephemeral oospore-derived sexual populations. The frequency occurrence of the clonal lineage EU_13_A2 (blue-13) and EU_1_A1 has dropped to 9.3% and 0.4%, respectively, whereas the frequency of EU_6_A1 increased to 20.4% due to severe outbreaks in parts of Britain. A progressive displacement of these three lineages is occurring (Cooke et al. 2019). Clone EU_36_A2, which was first sampled at a low frequency in Germany and the Netherlands in 2014, has spread rapidly in Europe to the frequency of 26% in 2019. Clone EU_41_A2, first recorded in Denmark in 2013, has now spread to neighboring states, and its frequency has also increased from 4.6 to 5.7% of the European population in 2019 (Schepers 2019).

In eastern Africa, the first late blight epidemic occurred in Kenya in 1941 and the pathogen was thought to be introduced via potato seed tubers from the United Kingdom. After 1 year of the epidemic, the disease was also noticed in Uganda, Democratic Republic of Congo, and Tanzania (Natraas 1944). The US-1 was probably dominant in Europe at the time of the introduction of *P. infestans* in eastern Africa and is assumed to be the only lineage introduced into the region. In eastern Africa, only the A1 mating type has been detected so far, thereby signifying the persistence of a clonal population (Njorog et al. 2016). The US-1 had been the only lineage reported in the eastern African region, apart from RW-1 and RW-2 genotypes in Rwanda in the mid-1980s (Forbes et al. 1998; Goodwin et al. 1994). However, these two genotypes (RW-1 and RW-2) were not detected in a later study in 2007 that reported all isolates from Rwanda to be US-1 (Pule et al. 2013). They further reported that US-1 was still the only lineage in central and eastern Africa apart from Kenya, where US-1 and a new lineage KE-1 were found. The new genotype KE-1 was first reported from Kenya in 2007 and later from Uganda in 2011 and found to be the only lineage on potato in Kenya (Njoroge et al. 2016). The recent population of *P. infestans* infecting potato in the eastern African region is dominated by KE-1 lineage, which had similar SSR fingerprints to that of EU_2_A1 (Njoroge et al. 2019). They further found decline in US-1 lineage but still present on potato in Uganda, Rwanda, Burundi, and Tanzania. Besides, a tomato-adapted US-1 sub-population is also still present in all the countries. Two new European lineages (EU_33_A2 and EU_13_A2) have been emerged recently in Nigeria and Senegal on potato and are a cause for concern for potato production in sub-Saharan Africa (Schepers 2019).

Although there have been a number of publications on the late blight population structure of *P. infestans* in Asian countries, a very few have used markers for comparative analysis (Forbes 2015). The Indian population of *P. infestans* has been characterized for phenotypic and genotypic characters (Chimote et al. 2010; Sharma et al. 2016, 2017) and ploidy status (Sharma et al. 2018). The findings of these studies have shown that population is possessing complex virulence genes, resistance to metalaxyl, Ia mtDNA haplotype, and varied allele size for SSR markers. There are records of at least four migrations of *P. infestans* into India over the past 100 years. The oldest samples of *P. infestans* collected from Bagalpur (Bihar) in 1913 by J.F. Dastur were the Ia mtDNA haplotype (Ristaino and Hu 2009), and the US-1 clonal lineage (Ib mt DNA haplotype) was present in India by the 1960s (Ristaino and Hu 2009). The occurrence of the A2 mating type in the 1990s in the northern hills provided the additional evidence of migration from an outside source (Singh et al. 1994), and more recently, the European 13_A2 genotype was intercepted in southern India (Chowdappa et al. 2013, 2015). Dey et al. (2018) found that mutations have generated substantial sub-clonal variation in EU_13_A2 genotype, having 19 out of 24 unique variants not yet reported elsewhere globally. Nevertheless, the Asian population of *P. infestans* has also been genotyped using markers, and findings revealed the widespread occurrence of aggressive genotype 13_A2 in many parts of Asia as reported from China (Li et al. 2013), India (Chowdappa et al. 2013, 2015; Dey et al. 2018), Bangladesh (Kessel et al. 2017), and Pakistan (Raza et al. 2020). In Vietnam, the *P. infestans* population is still the "old" US-1 (Le et al. 2008). The *P. infestans* population in northwestern China is genetically distant from European lineages, including the recently identified 13_A2 lineage (Tian et al. 2016), though its presence (13_A2) was reported in Sichuan and Yunnan provinces, south western China (Li et al. 2013). Four clonal lineages, viz., KR_1_A1, KR_2_A2, SIB-1, and US-11, have been reported from South Korea. KR_2_A2 was confined to Gyeongnam Province, whereas SIB-1 was dominant until 2013 and thereafter its frequency declined gradually. US-11 was first found in 2014, and its frequency has increased to become co-dominant with KR_1_A1. The EU_13_A2 genotype was not found in South Korea (Choi et al. 2020). The Indonesian population is dominated by EU_2_A1 (60%), EU_4_A1, and EU_13_A2 (1.5%) (Dangi et al. 2021).

7.4 Symptoms

P. infestans adopts a two-step infection style typical of hemibiotrophs. Infection generally starts when sporangia lands on a plant surface and release zoospores that encyst, germinate, and penetrate the host tissue or sporangia directly germinate and initiate the infection. Germ tubes form an appressorium and then a penetration peg, which pierces the cuticle and penetrates an epidermal cell to form an infection vesicle. Branching hyphae with narrow, digit-like haustoria expand from the site of penetration to neighboring cells through the intercellular space. At this biotrophic stage, *P. infestans* requires living cells to obtain nutrients. However, this stage of

Fig. 7.1 Late blight symptoms (**a**) Foliar blight on upper surface (**b**) On lower surface (**c**) On stem (**d**) On tubers

infection remains unnoticed to the naked eye, but at cellular level a repertoire of molecular interactions takes place. The first visible symptoms appear within 2–3 days when the pathogen switches to the necrotrophic stage. Later on, the mycelium develops sporangiophores that emerge through the stomata to produce numerous asexual spores that initiate new infections (Judelson and Blanco 2005). In leaves, water-soaked irregular pale green lesions mostly near tip and margins that enlarge into brown to purplish black necrotic spots appear. A white mildew, which consists of sporangiophores and spores of the pathogen, can be seen on the lower surface of the infected leaves especially around the edges of the necrotic lesions under high humidity (Nowicki et al. 2012). On stems and petioles, light to dark brown lesions encircle the stems; as a result, the affected stems and petioles become weak at such points and may collapse. Affected tubers show irregular reddish brown to purplish areas which extend into internal tissues of the tubers (Fig. 7.1).

Pathogenesis involves the secretion of proteins and other molecules by *P. infestans* that participate in helping the pathogen attach to plant surface, breaking down physical barriers to infection and influence the host physiology by suppressing or inducing host-defense responses (Huitema et al. 2004). Gene expression profiling during asexual development of *P. infestans* revealed highly dynamic transcriptome. Differentially expressed genes encode potential cellular regulators, especially protein kinases; metabolic enzymes involved in glycolysis, gluconeogenesis, or the biosynthesis of amino acids or lipids; regulators of DNA synthesis; structural proteins; and pathogenicity factors like cell-wall degrading enzymes, RXLR effectors and enzymes protecting against plant defense responses (Tani et al. 2004; Judelson et al. 2008). A MADS-box protein (*PiMADS*) is required for sporulation of *P. infestans* but not for hyphal growth or host colonization as both mRNA and protein levels decline upon spore germination (Leesutthiphonchal and Judelson 2018). *P. infestans* possesses a large repertoire of phospholipase D (PLD) proteins which are essentially required for the promotion of virulence, possibly by executing membrane modifications to support the growth of *P. infestans* in the host (Meijer et al. 2019). Identification of these factors involved in pathogen growth and development and in pathogenesis would be of help in designing management strategies.

7.5 Management of the Disease

Management of this devastating pathogen is challenged by its remarkable speed of adaptation, with respect to emergence of virulence towards resistant cultivars and to fungicide resistance (Haas et al. 2009). One of the prerequisites for durable management of late blight is up-to-date knowledge on characteristics of local *P. infestans* population and its dynamics. Since the pathogen population is continually evolving, the emerging clonal lineages with new traits highlights the need to tailor management to the local pathogen population. No single approach is effective; hence, combination of approaches in an integrated manner is essential to combat this devastating disease.

7.5.1 Cultural Practices

These are an important part of an integrated disease management program as they reduce the incidence and severity of the disease epidemic thereby reducing yield losses and lowering the requirements of fungicides (Mizubuti and Forbes 2002). Reduction of primary source of inoculum is the first step, and this can be achieved by eliminating volunteers and cull piles, waste heaps, infected tubers, use of certified seed and resistant varieties, balanced fertilization, adequate space between rows and plants, rotation with non-host crops, adequate hilling, harvest in dry conditions, and when the tubers are mature (Garrett and Dendy 2001; Perez and Forbes 2010). Onset of epidemic can be delayed by 3–6 weeks if all primary infection from early potato is eliminated (Forrer et al. 2000). Covering of dumps with black plastic sheet throughout the season is an important step in reducing the primary inoculum as it prevents re-growth and the proliferation of spores on the piles thereby reducing the risk to nearby crops (Cooke et al. 2012). Infection usually starts early in fields which are not subjected to crop rotations. A sound crop rotation for 3–4 years is an effective way of reducing the risk of soil-borne inoculum as oospores can remain infectious up to 48 months in soil (Turkensteen et al. 2000; Bodker et al. 2006; Hannukkala et al. 2007). Choice of suitable cultivars, well-aerated fields, pre-sprouting of tubers, early planting, use of resistant varieties, and mixtures of potato varieties (resistant and susceptible) are some of the measures against foliar blight (Meinck and Kolbe 1999; Garrett and Mundt 2000; Pilet et al. 2006). Strip cropping of potatoes significantly reduced late blight severity in organic production when planted perpendicular to the wind neighbored by grass clover (Bounes and Finckh 2008). Avoiding excess nitrogen and use of moderate nitrogen fertilization is often recommended as a cultural practice to delay the development of late blight, whereas higher dose of phosphorus and potassium has been found to give a higher yield in a late blight year (Roy et al. 2001). High ridging is often used to reduce tuber contamination by blight. Another approach to reduce tuber blight is to destroy the canopy when blight reaches to 75% severity. Elimination of infected foliage reduces the likelihood of tuber infection. Intercropping with garlic has been found effective against potato late blight under Ethiopian condition (Kassa and Sommartya 2006).

7.5.2 Host Resistance

Host resistance is the most preferred environment and economic option globally for the management of late blight. With the use of host resistance, fungicide load can be reduced either by lowering the fungicide dose or increasing the application intervals (Kirk et al. 2005; Cooke et al. 2012; Haverkort et al. 2016). Durable resistant cultivars with multiple resistant genes are needed today, which can be developed by a blend of conventional and molecular approaches. So far resistant genes from the wild species *Solanum demissum* and *S. stoloniferum* and the cultivated *S. tuberosum* subsp. *andigena* and *S. phureja* have been utilized into common potato in different parts of the world (Bradshaw et al. 2006). Thus, it warrants the breeders to search for new sources of resistance in wild gene pools and their faster deployment into cultivars through modern techniques. Late blight resistance genes/QTLs and molecular markers for late blight resistance genes/QTLs in potato have been reviewed by Tiwari et al. (2013). Genetic engineering may also provide options for generating resistant cultivars. A resistance gene effective against most known strains of blight has been identified from a wild relative of the potato, *Solanum bulbocastanum*, and introduced by genetic engineering into cultivated varieties of potato (Song et al. 2003; Van der Vossen et al. 2003). Introgression of RB gene in Indian popular potato cultivars has demonstrated variable level of late blight resistance and generation of valuable genetic material for resistance breeding (Shandil et al. 2017).

7.5.3 Use of Fungicides

The chemical-based management still continues to be the most common method to supplement host resistance and to manage the late blight. Recent changes in the population structure of the pathogen have led to the advent of new genotypes that are more aggressive and resistant to previously effective fungicides (Fry et al. 2015). Sixteen classes of fungicides with different modes of action are available for the control of oomycete plant pathogens (FRAC n.d.). The three most important single-site compounds are phenylamides, quinone outside inhibitors, and carboxylic acid amides (Gisi and Sierotzki 2014). Products containing mefenoxam or metalaxyl (a.i. mefenoxam) have been the most widely used fungicides for control of *P. infestans*. However, more recent dominant lineages are largely sensitive to mefenoxam (Matson et al. 2015; Saville et al. 2015). The build-up of resistance to single-site oomycides has accelerated the research for anti-oomycete compounds with new modes of action. The development strategy for creating new fungicides consists of fungicides that are (1) effective at an extremely low dosages, (2) readily degradable and less residual in the environment, and (3) selective toxic agrochemicals (Umetsu and Shirai 2020). Many fungicides possessing various novel modes of action have been launched or are under development. Two such novel compounds are ametoctradin (Quinone QoSI inhibition of the respiratory chain) binding to the mitochondrial bc1 Complex III (Fehr et al. 2015) and oxathiapiprolin (inhibitor of oxysterol-binding protein) (Sweigard et al. 2014). Oxathiapiprolin binds in the oxysterol-binding protein (OSBP) domain of oomycetes and inhibits zoospore and sporangial germination, stops mycelia growth in the host plants before visible symptoms occur, and inhibits further lesion growth and spore production and viability. It belongs to the FRAC U49 group of fungicides (Cohen 2015).

7.5.4 Alternatives to Fungicides

Various chemicals other than fungicides have also been found effective against late blight; for example, ammonium molybdate, cupric sulfate, and potassium metabisulfite have been reported to partially inhibit the growth and spore germination of *P. infestans*, whereas ferric chloride, ferrous ammonium sulfate, and $ZnSO_4$ completely inhibited growth and spore germination (Bhat et al. 2006). The foliar application of $ZnSO_4$ and $CuSO_4$ (0.2%) micronutrients in combination with host resistance delayed the onset of late blight by 12 days and subsequently reduced disease severity with higher yield (Basu et al. 2003). Phosphites (Phi), derived from phosphorous acid, are fungitoxic chemicals that can be combined with different elements such as calcium, copper, manganese, magnesium, potassium, or zinc and are classified by the US Environmental Protection Agency (US-EPA) as biopesticides, specifically biochemical pesticides (http://www.epa.gov/pesticides/biopesticides/). Thus, they have low environmental impact (Guest and Grant 1991). Besides their fungistatic or fungicidal activity (Fenn and Coffey 1984; Lobato et al.

2008), Phi stimulate defense mechanisms in plants against diseases (Daniel and Guest 2006; Andreu et al. 2006; Lobato et al. 2011) and promote growth (Thao and Yamakawa 2009). Because of these attributes, the horticultural industry widely uses Phi for oomycete control (Pilbeam 2003). Cicore et al. (2012) evaluated the effect of calcium phosphite (Phica) and potassium phosphite (PhiK) on late blight control and found that PhiK had significantly lower damage and higher yields than PhiCa and untreated control. Sub-phytotoxic dose of boron with reduced rate of propineb + iprovalidicab has been found more effective than treated with fungicides alone (Frenkel et al. 2010). Similarly, application of potassium phosphate in combination with reduced doses of fungicides provided the same level of protection as full dose of fungicides. Thus, combined treatments could help to reduce the quantity of traditional fungicides and may also decrease the selection pressure for fungicide resistance development in the pathogen. β-aminobutyric acid (BABA) has been known as an inducer of disease resistance. Plant activators, viz., BABA and phosphoric acid, have been evaluated against late blight with combination of fungicides or alone (Tsai et al. 2009). A 20–25% reduction of the fungicide dose in combination with BABA gave the same result on late blight development as full dose of Shirlan alone in field condition, while reduced dose of Shirlan alone sometimes resulted in less effective protection. The partially resistant cultivars Ovatio and Superb reacted to lower concentrations of BABA where no effect was found in susceptible cv. Bintje (Liljeroth et al. 2010). The expression of the defense-related genes and *P. infestans* effector proteins β-1,3 glucanase, PR-1 protein, phytophthora inhibitor, protease inhibitor, xyloglucanase, thaumatin protein, steroid binding proteins, proline, endochitinase, and cyclophilin genes was upregulated with the SAR activator treatment compared to unsprayed (CPRI 2014). Better results than with copper were achieved with Phosfik® (Ph), a phosphonate-based product. Two to three applications with 2–3 L/ha of Ph would be feasible to not exceed a minimal risk level (MLR) of 20 mg/kg of phosphorous acid as proposed by the European Food Safety Authority (Forrer et al. 2017). Due to an excellent environmental profile and a complex mode of action counteracting *Phytophthora infestans* resistance, phosphonate-based products would be most suitable for sustainable late blight management in integrated disease management programs.

7.5.5 Biocontrol

New strategies to manage plant diseases without harming the environment are urgently needed. Biocontrol agents and bio-pesticides could be a safe option to the use of synthetic fungicides. Some workers have reported the use of *Trichoderma* isolates (Yao et al. 2016), *Chaetomium globosum* (Shanthiyaa et al. 2013), *Trichoderma viride*, and *Penicillium viridicatum* (Gupta 2016) and species of *Bacillus*, *Pseudomonas*, *Rahnella*, and *Serratia* (Daayf et al. 2003) as biocontrol agents in the management of late blight disease in potato. The bio-based products, viz., neem-based products and bio-agents (*T. viride* and *P. fluorescens*), have shown some efficacy against late blight under field conditions (Lal et al. 2021). The biocontrol agents in general have been found to be very effective under laboratory and glasshouse conditions but less effective under field conditions (Arora 2000). However, an integrated use of biocontrol agents along with fungicides could help to reduce the quantity of fungicides used in the management of late blight (Lal et al. 2017). Biosurfactants produced by microbes can be used as alternatives to chemical surfactants because of their low toxicity, high specificity, and biodegradability (Lima et al. 2011). Significant reduction in late blight development was observed when plants were treated with biosurfactant—*Pseudomonas koreensis* 2.74—and also, biosurfactants have the potential to induce resistance in potato to late blight (Bengtsson et al. 2015). The biosurfactant produced by *Pseudomonas aeruginosa* has shown high efficacy against *P. infestans* under in vitro and glass house conditions (Tomar et al. 2013, 2014). The rhamnolipid-based formulation prepared from *P. aeruginosa* biosurfactant was found effective against late blight when evaluated through detached leaf (Tomar et al. 2019) and could be used in field spray as green chemicals.

Plant-associated bacteria contribute to their host's health in diverse ways, among which the emission of disease inhibiting volatile organic compounds (VOCs) is one option. Volatile organic compounds (VOCs) produced by the plant microbiota have been demonstrated to elicit plant defenses and inhibit the growth and development of numerous plant pathogens. The inhibitory impact of volatiles emitted by *Pseudomonas* species against late blight has been shown by impeding mycelial growth and sporangia germination of *P. infestans* (Bailly and Weisskopf 2017). The VOCs containing sulfur compound S-methyl methane thiosulfonate (MMTS) had shown high *in planta* protective potential against late blight without phytotoxic effects. Short exposure times were sufficient to protect plants against infection. This protective activity of MMTS is not mediated by the plant immune system but is due to its anti-oomycete activity (Chinchilla et al. 2019). This provides new perspectives for plant protection by opening new research avenues on the role of VOCs in the interaction between plants and their microbiome and thus could help select for efficient biocontrol strategies and lead to a greener chemical disease management in the field.

In organic potato production, the only synthetic direct control measure allowed is the use of copper-based products despite its persistence in soil and toxicity to soil organisms (Buenemann et al. 2006). Based on such reports about the toxicity of copper, the EU proposed a ban of copper fungicides as early as 2002, though it was not imposed as of now, but this would have threatened the feasibility of organic potato production. This initiative led to intensified research for new approaches to reduce the risk of late blight attacks and for natural products to replace or reduce the use of copper (Leifert and Wilcockson 2005). Three promising botanicals, including bark of buckthorn (*Frangula alnus*, FA), roots of medicinal rhubarb (*Rheum palmatum*), and galls of the nutgall tree (*Galla chinensis*), have been reported effective under field conditions and could replace copper reaching a level close to that of 2–3 kg copper per hectare and year (Forrer et al. 2017).

7.5.6 Late Blight Forecasting

Currently, late blight management has been heavily based on numerous fungicide applications due to introduction of new, more aggressive genotypes of the pathogen (Schepers 2017, 2019). However, this strategy faces increasing concerns due to societal pressure for reducing pesticide use on crops and acreage of organically grown food crops. Innovative and effective control measures are needed if fungicide use is to be reduced or, as in the case of organic production, eliminated. One way of achieving this goal is through the use of forecasting models and decision support systems (DSSs). Forecasting allows a better control of a disease and a more efficient use of fungicides by making informed disease management decisions. Various late blight forecasting models and DSSs have been developed across the globe for the management of late blight in different agro-ecologies (Table 7.1). The DSS-based strategy can deliver general or site-specific information to the stakeholders via print and electronic media (Cooke et al. 2012) enabling them to take firm decisions on the management of late blight thereby resulting in economic gains and environment protection (Sekhon et al. 2017; Liu et al. 2017; Sharma 2019).

Table 7.1 List of forecasting models and decision support systems used for forecasting and management of late blight (Source: Singh and Sharma 2013; Schepers 2019)

Decision support system	Country	Decision support system	Country
BliteCast, SimCast, BlightPro	USA	Guentz-Divoux	Belgium, France
Blight-watch, plant plus, BlightCAST	England, Wales, Scotland	Estonian crop research institute	Estonia
Plant plus	Latvia	Mileos	France
Phytophthora model Weihenstephan, ISIP, Phytoprog, SimPhyt, ProPlant, ProGeb	Germany	Prophy, plant plus, Akkerwels (WUR model)	Netherlands
Met. Service based on Irish rules (Bourke)	Ireland	VIPS (Naerstad model)	Norway
Plant-plus, VNIIF blight, Agrodozor	Russia	Plant plus, blight management (DK), VIPS (no)	Sweden
Bio-PhytoPRE, PhyoPRE, PhytoPRE +2000	Switzerland	WISDOM, web-blight, NegFry	Denmark
IPI, MIP	Italy	Blight watch	UK
China-blight	China	Indo-Blightcast, Jhulsacast	India

7.6 Looking Forward

Phytophthora infestans is capable of overcoming host resistance and fungicides; hence, late blight would continue to be the main constraint in potato cultivation throughout the world. Nevertheless, the advances in molecular, sensor, computational, and electronics technologies would provide stable solutions for its management. New high-throughput methods (remote sensing, image processing, UAV, etc.) would be of significance in disease detection and surveillance. Robust, quick, and onsite detection methods are needed for early diagnosis of the pathogen and monitoring of population structure. Research is warranted on development of new oomycides having efficacy at very low dosages, highly degradable, and with novel mode of action. Besides, there is need to identify new molecules of biological origin that can be used under organic production. Smartphone-based systems can be of help in monitoring, forecasting/DSSs, and dissemination of the disease information to the stakeholders. Emerging research topics on *P. infestans* include genome editing for genetic improvement of plant disease resistance and the role of the pathogen–microbiota interaction in promotion or suppression of the disease.

References

Acuna I, Sagredo B, Gutierrez M et al (2012) Characterization of *Phytophthora infestans* population in Chile. In: Proceedings of the thirteenth EuroBlight workshop, St Petersburg, Russia, 9–12 October 2011. Praktijkonderzoek Plant & Omgeving, PPO, pp 145–150

Ahmed I, Mirza JI (1995) Occurrence of A2 mating type of *Phytophthora infestans*. In: Research and Development of Potato Production in Pakistan. Proceedings of the National Seminar held at NARC, Islamabad, Pakistan 23–25 April, 1995

Andreu AB, Guevara MG, Wolski EA et al (2006) Enhancement of natural disease resistance in potatoes by chemicals. Pest Manag Sci 62:162–170

Arora RK (2000) Bio-control of potato late blight. In: Paul KSM, Shekhawat GS, Singh BP, Pandey SK (eds) Potato global research and development. Indian Potato Association, Central Potato Research Institute, Shimla, pp 620–623

Arora RK, Shrama S, Singh BP (2014) Late blight disease of potato and its management. Potato J 41:16–40

Bailly A, Weisskopf L (2017) Mining the volatilomes of plant-associated microbiota for new biocontrol solutions. Front Microbiol 8:1638

Bakonyi J, Ersek T (1997) First report of A2 mating type of *Phytophthora infestans* on potato in Hungary. Plant Dis 81:1094

Baldauf SL, Roger AJ, Wenk-Siefert I, Doolittle WF (2000) A kingdom-level phylogeny of eukaryotes based on combined protein data. Science 290:972–977

Basu A, Hossain MM, Konar A, Chettri M (2003) Integrated management of late blight disease of potato in West Bengal. Ann Pl Protec Sci 1:64–66

Beakes GW, Glocking SL, Sekimoto S (2012) The evolutionary phylogeny of the oomycete "fungi". Protoplasma 249:3–19

Bengtsson T, Holefors A, Liljeroth E et al (2015) Biosurfactants have the potential to induce defence against *Phytophthora infestans* in potato. Potato Res 58:83–90

Bhat MN, Rani A, Singh BP (2006) Efficacy of inorganic salt against potato late blight. Potato J 34: 83–84

Bodker L, Pedersen H, Kristensen K et al (2006) Influence of crop history of potato on early occurrence and disease severity of potato late blight caused by *Phytophthora infestans*. In: Westerdijk CE, Schepers HTAM (eds) Proceedings of the 9th workshop of European network for development of an integrated control strategy of potato late blight, pp 53–56

Bounes H, Finckh MR (2008) Effects of strip intercropping of potatoes with non-hosts on late blight severity and tuber yield in organic production. Plant Pathol 57:916–927

Bradshaw JE, Bryan GJ, Ramsay G (2006) Genetic resources (including wild and cultivated *solanum* species) and progress in their utilisation in potato breeding. Potato Res 49:49–65

Brurberg MB, Elameen A, Le VH et al (2011) Genetic analysis of *Phytophthora infestans* populations in the Nordic European countries reveals high genetic variability. Fungal Biol 115:335–342

Buenemann EK, Schwenke GD, Van Zwieten L (2006) Impact of agricultural inputs on soil organisms—a review. Aust J Soil Res 44:379–406

Cardenas M, Grajales A, Sierra R et al (2011) Genetic diversity of *Phytophthora infestans* in the northern Andean region. BMC Genet 12:23

Casa-Coila VH, Lehner MDS, Hora Júnior BT et al (2017) First report of *Phytophthora infestans* self-fertile genotypes in southern Brazil. Plant Dis 101(9):1682

Cavalier-Smith T, Chao E (2006) Phylogeny and megasystematics of phagotrophic heterokonts (kingdom Chromista). J Mol Evol 62:388–420

Chimote VP, Kumar M, Sharma PK et al (2010) Characterization of changes in phenotype and genotype of *Phytophthora infestans* isolates from India. J Plant Pathol 92:669–677

Chinchilla D, Bruisson S, Meyer S et al (2019) A sulfur-containing volatile emitted by potato-associated bacteria confers protection against late blight through direct anti-oomycete activity. Sci Rep 9(1):18778

Choi JG, Hong SY, Kessel GJT et al (2020) Genotypic and phenotypic characterization of *Phytophthora infestans* in South Korea during 2009-2016 reveals clonal reproduction and absence of EU_13_A2 genotype. Plant Pathol 69:932–943

Chowdappa P, Kumar NBJ, Madhura S et al (2013) Emergence of 13_A2 blue lineage of *Phytophthora infestans* was responsible for severe outbreaks of late blight on tomato in south-west India. J Phytopathol 161(1):49–58

Chowdappa P, Nirmal Kumar BJ, Madhura S et al (2015) Severe outbreaks of late blight on potato and tomato in South India caused by recent changes in the *Phytophthora infestans* population. Plant Pathol 64:191–199

Chycoski CI, Punja ZK (1996) Characteristics of populations of *Phytophthora infestans* from potato in British Columbia and other regions of Canada during 1993-1995. Plant Dis 80:579–589

Cicore PL, Suorez PA, Andreu AB (2012) Phosphites effect on late blight control and physiological parameters in commercial potato (*Solanum tuberosum*) in Argentina. Pest Technol 6:27–31

CIP (1997) Annual report. The International Potato Centre, Lima, p 179

Cohen Y (2015) The novel oomycide oxathiapiprolin inhibits all stages in the asexual life cycle of *Pseudoperonospora cubensis*—causal agent of cucurbit downy mildew. PLoS One. https://doi.org/10.1371/journal.pone.0140015

Cooke DEL, Cano LM, Raffaele S et al (2012) Genome analyses of an aggressive and invasive lineage of the Irish potato famine pathogen. PLOS Pathogens 10:14

Cooke DEL, Kessel GJT, Lassen P, Hansen JG (2019) The European population of Phytophthora infestans in a global context. In: Seventeenth Euroblight workshop, York, 12–15 May, 2019. WUR-Special Rep 19, pp 35–36

Cooke LR, Swan RE, Currie TS (1995) Incidence of the A2 mating type of *Phytophthora infestans* on potato crop in Northern Ireland. Potato Res 38:23–29

CPRI (2014) Annual Progress report. Central Potato Research Institute, Shimla

Cristinzio G, Testa A (1997) Occurrence of A2 mating type and self isolates of *Phytophthora infestans* in Italy. J Plant Pathol 79:121–123

Daayf F, Adam L, Fernando WGD (2003) Comparative screening of bacteria for biological control of potato late blight (strain US-8), using *in-vitro*, detached-leaves, and whole-plant testing systems. Can J Plant Pathol 25(3):276–284

Dangi S, Wharton P, Ambarwati AD et al (2021) Genotypic and phenotypic characterization of *Phytophthora infestans* population on Java, Indonesia. Plant Pathol 70:61–73

Daniel R, Guest DI (2006) Defence responses induced by potassium phosphonates in *Phytophthora palmivora*-challenged Arabidopsis thaliana. Physiol Mol Plant Pathol 67:194–201

Danies G, Small IM, Myers K et al (2013) Phenotypic characterization of recent clonal lineages of *Phytophthora infestans* in the United States. Plant Dis 97:873–881

Deahl KL, Groth RW, Young R et al (1991) Occurrence of the A2 mating type of *Phytophthora infestans* in potato fields in the United States and Canada. Am Potato J 68:717–726

Deahl KL, Pagani MC, Vilaro FL et al (2003) Characteristics of *Phytophthora infestans* isolates from Uruguay. Eur J Plant Pathol 109(3):277–281

Dey T, Saville A, Myers K et al (2018) Large sub-clonal variation in *Phytophthora infestans* from recent severe late blight epidemics in India. Sci Rep 8(4429):12

Dick M (2001) Straminipilous fungi: systematics of the Peronosporomycetes including accounts of the marine straminipilous protists, the plasmodiophorids and similar organisms. Kluwer, Dordrecht, p 660

Domazakis E, Lin X, Aguilera-Galvez C et al (2017) Effectoromics-based identification of cell surface receptors in potato. Methods Mol Biol 2017:337–353

Dong S, Raffaele S, Kamoun S (2015) The two-speed genomes of filamentous pathogens: waltz with plants. Curr Opin Genet Dev 35:57–65

Drenth A, Goodwin S, Fry WE, Davidse L (1993a) Genotypic diversity of *Phytophthora infestans* in the Netherlands revealed by DNA polymorphism. Phytopathology 83:1087–1092

Drenth A, Turkensteen LJ, Grovers F (1993b) The occurrence of the A2 mating type of *Phytophthora infestans* in the Netherlands; significance and consequences. Netherland J Plant Pathol 99(Suppl.3):57–67

Fehr M, Wolf A, Stammler G (2015) Binding of the respiratory chain inhibitor Ametoctradin to mitochondrial bc 1 complex. Pest Manag Sci. https://doi.org/10.1002/ps.4031

Fenn ME, Coffey MD (1984) Studies on the *in vitro* and *in vivo* antifungal activity of fosetyl-Al and phosphorous acid. Phytopathology 74:606–611

Forbes GA (2015) Recent developments concerning the population biology and control strategies of Phytophthora infestans in Asia and Africa. In: Fifteenth Euroblight workshop, 10–13 May, 2015, Brasov. PPO-Special Report No. 17, pp 51–56

Forbes GA, Goodwin SB, Drenth A et al (1998) A global marker database for Phytophthora infestans. Plant Dis 82:811–818

Forrer HR, Hecker A, Stecnblock T et al (2000) Hot water treatment of potato seed tubers- a practicable means to prevent primary foci and delay epidemics of potato late blight. IFOAM 2000: the world grows organic. In: Proceedings of 13th International IFOAM Scientific Conference, Basel, 28–31 Aug, 2000

Forrer HR, Vogelgsang S, Musa T (2017) Botanicals and phosphonate show potential to replace copper for control of potato late blight. J Fungi 3:65. https://doi.org/10.3390/jof3040065

FRAC (n.d.) Fungicide Resistance Action Committee. www.frac.info. Accessed 15 May 2021

Frenkel O, Yermiyahu U, Forbes GA et al (2010) Restriction of potato and tomato late blight development by sub-phytotoxic concentrations of boron. Plant Pathol 59:626–633

Fry EW, Goodwin SB (1997) Re-emergence of potato and tomato late blight in the United States. Plant Dis 81:1349–1357

Fry WE, Birch PRJ, Judelson HS et al (2015) Five reasons to consider *Phytophthora infestans* a re-emerging pathogen. Phytopathology 105:966–981

Fry WE, Myers K, Roberts PD et al (2013) The 2009 late blight pandemic in the eastern United States- causes and results. Plant Dis 97:296–306

Garret KA, Mundt CC (2000) Host diversity can reduce potato late blight severity for focal and general patterns of primary inoculum. Phytopathology 90:1307–1312

Garrett KA, Dendy SP (2001) Cultural practices in potato late blight management. In: Complementing resistance to late blight in the Andes, 13–16 February 2001. International Potato Center, Cochabamba, Bolivia
Gavino PD, Smart CD, Sandrock RW et al (2000) Implications of sexual reproduction for *Phytophthora infestans* in the United States: generation of an aggressive lineage. Plant Dis 84:731–735
Gilet A (1996) Potatoes: a new strain of late blight in France. Cultivar Rueil Malmaison 401:18–21
Gisi U, Sierotzki H (2014) Mechanisms of resistance: oomycete fungicides-phenylamides, quinone outside inhibitors and carboxylic acid amides. In: Ishii H, Hollomon D (eds) Fungicide resistance in plant pathogens: principles and a guide to practical management. Springer, Tokyo, pp 145–174
Goodwin SB, Cohen BA, Fry WE (1994) Panglobal distribution of a single clonal lineage of the Irish potato famine fungus. Proc Natl Acad Sci USA 91:11591–11595
Goodwin SB, Smart CD, Sandrock RW et al (1998) Genetic changes within populations of *Phytophthora infestans* in the United States and Canada during 1994 to 1996: role of migration and recombination. Phytopathology 88:939–949
Goodwin SB, Sujkowski LS, Fry WE (1996) Widespread distribution and probable origin of resistance to metalaxyl in clonal genotypes of *Phytophthora infestans* in the United States and western Canada. Phytopathology 86:793–800
Goss EM, Tabima JF, Cooke DEL et al (2014) The Irish potato famine pathogen *Phytophthora infestans* originated in Central Mexico rather than the Andes. Proc Natl Acad Sci USA 111: 8791–8796
Guenthner JF, Michael KC, Nolte P (2001) The economic impact of potato late blight on US growers. Potato Res 44(2):121–125
Guest DI, Grant BR (1991) The complex action of phosphonates as antifungal agents. Biol Rev 66: 159–187
Guo J, Lee T, Qu DY et al (2009) Phytophthora infestans isolates from northern China show high virulence diversity but low genotypic diversity. Plant Biol 11(1):57–67
Gupta J (2016) Efficacy of biocontrol agents against *Phytophthora infestans* on potato. Int J Eng Sci Comp 6(9):2249–2251
Haas BJ, Kamoun S, Zody MC et al (2009) Genome sequence and analysis of the Irish potato famine pathogen Phytophthora infestans. Nature 461:393–398
Hannukkala AO, Kaukoranta T, Lehtinen A, Rahkonen A (2007) Late blight epidemics on potato in Finland 1933-2002; increased and earlier occurrence of epidemics associated with climate change and lack of rotation. Plant Pathol 56:167–176
Harbaoui K, Hamada W, Li Y et al (2014) Increased difficulties to control late blight in Tunisia are caused by a genetically diverse *Phytophthora infestans* population next to the clonal lineage NA-01. Plant Dis 98(7):898–908
Haverkort AJ, Boonekamp PM, Hutten R et al (2016) Durable late blight resistance in potato through dynamic varieties obtained by cisgenesis: scientific and societal advances in the DuRPh project. Potato Res 59(1):35–66
Haverkort AJ, Struik PC, Visser RGF, Jacobsen E (2009) Applied biotechnology to combat late blight in potato caused by Phytophthora infestans. Potato Res 52:249–264
Hohl HR, Iselin K (1984) Strains of *Phytophthora infestans* from Switzerland with A2 mating type behavior. Trans Br Mycol Soc 83:529–530
Hu CH, Perez FG, Donahoo R et al (2012a) Recent genotypes of *Phytophthora infestans* in the eastern United States reveal clonal populations and reappearance of mefenoxam sensitivity. Plant Dis 96:1323–1330
Hu T, Zhu J, Cao K (2012b) China-blight: a web based DSS on potato late blight management in China. PPO 15:157–164
Huitema E, Bos JIB, Tian M et al (2004) Linking sequence to phenotype in *Phytophthora*-plant interactions. Trends Microbiol 12:193–200

Ivanyuk VG, Konstantinovich (1992) Quoted from: late blight: a threat to global food security. Proceedings of global initiative on late blight conference, March 16–19, 1999, Quito, p 21

Jiang RH, Tripathy S, Govers F, Tyler BM (2008) RXLR effector reservoir in two *Phytophthora* species is dominated by a single rapidly evolving superfamily with more than 700 members. Proc Natl Acad Sci USA 105:4874–4879

Jiang RH, Tyler BM (2012) Mechanisms and evolution of virulence in oomycetes. Annu Rev Phytopathol 50:295–318

Johnson DA, Cummings TF, Hamm PB et al (1997) Potato late blight in the Columbia Basin: an economic analysis of the 1995 epidemic. Plant Dis 81:103–106

Judelson HS, Ah-Fong AMV, Aux G et al (2008) Gene expression profiling during asexual development of the late blight pathogen *Phytophthora infestans* reveals a highly dynamic transcriptome. Mol Plant Microbe Interact 21(4):433–447

Judelson HS, Blanco FA (2005) The spores of *Phytophthora*: weapons of the plant destroyer. Nat Rev Microbiol 3(1):47–58

Kassa B, Sommartya T (2006) Effect of intercropping on potato late blight, *Phytophthora infestans* (Mont.) de Bary development and potato tuber yield in Ethiopia. Kasetsart J (Nat Sci) 40:914–924

Kelaniyangoda DB (2011) Exotic strains of *Phytophthora infestans* in Sri Lanka. Potato J 38(2): 185–187

Kessel G, Moene A, Valkengoed EV et al (2017) Geodata to control potato late blight in Bangladesh. Sixteenth Euroblight workshop, 14-17 May, 2017, Aarhus, Denmark. PPO-Special Rep 18, pp 59–60

Kirk WW, Abu-El Samen FM, Muhinyuza JB et al (2005) Evaluation of potato late blight management utilizing host plant resistance and reduced rates and frequencies of fungicide applications. Crop Prot 24(11):961–970

Knaus BJ, Tabima JF, Davis CE (2016) Genomic analyses of dominant US clonal lineages of *Phytophthora infestans* reveals a shared common ancestry for clonal lineages US11 and US18 and a lack of recently shared ancestry among all other US lineages. Phytopathology 106:1393–1403

Labato MC, Olivieri FP, Gonzalez Altamiranda EA et al (2008) Phosphite compounds reduce disease severity in potato seed tubers and foliage. Eur J Plant Pathol 122:349–358

Le VH, Ngo XT, Brurberg MB, Hermansen A (2008) Characterization of *Phytophthora infestans* population from Vietnam. Australasian Plant Pathol 37:592–599

Leesutthiphonchal W, Judelson HS (2018) A MADS-box transcription factor regulates a central step in sporulation of the oomycete Phytophthora infestans. Mol Microbiol 110(4):562–575

Leifert C, Wilcockson SJ (2005) Blight-MOP: development of a systems approach for the management of late blight (caused by *Phytophthora infestans*) in EU organic potato production. Biol Agric Hortic 23:393–412

Li BJ, Chen QH, Lv X et al (2009) Phenotypic and genotypic characterization of *Phytophthora infestans* isolates from China. J Phytopathol 157(9):558–567

Li Y, Shen H, Zhou Q et al (2017) Changing ploidy as a strategy: the Irish potato famine pathogen shifts ploidy in relation to its sexuality. Mol Plant Microbe Interact 30(1):45–52

Li Y, van der Lee T, Zhu JH et al (2013) Population structure of *Phytophthora infestans* in China-geographic clusters and present of the EU genotype Blue_13. Plant Pathol 62:932–942

Liljeroth E, Bengtsson T, Wiik L, Anderson E (2010) Induced resistance in potato to *Phytophthora infestans* effects of BABA in greenhouse and field tests with different potato varieties. Eur J Plant Pathol 127:171–183

Lima TM, Procopio LC, Brandao FD et al (2011) Biodegradability of bacterial surfactants. Biodegradation 22:585–592

Liu YX, Langemeier MR, Small IM et al (2017) Risk management strategies using precision agriculture technology to manage potato late blight. Agron J 109(2):562–575

Lobato MC, Machinandiasena MF, Tambascio C (2011) Effect of foliar application of phosphate on post-harvest potato tubers. Eur J Plant Pathol 130(2):155–160

Matson MEH, Small IM, Fry WE, Judelson HS (2015) Metalaxyl resistance in *Phytophthora infestans*: assessing role of RPA190 gene and diversity within clonal lineages. Phytopathology 105:1594–1600

Mehi L, Saurabh Y, Sanjeev S et al (2017) Integrated management of late blight of potato. J Appl Nat Sci 9(3):1821–1824

Mehi L, Sorabh C, Sanjay R et al (2021) Evaluation of bio-agents and neem-based products against late blight disease (*Phytophthora infestans*) of potato. Indian Phytopathol. https://doi.org/10.1007/s42360-021-00330-6

Meijer HJG, Schoina C, Wang S et al (2019) Phytophthora infestans small phospholipase D-like proteins elicit plant cell death and promote virulence. Mol Plant Pathol 20(2):180–193

Meinck S, Kolbe H (1999) Control of leaf and tuber blight in ecological potato cultivation. Kartoffelbau 50:172–175

Mizubuti E, Forbes G (2002) Potato late blight IPM in the developing countries. Global initiative in late blight conference. late blight: managing the global threat. March 11–13, 2002. Hamburg, p 93–97

Myint MM (2002) Research on management of potato late blight in Myanmar. In: Late Blight: Managing the Global Threat. GILB2002 conference 11–13 July, 2002 Hamberg, Germany, p 52

Natrass RM (1944) Potato blight East Afr Agric J 10:18–21

Nishimura R, Sato K, Lee WH et al (1999) Distribution of *Phytophthora infestans* in seven Asian countries. Ann Phytopathol Soc Japan 65:163–170

Njoroge AN, Andersson B, Lees AK et al (2019) Genotyping of *Phytophthora infestans* in eastern Africa reveals a dominating invasive European lineage. Phytopathology 109:670–680

Njoroge AW, Tusiime G, Forbes GA, Yuen JE (2016) Displacement of US-1 clonal lineage by a new lineage of *Phytophthora infestans* on potato in Kenya and Uganda. Plant Pathol 65:587–592

Nowicki M, Foolad MR, Nowakowska M, Kozik EU (2012) Potato and tomato late blight caused by *Phytophthora infestans*: an overview of pathology and resistance breeding. Plant Dis 96:4–17

Oh SK, Young C, Lee M et al (2009) In planta expression screens of *Phytophthora infestans* RXLR effectors reveal diverse phenotypes, including activation of the *Solanum bulbocastanum* disease resistance protein Rpi-blb2. Plant Cell 21:2928–2947

Oyarzun PJ, Ordonez ME, Forbes GA, Fry WE (1997) First report of *Phytophthora infestans* A2 mating type in Eucaodr. Plant Dis 81:311

Perez W, Forbes G (2010) Potato late blight: technical manual. International Potato Center (CIP), Lima. http://www.cipotato.org/publications/pdf/005446.pdf

Pilbeam R (2003) Effects of phosphate on disease development and histological responses in *Eucalyptus marginata* infected with *Phytophthora cinnamomi*. Ph.D. thesis. School of Engineering, Murdoch University, p 176

Pilet F, Chacón G, Forbes GA, Andrivon D (2006) Protection of susceptible potato cultivars against late blight in mixtures increases with decreasing disease pressure. Phytopathology 96(7): 777–783

Plata G (1998) Fenotipos de virulencia en Morochata y tipo sexual de apareamiento en Bolivia de Phytophthora infestans que afectan al cultivo de la en la zona de Morochata y determinación del tipo sexual de apareamiento de Phytophthora infestans en Bolivia. Tesis Ing. Agr., Facultad de Ciencias Agrícolas Pecuarias y Forestales "Martín Cárdenas" Universidad Mayor de San Simón. Cochabamba

Pule BB, Meitz JC, Thompson AH et al (2013) Phytophthora infestans populations in central, eastern and southern African countries consist of two major clonal lineages. Plant Pathol 62: 154–165

Raza W, Ghazanfar MU, Sulvian L (2020) Mating type and aggressiveness of Phytophthora infestans (Mont.) de bary in potato-growing areas of Punjab, Pakistan, 2017-2018 and identification of genotype 13_A2 in 2019-20. Potato Res. https://doi.org/10.1007/s11540-020-09467-9

Rekad FZ, Cooke DEL, Puglisi I (2017) Characterization of *Phytophthora infestans* populations in northwestern Algeria during 2008-2014. Fungal Biol 121(5):467–477
Ristaino JB, Hu C (2009) DNA sequence analysis of the late blight pathogen gives clues to the world-wide migration. Acta Hortic 834:27–40
Roy SK, Sharma RC, Trehan SP (2001) Integrated nutrient management by using farmyard manure and fertilizers in potato-sunflower-paddy rice rotation in the Punjab. J Agric Sci 137:271–278
Runno-Paurson E, Fry WE, Myers KL et al (2009) Characterization of *Phytophthora infestans* isolates collected from potato in Estonia during 2002-2003. Eur J Plant Pathol 124(4):565–575
Sanjeev S, Singh BP, Jeevalatha A, Chakrabarti SK (2017) Molecular characterization and pathogenicity of Indian *Phytophthora infestans* isolates reveals no correlation between phenotypes and their geographic origin. J Mycol Plant Pathol 47(1):1–12
Saville A, Graham K, Grünwald N (2015) Fungicide sensitivity of US genotypes of *Phytophthora infestans* (Mont.) de Bary to six oomycete targeted compounds. Plant Dis 99:659–666
Saville A, Ristaino JB (2019) Genetic structure and subclonal variation of extant and recent US lineages of Phytophthora infestans. Phytopathology 109:1614–1627
Saville AC, Martin MD, Ristaino JB (2016) Historic late blight outbreaks caused by a widespread dominant lineage of *Phytophthora infestans* (Mont.) de Bary. PLoS One 11:e0168381
Schepers HTAM (2017) Proceedings of the sixteenth Euroblight workshop. PAGV Special Report No. 18, p 300
Schepers HTAM (ed) (2019) Proceedings of the seventeenth Euroblight workshop. WUR Special Report No 19, p 252
Schick R (1932) Uber das verhalten von *Solanum demissum, Solanum tuberosum* und ihren Basterden gegunber verschiedenen Herkunjten von Phytophthora infestans. Zuechter 4:233–237
Sekhon PS, Amarjit S, Sandeep J, Monica A (2017) Use of web-based decision support system for the management of late blight of potato in Punjab. Indian J Econ Dev 13(3):576–580
Shandil RK, Chakrabarti SK, Singh BP et al (2017) Genotypic background of the recipient plant is crucial for conferring RB gene mediated late blight resistance in potato. BMC Genet 18:22. https://doi.org/10.1186/s12863-017-0490-x
Shanthiyaa V, Saravanakumar D, Rajendran L et al (2013) Use of *Chaetomium globosum* for biocontrol of potato late blight disease. Crop Prot 52:33–38
Sharma S (2019) Scheduling fungicide application for economic management of late blight using indo-Blightcast model. Potato J 46(2):124–131
Sharma S, Guleria A, Lal M et al (2018) Cataloguing variability in *Phytophthora infestans* with respect to ploidy status and response of different polyploids to temperature. Indian Phytopathol 71(1):183–189
Sharma S, Singh BP, Sumit S, Patil VU (2016) Phenotypic and genotypic characterization of *Phytophthora infestans* population of Himachal Pradesh. Indian Phytopathol 69(4):391–395
Singh BP, Roy S, Bhattacharyya SK (1994) Occurrence of A2 mating type of Phytophthora infestans in India. Potato Res 37:227–231
Singh BP, Sharma S (2013) Forecasting of potato late blight. Intern J Innov Hort 2(2):1–11
Sjoholm L, Andersson B, Hogberg N et al (2013) Genotypic diversity and migration patterns of *Phytophthora infestans* in the Nordic countries. Fungal Biol 117(10):722–730
Song J, Bradeen JM, Naess SK et al (2003) Gene *RB* cloned from *Solanum bulbocastanum* confers broad spectrum resistance to potato late blight. Proc Natl Acad Sci U S A 100:9128–9133
Sonica T, Singh BP, Mehi L et al (2014) Screening of novel microorganisms for biosurfactant and biocontrol activity against Phytophthora infestans. J Environ Biol 35(5):893–899
Sweigard JA, Andeassi J, Pember S et al (2014) Discovery of the target site of oxathiapiprolin (DuPont Zorvec disease control). 13th IUPAC congress of pesticide chemistry, San Francisco
Tani S, Yatzkan E, Judelson HS (2004) Multiple pathways regulate the induction of genes during zoosporogenesis in Phytophthora infestans. Mol Plant Microbe Interact 17(3):330–336
Thao HTB, Yamakawa T (2009) Phosphite (phosphorous acid): fungicide, fertilizer or biostimulator? Soil Sci Plant Nutr 55:228–234

Tian YE, Yin JL, Sun JP et al (2016) Population genetic analysis of *Phytophthora infestans* in northwestern China. Plant Pathol 65:17–25

Tiwari JK, Siddappa S, Singh BP et al (2013) Molecular markers for late blight resistance breeding of potato: an update. Plant Breed 132:237–245

Tomar S, Lal M, Khan MA et al (2019) Characterization of glycolipid biosurfactant from *Pseudomonas aeruginosa* PA1 and its efficacy against Phytophthora infestans. J Environ Biol 40(4): 725–730

Tomar S, Singh BP, Khan MA et al (2013) Identification of *Pseudomonas aeruginosa* strain producing biosurfactant with antifungal activity against Phytophthora infestans. Potato J 40: 155–163

Tooley PW, Fry WE, Villarreal Gonzalez MJ (1985) Isozyme characterization of sexual and asexual *Phytophthora infestans* populations. J Hered 76:431–435

Torto TA, Li S, Styer A (2003) EST mining and functional expression assays identify extracellular effector proteins from the plant pathogen Phytophthora. Genome Res 13:1675–1685

Tsai JN, Ann PJ, Wang IT et al (2009) Control of *Phytophthora* late blight of potato and tomato with neutralized phosphorous acid. J Taiwan Agric Res 58:185–195

Turkensteen J, Flier WG, Wanningen R, Mulder A (2000) Production survival and infectivity of oospores of Phytophthora infestans. Plant Pathol 49:688–696

Umetsu N, Shirai Y (2020) Development of novel pesticides in the 21st century. J Pestic Sci 45(2): 54–74

van der Vossen EAG, Sikkema ABL, Hekkert J et al (2003) An ancient R gene from the wild potato species *Solanum bulbocastanum* confers broad-spectrum resistance to *Phytophthora infestans* in cultivated potato and tomato. Plant J 36:867–882

Vargas AM, Quesado-Ocampo LM, Caspedes MC et al (2009) Characterization of *Phytophthora infestans* populations in Colombia: first report of the A2 mating type. Phytopathology 90(1): 82–88

Vorobyeva YV, Gridnev VV, Bashaeva EG et al (1991) On the occurrence of the A2 mating type isolates of Phytophthora infestans (Mont.) de Bary in the USSR. Mikologija i fitopatologija 1991:62–67

Wang J, Fernandez-Pavia SV, Larsen MM et al (2017) High levels of diversity and population structure in the potato late blight pathogen at the Mexico Centre of origin. Mol Ecol 26(4): 1091–1107

Whisson SC, Boevink PC, Moleleki L et al (2007) A translocation signal for delivery of oomycete effector proteins into host plant cells. Nature 450:115–118

Widmark AK, Anderson B, Cassel Lundhagen A et al (2007) Phytophthora infestans in a single field in Southwest Sweden early in spring: symptoms, spatial distribution and genotypic variation. Plant Pathol 56(4):573–579

Yao Y, Li Y, Chen Z et al (2016) Biological control of potato late blight using isolates of Trichoderma. Am J Potato Res 93:33–42

Yoon HS, Hackett JD, Pinto G, Bhattacharya D (2002) The single, ancient origin of chromist plastids. Proc National Acad Sci USA 99:15507–15512

Yoshida K, Schuenemann VJ, Cano LM, Pais M, Mishra B, Sharma R et al (2013) The rise and fall of the *Phytophthora infestans* lineage that triggered the Irish potato famine. eLife 2:e00731. https://doi.org/10.7554/eLife.00731.001

Yuen JE, Andersson B (2013) What is the evidence for sexual reproduction of *Phytophthora infestans* in Europe? Plant Pathol 62:485–491

Zhiming Z, Yuqin L, Shimin T et al (1996) The occurrence of potato late blight pathogen *Phytophthora infestans* A2 mating type in China. J Hebei Agril Univ 19:61–65

Zhu W, Shen L, Fang Z et al (2016) Increased frequency of self-fertile isolates in Phytophthora infestans may attribute to their higher fitness relative to the A1 isolates. Sci Rep 6:29428. https://doi.org/10.1038/srep29428

Role of Genetic Resources in Management of Potato Pests and Diseases

8

Vikas Mangal, Salej Sood, Dalamu, Vinod Kumar, and Vinay Bhardwaj

Abstract

Potato (*Solanum tuberosum* L.) is the fourth most essential staple crop after rice, wheat, and maize. It is a new world crop that was not known to other parts of the world until the sixteenth century. The cultivated potato and its wild relatives belong to the genus *Solanum*, the largest genus with 1500–2000 species. It provides a substantial part of the world's food supply, but vulnerable to many pests and diseases. Many biotic stresses (diseases and pests) affect potato plants in the farming fields. In particular, late blight, potato cyst nematode (*Globodera pallida* and *Globodera rostochiensis*), bacterial wilt (*Ralstonia solanacearum*), common scab (*Streptomyces scabies*), viral diseases (mainly *Potato virus X* (PVX) and *Potato virus Y* (PVY)), Colorado potato beetle, and potato aphids have become the main focuses of resistance breeding. Since the genetic base of present cultivated potatoes is very narrow, landraces and wild relatives are considered to be valuable sources of variation for genetic enhancement and crop improvement because they harbor an enormous amount of genetic diversity. In more than 150 years of potato breeding, wild potato species have made significant contributions to potato improvement in terms of resistance to diseases and insect pests. Numerous wild species have been used for variety development as a parent, because of their disease resistance traits like *S. acaule* (PVX, potato spindle tuber viroid, *Potato leaf roll virus* (PLRV), wart, and *Globodera*), *S. demissum* (late blight and PLRV), *S. chacoense* (*potato virus A* (PVA), PVY, late blight, Colorado beetle, tuber moth), *S. spegazzinii* (*Fusarium*, wart, *Globodera*), *S. stoloniferum* (PVA, PVY), and *S. vernei* (*Globodera*). *S. microdontum*, *S. phureja*, *S. sparsipilum*, *S. commersonii*, *S. maglia*,

V. Mangal · S. Sood · Dalamu · V. Kumar · V. Bhardwaj (✉)
Division of Crop Improvement, ICAR-Central Potato Research Institute, Shimla, Himachal Pradesh, India

S. Kumar Chakrabarti et al. (eds.), *Sustainable Management of Potato Pests and Diseases*, https://doi.org/10.1007/978-981-16-7695-6_8

S. tuberosum subsp. andigena, and *S. verrucosum* have also been used as breeding materials. So far, auspicious genetic resources for disease and pest resistance exist over the wide range of primitive cultivars and wild relatives of potato. The potential for using available genetic resources in resistance breeding program depends on their crossability with the cultivated potato (*S. tuberosum*).

Keywords

Biotic stress · Disease resistance · Wild species · *Tuberosum* · Andigena · *R1* gene · Germplasm · *Solanum phureja*

8.1 Introduction

Potato (*Solanum tuberosum* L.) is one of the important staple crops in the world. It is a new world crop that was not known to other parts of the world until the sixteenth century. Over the next six centuries, potato cultivation expanded from the Andes highlands in South America, (center of origin) to other parts of the world. The basic number of chromosomes of cultivated potato is $n = 12$ and varies from diploid ($2n = 2x = 24$), triploid ($2n = 3x = 36$), tetraploid ($2n = 4x = 48$) to pentaploid ($2n = 5x = 60$). Potato is currently the fourth most essential staple crop after rice, wheat, and maize (De Haan and Rodriguez 2016) and provides a substantial part of the world's food supply, but it is vulnerable to many pests and diseases. The Solanaceae family consists of 3000–4000 species in around 90 genera. The cultivated potato and its wild relatives belong to the genus *Solanum*, the largest genus with 1500–2000 species. In this genus, more than a thousand species have been identified. Generally, tuber-bearing *Solanum* species are included in the Petota section, and this section is divided into two subsections, Potatoe and Estolonifera (Hawkes 1990). The subsection Potatoe includes all tuber-bearing potatoes, also containing common potato (*Solanum tuberosum*, series *tuberosa*). Two non-tuber-bearing series, Etuberosa and Juglandifolia, were enlisted in subsection Estolonifera, but the classification of cultivated potatoes is still under discussion. Bukasov (1971) and Lechnovich (1971) recognized 21 species. Hawkes (1990) pointed 7 cultivated potato species, while Ochoa (1999) only identified 9 species and 141 intraspecific taxa. Spooner et al. (2007) proposed to reclassify cultivated potatoes into four species: (1) *S. tuberosum*, (2) *S. ajanhuiri* (diploid), (3) *S. juzepczukii* (triploid), and (4) *S. curtilobum* (pentaploid). However, a sensible concern is that many web searchable databases of potato germplasm resources in the world still use the classification and description of Hawkes (1990). Hawkes (1990) separated *S. tuberosum* into two subspecies: tuberosum and andigena, which are tetraploid ($2n = 4x = 48$). The *S. tuberosum* is the cultivated potato used worldwide, while the andigena subspecies is limited to Central and South America.

Cultivated potatoes can be classified as native varieties, landraces that were still grown in South America today, or improved varieties, grown worldwide. Potato landraces are highly diverse, with various tuber shapes, skins, and flesh colors. Since the genetic base of modern cultivated potatoes is very narrow (Jansky et al. 2013, 2016), landraces and wild relatives are considered to be valuable sources of variation for genetic enhancement and crop improvement because they harbor a large amount of genetic diversity. Many expeditions collected important potato genetic resources and kept them in the gene bank. Their effective collection, characterization, conservation, and use will be an important key for adaptation under future climate change scenarios. Genetic resources are strategic resources for sustainable crop production. Their effective protection and use are essential to continue to feed the growing world population. Gene banks play a key role in the conservation and distribution of germplasm for crop improvement and sustainable food production research. They are the starting point of any plant improvement program. Globally, 20% of potato germplasm is kept in medium-term storage, 11% is kept in short-term storage for immediate use, and 69% is kept under unknown storage conditions.

Potato germplasm, including wild and cultivated potatoes, is conserved in gene banks throughout the world. The report classified the collection into four categories: (1) wild relatives, (2) native cultivars, (3) modern cultivars of the common potato (*Solanum tuberosum* subsp. tuberosum), and (4) other germplasm (e.g., interspecific hybrids, breeding clones, etc.). Many gene banks around the world are maintaining potato genetic resources including wild types. The current accession number of potato in these gene banks is as follows (Machida-Hirano and Niino 2017):

- International Potato Center (CIP), Peru, 6768
- Leibniz Institute of Plant Genetics and Crop Plant Research (IPK)/The Groß Lüsewitz Potato Collection (GLKS), Germany, 6124
- Northern Region 6 (NR6), USA, 5808
- Vavilov Institute of Plant Industry, Russia, 9000
- Central Potato Research Institute (CPRI), India, 4552
- Potato Research Institute, Czechoslovakia, 2225

In India at CPRI, Shimla, the potato germplasm is being maintained in the field gene bank, true seeds, and in vitro conservation. Potato landraces and their wild relatives are diverse sources of genetic variation owing to their great range of ecological adaptation and wide geographical distribution which could be a source for the development of disease and insect-resistant variety. In more than 150 years of potato breeding, wild potato species have made significant contributions to potato improvement in terms of resistance to diseases and insect pests (Bradshaw and Ramsay 2005; Bradshaw 2009). The potato is considered to be a crop that primarily uses wild relatives for crop improvement. Different wild potato species and traits identified in them are mentioned in Tables 8.1 and 8.2. The huge diversity within wild species and even in germplasm requires careful screening to identify individual clones with resistance genes.

Table 8.1 Solanum species and reported useful traits for insect and pest resistance

Name of the species	Fungus resistance			Bacterial resistance		Virus resistance				Insect resistance		Nematode resistance	
	Late blight	Wart	Common scab	Bacterial wilt	Soft rot: black leg	Potato virus X	Potato virus Y	Potato leaf roll virus	Spindle tuber viroid	Colorado beetle	aphids	Potato cyst nematode	Root knot nematode
S. acaule		**		*		**	*	**	*	*		*	*
S. andreanum											*		
S. berthaultii	*	*				*			*	*	*	*	*
S. boliviense												*	*
S. brevicaule		*						*		*		*	*
S. brevidens							*	*					
S. bulbocastanum	**	*		*	*						**	*	*
S. candolleanum		*					*			*	*		*
S. cardiophyllum	**	*					*			*	*	*	*
S. chacoense	*	*	*	*	*	*	**			*			*
S. chomatophilum										*	*		
S. clarum				*						*	*	*	
S. colombianum	*												
S. commersonii	*	*	*		*					*			*
S. contumazaense													
S. demissum	**	*			*	*	**	*		*			*
S. edinense	*									*			
S. guerreroense									*				
S. hjertingii					*								
S. infundibuliforme		*						*		*	*		
S. iopetalum		*				*							*
S. jamesii		*						*		*			*
S. kurtzianum		*				*		*		*		*	*
S. lignicaule											*		

S. medians											*		
S. microdontum	*	*		*	*			*			*		*
S. morelliforme	*									*			
S. oxycarpum	*									*			
S. phureja				*									
S. pinnatisectum	**	*		*	*		*			*			*
S. polyadenium	*	*		*				*		*	*	*	*
S. raphanifolium		*						*					*
S. sparsipilum												*	
S. stenotomum				*									
S. stipuloideum	*	*								*	*	*	*
S. stoloniferum	**	*			*		**	*		*	*	*	*
S. tarijense										*			
S. trifidum	*	*								*	*		
S. vernei	*			*						*		*	*
S. verrucosum	*					*				*			*
S. violaceimarmoratum										*			
S. tuberosum Andigenum group	*	*	*	*	*	**	*	*				*	*
S. curtilobum						*							*
S. juzepczukii						*							

Adapted from Sinden et al. (1986), Kasai et al. (2000), Fock et al. (2001), Hosaka et al. (2001), Song et al. (2003), Bradshaw and Ramsay (2005), Kim-Lee et al. (2005), Bradshaw et al. (2006b), Brown et al. (2006), Dalamu et al. (2012), Chen et al. (2013), Spooner et al. (2014), Machida-Hirano (2015), Sood et al. (2017).
* species include individuals who possess resistance, tolerance, or good quality. The resistance/tolerance mentioned in the table is based on the information available for the particular species; ** species include individuals who possess immunity, high resistance, high tolerance, or high quality.

Table 8.2 Potato somatic hybrids developed using different wild species for disease and pest resistance

Target trait (disease/pest)	Somatic hybrids	Reference
Late blight	*S. cardiophyllum* (+) *S. tuberosum*	Chandel et al. (2015)
	S. pinnatisectum (+) *S. tuberosum*	Singh et al. (2016), Tiwari et al. (2016)
	S. × *michoacanum* (+) *S. tuberosum*	Smyda et al. (2013)
	S. chacoense (+) *S. tuberosum*	Rakosy-Tican and Aurori (2015)
	S. bulbocastanum (+) *S. tuberosum*	Iovene et al. (2012)
	S. villosum (+) *S. tuberosum*	Tarwacka et al. (2013)
Bacterial wilt	*Solatium brevidens* (+) *S. tuberosum*	Austin et al. (2021)
	S. chacoense (+) *S. tuberosum*	Guo et al. (2010), Chen et al. (2016)
Potato virus Y (PVY)	*S. berthaultii* (+) *S. tuberosum*	Nouri-Ellouz et al. (2016)
	S. etuberosum (+) *S. tuberosum*	Tiwari et al. (2010, 2015)
Potato virus X (PVX)	S. acaule (+) S. tuberosum	Yamada et al. (1997)
Potato leaf roll virus (PLRV)	*S. etuberosum* (+) *S. tuberosum* haploid x *S. berthaultii*	Webber Iii et al. (2017)
Common scab	*S. brevidens* (+) *S. tuberosum*	Ahn and Park (2013)
Colorado potato beetle	*S. cardiophyllum* (+) *S. tuberosum*	Thieme et al. (2010)
	S. chacoense (+) *S. tuberosum*	Rakosy-Tican and Aurori (2015), Molnár et al. (2017)
Columbia root-knot nematode	*S. bulbocastanum* (+) *S. hougasii*	Brown et al. (2006)
Potato cyst nematode (PCN)	*S. tuberosum* (+) *S. sanctae-rosae*	Harding and Millam (2000)

8.2 Uses of Wild Species in Potato Breeding

The value of germplasm depends on its genetic diversity, availability, and practicality. In this sense, this crop stands out among all other crops (Bamberg and Del Rio 2005). For more than 100 years, wild potatoes have been used for disease resistance in breeding programs. Numerous wild species have been used for variety development as parentage in Europe and North America, because of their disease resistance traits like *S. acaule* (potato virus X, potato spindle tuber viroid (PSTV), PLRV, wart, and *Globodera*), *S. demissum* (late blight and potato leaf roll virus (PLRV)), *S. chacoense* (potato virus A, potato virus Y, late blight, Colorado beetle, tuber moth), *S. spegazzinii* (*Fusarium*, wart, *Globodera*), *S. stoloniferum* (potato virus A, potato virus Y), and *S. vernei* (*Globodera*). *S. microdontum*, *S. phureja*, *S. sparsipilum*, *S. commersonii*, *S. maglia*, *S. tuberosum* subsp. andigena, and *S. verrucosum* have also been used as breeding materials.

Plant diseases caused by fungi, viruses, bacteria, and nematodes cause massive yield losses annually (Li et al. 2020). A recent survey disclosed that worldwide crop losses in potato caused by pathogens and pests range from 8.1% to 21.0% (Savary et al. 2019). Potato late blight caused by pathogen *Phytophthora infestans* caused the Irish famine in the 1840s and is still widespread in most potato-producing areas. The disease causes annual economic losses of up to 5 billion US dollars worldwide (Judelson and Blanco 2005) and nearly 16% of production losses (Haverkort et al. 2009). Cultivating disease-resistant varieties and frequent use of fungicides are the only control measures. Wild potato relatives carry biotic and abiotic stress resistance genes that have not been found in cultivated potatoes (Machida-Hirano 2015; Machida-Hirano and Niino 2017). The introgression of genes from wild species to cultivated backgrounds began with crosses between *S. tuberosum*, *S. commersonii*, and *S. maglia* between 1824 and 1909 but was unsuccessful. This may be due to the highly sterile hybrid triploid clone. Later on, late blight pathogen *Phytophthora infestans*-resistant hybrids of potato with *Solanum demissum* and *S. edinense* were identified. Late blight resistance breeding focuses on the use of the main dominant 11 R gene of *S. demissum*, but potato breeding is complexed due to its autotetraploid genome, asexual propagation, and breeding principles and practices that are quite different from those employed for the majority of diploid (or allopolyploid), seed-propagated crops.

The first commercially successful example of using wild potatoes in variety development was the cultivar Pentland Ace (with the resistance gene *R3*), which was produced by crossing *S. phureja* and *S. demissum* and backcrossed with *S. tuberosum* three times (Bradshaw et al. 2006b). In due course of time cultivar, Pentland Dell (having resistances gene *R1*, *R2*, and *R3*) was released in 1963. This marked the beginning of gene introgression breeding, followed by gene introgression of resistance genes to viruses, cyst nematodes, and other traits also. However, the overall use of wild relatives to incorporate traits into new varieties is very limited (Bethke et al. 2017). Genetic diversity, availability, and usefulness of germplasm have always been the driving force for the integration of wild genes into cultivated genes (Bamberg and Rio 2005). This happened because potato has an extremely large secondary gene pool and rich genetic reservoir (Mihovilovich et al. 2015) consisting of related wild species which provides a unique, rich, and diverse source of genetic variation.

Potato plants are affected by many biotic stresses (diseases and pests) in the farming fields. In particular, late blight, potato cyst nematode (*Globodera pallida* and *Globodera rostochiensis*), bacterial wilt (*Ralstonia solanacearum*), common scab (*Streptomyces scabies*), and viral diseases (mainly PVX and PVY) have become the main focuses of resistance breeding (Asano and Tamiya 2016). These plant pathogens represent a continuous and serious threat towards achieving food security goals and significantly reduce crop yields (Armstrong et al. 2019).

8.3 Genetic Resources for Disease and Pest Management

8.3.1 Late Blight of Potato

Late blight (LB) caused by the pathogen *Phytophthora infestans* is considered to be the most devastating disease of cultivated potato (Chen et al. 2017; Karki et al. 2021). This pathogenic *Phytophthora infestans* is a specialized airborne pathogen that infects potato leaves, fruits, and stems. Airborne transmission, aggressiveness, and extraordinary adaptability make it a pathogen with high evolutionary potential. This pathogen is a diploid, heterothallic fungus with two mating types (A1 and A2). Many wild species are the sources of late blight resistance that has been introgressed into the cultivated potato (Pavek and Corsini 2001). Many R genes have been identified (Table 8.3) that confer late blight resistance in various potato species, and many of these R genes have been used in potato breeding. Resistance genes (R) have been searched for species within the Petota section of *Solanum* that originated in Mexico (Vleeshouwers et al. 2011b) and also a proposed origin of the pathogen *P. infestans*. The discovery of resistance genes, effector proteins, and their specific mode of action are being developed for durable resistance breeding.

LB is a major potato disease worldwide. Even for the most resistant varieties, the use of fungicides to control LB is still the norm. *Solanum demissum* was the first wild potato species that was successfully used in the breeding of late blight-resistant cultivars. So far, 11 major resistance genes (Table 8.3) have been identified in *Solanum demissum* species (Huang et al. 2005; Hein et al. 2009), conferring dominant resistance to this pathogen. *R1*, *R2*, *R3*, *R4*, and *R10* have been introgressed into cultivated potato (Vleeshouwers et al. 2011a; Rodewald and Trognitz 2013). However, their resistance is race-specific, and for each R gene, it has been quickly overcome due to the rapid adaptation of the pathogen. Breeding programs have also used broad-spectrum and durable resistance genes derived from another wild species, such as *S. bulbocastanum* ($2n = 2x = 24$) which contains several R genes. The R genes identified (*Rpi-blb1* to *Rpi-blb3*, *Rpi-bt1*, *Rpi-apbt*) (Lokossou et al. 2009) likely target effector molecules that are

widespread among different pathogen clonal lineages (Rodewald and Trognitz 2013). So far, some potentially more long-lasting, broad-spectrum R genes have been identified and cloned (Table 8.3). Thus in 2004, the first potato cultivar Biogold (Van Rijn BV) was released which carried late blight resistance from *S. bulbocastanum* via the ABPT bridging clones produced by Hermsen and Ramanna in the early 1970s (Huang et al. 2005). One of the most successful cultivars to be introduced into China by CIP, CIP-24, had *S. acaule*, *S. stoloniferum*, and *S. demissum* in its pedigree (Ortiz 2001). At ICAR-CPRI, Shimla breeding for late blight resistance was initiated using *S. verrucosum* in the year 1975 with *S. phureja* acting as a bridge species to enhance crossability. Parental lines sharing gene pool from wild and semi-cultivated *Solanum* species like *S. demissum*, *S. acaule*, *S. microdontum*, *S. chacoence*, *S. hougassi*, and *S. stoloniferum* were developed. In India, the late blight screening of germplasm and advanced potato hybrids is routinely done under laboratory conditions through detached leaf assay and tuber

Table 8.3 Mapping of different late blight (LB) resistance gene in different wild species

Resistant gene/locus	Species	Reference
R1, R2, R3 (*R3a* and *R3b*), *R4, R5, R6, R7, R8, R9, R10, R11, Rpi-dmsf1*	*S. demissum*	Black et al. (1953), Huang et al. (2005), Bradshaw et al. (2006a), Hein et al. (2009)
RB/Rpi-blb1, Rpi-blb2, Rpi-blb3, Rpi-bt1, Rpi-apbt	*S. bulbocastanum*	Song et al. (2003), Van Der Vossen et al. (2003, 2005), Park et al. (2005a, b), Lokossou et al. (2009), Oosumi et al. (2009)
Rpi-amr1, Rpi-amr3	*S. americanum*	Witek et al. (2016, 2020)
Rpi1, Rpi2	*S. pinnatisectum*	Kuhl et al. (2001), Yang et al. (2017)
Rpi-ver1	*S. verrucosum*	Chen et al. (2018)
Rpi-mch1	*S. michoacanum*	Śliwka et al. (2012b)
Rpi-rzc1	*S. ruizceballosii*	Śliwka et al. (2012a)
Rpi-qum1	*S. circaeifolium*	Verzaux et al. (2012)
Rpi-edn1.1, Rpi-edn2	*S. edinense*	De vetten et al. (2011), Champouret (2010)
Rpi-sto1, Rpi-sto 2	*S. stoloniferum*	Vleeshouwers et al. (2008), Wang et al. (2008)
Rpi-pta1, Rpi-pta2	*S. papita*	Vleeshouwers et al. (2008), Wang et al. (2008), Champouret (2010)
Rpi-vnt1.1, Rpi-vnt1.2, Rpi-vnt1.3	*S. venturii*	Foster et al. (2009), Pel et al. (2009)
Rpi-ber, Rpi-ber1, Rpi-ber1	*S. berthaultii*	Ewing et al. (2000), Rauscher et al. (2006), Park et al. (2009)
Rpi-cap1	*S. capsicibaccatum*	Jacobs et al. (2010)
Rpi-mcd1	*S. microdontum*	Sandbrink et al. (2000), Tan et al. (2008)
Rpi-moc1	*S. mochiquense*	Smilde et al. (2005)
Rpi-pcs	*S. paucissectum*	Villamon et al. (2005)
Rpi-phu1	*S. phureja*	Śliwka et al. (2006)
Rpi-snk1.1, Rpi-snk1.2	*S. schenckii*	Champouret (2010), Jacobs et al. (2010)
Rpi-plt1	*S. polytrichon*	Wang et al. (2008)
Rpi-dlc1	*S. dulcamara*	Golas et al. (2010)
Rpi-chc1	*S. chacoense*	Vossen et al. (2009)
Rpi-hjt1.1, Rpi-hjt1.2, Rpi-hjt1.3	*S. hjertingii*	Champouret (2010)
Rpi-bst1	*S. brachistotrichum*	Hein et al. (2009)

slice or whole tuber methods and under natural epiphytotic conditions in hills (Fig. 8.1).

In India, wild species *S. microdontum* and *S. verrucosum* have been used as donors of durable resistance to late blight. Varieties like Kufri Jeevan, Kufri Khasigaro, Kufri Jyoti, Kufri Naveen, Kufri Muthu, Kufri Neelamani, Kufri Badshah, Kufri Megha, and Kufri Jawahar were developed by incorporating

Fig. 8.1 Screening of genotypes against late blight under natural epiphytotic conditions (**a**) Susceptible (**b**) Resistant

resistance genes in indigenously developed material, through exotic material from UK having *S. demissum* genes. The genetic base of Indian potato varieties can be traced to just 49 ancestors of which ten from the UK account for 41% of the total genomic constitution (Gopal and Oyama 2005). The most common ancestors were two clones from the UK, 2814a1, and 3069d4, which can be traced back to the cross between *S. rybinii* (a variant of *S. phureja*) and *S. demissum*. ICAR-CPRI, Shimla has developed two interspecific somatic hybrids using androgenic dihaploid clone 'C-13' (regenerated from *S. tuberosum* cv. Kufri Chipsona-2) and wild species *S. pinnatisectum* (1EBN) and *S. cardiophyllum* (1EBN) for late blight resistance. Furthermore, somatic fusion has allowed the production of hybrid between *S. tuberosum* (tetraploid, 4EBN) and diploid 1EBN species, for example, the non-tuber-bearing species *S. brevidens* that has tuber early blight and soft rot resistances (Tek et al. 2004) and *S. bulbocastanum* that has a major gene (*Rpi-blb1* to *Rpi-blb3*, *Rpi-bt1*, *Rpi-apbt*) for broad-spectrum and durable resistance to late blight (Naess et al. 2001). The transgenic Rx-mediated resistance was indistinguishable from the Rx-mediated phenotype in cultivar Cara. The *R1* resistance gene has been cloned and introduced into the susceptible cultivar Desiree and shown to give a typical hypersensitive response, similar to the resistant line hosting *R1* (Ballvora et al. 2002), but *R1* has failed to give durable resistance. The *RB* and *Rpi-blb1* genes (which are allelic) from *S. bulbocastanum* have been cloned and introduced into the susceptible cultivars Katahdin and Impala, respectively, and shown to confer broad-spectrum resistance to late blight (Song et al. 2003; Van Der Vossen et al. 2003).

Several late blight resistance genes were mapped in different wild species for late blight resistance which are mentioned in Table 8.3.

Dr. Black, at the Scottish Plant Breeding Station (SPBS), crossed *S. demissum* (6x) with the Alness (4x) and secured a pentaploid (5x) clone, where he was able to introgress major dominant R-genes. Later, he secured a few artificial tetraploid seedlings by hybridizing *S. phureja* (2x) with *S. demissum* (6x) and used them in genetic studies and his breeding program (Black 1970). Bachmann-Pfabe et al. (2019) after screening wild population tuber blight resistance were identified in accessions of less investigated species such as *S. acaule*, *S. fendleri*, *S. trifidum*, *S. megistracrolobum*, *S. polytrichon*, *S. jamesii*, and *S. tarnii*.

8.3.2 Viruses

The complex of viruses leads to the degeneration of vegetative plant material and greater yield losses. Many viruses are common in cultivated potato. Nearly 37 viruses naturally infect potato and about one-third of them cause economically important diseases. Potato leaf roll virus (PLRV), potato virus X (PVX), potato virus Y (PVY), potato virus A (PVA), potato virus S (PVS), and potato spindle tuber viroid (PSTVd), especially in warmer climate countries, cause serious damage to potato crop. Seed-borne viral diseases caused by PLRV and PVY lead to more than 50% yield reduction due to seed degeneration. The effective means to control the virus is through resistance breeding. Wild potato species possess multiple genes that confer resistance to different potato viruses, some of which have been introgressed into commercial cultivars through conventional breeding and marker-assisted selection (MAS) (Barker and Dale 2006).

Salaman and Pethybridge (1921) recognized that the degradation of potato crops in successive vegetative generations was the result of viral infections and led to targeted breeding for resistance starting in the 1930s (Solomon-Blackburn and Barker 2001). This involved the screening of germplasm collections for sources of resistance. Genes conferring resistance in a non-specific manner were preferred, and the following proved particularly useful: Rx genes for resistance to PVX from andigena and *S. acaule*; Ny genes for hypersensitive resistance to PVY from *S. demissum* and *S. microdontum*, both in a background of field resistance from *S. phureja*, and from *S. chacoense*; and Ry gene for resistance to PVY from *S. stoloniferum*. The genes have provided durable resistance. Most viruses that infect potato crops, such as PVY, are transmitted by aphids in a non-persistent manner, mainly by species that do not colonize potato plants. PVY, one of the important potato diseases, can reduce yield by 80%. Wild species are rich in genes for hypersensitive resistance (HR) to potato viruses. Many genes for HR against viruses were introduced in the cultivated potato gene pool from wild species and used in resistance breeding (Zimnoch-Guzowska et al. 2013; Valkonen 2015). The *Ryadg* gene is highly resistant to all known PVY strains and has been mapped and cloned from *S. andigena* (Hämäläinen et al. 1998). The major gene resistances to PVY have proved durable, and the hybrid parent MPI 61.303/34 from the Max Planck Institute

Fig. 8.2 Screening of genotypes against *Apical leaf curl New Delhi virus* (ALCNDV) (**A**) Susceptible (Kufri Pukhraj) (**B**) Resistant (Kufri Karan)

in Koln has been widely used in European breeding as the source of the Ry gene for resistance from *S. stoloniferum*. Several genes for resistance to PVY have been found, and markers have also been developed for their detection (Fulladolsa et al. 2015). These include molecular markers ADG2 BbvI, RYSC4, and RYSC3 for detection of Ry_{adg} from andigena, on chromosome XI (Kasai et al. 2000); 38–530 and CT220 for Ry_{chc} from *Solanum chacoense* on chromosome IX (Sato et al. 2006); and GP122, YES3-3B, and STM003 for Rysto from *S. stoloniferum*, on chromosome XII (Song and Schwarzfischer 2008; Valkonen et al. 2008). ICAR-CPRI, Shimla has produced a triplex clone YY-6/3 C-11 carrying extreme resistance gene to Ryadg in triplex dose to PVY using marker-assisted selection (MAS) (Kaushik et al. 2013). In India, the screening of potato germplasm against viruses like PVA, PVM, PVS, PVY, PVX, PLRV, and ALCNDV is done under lab and field conditions (Fig. 8.2). Likewise, the major gene resistances to PVX have proved durable despite the occurrence of resistance-breaking strains.

8.3.3 Bacterial Wilt

Bacterial wilt (BW) caused by pathogen *Ralstonia solanacearum* is considered one of the most crucial potato diseases in tropical hot and humid regions. Resistance to

Fig. 8.3 In vitro evaluation of potato germplasm for resistance to *Ralstonia solanacearum* (*a* = 04 days after inoculation; *b* = 06 days after inoculation

BW has been found in many wild species mentioned in Table 8.1. Chemical controls are not available for bacterial wilt. Five to seven years of crop rotation with non-susceptible crops are effective methods for managing soil-borne inoculants. Somatic hybrids (Fock et al. 2000; Chen et al. 2013) or protoplast fusion (Fock et al. 2001; Kim-Lee et al. 2005) techniques were used for the transfer of resistance genes. Some resistance is available with a *Solanum demissum* and *S. phureja* background.

The result of the introgression from wild species, *Solanum brevidens*, was a high-yielding clone, C75–5 + 297, with resistances to tuber soft rot. Using both cytogenetic and molecular approaches, Tek et al. (2004) showed that C75–5 + 297 had 47 chromosomes, including four copies of chromosome 8, three from potato, and one from *Solanum brevidens* which was the only portion of the wild species genome present. In contrast, Barone et al. (2001) did find 48 chromosomes and evidence of recombination between *S. commersonii* (2x 1EBN) and *Solanum tuberosum* chromosomes in their marker-assisted introgression of bacterial wilt resistance. Recently (unpublished data), ICAR-CPRI, Shimla has identified few cultivated potato genotypes tolerant to bacterial wilt (up to 15 DAI) under in vitro studies (Fig. 8.3).

8.3.4 Common Scab

Common scab caused by soil-borne *Streptomyces* spp. bacteria is a serious problem for potato growers. Common scab mostly affects the marketable yield and

Fig. 8.4 Screening of potato varieties/germplasm for sensitivity to thaxtomins produced by *Streptomyces scabies*

occasionally total yield of potato. On potato, common scab symptoms range from a superficial/raised brown spots on the skin to dark pits which extends several millimeters into the tuber tissue (Navarro et al. 2015). Wild species of solanum have made significant contributions to variety development (Maxted et al. 2012) because they provide a potential source of resistance for common scab (Braun et al. 2017). Hosaka et al. (2000) screened 100 accessions of 18 diploid wild potato species and chose several resistant/tolerant genotypes. The wild species *S. canasense*, *S. bukasovii*, *S. multidissectum*, and *S. chacoense* produced the resistant clones. Resistance has also been detected in the cultivated diploid *S. tuberosum* Phureja Group and dihaploids of *S. tuberosum*, in addition to some historic russet cultivars, like Russet Burbank and Russet Rural. Resistance has also been reported in the French cultivar Belle de Fontenay and Dutch cultivars Monalisa and Sirtema (Pasco et al. 2005). New cultivars with resistance include GemStar Russet, Alta Crown, Freedom Russet, Kalkaska, Liberator, Megachip, Marcy, McBride, Millennium Russet, Owyhee Russet, Premier Russet, Teton Russet, Summit Russet, and Western Russet (Yilma et al. 2012; Novy et al. 2014). Germplasm and population offspring evaluation are also important to identify and improve resistance in varieties available to growers. ICAR-CPRI, Shimla has identified few cultivated potato genotypes tolerant to common scab under laboratory conditions (Fig. 8.4).

8.3.5 Nematodes

Nematodes are an important potato pest worldwide, with an average loss of 10% in infected areas. The pest also contributes, with *Verticillium* wilt, to a syndrome called early dying which is more serious than either of the separate diseases (Evans and Brodie 1980). The use of new resistant varieties derived from *S. tuberosum* ssp. *andigena* and strict quarantine practices can effectively control this pathotype. Other races endemic throughout Latin America and Europe are controlled with varieties deriving resistance from *S. tuberosum* ssp. *andigena* and *S. vernei* (Brodie et al. 1991).

Potato cyst nematode is a major constraint in potato production. *Globodera pallida* and *G. rostochiensis*, collectively known as potato cyst nematodes (PCN), have a narrow host range, reproducing primarily on solanaceous crops (Whitworth et al. 2018). PCN has a highly restricted host range, and hatching happens only in the presence of roots diffusates of a suitable host. Under unmanaged conditions, PCN can cause up to 75% yield loss in potato. The natural resistance and use of nematicides are the control options for PCN. The wild species mostly exploited in PCN resistance breeding were *S. tuberosum* ssp. *andigena*, *S. sparsipilum*, *S. vernei*, *S. spegazzinii*, and *S. gourlayi*. The commonly used *H1* gene from andigena potato (Bakker et al. 2004) has remained effective against *Globodera rostochiensis* in Britain because Ro1 was the main pathotype, but the widespread deployment of *H1* gene and extensive use of *G. rostochiensis*-resistant cultivars have encouraged the spread of another species *G. pallida* (Bradshaw and Ramsay 2005). Monogenic dominant *H1* gene was effective against pathotypes Ro1 and Ro4 of *Globodera rostochiensis*. PCN populations capable of overcoming the *H1* resistance gene were soon found due to emerging of new pathotype Pa2/3 of *G. pallida*. Quantitative resistance to both PCN (*G. pallida* and *G. rostochiensis*) was found in a diploid wild species, *S. vernei*. Colchicine treatment of *S. vernei* species produced tetraploid plants which were crossed with cultivated potatoes in 1957 and 1958. These resulting hybrids were again intercrossed and outcrossed to other cultivars/variety, and after PCN screening, cultivars Morag and Glenna were released for cultivation in 1985 and 1987, respectively. Sant'e from the Netherlands and Nadine from Caithness Breeders are other examples of cultivars with PCN resistance from *S. vernei*.

Another source of resistance that was effectively incorporated into the European potato cultivars was quantitative resistance to *G. pallida*, from andigena germplasm (CPC 2802), and is known as *H3*. In the year 1969, after selfing CPC 2802, the resulting genotype was crossed with Maris Piper, followed by three backcrosses to Tuberosum to give clones 12601ab1, 14069a4, and 12674ab1. Paal et al. (2004) have cloned the Gro1 gene for resistance to *G. rostochiensis* from *S. spegazzinii* and introduced it into the susceptible cultivar Desire and shown that it confers resistance to pathotype Ro1. Along with the above genes, several other genes or QTLs are mapped which were involved in PCN resistance (Table 8.4).

Turner (1989) screened 35 species for new sources of resistance to *G. rostochiensis* and *G. pallida* and found resistance to Ro1–5 and Pa1–3 in

Table 8.4 Genes and QTLs involved in PCN resistance

Genes/ QTLs	Nematode species	Potato species	Resistant to pathotype	References
H1	*G. rostochiensis*	*S. tuberosum* ssp. *andigena*	Ro1 and Ro 4	Ellenby (1948), Bakker et al. (2004)
H2	*G. pallida*	*S. multidissectum*	Pa1 (highly resistant), Pa2/3 (moderately resistant)	Dunnett (1961), Strachan et al. (2019)
H3, Gpa2	*G. pallida*	*S. tuberosum* ssp. *andigena*	Pa2/3	Van Der Voort et al. (1997), Bryan et al. (2004)
Gpa5, Gpa6	*G. pallida*	*S. vernei*	Pa2/3	Van Der Voort et al. (2000)
Gro 6	*G. rostochiensis*	*S. vernei*	Ro1 and Ro 4	Jacobs et al. (1996)
Gro1–4, Gro1.2, Gro1.3, Gro1.4	*G. rostochiensis*	*S. spegazzinii*	Ro1	Kreike et al. (1993, 1996), Paal et al. (2004)
$GpaV^{s}_{sp}$, $GpaXI^{s}_{sp}$	*G. pallida*	*S. sparsipilum*	Pa2/3	Caromel et al. (2005)
Gpa, GpaM1, GpaM2, GpaM3	*G. pallida*	*S. spegazzinii*	Pa2/3	Kreike et al. (1994), Caromel et al. (2003)
Grp1	*G. rostochiensis, G. pallida*	*S. tuberosum, S. oplocense, S. vernei,* and *S. tuberosum* ssp. *andigena*	Ro5, Pa2/3	Van Der Voort et al. (1998)
$GpaXI^{l}_{tar}$	*G. pallida*	*S. tarijense*	Pa3	Adillah Tan et al. (2009)
Ro2_A, Ro2_B	*G. rostochiensis*	*S. tuberosum* ssp. *andigena* and *S. vernei*	Ro2	Park et al. (2019)
Pa2/3_A, Pa2/3_B, GpaIV	*G. pallida*	*S. tuberosum* ssp. *andigena* and *S. vernei*	Pa2/3	Bradshaw et al. (1998), Park et al. (2019)

S. kurtzianum, *S. stenotomum*, *S. sparsipilum*, and *S. stenotomum* × *S. spegazzinii*. Castelli et al. (2003) screened 198 accessions from 63 species in the previously untested germplasm in the CPC which came from Hawkes' collection. With *G. pallida* (Pa2/3), roughly equal distributions of resistant and susceptible accessions were found throughout South America and Mexico, whereas with *G. rostochiensis* (Ro1), the majority of resistant accessions originated from the southern part of South America, mainly Argentina (the origin of *G. rostochiensis*). A high proportion (37%) of accessions were resistant to both species of nematode from five *Solanum* species like *S. palustre*, *S. mochiquense*, *S. okadae*, *S. neocardenasii*, and *S. semidemissum*. The accessions with resistance to a number of populations of both nematode species, and hence the most promising for use in breeding programs, were *S. canasense*, *S. gourlayi*, *S. okadae*, *S. spegazzinii*, and *S. verrucosum* (Castelli et al. 2003). When Van Soest et al. (1983) screened the German-Netherlands Potato Collection, they found resistance to several pathotypes of cyst nematodes in *S. gourlayi* (Ro5, Pa1–3), *S. oplocense* (Pa2/3), *S. multidissectum* (Ro1–3,5, Pa2,3), *S. spegazzinii* (Ro1,3,5, Pa2,3), *S. sucrense* (Ro1,2,5, Pa1–3), and *S. vernei* (Ro1–3,5, Pa1–5). Limited screenings from 1986 to 1995 identified high levels of foliage resistance in seven Mexican (*S. pinnatisectum*, *S. polyadenium*, *S. verrucosum*, *S. papita*, *S. polytrichon*, *S. stoloniferum*, *and S. brachycarpum*) and one Bolivian (*S. circaeifolium*) species (Bradshaw et al. 1995). In India, the breeding for PCN resistance mostly uses sources from *S. vernei* (Dalamu et al. 2012). The first cyst nematode-resistant variety, Kufri Swarna, was released in the year 1985. Another *S. vernei*-derived resistant hybrid Kufri Neelima was released in the year 2012 for Nilgiri hills. Recently, a PCN immune variety, Kufri Sahyadri, and moderately tolerant variety, Kufri Karan, were released in 2019 and 2020, respectively. The germplasm is screened under controlled conditions against both species of PCN (Fig. 8.5), and many tolerant accessions have been identified for use in resistance breeding program.

8.3.6 Pest Resistance

Potato (*S. tuberosum*) crop is attacked by a diversity of insect pests. A wide range of insect-pest resistance has been detected in wild species. In the past many years, large numbers of species were evaluated for resistance to one or more of the following insect pests: *Myzus persicae* (aphid), potato tuber moth, Colorado potato beetle (CPB), potato flea beetle, and potato leafhopper. Numerous studies showed that resistance to insect pests is due to glandular trichomes, glycoalkaloids, and other unidentified mechanisms (Pelletier et al. 2013). Among the pests, Colorado potato beetle, potato tuber moth (PTM), and potato weevil are the primary problems of global scale. Flanders et al. (1992) evaluated 100 wild potato species for resistance to various insect pests and reported that resistance mechanism was associated with glandular trichomes, glycoalkaloid tomatine, and dense hairs. Jansky et al. (2009) revealed resistance to CPB (Colorado potato beetle) in some wild species which is characterized by dense glandular trichomes (*Solanum polyadenium* and *S. tarijense*) and high levels of glycoalkaloids (*S. chacoense*). Wild species *S. hougasii* showed resistance to Columbia root-knot nematode (Brown et al. 1991). The glandular trichomes of *S. tarijense*, *S. berthaultii*, *and S. neocardenasii* have been linked to CPB resistance (Maharijaya and Vosman 2015). *S. cardiophyllum* and *S. circaeifolium* resistance to *Myzus persicae* and *S. chomatophilum* resistance to *Macrosiphum euphorbiae* have been attributed to antibiosis mechanisms (Pelletier

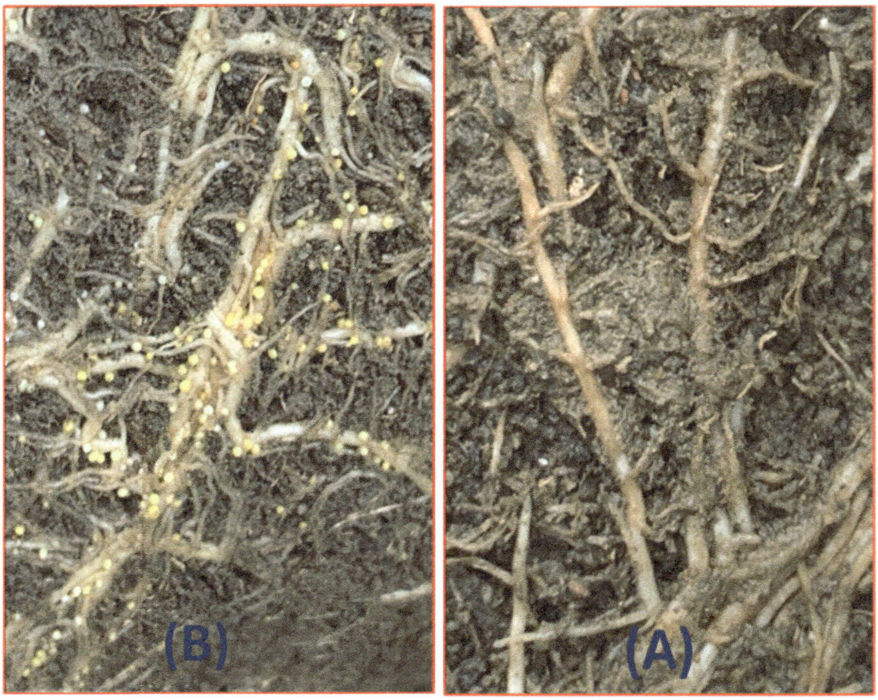

Fig. 8.5 Screening of genotypes against potato cyst nematode (PCN) (**a**) Resistant (SM/11–120) (**b**) Susceptible (Kufri Himalini)

et al. 2011). However, so far, progress has been limited mainly to the introgression of a few trichome-bearing species and, in particular, *S. berthaultii*. Many other wild potato species show promise as new sources of resistance, particularly against foliage herbivores.

8.4 Conclusion and Future Outlook

One of the important problems of people researching potato around the world is to assure food security and increasing productivity sustainably in changing climatic conditions. Wild relatives and primitive cultivars of potato have been mostly used as donor parents or sources of resistance to diseases and pests for potato breeding. Techniques have been advanced to integrate beneficial alleles/genes from its wild species into cultivated potato so that a wider gene pool can be used more effectively. Currently, there are a huge number of potato germplasm resources containing useful alleles in different gene banks all over the world; however, further collection may add new genes/alleles. In situ conservation of wild species is important to maintain the integrity of the gene pool and allowing natural evolution to occur in populations.

Accurate identification and characterization of species are requisite for the effective use of germplasm resources; therefore, taxonomic research and update taxonomic descriptions collected from potato gene banks are essential.

Genetic complexity, unpredictable expression in adapted backgrounds, and inbreeding depression hindered the proper introgression of resistance or tolerance traits from wild species into the cultivated potato. Traditional potato breeding approaches for introgression of disease resistance genes require years or decades as the resultant progeny must contain some undesirable traits along with the new trait of interest. Also, the time involved in the elimination of undesirable wild germplasm traits may lead to the evolution of pathogen populations. To overcome this, quick methods for characterization of emerging strains and approaches for quick deployment of resistance are needed. The use of molecular markers and cisgenic approach involving the introduction of genetic material, derived from its wild relatives, and lacking any selectable markers for antibiotic resistance or genomic selection for quantitatively inherited traits is required. Identification of numerous sequences involved in multigenic traits and simultaneously introgression them into new cultivars are possible with advances in genetic and genomics tools and offer a leading hand for trait improvement in crop plants particularly potato.

The genetic base of cultivated potato is still narrow due to selection during domestication, adaptation to new growing environments, and disease pressure. A wide gap exists between a large number of wild species evaluated that show promise and the actual number used in breeding. Wild species that carry disease resistance genes, they are easily accessible through the biotechnological approaches which allow the use of new species both within and outside of section Petota that have never been used before in breeding programs. By manipulation of ploidy, irrespective of Endosperm Balance Number, any potato species (donor) can be used for the introgression of desired genes into *S. tuberosum*. As discussed above, auspicious genetic resources for disease and pest resistance exist over the wide range of primitive cultivars and wild relatives of potato. The potential for using available genetic resources in resistance breeding program depends on their crossability with the cultivated potato (*S. tuberosum*).

References

Adillah Tan MY, Park TH, Alles R, Hutten RCB, Visser RGF, Van Eck HJ (2009) GpaXItarl originating from *solanum tarijense* is a major resistance locus to *Globodera pallida* and is localised on chromosome 11 of potato. Theor Appl Genet 119:1477–1487

Ahn YK, Park TH (2013) Resistance to common scab developed by somatic hybrids between *Solanum brevidens* and *Solanum tuberosum*. Acta Agric Scand Sect B Soil Plant Sci 63:595–603

Armstrong MR, Vossen J, Lim TY, Hutten RCB, Xu J, Strachan SM, Harrower B, Champouret N, Gilroy EM, Hein I (2019) Tracking disease resistance deployment in potato breeding by enrichment sequencing. Plant Biotechnol J 17:540–549

Asano K, Tamiya S (2016) Breeding of pest and disease resistant potato cultivars in Japan by using classical and molecular approaches. Japan Agric Res Q 50:1–6

Austin S, Lojkowska E, Ehlenfeldt MK, Kelman A, Helgeson JP (2021) Fertile interspecific somatic hybrids of Solatium: a novel source of resistance to Erwinia soft rot. Phytopathology 78:1216–1220

Bachmann-Pfabe S, Hammann T, Kruse J, Dehmer KJ (2019) Screening of wild potato genetic resources for combined resistance to late blight on tubers and pale potato cyst nematodes. Euphytica 215:1–15

Bakker E, Achenbach U, Bakker J, Van Vliet J, Peleman J, Segers B, Van Der Heijden S, Van Der Linde P, Graveland R, Hutten R, Van Eck H, Coppoolse E, Van Der Vossen E, Bakker J, Goverse A (2004) A high-resolution map of the H1 locus harbouring resistance to the potato cyst nematode *Globodera rostochiensis*. Theor Appl Genet 109:146–152

Ballvora A, Ercolano MR, Weiß J, Meksem K, Bormann CA, Oberhagemann P, Salamini F, Gebhardt C (2002) The R1 gene for potato resistance to late blight (*Phytophthora infestans*) belongs to the leucine zipper/NBS/LRR class of plant resistance genes. Plant J 30:361–371

Bamberg JB, Del Rio A (2005) Conservation of genetic resources. In: Genetic improvement of solanaceous crops, vol 1: potato. Science, p 451

Barker H, Dale MFB (2006) Resistance to viruses in potato. In: Natural resistance mechanisms of plants to viruses. Springer, Amsterdam, pp 341–366

Barone A, Sebastiano A, Carputo D, Della Rocca F, Frusciante L (2001) Molecular marker-assisted introgression of the wild *Solanum commersonii* genome into the cultivated S. tuberosum gene pool. Theor Appl Genet 102:900–907

Bethke PC, Halterman DA, Jansky S (2017) Are we getting better at using wild potato species in light of new tools? Crop Sci 57:1241–1258

Black W (1970) The nature and inheritance of field resistance to late blight (*Phytophthora infestans*) in potatoes. Am Potato J 47:279–288

Black W, Mastenbroek C, Mills WR, Peterson LC (1953) A proposal for an international nomenclature of races of *Phytophthora infestans* and of genes controlling immunity in *Solanum demissum* derivatives. Euphytica 2:173–179

Bradshaw JE (2009) Potato breeding at the Scottish plant breeding station and the Scottish crop research institute: 1920-2008. Potato Res 52:141–172

Bradshaw JE, Bryan GJ, Lees AK, McLean K, Solomon-Blackburn RM (2006a) Mapping the R10 and R11 genes for resistance to late blight (*Phytophthora infestans*) present in the potato (*Solanum tuberosum*) R-gene differentials of Black. Theor Appl Genet 112:744–751

Bradshaw JE, Bryan GJ, Ramsay G (2006b) Genetic resources (including wild and cultivated solanum species) and progress in their utilisation in potato breeding. Potato Res 49:49–65

Bradshaw JE, Hackett CA, Meyer RC, Milbourne D, McNicol JW, Phillips MS, Waugh R (1998) Identification of AFLP and SSR markers associated with quantitative resistance to Globodera pallida (stone) in tetraploid potato (Solanum tuberosum subsp. tuberosum) with a view to marker-assisted selection. Theor Appl Genet 97:202–210

Bradshaw JE, Ramsay G (2005) Utilisation of the commonwealth potato collection in potato breeding. Euphytica 146:9–19

Bradshaw JE, Stewart HE, Wastie RL, Dale MFB, Phillips MS (1995) Use of seedling progeny tests for genetical studies as part of a potato (*Solanum tuberosum* subsp. *tuberosum*) breeding programme. Theor Appl Genet 90:899–905

Braun S, Gevens A, Charkowski A, Allen C, Jansky S (2017) Potato common scab: a review of the causal pathogens, management practices, varietal resistance screening methods, and host resistance. Am J Potato Res 94:283–296

Brodie BB, Plaisted RL, de Scurrah MM (1991) The incorporation of resistance to *Globodera pallida* into *Solanum tuberosum* germplasm adapted to North America. Am Potato J 68:1–12

Brown CR, Mojtahedi H, James S, Novy RG, Love S (2006) Development and evaluation of potato breeding lines with introgressed resistance to Columbia root-knot nematode (*Meloidogyne chitwoodi*). Am J Potato Res 83:1–8

Brown CR, Mojtahedi H, Santo GS (1991) Resistance to Columbia root-knot nematode in Solanum ssp. and in hybrids of *S. hougasii* with tetraploid cultivated potato. Am Potato J 68:445–452

Bryan G, McLean K, Pande B, Purvis AA, Hackett CE, Bradshaw J, Waugh R (2004) Genetical dissection of H3-mediated polygenic PCN resistance in a heterozygous autotetraploid potato population. Mol Breed 14:105–116

Bukasov SM (1971) Cultivated potato species. In: Bukasov SM (ed) Flora of cultivated plants, vol IX. Kolos, Leningrad, pp 5–40

Caromel B, Mugniéry D, Kerlan MC, Andrzejewski S, Palloix A, Ellissèche D, Rousselle-Bourgeois F, Lefebvre V (2005) Resistance quantitative trait loci originating from *solanum sparsipilum* act independently on the sex ratio of *Globodera pallida* and together for developing a necrotic reaction. Mol Plant Microbe Interact 18:1186–1194

Caromel B, Mugniéry D, Lefebvre V, Andrzejewski S, Ellissèche D, Kerlan MC, Rousselle P, Rousselle-Bourgeois F (2003) Mapping QTLs for resistance against *Globodera pallida* (stone) Pa2/3 in a diploid potato progeny originating from *Solanum spegazzinii*. Theor Appl Genet 106: 1517–1523

Castelli L, Ramsay G, Bryan G, Neilson SJ, Phillips MS (2003) New sources of resistance to the potato cyst nematodes *Globodera pallida* and *G. rostochiensis* in the commonwealth potato collection. Euphytica 129:377–386

Champouret N (2010) Functional genomics of Phytophthora infestans effectors and solanum resistance genes. PhD thesis, Wageningen University, p 154

Chandel P, Tiwari JK, Ali N, Devi S, Sharma S, Sharma S, Luthra SK, Singh BP (2015) Interspecific potato somatic hybrids between *Solanum tuberosum* and *S. cardiophyllum*, potential sources of late blight resistance breeding. Plant Cell Tiss Org Cult 123:579–589

Chen L, Guo X, Wang H, Xie C, Cai X, He L, Zhou J, Liu J (2016) Tetrasomic inheritance pattern of the pentaploid *Solanum chacoense* (+) *S. tuberosum* somatic hybrid (resistant to bacterial wilt) revealed by SSR detected alleles. Plant Cell Tiss Org Cult 127:315–323

Chen L, Guo X, Xie C, He L, Cai X, Tian L, Song B, Liu J (2013) Nuclear and cytoplasmic genome components of *Solanum tuberosum* + *S. chacoense* somatic hybrids and three SSR alleles related to bacterial wilt resistance. Theor Appl Genet 126:1861–1872

Chen S, Borza T, Byun B, Coffin R, Coffin J, Peters R, Wang-Pruski G (2017) DNA markers for selection of late blight resistant potato breeding lines. Am J Plant Sci 08:1197–1209

Chen X, Lewandowska D, Armstrong MR, Baker K, Lim TY, Bayer M, Harrower B, McLean K, Jupe F, Witek K, Lees AK, Jones JD, Bryan GJ, Hein I (2018) Identification and rapid mapping of a gene conferring broad-spectrum late blight resistance in the diploid potato species *Solanum verrucosum* through DNA capture technologies. Theor Appl Genet 131:1287–1297

Dalamu BV, Umamaheshwari R, Sharma R, Kaushik SK, Joseph TA, Singh BP, Gebhardt C (2012) Potato cyst nematode (PCN) resistance: genes, genotypes and markers- an update. Sabrao J Breed Genet 44:202–228

De Haan S, Rodriguez F (2016) Potato origin and production. In: Advances in potato chemistry and technology, 2nd edn. Elsevier, Amsterdam, pp 1–32

Dunnett JM (1961) Inheritance of resistance to potato root eelworm in a breeding line stemming from *solanum multidissectum* Hawkes. Scottish Plant Breed Stn Rep 1961:39–46

Ellenby C (1948) Resistance to the potato-root eelworm. Nature 162:704

Evans K, Brodie BB (1980) The origin and distribution of the golden nematode and its potential in the U.S.a. Am Potato J 57:79–89

Ewing EE, Šimko I, Smart CD, Bonierbale MW, Mizubuti ESG, May GD, Fry WE (2000) Genetic mapping from field tests of qualitative and quantitative resistance to *Phytophthora infestans* in a population derived from *Solanum tuberosum* and *Solanum berthaultii*. Mol Breed 6:25–36

Flanders KL, Hawkes JG, Radcliffe EB, Lauer FI (1992) Insect resistance in potatoes: sources, evolutionary relationships, morphological and chemical defenses, and ecogeographical associations. Euphytica 61:83–111

Fock I, Collonnier C, Luisetti J, Purwito A, Souvannavong V, Vedel F, Servaes A, Ambroise A, Kodja H, Ducreux G, Sihachakr D (2001) Use of *solanum stenotomum* for introduction of resistance to bacterial wilt in somatic hybrids of potato. Plant Physiol Biochem 39:899–908

Fock I, Collonnier C, Purwito A, Luisetti J, Souvannavong V, Vedel F, Servaes A, Ambroise A, Kodja H, Ducreux G, Sihachakr D (2000) Resistance to bacterial wilt in somatic hybrids between *Solanum tuberosum* and *Solanum phureja*. Plant Sci 160:165–176

Foster SJ, Park TH, Pel M, Brigneti G, Sliwka J, Jagger L, Van Der Vossen E, Jones JDG (2009) Rpi-vnt1.1, a tm-22 homolog from *Solanum venturii*, confers resistance to potato late blight. Mol Plant Microbe Interact 22:589–600

Fulladolsa AC, Navarro FM, Kota R, Severson K, Palta JP, Charkowski AO (2015) Application of marker assisted selection for potato virus y resistance in the university of Wisconsin potato breeding program. Am J Potato Res 92:444–450

Golas TM, Sikkema A, Gros J, Feron RMC, van den Berg RG, van der Weerden GM, Mariani C, Allefs JJHM (2010) Identification of a resistance gene Rpi-dlc1 to Phytophthora infestans in European accessions of *Solanum dulcamara*. Theor Appl Genet 120:797–808

Gopal J, Oyama K (2005) Genetic base of Indian potato selections as revealed by pedigree analysis. Euphytica 142:23–31

Guo X, Xie C, Cai X, Song B, He L, Liu J (2010) Meiotic behavior of pollen mother cells in relation to ploidy level of somatic hybrids between *Solanum tuberosum* and *S. chacoense*. Plant Cell Rep 29:1277–1285

Hämäläinen JH, Sorri VA, Watanabe KN, Gebhardt C, Valkonen JPT (1998) Molecular examination of a chromosome region that controls resistance to potato Y and a potyviruses in potato. Theor Appl Genet 96:1036–1043

Harding K, Millam S (2000) Analysis of chromatin, nuclear DNA and organelle composition in somatic hybrids between *Solanum tuberosum* and *Solanum sanctae-rosae*. Theor Appl Genet 101:939–947

Haverkort AJ, Struik PC, Visser RGF, Jacobsen E (2009) Applied biotechnology to combat late blight in potato caused by *Phytophthora infestans*. Potato Res 52:249–264

Hawkes JG (1990) The potato: evolution, biodiversity and genetic resources. Choice Rev 28:2137

Hein I, Birch PRJ, Danan S, Lefebvre V, Odeny DA, Gebhardt C, Trognitz F, Bryan GJ (2009) Progress in mapping and cloning qualitative and quantitative resistance against *Phytophthora infestans* in potato and its wild relatives. Potato Res 52:215–227

Hosaka K, Hosaka Y, Mori M, Maida T, Matsunaga H (2001) Detection of a simplex RAPD marker linked to resistance to potato virus Y in a tetraploid potato. Am J Potato Res 78:191–196

Hosaka K, Matsunaga H, Senda K (2000) Evaluation of several wild tuber-bearing solanum species for scab resistance. Am J Potato Res 77:41–45

Huang S, Van Der Vossen EAG, Kuang H, Vleeshouwers VGAA, Zhang N, Borm TJA, Van Eck HJ, Baker B, Jacobsen E, Visser RGF (2005) Comparative genomics enabled the isolation of the R3a late blight resistance gene in potato. Plant J 42:251–261

Iovene M, Aversano R, Savarese S, Caruso I, Di Matteo A, Cardi T, Frusciante L, Carputo D (2012) Interspecific somatic hybrids between *Solanum bulbocastanum* and *S. tuberosum* and their haploidization for potato breeding. Biol Plant 56:1–8

Jacobs JME, Van Eck HJ, Horsman K, Arens PFP, Verkerk-Bakker B, Jacobsen E, Pereira A, Stiekema WJ (1996) Mapping of resistance to the potato cyst nematode *Globodera rostochiensis* from the wild potato species *Solanum vernei*. Mol Breed 2:51–60

Jacobs MMJ, Vosman B, Vleeshouwers VGAA, Visser RGF, Henken B, van den Berg RG (2010) A novel approach to locate *Phytophthora infestans* resistance genes on the potato genetic map. Theor Appl Genet 120:785–796

Jansky SH, Charkowski AO, Douches DS, Gusmini G, Richael C, Bethke PC, Spooner DM, Novy RG, De Jong H, De Jong WS, Bamberg JB, Thompson AL, Bizimungu B, Holm DG, Brown CR, Haynes KG, Sathuvalli VR, Veilleux RE, Creighton Miller J, Bradeen JM, Jiang J (2016) Reinventing potato as a diploid inbred line-based crop. Crop Sci 56:1412–1422

Jansky SH, Dempewolf H, Camadro EL, Simon R, Zimnoch-Guzowska E, Bisognin DA, Bonierbale M (2013) A case for crop wild relative preservation and use in potato. Crop Sci 53:746–754

Jansky SH, Simon R, Spooner DM (2009) A test of taxonomic predictivity: resistance to the Colorado potato beetle in wild relatives of cultivated potato. J Econ Entomol 102:422–431

Judelson HS, Blanco FA (2005) The spores of Phytophthora: weapons of the plant destroyer. Nat Rev Microbiol 3:47–58

Karki HS, Jansky SH, Halterman DA (2021) Screening of wild potatoes identifies new sources of late blight resistance. Plant Dis 105:368–376

Kasai K, Morikawa Y, Sorri VA, Valkonen J, Gebhardt C, Watanabe KN (2000) Development of SCAR markers to the PVY resistance gene Ry adg based on a common feature of plant disease resistance genes. Genome 43:1–8

Kaushik SK, Sharma R, Garg ID, Singh BP, Chakrabarti SK, Bhardwaj V, Pandey SK (2013) Development of a triplex (YYYy) parental potato line with extreme resistance to potato virus Y using marker assisted selection. J Hortic Sci Biotech 88:580–584

Kim-Lee H, Moon JS, Hong YJ, Kim MS, Cho HM (2005) Bacterial wilt resistance in the progenies of the fusion hybrids between haploid of potato and *Solanum commersonii*. Am J Potato Res 82: 129–137

Kreike CM, de Koning JRA, Vinke JH, van Ooijen JW, Gebhardt C, Stiekema WJ (1993) Mapping of loci involved in quantitatively inherited resistance to the potato cyst-nematode *Globodera rostochiensis* pathotype Ro1. Theor Appl Genet Int J Plant Breed Res 87:464–470

Kreike CM, de Koning JRA, Vinke JH, van Ooijen JW, Stiekema WJ (1994) Quantitatively-inherited resistance to *Globodera pallida* is dominated by one major locus in *Solanum spegazzinii*. Theor Appl Genet 88:764–769

Kreike CM, Kok-Westeneng AA, Vinke JH, Stiekema WJ (1996) Mapping of QTLs involved in nematode resistance, tuber yield and root development in solanum sp. Theor Appl Genet 92: 463–470

Kuhl JC, Hanneman RE, Havey MJ (2001) Characterization and mapping of Rpi1, a late-blight resistance locus from diploid (1EBN) Mexican *Solanum pinnatisectum*. Mol Genet Genomics 265:977–985

Lechnovich VS (1971) Cultivated potato species. In: Bukasov SM (ed) Chapter 2: Flora of cultivated plants, vol IX. Kolos, Leningrad, pp 41–304

Li W, Deng Y, Ning Y, He Z, Wang GL (2020) Exploiting broad-spectrum disease resistance in crops: from molecular dissection to breeding. Annu Rev Plant Biol 71:575–603

Lokossou AA, Park T, van Arkel G, Arens M, Ruyter-Spira C, Morales J, Whisson SC, PRJ B, RGF V, Jacobsen E, van der Vossen EAG (2009) Exploiting knowledge of R/Avr genes to rapidly clone a new LZ-NBS-LRR family of late blight resistance genes from potato linkage group IV. Mol Plant Microbe Interact 22:630–641

Machida-Hirano R (2015) Diversity of potato genetic resources. Breed Sci 65:26–40

Machida-Hirano R, Niino T (2017) Potato genetic resources. CABI, Wallingford, pp 11–30

Maharijaya A, Vosman B (2015) Managing the Colorado potato beetle; the need for resistance breeding. Euphytica 204:487–501

Maxted N, Kell S, Ford-Lloyd B, Dulloo E, Toledo Á (2012) Toward the systematic conservation of global crop wild relative diversity. Crop Sci 52:774–785

Mihovilovich E, Sanetomo R, Hosaka K, Ordoñez B, Aponte M, Bonierbale M (2015) Cytoplasmic diversity in potato breeding: case study from the international potato center. Mol Breed 35:1–10

Molnár I, Besenyei E, Thieme R, Thieme T, Aurori A, Baricz A, Banciu HL, Rakosy-Tican E (2017) Mismatch repair deficiency increases the transfer of antibiosis and antixenosis properties against Colorado potato beetle in somatic hybrids of *Solanum tuberosum* + *S. chacoense*. Pest Manag Sci 73:1428–1437

Naess S, Bradeen J, Wielgus S, Haberlach G, McGrath J, Helgeson J (2001) Analysis of the introgression of *Solanum bulbocastanum* DNA into potato breeding lines. Mol Genet Genomics 265:694–704

Navarro FM, Rak KT, Banks E, Bowen BD, Higgins C, Palta JP (2015) Strategies for selecting stable common scab resistant clones in a potato breeding program. Am J Potato Res 92:326–338

Nouri-Ellouz O, Triki MA, Jbir-Koubaa R, Louhichi A, Charfeddine S, Drira N, Gargouri-Bouzid R (2016) Somatic hybrids between potato and *S. berthaultii* show partial resistance to soil-borne fungi and potato virus Y. J Phytopathol 164:485–496

Novy RG, Whitworth JL, Stark JC, Charlton BA, Yilma S, Knowles NR, Pavek MJ, Spear RR, Brandt TL, Olsen N, Thornton M, Brown CR (2014) Teton russet: an early-maturing, dual-purpose potato cultivar having higher protein and vitamin C content, low asparagine, and resistances to common scab and fusarium dry rot. Am J Potato Res 91:380–393

Ochoa CM (1999) The potatoes of South America: Peru – part I. Allen Press, Kansas City, p 1036

Oosumi T, Rockhold DR, Maccree MM, Deahl KL, McCue KF, Belknap WR (2009) Gene Rpi-bt1 from *Solanum bulbocastanum* confers resistance to late blight in transgenic potatoes. Am J Potato Res 86:456–465

Ortiz R (2001) The state of the use of potato genetic diversity. In: Broadening the genetic base of crop production. CABI Publishing, Wallingford, pp 181–200

Paal J, Henselewski H, Muth J, Meksem K, Menéndez CM, Salamini F, Ballvora A, Gebhardt C (2004) Molecular cloning of the potato Gro1-4 gene conferring resistance to pathotype Ro1 of the root cyst nematode *Globodera rostochiensis*, based on a candidate gene approach. Plant J 38:285–297

Park J, Hackett CA, Dandurand LM, Wang X, De Jong WS (2019) QTL for resistance to *Globodera rostochiensis* Pathotype Ro2 and G. pallida Pathotype Pa2/3 in autotetraploid potato. Am J Potato Res 96:552–563

Park TH, Foster S, Brigneti G, Jones JDG (2009) Two distinct potato late blight resistance genes from *Solanum berthaultii* are located on chromosome 10. Euphytica 165:269–278

Park TH, Gros J, Sikkema A, Vleeshouwers VGAA, Muskens M, Allefs S, Jacobsen E, Visser RGF, Van Der Vossen EAG (2005a) The late blight resistance locus Rpi-blb3 from *Solanum bulbocastanum* belongs to a major late blight R gene cluster on chromosome 4 of potato. Mol Plant Microbe Interact 18:722–729

Park TH, Vleeshouwers VGAA, Hutten RCB, Van Eck HJ, Van Der Vossen E, Jacobsen E, Visser RGF (2005b) High-resolution mapping and analysis of the resistance locus Rpi-abpt against *Phytophthora infestans* in potato. Mol Breed 16:33–43

Pasco C, Jouan B, Andrivon D (2005) Resistance of potato genotypes to common and netted scab-causing species of Streptomyces. Plant Pathol 54:383–392

Pavek JJ, Corsini DL (2001) Utilization of potato genetic resources in variety development. Am J Potato Res 78:433–441

Pel MA, Foster SJ, Park TH, Rietman H, Ven Arkel G, Jones JDG, Van Eck HJ, Jacobsen E, Visser RGF, Van Der Vossen EAG (2009) Mapping and cloning of late bright resistance genes from *Solanum venturii* using an interspecific candidate gene approach. Mol Plant Microbe Interact 22: 601–615

Pelletier Y, Horgan FG, Pompon J (2013) Potato resistance against insect herbivores: resources and opportunities. In: Insect pests of potato: global perspectives on biology and management, pp 439–462

Pelletier, Y., Horgan, F. G., & Pompon, J. (2013). Potato resistance against insect herbivores: Resources and opportunities. Insect Pests of Potato: Global Perspectives on Biology and Management, 439-462.

Rakosy-Tican E, Aurori A (2015) Green fluorescent protein (GFP) supports the selection based on callus vigorous growth in the somatic hybrids *Solanum tuberosum* L. + *S. chacoense* bitt. Acta Physiol Plant 37:1–5

Rauscher GM, Smart CD, Simko I, Bonierbale M, Mayton H, Greenland A, Fry WE (2006) Characterization and mapping of RPi-ber, a novel potato late blight resistance gene from *Solanum berthaultii*. Theor Appl Genet 112:674–687

Rodewald J, Trognitz B (2013) Solanum resistance genes against *Phytophthora infestans* and their corresponding avirulence genes. Mol Plant Pathol 14:740–757

Rouppe Van Der Voort J, Lindeman W, Folkertsma R, Hutten R, Overmars H, Van Der Vossen E, Jacobsen E, Bakker J (1998) A QTL for broad-spectrum resistance to cyst nematode species (*Globodera* spp.) maps to a resistance to gene cluster in potato. Theor Appl Genet 96:654–661

Rouppe Van Der Voort J, Wolters P, Folkertsma R, Hutten R, Van Zandvoort P, Vinke H, Kanyuka K, Bendahmane A, Jacobsen E, Janssen R, Bakker J (1997) Mapping of the cyst nematode resistance locus Gpa2 in potato using a strategy based on comigrating AFLP markers. Theor Appl Genet 95:874–880

Salaman RN, Pethybridge G (1921) Some recent work on the potato blight. In: Rep. Intern. Potato Conf. Roy. Hort. Soc, vol 122

Sandbrink JM, Colon LT, Wolters PJCC, Stiekema WJ (2000) Two related genotypes of *Solanum microdontum* carry different segregating alleles for field resistance to *Phytophthora infestans*. Mol Breed 6:215–225

Sato M, Nishikawa K, Komura K, Hosaka K (2006) Potato virus Y resistance gene, Rychc, mapped to the distal end of potato chromosome 9. Euphytica 149:367–372

Savary S, Willocquet L, Pethybridge SJ, Esker P, McRoberts N, Nelson A (2019) The global burden of pathogens and pests on major food crops. Nat Ecol Evol 3:430–439

Sinden SL, Sanford LL, Cantelo WW, Deahl KL (1986) Leptine Glycoalkaloids and resistance to the Colorado potato beetle (Coleoptera: Chrysomelidae) in *Solanum chacoense*. Environ Entomol 15:1057–1062

Singh R, Tiwari JK, Rawat S, Sharma V, Singh BP (2016) In silico identification of candidate microRNAs and their targets in potato somatic hybrid *Solanum tuberosum* (+) S. pinnatisectum for late blight resistance. https://search.informit.org/doi/10.3316/INFORMIT.888051540115882. Accessed 12 Jun 2021

Śliwka J, Jakuczun H, Chmielarz M, Hara-Skrzypiec A, Tomczyńska I, Kilian A, Zimnoch-Guzowska E (2012a) A resistance gene against potato late blight originating from *solanum* × *michoacanum* maps to potato chromosome VII. Theor Appl Genet 124:397406

Śliwka J, Jakuczun H, Chmielarz M, Hara-Skrzypiec A, Tomczyńska I, Kilian A, Zimnoch-Guzowska E (2012b) Late blight resistance gene from *solanum ruiz-ceballosii* is located on potato chromosome X and linked to violet flower colour. BMC Genet 13:1–12

Śliwka J, Jakuczun H, Lebecka R, Marczewski W, Gebhardt C, Zimnoch-Guzowska E (2006) The novel, major locus Rpi-phu1 for late blight resistance maps to potato chromosome IX and is not correlated with long vegetation period. Theor Appl Genet 113:685–695

Smilde WD, Brigneti G, Jagger L, Perkins S, Jones JDG (2005) Solanum mochiquense chromosome IX carries a novel late blight resistance gene Rpi-moc1. Theor Appl Genet 110:252–258

Smyda P, Jakuczun H, Dębski K, Śliwka J, Thieme R, Nachtigall M, Wasilewicz-Flis I, Zimnoch-Guzowska E (2013) Development of somatic hybrids *solanum* × *michoacanum* bitter. (Rydb.) (+) *S. tuberosum* L. and autofused 4x *S.* × *michoacanum* plants as potential sources of late blight resistance for potato breeding. Plant Cell Rep 32:1231–1241

Solomon-Blackburn RM, Barker H (2001) A review of host major-gene resistance to potato viruses X, Y, a and V in potato: genes, genetics and mapped locations. Heredity (Edinb) 86: 8–16

Song J, Bradeen JM, Naess SK, Raasch JA, Wielgus SM, Haberlach GT, Liu J, Kuang H, Austin-Phillips S, Buell CR, Helgeson JP, Jiang J (2003) Gene RB cloned from *Solanum bulbocastanum* confers broad spectrum resistance to potato late blight. Proc Natl Acad Sci USA 100:9128–9133

Song YS, Schwarzfischer A (2008) Development of STS markers for selection of extreme resistance (Ry sto) to PVY and maternal pedigree analysis of extremely resistant cultivars. Am J Potato Res 85:159–170

Sood S, Bhardwaj V, Pandey SK, Chakrabarti SK (2017) History of potato breeding: improvement, diversification, and diversity. In: The Potato genome. Springer, Cham, pp 31–72

Spooner DM, Ghislain M, Simon R, Jansky SH, Gavrilenko T (2014) Systematics, diversity, genetics, and evolution of wild and cultivated potatoes. Bot Rev 80:283–383

Spooner DM, Núñez J, Trujillo G, Herrera MDR, Guzmán F, Ghislain M (2007) Extensive simple sequence repeat genotyping of potato landraces supports a major re-evaluation of their gene pool structure and classification. Proc Natl Acad Sci U S A 104:19398–19403

Strachan SM, Armstrong MR, Kaur A, Wright KM, Lim TY, Baker K, Jones J, Bryan G, Blok V, Hein I (2019) Mapping the H2 resistance effective against *Globodera pallida* pathotype Pa1 in tetraploid potato. Theor Appl Genet 132:1283–1294

Tan MYA, Hutten RCB, Celis C, Park TH, Niks RE, Visser RGF, Van Eck HJ (2008) The RPi-mcd1 locus from *Solanum microdontum* involved in resistance to *Phytophthora infestans*, causing a delay in infection, maps on potato chromosome 4 in a cluster of NBS-LRR genes. Mol Plant Microbe Interact 21:909–918

Tarwacka J, Polkowska-Kowalczyk L, Kolano B, Śliwka J, Wielgat B (2013) Interspecific somatic hybrids *Solanum villosum* (+) S. tuberosum, resistant to *Phytophthora infestans*. J Plant Physiol 170:1541–1548

Tek AL, Stevenson WR, Helgeson JP, Jiang J (2004) Transfer of tuber soft rot and early blight resistances from *Solanum brevidens* into cultivated potato. Theor Appl Genet 109:249–254

Thieme R, Rakosy-Tican E, Nachtigall M, Schubert J, Hammann T, Antonova O, Gavrilenko T, Heimbach U, Thieme T (2010) Characterization of the multiple resistance traits of somatic hybrids between *Solanum cardiophyllum* Lindl. And two commercial potato cultivars. Plant Cell Rep 29:1187–1201

Tiwari JK, Devi S, Chandel P, Ali N, Bhardwaj V, Singh BP (2016) Organelle genome analysis in somatic hybrids between *Solanum tuberosum* and *S. pinnatisectum* revealed diverse cytoplasm type in potato. Agric Res 5:22–28

Tiwari JK, Sarkar PD, Pandey SK, Gopal J, Kumar SR (2010) Molecular and morphological characterization of somatic hybrids between *Solanum tuberosum* L. and *S. etuberosum* Lindl. Plant Cell Tiss Org Cult 103:175–187

Tiwari JK, Saurabh S, Chandel P, Devi S, Ali N, Bist CM, Singh BP (2015) Analysis of genetic and epigenetic changes in potato somatic hybrids between *Solanum tuberosum* and *S. etuberosum* by AFLP and MSAP markers. Agric Res 4:339–346

Turner SJ (1989) New sources of resistance to potato cyst-nematodes in the commonwealth potato collection. Euphytica 42:145–153

Valkonen JPT (2015) Elucidation of virus-host interactions to enhance resistance breeding for control of virus diseases in potato. Breed Sci 65:69–76

Valkonen JPT, Wiegmann K, Hämäläinen JH, Marczewski W, Watanabe KN (2008) Evidence for utility of the same PCR-based markers for selection of extreme resistance to potato virus Y controlled by Rysto of *Solanum stoloniferum* derived from different sources. Ann Appl Biol 152:121–130

Van Der Voort JR, Van Der Vossen E, Bakker E, Overmars H, Van Zandvoort P, Hutten R, Lankhorst RK, Bakker J (2000) Two additive QTLs conferring broad-spectrum resistance in potato to *Globodera pallida* are localized on resistance gene clusters. Theor Appl Genet 101: 1122–1130

Van Der Vossen E, Sikkema A, Te Lintel HB, Gros J, Stevens P, Muskens M, Wouters D, Pereira A, Stiekema W, Allefs S (2003) An ancient R gene from the wild potato species *Solanum bulbocastanum* confers broad-spectrum resistance to Phytophthora infestans in cultivated potato and tomato. Plant J 36:867–882

Van Der Vossen EAG, Gros J, Sikkema A, Muskens M, Wouters D, Wolters P, Pereira A, Allefs S (2005) The Rpi-blb2 gene from *Solanum bulbocastanum* is an mi-1 gene homolog conferring broad-spectrum late blight resistance in potato. Plant J 44:208–222

Van Soest LJM, Rumpenhorst HJ, Huijsman CA (1983) Resistance to potato cyst-nematodes in tuber-bearing solanum species and its geographical distribution. Euphytica 32:65–74

Verzaux E, van Arkel G, Vleeshouwers VGAA, van der Vossen EAG, Niks RE, Jacobsen E, Vossen J, Visser RGF (2012) High-resolution mapping of two broad-spectrum late blight resistance genes from two wild species of the *Solanum circaeifolium* group. Potato Res 55: 109–123

Villamon FG, Spooner DM, Orrillo M, Mihovilovich E, Pérez W, Bonierbale M (2005) Late blight resistance linkages in a novel cross of the wild potato species *Solanum paucissectum* (series Piurana). Theor Appl Genet 111:1201–1214

Vleeshouwers VGAA, Finkers R, Budding D, Visser M, Jacobs MMJ, van Berloo R, Pel M, Champouret N, Bakker E, Krenek P, Rietman H, Huigen DJ, Hoekstra R, Goverse A, Vosman B, Jacobsen E, Visser RGF (2011a) SolRgene: an online database to explore disease resistance genes in tuber-bearing solanum species. BMC Plant Biol 11:116

Vleeshouwers VGAA, Raffaele S, Vossen JH, Champouret N, Oliva R, Segretin ME, Rietman H, Cano LM, Lokossou A, Kessel G, Pel MA, Kamoun S (2011b) Understanding and exploiting late blight resistance in the age of effectors. Annu Rev Phytopathol 49:507–531

Vleeshouwers VGAA, Rietman H, Krenek P, Champouret N, Young C, Oh SK, Wang M, Bouwmeester K, Vosman B, Visser RGF, Jacobsen E, Govers F, Kamoun S, Van der Vossen EAG (2008) Effector genomics accelerates discovery and functional profiling of potato disease resistance and *Phytophthora infestans* avirulence genes. PLoS One 3:e2875

Vossen JH, Nijenhuis M, Arens-de Reuver MJB et al (2009) Cloning and exploitation of a functional R gene from Solanum chacoense US patent WO 2011034433 A1. Published by the world intellectual property organization 18 September 2009

Wang M, Allefs S, Van Den Berg RG, Vleeshouwers VGAA, Van Der Vossen EAG, Vosman B (2008) Allele mining in solanum: conserved homologues of Rpi-blb1 are identified in *Solanum stoloniferum*. Theor Appl Genet 116:933–943

Webber Iii C, Harris M, Sherefler J, Durnova M, Christopher C (2017) USDA ARS. https://www.ars.usda.gov/research/publications/publication/?seqNo115=156337%0Ahttps://www.ars.usda.gov/research/publications/publication/?seqNo115=147392%0Ahttps://www.ars.usda.gov/research/publications/publication/?seqNo115=192159%0Ahttps://www.ars.usda. Accessed 12 Jun 2021

Whitworth JL, Novy RG, Zasada IA, Wang X, Dandurand LM, Kuhl JC (2018) Resistance of potato breeding clones and cultivars to three species of potato cyst nematode. Plant Dis 102:2120–2128

Witek K, Jupe F, Witek AI, Baker D, Clark MD, Jones JDG (2016) Accelerated cloning of a potato late blight-resistance gene using RenSeq and SMRT sequencing. Nat Biotechnol 34:656–660

Witek K, Lin X, Karki H, Jupe F, Witek A, Steuernagel B, Stam R, van Oosterhout C, Fairhead S, Cocker J, Bhanvadia S, Barrett W, Wu C-H, Adachi H, Song T, Kamoun S, Vleeshouwers VG, Tomlinson L, Wulff BB, Jones JD (2020) A complex resistance locus in *Solanum americanum* recognizes a conserved Phytophthora effector. bioRxiv 2020.05.15.095497

Yamada T, Misoo S, Ishii T, Ito Y, Takaoka K, Kamijima O (1997) Characterization of somatic hybrids between tetraploid *Solanum tuberosum* L. and Dihaploid *S. acaule*. Breed Sci 47:229–236

Yang L, Wang D, Xu Y, Zhao H, Wang L, Cao X, Chen Y, Chen Q (2017) A new resistance gene against potato late blight originating from *Solanum pinnatisectum* located on potato chromosome 7. Front Plant Sci 8:1729

Yilma S, Vales MI, Charlton BA, Hane DC, James SR, Shock CC, Mosley AR, Culp D, Feibert E, Leroux L, Karaagac E, Knowles NR, Pavek MJ, Stark JC, Novy RG, Whitworth JL, Pavek JJ, Corsini DL, Brandt TL, Olsen N, Brown CR (2012) Owyhee russet: a variety with high yields of U.S. no. 1 tubers, excellent processing quality, and moderate resistance to fusarium dry rot (*Fusarium solani* var. *coeruleum*). Am J Potato Res 89:175–183

Zimnoch-Guzowska E, Yin Z, Chrzanowska M, Flis B (2013) Sources and effectiveness of potato pvy resistance in IHAR's breeding research. Am J Potato Res 90:21–27

9 Biology and Management of Aphids Infesting Potato

Mohd Abas Shah, S. Subhash, Kailash C. Naga, and Sanjeev Sharma

Abstract

Aphids are the most important pests of potato worldwide. They are sap-feeding insects, but the major damage inflicted by aphids in potato crops is by transmission of numerous potato viruses limiting disease-free seed production with a progressive decline in yield. Potato crops are infested by a number of colonizing and noncolonizing species of aphids; the noncolonizing aphids are more important for the spread of nonpersistent viruses like *potato virus Y* (PVY), and the persistent viruses like *potato leaf roll virus* (PLRV) are mainly spread by colonizing aphids. More than 22 species of aphids are recorded worldwide that colonize potato plants, and more than 110 species are known to transiently visit the crops. Various attributes of aphid biology and ecology have contributed to their success as crop pests. The host-finding and feeding behavior of aphids predisposes them to being the predominant vectors of various viruses. Controlling the spread of PVY remains a challenge to the potato industry worldwide because of its nonpersistent mode of transmission and the evolution of new strains and variants. Various countries operate networks of traps to monitor the flight activity of aphid species in seed potato. It has been reported that aphids other than *M. persicae* are more important for the early-season spread of viruses like PVY. Currently, the aphid management

M. A. Shah (✉)
ICAR-Central Potato Research Institute-Regional Station, Jalandhar, Punjab, India
e-mail: mabas.shah@icar.gov.in

S. Subhash
ICAR-Central Potato Research Institute-Regional Station, Modipuram, Meerut, Uttar Pradesh, India

K. C. Naga · S. Sharma
Division of Plant Protection, ICAR-Central Potato Research Institute, Shimla, Himachal Pradesh, India

S. Kumar Chakrabarti et al. (eds.), *Sustainable Management of Potato Pests and Diseases*, https://doi.org/10.1007/978-981-16-7695-6_9

methods in potato are mostly reliant on the use of various insecticides and mineral oils. Moreover, the use of infection-free seed, roguing, and use of cultural practices such as manipulation of planting and haulm-cutting dates are the most useful to keep the incidence of virus under control.

Keywords

Nonpersistent virus · Vector pressure · Aphid monitoring · Virus transmission · Host alternation · Parthenogenesis · Noncolonizing aphids

9.1 Introduction

Potato originated in the Andean highlands of South America and is now cultivated in a major part of the world across the temperate, subtropical, and tropical agroecologies. Its wide geographical distribution also exposes it to a plethora of diverse phytophagous arthropods. Kroschel et al. (2020) described a total of 49 species of insect pests infesting potato crops in different parts of the world. Out of these, 6 major and 32 minor species are prevalent throughout the temperate, tropical, and subtropical regions; 9 major species are prevalent in the tropical and subtropical regions; 2 major species affect potato crops in the temperate regions. Among the global pests of potato, aphids are the most important. Aphids are sap-feeding insects, but the major damage inflicted by aphids in potato crops is by transmission of numerous potato viruses. The resulting viral disease leads to considerable yield reductions, limits the production of disease-free seed potatoes, and causes a progressive degeneration of seed stocks.

Aphids (Aphididae: Hemiptera) are a diverse group of insects with more than 5000 species reported worldwide (Remaudière and Remaudière 1997; Favret 2014). They are distributed worldwide but are most abundant and most diverse in the temperate areas. Although many of these can infest the crop plants, only around 100 of them are of economic importance (Blackman and Eastop 2017). Aphids are generally recognized by a number of common morphological characteristics, e.g., soft body with head, thorax, and abdomen; siphunculi (secretory organs); five- or six-segmented antennae composed of two basal segments and a segmented flagellum with a terminal process; two-segmented tarsi, with the first segment much shorter than the second; and a cauda, which is often used for flicking away droplets of honeydew from the anus. These features have been modified, reduced, or secondarily lost in some species (Blackman and Eastop 2017).

Due to their remarkable ability to adapt and colonize diverse ecological situations, aphids are major pests of various crops, including potato. The cyclic parthenogenesis enables aphids to alternate sexual and asexual generations. The asexual reproduction leads to faster multiplication rates and quick colonization of the secondary hosts where they can cause severe crop damage. Aphids characteristically exhibit polyphenism, which is the production of different phenotypes from the same genotype. Polyphenism is the major reason for the success of the insects in general

(Simpson et al. 2011), allowing them to partition their life-history stages (larvae dedicated to feeding and growth and adults dedicated to reproduction and dispersal), to adopt different phenotypes in response to environmental change (seasonal morphs), and to cope with temporally heterogeneous environments (dispersal morphs) (Field et al. 2017). Aphids exhibit a range of continuous morphological variation, wider than in many other insect groups. Increases or decreases in size due to nutritional effects, for example, can accumulate over several generations, because the size of the mother can affect the size of her offspring. There may be large seasonal differences, with some species producing dwarf individuals when food quality is poor in midsummer.

Aphids can damage potato cops directly by feeding on sap and indirectly by transmitting various viral diseases. Although the direct damage inflicted by aphids is rarely of much significance, sap sucking by a large number of aphids can considerably weaken the plant, slow down the rate development, and reduce the tuber yield. Leaf deformation due to aphid feeding is also possible. Production of honey dew can promote the growth of sooty molds on foliage, potentially leading to reduced photosynthetic area and reduced yield. The most important damage caused by aphids in potato crops is due to the spread of viruses, which leads to reduced tuber yield and degeneration of seed stocks (Kroschel et al. 2020). The most important potato viruses transmitted by aphids are *potato virus Y* (PVY) and *potato leaf roll virus* (PLRV), which can cause losses worth millions of rupees (Loebenstein et al. 2001).

In this chapter, we provide an overview of the biology and ecology of the aphids with discussion in the context of potato. Separate sections dealing with virus transmission characteristics of aphids with emphasis on potato viruses are given, and the state of art with respect to transmission of potato viruses by aphids is provided. Finally, we provide a summary of the management methods generally adopted by potato farmers with concluding remarks.

9.2 Species Composition and Colonization

Potato crops are infested by a large number of colonizing and noncolonizing aphids. The colonizing species feed and breed on potato plants whereas the noncolonizing species are occasional transient visitors. More than 22 species of aphids are recorded worldwide that colonize potato plants (Blackman and Eastop 1994, 2000a, b, 2006) (Table 9.1; Fig. 9.1). Most of these aphids are polyphagous with worldwide distribution.

A large number of aphid species are reported on potato crops from different parts of India. Earlier, five major species infesting potato under Indian conditions were known, viz., *Myzus persicae* (peach potato aphid or green peach aphid), *Aphis gossypii* (melon aphid or cotton aphid), *A. fabae* (black bean aphid), *Rhopalosiphoninus latysiphon* (bulb and potato aphid), and *Rhopalosiphum rufiabdominale* (rice root aphid), in addition to two minor species *Rhopalosiphum nymphaeae* (water lily aphid) and *Tetraneura nigriabdominalis* (rice root aphid) (Pushkarnath 1959; Bindra and Sekhon 1971; Verma 1977; Sekhon and Bindra

Table 9.1 List of aphid species colonizing potato (after Blackman and Eastop 1994, 2000a, b, 2006)

S. No.	Species	Common name	Life cycle
1.	*Acyrthosiphon malvae* (Mosley)	Geranium aphid; pelargonium aphid	Autoecious holocyclic
2.	*Aphis craccivora* Koch	Cowpea aphid, black legume aphid	Anholocyclic, sexual morphs recorded from India and Germany
3.	*Aphis fabae* Scopoli	Black bean aphid	Heteroecious holocyclic
4.	*Aphis frangulae* ssp. *beccabungae*	Alder buckthorn-potato aphid	Heteroecious holocyclic
5.	*Aphis gossypii* Glover	Melon aphid; cotton aphid	Anholocyclic/holocyclic
6.	*Aphis nasturtii* Kaltenbach	Buckthorn aphid; buckthorn-potato aphid	Heteroecious holocyclic
7.	*Aphis solanella* Theobold	Black bean aphid	Heteroecious holocyclic
8.	*Aphis spiraecola* Patch	Spiraea aphid; green citrus aphid	Anholocyclic/holocyclic
9.	*Aulacorthum solani* (Kaltenbach)	Glasshouse potato aphid; foxglove aphid	Anholocyclic/holocyclic
10.	*Brachycaudus helichrysi* (Kaltenbach)	Leaf-curling plum aphid	Heteroecious holocyclic/ anholocyclic
11.	*Macrosiphum euphorbiae* (Thomas)	Potato aphid	Heteroecious holocyclic/ anholocyclic
12.	*Myzus antirrhinii* (Macchiati)	–	Anholocyclic
13.	*Myzus ascalonicus* Doncaster	Shallot aphid	Anholocyclic
14.	*Myzus ornatus* Laing	Violet aphid	Anholocyclic, males recorded from India
15.	*Myzus persicae* (Sulzer)	Peach potato aphid; green peach aphid	Heteroecious holocyclic/ anholocyclic
16.	*Neomyzus circumflexus* (Buckton)	Mottled arum aphid	Anholocyclic
17.	*Pemphigus* sp.	–	Not clear
18.	*Pseudomegoura magnoliae* (=*Aulacorthum magnoliae*) (Essig and Kuwana)	–	Mainly anholocyclic
19.	*Rhopalosiphoninus latysiphon* (Davidson)	Bulb and potato aphid	Anholocyclic
20.	*Rhopalosiphum rufiabdominale* (Sasaki)	Rice root aphid	Heteroecious Holocyclic/ anholocyclic
21.	*Smynthurodes betae* Westwood	Bean root aphid	Heteroecious Holocyclic/ anholocyclic
22	*Uroleucon compositae* (Theobald)	Artichoke aphid	Anholocyclic

Aphis craccivora (Apterae)

Aphis spiraecola (Alate and nymphs)

Aphis fabae (Apterae)

Aulacorthum solani (Alate)

Macrosiphum euphorbiae (Apterous)

Myzus persicae (Alate)

Fig. 9.1 Common aphids infesting potato

1979; Kashyap and Verma 1982; Misra and Agarwal 1987; Kumara et al. 2017). Later, Bhatnagar et al. (2017) compiled information on 13 species of aphids recorded on potato crops in India, viz., *M. persicae*, *A. gossypii*, *A. fabae*, *A. spiraecola*

Table 9.2 A summary of studies on noncolonizing aphid species visiting potato crops from across the world

S. No.	Location	No. of species/ taxa reported	Period of study	Reference
1.	Harpenden, England	119	1984	Harrington et al. (1986)
2.	Wageningen, Netherlands	105	1983–1985	Piron (1986)
3.	Sweden	80	1976–1984	Sigvald (1987)
4.	New Brunswick, Canada	62	1984–1987	Boiteau et al. (1988)
5.	Southern Sweden	>20	1975–1979	Sigvald (1989)
6.	Southern and central Sweden	21	1975–1980	Sigvald (1990)
7.	Netherlands	122	1983–1987	De Bokx and Piron (1990)
8.	Minnesota and North Dakota, USA	34	1992–1994	DiFonzo et al. (1997)
9.	Hungary	28	1982–2001	Kuroli and Lantos (2006)
10.	Tunisia	103	2002–2004	Boukhris-Bouhachem et al. (2007)
11.	Tunisia	15	2001–2006	Boukhris-Bouhachem et al. (2010)
12.	Northern Finland	83	2007–2010	Kirchner et al. (2013)
13.	Idaho, USA	46	2012–2013	Mondal et al. (2016)
14.	Hokkaido, Japan	19	2016	Sano et al. (2019)
15.	Northwest Russia	43	2013–2017	Sukhoruchenko et al. (2019)

(spiraea aphid; green citrus aphid), *A. nerii* (oleander aphid), *A. craccivora* (cowpea aphid or groundnut aphid or black legume aphid), *Macrosiphum euphorbiae* (potato aphid), *Brevicoryne brassicae* (cabbage aphid or mealy cabbage aphid), *Aulacorthum solani* (glasshouse potato aphid), *Lipaphis erysimi* (mustard aphid or turnip aphid), *Hyadaphis coriandri* (coriander aphid), *Rhopalosiphum rufiabdominalis*, and *Rhopalosiphum maidis* (corn leaf aphid). In addition to these, *Myzus ornatus* Laing (ornate Aphid or violet aphid) and *Macrosiphum rosae* (Linn.) (rose aphid) are reported from potato in India.

A large number of aphid species are reported worldwide to transiently visit the potato plants while searching for their own host plant(s). The species composition is studied either based on sampling from potato foliage or with the help of Moericke yellow water pan traps or other types of impaction traps. In some of the studies, more than 120 species/specie groups have been collected from traps in potato fields (De Bokx and Piron 1990). In Table 9.2, a summary of studies on species composition of aphids visiting potato crops is given. Most of these aphids are the pests of other crops or originate from a large number of weed flora. Noncolonizing aphids are the major spreaders of nonpersistent viruses like PVY under filed conditions. Further discussion on the importance of noncolonizing aphids is given in other relevant sections.

9.3 Life Cycles and Dispersal

Aphids have complex life cycles characterized by host alternation and facultative parthenogenesis (Blackman and Eastop 2000a, b). Depending on their ability to host alternate, the life cycle may be heteroecious or autoecious. Aphids that practice host alternation are heteroecious; they live and sexually reproduce on a primary host, mostly woody perennials, during winter and colonize secondary hosts during the rest of the year before coming back to their primary host. Although heteroecy is considered a primitive life strategy in aphids, only about 10% of the modern-day aphid species are heteroecious. In contrast, majority of the species of aphids live on the same or a group of closely related herbaceous hosts throughout the year, commonly referred to as auto−/monoecious species (Williams and Blackman 2007).

Depending on their ability to undergo sexual reproduction, the aphid life cycle may be holocyclic or anholocyclic (Blackman and Eastop 2000a, b). Most of the aphid species alternate parthenogenesis and sexual reproduction and are called holocyclic. Such aphid species switch over to parthenogenesis from the first generation in spring to the appearance of sexual morphs in autumn. The asexual phase is spent partly on the primary host and mainly on the secondary hosts. The appearance of sexual morphs is induced by seasonal changes in temperature and photoperiod. In contrast, some species are anholocyclic; they do not produce sexual morphs or eggs and only reproduce by parthenogenesis (Fenton et al. 1998; Williams and Dixon 2007). Such species continue to utilize herbaceous hosts throughout the year, including winters. Although some species are strictly holocyclic or anholocyclic, certain populations in some holocyclic aphids can lose their sexual phase and become anholocyclic or generate only male populations (androcycly) during winter, mostly leading to production of abortive eggs (Blackman 1971; Margaritopoulos et al. 2002).

The viviparous mode of reproduction in aphids confers a rapid reproduction rate with short developmental times, resulting in population growth that is atypically high, even for insects. For instance, Dixon (1971) estimated that aphid populations in potato fields can reach densities of 2×10^9 individuals per hectare. Douglas (2003) suggested that such rates of population increase reflect nutrient allocation to the reproductive system. Energy is preferentially invested in embryo biomass and larval development rather than in maternal tissues. Aphids have telescoping generations, i.e., ovarian development and embryo formation start at the same time in embryonic mothers (Powell et al. 2006). Parthenogenetic reproduction results in clonal aphid colonies that have the same genotype. With this reproduction mode, an atypical characteristic can be amplified and become predominant in a given population after several generations. This can explain why aphids are able to quickly adapt to disturbances in their environment. Aphid populations may crash depending on the weather (Barlow and Dixon 1980), deteriorating resources, or pesticide treatments. However, parthenogenesis rapidly generates new populations that are adapted to their environment and, in some cases, resistant to pesticides.

Parthenogenesis generally occurs during the warmer months of the year and maximizes offspring production. In fall, it is interrupted and followed by sexual reproduction that produces overwintering eggs. Aphids produce both apterous (wingless) and alate (winged) morphs. Production of alate morphs is energetically costly (Dixon et al. 1993). Alates appear at different times during the year. They are considered to be colonizers and use winds to disperse and locate new hosts. Wingless fundatrices emerge from eggs laid on the primary host. Their alate progeny are the spring migrants. Alate production is completed within a 2-week period (Radcliffe 1982). These individuals fly to secondary hosts (e.g., potato) and, when conditions are favorable, generate apterous and parthenogenetic populations. During summer, overpopulation of aphids, degradation of host-plant nutritional suitability, or variations in light intensity, temperature, and precipitation induce the decline in aphid populations and the appearance of winged morphs that move to more suitable host habitats. In autumn, as day length and temperature decrease, the quality of secondary host plants is altered. These factors generate the appearance of a new generation of virginoparous alates that migrate to the primary host. After the second generation on the primary host, oviparous females appear and are fertilized by winged males (Radcliffe 1982). After reproduction, oviparous females lay their eggs on the primary host for overwintering (Powell et al. 2006). Timing of flight and the number of migrants is important for colonization, clonal fitness, and overwintering success. Aphids that colonize potato are mainly heteroecious and holocyclic, whereas as others switch from other herbaceous hosts to colonize potato or visit it transiently.

9.4 Ecology and Chemical Interactions

Other than rapid reproduction, alternation of sexual and asexual phases, and long-range migration, the most noteworthy feature of aphids is the adaptation to host-plant ecology and physiology. This includes host-plant and feeding-site discrimination using sensitive chemosensory cues, role of endosymbionts and chemical communication among the members of the species and between species.

9.4.1 Host-Plant Selection and Feeding

Host-plant selection in insects includes a sequence of behavioral responses. The sequence includes habitat location, host-plant location, host-plant acceptance, and host use. In general, a number of sensory cues, such as visual, olfactory, gustatory, and tactile stimuli as well as humidity and light intensity (Bernays and Chapman 1994), are used by insects during host selection. To locate a suitable host plant, winged aphids are confronted by various challenges, particularly depending on their host-plant range. Among the aphid species, only 5% are considered as polyphagous (Blackman and Eastop 2000a, b), and many others exploit not more than one or few closely related plant species and are highly specialized in their feeding preference (Dixon 1998).

A series of complex behaviors is involved in host-finding behavior by alate morphs of aphids, and these are closely linked with migration and function of dispersal. The sequence of host-selection behavior in aphids can be broadly categorized into three steps, (a) approach and landing on the plant, (b) leaf-surface exploration and brief probes, and (c) host acceptance, after assessment of the phloem sap, which leads to sustained sap ingestion (Niemeyer 1990; Caillaud 1999; Powell et al. 2006). The discrimination between host and nonhost plants involves perception of visual and volatile cues before landing (Nottingham and Hardie 1993; Powell et al. 1999) but also gustatory cues perceived during brief plant subepidermal probes (Bernays and Funk 2000; Caillaud and Via 2000; Powell and Hardie 2000; Funk and Bernays 2001) and during phloem sap ingestion (Van Helden and Tjallingii 1993). The relative importance of each of these steps is influenced by the aphid specialization with respect to the plant (Bernays and Funk 1999; Funk and Bernays 2001) and according to the aphid species (Tosh et al. 2003). Stylet penetration in the epidermis allows aphids to evaluate the phytochemistry of the plant and to detect antifeedant compounds, providing aphids with the information to decide whether to accept or reject the plant. Saguez et al. (2013) and Pettersson et al. (2007) have discussed the host-finding behavior, feeding, and nutrition in aphids in detail.

Since the past few decades, the research revealed that the host-finding and host-selection behavior of aphids are influenced by naturally occurring chemical compounds (Pickett et al. 1992; Pickett and Glinwood 2007; Webster 2012; Pickett et al. 2013). These comprise of (a) volatile organic compounds (VOCs) emitted by host and nonhost plants and (b) volatiles emitted by aphids (pheromones). The aphids' sensory receptor organs called rhinaria (Park and Hardie 2004), circular or oval structures located on the antennae (Shambaugh et al. 1978), perceive these small-molecular-weight lipophilic compounds (Pickett et al. 2013). A third method of chemical stimuli influencing the aphid host-finding and host-selection behavior is at the point when the aphid is making contact with the plant (Backus 1988; Powell et al. 1999, 2006; Alfaro-Tapia et al. 2007). In certain cases, specific VOCs are used by aphids as host cues. In some other cases, individual VOCs act as nonhost cues during host finding; further it depends on the host range of the aphid species. Apart from the effects of individual compounds, there are also specific effects of VOC blends (relative concentration of chemicals in a mixture of VOCs) on aphid choice behavior (Bruce et al. 2005). For example, VOCs that act as host cues in a blend can become nonhost cues when presented individually (Webster et al. 2010).

According to Powell and Hardie (2001), it is common that aphid species are able to respond to their primary host plant volatile cues, but there is variation in response to volatile cues by individuals from different developmental stages/phenotypes. Summer female aphids (virginoparae) do not show host-plant selectivity, whereas autumn return migrants (gynoparae and males) show olfactory responses to their primary host plant (Powell and Hardie 2001). Phenotypic differences were also identified by electroantennography among the different female phenotypes (virginoparae and gynoparae) and males of *A. fabae* (Powell and Hardie 2001). Wingless aphids of *Macrosiphum euphorbiae* is attracted to potato foliage, while winged aphids are not (Narayandas et al. 2006). A synergism between host-plant volatile and pheromone component has also been shown for aphids, for example, *A. fabae* primary host plant odors increase the response to the sex pheromone, released by mature oviparae aphids, when they return to their host in the autumn (Powell and Hardie 2001).

9.4.2 Endosymbiosis

Symbioses have evolved independently between various insect groups and microorganisms. Almost all of the insects harboring endosymbionts live through the life cycle on nutritionally unbalanced or poor diets. Majority of aphid species possess intracellular bacteria of the genus *Buchnera*, including the ones that colonize potato, namely, *Myzus persicae* and *Aphis gossypii*. *Buchnera* has an obligate association with aphids and are vertically transmitted via the aphid ovary. The *Buchnera aphidicola* benefits *M. persicae* by providing essential amino acids and vitamins that it cannot obtain in sufficient quantities from its diet (Douglas 1998; Prosser and Douglas 1991). Hence the presence of *B. aphidicola* is necessary for the survival and reproduction of the aphids, and the *Buchnera*-free aphids develop poorly and produce no or a few offspring. Disrupting this endosymbiotic bacterium of *M. persicae* can also change the feeding behavior, resulting in delayed host-plant acceptance (Machado-Assefh et al. 2015). The association of *Buchnera* also provides nonnutritional benefits like thermal tolerance and protection from the natural enemies to the aphids. The obligate endosymbionts limit the thermal tolerance of the host species. For example, the exposure of *Aphis gossypii* to elevated heat did not change *Buchnera* titer, resulting in enhanced fecundity. In contrast, heat suppressed the *Buchnera* titer in *A. fabae*; hence they suffered enhanced mortality, delayed development, and reduced fecundity (Zhang et al. 2019). Endosymbiotic bacteria also help *M. persicae* in the circulative transmission of PLRV. The endosymbiotic bacteria synthesize a predominant protein called symbionin and release it in the hemolymph. The symbionin interact with the coat protein of the virus and protect it from enzymatic breakdown in the vector hemolymph (Van den Heuvel et al. 1994).

Apart from the primary obligate bacteria, aphids harbor many facultative bacteria that are not necessarily required for aphid survival or reproduction but may give fitness advantages. Facultative secondary symbionts inhabit bacteriocytes, sheath cells, or hemocoel and are maternally or horizontally transmitted. Seven facultative endosymbionts have been reported from *M. persicae*, namely, *Hamiltonella defensa*, *Serratia symbiotica*, *Regiella insecticola*, *Wolbachia*, *Rickettsia*, *Arsenophonus*, and *Spiroplasma* (Vorburger et al. 2010; Xu et al. 2021). Among these facultative endosymbionts, *Regiella insecticola* have been reported in *M. persicae* to give protection against its two major parasitoids *Aphidius colemani* and *Diaeretiella rapae* (Von Burg et al. 2008; Vorburger et al. 2010).

9.4.3 Semiochemicals

Intraspecific communication in aphids is meant for attracting mates, aggregation, avoidance of competition, and warning against threats, like most other insects. Such signals are pivotal at different stages of the complex aphid life cycles, such as finding of correct primary and secondary hosts, finding mates before ensuing sexual reproduction, and evading predators and parasitoids who are able to respond to some of such cues. Therefore, aphids make extensive use of various semiochemicals at different stages of the life cycles.

The sex pheromones are produced in glandular epidermal cells on the tibiae of the hind legs of the sexual females and perceived by placoid sensilla, in the secondary rhinaria on the antennae of male aphids. During pheromone release, the female engages in typical "calling" behavior, with the hind legs raised (Hardie et al. 1991; Dewhirst et al. 2010). The pheromones usually comprise (4aS,7S,7aR)-nepetalactone and (1R,4aS,7S,7aR)-nepetalactol, monoterpenoids in the cyclopentanoid or iridoid series (Campbell et al. 2003). A further compound, (1S,2R,3S)-dolichodial, has been identified from oviparae of *Dysaphis plantaginea* (rosy apple aphid) (Dewhirst et al. 2008). Most aphids examined so far employ a limited range of pheromone components, but there are differences in relative and absolute compositions.

The asexual forms, and most often the apterae, release an alarm pheromone when disturbed. Nearby aphids exhibit a variety of behaviors, ranging from the removal of mouthparts from the plant and moving away to running, dropping off the plant, and even attacking the predator. Moreover, exposure to alarm pheromone can lead to an increase in the production of winged morphs in an aphid colony (Hardie et al. 1991; Vandermoten et al. 2012). The alarm pheromone is secreted along with the honeydew through siphunculi. The main component of the alarm pheromone of many aphids is the sesquiterpene hydrocarbon (E)-β-farnesene (Bowers et al. 1972; Edwards et al. 1973; Wientjens et al. 1973; Pickett and Griffiths 1980). Other components may also be present. For example, the alarm pheromone of *Megoura viciae* (vetch aphid) contains the monoterpenes α-pinene, β-pinene, (Z, E)-α-farnesene, and (E,E)-α-farnesene, in addition to (E)-β-farnesene, and these can synergize the activity of the latter.

A series of chemicals and their combinations have been demonstrated to have a role in the aggregation of aphid colonies and regulation of overcrowding. Similarly, semiochemicals from host plants are being identified that help the aphids to locate primary and secondary hosts (Pickett et al. 2017).

9.5 Virus Transmission by Aphids

9.5.1 Aphid Characteristics

Majority of the plant viruses are transmitted by arthropod, nematode, or fungal vectors, and among these, aphids are the most important family transmitting more viruses than any other group. More than 5000 aphid species have been described, and of these, over 190 have been reported to transmit plant viruses with many species able to transmit more than one virus (Remaudière and Remaudière 1997; Nault 1997; Hull 2002). Potato is infected by more than 30 RNA viruses (Salazar 1996), among which 13 are transmitted by aphids (Brunt and Loebenstein 2001). The two most important potato viruses transmitted by aphids are the PLRV and PVY. Other than these, *potato virus M* (PVM), *potato virus S* (PVS), *potato latent virus* (PLV), and *potato yellowing virus* (PYV) can become sporadically important (Brunt and Loebenstein 2001).

The virus transmission by an aphid consists of acquiring a virion from an infected plant, its retention in or on the aphid, and its inoculation in another plant to establish infection. The aphid may not be able to immediately release the virus and can do that only after some time has elapsed—the "latent period." Depending on the time for which the aphid can retain a virus in or on it to remain viruliferous, the modes of transmission are generally classified as nonpersistent, semi-persistent, or persistent. In nonpersistent transmission, virus acquisition and inoculation require few seconds to minutes, and there is no latent period involved in between. Such viruses are carried on the stylets of aphids and are retained for a very short time, e.g., potyviruses (*potato viruses A*, *Y*, and *V*), PVM (some strains), PVS (some strains), etc. The most important nonpersistent potato virus is PVY. For semi-persistent viruses, acquisition and inoculation take longer (usually 15 min), and there is no latent period in between. The aphids remain viruliferous for about 2 days. The persistent viruses take much longer for acquisition and inoculation, and there is a significant latent period involved. The aphids remain viruliferous for the lifetime after the latent period has passed, e.g., PLRV.

Several characteristics of aphids predispose them to being efficient virus vectors. Among the most important factors is the feeding behavior of aphids. After landing and tarsal contact with green surfaces, aphids tend to make brief stylet insertions ("probes") into the epithelial or parenchymal tissues. Probing behavior is a particularly important feature of host-plant selection by aphids, which provides information about host quality (Powell and Hardie 2000; Powell et al. 2006). Due to apparent lack of chemosensillae on the stylets, aphids need to ingest plant sap into the pharyngeal area of the foregut for chemosensory assessment. During the probing, the stylets puncture the epidermal cells for a very brief period of time and during which the virion of the nonpersistent virus are acquired. Aphids make several such probes on a plant before actual feeding on the phloem sap or rejection of the plant and moving on to the next plant. This phenomenon continues, and aphids tend to probe several plants before settling for feeding. This is perhaps the most important reason for quick spread of nonpersistent viruses by aphids and failure of chemical control to check such spread of viruses.

Molecular interaction of the aphid-virus-plant complex indicates a complex plethora of pathogenesis and defense reactions. For examples, the gelling saliva of aphids is known to contain phenoloxidases, peroxidases, pectinases, and glucosidases (Cherqui and Tjallingii 2000; Tjallingii 2006) whereas the watery saliva is a complex mixture of enzymes, e.g., those capable of degrading plant cell walls or preventing occlusion of sieve tubes and others capable of eliciting plant defense responses (Will et al. 2009, 2012; Bak et al. 2013). Proteome analysis of the saliva identified a wide range of secreted effectors with complex roles (Elzinga and Jander 2013; Pitino and Hogenhout 2013).

Virus infection of plants has been shown to increase the fitness of the aphids feeding on such plants. Viruses possibly affect the aphids directly or by manipulating the host plants to their advantage. This is further discussed in other sections of this chapter. Other than the host selection and feeding behavior of aphids, other biological characteristics help them spread viruses at alarming rates in crop plants, e.g., life cycle and dispersal, and host range, which are further discussed in other sections of this chapter.

9.5.2 Role of Colonizing and Noncolonizing Aphids

Broadbent (1948) was first to suggest that alatae of species that did not colonize potato could be potential vectors of PVY because of the brief probes they make when visiting potato crops. Till the 1990s, hundreds of noncolonizing aphids were evaluated for their ability to transmit potato viruses, PVY in particular. Among these, around 65 species are now established as vectors of PVY strains (Table 9.3). Although nonpersistent viruses are retained for a few seconds to minutes in their vectors, the retention times for PVY^N in its vectors can range from 4 h (Proeseler and Weidling 1975) to 17 h (Kostiw 1975). Therefore, it is to be expected that the noncolonizing aphids originating either from nearby or far locations can bring the viruses along and inoculate potato plants. In spite of this, the sources of virus within the crop fields (infected seed) are demonstrated to be more important in the spread of viruses in seed potato crops.

Although the colonizing species are more efficient at virus transmission compared with the noncolonizing species, the latter are the most important vectors of nonpersistent viruses because of their huge numbers (Halbert et al. 2003). Opposite to this, the spread of persistent viruses like PLRV is mainly accomplished by the colonizing species (Table 9.4). Persistent viruses are acquired when the aphids finally feed on phloem sap. Since the process of accepting a plant as host and locating the phloem takes a while, therefore, noncolonizing aphids are theoretically incapable of spreading persistent viruses like PLRV.

Table 9.3 List of aphid species known to transmit PVY (modified after Al-Mrabeh 2010; Lacomme et al. 2017)

S. No	Aphid species	Major host plants	PVY strain	Transmission efficiency (%)
1.	*Acyrthosiphon pisum*	Fabaceae, important pest of peas and alfalfa	PVY^N	14.0
2.	*Acyrthosiphon primulae*	*Primula* spp.	PVY^N	15.0
3.	*Anoecia corni*	Host alternation between *Cornus sanguinea* and roots of Poaceae	PVY^O	–
4.	*Aphis citricola* (=*Aphis spiraecola*)	Caprifoliaceae, Compositae, Rosaceae, Rubiaceae and Rutaceae, major pest of *Citrus*	PVY (pepper)	6.2
5.	*Aphis craccivora*	Fabaceae, major pest of leguminous crops	PVY (pepper)	4.0
6.	*Aphis fabae*	Host alternation between *Euonymus europaeus* and a variety of plants; *Aphis fabae* s. str. Colonizes *Vicia faba*	PVY^O, PVY^N	24.0
7.	*Aphis fabae cirsiacanthoides*	Host alternation between *Euonymus europaeus* and *Cirsium arvense*	PVY^O, PVY^N	39.3 for PVY^O, 80 for PVY^N
8.	*Aphis frangulae*	Sexual phase in Europe on *Rhamnus frangula*, host alternates to a wide range of plants depending on the subspecies		–
9.	*Aphis glycines*	Fabaceae, particularly *Glycine* spp., a major pest of soybean	PVY^O, PVY^N, PVY^{NTN}	14–75
10.	*Aphis gossypii*	On a very wide range of host plants, major pest of cotton and cucurbits, and in glasshouses in cold temperate regions	PVY^O	31
11.	*Aphis helianthi* (=*Aphis asclepiadis*, *A. carduella*)	Compositae/Asteraceae and Umbelliferae/Apiaceae		–
12.	*Aphis nasturtii*	Sexual phase on *Rhamnus* spp., on *Nasturtium officinale*, potato, *Veronica beccabunga*, *Drosera rotundifolia*, and *Rumex* spp.	PVY^O	7.1
13.	*Aphis pomi*	Rosaceae including *Chaenomeles*, *Cydonia*, *Malus*, and *Pyracantha*	PVY^O, PVY^N	2–9
14.	*Aphis rumicis*	On *Rumex* spp. and *Rheum* spp.		–

(continued)

Table 9.3 (continued)

S. No	Aphid species	Major host plants	PVY strain	Transmission efficiency (%)
15.	*Aphis sambaci*	*Sambucus* spp.; host alternation occurs in roots of plants such as *Cerastium*, *Dianthus*, *Silene*, *Melandrium*, *Moehringia*, and *Spergula* and also often on *Rumex*, *Capsella*, *Oenothera*, and *Saxifraga*	PVY^O, PVY^N	4.3–12
16.	*Aphis spiraecola*	See *Aphis citricola*		–
17.	*Aulacorthum solani*	Foxglove, extremely polyphagous	PVY^O, PVY^N	5
18.	*Brachycaudus cardui*	Compositae, e.g., *Arctium*, *Carduus*, *Cirsium*, *Cynara*, *Chrysanthemum*, *Tanacetum*, *Matricaria*), and Boraginaceae, e.g., *Borago*, *Cynoglossum*, *Echium*, *Symphytum*		–
19.	*Brachycaudus helichrysi*	Sexual phase on *Prunus* spp., host alternates to Compositae/Asteraceae and Boraginaceae	PVY^O, PVY^N	7.2 for PVY^O 0.9 to 5.9 for PVY^N
20.	*Brevicoryne brassicae*	Brassicaceae		–
21.	*Capitophorus elaeagni*	*Elaeagnus* spp. and sometimes on *Hippophae* migrate to Compositae (*Arctium*, *Carduus*, *Cirsium*, *Cynara*, *Gerbera*, *Silybum*)	PVY^O	2
22.	*Capitophorus hippophaes*	Elaeagnaceae (*Elaeagnus* spp., *Hippophae* spp.) migrate to Polygonaceae such as *Polygonum* and *Persicaria* spp.	PVY^N	3
23.	*Caveriella aegopodii*	Numerous genera and species of Umbelliferae, sexual phase on various *Salix* spp.	PVY, PVY^N	0.2–0.4
24.	*Caveriella pastinacae*	Host alternates from *Salix* to *Heracleum*, less commonly to *Pastinaca*		–
25.	*Cryptomyzus ballotae*	*Ballota nigra*	PVY	100
26.	*Cryptomyzus galeopsidis*	*Ribes* spp., migrating to *Lamium* and *Galeopsis*	PVY^N	17.4
27.	*Cryptomyzus ribis*	On *Ribes* spp., migrating to *Stachys* spp.	PVY^N	15.4

(continued)

Table 9.3 (continued)

S. No	Aphid species	Major host plants	PVY strain	Transmission efficiency (%)
28.	*Diuraphis noxia*	On grasses and cereals *Agropyron*, *Anisantha*, *Andropogon*, *Bromus*, *Elymus*, *Hordeum*, *Phleum*, *Triticum*	PVY pepper	4–7
29.	*Drepanosiphum platanoidis*	*Acer pseudoplatanus*, common on sycamores	PVY^N	0.6
30.	*Dysaphis plantaginea*	*Malus* spp., *Pyrus* spp., *Plantago* spp.		–
31.	*Dysaphis aucuparie*	On *Sorbus torminalis*, migrating to *Plantago* spp.		–
32.	*Hayhurstia atriplicis*			–
33.	*Hyadaphis foeniculi*	On *Lonicera* spp., migrating to various Umbelliferae	PVY^N	14.7
34.	*Hyalopterus pruni*	On *Prunus domestica*, migrating to *Phragmites* or sometimes to *Arundo donax*	PVY^N	13.9
35.	*Hyperomyzus lactucae*	On *Ribes* spp., migrating to *Sonchus* spp. and occasionally other Asteraceae	PVY^N	17.4
36.	*Hyperomyzus pallidus*	On Amaranthaceae, usually *Atriplex* and *Chenopodium* spp.		–
37.	*Lipaphis erysimi*	On various Brassicaceae (*Arabis*, *Capsella*, *Coronopus*, *Erysimum*, *Isatis*, *Lepidium*, *Matthiola*, *Sinapis*, *Sisymbrium*, *Thlaspi*, etc.) but not usually on field *Brassica* crops		–
38.	*Macrosiphum euphorbiae*	Sexual phase on *Rosa*, secondary hosts in more than 20 different plant families	PVY^N	29
39.	*Macrosiphum rosae*	On *Rosa* spp. in spring, migrating to Dipsacaceae (*Dipsacus*, *Knautia*, *Succisa*) and Valerianaceae (*Centranthus*, *Valeriana*)		–
40.	*Metopolophium albidum*	On grasses such as *Arrhenatherum elatius*	PVY^N	11
41.	*Metopolophium dirhodum*	On *Rosa* spp. in spring, migrating to numerous species of Poaceae and Cyperaceae.	PVY^N	3

(continued)

Table 9.3 (continued)

S. No	Aphid species	Major host plants	PVY strain	Transmission efficiency (%)
42.	*Metopolophium festucae*	Poaceae	PVY^O	0.5
43.	*Myzaphis rosarum*	Wild and cultivated *Rosa*	PVY^O	10
44.	*Neomyzus circumflexus*	*Sonchus oleraceus*	PVY^O, PVY^N	–
45.	*Myzus ascalonicus*	Polyphagous, Alliaceae, Caryophyllaceae, Compositae, Brassicaceae, Liliaceae, and Rosaceae		–
46.	*Myzus cerasi*	On *Prunus* spp., migrating to secondary hosts in Rubiaceae (*Asperula*, *Gallium*), Orobanchaceae (*Euphrasia*, *Rhinanthus*), Plantaginaceae (*Veronica*), and certain Brassicaceae (*Capsella*, *Cardamine*, *Coronopus*, *Lepidium*)	PVY^O, PVY^N	3.2
47.	*Myzus certus*	On Caryophyllaceae (*Cerastium*, *Dianthus*, *Stellaria*)	PVY^N	71.0
48.	*Myzus ligustri*	Privet hedges (*Ligustrum ovalifolium*, *L. vulgare*)	PVY^O, PVY^N	30.0 for PVY^O, 76.3 for PVY^N
49.	*Myzus myosotidis*	*Myosotis scorpioides* (=*palustris*)	PVY^O	100.0
50.	*Myzus persicae nicotianae*	Host alternates from *Prunus* to tobacco	PVY, PVY^N	15.3
51.	*Myzus persicae*	Host alternates from *Prunus* to a wide variety of plants.	PVY^O, PVY^N	50.0–71.0
52.	*Phorodon humuli*	On *Prunus* spp., migrating to *Humulus lupulus*	PVY^N	35
53.	*Rhopalosiphum oxyacanthae* (=*R. insertum*)	On Pyroideae (*Cotoneaster*, *Crataegus*, *Malus*, *Pyrus*, *Sorbus*) migrating to Poaceae (*Agropyron*, *Agrostis*, *Alopecurus*, *Dactylis*, *Festuca*, *Glyceria*, *Phalaris*, *Poa*, *Triticum*)	PVY^N	50
54.	*Rhopalosiphum maidis*	On *Avena*, *Hordeum*, *Oryza*, *Saccharum*, *Secale*, *Sorghum*, *Triticum*, and *Zea*, migrating to *Prunus* spp.	PVY^O	1.5
55.	*Rhopalosiphum padi*	On *Prunus* spp., migrating to numerous grasses and cereals	PVY^O, PVY^N	2–11.5

(continued)

Table 9.3 (continued)

S. No	Aphid species	Major host plants	PVY strain	Transmission efficiency (%)
56.	*Rhopalosiphum pseudobrassicae* (=*Lipaphis pseudobrassicae*)	Brassicaceae, including *Barbarea*, *Brassica*, *Capsella*, *Iberis*, *Raphanus*, and *Rorippa*		–
57.	*Schizaphis graminum*	Various species of Poaceae		
58.	*Sitobion avenae*	On numerous species of Poaceae, including all the cereals and pasture grasses	PVY^O, PVY^N	0.1–1.8
59.	*Sitobion fragariae*	Apterae on *Rubus* and other Rosaceae, migrating to Poaceae	PVY^O, PVY^N	0.5–10.1
60.	*Sitobion graminum*	Most probably *Schizaphis graminum*		–
61.	*Staphylae tulipaellus* (=*Rhopalosiphoninus staphyleae* ssp. *tulipaellus* Theobald)	*Beta vulgaris*, also recorded from the roots *Galium*, *Lycopersicon*, *Rumex*, *Tulipa*, and *Viola*		–
62.	*Therioaphis trifolii*	On many plants of Leguminosae/Fabaceae in the genera *Astragalus*, *Lotus*, *Medicago*, *Melilotus*, *Onobrychis*, *Ononis*, and *Trifolium*		–
63.	*Tetraneura ulmi*	Poaceae	PVY^N	–
64.	*Uroleucon* spp.	Compositae/Asteraceae	PVY^O, PVY^N	0.5–8.3
65.	*Uroleucon sonchi*	Mainly on *Sonchus* spp. and other genera in the tribe Cichoriaceae (*Lactuca*, *Cichorium*, *Hieracium*, *Ixeridium*, *Picris*, *Reichardia*)	PVY	–

9.5.3 Role of Apterae

The relative role of apterae in within-field spread of potato viruses continues to be a controversial topic. However, evidence has been slowly accumulating, which shows that apterae leave their host plants quite readily and can then play an important role in the local spread of virus within crops (Hodgson 1991). The voluntary movement of apterae could be particularly significant along the leaf blades of canopies of adjacent plants in touch, or by walking across soil from one plant to another (Ferrar 1969; Alyokhin and Sewell 2003). Major factors thought to affect dispersal by apterae include climatic effects (wind, rain), parasitoids and predators, host-plant

Table 9.4 List of reported aphid vectors of *potato leaf roll virus* (PLRV) (Source: https://aphmon.fera.co.uk/plrv_vector_info.cfm)

S. No.	Species	Relative efficiency factor
1.	*Aphis fabae*	0.30
2.	*Aphis gossypii*	0.50
3.	*Aphis nasturtii*	0.25
4.	*Aulacorthum circumflexum*	0.90
5.	*Aulacorthum solani*	0.30
6.	*Macrosiphum euphorbiae*	0.15
7.	*Myzus ascalonicus*	0.30
8.	*Myzus ornatus*	0.30
9	*Myzus persicae*	1.00
10.	*Phorodon humuli*	0.12
11.	*Rhopalosiphoninus latysiphon*	0.30
12.	*Rhopalosiphoninus staphyleae*	0.10

quality, and intra- and interspecific population interactions (summarized in Hodgson 1991). Hodgson (1991) found that apterous dispersal is frequent in *Myzus persicae* (Sulzer), *Brevicoryne brassicae* (Linnaeus), and *Megoura viciae* Buckton and arguably in other aphid species; the movement occurs at low population densities, mainly due to a reduction in the host-plant quality, and the main emigrants are young adults or fourth instar apteriform nymphs. Narayandas and Alyokhin (2006) reported that regardless of canopy overlap, most apterae of *Macrosiphum euphorbiae* (Thomas) moved within the rows of potato plants. Wind, rain, and mechanical raking significantly encouraged aphid movement between plants with overlapping canopies. Therefore, it is plausible to conclude that movement of apterae could have implications for within-field and along the row spread of viruses; however, the exact role needs to be ascertained for specified conditions (Narayandas and Alyokhin 2006).

9.5.4 Virus Induced Changes in Host Plant and Aphid

Plant viruses depend on both host plant and vectors for a successful infection and survival. Such vector-borne pathogens can modify their hosts and vectors in such ways that shape the frequency and nature of interactions between them, resulting in significant implications on transmission and spread of disease. In virus-induced host-plant manipulation, host odors are particularly probable targets for manipulation for the insect-borne pathogens as the insect uses host-released volatile compounds as key foraging cues, particularly host recognition and acceptance. *Cucumber mosaic virus* significantly increases the attractiveness of infected host plants by inducing elevated emissions of a plant volatile blend for *M. persicae* and *A. gossypii* (Mauck et al. 2010). Similarly *bean common mosaic virus* (BCMV), *bean common mosaic necrosis virus* (BCMNV), and *cucumber mosaic virus* (CMV) considerably reduce host-plant quality, inducing dispersal of *M. persicae* and *A. gossypii* from such plants but increasing the attractiveness of infected host plants to aphids via increased emissions of a plant volatile blend (Wamonje et al. 2020). Thus, these viruses appear to attract insect vectors deceptively to infected plants from which they then disperse rapidly; this is a pattern highly conducive to the nonpersistent transmission.

Viruses can also alter the host-plant metabolism or plant defense pathways that favor vector's attraction, settling, or feeding which, in turn, can be favorable for virus propagation and spread. Bak et al. (2019) reported that PVY and *turnip mosaic virus* manipulate host physiology by induction of ethylene signaling, which mediates *M. persicae* attraction to infected plants and hence virus spread. Similarly, PLRV infection attenuates the induction of jasmonic acid and ethylene using transient expression of three PLRV proteins (P0, P1, and P7) in potato and *Nicotiana benthamiana*. Attenuated induction of aphid-induced phytohormones manifests to alter host physiology and, in turn, aphid behavior and fecundity (Patton et al. 2020).

To understand the direct effect of the plant viruses on their vectors, Rajabaskar et al. (2014) carried out a study using *M. persicae*-PLRV pathosystem and observed that the viruliferous aphids prefer to settle on the healthy potato plants, whereas the non-viruliferous aphids preferred potato plants infected with PLRV. The direct effects on the vector upon acquisition of virus in terms of vector performance, behavior, or fecundity and longevity are also documented, which, in turn, could have implications for multiplication and spread of the viruses (Rajabaskar et al. 2014; Eigenbrode et al. 2018).

9.5.5 Virus Transmission Efficiency of Aphids and Vector Pressure

Numerous species of aphids visit potato crops transiently, and a number of species can breed on potato plants. Among these, the number of species that are physically capable of transmitting nonpersistent viruses like PVY is much higher compared with those that can transmit the persistent viruses like PLRV. The vectors are able to transmit PVY with variable efficacy (Kostiw 1979; Van Hoof 1980; Sigvald 1984; Harrington and Gibson 1989; De Bokx and Piron 1990). For instance, if a particular aphid species was found to transmit PVY 50 times out of the 100 times it fed, we would say that that species has a transmission efficiency of 50%. The peach potato aphid, *Myzus persicae*, is generally accepted as the most efficient vector of PVY. The virus transmission efficiency of all other species of aphids are expressed relative to the transmission efficiency of *M. persicae*, generally referred to as relative efficiency factor (REF); *M. persicae* is assigned an efficiency factor of 1. These REFs for the different aphid species are used to calculate the cumulative vector pressures of all the vector species present and contribute to PVY forecasting or control systems. Vector pressure is given by the product of the count of individuals of a particular species caught in traps in a particular period of time, mostly 1 week, and its corresponding REF. Vector pressure is considered as an important measure of estimating the risk the PVY spread in seed potato crops.

The virus transmission efficiency of aphids has been evaluated since the 1980s using different methods, mainly in Europe. In one method, the aphids were caught alive from potato fields, allowed to probe PVY-infected plants, and subsequently transferred to healthy potato plants. The resulting percentage of infected plants gave a measure of virus transmission efficiency of the aphids (Ryden et al. 1983; Sigvald 1984, 1986; De Bokx and Piron 1990). In the alternative method, the aphids caught alive from the potato fields were directly transferred to healthy plants (mostly tobacco) to determine their transmission efficiency (Harrington et al. 1986; Kostiw 1979; Katis and Gibson 1985; Woodford 1992; Boiteau et al. 1998; Halbert et al. 2003). Lately, the apterae from aphid cultures were used to assess their efficiency at transmitting PVY strains (Verbeek et al. 2010). The results most often differ among the studies mainly due to the use of different methods, biotypes of aphids, and host plants used (Verbeek et al. 2010). Earlier studies evaluated the transmission efficiency for strains like PVY^{O} and PVY^{N}; the prevalence of recombinant strains like PVY^{NTN} and $PVY^{N\text{-}Wi}$ has necessitated a fresh evaluation of the virus transmission efficiencies. It is reported that strains like PVY^{NTN} and $PVY^{N\text{-}Wi}$ are transmitted at a higher rate than PVY^{O} or PVY^{N} (Verbeek et al. 2010; Mondal et al. 2016).

The REFs and the vector pressure are used for forecasting incidence of PVY and to take management decisions, particularly the timing of insecticide application, selection of the kind of pesticide to be applied, and decision on the time for cutting of haulms in seed potato crops. Many countries or regions producing seed potato operate trapping networks to monitor the flights of aphids and to alert farmers about the risk of virus spread in the current crop season. The transmission risk is mostly evaluated in terms of vector pressure (calculated by multiplying the abundance of each aphid species by its corresponding relative transmission efficiency factor (REF value) (van Harten 1983; Verbeek et al. 2010) and summing over the species (Basky 2002, 2006; Northing 2009; Kirchner et al. 2011).

During the early years, the population counts of aphids on potato plants were the determinants. However, from 1951 the flight activity of *M. persicae* became the main criterion, and this was recorded by using many Moericke (yellow water) traps. When an average of two or more *M. persicae* were caught in the yellow traps of one region on 1 day, this was taken as an indication that the summer flight of this species had started. The haulms of basic seed fields were usually destroyed within 10 days of that particular day (Hille Ris Lambers 1972). As long as PLRV was the most important virus disease in the Netherlands, this system functioned satisfactorily. However, in the 1950s, a new strain of PVY^{N} invaded Europe, and the symptoms caused by it were mostly overlooked. As a result, roguing, which had been a good way to control other long known strains of PVY, was less effective, and there was a rapid spread of PVY^{N}. Since then PVY^{N} has had to be taken into account in seed potato production (Van Harten 1983). Since 1976, much information has been published on early spread of PVY^{N} in the Netherlands (van Hoof 1977, 1979) and on the ability and efficiency of many aphid species to transmit it (Kostiw 1979; Ryden 1979; van Hoof 1980).

By attributing relative efficiency factors to predominant vector species and considering their flights as recorded with suction traps in the Netherlands, values of vector pressure were obtained that correlate well with weekly infection of bait plants (Van Harten 1983). In Sweden, the relationship between occurrence of alate aphids and the proportion of PVY-infected progeny tubers has been studied since 1975. A dynamic simulation model for PVY has been designed for predicting the incidence of PVY. The simulation model describes a system which includes, e.g., healthy and PVY diseased potato plants, different aphid species as virus vectors and their efficiency as virus vectors, the susceptibility of the potato crop according to mature plant resistance, and date of haulm destruction. There was a good correlation between model output and samples of progeny tubers tested for PVY (Sigvald 1992). Basky (2002) conducted an aphid and virus survey in Hungary yearly between 1993 and 2000. Aphid flight was monitored using yellow pan traps, and virus infection in seed potato progeny tubers was tested with double-antibody sandwich ELISA and varied between 0.75% and 31.8% (PVY) and 0% and 13.25% (PLRV). A simple linear regression analysis showed that the factors examined, i.e., total aphid number, vector number, cumulative vector intensity, and age-corrected vector intensity, had significant effects on the proportion of PVY- and PLRV-infected progeny tubers in seed potato fields. Kirchner et al. (2011) modeled the seasonal increase in PVY incidence using aphid counts in traps, the relative vector efficiencies of the aphids, virus resistance of cultivars, and the initial infection rate of the seed tubers as explanatory variables in generalized linear mixed modeling in Finland. Results of this modeling approach showed that the incidence of seed-borne PVY infection and the early-season vector flights are the most important factors contributing to the incidence of PVY in the yield. Steinger et al. (2015) used a linear regression model including the cumulative sums (until mid-June) of two aphid species (*Brachycaudus helichrysi* and *Phorodon humuli*) as predictor variables for virus disease, which was remarkably well supported by the data ($R^2 = 0.86$). Remarkably, the abundance of *M. persicae*, often considered the main vector of PVY, was not correlated with virus incidence. Taken together, the analysis suggests that the early migrating aphid *B. helichrysi*, rather than *M. persicae*, is the main vector of PVY in Switzerland and that suction trap data are useful for the design of decision-support systems aimed at optimizing virus control in seed potato production.

Extensive aphid monitoring programs using suction traps have been running successfully in European countries, the USA, and New Zealand, for example. The oldest network is in the UK, which has been running for more than 50 years. In the United Kingdom, aphids relevant to seed potato protection are monitored by the Rothamsted/SASA suction-trap network and the FERA yellow water-pan trap network. Suction trap aphid data and weather data are used to forecast the start of aphid flights. Each week, results of trap catch (species composition and abundance) with a cumulative vector pressure index are published and made available to the farmers and others involved with this sector. This index is designed to give the user an assessment of the risk to their crop of PVY spread and helps in decision-making processes when considering the need for insecticide treatments and in deciding the best time to burn down/cut haulms of potato crops (https://secure.fera.defra.gov.uk/aphmon/index.cfm).

On similar lines, to monitor aphid flight, a national aphid-monitoring suction trap network has been established in South Africa in 2005. The network consists of nine 12.2-m-high Rothamsted-type suction traps, which are situated throughout major seed potato-growing regions. Each trap represents aphid samples over a radius of approximately 80 km. The aim of the South African network is to provide seed potato growers with aphid abundance data on a regional level to assess virus risk. The monitoring network and associated web-based database are to serve as an early warning system to assist growers in making management decisions regarding the location and timing of aphid control measures. To view long- and short-term trends in aphid abundance and keep track of aphid numbers and vector pressure, seed growers can apply for user registration on the website of Potatoes South Africa (Kruger and Laebscher 2012) (www.potatoes.co.za).

9.6 Management of Aphids in Potato Crops

Since the managements of aphids is the most important way to manage the incidence of aphid-transmitted viruses in potatoes, various tactics are adopted for the management of aphid-virus complex in seed potatoes. In ware potatoes, a comparatively less stringent pest management regime is adopted. Dupuis et al. (2017) and Pickup and Lacomme (2017) have discussed the subject in detail. The various aspects of the integrated management of aphids in potato cops are discussed as follows.

9.6.1 Monitoring of Aphids

The management of aphids in potato is principally the management of aphid-transmitted potato viruses. Aphids spread viruses when they move from an infected plant to a healthy one. Therefore, it is imperative to monitor the flight activity of aphids to assess the risk of virus spread under filed conditions. As described before, the flight activity of aphids is monitored using the yellow water pan traps or the section traps; each of these has its own merits and demerits. The information on abundance, species composition and flight activity, and ensuing risk of virus spread is made available to the farmers to decide the timings of pesticide/mineral oil applications or the timings for cutting of haulms (Pickup and Lacomme 2017). Various networks of suction traps and water pan traps are being operated in different parts of the world, as described in the earlier section.

9.6.2 Chemical Control

Various contact and systemic insecticides are used worldwide for the management of aphids in potato crops. Among the most commonly used ones are the neonicotinoids including imidacloprid, clothianidin, and thiamethoxam as seed treatment and foliar sprays. Other than these, dinotefuran and nitenpyram are also recommended. Due to their systemic ability and persistence, these are very popular among farmers (Dewar and Denholm 2017). Among the synthetic pyrethroids, cypermethrin, deltamethrin, esfenvalerate, lambda-cyhalothrin, and beta-cyfluthrin are effective due to their knockdown ability and ability to control nonpersistent viruses (Bedford et al. 1998). Among the new-chemistry insecticides, pymetrozine and flonicamid exert similar effects against aphids, causing irreversible cessation of feeding within a few hours of application, followed eventually by starvation and death, and are highly effective against aphids (Schwinger et al. 1994; Morita et al. 2007). Spirotetramat among the novel classes—the tetronic and tetramic acid derivatives—has shown promising results against aphids (Bruck et al. 2009). However, many aphids became resistant to insecticides (Radcliffe 1982; Devonshire et al. 1998; Foster et al. 2000). Various mechanisms have been shown to confer resistance to organophosphorus, carbamates, and pyrethroid compounds (Radcliffe 1982; Wheelock et al. 2005). Therefore, the use of insecticides should be strictly as per the resistance management guidelines, e.g., those of Insecticide Resistance Action Committee (https://irac-online.org/) (Nauen et al. 2019).

Petroleum-derived spray oils are long known to possess insecticidal activity. Mineral oils have been demonstrated to reduce the spread of PVY by more than 50% in comparison with untreated control on many occasions. The usual practice is to apply 5–10 L/ha with season-long spraying program at weekly intervals. Mineral oils possess direct toxicity toward the vector aphids, interfere with feeding behavior and binding of virions within the stylets of aphids, and impede the infection process post-inoculation. All these together or alone contribute to reducing the spread of nonpersistent viruses like PVY in field. Perhaps the most important limitation of mineral oils is the necessity for complete coverage of the foliage. Therefore, fresh foliage after treatment continues to be susceptible to probing by aphids and the consequent virus transmission. Therefore, mineral oils are applied more frequently in the early season and also when the aphid flight activity is higher (Yang et al. 2019; Shah et al. 2021).

9.6.3 Cultural Control

Weeding and general cleanliness in and around crop fields and removal of overwintering hosts can help reduce the incidence of aphids. Mulches including plastic reflective mulches and straw mulches have been demonstrated to considerably reduce the landing rate and population growth of aphids on potato (Summers et al. 2004; Shah et al. 2020). Similarly, intercropping with onion, garlic, or coriander is known to reduce aphid population (Lehmhus et al. 1996; Vidal 1997). Manipulation of planting and haulm-cutting dates to evade the periods of high aphid activity are practiced worldwide to reduce the incidence of aphid-borne viruses in seed potatoes (Pushkarnath 1959, 1967). Chang et al. (2017) have discussed the subject at length.

9.6.4 Natural Enemies and Microbials

Natural enemies of aphids belong to diverse taxonomic groups, from entomopathogenic fungi to parasitoids, and include generalist and specialist predators, many of which are commercially available (Hance et al. 2017). Most common among these are the braconid and aphelinid parasitoids, coccinellids, predatory bugs, lacewings, and syrphids. The natural enemies work better if their populations are conserved under field conditions by provision of food and refugia. Since the activity of natural enemies is slow, therefore, their role in the management of aphid-virus complex is limited.

Numerous biological control products that use one or more species of entomopathogenic fungi, e.g., *Beauveria bassiana*, *Lecanicillium* spp., are commercially available for aphid control. Proper timing of application is very important when these products are used because fungal spores are strongly influenced by environmental conditions, such as temperature and relative humidity (Kim et al. 2013).

Other than these, the extracts of many plants are known to reduce the aphid population through lethal or sublethal effects, e.g., garlic, neem, red chilli. Insecticidal soaps are used as a safer alternative in some occasions. Potassium silicate foliar sprays have been demonstrated to reduce the population of aphids by at least 60% with considerable reduction in the incidence of viruses in potato crops (Shah et al. 2019).

9.7 Conclusion and Future Outlook

The most significant type of damage inflicted by aphids in potato crops is through the spread of various potato viruses. Potato crops are infested by a number of colonizing and noncolonizing species of aphids, the noncolonizing aphids being more important for the spread of nonpersistent viruses like PVY. Various attributes of aphid biology and ecology have contributed to their success as crop pests. The host-finding and feeding behavior of aphids predisposes them to being the predominant vectors of various viruses. Controlling the spread of PVY remains a challenge to the potato industry worldwide because of its nonpersistent mode of transmission and the evolution of new strains and variants. The control strategies help reduce PVY transmission by aphids; however, each individual control strategy has its own limitations. Various countries operate networks of traps to monitor the flight activity of aphid species in seed potato. It has been reported that aphids other than *M. persicae* are more important for the early-season spread of viruses like PVY. Currently, the aphid management methods in potato are mostly reliant on the use of various insecticides and mineral oils. Besides, the use of infection-free seed, roguing, and use of cultural practices such as manipulation of planting and haulm-cutting dates are the most useful to keep the incidence of virus under control. Resistant cultivars (resistant to aphids and the viruses) with good agronomic traits and customer acceptance could go a long way in the sustainable management of vector-virus complex in potato.

Acknowledgments Table 9.3 was modified from Lacomme et al. (2017) with kind permission from Springer Nature.

References

Al Mrabeh A, Anderson E, Torrance L, Evans A, Fenton B (2010) A literature review of insecticide and mineral oil use in preventing the spread of nonpersistent viruses in potato crops. Agriculture and Horticulture Development Board, London. http://www.potato.org.uk/sites/default/files/publication_upload/Review%20of%20Spray%20Oils%20Insecticides%20and%20Potyvirus%20Transmission.pdf. Accessed 20 Dec 2020

Alfaro-Tapia A, Verdugo JA, Astudillo LA, Ramiırez CC (2007) Effect of epicuticular waxes of poplar hybrids on the aphid *Chaitophorus leucomelas* (Hemiptera: Aphididae). J Appl Entomol 131:486–492

Alyokhin A, Sewel G (2003) On soil movement and plant colonization by walking wingless morphs of three aphid species (Homoptera: Aphididae) in greenhouse arenas. Environ Entomol 32: 1393–1398

Backus EA (1988) Sensory systems and behaviours which mediate hemipteran plant-feeding: a taxonomic overview. J Insect Physiol 34:151–165

Bak A, Martinière A, Blanc S, Drucker M (2013) Early interactions during the encounter of plants, aphids and arboviruses. Plant Signal Behav 8(6):e24225. https://doi.org/10.4161/psb.24225

Bak A, Patton MKF, Perilla-Henao LM, Aegerter BJ, Casteel CL (2019) Ethylene signaling mediates potyvirus spread by aphid vectors. Oecologia 190(1):139–148. https://doi.org/10.1007/s00442-019-04405-0

Barlow ND, Dixon AFG (1980) Simulation of lime aphid population dynamics. Pudoc, Wageningen, p 165

Basky Z (2002) The relationship between aphid dynamics and two prominent potato viruses (PVY and PLRV) in seed potatoes in Hungary. Crop Prot 21:823–827

Basky Z (2006) Cumulative vector intensity and seed potato virus infection in Hungary. Int J Hortic Sci 12:61–64

Bedford ID, Kelly A, Banks GK, Fuog D, Markham PG (1998) The effect of pymetrozine, a feeding inhibitor of Homoptera, in preventing transmission of cauliflower mosaic caulimovirus by the aphid species *Myzus persicae* (Sulzer). Ann Appl Biol 132:453–462

Bernays EA, Chapman RF (1994) Host-plant selection by phytophagous insects. Chapman & Hall, London

Bernays EA, Funk DJ (1999) Specialists make faster decisions than generalists: experiments with aphids. Proc R Soc B 266:151–156

Bernays EA, Funk DJ (2000) Electrical Penetration Graph analysis reveals population differentiation of host-plant probing behaviors within the aphid species *Uroleucon ambrosiae*. Entomol Exp Appl 97:183–191

Bhatnagar A, Jandrajupalli S, Venkateswarlu V, Malik K, Shah MA, Singh BP (2017) Mapping of aphid species associated with potato in India using morphological and molecular taxonomic approaches. Potato J 44(2):126–134

Bindra OS, Sekhon SS (1971) Survey of aphid vectors of potato viruses in the plains of the Punjab. Indian J Hortic 28(2):161–166

Blackman RL (1971) Variation in the photoperiodic response within natural populations of *Myzus persicae* (Sulz.). Bull Entomol Res 60:533–546. https://doi.org/10.1017/S0007485300042292

Blackman RL, Eastop VF (1994) Aphids on the World's trees. CAB International, Wallingford, p 987

Blackman RL, Eastop VF (2000a) Aphids on the world's crops: An identification and information guide, 2nd edn. Wiley, Chichester, p 466

Blackman RL, Eastop VF (2000b) Aphids on the world's crop: an identification and information guide, 2nd edn. Wiley, New York

Blackman RL, Eastop VF (2006) Aphids on the world's herbaceous plants and shrubs, 2nd edn. Wiley, Chichester, p 1439

Blackman RL, Eastop VF (2017) Taxonomic issues. In: van Emden HF, Harrington R (eds) Aphids as crop pests, 2nd edn. CABI, Oxfordshire, pp 1–27

Boiteau G, Singh M, Singh RP, Tai G, Turner T (1998) Rate of spread of PVY^N by alate *Myzus persicae* (Sulzer) from infected to healthy plants under laboratory conditions. Potato Res 41: 335–344

Boiteau G, Singh RP, Parry RH, Pelletier Y (1988) The spread of PVY^o in New Brunswick potato fields: timing and vectors. Am Potato J 65:639–649

Boukhris-Bouhachem S, Djilani-Khouadja F, Fakhfakh H, Glais L, Tribodet M, Kerlan C (2010) Incidence and characterization of Potato virus Y in seed potatoes in Tunisia. Potato Res 53:151–166. https://doi.org/10.1007/s11540-010-9159-9

Boukhris-Bouhachem S, Souissi R, Turpeau E, Rouzé-Jouan J, Fahem M, Brahim NB, Hulle M (2007) Aphid (Hemiptera: Aphidoidea) diversity in Tunisia in relation to seed potato production. Ann Soc Entomol Fr 43(3):311–318

Bowers WS, Nault LR, Webb RE, Dutky SR (1972) Aphid alarm pheromone: isolation, identification, synthesis. Science 177:1121–1122

Broadbent L (1948) Aphid migration and the efficiency of the trapping method. Ann Appl Biol 35: 379–394

Bruce TJA, Wadhams LJ, Woodcock CM (2005) Insect host location: a volatile situation. Trends Plant Sci 10:269–274

Bruck E, Elbert A, Fischer R, Krueger S, Kuhnhold J et al (2009) Movento©, an innovative ambimobile insecticide for sucking insect pest control in agriculture: biological profile and field performance. Crop Prot 28:838–844

Brunt AA, Loebenstein G (2001) The main viruses infecting potato crops. In: Loebenstein G, Berger PH, Brunt AA, Lawson RH (eds) Virus and virus like diseases of potatoes and production of seed-potatoes. Kluwer, Dordrecht, pp 65–134

Caillaud MC (1999) Behavioural correlates of genetic divergence due to host specialization in the pea aphid, *Acyrthosiphon pisum*. Entomol Exp Appl 91:227–232

Caillaud MC, Via S (2000) Specialized feeding behavior influences both ecological specialization and assortative mating in sympatric host races of pea aphids. Am Nat 156:606–621

Campbell CAM, Cook FJ, Pickett JA, Pope TW, Wadhams LJ, Woodcock CM (2003) Responses of the aphids *Phorodon humuli* and *Rhopalosiphum padi* to sex pheromone stereochemistry in the field. J Chem Ecol 29:2225–2234

Chang MG, Gurr GM, Tylianakis JM, Wratten SD (2017) Cultural control. In: van Emden HF, Harrington R (eds) Aphids as crop pests, 2nd edn. CABI, Oxfordshire, pp 494–514

Cherqui A, Tjallingii WF (2000) Salivary proteins of aphids, a pilot study on identification, separation and immunolocalisation. J Insect Physiol 46:1177–1186

De Bokx JA, Piron PGM (1990) Relative efficiency of a number of aphid species in the transmission of Potato virus YN in the Netherlands. Neth J Plant Pathol 96:237–246

Devonshire AL, Field LM, Foster SP, Moores GD, Williamson MS, Blackman RL (1998) The evolution of insecticide resistance in the peach–potato aphid, *Myzus persicae*. Philos Trans R Soc Lond B Biol Sci 353:1677–1684

Dewar AM, Denholm I, van Emden HF, Harrington R (2017) Chemical control. In: van Emden HF, Harrington R (eds) Aphids as crop pests, 2nd edn. CABI, Oxfordshire, pp 398–425

Dewhirst SY, Birkett MA, Fitzgerald JD, Stewart-Jones A, Wadhams LJ et al (2008) Dolichodial: a new aphid sex pheromone component? J Chem Ecol 34:1575–1583

Dewhirst SY, Pickett JA (2010) Production of semiochemical and allelobiotic agents as a consequence of aphid feeding. Chem 20:89–96

DiFonzo CD, Ragsdale DW, Radcliffe EB, Gudmestad NC, Secor GA (1997) Seasonal abundance of aphid vectors of Potato virus Y in the Red River Valley of Minnesota and North-Dakota. J Econ Entomol 90:824–831

Dixon AFG (1971) The role of intra-specific mechanisms and predation in regulating the numbers of the lime aphid *Eucallipterus tiliae* L. Oecologia 8:179–193. https://doi.org/10.1007/BF00345812

Dixon AFG (1998) Aphid ecology, an optimization approach. Chapman & Hall, London

Dixon AFG, Horth S, Kindlmann P (1993) Migration in Insects: cost and strategies. J Anim Ecol 62:182–190. https://doi.org/10.2307/5492

Douglas AE (1998) Nutritional interactions in insect-microbial symbioses: aphids and their symbiotic bacteria Buchnera. Annu Rev Entomol 43:17–37. https://doi.org/10.1146/annurev.ento.43.1.17

Douglas AE (2003) Nutritional physiology of aphids. Adv Insect Physiol 31:73–140. https://doi.org/10.1016/S0065-2806(03)31002-1

Dupuis B, Bragard C, Carnegie S, Kerr J, Glais L, Singh M, Nolte P, Rolot JL, Demeulemeester K, Lacomme C (2017) Potato virus Y: control, management and seed certification programmes. In: Lacomme C, Glais L, Bellstedt DU, Dupuis B, Karasev AV, Jacquot E (eds) Potato virus Y: biodiversity, pathogenicity, epidemiology and management. Springer, Cham, pp 177–206

Edwards LJ, Siddall JB, Dunham LL, Uden P, Kislow CJ (1973) Trans-β-farnesene, alarm pheromone of the green peach aphid, *Myzus persicae* (Sulzer). Nature 241:126–127

Eigenbrode SD, Bosque-Pérez NA, Davis TS (2018) Insect-borne plant pathogens and their vectors: ecology, evolution, and complex interactions. Annu Rev Entomol 63:169–191. https://doi.org/10.1146/annurev-ento-020117-043119

Elzinga DA, Jander G (2013) The role of protein effectors in plant–aphid interactions. Curr Opin Plant Biol 16:451–456

Favret C (2014) Aphid species file. http://aphid.speciesfile.org/HomePage/Aphid/HomePage.aspx. Accessed 20 Dec 2020

Fenton B, Woodford JAT, Malloch G (1998) Analysis of clonal diversity of the peach–potato aphid, *Myzus persicae* (Sulzer), in Scotland, UK and evidence for the existence of a predominant clone. Mol Ecol 7:1475–1487

Ferrar P (1969) Interplant movement of apterous aphids with special reference to *Myzus persicae* (Sulz.) (Hem., Aphididae). Bull Entomol Res 58(3):653–660. https://doi.org/10.1017/S0007485300057370

Field LM, Bass C, Davies TGE, Williamson MS, Zhou JJ (2017) Aphid genomics and its contribution to understanding aphids as crop pests. In: van Emden HF, Harrington R (eds) Aphids as crop pests, 2nd edn. CABI International, Wallingford, pp 37–49

Foster SP, Denholm I, Devonshire AL (2000) The ups and downs of insecticide resistance in peach–potato aphids (*Myzus persicae*) in the UK. Crop Prot 19:873–879

Funk DJ, Bernays EA (2001) Geographic variation in host specificity reveals host range evolution in *Uroleucon ambrosiae* aphids. Ecology 82:726–739

Halbert SE, Corsini DL, Wiebe MA (2003) Potato virus Y transmission efficiency for some common aphids in Idaho. Am J Potato Res 80(2):87–91

Hance T, Kohandani-Tafresh F, Munaut F (2017) Biological control. In: van Emden HF, Harrington R (eds) Aphids as crop pests, 2n edn. CABI, Oxfordshire, p 493

Hardie J, Nottingham SF, Powell W, Wadhams LJ (1991) Synthetic aphid sex pheromone lures female aphid parasitoids. Entomol Exp Appl 61:97–99

Harrington R, Gibson RW (1989) Transmission of potato virus Y by aphids trapped in potato crops in southern England. Potato Res 32(2):167–174

Harrington R, Katis N, Gibson RW (1986) Field assessment of the relative importance of different aphid species in the transmission of potato virus Y. Potato Res 29(1):67–76. https://doi.org/10.1007/BF02361982

Hille Ris Lambers D (1972) Aphids: their life cycles and their role as virus vectors. In: de Bokx JA (ed) Viruses of potatoes and seed potato production. Pudoc, Wageningen, pp 36–56

Hodgson C (1991) Dispersal of apterous aphids (Homoptera: Aphididae) from their host plant and its significance. Bull Entomol Res 81:417–427
Hull R (2002) Matthews' plant virology, 4th edn. Academic Press, London, p 1001
Kashyap RK, Verma AN (1982) New record of aphids infesting seed crop of potato. J Indian Potato Assoc 9:157–158
Katis N, Gibson RW (1985) Transmission of potato virus Y by cereal aphids. Potato Res 28(1): 65–70
Kim JJ, Jeong G, Han JH, Lee S (2013) Biological control of aphid using fungal culture and culture filtrates of *Beauveria bassiana*. Mycobiol 41(4):221–224
Kirchner SM, Döring TF, Hiltunen LH, Virtanen E, Valkonen JPT (2011) Information-theory-based model selection for determining the main vector and period of transmission of Potato virus Y. Ann Appl Biol 159(3):414–427
Kirchner SM, Hiltunen L, Döring TF, Virtanen E, Palohuhta JP, Valkonen JPT (2013) Seasonal phenology and species composition of the aphid fauna in a northern crop production area. PLoS One 8:e71030
Kostiw M (1975) Investigation on the retention of potato viruses M and Y in two species of aphids (*Myzus persicae* Sulz. and *Aphis nasturtii* Kalt.). Potato Res 18(4):637–640
Kostiw M (1979) Transmission of potato virus Y by *Rhopalosiphum padi* L. Potato Res 22:237–238
Kroschel J, Mujica N, Okonya J, Alyokhin A (2020) Insect pests affecting potatoes in tropical, subtropical, and temperate regions. In: Campos H, Ortiz O (eds) The potato crop-its agricultural, nutritional and social contribution to humankind. Springer, Cham, pp 251–306
Krüger K, Laubscher K (2012) The South African aphid monitoring network. *Technical News*, May–June 2012, pp. 24–25. Potatoes South Africa
Kumara BB, Kalleshwaraswamy CM, Ali S, Kadian MS, Venkataravanappa V (2017) Species composition and population dynamics of aphids influencing potato virus Y (PVY) incidence in Karnataka. J Entomol Zool Stud 5(6):1242–1246
Kuroli G, Lantos ZS (2006) Long-term study of alata aphid flight activity and abundance of potato colonizing aphid species. Acta Phytopathol Entomol Hung 41(3–4):261–273
Lacomme C, Pickup J, Fox A, Glais L, Dupuis B, Steinger T, Rolot JL, Valkonen JP, Kruger K, Nie X, Modic S (2017) Transmission and epidemiology of Potato virus Y. In: Lacomme C, Glais L, Bellstedt D, Dupuis B, Karasev A, Jacquot E (eds) Potato virus Y: biodiversity, pathogenicity, epidemiology and management. Springer, Cham, pp 141–176
Lehmhus J, Vidal S, Hommes M (1996) Population dynamics of herbivorous and beneficial insects found in plots of white cabbage undersown with clover. Bull OILB/SROP 19:115–121
Loebenstein G, Berger PH, Brunt AA, Lawson RH (eds) (2001) Virus and virus-like diseases of potatoes and production of seed potatoes. Kluwer Academic Publisher, Dordrecht
Machado-Assefh CR, Lopez-Isasmendi G, Tjallingii WF, Jander G, Alvarez AE (2015) Disrupting *Buchnera aphidicola*, the endosymbiotic bacteria of *Myzus persicae*, delays host plant acceptance. Arthropod-Plant Int 9(5):529–541. https://doi.org/10.1007/s11829-015-9394-8
Margaritopoulos JT, Tsitsipis JA, Goudoudaki S, Blackman RL (2002) Life cycle variation of *Myzus persicae* (Hemiptera: Aphididae) in Greece. Bull Entomol Res 92(4):309–319
Mauck KE, De Moraes CM, Mescher MC (2010) Deceptive chemical signals induced by a plant virus attract insect vectors to inferior hosts. PNAS 107(8):3600–3605. https://doi.org/10.1073/pnas.0907191107
Misra SS, Agrawal HO (1987) Potato aphids: a review of the species, their identification, importance, control and pesticide residues in potatoes in India. Int J Pest Manage 33(1):39–43
Mondal S, Wenninger EJ, Hutchinson PJ, Weibe MA, Eigenbrode SD, Bosque-Pérez NA (2016) Contribution of noncolonizing aphids to Potato virus Y prevalence in potato in Idaho. Environ Entomol 45(6):1445–1462
Morita M, Ueda T, Yoneda T, Koyanagi T, Haga T (2007) Flonicamid, a novel insecticide with a rapid inhibitory effect on aphid feeding. Pest Manag Sci 63:969–973

Narayandas GK, Alyokhin AV (2006) Interplant movement of potato aphid (Homoptera: Aphididae) in response to environmental stimuli. Environ Entomol 35:733–739

Narayandas GK, Alyokhin AV, Alford R, Weber D, Dickens JC (2006) Response of potato aphid (Homoptera: Aphididae) to synthetic potato derived Colorado potato beetle (Chrysomelidae: Coleoptera) attractant and natural potato odor. J Econ Entomol 99(1203):1208

Nauen R, Slater R, Sparks TC, Elbert A, Mccaffery A (2019) IRAC: insecticide resistance and mode-of-action classification of insecticides. In: Jeschke P, Witschel M, Krämer W, Schirmer U (eds) Modern crop protection compounds, 3rd edn. Wiley, Weinheim, pp 995–1012

Nault LR (1997) Arthropod transmission of plant viruses: a new synthesis. Ann Entomol Soc Am 90:521–541

Niemeyer HM (1990) The role of secondary plant compounds in aphid-host interactions. In: Campbell RK, Eikenbary RD (eds) Aphid–plant genotype interactions. Elsiever, Amsterdam, pp 187–205

Northing P (2009) Extensive field based aphid monitoring as an information tool for the UK seed potato industry. Asp Appl Biol 94:31–34

Nottingham SF, Hardie J (1993) Flight behaviour of the black bean aphid, *Aphis fabae*, and the cabbage aphid, *Brevicoryne brassicae*, in host and non-host plant odour. Physiol Entomol 18: 389–394

Park KC, Hardie J (2004) Electrophysiological characterization of olfactory sensilla in the black bean aphid, *Aphis fabae*. J Insect Physiol 50:647–655

Patton MF, Bak A, Sayre JM, Heck ML, Casteel CL (2020) A polerovirus, Potato leafroll virus, alters plant–vector interactions using three viral proteins. Plant Cell Environ 43(2):387–399. https://doi.org/10.1111/pce.13684

Pettersson J, Tjallingii WF, Hardie J (2007) Host-plant selection and feeding. In: van Emden HF, Harrington R (eds) Aphids as crop pests. CABI, Oxfordshire, pp 87–113

Pickett JA, Allemann RK, Birkett MA (2013) The semiochemistry of aphids. Nat Prod Rep 30: 1277–1283

Pickett JA, Bruce TJA, Glinwood RT (2017) Chemical ecology. In: van Emden HF, Harrington R (eds) Aphids as crop pests, 2nd edn. CABI, Oxfordshire, pp 148–172

Pickett JA, Glinwood RT (2007) Chemical ecology. In: van Emden HF, Harrington R (eds) Aphids as crop pests. CABI Publishing, Wallingford, pp 235–260

Pickett JA, Griffiths DC (1980) Composition of aphid alarm pheromones. J Chem Ecol 6:349–360

Pickett JA, Wadhams LJ, Woodcock CM, Hardie J (1992) The chemical ecology of aphids. Annu Rev Entomol 37:67–90

Pickup J, Lacomme C (2017) IPM case studies: seed potato. In: van Emden HF, Harrington R (eds) Aphids as crop pests, 2nd edn. CABI, Oxfordshire, pp 597–606

Piron PGM (1986) New aphid vectors of potato virus Y^N. J Plant Pathol 92(5):223–229

Pitino M, Hogenhout SA (2013) Aphid protein effectors promote aphid colonization in a plant species-specific manner. Mol Plant Microbe Interact 26:30–139

Powell G, Hardie J (2000) Host-selection behaviour by genetically identical aphids with different plant preferences. Physiol Entomol 25:54–62

Powell G, Hardie J (2001) The chemical ecology of aphid host alteration: How do return migrants find the primary host plant? Appl Entomol Zool 36:259–267

Powell G, Maniar SP, Pickett JA, Hardie J (1999) Aphid responses to non-host epicuticular lipids. Entomol Exp Appl 91:115–123

Powell G, Tosh CR, Hardie J (2006) Host plant selection by aphids: behavioral, evolutionary, and applied perspectives. Annu Rev Entomol 51:309–330. https://doi.org/10.1146/annurev.ento.51.110104.151107

Proeseler G, Weidling H (1975) Die Retentionszeit von Stämmen des Kartoffel-Y-Virus in verschiedenen Aphidenarten und Einfluß der Temperatur. Arch Phytopathol Pflanzenschutz 11(5):335–345

Prosser WA, Douglas AE (1991) The aposymbiotic aphid: an analysis of chlortetracycline-treated pea aphid, *Acyrthosiphon pisum*. J Insect Physiol 37(10):713–719. https://doi.org/10.1016/0022-1910(91)90104-8
Pushkarnath (1959) Producing healthy seed potatoes in the plians: a new approach. Indian Potato J 1:63–72
Pushkarnath (1967) Seed potato production in the sub-tropical plains of India. Am Potato J 44:429–441
Radcliffe EB (1982) Insect pests of potato. Annu Rev Entomol 127:173–204. https://doi.org/10.1146/annurev.en.27.010182.001133
Rajabaskar D, Bosque-Pérez NA, Eigenbrode SD (2014) Preference by a virus vector for infected plants is reversed after virus acquisition. Virus Res 186:32–37. https://doi.org/10.1016/j.virusres.2013.11.005
Remaudière G, Remaudière M (1997) Catalogue des Aphididae du Monde. INRA, Paris, p 473
Ryden K (1979) Havrebladlusen, Rhopalosiphum padi, kan sprida potatisvirus Y. Vaxtskyddsnotiser 43:51–53
Ryden K, Brishammar S, Sigvald R (1983) The infection pressure of Potato virus Y^O and the occurrence of winged aphids in potato fields in Sweden. Potato Res 26:229–235
Saguez J, Giordanengo P, Vincent C (2013) Aphids as major potato pests. In: Giordanengo P, Vincent C, Alyokhin A (eds) Insect pests of potato: global perspectives on biology and management. Elsevier, San Diego, pp 31–63. https://doi.org/10.1016/B978-0-12-386895-4.00003-X
Salazar LF (1996) Potato viruses and their control. International Potato Center, Peru
Sano M, Ohki T, Takashino K, Toyoshima S, Maoka T (2019) Species composition of alate aphids (Hemiptera: Aphididae) harboring potato virus Y and the harbored virus strains in Hokkaido, northern Japan. J Econ Entomol 112(1):85–90
Schwinger M, Harrewijn P, Kayser H (1994) Effect of pymetrozine (CGA 15′944), a novel aphicide, on feeding behaviour of aphids. In: Proceedings of 8th IUPAC international congress of pesticide chemistry, Washington DC, July 1994, p 230
Sekhon SS, Bindra OS (1979) Survey for aphid vectors of potato viruses in Kulu valley. Indian J Hortic 36:208–211
Shah MA, Naga KC, Subhash S, Sharma S, Kumar R (2021) Use of petroleum-derived spray oils for the management of vector-virus complex in potato. Potato Res. https://doi.org/10.1007/s11540-021-09505-0
Shah MA, Sharma S, Kumar R, Singh RK (2020) Evaluation of polypropylene row covers for excluding virus vectors and their effect on the incidence of diseases and yield in potato. Indian Phytopathol 73(4):751–757
Shah MA, Sharma S, Sharma J (2019) Bio-efficacy of potassium silicate against aphids and whitefly in potato. Potato J 46(2):132–137
Shambaugh GF, Frazier JL, Castell AEM, Coons LB (1978) Antennal sensilla of seventeen aphid species (Homoptera: Aphidinae). Int J Insect Morphol Embryol 7:389–404
Sigvald R (1984) The relative efficiency of some aphid species as vectors of Potato Virus Y^o (PVY^o). Potato Res 27(3):285–290
Sigvald R (1986) Forecasting the incidence of potato virus YO. In: McLean GD, Garrett RG, Ruesink WG (eds) Plant virus epidemics, monitoring, modelling and predicting outbreaks. Academic Press, Sydney, pp 419–441
Sigvald R (1987) Forecasting Potato virus Y. Waxtskyddsnotiser 51:131–137
Sigvald R (1989) Relationship between aphid occurrence and spread of potato virus Y°(PVY°) in field experiments in southern Sweden. J App Entomol 108(1–5):35–43
Sigvald R (1990) Aphids on potato foliage in Sweden and their importance as vectors of potato virus Y^0. Acta Agric Scand 40(1):53–58. https://doi.org/10.1080/00015129009438547
Sigvald R (1992) Progress in aphid forecasting systems. Neth J Plant Pathol 98(2):55–62
Simpson SJ, Sword GA, Lo N (2011) Polyphenism in insects. Curr Biol 21:R738–R749

Steinger T, Goy G, Gilliand H, Hebeisen T, Derron J (2015) Forecasting virus disease in seed potatoes using flight activity data of aphid vectors. Ann Appl Biol 166(3):410–419

Sukhoruchenko GI, Ivanova GP, Volgarev SA, Berim MN (2019) Species composition of aphids (hemiptera, aphididae) on seed potato plantings in Northwest Russia. Entomol Rev 99(8): 1113–1124

Summers CG, Mitchell JP, Stapleton JJ (2004) Management of aphid-borne viruses and *Bemisia argentifolii* (Homoptera: Aleyrodidae) in zucchini squash by using UV reflective plastic and wheat straw mulches. Environ Entomol 33:447–1457

Tjallingii WF (2006) Salivary secretions by aphids interacting with proteins of phloem wound responses. J Exp Bot 57:739–745

Tosh CR, Powell G, Hardie J (2003) Decision making by generalist and specialist aphids with the same genotype. J Insect Physiol 49:659–669

Van den Heuvel JF, Verbeek M, Van der Wilk F (1994) Endosymbiotic bacteria associated with circulative transmission of potato leafroll virus by *Myzus persicae*. J Gen Virol 75(10): 2559–2565. https://doi.org/10.1099/0022-1317-75-10-2559

Van Harten A (1983) The relation between aphid flights and the spread of Potato virus Y^N (PVY^N) in the Netherlands. Potato Res 26:1–15

Van Helden M, Tjallingii WF (1993) Tissue localisation of lettuce resistance to the aphid *Nasonovia ribisnigri* using electrical penetration graphs. Entomol Exp Appl 68:269–278

Van Hoof H (1980) Aphid vectors of potato virus Y^N. Eur J Plant Pathol 86:159–162

van Hoof HA (1977) Determination of the infection pressure of potato virus YN. Eur J Plant Pathol 83:123–127

Van Hoof HA (1979) Infection pressure of potato virus Y^N. Neth J Plant Pathol 85:31–37. https://doi.org/10.1007/BF01976717

Vandermoten S, Mescher MC, Francis F, Haubruge E, Verheggen FJ (2012) Aphid alarm pheromone: an overview of current knowledge on biosynthesis and functions. Insect Biochem Mol Biol 42:155–163

Verbeek M, Piron P, Dullemans A, Cuperus C, van der Vlugt R (2010) Determination of aphid transmission efficiencies for N, NTN and Wilga strains of Potato virus Y. Ann Appl Biol 156: 39–49

Verma KD (1977) Aphids and their role in potato cultivation. In: Nagaich BB (ed) Recent technologies in potato improvement and production. CPRI, Shimla, pp 256–260

Vidal S (1997) Factors influencing the population dynamics of *Brevicoryne brassicae* in undersown Brussels sprouts. Biol Agric Hortic 15:285–295

Von Burg S, Ferrari J, Müller CB, Vorburger C (2008) Genetic variation and covariation of susceptibility to parasitoids in the aphid *Myzus persicae*: No evidence for trade-offs. Proc Royal Soc B: Biol Sci 275(1638):1089–1094. https://doi.org/10.1098/rspb.2008.0018

Vorburger C, Gehrer L, Rodriguez P (2010) A strain of the bacterial symbiont *Regiella insecticola* protects aphids against parasitoids. Biol Lett 6:109–111. https://doi.org/10.1098/rsbl.2009.0642

Wamonje FO, Donnelly R, Tungadi TD, Murphy AM, Pate AE, Woodcock C, Caulfield J, Mutuku JM, Bruce TJA, Gilligan CA, Pickett JA, Carr JP (2020) Different plant viruses induce changes in feeding behavior of specialist and generalist aphids on common bean that are likely to enhance virus transmission. Front Plant Sci 10:1–13. https://doi.org/10.3389/fpls.2019.01811

Webster B (2012) The role of olfaction in aphid host location. Physiol Entomol 37:10–18

Webster B, Bruce T, Pickett J, Hardie J (2010) Volatiles functioning as host cues in a blend become nonhost cues when presented alone to the black bean aphid. Anim Behav 79:451–457

Wheelock CE, Shan G, Ottea J (2005) Overview of carboxylesterases and their role in the metabolism of insecticides. J Pestic Sci 30:75–83

Wientjens WHJM, Lakwijk AC, van der Marel T (1973) Alarm pheromone of grain aphids. Experientia 29:658–660. https://doi.org/10.1007/BF01944756

Will T, Kornemann SR, Furch ACU, Tjallingii WF, van Bel AJE (2009) Aphid watery saliva counteracts sieve-tube occlusion: a universal phenomenon? J Exp Biol 212:3305–3312

Will T, Steckbauer K, Hardt M, van Bel AJE (2012) Aphid gel saliva: sheath structure, protein composition and secretory dependence on stylet-tip milieu. PLoS One 7:e46903

Williams IS, Dixon AFG (2007) Life cycles and polymorphism. In: van Emden H, Harrington R (eds) Aphids as crop pests. CAB International, Wallingford, pp 69–86

Woodford JAT (1992) Virus transmission by aphids in potato crops. Neth J Plant Pathol 98:47–54

Xu S, Jiang L, Qiao G, Chen J (2021) Diversity of bacterial symbionts associated with *Myzus persicae* (Sulzer) (Hemiptera: Aphididae: Aphidinae) revealed by 16S rRNA Illumina sequencing. Microb Ecol 81:784–794. https://doi.org/10.1007/s00248-020-01622-6

Yang Q, Arthurs S, Lu Z, Liang Z, Mao R (2019) Use of horticultural mineral oils to control potato virus Y (PVY) and other non-persistent aphid-vectored viruses. Crop Prot 118:97–103

Zhang B, Leonard SP, Li Y, Moran NA (2019) Obligate bacterial endosymbionts limit thermal tolerance of insect host species. PNAS USA 116(49):24712–24718. https://doi.org/10.1073/pnas.1915307116

Biology and Management of Whiteflies in Potato Crops

10

Mohd Abas Shah, S. Subhash, R. Ellango, R. Asokan, Sanjeev Sharma, and S. K. Chakrabarti

Abstract

The sweetpotato whitefly (*Bemisia tabaci*) is a major pest of many crops. *B. tabaci* is characterized by a huge genetic variation among its populations such that dozens of biotypes/genetic groups are recognized. Other than their ability to quickly develop resistance to insecticides, some of the biotypes like B and Q are highly invasive. In potato crops, *B. tabaci* is a major problem in tropical and subtropical countries like India due to the transmission of *Begomovirus*, *Tomato leaf curl New Delhi virus* [potato], which can lead to huge yield losses and degeneration of seed stocks. *B. tabaci* is known to exhibit various patterns of population dynamics on potato crops in the Indo-Gangetic plains. Although potato is not the best host for whiteflies, the crops are infested by immigrating populations of *B. tabaci*. On the other hand, greenhouse whitefly (*Trialeurodes vaporariorum*) is currently considered an occasional, minor pest of potato crops, which very rarely requires control measures unless it is associated with a Potato yellow vein disease (PYVD) epidemic. PYVD caused by the *Crinivirus, Potato*

M. A. Shah (✉)
ICAR-Central Potato Research Institute-Regional Station, Jalandhar, Punjab, India
e-mail: mabas.shah@icar.gov.in

S. Subhash
ICAR-Central Potato Research Institute-Regional Station, Modipuram, Meerut, Uttar Pradesh, India

R. Ellango · R. Asokan
Division of Biotechnology, Indian Institute of Horticultural Research, Bangalore, Karnataka, India
e-mail: asokan.r@icar.gov.in

S. Sharma · S. K. Chakrabarti
Division of Plant Protection, ICAR-Central Potato Research Institute, Shimla, Himachal Pradesh, India

S. Kumar Chakrabarti et al. (eds.), *Sustainable Management of Potato Pests and Diseases*, https://doi.org/10.1007/978-981-16-7695-6_10

yellow vein virus (PYVV), an important disease of potato in Colombia, Ecuador, and northern Peru, leads to appreciable yield reductions. As of now, the management of whitefly-virus complex in potato crops is mainly dependent on insecticidal applications.

Keywords

Begomovirus · Seed degeneration · Vector · Insecticides · Persistent virus · Sweetpotato whitefly · Greenhouse whitefly · Biotypes · Bionomics

10.1 Introduction

The sweetpotato whitefly, *Bemisia tabaci* (Gennadius) (Hemiptera: Sternorrhyncha: Aleyrodidae), is a phloem-feeding insect that lives predominantly on herbaceous species. It is a major pest of ornamental, vegetable, grain legume, and cotton production, causing damage directly through feeding and indirectly through the transmission of plant pathogenic viruses, primarily *Begomoviruses* (Jones 2003). It has a global distribution (Dinsdale et al. 2010). *B. tabaci* is a major problem for quality seed potato production in countries like India where potato is predominantly cultivated in the subtropics.

The whitefly, *B. tabaci*, is a species complex that has attracted attention because of its unusually plastic or variable phenotypic traits, including host range, environmental adaptation, fecundity, and variable dispersal behaviors (Boykin et al. 2007). Also, the *B. tabaci* complex has long confounded systematists owing to the lack of morphological characters that can be linked to diverse phenotypes. Members of the *B. tabaci* complex further vary with respect to composition of secondary endosymbionts that are thought to contribute to certain aspects of fitness. The suite of phenotypic characters of certain biological types closely aligns with recognized invasive behaviors. Particularly invasive *B. tabaci* often exhibit resistance to certain insecticides used in agricultural production systems, which similarly seems related to the inherent plasticity of the species (De Barro et al. 2011).

The whitefly, *B. tabaci*, retains the status as serious pest not only because of its nature as a cryptic species but also the damage it causes in agricultural crops by feeding on the phloem sap and its ability to serve as vector for hundreds of plant viruses. Gennadius (1889) first described this insect from tobacco in Greece, named as *Aleyrodes tabaci* and now known as *Bemisia tabaci*. In the last three decades, the dramatic increase in research interest in the whitefly was due to its cryptic nature, wherein the members vary greatly in their biology, mainly host range, fecundity, and insecticide resistance, and in their ability to transmit plant viruses and induce plant disorders.

The worst damage from *B. tabaci* infestation is usually a consequence of its role as a virus vector. The whiteflies transmit plant viruses belonging to the genera *Begomoviruses* (family Geminiviridae), *Crinivirus* and *Closterovirus* (Closteroviridae), and *Ipomovirus* (Potyviridae) (Hadjistylli et al. 2016). The *B. tabaci* transmit *Begomoviruses* in a persistently circulative manner, while the

Criniviruses and *Closteroviruses* are transmitted semipersistently (Brown and Bird 1995). In addition to this, the greenhouse whitefly (GHW) can be a serious problem for potato cultivation in the Andean region due to the transmission of *potato yellow vein virus* (PYVV). In this chapter, the major characteristics of the sweetpotato whitefly, which make it the major pest of many economic crops, are discussed, with special reference to potato. Also, an overview of the importance of GHW for the potato production system is discussed. A summary of major tactics adopted for the management of the whitefly virus complex is also given.

10.2 *Bemisia tabaci* Genetic Diversity: Biotypes/Genetic Groups

The *B. tabaci* complex is a "cryptic species" in that its members exhibit a range of genetic variation and are collectively considered a sibling species group, although the morphological characters in the pupal case that are useful for species identification lack variations sufficient for finer-scale taxonomic purposes. Variants of *B. tabaci* for which biological (phenotypic) differences are recognized have been referred to as "biotypes" and previously as races (Bird 1957; Bird and Sanchez 1971; Bird and Maramorosch 1975, 1978). More than 30 biotypes have been characterized to varying degrees. For *B. tabaci*, variants have been distinguished based on host range, life history traits, and other evidence for phenotypic variation, including differential virus transmission, which influences the epidemiology of viral outbreaks and their control (Brown 2010). The first major outbreak in the Americas of a then-uncharacterized race or biotype of *B. tabaci* occurred in 1981 in the southwestern USA, later called the Arizona "A" biotype (AZ-A), which represented the local or endemic haplotype, soon to be followed by the invasive, pesticide-resistant type with more efficient virus transmission ability, the B biotype (Brown 1990).

For *B. tabaci*, the greatest numbers of populations have been evaluated using allozyme electrophoresis, random amplification of polymorphic DNA (RAPD), polymerase chain reaction (PCR), mitochondrial cytochrome oxidase 1 (CO1), and ribosomal internal transcribed spacer (ITS)1, with biotypes being morphologically indistinguishable (Rosell et al. 1997). Perring (2001) summarized 41 distinct populations of *B. tabaci*; 24 of these populations were given a specific biotype designation, while the remaining 17 populations have not been labeled, the number of designated biotypes being at least 32 now (Simon et al. 2003; Zang et al. 2006). Previous studies acknowledged that *B. tabaci* is genetically complex and composed of numerous well-defined genetic groups (Maruthi et al. 2004; De Barro et al. 2005). However, in all these studies, the point at which group structure has been applied has varied considerably and inconsistently both among and within individual studies.

Based on phylogenetic analysis and pairwise comparisons of genetic distance using mitochondrial cytochrome oxidase 1 gene (*CO1*) between genetic groups of *B. tabaci* worldwide, Dinsdale et al. (2010) provided a framework to suggest that *B. tabaci* is a cryptic (or sibling) species complex containing 11 higher-genetic groups and at least 24 morphologically indistinguishable species. Following the "phylogenetic" species bounds proposed by Dinsdale et al. (2010), Hu et al. (2011)

added four more species, increasing the total number of cryptic species to 28. The field surveys conducted in India indicated that more putative species may be added to the list (Chowda-Reddy et al. 2012). Currently, at least 47 *B. tabaci* sister clades (species) have been proposed based on the differences observed in the sequence of the mitochondrial cytochrome c oxidase subunit I gene (mtCOI) fragment (Alemandri et al. 2015; Firdaus et al. 2013; Hu et al. 2011, 2018; Boykin et al. 2017; Mugerwa et al. 2018). Divergence of at least 3.5% in mtCOI sequences is accepted as a criterion for separation of *B. tabaci* species (De Barro 2012; Boykin and De Barro 2014).

Two invasive *B. tabaci* species, Middle East-Asia Minor 1 (MEAM1; formerly known as B biotype) and Mediterranean (MED; formerly known as Q biotype), have spread to a significant number of countries in the world in recent decades. *B. tabaci* has a well-deserved reputation for being invasive, mostly because biotype B has spread from its origins in the Middle East–Asia Minor region to at least 50 countries in Africa, the Americas, Asia, Australia, and Europe via the trade in ornamentals (Cheek and Macdonald 1994; Dalton 2006). The Q biotype has also begun to invade from its origin in countries bordering the Mediterranean Basin to at least ten countries in Africa, the Americas, Asia, and Europe (Zhang et al. 2005; Dalton 2006) again via the trade in ornamental species. The invasive ability and damage potential of the B biotype have earned it a place as one of the world's top 100 insidious species (http://www.issg.org) of global agriculture.

Ellango et al. (2015) reported six (06) genetic groups of *B. tabaci* from India, viz., Asia I, Asia II 1, Asia II 5, Asia II 7, Asia II 8, Asia II 11, and MEAM I. Asia I is the most widely distributed genetic group in India, followed by Asia II-1. The possible taxonomic confusion for Asia II 1 (Dinsdale et al. 2010; De Barro et al. 2011), which includes biotypes K, P, PCG-1, ZHJ2, PK1, and SY, has been explained as in the case of biotypes K and P, there is a slight difference in esterase banding patterns, and subsequent examination of their mtCO1 showed <2% sequence variation (Bedford et al. 1994; De Barro et al. 2011). ZHJ2 was identified as another biotype, but without comparison to material from Pakistan and Nepal where P and K were obtained. ZHJ2 and K had an identical mtCOI (De Barro et al. 2011). Similarly, biotypes PCG-1, PK1, and SY were all raised without reference to K, P, and ZHJ2 and again have mtCO1 that were either identical to K or show <2% mtCOI sequence divergence. In all cases there is no data showing biological differences, and the identifications have been based solely on molecular data of one form or another (De Barro et al. 2011).

The distribution of *B. tabaci* species across India is shown in Table 10.1. Asia I is the most widely distributed species recorded from 61 locations, followed by Asia II-1, which was found in 31 locations. Asia II-8, Asia II-7, and Asia II-5 are localized in 12, 3, and 8 locations, respectively. The presence of MEAM1 (previously "B" biotype) was recorded only in Karnataka southeast region. The newly identified genetic group Asia II-11 was located in Karnataka northwest region (three locations) (Ellango et al. 2015). Among the biotypes/genetic groups of *B. tabaci*, Asia II-1, Asia 1, and Asia II-5 are common in the Indo-Gangetic plains (Chaubey et al. 2015;

Table 10.1 Details of the distribution of *B. tabaci* genetic groups across India (after Ellango et al. 2015)

Asia I	Asia II 1	Asia II 5	Asia II 7	Asia II 8	Asia II 11	MEAM I
Haryana	Jammu and Kashmir	Karnataka	New Delhi	Tamil Nadu	Karnataka	Karnataka
Utter Pradesh	Himachal Pradesh	Kerala	Kerala	Karnataka		
Gujarat	Punjab			Kerala		
Maharashtra	Rajasthan					
Tamil Nadu	Haryana					
Karnataka	New Delhi					
Kerala	Utter Pradesh					
	Andhra Pradesh					

Ellango et al. 2015; Hashmi et al. 2017; CPRI 2019) and frequently reported on potato crops.

10.3 Bio-ecology of Sweetpotato Whitefly, *B. tabaci*

10.3.1 Host Plants and Nature of Damage

B. tabaci has a very wide host range. There are more than 600 hosts worldwide (Dhawan et al. 2007; Nombela and Muñiz 2009; Chandrashekar and Shashank 2017). The major plant families that serve as host for *B. tabaci* are Malvaceae, Curcubitaceae, Euphorbiaceae, Convolvulaceae, and Solanaceae. The major damage inflicted by whitefly is in crops like sweet potato, cucumber, water melon, squash, eggplant, pepper, tomato, potato, lettuce, broccoli, cotton, soybean, and, to a lesser degree, alfalfa, and many other crops are hosts, but their suitability varies. Various weeds and field crops may favor survival of white flies during the vegetable-free period. *Lantana camara*, *Hibiscus esculentus*, *Solanum nigrum*, and *Datura* sp. are examples of suitable weed hosts (Dhawan and Simwat 1997).

Adults and nymphs of whitefly use their piercing-sucking mouth parts to feed on the phloem of host plants. This results in direct damage, which is manifested in localized spotting, yellowing, or leaf drop (Broad and Puri 1993). Under heavy feeding pressure, wilting and severe growth reduction may occur (Malik et al. 2005). Systemic effects may occur, with uninfested leaves and other tissues being severely damaged as long as feeding whiteflies are present on the plant (Butter and Kular 1999). Dhawan and Mandal (2008) reported that potato plants infected with PALCV showed stunting, crinkling, vein thickening, curling, waviness of leaf margins, and leaf distortion. There can be contamination of leaves by honeydew and sooty molds, which adversely affects the photosynthesis of plant, leading to reduction in yield (Reddy and Rao 1989). Gerling (2002) suggested that nymphs, but not adults, produced a translocable toxicogenic secretion. In addition to direct damage, *B. tabaci* also causes damage indirectly by transmitting viruses. The whitefly transmits a *Begomovirus*, *tomato leaf curl New Delhi virus* (ToLCNDV) [potato], the pathogen of potato apical leaf curl disease, which is now a major constraint for quality seed potato production in India (Usharani et al. 2004a; Chandel et al. 2010; Jeevalatha et al. 2016; Sridhar et al. 2016; Jeevalatha et al. 2017a; Bhatnagar et al. 2017a, b). A disease-free potato seed production cannot be successful in the presence of *B. tabaci*.

10.3.2 Bionomics and Dispersal of *B. tabaci*

B. tabaci can complete a generation in about 20–30 days under favorable weather conditions (Saini 1998). Whiteflies produce many generations in a year and reach high populations. At least three generations are completed on a potato crop (CPRI 2004). Temperature in the range of 26–32 °C and RH of 60–70% is optimal for whitefly development (Traboulsi 1995). The lower and upper developmental thresholds are about 10 and 30 °C (Gerling 2002). In addition, *B. tabaci* can protect themselves from heat damage in extreme environments by residing under the leaves, producing heat shock proteins, and raising the levels of sorbitol in their blood by 15–27-fold under increased temperature (Gerling 2002). Adults typically live 10–20 days and may produce 50–150 eggs or even up to 300 eggs (Reddy and Rao 1989). Female whiteflies are diploid and emerge from fertilized eggs, whereas male whiteflies are haploid and emerge from unfertilized eggs. Eggs are initially whitish in color and change to a brown color near hatching, within 5–7 days. After hatching, the whitefly nymph develops through four instar stages. After the fourth instar, the nymph transforms into a pupa during which the eyes become deep red and the body becomes yellow. Adult whiteflies have light yellow bodies and white wings, which is attributed to the secretion of wax across its wings and body (Brown et al. 1995) (Fig. 10.1).

Dispersal is an integral component of the ecology of *B. tabaci* that enables host finding and colonization in a constantly shifting environment. Dispersal is also a critical mechanism enabling *B. tabaci* to spread plant viruses, distribute insecticide

Fig. 10.1 *Bemisia tabaci* on potato: pupal stage (left), adults (center), adults on potato leaf (right)

resistance genes, and escape natural enemies. Although *B. tabaci* may be a weak flyer, it is nonetheless highly adaptable at moving considerable distances within its environment. Large swarms of flying *B. tabaci* in Brazil were observed in small towns several kilometers from soybean fields where they originated, prompting Costa (1976) to suggest that *B. tabaci* moves farther than is generally recognized.

Insights gained from decades of whitefly flight and migration studies have revealed various strategies of dispersal in terms of migratory vs. foraging flight (Kennedy 1985). Mark and recapture studies have demonstrated a strong directional component to dispersal by *B. tabaci.* Whiteflies marked the evening before with DayGlo^{R} dust were consistently trapped the following day as far as 2.7 km from the source field in Yuma, AZ, USA (Byrne et al. 1996). Whether individuals trapped at this distance represented foraging or migratory flyers was uncertain, but the regularity in which at least some individuals attained the 2.7-km distance prompted the suggestion that migrating whiteflies were capable of exceeding this distance (Isaacs and Byrne 1998). The conclusion was supported by Cohen et al. (1988) who found that whiteflies marked on the weed *Cynanchum acutum* were subsequently recaptured 6 days later on traps near tomato fields some 7 km away. In another field study, a series of traps was positioned at four heights from the ground level up to 7.2 m at six equidistant locations out to 100 m from the source field. A gradual decline in the numbers of whiteflies caught with increasing distance out to 100 m was observed for both males and females.

Gerling and Horowitz (1984) reported that whiteflies within the cotton field flew near the ground, whereas in the open air, they flew >2 m. The airborne populations above 2 m land on the ground, and the whiteflies do not recognize the host plant before their descent. If they happen to reach a plant canopy, they disperse on the plants and search for suitable sites (Prokopy and Owen 1983). However, if the whiteflies reach the bare ground, they fly about looking for the right substrate to land on (Naranjo et al. 2010). Using yellow sticky traps placed 0 to 5 m above ground level in a fallow field, Gerling and Horowitz (1984) reported that the principal whitefly catch is at ground level, but a definite fraction was caught as high as 5 m above ground level (3–5%). Byrne et al. (1986) found that trap height is as important as the design, and cylindrical traps at ground level caught over sevenfold as many whiteflies as at 50 cm, and more than 12-fold as many as at 100 cm. In another field study, a series of traps was positioned at four heights from ground level up to 7.2 m. Whiteflies started to disperse in the direction of the prevailing wind at dawn, and most of the individuals (70%) were observed flying near the ground. The mean proportion of the aerial population trapped decreased exponentially with increasing height above the ground. Less than 5% of marked whiteflies were caught at the 7.2-m height (Isaacs and Byrne 1998). The pattern of vertical distribution was reported to be different in and around the crop as compared with larger distances from the source fields. Dispersal flights are most conspicuous during the first few hours of daylight on summer days before temperatures reach prohibitive levels.

10.4 Importance of Sweetpotato Whitefly, *B. tabaci*, in Seed Production Systems

Bemisia tabaci is a common pest in tropical and subtropical regions but is less prominent in temperate habitats. It is a major pest of potato in India (Shah et al. 2021a) and theoretically all the agro-ecologies where potato is cultivated in the subtropics or tropics. It is reported as a minor pest of potato in parts of China and Australia (Xu et al. 2013; Kroschel et al. 2020).

The whitefly transmits a *Begomovirus*, ToLCNDV [potato], the pathogen of potato apical leaf curl disease, which is now a major constraint for quality seed potato production in India (Usharani et al. 2004a, b; Chandel et al. 2010; Jeevalatha et al. 2016; Sridhar et al. 2016; Jeevalatha et al. 2017a; Bhatnagar et al. 2017a, b). In 2001, the association of a *Begomovirus* with this disease was proved by Garg et al. (2001) using immune electron microscopy, which was named as potato apical leaf curl disease (PALCD). Later, the cause of this disease was confirmed as a variant of ToLCNDV by Usharani et al. (2004a) and identified as a new strain of ToLCNDV and given the name ToLCNDV-potato. Apical leaf curl symptoms in potato comprised curling and bunchiness of apical leaves along with mosaic and chlorosis (Garg et al. 2001). The primary infection with PALCV in the field appears within about 40–45 days from the date of planting.

The incidence of this virus has immensely increased year by year, and it captured the first position in Indian potato viruses in the last two decades (Kumar et al. 2021). The disease incidence was recorded higher particularly in Indo-Gangetic plains (40–100% infection), which caused significant yield losses in susceptible varieties (Lakra 2002). Up to 40% incidence of PALCD was reported from West Bengal (Saha et al. 2014). The yield losses were observed up to 60.8%, with a significant reduction in size and number of tubers per plant in most prominent potato cultivars, i.e., Kufri Pukhraj and Kufri Khyati, in India (Lakra 2002, 2003; Chandel et al. 2010). The incidence of the PALCV has been observed to be higher in early planted crops when the temperature is high in October than in November-planted potato crops.

A successful potato production needs seed tubers in which virus infection levels are held below the critical thresholds. Because of the critical role of whitefly in the transmission of geminivirus, whitefly needs special attention in potato seed production.

Geminiviruses are transmitted by *B. tabaci* in a persistent circulative manner (Brown 1994; Duffus 1994). An acquisition-access feeding of 2–24 h followed by an inoculation-access period of 2–3 days is optimum for a successful transmission of the virus (Khurana and Singh 2003). Generally, the feeding period required for a whitefly to become infected by geminiviruses is around 1 day (Schuster et al. 2009). Once an adult has acquired the virus by feeding on an infected plant, it may retain the virus for a long period and transmit it to healthy plants (Brown 1994). After acquisition, whiteflies can transmit the virus up to 5–20 days (Khurana and Singh 2003). Transmission occurs only after a latent period of 4–10 h. The females are more efficient in transmitting the virus than the males (Boulehya et al. 1997).

ToLCNDV-potato consists of two circular ssDNA components, namely, DNA-A and DNA-B of 2.7 kb and 2.6 kb, respectively (Jeevalatha et al. 2017a). DNA-A encodes all information for viral encapsidation (AV1; CP) and replication (AC1; Rep), replication enhancer protein (AC2; REn), AV2, a pathogenicity determinant, and a transcriptional activator protein (AC3; TrAP) and can replicate autonomously. DNA-B is dependent on DNA-A for its replication, and it is required for systemic infection and symptom expression, nuclear localization, and systemic movement (Yadava et al. 2010; Nash et al. 2011; Jeevalatha et al. 2017a). The function of AC4 and AC5 is still not clear. The DNA-B encodes the nuclear shuttle protein (BV1; NSP) and the movement protein (BC1; MP) (Fondong 2013). ToLCNDV is occasionally reported along with betasatellites (Sivalingam et al. 2010; Jyothsna et al. 2013). Betasatellite has been reported to enhance the symptom severity in host plants (Kumar et al. 2010; Shahid 2020). The association between betasatellite and ToLCNDV-potato has also been reported (Usharani et al. 2004b; Jyothsna et al. 2013).

Phylogenetic analysis revealed that the DNA-A and DNA-B components of eight different isolates shared 94.6–99.4% and 97.2–99.5% of homology within the isolates, respectively. An identical grouping was also observed in the AC1 and AC4 genes in eight isolates. However, the DNA-A component of ToLCNDV-potato shared greater than 90.0% similarity with the DNA-A of ToLCNDV isolates of tomato, bhindi, and other cucurbitaceous crops, 89.0–90.0% similarity with ToLCNDV-papaya isolates, and 70.4–74.0% similarity with other tomato leaf curl viruses (Jeevalatha et al. 2017a). Moreover, the DNA-B component of this virus has shown 86.6–91.7% similarity with the ToLCNDV isolates of other crops. Evidence showed that all the eight Indian isolates of ToLCNDV-potato are closely related to other ToLCNDV. There are reports highlighting that ToLCNDV-potato originated from the genetic recombination between ToLCNDV and some other *Begomoviruses* (Moriones et al. 2017).

10.5 Seasonal Abundance and Population Dynamics of *B. tabaci*

More than 85% of potato production in India is realized from the subtropical plains (Indo-Gangetic plains) where potatoes are cultivated during winter (Khurana and Naik 2003). Cultivation in subtropics leads to the infestation of sweetpotato whitefly or cotton whitefly, *B. tabaci*, in the potato crops. The incidence of apical leaf curl disease, transmitted by whitefly, to the extent of 40–100% with significant yield losses and degeneration of seed stocks are reported (Lakra 2002). For the proper management of the whitefly-virus complex, the study of the population dynamics of the pest is of pivotal importance. The population dynamics of whitefly infesting potato crops in India has been attempted at various locations, e.g., Gwalior (Bhatnagar 2007, 2009), Hair (Lakra 2003, 2005; Kumar and Gupta 2016), Modipuram (Kishore et al. 2005; Malik and Singh 2007), and Nadia (West Bengal) (Amitava et al. 2010). These studies evaluated the location-specific trends of the population dynamics of whitefly with correlation with local weather parameters.

In a detailed study of the population dynamics of whitefly in the Jalandhar region of Punjab (India), Shah et al. (2021a) found that the whitefly adults appeared on the crop immediately after emergence and peaked (1.5–2.5 per plant) in the first week of November. The adults remained on the crops for 85.58 ± 4.95 days. The average daily temperature emerged as the strongest predictor for the population fluctuation of adult whiteflies on potato plants. The trap catch was highest in the first 2–3 weeks after crop emergence (15–47 per trap) and decreased abruptly afterward, and for the remainder of the crop season, very few whiteflies were trapped. The first 2–3 weeks represent the phase when maximum immigration of adults occurs in the potato crops. The flight activity of the adults continued until the maximum daily temperature did not fall below 13 °C.

The populations of *B. tabaci*, being a multivoltine insect that has no diapause or quiescent stage, are sustained through the continual exploitation of multiple host resources, both wild and cultivated, over the annual cycle. Therefore, in addition to temperature, cropping sequence also shapes the pattern of whitefly infestation in crops (Murugan and Uthamasamy 2001; Naranjo et al. 2009). Shah et al. (2019a) reported that the whitefly, *B. tabaci*, exhibits different patterns of population dynamics across the Indo-Gangetic plains on potato crops depending on the cropping sequence adopted by the farmers and the daily temperature during the winter months, i.e., December and January. The whitefly incidence was higher at locations where potato is preceded by crops preferred by whitefly, such as cotton, broad beans, groundnut, etc. (Fig. 10.2). Potato crops do not sustain a high whitefly population on their own, and the whitefly assemblage on potato crops at most of the locations is the result of immigration from adjoining and preceding crops (Shah et al. 2019a, 2021a).

10.6 Sampling Scheme for Sweetpotato Whitefly in Potato

Reliable and cost-effective sampling methods are central to the study of biology and ecology of *B. tabaci* and are critical to the development of monitoring programs for pest management applications (Naranjo and Flint 1995). The development of sampling plans, including the size of the sample unit, the number of samples to be taken, and the allocation of samples within the sample universe, depends on the understanding of the underlying spatial distribution of the target pests (Southwood 1978). Since adults are relatively easy to monitor, a variety of methods have been developed and used worldwide for sampling the mobile stage of whitefly (Butler et al. 1986; Ohnesorge and Rapp 1986; Ekbom and Xu 1990).

For efficient monitoring and management of *B. tabaci*, the within-plant distribution of *B. tabaci* was explored, and a sequential sampling plan was developed for potato crops by Shah et al. (2020a). The highest proportion of *B. tabaci* adults was found on leaf numbers 3–7, with leaf numbers beginning at the apical meristem (Fig. 10.3). The count of adults from three leaves (nodes 4, 5, and 6) from the top stratum in a plant was proposed as a sample unit for adult *B. tabaci* in potato (Shah et al. 2020a). Based on the proposed sample unit, Green and Kuno's methods were used to develop fixed-precision sequential sampling plans (Kuno 1969; Green 1970).

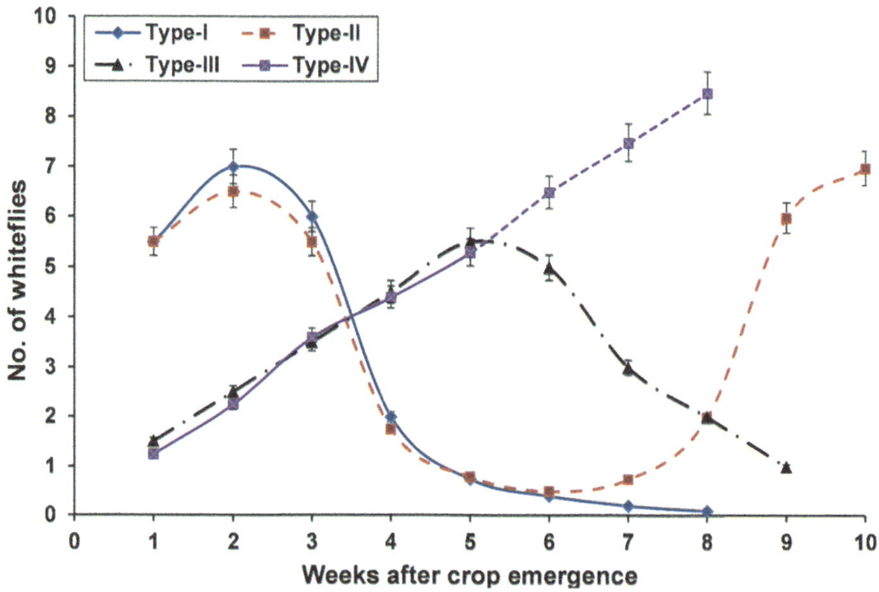

Fig. 10.2 Theoretical framework of the population dynamics of cotton whitefly, *B. tabaci*, infesting potato crops in the Indo-Gangetic plains (after Shah et al. 2019a). Types I and II are found at locations where the minimum daily temperature falls below 10–12 °C during December to January, while Types III and IV are found at locations where the temperature continues to remain suitable for whitefly growth and development. The actual number of whiteflies is determined by the extent of carryover from preceding crops

Both the plans yielded a similar average required sample size, except at lower densities (≤ 1 whitefly per plant) (Table 10.2). At the precision level of 0.1, Green's plan suggested a sample size of 122 for a mean density of 0.5 whitefly adults per plant and 161 samples as per Kuno's plan. At a precision level of 0.25 (normally used for making pest management decision), both the plans suggested a maximum number of 20–26 samples for the lowest density of whitefly noted.

10.7 Greenhouse Whitefly, *Trialeurodes vaporariorum* Infesting Potato Crops

The GWF, *Trialeurodes vaporariorum* Westwood (Hemiptera: Aleyrodidae), is a serious pest of many fruits, vegetables, and ornamental crops in subtropical regions and in greenhouses worldwide. Adults and nymphs typically cause reduction in plant vigor by sucking sap, and through the production of honeydew, which serves as a substrate for sooty molds (Byrne and Bellows Jr. 1991). However, the GHW is currently considered an occasional minor pest of potato crops, which very rarely requires control measures (CIP 2016; DAFWA 2017; Godfrey and Haviland 2017),

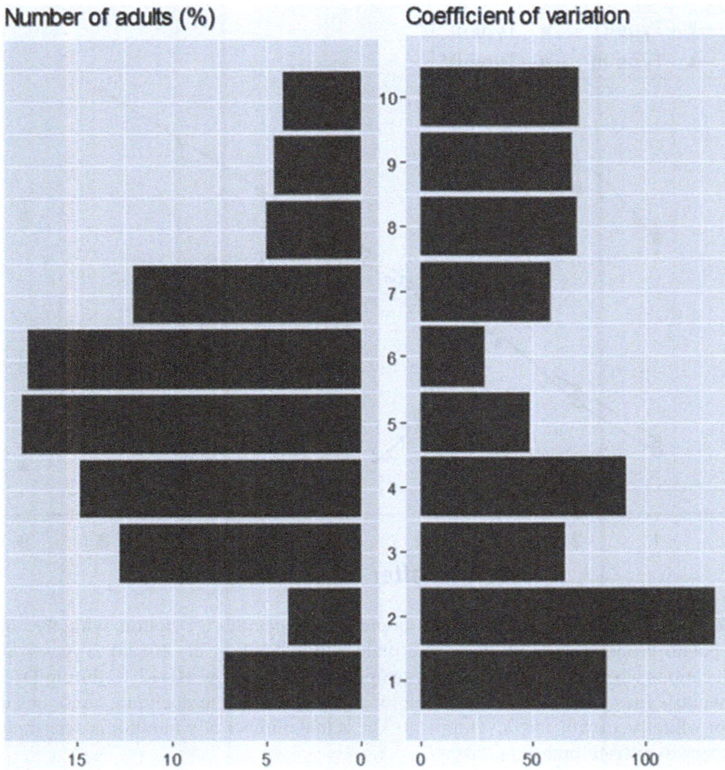

Fig. 10.3 Distribution and associated coefficients of variation for adult *Bemisia tabaci* along the main stem of potato in field (leaf numbers 1–10, counting down from terminal leaf) (after Shah et al. 2020a)

unless it is associated with a potato yellow vein disease (PYVD) epidemic. González-Dufau et al. (2018) determined the demographic parameters of the GHW on potatoes (cultivar Puren) and tomato (cultivar Tropic) and reported that the tomato plant was the best host with the highest intrinsic rate of development and the lowest mortality of GWF compared with the potato plant. Potato plants are not the preferred host for GWF as are beans, red kidney beans, and tomato (Saldarriaga et al. 1988).

10.7.1 Symptoms, Etiology, Epidemiology, and Economic Importance of PYVD

PYVD caused by PYVV, a member of the genus *Crinivirus* (family Closteroviridae), is an important disease of potato in Colombia, Ecuador, and northern Peru, causing up to 50% yield reductions (Guzmán-Barney et al. 2012;

Table 10.2 Estimates of average sample numbers required at prefixed-precision levels ($D = 0.1$ and 0.25) from sampling stop lines as per Green and Kuno's sampling plan (based on Shah et al. 2020a)

Sampling plan	Density (insects per plant)	Precision	
		$D = 0.1$	$D = 0.25$
Green's	0.5	122	20
	1	79	13
	2	51	8
	3	40	6
	4	33	5
	5	29	5
	6	26	4
Kuno's	0.5	161	26
	1	88	14
	2	51	8
	3	39	6
	4	33	5
	5	29	5
	6	26	4

Salazar et al. 2005). PYVD symptoms are characterized by a yellowing of secondary veins that often begins in older leaves but slowly progresses to most plant foliage causing early senescence as well as reduction of photosynthetic capacity and plant vigor (Fig. 10.4) (Salazar et al. 2005). PYVD was first reported in Ecuador in 1943. Since then, it has been reported in Venezuela, Colombia, Ecuador, and Peru (Salazar 2006). In Colombia, PYVD had rarely been considered a limiting disease, until 2014, when the country's agricultural authority (Instituto Colombiano Agropecuario (ICA)) declared a sanitary emergency because of a serious PYVD re-emergence in potato-producing areas in Colombia (ICA 2014). The regional PYVD re-emergence is part of a worldwide emergence of *Criniviruses*, which has been associated with the outbreak of whitefly populations in areas where they regularly or persistently occur (Tzanetakis et al. 2013).

PYVV is considered a quarantine pathogen by various agencies, such as the European and Mediterranean Plant Protection Organization (EPPO) and the Animal and Plant Health Inspection Service (APHIS) (López et al. 2006).

PYVV is transmitted in a semipersistent manner by the GHW *T. vaporariorum* (Lemma and Pulgarín Navarro 1989; Salazar et al. 2005; Tamayo and Navarro Alzate 1984) and through seed tuber and underground stem grafts (Alba 1952; Salazar 1996). PYVV is also known to asymptomatically infect weed species of the genus *Polygonum*. Potatoes, tomatoes, and various weeds of *Polygonum* sp. (Polygonaceae) can act as reservoirs for PYVV (Salazar et al. 2000). PYVV has also been reported to occur in the field in mixed infections with *Potyviruses* in potato (Villamil-Garzón et al. 2014). Gamarra et al. (2020a) evaluated the effect of temperature on the efficiency of PYVV transmission by the GHW. The vector

Fig. 10.4 Symptoms of potato yellow vein disease (PYVD) (reused from Nino et al. (2021) under CC BY 4.0). Upper panel, characteristic PYVD symptoms caused by PYVV in a *Solanum phureja* plant at the forefront of the image, consisting of mild yellowing of the leaf blade but confined around foliar veins; lower panel to the left, stronger PYVD symptoms characterized by extended and intense yellowing in the sink areas of leaves in a *S. phureja* plant; lower panel to the right, atypical strong symptoms characterized by an extended mosaic of green and intense areas of yellow in *Solanum tuberosum*

capacity to transmit the virus was highest at 15 °C (about 70% probability of infection) but decreased radically as temperature deviated from this optimum temperature to <10% at temperatures of 10 and 20 °C, respectively. In another study,

Gamarra et al. (2020b) reported that *T. vaporariorum* completed its life cycle at constant temperatures above 15 °C and below 32 °C, although the cycle was completed at daily fluctuating temperatures between 5 °C and 35 °C on potato leaves. The overall nonlinear modeling of the development rate data portrayed population development within the temperature range of 14° to 32 °C with a maximum finite rate of population increase (=1.14) at 23 °C.

10.7.2 Origin, Distribution, and Genetic Structure of PYVV

The origin of PYVV has been traced to Northern Ecuador and the Central West Colombia region (Alba 1952; Tamayo and Navarro Alzate 1984), and since then the virus has spread throughout the Central Andes, particularly to the important potato-producing areas of northern Peru (Salazar 1996) and all potato-growing regions in the Andean highlands of Colombia (Franco-Lara et al. 2013; Guzmán-Barney et al. 2012; Guzmán et al. 2006; Rodríguez et al. 2015) and Venezuela. Nasruddin and Mound (2016) reported GWF for the first time in the South Sulawesi Province of Indonesia, which caused significant damage to field-grown potato crops. The infested plants had an average number of 68 adult whiteflies per leaflet, which inhibited plant growth and reduced yield by 39%.

The first genome of PYVV was fully sequenced by Livieratos et al. (2004) and was found to comprise three separate RNAs: RNA 1, RNA 2, and RNA 3, of ~8, 5.3, and 3.8 kb size, respectively. RNA 1 contains the replication module and codes for three ORFs. In RNA 2, five ORFs were predicted, Hsp70h, p7, p60, p10, and CP, which correspond to the conserved gene array of *Closteroviruses*. RNA 3 encodes three potential ORFs, p4, CPm, and p26, also consistent with the *Closterovirus* gene array (Livieratos et al. 2004). In addition to the initial Peruvian isolate sequenced (Livieratos et al. 2004), full PYVV sequences were later obtained in Colombia from infected potatoes (*Solanum phureja*) (Álvarez et al. 2017), tomatoes (*Solanum lycopersicum*) (Muñoz Baena et al. 2017), and lulo plants (*Solanum quitoense*) (Gallo et al. 2018; Niño et al. 2021).

10.7.3 Sampling and Economic Threshold Level

Jumardi et al. (2020) determined the vertical distribution of GWF in potato plants. They reported that about 81, 18, and 1% of the eggs were laid on the upper, middle, and lower parts of the canopy, respectively. Similarly, about 80, 17, and 3% of adults were found on the upper, middle, and lower parts, respectively. In contrast, no nymphs were found in the upper part of the canopy, but about 39 and 61% were found in the middle and lower parts, respectively. They suggested the sampling of respective strata of potato foliage for efficient sampling. Rincon et al. (2019) determined the economic injury levels (EIL) of GHW incidence and PYVV infection in potato crops. It was found that the direct injury caused by GWF feeding does not affect potato yield; however, an exponential reduction in crop yield was observed with the increase in PYVD incidence. Rincon et al. (2019) recommended the threshold between 100 and 1200 infected plants/ha for initiation of control measures, depending mainly on potato market price. Further, they suggested that disease incidence should be used to calculate the EIL for PYVD management, instead of focusing on estimations of vector population size.

10.8 Management of Whiteflies and Whitefly-Transmitted Viruses in Potato

The management of vector-virus complex in crops like potato usually entails a combination of methods to ensure that the produce is virus-free. A comprehensive management strategy involves the strengthened seed certification system, good phytosanitary measures, use of host plant resistance, need-based pesticide application for vector control, and management of the tubers during harvest and storage. Strategies like production of virus-free planting material, tuber indexing and clonal multiplication, seed testing, etc. have been discussed by Kumar et al. (2021) (this book). In this section, the strategies for the management of vector populations are discussed.

10.8.1 Monitoring Pest Populations

Direct observation and use of yellow sticky traps are useful methods for monitoring whiteflies and for early detection and documentation of relative whitefly abundance over time. The detailed sampling plans for *B. tabaci* for monitoring incidence in potato crops have been developed (Shah et al. 2020a). Similarly, guidelines for monitoring of *T. vaporariorum* are being developed (Jumardi et al. 2020). In India, the ETL for *B. tabaci* has been set at two adult whiteflies per plant, whereas for *T. vaporariorum*, the ETL based on virus incidence rather than the vector population has been developed (Rincon et al. 2019; Shah et al. 2020a). The subject is dealt with in the earlier sections of this chapter. Whitefly adults are monitored for timing applications of adulticides and nymphs for timing IGR applications. Whitefly adults are highly mobile, and inter-crop movement may be of concern due to viral disease. Thus, monitoring adult populations and their movement and the percentage of adults carrying a virus is important for an area-wide whitefly management program or for viral disease management programs (Salati et al. 2002).

10.8.2 Chemical Control

Although some biological and physical control methods as well as other approaches have been useful in the management of *B. tabaci*, the use of insecticides remains the primary means of control. Several comprehensive reviews of chemical control against *B. tabaci* and insecticide resistance in this pest have been published (e.g.,

Denholm et al. 1996; Horowitz and Ishaaya 1996; Palumbo et al. 2001; Horowitz et al. 2007; Castle et al. 2010; Naveen et al. 2017). The most extensively used insecticide classes—organochlorines, organophosphates (OPs), carbamates, and pyrethroids—have generally been the most seriously threatened by resistance; in addition, there is a tendency to ban their use because of their detrimental effect on humans and the environment. Early reports (e.g., Sharaf 1986; Dittrich et al. 1990; Horowitz and Ishaaya 1996) presented data on more than 50 conventional insecticides for controlling populations and suppressing virus transmission by *B. tabaci*. The most common conventional insecticides used were carbamates, Ops, and pyrethroids. Until the mid-1990s, spray mixtures of synergized pyrethroids had been the most effective combination for controlling *B. tabaci* populations (Horowitz and Ishaaya 1996; Prabhaker et al. 1998; Palumbo et al. 2001; Castle et al. 2010). It is difficult to achieve comprehensive control of *B. tabaci* using conventional insecticides because of the underleaf habitat of immature stages and adults, the presence of older larvae in the lower canopy of the crop, the pest's highly polyphagous nature, and the frequent dispersion of adults (Horowitz and Ishaaya 1996; Palumbo et al. 2001).

Most of the new chemistry insecticides are preferable because of their specificity to target pests, their effectiveness at low rates, and their nonpersistent characteristics in the environment. However, the rate of success for the management of vector-virus complexes is variable. The most promising new chemistry insecticides recommended for use against whitefly are described below. This section is largely based on the account put forth by Horowitz et al. (2011). The detailed discussion on chemistry, mode of action, and safety of newer insecticides is given in a separate chapter of this volume.

10.8.2.1 Neonicotinoid Insecticides

The neonicotinoids exhibit systemic and translaminar properties and high residual activity, especially against sucking insects, such as whiteflies, aphids, and leafhoppers. Imidacloprid was the first commercial neonicotinoid successfully used for controlling agricultural pests. The major neonicotinoids currently in use include imidacloprid, thiamethoxam, clothianidin, acetamiprid, thiacloprid, and dinotefuran.

10.8.2.2 Insect Growth Regulators (IGRs)

The most important IGRs used for the control of whitefly are buprofezin (Ishaaya 1990) and pyriproxyfen (Ishaaya and Horowitz 1992). Another IGR, novaluron, an inhibitor of chitin synthesis, also has some effect on *B. tabaci* (Ishaaya et al. 2003). Buprofezin is a thiadizine-like compound with long residual activity and has both contact and vapor activity; it also affects the nymphal stages of sucking insects, especially whiteflies (Ishaaya et al. 1988; De Cock and Degheele 1998). Pyriproxyfen is a potent juvenile hormone (JH) mimic and is considered a leading insecticide for controlling whiteflies (Ishaaya and Horowitz 1995; Horowitz et al. 2005; Crowder et al. 2008; Castle et al. 2010), especially biotype B.

10.8.2.3 Diafenthiuron, a Thiourea Derivative

Diafenthiuron is an effective whitefly-controlling compound that has been used particularly in Europe and Israel as an alternative to pyriproxyfen for *B. tabaci* control in cotton since 1998 (Horowitz et al. 1999; Palumbo et al. 2001). Diafenthiuron suppresses the formation of whitefly progeny when adult females are exposed to treated plants (Ishaaya et al. 1993); it is also more potent against nymphs than against pupae or eggs. It is considered as one of the few whitefly adulticides still used effectively.

10.8.2.4 Pyridine Insecticides (Pymetrozine)

Pymetrozine is highly specific against sucking insect pests (Flückiger et al. 1992a, b; Fuog et al. 1998). It affects the nerves controlling the salivary pump and causes immediate and irreversible cessation of feeding due to an obstruction of stylet penetration, followed by starvation and insect death (Kayser et al. 1994). Pymetrozine has systemic and translaminar activities and can be used as a drench or in foliar application (Flückiger et al. 1992a, b).

10.8.2.5 The Ketoenols: Spiromesifen and Spirotetramat

Spiromesifen belongs to a new class of pesticides that are derivatives of spirocyclic tetronic acid, which affects mainly whiteflies and mites. Spiromesifen acts effectively on the egg and early nymphal stages of *B. tabaci* (both biotypes B and Q), but adults and late nymphal stages are only moderately affected (Prabhaker et al. 2008; Kontsedalov et al. 2009). Another insecticide belonging to the keto-enol group is spirotetramat, a novel spirocyclic tetramic acid derivative and also a lipid biosynthesis inhibitor. Spirotetramat is a systemic insecticide with phloem and xylem mobility for the control of sucking insects, including aphids, whiteflies, psyllids, and scales. It is particularly effective against juvenile stages of sucking pests, and it significantly reduces the fecundity and fertility of *B. tabaci* females (Brück et al. 2009).

10.8.2.6 Ryanodine Receptor Insecticides (the Diamides)

Ryanodine is a plant alkaloid used as a natural botanical insecticide. Recently, two classes of synthetic agents have been developed for commercial compounds that target insect ryanodine receptors. So far, two insecticides are being studied and registered: Rynaxypyr® (Chlorantraniliprole), which is more potent against lepidopteran pests, and Cyazypyr™ (Cyantraniliprole), which targets sucking pests, such as whiteflies and aphids, as well as other types of insect pests (Sattelle et al. 2008; Lahm et al. 2009).

Suppression of whitefly populations using insecticides, especially in areas with regular incidence of whiteflies, can be an important component of a successful IPM package. Insecticides are most commonly applied as foliar sprays or injected into the soil but may also be applied via chemigation through drip irrigation. Soil applications are typically systemic insecticides, mostly in the neonicotinoid chemical class. The prophylactic use of soil-applied systemic insecticides has been reported to slow down, reduce, or delay virus transmission by whiteflies; however, the use of insecticides alone often does not deliver sufficient protection from viruses to prevent economically important crop damage. In India, seed treatment with imidacloprid 17.8 SL (0.04%) for 10 min and foliar application at 0.03% at 75% crop emergence followed by thiamethoxam 25WG (0.05%) after 15 days is recommended for the management of whitefly in potato crops. The sprays can be repeated as per requirement. Various new chemistry molecules like spiromesifen, IGRs, and knockdown insecticides are also being used across the locations (Shah et al. 2020b).

10.8.2.7 Nonconventional Insecticides

Shah et al. (2019b) reported that the whitefly population reduced by 53.27% and 61.42% after 3 and 7 days, respectively, of foliar application with potassium silicate (0.3%) in potato. Potassium silicate sprays increased the leaf silicon concentration significantly, and the virus incidence (mild and severe mosaic, leaf roll, and apical leaf curl) in potassium silicate-treated plots was at par with those of the recommended insecticides but much lower than the control.

Significant repellency of mineral oils to whitefly adults is reported on many crops like cotton, melon, squash, chrysanthemum, and tomato (summarized in Liang and Liu 2002). Xue et al. (2002) reported that mineral oil concentration significantly affected the number of eggs deposited on tomato leaves. Schuster et al. (2009) demonstrated that oil reduces the settling of *B. tabaci* adults sufficiently to interfere with *Begomovirus* transmission. This reduction of settling could be one of the mechanisms by which field applications can result in fewer virus-infected plants. Malik et al. (2020) reported that horticultural mineral oil spraying at 50 mL/10 L after a series of insecticidal applications achieved a maximum percent reduction in *B. tabaci* population (74.5%) as comapred to untreated control, with a maximum percent reduction in viral infection (93.0%) over untreated control in potato. A summary of studies on the application of mineral oils against whiteflies is given by Shah et al. (2021b). The attempts to use botanicals and microbial secondary metabolites against whiteflies are summarized in other chapters of this volume.

10.8.3 Physical Control

10.8.3.1 Yellow Sticky Traps

Stationary yellow sticky traps are recommended to be installed when planting around fields to capture whitefly adults migrating from other crops. The use of mobile and stationary yellow sticky traps can effectively reduce whitefly adult populations in potato (Cisneros and Mujica 1999; Mujica 1998). In India, yellow sticky traps (15×30 cm^2) placed just above the canopy height at 60 traps/ha at equidistance from each other are recommended for mass trapping in potato crops (Malik et al. 2021).

10.8.3.2 Protected Culture in Greenhouses and Screen Houses

Crops can be protected from whitefly damage and virus infection by physical means, i.e., preventing the insects from coming in contact with susceptible plants. In the most extreme case, the entire crop is grown in a greenhouse or screen house, and plants are protected from whiteflies for the entire production cycle, e.g., vegetables. When these structures are kept free of whiteflies (e.g., through the use of glass, plastic, or screen; vents covered with screen and double doors with a positive pressure), excellent management of whiteflies and virus can be achieved (Ausher 1997; Berlinger et al. 2002). In India, it is a common practice to cultivate early-stage seed potato crops inside net houses. Temporary net-house structures, which can be dismantled during crop harvest, are recommended by CPRI, Shimla, to produce quality seed potatoes like the early-generation minitubers and tissue culture-based plants.

10.8.3.3 Floating Row Covers

The other commonly used method of exclusion is the covering of young plants, either those emerging from seeds or those that have been transplanted, with protective netting. This netting is a spun-bonded polyester material (commercially available as Agribon or Agril) and is placed directly over the rows of emerging seedlings or transplants. Spun-bonded or nonwoven fabrics protect plants from insect-vectored viruses and their aphid and whitefly vectors in many crops, for example, tomato (Berlinger et al. 2002; Al-Shihi et al. 2016), cucurbits (Natwick et al. 1988; Conway et al. 1989; Perring et al. 1989; Webb and Linda 1992; Walters 2003), zucchini (Costa et al. 1994), etc. Shah et al. (2020c) evaluated the 25 GSM spun-bonded row covers for exclusion of vectors from potato crops. Aphids and whiteflies were completely excluded from the covered plots, while a significantly higher number was recorded in the uncovered control. However, the terminal intensity of late blight was higher and the total tuber yield was lower in the covered plots as compared with the uncovered control.

10.8.3.4 Barrier Crops and Mulches

Physical barriers can be designed to prevent the movement of whiteflies into fields of susceptible crops. Barriers may be nonliving, such as plastic (yellow plastic with sticky material to trap insects) or screen, or living, such as the planting of a tall-growing plant species (nonhosts of the whitefly and viruses) in between fields of susceptible crops. The best barrier plants for viruses are monocots, such as corn, sorghum, and elephant grass. However, there is little evidence that barriers effectively reduce whitefly migration or virus spread because whiteflies can fly or be carried by wind over barriers and transmit viruses for long periods of time due to the persistent nature of transmission (Hilje et al. 2001). Thus, barriers are generally not an essential component of the IPM package for whitefly-transmitted viruses.

Mulches are designed to prevent insects from recognizing and landing on a crop that is susceptible to virus infection. Like barriers, mulches can be nonliving (plastic or some other material) or living (plants grown among the susceptible crop). In terms of nonliving mulches, the most effective materials are colored or UV-reflective plastic. These have been reported to be successful in reducing whitefly population densities and the incidence of viruses (Antignus 2000). In Florida, a UV-reflective mulch treatment reduced the incidence of CuLCrV in zucchini squash (Nyoike et al.

2008) and Tomato mottle virus (ToMoV) and TYLCV in Florida and Jordan, respectively (Csizinszky et al. 1995; Suwwan et al. 1988). These mulches can also result in improved crop growth. On the other hand, mulches are expensive and labor-intensive and can be deleterious to the environment. The use of mulches is being encouraged in high-value seed potato crops.

10.8.3.5 Roguing

This strategy involves the physical removal of virus-infected plants over the course of the growing season. Roguing needs to be done soon after plots are established and is most helpful if the incidence of the virus is low (<5%). If whitefly populations are high, plants should be treated with an insecticide to kill whitefly adults prior to roguing. If nymphs are present, rogued plants should be removed and disposed of well away from production fields. Ideally, fields should be monitored frequently and symptomatic plants removed. In India, rouging thrice is recommended for seed potato crops to remove the symptomatic plants: first after the establishment of the crop stand, second during the mid-season, and third before the haulms are cut (before senescence sets in) (Venkatasalam et al. n.d.).

10.8.4 Biological Control

Although many natural enemies of whiteflies are known to interact with whitefly populations, only few of them have been attempted at the field level (Arnó et al. 2010). Three species of entomopathogenic fungi active against *B. tabaci* are available commercially: *Paecilomyces fumosoroseus* (=*Isaria fumosorosea*), *Verticillium lecanii*, and *Beauveria bassiana*. The first two are naturally found to infect whiteflies, whereas *B. bassiana* is only seen infecting whiteflies when applied as part of a formulation. Entomopathogenic fungi are easy to apply, although good coverage is required on the abaxial foliar surfaces where whiteflies reside. These fungi present essentially no risk to human health, and most studies show that they are relatively innocuous to other natural enemies (Goettel et al. 2001; Vestergaard et al. 2003; Zimmerman 2008). The use of fungal products is compatible with many insecticides, and resistance to mycopesticides has not yet been reported. However, fungi are slow-acting compared with chemical insecticides, exhibit poor adulticidal activity, and are incompatible with many commonly used fungicides. In addition, they are relatively expensive, have limited shelf life, and are dependent on favorable environmental conditions (Inglis et al. 2001; Faria and Wraight 2001; Vidal et al. 2003). The use of entomopathogenic fungi against whiteflies in potato crops seems to hold some potential, especially in the case of organic cultivation.

10.8.5 Host Resistance

The genetic resistance source for ToLCNDV in potato is not identified so far. However, the lowest seed degeneration was observed in Kufri Bahar (4.5% yield reduction) even under a high whitefly population pressure and with repeated use of the same seed stock, while other varieties showed faster degeneration under field conditions (Lakra 2003). The mechanism behind the resistance is not clearly understood (Bhatnagar et al. 2017b; Jeevalatha et al. 2017b). Konar et al. (2013) reported that nine germplasm lines, viz., Kufri Chandramukhi, Kufri Jawhar, Kufri Ashoka, Kufri Pukhraj, Atlantic, Kufri Bahar, J/97–165, J/97–168, and J/95–144, were more susceptible to the whitefly, while four lines, viz., Kufri Chipsona-1, Kufri Chipsona-2, DSP-7, and MS/99–1871, were comparatively less susceptible. J/96–84 and J/96–149 showed moderate level of susceptibility.

Silva et al. (2008) evaluated the resistance of 24 potato genotypes to *B. tabaci* biotype B. In the free-choice test, potato genotypes NYL 235–4 and IAC-1966 were the most attractive to adults, while cultivars Achat, Aracy Ruiva, and Monte Bonito had the lowest number of adults. Also in this assay, cultivars Achat, Ibituaçu, Panda, IAC-1966, and Agata presented the lowest number of eggs, while in the no-choice test, only cultivar Achat and IAC-1966 remained resistant. Consequently, for these two genotypes, non-preference is the oviposition resistance mechanism. The clone NYL 235–4 had the greatest number of simple trichome (ST) and glandular trichome (GT), while the clone IAC-1966 had the lowest number of ST and the clone IAC-6290 the lowest number of GT. There were significant correlations between adult attractiveness and oviposition preference, between oviposition preference and ST density, and between oviposition preference and GT density. Considering all characteristics, the cultivar Achat was the most resistant to *B. tabaci* biotype B among all the potato genotypes studied.

Boiteau and Singh (1988) reported that a clone of the wild potato *Solanum berthaultii* Hawkes can trap adult greenhouse whiteflies in the exudate from its glandular trichomes and reduce the whitefly incidence on the wild plants by more than 50%. Guzmán and Rodríguez (2010) evaluated 62 two virus-free accessions of *Solanum phureja* from the Colombian central collection for their susceptibility to infection with PYVV. Twelve accessions were free of virus infection as judged by the absence of symptoms and negative RT-PCR assay results over two cycles of tuber setting and germination. These accessions, which may contain PYVV resistance genes for future use in potato breeding programs, were Col 39, Col 59, Col 70, Col 77, Col 87, Col 90, Col 92, and Col 97; PI 225669 and PI 275110; and Phuc 8 and Phuc 12.

Genetically engineered resistance against various *Begomoviruses* has been attempted in various crops (Horowitz et al. 2011). However, no major progress has yet been made in potato crop.

10.9 Conclusion and Future Outlook

The sweetpotato whitefly continues to be one of the major agricultural pests worldwide. The emergence of the whitefly-virus complex as a major problem in potato crops is comparatively recent. The management of whiteflies is complicated by the huge genetic variation that exists in the pest populations, their ability to quickly evolve insecticide resistance, and the dispersal and migration from crop to crop with year-round activity. The use of insecticides continue to be the major approach for managing the whitefly-virus complex; however, alternative control methods are also being continuously explored. Exploring the potato germplasm for resistance/tolerance against the whitefly-vector complex is one promising line of work. The genetically engineered resistance opens up the way to broaden and enrich the pool of natural resistance genes against viral diseases. The use of transgenes based on pathogen-derived resistance is being attempted in many crops. The potential candidates include replication (*Rep*)-associated proteins, movement proteins (*mp*), and gene silencing, e.g., the susceptibility genes, although RNAi and CRISPR/Cas9 hold great promise. For now, the whitefly-virus complex will continue to be a problem for growers and other associated stakeholders.

Acknowledgments Table 10.1 was modified from Ellango et al. (2015); Table 10.2 and Fig. 10.3 were modified from Shah et al. (2020a) and Table 10.2 from Shah et al. (2019a). Figure 10.4 was reused from Nino et al. (2021) under CC BY 4.0.

References

Alba V (1952) El Amarillamiento De Las Venas De La Papa, Una Enfermedad Causada Por Virus, Tesis. Universidad Nacional de Colombia, Medellin, Colombia

Alemandri V, Vaghi Medina CG, DumÓn AD, Argüello Caro EB, Mattio MF, Garcia Medina S, López Lambertini PM, Truol G (2015) Three members of the *Bemisia tabaci* (Hemiptera: Aleyrodidae) cryptic species complex occur sympatrically in Argentine horticultural crops. J Econ Entomol 108(2):405–413

Al-Shihi AA, Al-Sadi AM, Al-Said FA, Ammara U, Deadman ML (2016) Optimizing the duration of the floating row cover period to minimize the incidence of tomato yellow leaf curl disease and maximize the yield of tomato. Ann Appl Biol 168(3):328–336

Álvarez D, Gutiérrez P, Marín M (2017) Secuenciación del genoma del *Potato yellow vein virus* (PYVV) y desarrollo de una prueba molecular para su detección. Bioagro 29:3–14

Amitava K, Singh NJ, Rajat P (2010) Influence of intercropping on population dynamics of major insect pests and vectors of potato. J Entomol Res 34(2):151–154

Antignus Y (2000) Manipulation of wavelength dependent behavior of insects: an IPM tool to impede epidemics and restrict spread of insect-borne viruses. Virus Res 71:213–220

Arnó J, Gabarra R, Liu TX, Simmons AM, Gerling D (2010) Natural enemies of *Bemisia tabaci*: predators and parasitoids. In: Stansly PA, Naranjo SE (eds) Bemisia: bionomics and management of a global pest. Springer, Dordrecht, pp 385–421

Ausher R (1997) Implementation of integrated pest management in Israel. Phytoparasitica 25:119–141

Bedford ID, Briddon RW, Brown JK, Rosell RC, Markham PG (1994) Geminivirus transmission and biological characterisation of *Bemisia tabaci* (Gennadius) biotypes from different geographic regions. Ann Appl Biol 125:311–325

Berlinger MJ, Taylor RAJ, Lebiush-Mordechi J, Shalhevet S, Spharim I (2002) Efficiency of insect exclusion screens for preventing whitefly transmission of tomato yellow leaf curl virus of tomatoes in Israel. Bull Entomol Res 92:367–373

Bhatnagar A (2007) Incidence and succession of thrips, leafhoppers and whitefly in combination of planting dates and potato varieties. Ann Plant Prot Sci 15(1):101–105

Bhatnagar A (2009) Effect of two pesticides and nitrogen application on insect vectors population of potato crop. Potato J 36(1–2):75–77

Bhatnagar A, Pant R, Sridhar J, Chakrabarti S, Lal M (2017b) Incidence of apical leaf curl disease (ToLCNDV), and economics and reaction of potato (*Solanum tuberosum*) cultivars against whitefly, *Bemisia tabaci* in northern India. Indian J Agric Sci 87(12):1673–1678

Bhatnagar A, Shridhar J, Venkateswarlu V, Malik K, Shah MA, Singh BP (2017a) Mapping of aphid species associated with potato in India using morphological and molecular taxonomic approaches. Potato J 44(2):126–134

Bird J (1957) A whitefly-transmitted mosaic of *Jatropha gossypiifolia*. Tech Paper PR Agric Exp Stn 22:1–35

Bird J, Maramorosch K (1975) Tropical diseases of legumes. Academic Press, New York

Bird J, Maramorosch K (1978) Viruses and virus diseases associated with whiteflies. Adv Virus Res 22:55–110

Bird J, Sanchez J (1971) Whitefly-transmitted viruses in Puerto Rico. Tech Paper PR Agric Exp Stn 55:461–467

Boiteau G, Singh RP (1988) Resistance to the greenhouse whitefly, Trialeurodes vaporariorum (Westwood)(Homoptera: Aleyrodidae), in a clone of the wild potato Solanum berthaultii Hawkes. Ann Entomol Soc Am 81(3):428–431

Boulehya S, Najar A, Sghairi R, Jarraya A (1997) Whitefly report of Tunisia. In: Ioannou N (ed) Management of the whitefly-virus complex. Proceedings of the FAO workshop on management of the whitefly-virus complex in vegetable and cotton production in the Near East, 2-6 October, 1995, Larnaca, Cyprus. FAO, Rome, p 71–75

Boykin LM, De Barro PJ (2014) A practical guide to identifying members of the *Bemisia tabaci* species complex: and other morphologically identical species. Front Ecol Evol 2:45

Boykin LM, Savill A, De Barro P (2017) Updated mtCOI reference dataset for the *Bemisia tabaci* species complex. F1000Res 6:1835. https://doi.org/10.12688/f1000research.12858.1

Boykin LM, Shatters RGJ, Rosell RC, McKenzie CL, Bagnall RA, De Barro PJ, Frohlich DR (2007) Global relationships of *Bemisia tabaci* (Hemiptera: Aleyrodidae) revealed using Bayesian analysis of mitochondrial COI DNA sequences. Mol Phylogenet Evol 44:1306–1319

Broad VK, Puri SN (1993) Some studies on the behaviour of whitefly. J Maharashtra Univ 18:101–103

Brown JK (1990) An update on the whitefly-transmitted geminiviruses in the Americas and the Caribbean Basin. FAO Plant Prot Bull 39:5–23

Brown JK (1994) Current status of *Bemisia tabaci* as a plant pest and virus vector in agroecosystems worldwide. FAO Plant Prot Bull 42:3–32

Brown JK (2010) Phylogenetic biology of the *Bemisia tabac*i sibling species group. In: Stansly PA, Naranjo SE (eds) Bemisia: bionomics and management of a global pest. Springer, Dordrecht, pp 31–67

Brown JK, Bird J (1995) Variability within the *Bemisia tabaci* species complex and its relation to new epidemics caused by geminiviruses. CEIBA 36:73–80

Brown JK, Frohlich D, Rosell R (1995) The sweet potato/silver leaf whiteflies: biotypes of *Bemisia tabaci* (Genn.), or a species complex? Annu Rev Entomol 40:511–534

Brück E, Elbert A, Fischer R et al (2009) Movento®, an innovative ambimobile insecticide for sucking insect pest control in agriculture: biological profile and field performance. Crop Prot 28: 838–844

Butler GD Jr, Henneberry TJ, Hutchison WD (1986) Biology, sampling and population dynamics of *Bemisia tabaci*. In: Russell GE (ed) Agricultural zoology reviews. Hants Intercept, Andover, pp 167–195

Butter NS, Kular JS (1999) Resurgence of whitefly in cotton and its management. Indian J Entomol 61(1):85–90

Byrne DN, Bellows TS Jr (1991) Whitefly biology. Annu Rev Entomol 36:431–457

Byrne DN, Rathman RJ, Orum TV, Palumbo JC (1996) Localized migration and dispersal by the sweet potato whitefly, Bemisia tabaci. Oecologia 105:320–328

Byrne DN, Von Bretzel PK, Hoffman CJ (1986) Impact of trap design and placement when monitoring for the bandedwinged whitefly and the sweetpotato whitefly (Homoptera: Aleyrodidae). Environ Entomol 15(2):300–304

Castle SJ, Palumbo JC, Prabhaker N, Horowitz AR, Denholm I (2010) Ecological determinants of *Bemisia tabaci* resistance to insecticides. In: Stansly PA, Naranjo SE (eds) Bemisia: bionomics and management of a global pest. Springer, Dordrecht

Chandel RS, Banyal DK, Singh BP, Malik K, Lakra BS (2010) Integrated management of whitefly, *Bemisia tabaci* (Gennadius) and potato apical leaf curl virus in India. Potato Res 53(2):129–139

Chandrashekar K, Shashank PR (2017) Indian contribution to whitefly (*Bemisia tabaci*) research. In: Mandal B et al (eds) A century of plant virology in India. Springer, Singapore, pp 563–580

Chaubey R, Andrew RJ, Naveen NC, Rajagopal R, Ahmad B, Ramamurthy VV (2015) Morphometric analysis of three putative species of *Bemisia tabaci* (Hemiptera: Aleyrodidae) species complex from India. Ann Entomol Soc Am 108(4):600–612. https://doi.org/10.1093/aesa/sav028

Cheek S, MacDonald O (1994) Management of Bemisia tabaci. Pest Sci 42:135–137

Chowda-Reddy R, Kirankumar M, Seal SE, Muniyappa V, Valand GB, Govindappa MR, Colvin J (2012) Bemisia tabaci phylogenetic groups in India and the relative transmission efficacy of Tomato Leaf Curl Bangalore Virus by an indigenous and an exotic population. J Integr Agric 11:235–248

CIP (2016) Potato pests and diseases. International Potato Center, Lima

Cisneros F, Mujica N (1999) Biological and selective control of the sweet potato whitefly, Bemisia tabaci (Gennadius) (Hom.: Aleyrodidae). International Potato Center, Lima, pp 255–264

Cohen S, Kern J, Harpaz I, Ben-Joseph R (1988) Epidemiological studies of the tomato yellow leaf curl virus (TYLCV) in the Jordan Valley. Israel Phytoparasitica 16(3):259–270

Conway KE, McGraw BD, Motes JE, Sherwood JL (1989) Evaluations of mulches and row covers to delay virus diseases and their effects on yield of yellow squash. Appl Agric Res 4:201–207

Costa AS (1976) Whitefly-transmitted plant diseases. Annu Rev Phytopathol 14(1):429–449

Costa HS, Johnson MW, Ullman DE (1994) Row covers effect on sweet potato whitefly (Homoptera: Aleyrodidae) densities, incidence of silver leaf, and crop yield in zucchini. J Econ Entomol 87(6):1616–1621

CPRI (2004) Annual progress report 2003–04. Central Potato Research Institute, Shimla

CPRI (2019) Annual report 2018–19. ICAR-Central Potato Research Institute, Shimla. https://cpri.icar.gov.in//content/Index/?qlid=4121&Ls_is=4233&lngid=1. Accessed 10 May 2020

Crowder DW, Ellsworth PC, Tabashnik BE, Carriére Y (2008) Effects of operational and environmental factors on evolution of resistance to pyriproxyfen in the sweetpotato whitefly (Hemiptera: Aleyrodidae). Environ Entomol 37(6):1514–1524

Csizinszky AA, Schuster DJ, Kring JB (1995) Color mulches influence yield and insect pest Populations in tomatoes. J Am Soc Hort Sci 120:778–784

DAFWA (2017) Whitefly: potato and cabbage Pest in Indonesia and Western Australia. Department of Agriculture and Food, Western Australia, South Perth. https://www.agric.wa.gov.au/cabbage/whitefly-potato-andcabbage-pest-indonesia-and-western-australia. Accessed 2 Feb 2017

Dalton R (2006) The Christmas invasion. Nature 443:898–900

De Barro PJ (2012) The Bemisia tabaci species complex: questions to guide future research. J Integr Agric 11(2):187–196

De Barro PJ, Liu SS, Boykin LM, Dinsdale AB (2011) Bemisia tabaci: a statement of species status. Annu Rev Entomol 56:1–19

De Barro PJ, Trueman JWH, Frolich DR (2005) Bemisia argentifolii is a race of *B. tabaci* (Hemiptera: Aleyrodidae): the molecular differentiation of *B. tabaci* populations around the world. Bull Entomol Res 95:193–203

De Cock A, Degheele D (1998) Buprofezin: a novel chitin synthesis inhibitor affecting specifically planthoppers, whiteflies and scale insects. In: Ishaaya I, Degheele D (eds) Insecticides with novel modes of action: mechanism and application. Springer, Berlin/Heidelberg/New York, pp 74–91

Denholm I, Cahill M, Byrne FJ, Devonshire AL (1996) Progress with documenting and combating insecticide resistance in Bemisia. In: Gerling D, Mayer RT (eds) Bemisia: 1995 taxonomy, biology, damage, control and management. Intercept Ltd, Andover, pp 577–603

Dhawan AK, Butter NS, Narula AM (2007) The cotton whitefly Bemisia tabaci (Gennadius). Technical Bulletin, Department of Entomology, Punjab Agricultural University, Ludhiana

Dhawan AK, Simwat GS (1997) Insecticide induced resurgence in sucking pests in upland cotton in Punjab, India. In: Third international congress of entomological sciences, March 18–20, 1997, National Agricultural Research Centre, Islamabad, p 12

Dhawan P, Mandal RB (2008) Effect of apical leaf curl Begomovirus disease on growth and yield parameters of potato. In: Pandey SK (ed) Proceedings of the global potato conference–opportunities and challenges in the new millennium. Central Potato Research Institute, Shimla, pp 159–160

Dinsdale A, Cook L, Riginos C, Buckley Y, De Barro PJ (2010) Refined global analysis of *Bemisia tabaci* (Gennadius) (Hemiptera: Sternorrhyncha: Aleyrodoidea) mitochondrial CO1 to identify species level genetic boundaries. Ann Entomol Soc Am 103:196–208

Dittrich V, Ernst GH, Ruesh O, Uk S (1990) Resistance mechanisms in sweetpotato whitefly (Homoptera: Aleyrodidae) populations from Sudan, Turkey, Guatemala, and Nicaragua. J Econ Entomol 83:1665–1670

Duffus JE (1994) Diseases vectored by whiteflies: etiology, ecology, geographical distribution and possible control measures. In: Proceedings of 5th Arab congress of plant protection, November 27- December 2, 1994, FES

Ekbom BS, Xu R (1990) Sampling and spatial patterns of whiteflies. In: Gerling D (ed) Whiteflies: their bionomics, pest status and management. Intercept Ltd., London, pp 107–121

Ellango R, Singh ST, Rana VS, Priya NG, Raina H, Chaubey R, Naveen NC, Mahmood R, Ramamurthy VV, Asokan R, Rajagopal R (2015) Distribution of *Bemisia tabaci* genetic groups in India. Environ Entomol 44(4):1258–1264

Faria M, Wraight SP (2001) Biological control of *Bemisia tabaci* with fungi. Crop Prot 20:767–778

Firdaus S, Vosman B, Hidayati N, Jaya Supena ED, Visser GF, van Heusden AW (2013) The *Bemisia tabaci* species complex: additions from different parts of the world. Insect Sci 20(6): 723–733

Flückiger CR, Kristinsson H, Senn R, Rindlisbacher A, Buholzer H, Voss G (1992a) CGA 215¢ 944—a novel agent to control aphids and whiteflies. In: Brighton crop protection conference—pests and diseases, vol 1, Brighton, pp 43–50

Flückiger CR, Senn R, Buholzer H (1992b) CGA 215′944—opportunities for use in vegetables. In: Proceedings of 1992 Brighton crop protection conference—pests and diseases, vol 3, Brighton, pp 1187–1192

Fondong VN (2013) Geminivirus protein structure and function. Mol Plant Pathol 14:635–649

Franco-Lara L, Rodríguez D, Guzmán-Barney M (2013) Prevalence of potato yellow vein virus (PYVV) in *Solanum tuberosum* Group Phureja fields in three states of Colombia. Am J Potato Res 90(4):324–330

Fuog D, Fergusson SJ, Flückiger C (1998) Pymetrozine: a novel insecticide affecting aphids and whiteflies. In: Ishaaya I, Degheele D (eds) Insecticides with novel modes of action: mechanism and application. Springer, Berlin/Heidelberg/New York, pp 40–49

Gallo Y, Toro LF, Jaramillo H, Gutiérrez PA, Marín M (2018) Identificación y caracterización molecular del genoma completo de tres virus en cultivos de lulo (*Solanum quitoense*) de Antioquia (Colombia). Rev Colomb Cienc Hortíc 12:281–292

Gamarra H, Carhuapoma P, Cumapa L, González G, Muñoz J, Sporleder M, Kreuze J (2020a) A temperature-driven model for potato yellow vein virus transmission efficacy by *Trialeurodes vaporariorum* (Hemiptera: Aleyrodidae). Virus Res 289:198109

Gamarra H, Sporleder M, Carhuapoma P, Kroschel J, Kreuze J (2020b) A temperature-dependent phenology model for the greenhouse whitefly *Trialeurodes vaporariorum* (Hemiptera: Aleyrodidae). Virus Res 289:198107

Garg ID, Khurana SMP, Kumar S, Lakra BS (2001) Association of a geminivirus with potato apical leaf curl in India and its immunized electron microscopic detection. J Indian Potato Assoc 28: 227–232

Gennadius P (1889) Disease of the tobacco plantations in the Trikonia. The aleurodidd of tobacco. Ellenike Georgia 5:1–3

Gerling D (2002) Whiteflies revisited Manejo Integrado de Plagas 63:13–21

Gerling D, Horowitz AR (1984) Yellow traps for evaluating the population levels and dispersal patterns of *Bemisia tabaci* (Gennadius) (Homoptera: Aleyrodidae). Ann Entomol Soc Am 77: 753–759

Godfrey LD, Haviland DR (2017) UC IPM pest management guidelines: potato pest management guidelines. Agriculture and Natural Resources. University of California. http://ipm.ucanr.edu/PMG/selectnewpest.potatoes.html. Accessed 5 Jan 2017

Goettel MS, Hajek AE, Siegel JP, Evans HC (2001) Safety of fungal biocontrol agents. In: Butt TM, Jackson CW, Magan N (eds) Fungi as biocontrol agents: progress, problems and potential. CABI, Wallingford, pp 347–376

González-Dufau GI, Santamaría-Guerra J, Castrejon K, Herrera I, Monzón A (2018) Parámetros demográficos de Trialeurodes vaporariorum (Hemiptera: Aleyrodidae) en los cultivos de papa y tomate. Ciencia Agropecuaria 10(28):37–55

Green RH (1970) On fixed precision sequential sampling. Res Popul Ecol 12:249–251

Guzmán M, Rodríguez P (2010) Susceptibility of *Solanum phureja* (Juz. et Buk.) to Potato yellow vein virus. Agron Colomb 28(2):219–224

Guzmán M, Ruiz E, Arciniegas N, Coutts RH (2006) Occurrence and variability of Potato yellow vein virus in three departments of Colombia. J Phytopathol 154:748–750

Guzmán-Barney M, Franco-Lara L, Rodríguez D, Vargas L, Fierro JE (2012) Yield losses in *Solanum tuberosum* Group Phureja cultivar Criolla Colombia in plants with symptoms of PYVV in field trials. Am J Potato Res 89(6):438–447

Hadjistylli M, Roderick GK, Brown JK (2016) Global population structure of a worldwide pest and virus vector: genetic diversity and population history of the *Bemisia tabaci* sibling species group. PLoS One 11:e0165105

Hashmi TR, Dey D, Meshram NM, Devi SR, Ajit V, Narendra T, Ram P (2017) Records on genetic group of *Bemisia tabaci* (Gennadius) and associated endosymbionts feeding on potato plants. Environ Ecol 35(3C):2403–2408

Hilje L, Costa HC, Stansly PA (2001) Cultural practices for managing Bemisia tabaci and associated viral diseases. Crop Prot 20:801–812

Horowitz AR, Antignus Y, Gerling D (2011) Management of *Bemisia tabaci* whiteflies. In: Thompson WMO (ed) The whitefly, *Bemisia tabaci* (Homoptera: Aleyrodidae) interaction with geminivirus-infected host plants. Springer, Dordrecht, pp 293–322

Horowitz AR, Denholm I, Morin S (2007) Resistance to insecticides in the TYLCV vector, Bemisia tabaci. In: Czosnek H (ed) Tomato yellow leaf curl virus disease. Springer, Dordrecht, pp 305–325

Horowitz AR, Ishaaya I (1996) Chemical control of *Bemisia tabaci* – management and application. In: Gerling D, Mayer RT (eds) Bemisia: taxonomy, biology, damage, control and management. Intercept Ltd, Andover, pp 537–556

Horowitz AR, Kontsedalov S, Khasdan V, Ishaaya I (2005) Biotypes B and Q of Bemisia tabaci and their relevance to neonicotinoid and pyriproxyfen resistance. Arch Insect Biochem Physiol 58: 216–225

Horowitz AR, Mendelson Z, Cahill M, Denholm I, Ishaaya I (1999) Managing resistance to the insect growth regulator pyriproxyfen in Bemisia tabaci. Pestic Sci 55:272–276

Hu J, DeBarro PJ, Zhao H, Wang J, Nardi F, Liu SS (2011) An extensive field survey combined with a phylogenetic analysis reveals rapid and widespread invasion of two alien whiteflies in china. PLoS One 6:e16061

Hu J, Zhang X, Jiang Z, Zhang F, Liu Y, Li Z, Zhang Z (2018) New putative cryptic species detection and genetic network analysis of *Bemisia tabaci* (Hemiptera: Aleyrodidae) in China based on mitochondrial COI sequences. Mitochondrial DNA Part A 29(3):474–484

ICA (2014) Resolucion 4213 de 2014. Instituto Colombiano Agropecuario, Bogota, Colombia. http://extwprlegs1.fao.org/docs/pdf/col144725.pdf. Accessed 20 Aug 2020

Inglis GD, Goettel MS, Butt TM, Strasser H (2001) Use of hyphomycetous fungi for managing insect pests. In: Butt TM, Jackson CW, Magan N (eds) Fungi as biocontrol agents. progress, problems and potential. CABI, Wallingford, pp 1–8

Isaacs R, Byrne DN (1998) Aerial distribution, flight behaviour and eggload: Their inter-relationship during dispersal by the sweetpotato whitefly. J Anim Ecol 67(5):741–750

Ishaaya I (1990) Buprofezin and other IGRs for controlling cotton pests. Pestic Outlook 1(2):30–33

Ishaaya I, Horowitz AR (1992) A novel phenoxy juvenile hormone analog (pyriproxyfen) suppresses embryogenesis and adult emergence of the sweetpotato whitefly (Homoptera: Aleyrodidae). J Econ Entomol 85:2113–2117

Ishaaya I, Horowitz AR (1995) Pyriproxyfen, a novel insect growth regulator for controlling whiteflies: mechanism and resistance management. Pestic Sci 43:227–232

Ishaaya I, Kontsedalov S, Horowitz AR (2003) Novaluron (Rimon), a novel IGR: potency and cross-resistance. Arch Insect Biochem Physiol 54:157–164

Ishaaya I, Mendelson Z, Horowitz AR (1993) Toxicity and growth-suppression exerted by diafenthiuron in the sweetpotato whitefly, Bemisia tabaci. Phytoparasitica 21:199–204

Ishaaya I, Mendelson Z, Melamed-Madjar V (1988) Effect of buprofezin on embryogenesis and progeny formation of sweetpotato whitefly (Homoptera: Aleyrodidae). J Econ Entomol 81:781–784

Jeevalatha A, Chakrabarti SK, Sharma S, Sagar V, Malik K, Raigond B, Singh BP (2017a) Diversity analysis of Tomato leaf curl New Delhi virus-potato causing apical leaf curl disease of potato in India. Phytoparasitica 45:33–43

Jeevalatha A, Kaundal P, Kumar A, Guleria A, Sundaresha S, Pant RP, Sridhar J, Venkateswarlu V, Singh BP (2016) SYBR green based duplex RT-qPCR detection of a *Begomovirus*, tomato leaf curl New Delhi virus-[potato] along with potato virus X and potato leaf roll virus in potato. Potato J 43:125–137

Jeevalatha A, Siddappa S, Kumar A, Kaundal P, Guleria A, Sharma S, Nagesh M, Singh BP (2017b) An insight into differentially regulated genes in resistant and susceptible genotypes of potato in response to tomato leaf curl New Delhi virus-[potato] infection. Virus Res 232:22–33

Jones DR (2003) Plant viruses transmitted by whiteflies. Eur J Plant Pathol 109:195–219

Jumardi J, Melina M, Nasruddin A (2020) Vertical distribution of the greenhouse whitefly, Trialeurodes vaporariorum Westwood, within potato plant canopy. IOP Conf Ser: Earth Environ Sci 486:012144

Jyothsna P, Haq QMI, Singh P, Sumiya KV, Praveen S, Rawat R, Briddon RW, Malathi VG (2013) Infection of *Tomato leaf curl New Delhi virus* (ToLCNDV), a bipartite *Begomovirus* with betasatellites, results in enhanced level of helper virus components and antagonistic interaction between DNA B and betasatellites. Appl Microbiol Biotechnol 97:5457–5471. https://doi.org/10.1007/s00253-012-4685-9

Kayser H, Kaufmann L, Schürmann F (1994) Pymetrozine (CGA 215′944): a novel compound for aphid and whitefly control. An overview of its mode of action. In, Proceedings of 1994 Brighton crop protection conference—pests and diseases, vol 2, Brighton, pp 737–742

Kennedy JS (1985) Migration, behavioral and ecological. In: Rankin MA (ed) Contributions in marine science. Austin Marine Science Institute, University of Texas, Austin

Khurana SMP, Naik PS (2003) The potato: an overview. In: Khurana SMP, Minhas JS, Pandey SK (eds) The potato – production and utilization in subtropics. Mehta Publishers, New Delhi, pp 1–14

Khurana SMP, Singh MN (2003) Vector and true seed transmitted viruses. In: Khurana SMP, Minhas JS, Pandey SK (eds) The potato–production and utilization in sub-tropics. Mehta Publishers, New Delhi, pp 221–229

Kishore R, Singh BP, Parihar SBS (2005) Population dynamics of whitefly (*Bemisia tabaci* Genn.) on potato crop in relation to weather factors. Pro Indian Nat Sci Acad 75(4):257–260

Konar A, Singh NJ, Mandal P, More KA (2013) Screening of potato germplasm against aphid and whitefly in Gangetic plains of West Bengal. J Entomol Res 37(3):229–232

Kontsedalov S, Gottlieb Y, Ishaaya I, Nauen R, Horowitz AR, Ghanim M (2009) Toxicity of spiromesifen to the developmental stages of *Bemisia tabaci* biotype B. Pest Manag Sci 65:5–13

Kroschel J, Mujica N, Okonya J, Alyokhin A (2020) Insect pests affecting potatoes in tropical, subtropical, and temperate regions. In: Campos H, Ortiz O (eds) The potato crop. Springer, Cham, pp 251–306

Kumar M, Gupta A (2016) Effect of weather variables on whitefly (*Bemisia tabaci* Gennadius) population in development of potato apical leaf curl virus disease. J Agromet 18(2):288–291

Kumar P, Singh AK, Chattopadhyay B, Chakraborty S (2010) Molecular characterization of a new species of *Begomovirus* and betasatellite causing leaf curl disease of tomato in India. Virus Res 152:19–29

Kumar R, Tiwari RK, Jeevalatha A et al (2021) Potato apical leaf curl disease: current status and perspectives on a disease caused by tomato leaf curl New Delhi virus. J Plant Dis Prot 128:897–911. https://doi.org/10.1007/s41348-021-00463-w

Kuno E (1969) A new method of sequential sampling to obtain the population estimates with a fixed level of precision. Res Popul Ecol 11:127–136

Lahm GP, Cordova D, Barry JD (2009) New and selective ryanodine receptor activators for insect control. Bioorg Med Chem 17:4127–4133

Lakra BS (2002) Leaf curl: a threat to potato crop in Haryana. J Mycol Plant Pathol 32:367

Lakra BS (2003) Effect of date of planting on whitefly population, leaf curl incidence and yield of potato cultivars. J Indian Potato Assoc 30:115–116

Lakra BS (2005) Monitoring of foliar diseases and insect pests of potato at different dates of sowing and their impact on tuber yield in Haryana. Potato J 32(3–4):189–190

Lemma D, Pulgarín Navarro JM (1989) Relaciones insecto patógeno en el problema del amarillamiento de las venas de la papa. Medellín, Colombia, p 94

Liang G, Liu TX (2002) Repellency of a kaolin particle film, Surround, and a mineral oil, sunspray oil, to silverleaf whitefly (Homoptera: Aleyrodidae) on melon in the laboratory. J Econ Entomol 95(2):317–324

Livieratos IC, Eliasco E, Müller G, Olsthoorn RCL, Salazar LF (2004) Analysis of the RNA of Potato yellow vein virus: Evidence for a tripartite genome and conserved 3′-terminal structures among members of the genus *Crinivirus*. J Gen Virol 85:2065–2075

Lopez R, Asensio C, Guzman M, Boonham N (2006) Development of real-time and conventional RT-PCR assays for the detection of potato yellow vein virus (PYVV). J Virol Methods 136:24–29. https://doi.org/10.1016/j.jviromet.2006.03.026

Malik K, Chandel RS, Singh BP, Chandla VK (2005) Studies on potato apical leaf curl virus disease and its whitefly vector Bemisia tabaci. In: Proceedings of the annual meeting of Indian society of plant pathologists and centenary symposium on plant pathology. Central Potato Research Institute, Shimla, p 17

Malik K, Raghavendra KV, Kumar M (2020) Vector management strategies against *Bemisia tabaci* (Gennadius) transmitting potato apical leaf curl virus in seed potatoes. Potato Res 64:167. https://doi.org/10.1007/s11540-020-09470-0

Malik K, Shah MA, Naga K (2021) Aphid-vector surveillance. In: Chakrabarti SK, Bhardwaj V (eds) Potato in sub-tropics: a saga of success. IBPSS, New Delhi, pp 61–67

Malik K, Singh BP (2007) Studies on *Bemisia tabaci* and its management. Potato J 34(1–2):91–92

Maruthi MN, Colvin J, Thwaites RM, Banks GK, Gibson G, Seal SE (2004) Reproductive incompatibility and cytochrome oxidase I gene sequence variability among host-adapted and geographically separate *Bemisia tabaci* populations (Hemiptera: Aleyrodidae). Syst Entomol 29:560–568

Moriones E, Praveen S, Chakraborty S (2017) Tomato Leaf Curl New Delhi Virus: An emerging virus complex threatening vegetable and fiber crops. Viruses 9:264

Mugerwa H, Seal S, Wang HL et al (2018) African ancestry of New World, *Bemisia tabaci*-whitefly species. Sci Rep 8:2734. https://doi.org/10.1038/s41598-018-20956-3

Mujica N (1998) Uso de trampas amarillas en el control de adultos de mosca minadora y mosca blanca. International Potato Center, Lima

Muñoz Baena L, Gutiérrez Sánchez PA, Marín Montoya M (2017) Secuenciación del genoma completo del Potato yellow vein virus (PYVV) en tomate (*Solanum lycopersicum*) en Colombia. Acta Biol Colomb 22:5

Murugan M, Uthamasamy S (2001) Dispersal behaviour of cotton whitefly *Bemisia tabaci* (Genn.) under cotton based gardenland agro ecosystem of Coimbatore, Tamil Nadu. Madras Agric J 88(1/3):1–6

Naranjo SE, Castle SJ, De Barro PJ, Liu SS (2010) Population dynamics, demography, dispersal and spread of *Bemisia tabaci*. In: Stansly PA, Naranjo SE (eds) Bemisia: bionomics and management of a global pest. Springer, Dordrecht, pp 185–226

Naranjo SE, Ellsworth PC (2009) Fifty years of the integrated control concept: moving the model and implementation forward in Arizona. Pest Manag Sci 65(12):1267–1286

Naranjo SE, Flint HM (1995) Spatial distribution of adult *Bemisia tabaci* (Homoptera: Aleyrodidae) in cotton and development and validation of fixed-precision sampling plans for estimating population density. Environ Entomol 24(2):261–270

Nash TE, Dallas MB, Reyes MI, Buhrman GK, Ascencio-Ibanez JT, Hanley-Bowdoin L (2011) Functional analysis of a novel motif conserved across geminivirus rep proteins. J Virol 85: 1182–1192

Nasruddin A, Mound LA (2016) First record of *Trialeurodes vaporariorum* Westwood (Hemiptera: Aleyrodidae) severely damaging field grown potato crops in South Sulawesi, Indonesia. J Plant Protect Res 56:199–202. https://doi.org/10.1515/jppr-2016-0023

Natwick ET, Durazo AF III, Laemmlen F (1988) Direct row covers for insect and virus disease protection in desert agriculture. Plasticulture 78:35–46

Naveen NC, Chaubey R, Kumar D, Rebijith KB, Rajagopal R, Subrahmanyam B, Subramanian S (2017) Insecticide resistance status in the whitefly, *Bemisia tabaci* genetic groups Asia-I, Asia-II-1 and Asia-II-7 on the Indian subcontinent. Sci Rep 7:40634. https://doi.org/10.1038/srep40634

Niño Á, Del Toro FJ, Tenllado F, Canto T, Franco-Lara L (2021) Molecular insights on potato yellow vein *Crinivirus* infections in the highlands of Colombia. J Gen Virol 102(6):001604

Nombela G, Muñiz M (2009) Host plant resistance for the management of *Bemisia tabaci*: a multi-crop survey with emphasis on tomato. In: Stansly PA, Naranjo SE (eds) Bemisia: bionomics and management of a global pest. Springer, Dordrecht, pp 357–383

Nyoike TW, Liburd OE, Webb SE (2008) Suppression of whiteflies, *Bemisia tabaci* (Hemiptera: Aleyrodidae) and incidence of Cucurbit leaf crumple virus, a whitefly-transmitted virus of zucchini squash new to Florida, with mulches and imidacloprid. Fla Entomol 91:460–465

Ohnesorge B, Rapp G (1986) Methods for estimating the density of whitefly nymphs (*Bemisia tabaci* Genn.) in cotton. Trop Pest Manag 32:207–211

Palumbo JC, Horowitz AR, Prabhaker N (2001) Insecticidal control and resistance management for *Bemisia tabaci*. Crop Prot 20:739–765

Perring TM (2001) The *Bemisia tabaci* species complex. Crop Prot 20:725–737

Perring TM, Royalty RN, Farrar CA (1989) Floating row covers for the exclusion of virus vectors and the effect on disease incidence and yield of cantaloupe. J Econ Entomol 82(6):1709–1715

Prabhaker N, Castle SJ, Buckelew L, Toscano NC (2008) Baseline susceptibility of *Bemisia tabaci* B biotype (Hemiptera: Aleyrodidae) populations from California and Arizona to spiromesifen. J Econ Entomol 101:174–181

Prabhaker N, Toscano NC, Henneberry TJ (1998) Evaluation of insecticide rotations and mixtures as resistance management strategies for *Bemisia argentifolii* (Homoptera: Aleyrodidae). J Econ Entomol 91:820–826

Prokopy RJ, Owens ED (1983) Visual detection of plants by herbivorous insects. Annu Rev Entomol 28:337–364

Reddy AS, Rao NV (1989) Cotton whitefly (*Bemisia tabaci* Genn.). Indian J Plant Prot 17:171–179

Rincon DF, Vasquez DF, Rivera-Trujillo HF, Beltrán C, Borrero-Echeverry F (2019) Economic injury levels for the potato yellow vein disease and its vector, *Trialeurodes vaporariorum* (Hemiptera: Aleyrodidae), affecting potato crops in the Andes. Crop Prot 119:52–58. https://doi.org/10.1016/j.cropro.2019.01.002

Rodríguez PA, Franco Lara L, Guzmán Barney M (2015) Inter and intra variation of potato yellow vein virus in three potato species from Colombia. Rev Facult Nacional Agron Medell 68(1): 7387–7398

Rosell RC, Bedford ID, Frolich DR, Gill RJ, Brown JK, Markham PG (1997) Analysis of morphological variation in distinct populations of *Bemisia tabaci* (Homoptera: Aleyrodidae). Ann Entomol Soc Am 90:575–589

Saha A, Saha B, Saha D (2014) Molecular detection and partial characterization of a *Begomovirus* causing leaf curl disease of potato in sub-Himalayan West Bengal, India. J Environ Biol 35: 601–606

Saini HK (1998) Effect of synthetic pyrethroids on biology of whitefly Bemisia tabaci (Gennadius) on Gossypium hirsutum (Linn.). M.Sc. thesis, Punjab Agricultural University, Ludhiana

Salati R, Nahkla MK, Rojas MR, Guzman P, Jaquez J, Maxwell DP, Gilbertson RL (2002) Tomato yellow leaf curl virus in the Dominican Republic: characterization of an infectious clone, virus monitoring in whiteflies, and identification of reservoir hosts. Phytopathology 92(5):487–496

Salazar LF (1996) Potato viruses and their control. International Potato Center (CIP), Lima, p 214

Salazar LF (2006) Virus diseases of potatoes. Wiley, Chichester. https://doi.org/10.1038/npg.els.0004300

Salazar LF, Müller G, Querci M, Zapata JL, Owens RA (2000) Potato yellow vein virus: its host range, distribution in South America and identification as a crinivirus transmitted by *Trialeurodes vaporariorum*. Ann Appl Biol 137(1):7–19

Salazar LF, Müller G, Querci M, Zapata JL, Owens RA (2005) Potato yellow vein virus: its host range, distribution in South America and identification as a *Crinivirus* transmitted by *Trialeurodes vaporariorum*. Ann Appl Biol 137:7–19. https://doi.org/10.1111/j.1744-7348.2000.tb00052.x

Saldarriaga VA, Alvarez AM (1988) Jaramillo JE (1988) Effect of the potato yellow vein transmitted by *Trialeurodes vaporariorum* (Westwood) in potato. Rev Colomb Entomol 14(2):3–8

Sattelle DB, Cordova D, Cheek TR (2008) Insect ryanodine receptors: molecular targets for novel pest control chemicals. Invert Neurosci 8:107–119

Schuster DJ, Thompson S, Ortega LD, Polston JE (2009) Laboratory evaluation of products to reduce settling of sweetpotato whitefly adults. J Econ Entomol 102(4):1482–1489

Shah MA, Aarti B, Naga KC, Subhash S et al (2020b) Important potato pests and their management. In: Singh AK, Chakrabarti SK, Singh B, Sharma J, Dua VK (eds) Potato science and technology for sub-tropics. NIPA, New Delhi, pp 295–326

Shah MA, Malik K, Bhatnagar A, Katare S, Sharma S, Chakrabarti SK (2019a) Effect of temperature and cropping sequence on infestation pattern of *Bemisia tabaci* in potato. Indian J Agric Sci 89(11):50–55

Shah MA, Naga KC, Subhash S et al (2021a) Use of petroleum-derived spray oils for the management of vector-virus complex in potato. Potato Res. https://doi.org/10.1007/s11540-021-09505-0

Shah MA, Sharma S, Kumar R et al (2021b) Seasonal population dynamics and flight activity of sweetpotato whitefly, *Bemisia tabaci* on potato in the sub-tropical plains of India. Arthropod-Plant Interact 15:605–613. https://doi.org/10.1007/s11829-021-09835-3

Shah MA, Sharma S, Kumar R, Singh RK (2020c) Evaluation of polypropylene row covers for excluding virus vectors and their effect on the incidence of diseases and yield in potato. Indian Phytopathol 73:751–757

Shah MA, Sharma S, Sharma J (2019b) Bio-efficacy of potassium silicate against aphids and whitefly in potato. Potato J 46(2):132–137

Shah MA, Sridhar J, Sharma S (2020a) Studies on the within-plant distribution of whitefly, *Bemisia tabaci* (Gennadius) in potato and development of a sequential sampling plan. Int J Pest Manage. https://doi.org/10.1080/09670874.2020.1823520

Shahid MS (2020) Tomato yellow leaf curl virus interaction with betasatellites–a global threat to tomato production. Plant Pathol 69(7):1191–1192. https://doi.org/10.1111/ppa.13201

Sharaf N (1986) Chemical control of *Bemisia tabaci*. Agric Ecosyst Environ 17:111–127

Silva MS, Lourenção A, Souza-Dias JAC, Miranda Filho HS, Ramos VJ, Schammass EA (2008) Resistance of potato genotypes (*Solanum* spp.) to *Bemisia tabaci* biotype B. Hort Brasil 26:221–226

Simon B, Cenis JL, Beitia F, Saif K, Moreno IM et al (2003) Genetic structure of field populations of *Begomoviruses* and of their vector *Bemisia tabaci* in Pakistan. Phytopathology 93:1422–1422

Sivalingam PN, Malathi VG, Varma A (2010) Molecular diversity of the DNA-β satellites associated with tomato leaf curl disease in India. Arch Virol 757:764

Southwood TRE (1978) Ecological methods, 2nd edn. Chapman & Hall, London

Sridhar J, Venkateswarlu V, Jeevalatha A, Malik K, Bhatnagar A, Singh BP (2016) Squash and tissue print protocols for quick detection of tomato leaf curl New Delhi virus-potato in fresh and ethanol preserved single whitefly. Potato J 43(1):62–69

Suwwan MA, Akkawi M, Al-Musa AM, Mansour A (1988) Tomato performance and incidence of tomato yellow leaf curl (TYLC) virus as affected by type of mulch. Sci Hortic 37:39–45

Tamayo PJ, Navarro R (1984) Aumenta la incidencia del virus del amarillamiento de venas de la papa en Antioquia. ASCOLFI Informa (Bogota) 10:40–42

Traboulsi (1995) Bemisia tabaci: a report on the pest status with particular reference to the near east. FAO Plant Prot Bull 42:33–35

Tzanetakis IE, Martin RR, Wintermantel W (2013) Epidemiology of *Criniviruses*: an emerging problem in world agriculture. Front Microbiol 4:119. https://doi.org/10.3389/fmicb.2013.00119

Usharani KS, Srivastava A, Padmalatha KV, Malathi VG (2004b) First report of the association of a defective satellite DNA β molecule with a bipartite genome *Begomovirus* causing potato leaf curl disease in India. J Plant Pathol 86:177–180

Usharani KS, Surendranath B, Paul-Khurana SM, Garg ID, Malathi VG (2004a) Potato leaf curl—a new disease of potato in northern India caused by a strain of Tomato leaf curl New Delhi virus. Plant Pathol 253:235

Venkatasalam EP, Pandey KK, Singh V, Singh BP (n.d.) Seed potato production technology. Technical bulletin, published by ICAR-Central Potato Research Institute, Shimla, Himachal Pradesh, India, p 71

Vestergaard S, Cherry A, Keller S, Goettel M (2003) Safety of hyphomycete fungi as microbial control agents. In: Hokkanen HMT, Hajek AE (eds) Environmental impacts of microbial insecticides: needs and methods for risk assessment. Kluwer Academic, Dordrecht, pp 5–62

Vidal C, Fargues J, Rougier M, Smits N (2003) Effect of air humidity on the infection potential of hyphomycetous fungi as mycoinsecticides for *Trialeurodes vaporariorum*. Biocontrol Sci Technol 13:183–198

Villamil-Garzón A, Cuellar WJ, Guzmán-Barney M (2014) Natural co-infection of *Solanum tuberosum* crops by the Potato yellow vein virus and *Potyvirus* in Colombia. Agron Colomb 32:213–223

Walters SA (2003) Suppression of watermelon mosaic virus in summer squash with plastic mulches and row covers. Hort Technol 13:352–357

Webb SE, Linda SB (1992) Evaluation of spunbonded polyethylene row covers as a method of excluding insects and viruses affecting full-grown squash in Florida. J Econ Entomol 85(6): 2344–2352

Xu J, Liu N, Zhang R (2013) Other Pests – China. In: Giordanengo P, Vincent C, Alyokhin A (eds) Insect pests of potato. Elsevier, Amsterdam, pp 193–226

Xue YG, Watson DM, Nicetic O, Beattie GAC (2002) Impact of nC24 horticultural mineral oil deposits on oviposition by greenhouse whitefly *Trialeurodes vaporariorum* (Westwood) (Hemiptera: Aleyrodidae). Gen Appl Entomol 31:59–64

Yadava P, Suyal G, Mukherjee SK (2010) Begomovirus DNA replication and pathogenicity. Curr Sci 98:360–368

Zang LS, Jiang T, Xu J, Liu SS, Zhang YJ (2006) SCAR molecular markers of the B biotype and two non-B populations of the whitefly, *Bemisia tabaci* (Hemiptera: Aleyrodidae). J Agric Biotechnol 14:208–212

Zhang LS, Liu SS, Liu YQ, Ruan YM, Wan FH (2005) Competition between the B biotype and a non-B biotype of the whitefly, *Bemisia tabaci* (Homoptera: Aleyrodidae) in Zhejiang, China. Shengwu Duoyangxing 13:181–187

Zimmerman G (2008) The entomopathogenic fungi *Isaria farinosa* (formerly *Paecilomyces farinosus*) and the *Isaria fumosorosea* species complex (formerly *Paecilomyces fumosoroseus*): biology, ecology and use in biological control. Biocontrol Sci Tech 18:865–901

Biology and Management of Nematodes in Potato 11

Aarti Bairwa, E. P. Venkatasalam, Priyank H. Mhatre, Anuj Bhatnagar, Ashwani Kumar Sharma, Dalamu, Bhawna Dipta, S. Subhash, and Sanjeev Sharma

Abstract

Plant parasitic nematodes are a constraint in potato production, resulting in reduced productivity, abnormalities in tubers and malformations, all of which lead to loss of income to farmers. These tiny organisms result in a global yield loss of 12.3 percent ($157 billion), and in India alone, it accounts for $40.3 million. Because of their concealed nature, nematode damage is difficult to control and, as a result, is frequently disregarded. Nematode damage has symptoms that are similar to those of other diseases and abiotic stresses. Plant parasitic nematodes not only cause damage on their own, but they can form disease complexes with other micro-organisms, resulting in increased crop loss. A number of nematode species have been associated to potato; among them potato cyst nematodes and root-knot nematodes are economically important. In addition, false root-knot nematode, the potato rot nematode and root lesion nematode can also cause significant yield losses in potato. Use of certified nematode-free quality planting materials and resistant cultivars and crop rotations are recommended for preventing nematode infestations. Apart from these recent breakthroughs in RNAi and CRISPR/Cas biotechnological tools made nematode management effective and eco-friendly.

A. Bairwa (✉) · A. K. Sharma · Dalamu · B. Dipta · S. Sharma
Division of Plant Protection, ICAR-Central Potato Research Institute, Shimla, Himachal Pradesh, India

E. P. Venkatasalam · P. H. Mhatre
ICAR-Central Potato Research Station, Muthorai, Udhagamandalam, The Nilgiris, Tamil Nadu, India

A. Bhatnagar · S. Subhash
ICAR-Central Potato Research Institute-Regional Station, Modipuram, Meerut, Uttar Pradesh, India

S. Kumar Chakrabarti et al. (eds.), *Sustainable Management of Potato Pests and Diseases*, https://doi.org/10.1007/978-981-16-7695-6_11

Keywords

Potato · Nematodes · Species · Pathotype · Quarantine · Yield loss · Management · RNAi · CRISPR/Cas9

11.1 Introduction

The potato is a major horticultural crop all over the world, including India. Among the various root and tuber crops grown in India, the potato has the largest share and has attained production of 50.19 MT from an area of 2.17 million ha during 2020 (FAOSTAT 2017). Now, India is the second largest producer after China (91.88 MT). Potential yield of potato is determined by various biotic and abiotic factors. Among biotic factors, pathogens like fungi, bacteria, viruses, insects and nematodes play a crucial role leading to overall yield loss of 30–40%. There are a number of factors which enhance yield loss of potato tubers with nematode parasitism, viz. cultivar, atmosphere, soil composition, time of planting and nematode population (Noling 2016). Numerous potato plant parasitic nematode species are recorded; among which some cause significant yield losses, while others may cause negligible injuries. Potato cyst nematodes (PCN) (*Globodera* spp.) and root knot nematodes (RKN) (*Meloidogyne* spp.) are among the most economically important nematode pests of potato worldwide, including India. The initial record of cyst nematode infestation in the potato crop was recorded in the year 1881 by Julius Kuhn in Germany. Two species of PCN, viz. *Globodera rostochiensis* (Wollenweber) and *G. pallida* (Stone), are also popularly called golden nematodes, which hinder the sustainable production of potato. They are subjected to stringent quarantine and/or regulatory procedures, wherever they occur, and present a serious threat to domestic and international commerce in potatoes. RKN was recorded in 1889 by Neal from Florida in the USA, and ten species of RKN are reported to infect potatoes. RKN species such as *Meloidogyne incognita* (Kofoid and White), *M. javanica* (Treub), *M. arenaria* (Neal), *M. hapla* (Chitwood), *M. fallax* (Karssen), *M. thamesi* (Chitwood) and *M. chitwoodi* (Golden) have been associated with potato and have global significance. The false root-knot nematode *Nacobbus aberrans* (Thorne), the potato rot nematode *Ditylenchus destructor* (Thorne) and the root lesion nematode *Pratylenchus* spp. (Filipjev) can also cause significant yield losses in potato (Medina et al. 2017). Further, some nematode species can cause minor problems in potato, including the stubby-root nematodes, *Trichodorus* spp. (Cobb) and *Paratrichodorus* spp. (Siddiqi), the lance nematode *Hoplolaimus galeatus* (Cobb) and the dagger nematode *Xiphinema* spp. (Cobb).

11.2 Potato Cyst Nematodes (*Globodera* Species)

11.2.1 Origin and Distribution

PCN species, *G. pallida* and *G. rostochiensis*, are considered one of the globally important plant protection problems (CABI/EPPO 2020a; CABI/EPPO 2020b). The Andean Mountains of South America, which are the original home of potatoes, are also the origin of PCN. It was first introduced into Europe in the 1850s, along with the soil left on potato tubers which were sent for late blight resistance breeding, and it quickly spread across the world due to the introduction of European varieties. As a result, Europe has been labelled the "secondary node" for the PCN diaspora. The exact routes of PCN spread from South America to Europe, according to Franco et al. (1998), remain a matter of speculation. PCN likely spread from Europe to other countries through exported seed tubers of breeding materials (Evans and Stone 1977). PCN was considered to have been imported into Asian countries during World War II when human capital, medicine and military equipment were transported to many areas of Asia. However, PCN can be transmitted from Peru to Japan through polluted guano sacks and bird remains (Grenier and Benjamin 2017).

In India, Dr. F.G.W. Jones first detected the PCN in 1961 from Vijayanagaram farm in Udhagamandalam, The Nilgiri district of Tamil Nadu. Later on, their occurrence was reported from other parts of Nilgiri, Kodaikanal hills, adjoining hills of Karnataka and Idukki District in the Western Ghats of Kerala; accordingly the Tamil Nadu government imposed domestic quarantine during 1971. Recently, the occurrence of this pest has also been reported from some parts of the hilly regions of Himachal Pradesh, Jammu and Kashmir and Uttarakhand (Aarti et al. 2020a; Chandel et al. 2020). Accordingly, the Government of India in 2018 restricted the movement of potato seed tubers from infested areas. Other solanaceous crops like tomato (*Solanum lycopersicum*) and eggplant (*S. melongena*) and other members of Solanaceae like *Datura* spp., *Capsicum annuum* (chili pepper), *Hyoscyamus*, *Lycopersicon*, *Physalis* (husk tomatoes), *Physochlaina*, *Salpiglossis*, *Nicotiana acuminata*, *Saracha* and *Oxalis tuberosa* are also infested by both species of PCN (Sullivan et al. 2007). *S. sisymbriifolium* (Lam.), *S. mauritianum* and *S. nigrum* are reported as a potential trap crop for both species of PCN (Scholte and Vos 2000; Sullivan et al. 2007; Mhatre et al. 2021). As of now, PCN has turned into a major pest in the largest part of potato-growing areas affecting >80 countries in the temperate as well as cooler parts of the tropical and subtropical regions of the world.

11.2.2 Status of Species and Pathotypes

G. rostochiensis was first found in 1941 in the United States, in the 1960s in India and in the 1970s in Mexico (Grenier and Benjamin 2017). Presently, PCN has been recorded in 83 countries with *G. rostochiensis* (CABI/EPPO 2020a) and 64 countries with *G. pallida* (CABI/EPPO 2020b) in six continents, viz. Africa, North America, South America, Asia, Europe and Oceania. Kort et al. (1977) suggested an international scheme that designates five pathotypes of *G. rostochiensis* (Ro1 to Ro5) and three pathotypes of *G. pallida* (Pa1 to Pa3). *G. rostochiensis* (Ro1 race) populations in the United Kingdom have virulence against the H1 (ex-*andigena*) gene, which is close to that of *G. rostochiensis* populations in South America (Ro1 race). There are only a few pathotypes in England and Wales, namely, Ro1 (*G. rostochiensis*), Pa1, Pa2 and Pa3 (*G. pallida*). However, since it was difficult to differentiate between Pa2 and Pa3, the term Pa2/Pa3 was coined to characterize the broad range of virulence shown by European non-Pa1 *G. pallida*. A recent study in Scotland has revealed the presence of both Pa1 and Pa3 pathotypes, while populations in Northern Ireland have been found to be a mixture of Pa1, Pa3 and Pa2/Pa3 pathotypes. A significantly larger number of pathotypes occur in Europe, but are still incomparable to the range found in South America. These variations are thought to be a direct consequence of the few European introductions, but they may also have been influenced by the use of various cultivars and habitats, both of which would lead to the eventual expression of virulence. Ro1 has been documented in every country in Western Europe, including England and Wales, but pathotypes ranging from Ro2 to Ro5 have been reported in Germany, the Netherlands, Norway and Sweden (Turner and Evans 1998; Grenier and Benjamin 2017; Aarti et al. 2021).

The range of pathotypes of *G. pallida* in Europe, on the other hand, is thought to be identical to that in the United Kingdom. Less data is available in Central and Eastern Europe, where potato crop is also significant, but the widespread use of cultivars resistant only to Ro1 is likely to lead to the predominance of *G. pallida* pathotypes, as it has in England and Wales. Continuous breeding and selection of resistant potatoes in the United Kingdom, the Netherlands and Germany has revealed variation among species that were previously classified separately (Kort et al. 1977). In India, the differential host reactions of PCN populations revealed that the pathotypes Ro1 of *G. rostochiensis* and Pa2 of *G. pallida* are the most prevalent forms covering 75% of area. The other prevalent pathotypes are Ro2 (7%) and Ro5 (3%) of the *G. rostochiensis* and Pa1 (15%) and Pa3 (3%) of the *G. pallida* (Krishna Prasad 2006).

11.2.3 Biology

The hatching of cysts is stimulated by the chemical substances called hatching factors present in potato root diffusates (PRD) of the host plant roots. The second-stage juvenile (J_2) coming out of the cysts moves actively in soil and invades the roots by rupturing with its stylet. It enters through the epidermal cell walls and finally settles with its head towards the stele and feeds on cells in the pericycle, cortex or endodermis by forming a feeding tube. This induces enlargement of root cells and breakdown of their walls to form a large "syncytium" that provides nourishment for nematode development. The nematode moults and remains in the syncytium until its development is complete. The sex of the nematode is determined during J_3 stage (Fig. 11.1). Females become sedentary and swollen and remain attached to the roots and the posterior part of the body comes out by rupturing the root cells after the final

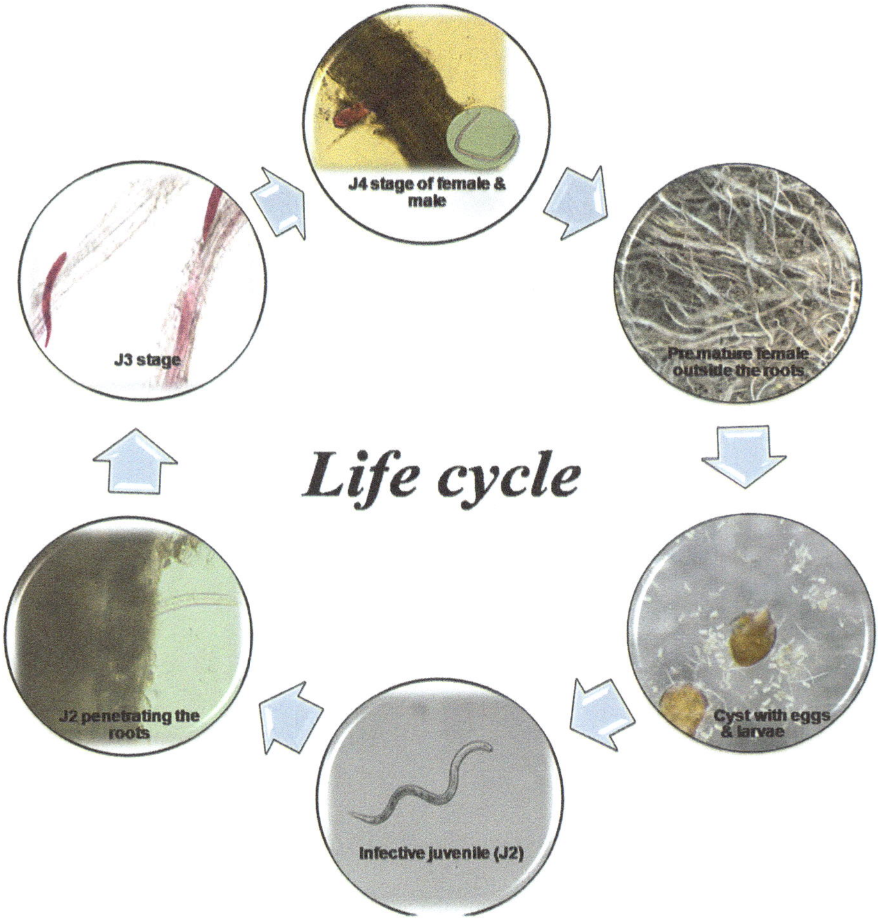

Fig. 11.1 Life cycle of potato cyst nematode

moult. Males retain their thread shape and come out of the roots to locate and mate with the females. The immature females of *G. rostochiensis* are golden yellow in colour, while *G. pallida* is white or cream in colour (Fig. 11.2).

The white PCN remain white or cream-coloured before finally turning brown, whereas the yellow PCN passes through a prolonged golden-yellow phase before and it also turns brown or leathery. After the female dies, the body wall thickens to form a hard brown cyst that is resistant to adverse weather conditions (Fig. 11.3). Each cyst contains 200–500 eggs and is easily dislodged in soil at harvest. The eggs inside the cysts can survive in soil for up to 30 years even in the absence of a suitable host. Mostly PCN complete the life cycle within 35–49 days. In general, one generation is completed in one crop season. High egg mortality in the cysts of

Fig. 11.2 Infection of *Globodera* species in potato root, *Globodera rostochiensis* (left), *Globodera pallida* (right)

Fig. 11.3 *Globodera* cyst (left), eggs and infective juveniles (J2s) (right)

Globodera spp. occurs in warmer climates, while in subtropical regions, its life cycle gets disturbed when the temperature goes beyond 28 °C (Caixeta et al. 2016). For the development on the host, *G. pallida* require 10–18 °C, while *G. rostochiensis* requires 15–25 °C (EPPO 2013). In the absence of a host, about 30–33% of eggs hatch spontaneously each year, subjected to environmental factors (Oostenbrink 1950; Aarti et al. 2021).

11.2.4 Spread

The PCN normally spreads by soil, water, compost and use of infested tubers in newer areas and also by the feet of animals and human beings and farm implements moving from infected to disease-free field.

Fig. 11.4 Potato field infested with *Globodera* spp.

11.2.5 Symptoms and Yield Loss

In case of low PCN population densities in soil, potato crop does not show any above-ground symptoms as most of the potato plants can tolerate nematode invasion. However, as the degree of invasion increases, the plant is unable to compensate and ultimately exhibits a range of symptoms. When the infestation is intense and localized, small patches of poorly growing plants appear in the field, and wilting may also occur during the hot sunny hours of the day (Fig. 11.4). As the season advances, the lower leaves turn yellow/brown and wither, leaving only the young leaves at the top. The entire plant shows a "tufted head" appearance, which ultimately causes the premature death of the plant. The browning and withering of the foliage gradually extend to withering. The root system is poorly developed, and depending upon the degree of infestation, the yield and size of the tubers are reduced.

The soil PCN population tolerance limit is 1.3–2.1 eggs per gram (Greco 1993), whereas the economic threshold is about 20 eggs per gram of soil (Evans and Stone 1977). Previously, Oerke et al. (1994) reported a yield loss of 30% worldwide, but Urwin et al. (2001) recorded estimated losses of more than 12%, whereas in India, Krishna Prasad (2008) recorded estimated yield loss ranging from 5 to 80% in high-infestation areas.

11.2.6 Management

PCN are extremely difficult to eliminate from contaminated soil once they have been established. Because no single control method is completely effective in achieving the desired level of nematode suppression, an integrated nematode management module combining a selective mixture of various options such as host resistance and chemical, biological and cultural methods is being proposed to reduce the PCN populace to levels that allow for cost-effective potato production.

11.2.6.1 Quarantine

Plants, plant products and goods are all subject to legal restrictions in almost every country in order to avoid pests and diseases introduced by humans from harming agriculture and the environment. Many pests and pathogens, such as nematodes, are widely distributed, but their biological range has not yet reached its full potential, and they may be absent from a country or geographic area (Taylor and Brown 1998). Because of the significant difficulty in eradicating PCN once it has been identified in the field, stringent quarantine regulations have been implemented in many parts of the world in order to control and prevent the spread of PCN. Within the EU, Council Directive 2000/29/EC allows member states to implement quarantine measures to prevent the spread of PCN, while EU Council Directive 2007/33/EC outlines PCN monitoring and management measures. This law was recently revised with the Plant Health Implementation Regulation 2016/2031. *G. rostochiensis* was discovered in New York in the 1940s, but an aggressive survey, quarantine and deployment of resistant cultivars carrying the H1 gene stopped the nematode from spreading further (Evans and Brodie 1980).

Strict local and national import controls have resulted in the localized eradication of PCN, though monitoring programs are still in place to keep an eye on this pest. Despite such strict measures, new outbreaks of PCN are recorded on a regular basis, including areas where potato production is heavily reliant. *G. pallida* was discovered in Idaho, one of the most significant potato-growing regions in the United States (Hafez et al. 2007), necessitating massive efforts to contain and eradicate the outbreak (Contina et al. 2020). The discovery of PCN in a number of sub-Saharan African countries, on the other hand, may be even more important (Mwangi et al. 2015; Niragire et al. 2019; Cortada et al. 2020).

Domestic quarantine prohibits seed tuber movement from infested to non-infested regions. The Ministry of Agriculture and Farmers Welfare in India has issued a notification under Section 4A of the Destructive Insects and Pests Act, 1914, prohibiting the movement of seed potatoes from PCN-infested areas to other states and union territories. Cysts are sedentary and unable to travel on their own. The most likely way for them to spread is through the movement of soil, rather than through the planting of an infested crop. The most successful way to avoid such a spread is to implement strict biosecurity measures. This involves ensuring the soil is not moved by farm equipment, tyres and shoe soles, planting materials and domestic and wild animals. Understanding the biology of PCN, including how they go through their life cycle, how they disperse and how likely they are to be detected at the county, local, farm and field levels, is critical to managing PCN populations. It's important to understand what lab findings mean and what growers can do to prevent PCN from spreading across the field.

11.2.6.2 Use of Certified Seed

A basic principle behind the EU PCN Directive 2007/33/EC is limiting the cultivation of certified seed potatoes to land that has been checked and found free of PCN. This helps avoid initial and subsequent introductions of PCN into fields.

11.2.6.3 Crop Rotation

Crop rotations, according to Urwin et al. (2001), hold PCN population densities below the damaging level. Maize and lima beans were found to be the best sequence for affecting PCN density, potato yield and profitability in Peru (Canto 1995), Ecuador (Ravelo 1984) and Cochabamba (Proinpa 1996). In Western Europe, a 7-year gap between potato crops of susceptible varieties is needed (Oostenbrink 1950; Jones 1970). Because of their limited host range, crop rotation with non-solanaceous crops is commonly recommended for PCN management. When a potato was grown at the end of a 4-year crop rotation involving potatoes, French beans, and peas, Menon and Thangaraju (1973) observed a 98.7–99.9% reduction in PCN in the fourth year and a yield increase of more than 90%. The use of resistant varieties alone in a 4-year crop rotation program increased yields by 67–78%.

PCN population density would be reduced by growing non-host crops in between host crops (Whitehead 1995). Crop rotation with PCN non-host crops such as radish, cabbage, cauliflower, turnip, garlic and carrot, and green manure crops such as lupin for 3 to 4 years, reduces cyst population by 50% (Krishna Prasad 1993). The number of cysts was reduced by 19.6–21.0%, and the number of eggs per cyst was reduced by 12.2–16.2% in radish, compared to other non-solanaceous crops. Garlic came in second, with a 15.9–17.7% and 10.3–11.6% reduction in cysts and eggs, respectively (Aarti et al. 2017). Crop rotation with barley has shown a reduction in *G. rostochiensis* up to 87% (Senasica 2013). Long rotations are often used to manage PCN, taking advantage of attrition caused by PCN's natural hatch and mortality. In the absence of a host plant, 20–30% of the population decline per year. But it is difficult to forecast since it is affected by variations in soil composition, soil type and other environmental factors such as aeration and moisture (Devine et al. 1999). However, farmers are typically hesitant to follow these suggestions because potatoes are a cash crop in hilly areas.

11.2.6.4 Inter-Cropping

When potatoes are intercropped with French beans (3:2 ratio), Manorama et al. (2005) found a higher potato equivalent yield and a lower cyst population. Potato and mustard inter-cropping in a 1:1 plant ratio, along with carbofuran application, reduced PCN infestation and increased potato yield (Devrajan and Balasubramanian 2008). Potato intercropped with radish at a 2:1 ratio was found to be successful in reducing the PCN population (Rf: 0.99) (Aarti et al. 2017).

11.2.6.5 Trap Cropping

The first form of trap crop is a potato crop, which must be uprooted 40 days after planting before the development of PCN females. This method has been used in the Netherlands to combat high infestations, but it necessitates the loss of a potential crop. The nematodes are caught and killed inside the plant prior to maturation, so highly efficient plant destruction is needed. In France, trap cropping decreased *G. pallida* populations by 80% per year and 98.5% with two trap crops and ethoprophos application. Cultivar Cara that is tolerant to *G. pallida* reduced the population by 75% when grown on complete ridges for 6 weeks in heavily infested soil (Whitehead 1977; Whitehead et al. 1994). However, in India, trap cropping with a susceptible potato cultivar attracted more juveniles than trap cropping with a resistant potato cultivar and decreased nematode population by 53%, but trap crops should be destroyed before the PCN life cycle is completed (Aarti et al. 2017). The second way of using a trap crop is to use a *S. tuberosum*-related crop that prevents PCN from completing its life cycle. There have been a number of candidate crops studied for this, but *S. sisymbriifolium* has shown the most promise so far (Dandurand et al. 2014). The use of the wild trap plant *S. sisymbriifolium* resulted in an 80% reduction in the PCN population in the region (Timmermans et al. 2007; Mhatre et al. 2021). Other species that have shown promise include *S. tuberosum*, *S. nigrum*, *S. dulcamara* and *D. stramonium* (Sparkes 2013). Growing potatoes to encourage PCN hatching and killing potato plants after nematode infestations in the potato roots can help reduce soil infestations (Webley and Jones 1981).

11.2.6.6 Host Plant Resistance

Globodera spp.-resistant varieties have been used successfully with a control rate of up to 95%. In addition, many breeding projects are underway around the world to find resistance genes for these nematodes (Sullivan et al. 2007). Resistance was found in 18 out of 22 *Solanum* accessions studied by Wolters et al. (1996), with the highest levels in *S. gourlayi* BGRC7180 and *S. neorossi* BGRC7211, as well as *S. sanctae rosae*, *S. sparsipilum* and *S. sucrense*. Resistance to *G. rostochiensis* R1A was observed in *S. andigena*, *S. gourlayi*, *S. spegazzini* and *S. vernei* in Germany, as well as resistance to *G. pallida* P4A/P5A in *S. gourlayi*, *S. spegazzini*, *S. sparsipilum* and *S. vernei*. All resistance was initially dependent on the H1 allele derived from *S. tuberosum* ssp. *andigena* CPC 1673, which was only successful against *G. rostochiensis* pathotypes R1A and R1B. Now that these pathotypes have spread widely and become virulent, the H1 allele is no longer successful against all *G. rostochiensis* populations (Phillips and Trudgill 1998). In comparison to *S. vernei*, Mulder (1994) found that resistant cultivars derived from *S. tuberosum* subsp. *andigena* had a high degree of tolerance.

According to Hockland et al. (2012), resistant potato varieties to *G. rostochiensis* (Ro1) exist in the United Kingdom and Europe, but no cultivars are resistant to all pathotypes of *G. pallida*. Some PCN-tolerant varieties, such as Cara, can produce robust growth while maintaining yield at moderately high PCN levels. High-resistance varieties of *G. rostochiensis*, such as Maris Piper, have become commonly cultivated, and the damage caused by this species decreased dramatically. Kishore et al. (1969) screened a large selection of germplasm against PCN in India in order to find resistance and incorporate it into commercial potato varieties. Dalamu et al. (2012) identified potato germplasm that was resistant to both PCN species found in *tuberosum* and *andigena* accessions. In India, to minimize PCN multiplication, the first *S. vernei*-derived resistant cultivar Kufri Swarna was launched in 1985 (Khan et al. 1985), followed by Kufri Neelima in 2012 (Joseph et al. 2012), Kufri Sahyadri in 2019 (Joseph et al. 2019) and Kufri Karan (Bhardwaj et al. 2019), all of which are suitable for the Nilgiri and Himalayan hills. However, in this area, conflict between breeder and nematode continues because of the development of virulence in both species of *Globodera*. The best way to handle PCN populations in the field is to develop and expand resistant varieties.

11.2.6.7 Antagonistic Plants

Antagonistic plants can withstand nematode infection at first, but later in their life cycle, plant factors can prevent them from developing further. *Crotalaria spectabilis*, *C. juncea*, *Tagetes patula*, *T. minuta*, *T. erecta* and *Estizolobium* spp. are used to combat root-knot nematode problems in potato fields in Brazil (Embrapa 2015) and may also be used to control *Globodera* spp.

11.2.6.8 Physical Control

Since only a few centimetres of soil in temperate areas reach lethal temperatures, soil solarization is best suited for small areas with long hot summers (Whitehead and Turner 1998). During the hot summer, *G. rostochiensis* eggs (97%) were unable to hatch in the top 10 cm layer of the soil (LaMondia and Brodie 1984). Solarization of the soil for 62 days decreased the population of *G. rostochiensis* by 95% (Mani et al. 1993).

11.2.6.9 Chemical Control

Nematicides are a reliable way to rapidly reduce the nematode population. The effectiveness of soil fumigation is highly dependent on the soil's condition and temperature. Soil can be fumigated above 5 °C with methyl bromide, 7 °C with 1,3-D or 10 °C with MITC fumigants (Whitehead and Turner 1998). Methyl bromide at 488–1464 kg/ha when applied under a gas-tight polythene board regulated PCN population in tomato and potato. However, since it is toxic to the ozone layer, it has been banned in several nations. Dazomet, a soil fumigant, was found to be more powerful than Telone in controlling nematodes in the United Kingdom (Whitehead et al. 1973). In the silty loam soil, Whitehead et al. (1994) found that ethoprophos at 11.2 kg/ha partially regulated *G. pallida*. The carbamate Vydate 10G (10% oxamyl) from DuPont and the organophosphates Nemathorin® (10% fosthiazate) from Syngenta® and Mocap 15G (15% ethoprophos) from Certis are the only granular nematicides currently available.

Organophosphates tend to be adsorbed onto organic matter, making such nematicides less effective; carbamate oxamyl is more likely to be effective in organic soils (Back et al. 2017). The liquid fumigant Telone II (1,3-dichloropropene) accounts for the majority of the nematicides used by weight. Although methyl bromide was the most effective fumigant at the time, other options are thought to be less effective. This is mostly due to the cyst's defence of the juveniles. Since good soil conditions at the time of fumigation treatment, including the right temperature and moisture content, are crucial to success, timing can mean applying the fumigation treatment in the rotation a year or two before the anticipated potato crop. Furthermore, applying the treatment is pointless if a proper surface seal cannot be achieved.

The granular carbamate and organophosphate forms are more accurately referred to as nematistats because their mode of action is to interrupt the juveniles' metabolism, feeding and movement without actually causing death. Different nematicides, such as DD, DBCP, Nemafos, V.C.13 and Dasanit 10G, have been tested in India. DD applied at 1000 l/ha in two 15-day split doses resulted in 98 to 100% reduction. Dasanit 10G was recommended for three crop seasons, with 300 kg/ha in the main season and 150 kg/ha in the second and third seasons (Gill 1974). After standardization, application of Furadan 3G at 2 kg a.i./ha at the time of planting is recommended for PCN as part of the potato package of practices in the Nilgiris (Krishna Prasad 2006). These chemicals, however, have recently been outlawed. The fumigant molecule Dazomet (Basamid 90G) at 40–50 g/m^2 was also found to be efficient in reducing PCN populations, but the soil must be covered with polythene sheet after application (Aarti et al. 2016). However, repeated use of nematicides is not only expensive but also hazardous to the environment. As a seed treatment, a calcium hypochlorite solution containing 9% available chlorine was found to be effective in reducing the PCN population (Manoharan et al. 1978). Potato tubers with cysts can be killed by immersing them in a sodium hypochlorite solution for 2 h and then rinsing them in water (Wood and Foot 1977). Soaking of PCN-infested un-sprouted seed potato tubers in 2% NaOCl solution (containing 4% usable chlorine) for 30 min resulted disintegration of all cysts without adversely affecting the tuber germinability after 2 months of storage (Aarti et al. 2020b).

11.2.6.10 Biofumigation

In the United Kingdom, biofumigation typically entails the cultivation of brassica green manure crops. Indian mustard (*Brassica juncea*), rocket (*Eruca sativa*) and oil radish (*Raphanus sativus*) are the most common species. Within a mid-July to early November time frame, the typical growth period is 8–14 weeks. During the growing season, several *Brassica* species, including Indian mustard, oil radish and rocket, have been shown to suppress PCN. Biofumigant crops are then incorporated as they reach early to mid-flowering (Back and William 2019). *B. juncea* (Indian/brown mustard) has a field efficacy of 15–95% in reducing the PCN population (Ngala et al. 2014). In India, biofumigation with 1 kg/m^2 radish leaves and polyethylene sheeting produced the highest yield (25.97 t/ha) and reduced the PCN reproduction factor (Rf: 1.21) (Umamaheswari et al. 2015).

11.2.6.11 Biocontrol Agents

Arbuscular mycorrhizal fungi have been shown to prevent PCN root invasion in laboratory experiments in the United Kingdom. *Pochonia chlamydosporia*, a fungus that parasitizes nematode larvae, has also been studied. It has performed well in some trials, but has not been scaled up to commercial quantities and may be susceptible to field fungicides. *Trichoderma harzianum*, *Plectosphaerella cucumerina* and *Penicillium oxalicum* are three other fungi that may be predators or rivals of PCN (Back et al. 2017).

Purpureocillium lilacinum (Thom), a fungus that parasitizes eggs; *Pseudomonas fluorescens*, a plant growth-promoting rhizobacteria; and *T. viride*, an antagonistic fungus, have all been shown to be potential biocontrol agents against PCN (Cronin et al. 1997; Devrajan et al. 2011). *P. lilacinum* was known to control PCN in the field by Davide and Zorilla (1983) and Seenivasan et al. (2007). Neem and talc powder formulations resulted in PCN control of 42.6 and 58.2%, respectively. However, in the recent past, several products with nematicidal effects have been introduced to the market. The majority of other possible biocontrol agents are still being studied to see whether they can solve implementation process issues. The use of biological control agents such as *P. fluorescens* and *P. lilacinus* (Seenivasan et al. 2007) as well as organic amendments such as neem cake (5 t/ha) combined with *T. viride* (5 kg/ha) resulted in a reduction in PCN population (Umamaheswari et al. 2012).

11.2.7 Integrated Management

The combined use of various management strategies under the IPM programme can help in keeping the populations below the economic threshold level. PCN can be effectively controlled when combined with 5 years of crop rotations with non-host crops, effective soil fumigation and the use of an effective trap crop (Whitehead and Turner 1998). A fumigant nematicide or a trap crop should be used to rapidly reduce large populations, accompanied by the planting of a potato crop covered by a granular nematicide (Phillips and Trudgill 1998).

G. rostochiensis was successfully managed using granular nematicides in combination with susceptible potato cultivars and crop rotations (Whitehead et al. 1991). When potato and mustard were intercropped at 1:1 plant ratio and carbofuran 3G (1 kg a.i./ha) was applied, PCN infestation was reduced, and potato yields were increased (Devrajan and Balasubramanian 2008). *P. fluorescens* (2.5 kg/ha) + neem cake (1 t/ha) + mustard intercrop (between potato rows) + carbofuran 3G (1 kg a.i./ha) increased tuber yield while lowering PCN population (Devrajan et al. 2004). There was a decrease in the PCN population after soil solarization (4 weeks) followed by application of neem cake (5 t/ha) in combination with *T. viride* (5 kg/ha) (Aarti et al. 2017). Manorama et al. (2016) recorded a 47% reduction in the PCN population in 2 years by rotating PCN-susceptible and PCN-resistant varieties and applying carbofuran at 2.0 kg a.i./ha.

Aside from IPM, the OEPP/EPPO has proposed several control steps for potato growers on an international basis. Growing resistant potato varieties, growing potato as a trap crop for 40 days, growing *S. sisymbriifolium* as a capture crop and soil fumigation are all potential control measures in the Netherlands. Growing resistant crops, crop rotation for at least 4 years and the elimination of volunteer potatoes have all been suggested in Slovenia. The use of PCN-resistant potato cultivars, crop rotation, chemical control, trap cropping with *S. sisymbriifolium*, green manures and fumigants are all practised in England and Wales. Usage of resistant potato cultivars for PCN pathotypes; application of nematicides such as metam sodium, metam potassium, ethoprophos, fosthiazate and oxamyl before planting susceptible cultivars; and crop rotation for ware (1 crop every 3 years) and seed (1 crop every 4 years) potato production are all part of Belgium's official control programme. Resistant varieties must be grown in Denmark for 2 years in a row. All equipment must be washed before being used in the fields. The harvested tubers of ware potatoes must not be planted at the same time as seed potatoes in infested fields, and the soil and other waste must be treated carefully to prevent further spread. Soil monitoring after 3 years of application is used to determine the effectiveness of the control programme. Since no nematicides are available in Germany, highly resistant varieties are used in a 6-year rotation. In France, seed potatoes are often tested before planting, and if PCN is found, growers are prohibited from growing potatoes for 6 years, and all volunteer potatoes must be destroyed. Plants such as grass, maize and cereals may be grown by growers to avoid the possibility of soil being exported in new fields (OEPP/EPPO 2014).

11.2.8 Novel Biotechnological Approaches for the Management of PCN (*Globodera* Spp.)

The most desirable and effective management method identified is host resistance. Resistant cultivars, on the other hand, are prone to the advent of novel virulence within the species. Farmers primarily rely on pesticides to regulate PCN, which can be extremely harmful. Due to the growing concern about the environment, it would be impossible to implement in the immediate future, necessitating the elimination of environmentally sustainable management strategies. As a result, a new gene-targeted management strategy based on RNAi (RNA interference), gene silencing, is an exciting and promising alternative. Silencing of known nematode effector proteins RNAi technology holds a great potential for plant resistance against various species of nematodes. The use of dsRNA in a drench/spray form not only helps resolve the issues associated with transgenic plants. In addition, this approach necessitates low-cost dsRNA generation, techniques to stabilize them for field delivery and dsRNA uptake by nematodes during feeding. In this area, ICAR-CPRI, Shimla also attempted to develop dsRNA based on pathogenicity genes such as flp-32c (Atkinson et al. 2013) and ams-1 (Jones et al. 2003) for the management of PCN (Aarti et al. 2020b). Apart from the above, CRISPR/Cas9 (Clustered Regularly Interspaced Short Repeats Palindromic Repeats/CRISPR-Associated Protein 9), a newly developed genome editing technique, allows researchers to create specific knockout mutants in 2 to 3 weeks, providing an alternative platform to investigate the role of a gene of interest in a cell-specific way while avoiding the embryonic mortality caused by some mutations. Furthermore, studies of potato transcriptomes in response to PCN infestation could be utilized to discover various genes that are up- or downregulated, followed by using CRIPSER/Cas9 techniques, so genes involved in nematode establishment can be downregulated and inhibited.

11.3 Root-Knot Nematode (RKN)

Worldwide, there are about 90 identified species of RKN. Out of these, *M. incognita*, *M. javanica*, *M. arenaria* and *M. hapla* are present in >95% of the soil (Hunt and Handoo 2009). Some of them have multiple races that parasitize >2000 plant species and cause significant damage to global agriculture (Carneiro et al. 2008; Moens et al. 2009). Among RKNs, *M. incognita* is common in the tropical regions followed by *M. javanica* and *M. arenaria*, while *M. hapla*, *M. chitwoodi*, *M. fallax* and *M. thamesi* are set up in cooler climates. *M. chitwoodi* is the most common and important RKN species affecting potato in temperate regions, such as North America, Europe and Australia. At temperatures below 6 °C, it seems to damage tubers. In India, the prominent RKN species, *M. incognita*, has been harming potato in both hills and plains, whereas *M. javanica* affects potato in northern India's mid-hills and plains. *M. hapla* is found in Uttarakhand, Himachal Pradesh, Jammu and Kashmir, Assam and Tamil Nadu, while *M. arenaria* is distributed in Uttar Pradesh's plains. The infection of RKN in potato tubers in India was firstly documented in Shimla by Thirumalachar in 1951.

11.3.1 Host Range

Many valuable grains, legumes, fruits, ornamentals and vegetables, including potato, have been reported to be infested by RKN species.

11.3.2 Biology

It infects the roots as well as tubers, but the early generation mostly affects the root system, with successive generations focusing on tubers. The vermiform second-stage juveniles emerge from the egg masses and begin feeding on the young potato roots. This leads to the formation of specialized, bigger cells known as "giant cells", which nourish the worms throughout their growth. Juveniles in the second stage (J2) moult and go through the J3 and J4 phases before becoming adult females or males. Males are migratory and vermiform or thread like, while adult females are sedentary and pear shaped. Males exit the root to find and mate with females. Females lay 300 to 400 eggs in a gelatinous matrix, which they commonly adhere to root galls. The juveniles frequently enter the tubers during tuber development since the root structure has begun to degrade. During the summer, the life cycle is completed in 25–30 days, and in the winter, it requires 65–100 days (Krishna Prasad 2008).

11.3.3 Symptoms and Yield Loss

Stunting and yellowing of plants with chlorotic leaves are among the above-ground symptoms caused by a reduction in water and nutrient intake by roots. The characteristic swellings known as "galls" are formed in the roots. Due to nematode infestation, warty "pimple-like" lesions on tubers diminish the commercial value and storage quality of potatoes. The presence of nematodes causes brown patches to appear in the flesh of the sliced tubers (Fig. 11.5). With a total tuber infestation of 100%, an initial inoculum of 200 juveniles (J2s) per 100 ml of soil resulted in a 40% production loss.

The RKN destroys 29–30% of vegetable harvests each year, developing root system that promotes plant growth as well as the invading nematode population. In India, the drop in potato output is greater than 12.2% (Sasser 1989). The nematodes can interact with other soil-borne fungus, bacteria and viruses, causing significant crop loss. The most important RKN-*Ralstonia solanacearum* interaction causes "pseudomonas wilt" in tomato, brinjal and potato (Vovlas et al. 2005; Gomes and Souza 2003). It's also been stated that potato tuber deformation is caused by *M. javanica*.

Fig. 11.5 Symptoms caused by root-knot nematode (*Meloidogyne* spp.) on tubers

11.3.4 Management of RKN

Cultural control RKN infestations can be reduced by using quality seed tubers, thorough ploughing during the summer months, maintaining high cleanliness and keeping the field weed-free. Crop rotation with non-host crops such as maize or wheat reduces nematode damage. A rotation should last at least 4 years and be accompanied by strong weed management. To reduce RKN damage, crops like oats, cotton and grasses that are resistant to RKN can be used in rotation with potatoes. Sorghum, maize and castor bean resistant to *M. javanica* have also been used as part of crop rotation strategies for this species. *Brassica* crops such as cabbage, cauliflower, mustard and Chinese cabbage are rotated with potato in general (Pinheiro et al. 2009).

Some proposed management techniques encompass regular and timely soil cultivation and drying, immediately destroying volunteer potato plants and tubers, planting certified potato tubers, selecting planting dates to avoid a high RKN population throughout tuber development, narrowing the potato cycle by using early maturing cultivars and, finally, the rational use of certified nematicides just before, during and after planting (Jones et al. 2017). Because of the lower temperature prevalent during the crop period, early planting of spring crops in the first week of January and late planting of fall crops in the second and third weeks of October minimize nematode infection in potatoes. In alternate rows with potato, growing antagonistic crop like French marigold, *T. patula*, to minimize nematode populations is suggested.

Host plant resistance Resistant genes from the wild potato species *Solanum sparsipilum* are being deployed in a breeding programme to produce *M. incognita*, *M. javanica* and *M. arenaria*-resistant potato cultivars. RMc1 (blb), from *Solanum* section Petota, is the encoded protein of the RKN resistant gene that is efficient against several races of *M. chitwoodi* (Brown et al. 2009). The wild potato *S. sparsipilum*, which is being used to develop resistant potato cultivars, has also been shown to be resistant to *M. incognita*, *M. javanica* and *M. arenaria* (The *International Potato Center-CIP*). Mc Cramick and Golden varieties of potatoes were found to be resistant to *M. chitwoodi*, whereas Oronek, ORA and Suzanna varieties were shown to be moderately resistant (Norshie et al. 2011).

Organic amendments Organic amendments are frequently used in agriculture to recycle energy and nutrients while also enhancing soil conditions for plant growth (Muchovej and Pacovsky 1997). Plant diseases are suppressed by some organic amendments, and plant parasitic nematodes are controlled by others (Ali et al. 2001). Some toxic plant components such as phenols, alkaloids, polyphenols and allochemicals inhibit phytopathogens and plant-parasitic nematodes indirectly by boosting soil microbiota (Shaukat et al. 2001). Powder made from *Avicennia marina* (mangrove) inhibited the knots of *M. javanica* and root-infecting fungus (Marium et al. 2008). The addition of organic matter from *T. minuta*, *Ricinus communis* and *D. stramonium* increased the parasitic activity of *P. lilacinus* against *M. javanica* eggs (Oduor-Owino et al. 1993).

Botanical amendments Phytochemicals or active components having nematicidal quality found in various plant portions have been proven to be useful in minimizing nematode infestation on plants (Saxena and Singh 2001; Rehman et al. 2012; Ojo and Umar 2013). Nath and Mukherjee (2000) found anti-inflammatory effects on egg hatching of *M. incognita* with the tuber extract of *Dioscorea floribunda*. Upadhyay et al. (2003) found enhanced juvenile mortality of potato RKN by azadirachtin, present in leaf and seed extracts of *Azadirachta indica*. Spanish cherry (*Mimusops elengi* L), *Lantana camara*, *Nicotiana tabacum*, *Syzygium aromaticum*, water hyacinth and devil pepper (*Rauvolfia tetraphylla*) leaf extracts were found to be useful in managing *Meloidogyne* populations (Ntalli et al. 2009; Ahmad et al. 2010; Umar and Mohammed 2013; Mandal and Nandi 2013). Marigolds (*Tagetes* species) that produce polythienyls have been found to suppress nematodes (Wang et al. 2007).

Chemical control Carbofuran 3G, applied at 1–2 kg a.i./ha, reduces nematode infestation and increases yield. Splitting the application, one at the time of planting and the other at the time of earthing up, will enhance the chemical's efficacy in managing RKNs (Jones et al. 2017).

Integrated management Because single control strategy is uneconomical and insufficient for better nematode management. Therefore, for better management, a careful balance of several strategies is always recommended. INM adoption for root-knot over a 2-year period results in an efficient and cost-effective production system.

11.4 Lesion Nematodes (*Pratylenchus* Spp.)

Pratylenchus spp., root lesion nematodes, are important plant parasites with a wide host range in tropical and subtropical climates, particularly in Brazil, the southern United States and Africa (Ferraz 1999; De Waele and Elser 2002). However, it is not a major pest in India. *P. andinus*, *P. brachyurus*, *P. coffeae*, *P. crenatus*, *P. minyus*, *P. penetrans*, *P. scribneri*, *P. thornei*, *P. vulnus*, *P. neglectus*, *P. mediterraneus* and *P. zeae* are some of the *Pratylenchus* species found in potato (Brodie et al. 1993; Mai et al. 1990). *Pratylenchus* species are smaller than 1 mm in length, and the nematode is a migratory endoparasitic nematode that lives inside and between the roots as well as in soil particles. Both males and females are wormlike, with the sexual characteristics being the only difference (Ferraz 1999).

The most common symptom is a root lesion; the lesions start out as little, elongated, water-soaked patches that develop brown to black over time. Patches of water stressed, less vigorous plants that turn yellow and die. On potato tubers in storage, the nematode develops a scabby appearance with sunken lesions or dark, wart-like lesions that develop purple (Davis and MacGuidwin 2000).

Pratylenchus spp. management strategies include the inclusion of crop rotation; the adoption of resistant cultivars; adequate physical, chemical and soil management; and weed control throughout the harvest and off-season. In a sandy soil in southern Alberta, Canada, the impacts of 3 to 6 years of crop rotation by rotating potatoes, dry beans, wheat, sugar beet and oats and by following soil management techniques such as reduced planting, cover crops and organic fertilizer applications decreased *P. neglectus* population densities in potatoes (Forge et al. 2015). Kimpinski et al. (2000) recorded lower population density of *P. penetrans* in marigold (*T. tenuifolia* cv. Nemakill and cv. Nemanon) as compared to other cover crops such as ryegrass, red clover, soybean and potato. The use of soil fumigation and resistant potato cultivars has been recommended by Dunn and Mai (1973). Treatment of infected tubers in hot water for 45–60 min at 50 °C helps reduce nematode spread (Koen 1969; Yokoo and Matsunobu 1975).

11.5 The False Root-Knot Nematode (*Nacobbus Aberrans*)

Nacobbus aberrans (Thorne & Allen) is a false root-knot nematode found mostly in Mexico and the United States, as well as in Argentina, Peru, Ecuador, Chile and Bolivia. This nematode species is a major problem on potato farms in Mexico and other South American nations. This species is a quarantined pest and is considered a significant pest to potatoes, with yield losses ranging from 55% to 90% (EPPO 1977; Vovlas et al. 2007). *N. aberrans* are also considered key nematode species severely harming potato productivity in the Andean region of Peru and Bolivia. It parasitizes several other commercially significant plant species such as *Solanum* and *Capsicum*, as well as carrots, lettuce, cabbage, peas, sugar beets, cucumbers and various weeds (Manzila-Lopes et al. 2002).

N. aberrans infection in potatoes has symptoms comparable to *Meloidogyne* spp., such as the creation of galls that are more distinct and rounder, whereas galls from RKN are elongated and produce swellings along the roots. Second-stage juveniles (J2s) hatch from eggs, travel across the soil in search of suitable roots and force penetration using their stylet and enzymes. They get into the vascular system and alter a group of cells that cause galls to form. The third, fourth and immature females of *N. aberrans* are migratory, unlike RKN species. In a gelatinous membrane that bulges from a fibre bundle, eggs are produced. Depending on the optimal temperature range of 14 to 25 °C, the nematode can accomplish 2–3 generations over the crop season (EPPO 1977; Manzila-Lopes et al. 2002). Nematicides, crop rotations (4–6 years), biological management with antagonist fungi and bacteria, the use of resistant or tolerant potato cultivars and quarantine rules for areas free of this nematode species are all options for controlling *N. aberrans* (Manzila-Lopes et al. 2002).

11.6 The Root-Rot Nematode (*Ditylenchus Destructor*)

It is an economically significant parasitic nematode, mainly when coupled with fungal diseases, and it is classified as a quarantine pest in various countries. In Europe, Russia, Asia, North America, Oceania and a few isolated parts of South America and South Africa, it is a serious pathogen of potatoes. However, this nematode is not found in India. Due to physiological and morphological abnormalities produced by the nematode infection, severely infected tubers result in a compromised plant, which may finally lead to plant mortality. Peeling off the tuber can expose little, white patches that indicate early infection (Mai et al. 1990). *D. dipsaci*, also known as the stem and bulb nematode, is more frequent in garlic, but it also harms other plant species, including potatoes, and attacks stalks, stolons and tubers. Gray to brownish lesions appear on affected potato tubers, and overall plant growth is weak (Asscheman et al. 1996). Due to their large host ranges, crop rotation cannot be used to control *D. destructor* and *D. dipsaci. D. dipsaci* resistance was also found in potato varieties, i.e. Innovator Aveka and Spunta. The potato cultivar Desiree was shown to be particularly vulnerable to both *D. destructor* and *D. dipsaci* (Mwaura et al. 2015).

11.7 The Stubby-Root Nematodes (*Trichodorus* and *Paratrichodorus* Spp.)

Stubby-root nematodes are ectoparasites that typically congregate at the root tips. They have an onchiostyle, which is a long, solid and curved stylet that they use to puncture plant cells while feeding, preferably meristem cells at the root tips (Decraemer 1995). Their direct feeding can cause significant damage, including enlarged roots and degeneration, premature withering and cessation of crop growth, a disorder termed "stubby root". Trichodorids have a wide distribution around the world, and they have features that aid in the transmission of specific viruses (Decraemer 1995). *Trichodorus* spp. can also be found in sandy soils around the world. Although they are more suited to monocotyledons, they occasionally parasitize dicots and are an important nematode of potato in the tropical and subtropical regions (Scurrah et al. 2005). *Trichodorus* spp. can spread viruses to potatoes, particularly the *Tobacco Rattle Virus* (TRV), a virus of the genus *Tobravirus* that induces the potato disease corky-ring spot (Pinheiro et al. 2015; Scurrah et al. 2005).

Trichodorus spp. juvenile and adult stages can transmit these viruses after eating on damaged plants. The viruses become lodged in the stylet region and do not circulate within the nematode's body. Yet, the worms might remain infected for up to 4 months. Potato plants infected with TRV exhibit symptoms such as necrosis, chlorosis and overall stunting. Necrotic lesions with brittle tissues may appear on affected potato tubers (Taylor and Brown 1998). Plant emergence is delayed, resulting in poor growth, lower tuber weight and potato output and low dry matter content (Decraemer 1995). Overall, stubby-root nematode control approaches include precautionary measures such as using certified seeds, cleaning farm machinery and equipment, preventing animal movement, crop rotation, cultural practices and the application of nematicides (Pinheiro et al. 2015).

11.8 Minor Nematode Species

Spiral (*Helicotylenchus* spp.), stunt (*Tylenchorhynchus* spp.), reniform (*Rotylenchulus reniformis*) and pin (*Paratylenchus* spp.) nematodes are small nematode parasites that feed on potato roots and cause significant production decreases (Krishna Prasad and Rajendran 1990).

11.9 Overall Strategies for Managing Nematodes in Potato Fields

Because of the biology of these plant parasites, effective control of potato nematodes is tricky and complex. They live in the soil, have a brief life cycle, reproduce quickly and have a high population. Only a few plant genotypes are resistant to them, and chemical nematicides have limited effect due to interactions with soil components or are avoided due to human and environmental adverse effects (Pinheiro et al. 2009). As a result, in order to succeed, nematode management techniques for potato should be carefully planned. In order to maximize control efficiency, it is recommended that many control strategies be used. Apart from that, RNAi and CRISPR techniques have quickly been adopted by researchers all over the world to modify the DNA sequences of organisms of concern.

11.10 Conclusion and Future Outlook

Nematode species have distinct biology and behaviour, making them difficult to manage or eradicate once established in a field. To pick the right management technique, accurate nematode identification at the species and pathotype/race levels is required. In general, nematode management in potatoes is usually accomplished when integrated management strategies are utilized, such as exclusion (quarantine rules, certified seed tubers and clean farm equipment), cultural methods (crop rotation, inter-cropping, antagonistic crops and cover crops), resistant varieties and, finally, pesticides. Growers, extension agencies and investigators must analyse these nematodes holistically, the damage they produce and whether these management measures are ecologically, legally and technically proficient for the sustainable cropping of potato cultivars.

References

Aarti B, Sharma S, Venkatasalam EP, Sharma A, Chakrabarti SK, Priyank HM, Singh RK (2020a) Decontamination of cyst (*Globodera* spp.) infested potato seed tubers with sodium hypochlorite (NaOCl). Indian J Agric Sci 90:1345–1347

Aarti B, Sundresha S, Venkatasalam EP, Bhawna D, Sharma S (2020b) Novel dsRNA based nematicide formulation for the management of potato cyst nematodes (*Globodera* spp.). CPRI Newsletter 80:1–2

Aarti B, Venkatasalam EP, Priyank HM, Sharma S (2021) Introduction of potato cyst nematodes, life cycle and their management through biobased amendments. In: Kaushal M, Prasad R (eds) Microbial biotechnology in crop protection. Springer, Singapore, pp 79–95

Aarti B, Venkatasalam EP, Sudha R, Singh BP (2016) Effect of new soil fumigant Basamid 90G on potato cyst nematode population and yield. CPRI Newsletter 65:2

Aarti B, Venkatasalam EP, Umamaheswari R, Sudha R, Singh BP (2017) Effect of cultural practices on potato cyst nematode population dynamics and potato tuber yield. Indian J Hort 74:91–96

Ahmad F, Rather MA, Siddiqui MA (2010) Nematicidal activity of leaf extracts from *Lantana camara* L. against *Meloidogyne incognita* (kofoid and white) Chitwood and its use to manage roots infection of *Solanum melongena* L. Braz Arch Biol Technol 53(3):543–548

Ali NI, Siddiqui IA, Zaki MJ, Shaukat SS (2001) Nematicidal potential of *Lantana camara* against *Meloidogyne javanica* in mung bean. Nematol Mediterr 29:99–102

Asscheman IE, Brinkman H, Bus CB, Van Delft M, Hotsma PH, Mejers CP, Mulder A, Turkensteen LJ, Wustman R (1996) Potato diseases (diseases, pests, defects), 1st edn. NIVAA, Holland, p 179

Atkinson LE, Stevenson M, McCoy CJ, Marks NJ, Fleming C (2013) Flp-32 ligand/receptor silencing phenocopy faster plant pathogenic nematodes. PLoS Path 29:1–16

Back M, William W (2019) Biofumigation for management of potato cyst nematodes (PCN). Produced by AHDB Stoneleigh Park Kenilworth Warwickshire

Back MA, Sosnowska D, Holgado R (2017) Management strategies: *Pochonia chlamydosporia* and IPM of plant-parasitic nematodes in temperate crops. In: Manzanilla-López R, Lopez-Llorca L (eds) Perspectives in sustainable nematode management through *Pochonia chlamydosporia* applications for root and rhizosphere health. Sustainability in plant and crop protection. Springer, Cham

Bhardwaj V, Salej S, Ashwani K, Vanishree G, Sharma S, Sundaresha S, Raigond B, Ravinder K, Aarti B, Lal M, Chakrabarti SK (2019) Efficiency and reliability of marker assisted selection for resistance to major biotic stresses in potato. Potato J 46(1):56–66

Brodie BB, Evans K, Franco J (1993) Nematode parasites of potato. In: Evans K, Trudgill DL, Webster JM (eds) Plant parasitic nematodes in temperate agriculture, 1st edn. CAB International, Wallingford, UK, pp 87–132

Brown CR, Mojtahedi H, Zhang LH, Riga E (2009) Independent resistant reactions expressed in root and tuber of potato breeding lines with introgressed resistance to *Meloidogyne chitwoodi*. Phytopathology 99:1085–1089

CABI/EPPO (2020a) *Globodera rostochiensis*. https://www.cabi.org/isc/datasheet/27034#todistributionDatabaseTable

CABI/EPPO (2020b) *Globodera pallida*. https://www.cabi.org/isc/datasheet/27033#todistributionDatabaseTable

Caixeta LB, Furlaneto C, Cares JE (2016) Gênero Heterodera. In: Oliveira CMG Santos MA

Canto M (1995) El nematode quiste de la papa en Latinoamerica. Congreso International de Nematologia Trophical, Rio Quente, Brazil

Carneiro RMDG, Dos Santos MFA, Almeida MRA, Mota FC, Gomes ACMM, Tigano MS (2008) Diversity of *Meloidogyne arenaria* using morphological, cytological and molecular approaches. Nematology 10:819–834

Chandel YS, Surendra SB, Rajan S, Sharmishtha T, Satya K, Vishal SS, Abhishek M, Walia RK (2020) Prevalence and spread of potato cyst nematodes, *Globodera* spp. in northern hilly areas of India. Curr Sci 118:1946. https://doi.org/10.18520/cs/v118/i12/1946-195

Contina JB, Dandurand LM, Knudsen GR (2020) A spatiotemporal analysis and dispersal patterns of the potato cyst nematode, *Globodera pallida*, in Idaho. Phytopathology 110:379–392

Cortada L, Omagwa J, Kisitu J, Adhiambo M, Haukeland S, Mburu H, Orr J, Jones JT, Wasukira A, Kisingiri JB, Tugume J, Birenge JB, Okonya JS, Coyne D (2020) First report of potato cyst nematode, *Globodera rostochiensis* infecting potato (*Solanum tuberosum* L.) in Uganda. Plant Dis 104:3082

Cronin D, Moenne-Loccoz Y, Fenton A, Dunne C, Dowling DN, O'gara F (1997) Role of 2,4-diacetylphloroglucinol in the interactions of the biocontrol pseudomonad strain F113 with the potato cyst nematode *Globodera rostochiensis*. Appl Environ Microbiol 63:1357–1361

Dalamu BV, Umamaheswari R, Sharma R, Kaushik SK, Joseph TA, Singh BP, Gerhardt C (2012) Potato cyst nematode (PCN) resistance: genes, genotypes and markers-an update. SABRAO J Breed Genet 44:202–228

Dandurand LM, Brown CR, Gajjar P (2014) Development of *Globodera pallida* in the trap crop *Solanum sisymbriifolium*. J Nematol 46:2

Davide RG, Zorilla RA (1983) Evaluation of fungus, *Paecilomyces lilacimus* (Thom) Samson, for the biological control of the potato cyst nematode *Globodera rostochiensis* Woll as compared with some nematicides. Philippine-Agric 66(4):397–404

Davis EL, Mac Guidwin AE (2000) Lesion nematode disease. Plant Health Instructor 2000:1030–1032. https://doi.org/10.1094/PHI-I-2000-1030-02

De Waele D, Elser A (2002) Migratory endoparasites: *Pratylenchus* and *Radopholus* species. In: Star JJ, Cook R, Bridge J (eds) Plant resistance to parasitic nematodes. CAB International, Wallingford, UK, pp 175–206

Decraemer W (1995) Species description. In Decraemer W The family Trichodoridae: stubby root and virus vector nematodes. 1st Rotterdam: Springer, pp 76–257

Devine KJ, Dunne C, O'Gara F, Jones PW (1999) The influence of in-egg mortality and spontaneous hatching on the decline of *Globodera rostochiensis* during crop rotation in the absence of the host potato crop in the field. Nematol 1:637–645

Devrajan K, Balasubramanian P (2008) Management of *Globodera rostochiensis* and *Globodera pallida* in potato by antagonistic crop, mustard in Nilgiris. Indian J Nematol 38:15–17

Devrajan K, Prabhu S, Seenivasan N, Sudha A, Ramakrishnan S, Anita B (2011) Occurrence of native microbial antagonists against potato cyst nematodes in the Nilgiri Hills of Tamil Nadu. Potato J 38:67–72

Devrajan K, Seenivasan N, Selvaraj N, Rajendran G (2004) An integrated approach for the management of potato cyst nematodes, *Globodera rostochiensis* and *G. pallida* in India. Nematol Mediterr 32:67–70

Dunn RA, Mai WF (1973) Reproduction of *Pratylenchus penetrans* in roots of seven cover crop species in greenhouse experiments. Plant Dis Reporter 57:728–730

Embrapa (2015) Nematoides na cultura da batata. Circular Tecnica 143:1–12

EPPO (1977) Data sheets on quarantine pests: *Nacobbus aberrans*. EPPO/CABI Bull 144:1–4

EPPO (2013) Globodera rostochiensis and G. Pallida PM 7-40. EPPO Bull 49:354–368

Evans K, Brodie BB (1980) The origin and distribution of the golden nematode and its potential in the USA. Am Potato J 57:79–89

Evans K, Stone AR (1977) A review of the distribution and biology of the potato cyst nematodes *Globodera rostochiensis* and *G. pallida*. PANS 23:178–189

FAOSTAT (2017). http://www.Fao.Org/faostat/en/#data/QCL/visualize

Ferraz LCCB (1999) Genero Pratylenchus–Os nematoides das lesões radiculares. In: Luz WC (ed) Revisão Anual de Patologia de Plantas, 1st edn. Passo Fundo, RAPP, pp 158–195

Forge TA, Larney FJ, Kawchuk LM, Pearson DC, Kock C, Blackshaw RE (2015) Crop rotation effects on *Pratylenchus neglectus* populations in the root zone of irrigated potato in southern Alberta. Can J Plant Pathol 37:1–12

Franco J, Oros R, Main G, Ortun N (1998) Potato cyst nematodes (*Globodera* species) in South America. In: Marks RJ, Brodie BB (eds) Potato cyst nematodes biology, distribution and control. CAB International, Wallingford, pp 239–270

Gill JS (1974) Advances made in potato nematology in India. J Indian Potato Assoc 1:51–54

Gomes CB, Souza RM (2003) Doenças causadas por nematoides. In: Pereira AS, Daniels J (eds) O cultivo da batata na região sul do Brasil, 1st edn. Embrapa Brasilia, Brazil, pp 321–349
Greco N (1993) Reviews: nematode problems affecting potato production in subtropical climates. Nematropica 23:213–220
Grenier E, Benjamin M (2017) Potato cyst nematodes around the world: all you wanted to knowabout their distribution and evolution history. Globodera Alliance Newslett 4:1–6
Hafez SL, Sundararaj P, Handoo ZA, Skantar AM, Carta LK, Chitwood DJ (2007) First report of the pale cyst nematode, *Globodera pallida*, in the United States. Plant Dis 91:3. https://doi.org/10.1094/PDIS-91-3-0325B
Hockland S, Niere B, Grenier E, Blok V, Phillips M, Den Nijs L, Anthoine G, Pickup G, Viaene N (2012) An evaluation of the implications of virulence in non-European populations of *Globodera pallida* and *G. rostochiensis* for potato cultivation in Europe. Nematology 14:1–13
Hunt D, Handoo Z (2009) Taxonomy, identification and principal species. In: Perry RN, Moens M, Star JL (eds) Root knot nematodes, 1st edn. CAB International, Wallingford, UK, pp 55–88
Jones FGW (1970) The control of the potato cyst nematode. J Royal Soc Arts 118:179–199
Jones JT, Furlanetto C, Bakker E, Banks B, Blok V, Chen Q, Phillips M, Prior A (2003) Characterization of a chorismate mutase from the potato cyst nematode *Globodera pallida*. Mol Plant Pathol 4:43–50
Jones RK, Storey SG, Knoetze R, Fourie H (2017) Nematode pests of potato and other vegetables: nematology in South Africa. A view from the 21st century, vol 1. Springer International Publishing, Cham, Switzerland, pp 231–260
Joseph TA, Latha M, Manorma K, Somasekhar N, Ravichandran G, Umamaheswari R (2012) Kufri Neelima: a new cyst nematode and late blight resistant potato variety suitable for Nilgiri hills. Paper presented in National consultation on potato research and development: way forward, OUAT, Orissa, India 26 Sept, 2012, p 80
Joseph TA, Sudha R, Venkatasalam EP, Aarti B, Umamaheswari R, Divya KL, Priyank HM (2019) Variety Kufri Sahyadri. Indian. J Genet 79:517–518
Khan IA, Gajaraja CP, Singh DB, Krishna Parsad KS (1985) Kufri Swarna: a high yielding late blight and cyst nematode resistant potato cultivar suitable for cultivation in Tamil Nadu hills. Extension bulletin no. 21 (E). CPRI, Shimla
Kimpinski J, Arsenaut WJ, Gallant CE, Sanderson JB (2000) The effects of marigolds (*Tagetts* spp) and other cover crops on *Pratylenchus* penetrans and on following potato crops. J Nematol 32: 531–536
Kishore H, Nirula KK, Vasudeva PP, Sharma AK (1969) All India nematology symposium, I. ARI, New Delhi, India, p 58
Koen H (1969) Thermotherapeutic control of *Pratylenchus brachyurus* in potato tubers. Phytophylactica 1:67–69
Kort J, Ross H, Rumpenhorst HJ, Stone AR (1977) An international scheme for identifying and classifying pathotypes of potato cyst nematodes *Globodera rostochiensis* and *G. pallida*. Nematologica 23:333–389
Krishna Prasad KS (1993) Nematodes: distribution, biology and management. In: Chadha KL, Grewal JS (eds) Advances in horticulture, vol 7. Malhotra Publishing House, New Delhi, India, pp 635–647
Krishna Prasad KS (2006) Potato cyst nematodes and their management in the Nilgiris (India). CPRI technical bulletin no. 77. Central Potato Research Institute, Shimla, India, p 20
Krishna Prasad KS (2008) Management of potato nematodes: an overview. J Hort Sci 3:89–106
Krishna Prasad KS, Rajendran TP (1990) Helicotylenchus dihystera: a potential pest of potato. J Ind Potato Assoc 17:204–205
LaMondia JA, Brodie BB (1984) Control of *Globodera rostochiensis* by solar heat. Plant Dis 68: 474–476
Mai WF, Brodie BB, Harisson MB, Jatala P (1990) Nematodes. In: Hooker WJ (ed) Compendium of potato diseases, 1st edn. The American Phytopathological Society, Saint Paul, MN, pp 93–101

Mandal TK, Nandi B (2013) Control of *Meloidogyne incognita* infestation in tomato by *Rauvolfia tetraphylla*. Indian J Sci Res 4(2):61

Mani A, Prakash KS, Zidgali TA (1993) Comparative effects of soil solarization and nematicides of three nematode species infecting potato. Curr Nematol 4:65–70

Manoharan AS, Vijayaraghavan S, Samboorharaman S (1978) Effect of treatment of seed tubers of potato to make potato cyst nematode non-viable, a new method of disinfestations. In: Proceedings of an international seminar on approaches towards increasing the potato production in developing countries. Abstracts 55 and 56, Indian Potato Association, Shimla

Manorama K, Lal SS, Prasad KKS (2005) Potato based intercropping system for Nilgiris (Abst.). In: national symposium on 'National Resources Management for eco development and livelihood security in southern India' held at Ootacamund, 24–25th Nov

Manorama K, Somasekhar N, Umamaheswari R (2016) Integrated potato cyst nematode (*Globodera* sp) management strategies for south Indian hills. Indian J Plant Prote 44:132–140

Manzila-Lopes RH, Costilla MA, Doucet M, Franco J, Inserra RN, Lehman PS, Cid del Prado-Vera I, Souza RM, Evans K (2002) The genus Nacobbus (Thorne & Allenn, 1944 Nematoda: Pratylenchidae): systematics, distribution, biology and management. Nematropica 32:149–227

Marium T, Shahnaz D, Fatima Sultana M, Muhammad JZ (2008) The effect of mangroves amendments to soil on root rot and root knot of potato (*Solanum tuberosum* L.). Acta Agrobot 61(1):115–121

Medina IL, Gomes CB, Correa VR, Mattos VS, Castagnone P, Carneiro RMDG (2017) Genetic diversity of *Meloidogyne* spp. parasitizing potato in Brazil and aggressiveness of *M. javanica* populations on susceptible cultivars. Nematol 19:69–80

Menon PPV, Thangaraju D (1973) Effect of crop rotation on the population of the golden nematode (*Heterodera rostochiensis* Woll.) in the Nilgiris. Indian J Agric Sci 43:304–306

Mhatre PH, Divya KL, Venkatasalam EP, Bairwa A, Sudha R, Saranya C, Guru-Pirasanna-Pandi G, Sharma S (2021) Evaluation of trap crop, *Solanum sisymbriifolium* and antagonistic crops against potato cyst nematodes, *Globodera* spp. S Afr J Bot 138:242–248

Moens M, Perry RN, Starr JL (2009) *Meloidogyne* species: a diverse group of novel and important plant parasites. In: Perry RN, Moens M, Starr JL, Root Knot Nematodes. Wallingford, UK, CAB International, 1–18

Muchovej RMC, Pacovsky RS (1997) Future directions of by-products and wastes in agriculture. In: Agricultural uses of by-products and wastes. (Rechcigl JE, Mac Kinnon HC), American Chemical Society Washington, DC 1-19

Mulder A (1994) Tolerance of the potato to stress associated with potato cyst nematodes, drought and pH. Agricultural University, Wageningen

Mwangi JM, Kariuki GM, Waceke JW, Grundler FMW (2015) First report of Globodera rostochiensis infesting potatoes in Kenya. New Dis Rep 31:18

Mwaura P, Niere B, Vidal S (2015) Resistance and tolerance of potato varieties to potato rot nematode (*Ditylenchus destructor*) and stem nematode (*Ditylenchus dipsaci*). Ann Appl Biol 166:257–270

Nath RC, Mukherjee B (2000) Dioscorea floribunda, a potential source of nematicides of plant origin. Nematol Medit 28(2):145–149

Ngala BM, Haydock PP, Woods S, Back MA (2014) Biofumigation with *Brassica juncea*, *Raphanus sativus* and *Eruca sativa*for the management of field populations of the potato cyst nematode *Globodera pallida*. Pest Manag Sci 71:759–769

Niragire I, Couvreur M, Karssen G, Uwumukiza B, Bert W (2019) First report of potato cyst nematode *Globodera rostochiensis* infecting potato (*Solanum tuberosum*) in Rwanda. Plant Dis 104:293. https://doi.org/10.1094/pdis-04-19-0891-pdn

Noling JW (2016) Nematode management in potatoes (Irish or white). University of Florida Agricultural Service-Extension-ENY 029(1):1–10

Norshie PM, Been TH, Chomaker CH (2011) Estimation of partial resistance in potato genotypes against *Meloidogyne chitwoodi*. Nematol 13(4):477–489

Ntalli NG, Menkissoglu-Spiroudi U, Giannakou IO, Prophetou-Athanasiadou DA (2009) Efficacy evaluation of a neem (*Azadirachta indica* a. Juss) formulation against root-knot nematodes *Meloidogyne incognita*. Crop Prot 28(6):489–494

Oduor-Owino P, Waudo SW, Makhatsa WL (1993) Effect of organic amendments on fungal parasitism of *Meloidogyne incognita* eggs and growth of tomato (*Lycopersicon esculentum* mill) cv. Moneymaker. Int J Pest Manag 39:459–461

OEPP/EPPO (2014) Report of the EPPO workshop on the management of *Globodera rostochiensis* and *G. pallida*. OEPP/EPPO Bull 44:21–26

Oerke EC, Dehne HW, Schonbeck F, Weber A (1994) Crop production and crop protectionestimated losses in major food and cash crops. Elsevier, Amsterdam, p 808

Ojo GT, Umar I (2013) Evaluation of some botanicals on root-knot nematode (*Meloidogyne javanica*) in tomato (*Lycopersicon esculentum* mill) in Yola Adamawa state. Nigeria Biological Forum 5(2):31–36

Oostenbrink M (1950) Het Aardappelaaltje (*Heterodera rostochiensis* Wollenweber), een gevaarlijke parasiet voor de eenzijdige aardappel-cultuur. In: Verslagen en Mededelingen von den Plantenziekten-Kundige Dienst te, vol 115. Wageningen, UK, H. Veenman and Zonen, p 230

Phillips MS, Trudgill DL (1998) Variation of virulence, in terms of quantitative reproduction of *Globodera pallida* populations, from Europe and South America, in relation to resistance from *Solanum vernei* and *S. tuberosum* ssp andigena CPC 2802. Nematologica 44:409–423

Pinheiro JB, Lopes CA, Henz GP (2009) Medidas gerais de controle de nematoides da batata. Circular Tecnica 76:1–9

Pinheiro JB, Silva FO, Pereira RB (2015) Nematoides na cultura da batata. Circular Tecnica 143:1–12

Proinpa (1996) Informe annual, Cochabamba, p 127

Ravelo J (1984) Dinamica poblacipnal de *Globodera pallida* (Stone, 1972) y combate mediante Manjeo Integrado de la Problacion. Memorias de la XII Reunion de ALAP, pp 461–474

Rehman B, Parihar K, Ganai MA, Siddiqui M (2012) Management of root-knot nematode, *Meloidogyne incognita* affecting chickpea, *Cicer arietinum* for sustainable production. Biosci Int 1(1):1–5

Sasser JN (1989) Plant parasitic nematode: the Farmer's hidden enemies. University Graphics, North Carolina State University, Raleigh, NC, p 115

Saxena R, Singh R (2001) Effectiveness of certain plant extracts for their nematicidal potebtialities. J Appl Zool Res 12(1):27–30

Scholte K, Vos J (2000) Effects of potential trap crops and planting date on soil infestation with potato cyst nematodes and root-knot nematodes. Ann Appl Biol 137:153–164

Scurrah MI, Niere B, Bridge J (2005) Nematodes parasite of *solanum* and sweet potatoes. In: Luc M, Sikora RA, Bridge J (eds) Plant parasitic nematodes in subtropical and tropical agriculture, vol 2205, 2nd edn. CAB Publishing, Wallingford, pp 193–220

Seenivasan N, Devrajan K, Selvaraj N (2007) Management of potato cyst nematodes, *Globodera* spp. through biological control. Indian J Nematol 37:27–29

Senasica (2013) Nematodo dorado de la papa- *Globodera rostochiensis* – Mexico–Segarpa. Ficha Técnica 19:1–24

Shaukat SS, Siddiqui IA, Ali NI, Zaki MJ (2001) Biological and chemical control of root infecting fungi and their effect on growth of mung bean. Pak J Biol Sci 4:1240–1242

Sparkes J (2013) Potential trap crops for the control of potato cyst nematodes (PCN). Agric Horticulture Dev Board 1:1–19

Sullivan MJ, Inserra RN, Franco J, Moreno-Leheude I, Greco N (2007) Potato cyst nematodes: plant host status and their regulatory impact. Nematropica 37:193–201

Taylor CE, Brown DJF (1998) Nematode vectors of plant viruses. CAB International, Wallingford, UK, p 286

Timmermans BGH, Vos J, Van Nieuwburg J, Stomph TJ, Van der Putten PEL, Molendijk PG (2007) Field performance of *Solanum sisymbriifolium*, a trap crop for potato cyst nematodes. I. Dry matter accumulation in relation to sowing time, location, season and plant density. Ann Appl Biol 150:89–97

Turner SJ, Evans K (1998) The origins, global distribution and biology of potato cyst nematodes *Globodera rostochiensis* (Woll.) and *Globodera pallida* (Stone). In: Marks RJ, Brodie BB (eds) Potato cyst nematodes: biology, distribution and control. CAB International, Wallingford, UK, pp 7–26

Umamaheswari R, Aarti B, Venkatasalam EP, Sudha R, Singh BP (2015) Effect of biofumigation on potato cyst nematodes. Potato J 42:124–129

Umamaheswari R, Somasekhar N, Manorama K, Joseph TA (2012) Eco-friendly management of potato cyst nematodes in the Nilgiris of Tamil Nadu. Potato J 39:185–190

Umar I, Mohammed BA (2013) Effect of water hyacinth *Eichhornia Crassipes* (Mart) leaf extract on the juvenile mortality of *Meloidogyne Incognita*. IOSR JAVS 4:46–48

Upadhyay KD, Dwivedi K, Uttam SK (2003) Effect of some plant extracts on the mortality and hatching of *Meloidogyne incognita* and *Heterodera cajani* infesting pigeonpea. Nematol Medit 31(1):29–31

Urwin PE, Troth KM, Zubko EI, Atkinson HJ (2001) Effective transgenic resistance to *Globodera pallida* in potato field trials. Mol Breed 8:95–101

Vovlas N, Mifsud D, Landa BB, Castillo P (2005) Pathogenicity of the root-knot nematode *Meloidogyne javanica* on potato. Plant Pathol 54:657–664

Vovlas N, Nico AI, De Luca F, De Giorgi C, Castillo P (2007) Diagnosis and molecular variability of an Argentinean population of *Nacobbus aberrans* with some observations on histopathology in tomato. J Nematol 39:17–26

Wang KH, Hooks C, Ploeg A (2007) Protecting crops from nematode pests: using marigold as an alternative to chemical nematicides. Plant Dis 2:1–6

Webley DP, Jones FGW (1981) Observations on *Globodera pallida* and *G. rostochiensis* on early potatoes. Plant Pathol 30:217–224

Whitehead AG (1977) Vertical distribution of potato, beet and pea cyst nematodes in some heavily infested soils. Plant Pathol 26:85–90

Whitehead AG (1995) Decline of potato cyst nematodes, *Globodera rostochiensis* and *G. pallida*, in spring barley microplots. Plant Pathol 44:191–195

Whitehead AG, Nichols AJF, Senior JC (1991) Control of potato pale cyst -nematode, *Globodera pallida*, with a granular nematicide and partially resistant potatoes. Ann Appl Biol 118:623–636

Whitehead AG, Nichols AJF, Senior JC (1994) The control of potato pale cyst-nematode (*Globodera pallida*) by chemical and cultural methods in different soils. J Agri Sci 123:207–218

Whitehead AG, Tite DJ, Fraser JE, French EM (1973) Control of potato cyst-nematode, *Heterodera rostochiensis*, in sandy, peaty and silt loam soils by dazomet and telone applied in different ways. Ann Appl Biol 75:257–265

Whitehead AG, Turner SJ (1998) Management and regulatory control strategies for potato cyst nematodes (*Globodera rostochiensis* and *Globodera pallida*). In: Marks RJ, Brodie BB (eds) Potato cyst nematodes: biology, distribution and control. CAB International, Oxford, UK, pp 135–152

Wolters PJCC, Vinke JH, Janseen R (1996) Resistance to *Globodera pallida* in eighteen wild *solanum* species. In: Abstracts of conference papers, posters and demonstrations, 13th Triennial Conference of the European Association for Potato Research. Veldhoven, The Netherlands, 425–426

Wood FH, Foot MA (1977) Decontamination of potato tubers grown in soil infested with potato cyst nematodes. N Z J Exp Agric 5:315–319

Yokoo T, Matsunobu K (1975) On thermal death point of the potato lesion nematode *Pratylenchus pratensis*. Agric Bull Saga Univ 5:97–102

Potato Viruses and Their Management 12

Ravinder Kumar, Rahul Kumar Tiwari, S. Sundaresha, Priyanka Kaundal, and Baswaraj Raigond

Abstract

Potatoes are vegetatively propagated, which can lead to the spread of viruses in the tubers. Aphid vectors are primarily responsible for the transmission of common viruses, such as PVY, PVS, PVM, PLRV, PVA, etc. The process of viral disease spread in the field has a big impact on chemical control approaches. Rapid degradation of seed stocks owing to viruses and associated infections needs frequent seed replacement in warm subtropical areas due to sufficient vector (aphids/thrips/whiteflies) population/activity. The growers will benefit from understanding potato viruses, their detection/elimination, and methods of indirect control to prevent their spread in potato crops, such as raising healthy seed crops from nucleus seed through isolation, sanitation, adjustment of planting and harvesting dates, haulms cutting, crop rotations, roguing, certification, etc. Furthermore, breeders have uncovered genetic resistance that can be introgressed into common cultivated types, providing a less expensive alternative to chemical control efforts. Viruses and recombinant viral strains have emerged in recent years, posing significant challenges to pathologists and breeders for seed certification and breeding. In this chapter, we review the different management approaches including use of resistance, seed systems, and cultural approaches. The newest concerns and challenges to potato production, including integrated management regimen for viruses, have been discussed in detail.

Keywords

Potato · Detection · Integrated management · Seed production · Tissue culture · Tuber · Virus

R. Kumar (✉) · R. K. Tiwari · S. Sundaresha · P. Kaundal · B. Raigond
ICAR-Central Potato Research Institute, Shimla, Himachal Pradesh, India

S. Kumar Chakrabarti et al. (eds.), *Sustainable Management of Potato Pests and Diseases*, https://doi.org/10.1007/978-981-16-7695-6_12

12.1 Introduction

Potato (*Solanum tuberosum* L.) area and production have consistently increased in tropical and subtropical regions of the world over the last few decades. The rate of potato production has surpassed the other major staple crops in developing countries (Devaux et al. 2020). The process of tropicalization of potatoes is expected to increase further due to the development of early maturing and heat-tolerant cultivars. However, this expansion of potatoes in nontraditional areas poses a major challenge of epidemics of viral diseases and the emergence of virus vectors in potato-growing regions of the world (Dahal et al. 2019). Throughout the world, potato is distressed by several viruses and their various strains belonging to different taxonomic groups (Table 12.1) resulting in severe decline in tuber quality and yield. Several factors govern the economic losses caused by potato viruses such as the strain of the virus, vector dynamics, variety grown in the area, growing conditions, etc. Due to the ever-changing vector dynamics under tropicalization in developing nations and the weak virus testing system, viral diseases are becoming serious threats causing up to 50% or more yield loss in potato (Harahagazwe et al. 2018). Besides, the practice of year-round potato cultivation in some tropical regions and the unavailability of suitable upland areas to produce high-quality seed potatoes where insect vector pressure is low enough aggravate potato viruses in these areas. The worldwide reported severe strains of *Potato virus Y* (PVY) and *Potato leafroll virus* (PLRV) can reduce 80% of tuber yield. The estimated loss of 30% occurs due to infection of *Potato virus X* (PVX), *Potato virus S* (PVS), and *Potato virus M* (PVM) (Khurana and Singh 1988). PVS is the latent virus that occurs in seed up to 90% level but causes a significant reduction in tuber yield only when combined with PVX and/or PVY. The secondary infection of PVX and PVS can cause a crop loss of 5–15%, while secondary infection of PVY^n and PLRV can reduce the yield by 15–30% and 40–70%, respectively (Singh and Somerville 1983). Similarly, whitefly transmitted *Begomovirus Tomato leaf curl New Delhi virus-potato* (ToLCNDV) known to cause apical leaf curl disease in India has become a serious problem in North Indian plains (Usharani et al. 2004). A *Tospovirus Groundnut bud necrosis virus* (GBNV) causing severe stem/leaf necrosis disease in plains/plateau of central/western India heavily infects early crops of potato (Kumar et al. 2019). The single reported viroid *Potato spindle tuber viroid* (PSTVd) is significantly devastating major potato-growing developed countries. Potato viruses also cause economic damage not only by crop losses but also by affecting seed quality and seed trade. The infestation of potato seeds with PLRV and PVY is the major cause for the rejection of the acreage entered in seed certification programs.

Sustainable potato production is feasible only if the viral diseases are kept under check especially in subtropics where the weather is highly conducive both for vectors and common viruses. The frequent spread of these viruses through tubers in susceptible cultivars causes rapid degeneration of potato crop in the field. It is very difficult to assess the direct losses caused by these viruses because of too many factors in a natural environment and the compensatory effect of non-infected and healthy plants in the vicinity of mildly or severely infected plants. The multiplication

Table 12.1 Important characteristics and genome structure of potato viruses (Modified after Kumar et al. 2020)

Virus (acronym)	Virus genus/group	Family	Morphology/ number of distinct particle size	Particle diameter	Vectors	Mode of transmission, spread	Geographical distribution
Potato leafroll virus (PLRV)	*Polerovirus* Group IV (+) ssRNA	*Luteoviridae*	Isometric/01	24	Aphid^{P}	TPS	Worldwide
Potato virus X (PVX)	*Potexvirus* Group IV (+) ssRNA	*Flexiviridae*	Filamentous/01	13	–	Contact, TPS	Worldwide
Potato virus Y (PVY)	*Potyvirus* Group IV (+) ssRNA	*Potyviridae*	Filamentous/01	11	Aphid^{NP}	TPS, mechanical	Worldwide
Potato virus A (PVA)	*Potyvirus* Group IV (+) ssRNA	*Potyviridae*	Filamentous/01	–	Aphid^{NP}	Mechanical	Worldwide
Potato virus M (PVM)	*Carlavirus* Group IV (+) ssRNA	*Flexiviridae*	Filamentous/01	12	Aphid^{NP}	Contact	Worldwide
Potato virus S (PVS)	*Carlavirus* Group IV (+) ssRNA	*Flexiviridae*	Filamentous/01	12	Aphid^{NP}	Contact	Worldwide
Tomato leaf curl New Delhi virus-potato (ToLCNDV-potato)	*Begomovirus* Group II (ssDNA)	*Geminiviridae*	Geminate particles	21–24 nm	Whitefly	–	India
Tomato spotted wilt virus (TSWV) or *Peanut bud necrosis virus*	*Tospovirus* Group IV (-) ssRNA	*Bunyaviridae*	Enveloped particle/01	70–110	Thrips^{P}	Mechanical	Hot climate, worldwide
Potato aucuba mosaic virus (PAMV)		*Flexiviridae*	Filamentous/01	11	Aphid^{HC}	TPS, contact	Worldwide (uncommon)

(continued)

Table 12.1 (continued)

Virus (acronym)	Virus genus/group	Family	Morphology/ number of distinct particle size	Particle diameter	Vectors	Mode of transmission, spread	Geographical distribution
	Potexvirus Group IV(+) ssRNA						
Alfalfa mosaic virus (AMV) *	*Alfamovirus* Group IV (+) ssRNA	*Bromoviridae*	Bacilliform/ 04-05	19	Aphid[NP]	TPS, pollen	Worldwide (uncommon)
Andean potato latent virus (APLV) *	*Tymovirus* Group IV (+) ssRNA	*Tymoviridae*	Isometric/01	28–30	Flea Beetle	TPS, pollen	South America
Andean potato mottle virus (APMV) *	*Comovirus* Group IV (+) ssRNA	*Comoviridae*	Isometric/01	28	Beetle	Contact	South America
Arracacha virus B – Oca strain (AVB-O) *	*Nepovirus* Group IV (+) ssRNA	*Sequiviridae*	Isometric/01	26	Unknown	TPS, pollen	Peru, Bolivia
Cucumber mosaic virus (CMV) *	*Cucumovirus* Group IV (+) ssRNA	*Bromoviridae*	Isometric/01	30	Aphid[NP]	Sap, TPS	Worldwide (uncommon)
Potato black ringspot virus (PBRSV) *	*Nepovirus* Group IV (+) ssRNA	*Comoviridae*	Isometric/01	26	Nematode[SP]	Soil-borne, TPS, pollen	Peru
Potato deforming mosaic virus (PDMV) *	*Begomovirus* Group II (ssDNA)	Geminiviridae	Segmented/02	18	Whitefly[SP]	TPS	Brazil
Potato latent virus (PotLV) *	*Carlavirus* Group IV (+) ssRNA	*Betaflexiviridae*	Filamentous/01	–	Aphid[NP]	Contact	North America

Tobacco rattle virus (TRV) *	*Tobravirus* Group IV (+) ssRNA	*Virgaviridae*	Rod or tubular/02	22	Nematode^{P}	Mechanically, TPS	Worldwide
Tobacco streak virus (TSV) *	*Ilarvirus* Group IV (+) ssRNA	*Bromoviridae*	Quasi-isometric/01	22–35	Thrips	Pollen, TPS, mechanical	South America
Potato yellow dwarf virus (PYDV) *	*Nucleorhabdovirus* Group V ((-) ssRNA)	*Rhabdoviridae*	Bacilliform	75	Leafhopper^{P}	Mechanical	North America
Potato yellow mosaic virus (PYMV) *	*Begomovirus* Group II (ssDNA)	*Geminiviridae*	Segmented/02	18–20	Whitefly^{SP}	–	Caribbean region
Potato mop-top virus (PMTV) *	Pomovirus Group IV (+) ssRNA	*Virgaviridae*	Rod or tubular/02	18–20	Fungus^{P}	Mechanical	western Europe and South America
Potato yellow vein virus (PYVV) *	*Crinivirus* Group IV (+) ssRNA	*Closteroviridae*	Filamentous	–	Whitefly^{P}	Infected tuber	South America
Potato yellowing virus (PYV) *	*Alfamovirus* Group IV ((+) ssRNA	Bromoviridae	Bacilliform	21	Aphid^{SP}	TPS	South America
Potato virus T (PVT) *	*Trichovirus* Group IV (+)ss RNA	*Flexiviridae*	Filamentous/01	12	–	Contact, TPS, pollen	South America
Potato virus U (PVU) *	*Nepovirus* Group IV (+) ssRNA	*Comoviridae*	Isometric/01	28	Nematode	Contact, TPS	Peru
Potato virus V (PVV) *	*Potyvirus* Group IV (+) ss RNA	*Potyviridae*	Filamentous/01	12–13	Aphid^{NP}	TPS	North Europe, South America

(continued)

Table 12.1 (continued)

Virus (acronym)	Virus genus/group	Family	Morphology/number of distinct particle size	Particle diameter	Vectors	Mode of transmission, spread	Geographical distribution
Solanum apical leaf curling virus (SALCV)*	*Begomovirus* Group II (ssDNA)	*Geminiviridae*	Segmented/03	18	Whitefly^{SP}	TPS	Peru
Tobacco mosaic virus (TMV)*	*Tobamovirus* Group IV (+) ssRNA	*Virgaviridae*	Rod or tubular/01	18	–	Contact, infected soil	Worldwide
Tobacco necrosis virus (TNV)*	*Necrovirus* Group IV (+) ssRNA	*Tombusviridae*	Isometric/01	26	Fungus^{P}	Soil-borne spores, mechanical	Europe, North America
Tomato black ring virus (TmBRV)*	*Nepovirus* Group IV (+) ssRNA	*Comoviridae*	Isometric/02	5–6	Nematode^{P}	Pollen, TPS	Europe
Tomato mosaic virus (ToMV)*	Tobamovirus Group IV (+) ssRNA	*Virgaviridae*	Rod or tubular/01	18	–	TPS, pollen, Contact	Hungary
Potato spindle tuber viroid (PSTVd)	*Pospiviroid* Circular (+)ssRNA	*Pospiviroidae*	Circular ssRNA only		Aphid^{CI}	TPS, pollen, contact	United States, Canada, South Africa, Russia

TPS true potato seed; *P/NP* persistently/non-persistently transmitted; *SP* semi-persistently transmitted; *HC* helper component involved for transmission, *CI* coinfection of PLRV essential for aphid transmission of viroid; * = viruses that are of quarantine importance in India or not reported in potato in India.

of seed stocks demands most care as the virus-free seed stock is liable to rapid infection from the infected volunteer plants, weeds, high vector pressure, and virus favoring cropping system. Management of potato virus and viroid diseases is a matter of vital importance and concern to farmers and scientists. To date, there is no direct method available to control the viral disease, and consequently, the current measures rely on indirect tactics to manage the potato viral disease. Only the possible strategies for potato virus disease management are (i) eradicating the source of infection to prevent the virus from reaching the potato crop, (ii) minimizing the spread of the virus by controlling its vector, (iii) utilizing virus-free potato seed material, and (iv) incorporating host-plant resistance to the cultivars. There are many technical aids available to scientists, field scouts, agronomists, and farmers to assist identification. This chapter includes the basic information on symptomatology, the economic importance of the disease, life cycle, and biology and management of major potato viruses.

12.2 Diversity and Distribution of Potato Viruses

There is immense diversity in viruses infecting potatoes (Table 12.1), and the greatest diversity has been reported from the Andean region which is also known as the place of origin for this wonder crop. The evolution of potatoes in the Andean region also facilitated the newer and rare virus strains which are not reported anywhere else. The nepoviruses *Potato Black Ringspot Virus*, *Potato Virus U*, and *Potato Virus B*, the tymoviruses *Andean potato latent virus* and *Andean potato mild mottle virus*, the Ilarvirus *Potato yellowing virus*, the crinivirus *Potato yellow vein virus*, the cheravirus *Arracacha virus B*, and the tepovirus *Potato virus T* are unique to the Andean region. In the Andean region, the most frequently intercepted viruses are PVX (30–82% incidence) and PVS (20–50%) followed by APMoV (4–15%) and APLV (2–6%). PVY and PLRV are the two most important viruses of worldwide economic importance. Together, these viruses cause as high as 80% yield loss in potatoes. Likewise, PVX is also reported across the world and causes 10–40% yield loss. PVS is also reported worldwide, but losses due to this virus are not severe until it comes as a mixed infection with PVX. PVA has sporadic occurrence but causes up to 40% yield loss. PLRV, PVX, PVY, PVA, PVM, PVS, GBNV, PAMV, and ToLCNDV are known to occur in India. PLRV and PVY are the most important and cause wide damage in all Indian varieties grown in different agroecological conditions. The viruses PVA and PVM and severe strains of PVX cause significant losses either singly or in different combinations. The reports of viruses infecting potatoes in Africa are limited. A study reported average incidences of 71%, 57%, 75%, and 41% of PLRV, PVY, PVA, and PVX, respectively. Besides these viruses, Chiunga and Valkonen (2013) reported the occurrence of PVM and PVS in two locations in Africa. A survey-based study in southwest Uganda reported PVX and PLRV as most frequent, followed by PVY and PVM in the year 2014 (CIP/IITA, unpublished). AMV and *Beet curly top virus* (BCTV) has been reported in Sudan, around Khartoum (Baldo et al. 2010). The most devastating and frequent viruses infecting potatoes in Europe include PLRV, PVY, PVA, PVM, and PVS. All of these viruses are aphid-borne. The PSTVd is also very damaging in potato cultivars prevalent in European countries. *Tomato spotted wilt virus* (TSWV) is emerging as a serious threat to potato cultivation in Europe particularly in changing climate scenarios. PMTV is also consistently reported in potato crops in Scotland, Northern Ireland, Denmark, the Czech Republic, and Austria. Recently, it has been detected in Poland (Santala et al. 2010). The important potato viruses in the Australian continent are PLRV, PVA, PVY, PVM, PVS, PAMV, CMV, AMV, TSWV, and *Lucerne Australian latent virus* (LALV), and the viroid PSTVd. The recombinant strain of PVY, viz., PVY^{NTN}, has created serious troubles in the potato seed chain and has been reported from eastern Australian states (Queensland, New South Wales, Victoria, South Australia).

12.3 Major Viruses and Viroid Infecting Potatoes

12.3.1 *Potato Virus Y* (PVY)

Potato virus Y (PVY) belonging to the genus *Potyvirus* within the family *Potyviridae* is a serious global threat in the potato production system and causes 10–100 percent yield losses in potato and other solanaceous crops worldwide. Over 20 years, PVY has caused heavy economic losses to potato industries especially in developing countries and parts of Europe and North America. The plethora of research on PVY during the last two decades revealed the occurrence of a huge number of variations and recombinations in this virus. The biological and phylogenetic basis of studies has classified PVY "strains" into 13 subgroups. (Kehoe and Jones 2016; Santillan et al. 2018). The strains of PVY are differentiated based on hypersensitive reactions on differential potato cultivars possessing strain-specific hypersensitive HR resistance genes. The hypersensitivity genes Nc, Ny, or Nz, respectively, correspond to strain groups PVY^{C}, PVY^{O}, and PVY^{Z} which elicit HR phenotypes. Two strains PVY^{N} and PVY^{E} can overcome all three hypersensitivity genes and develop completely different phenotypes in potatoes.

Venal necrosis is the most peculiar symptom of PVY^{N} (Chikh Ali et al. 2008; Green et al. 2020). PVY^{NTN} isolates cause severe superficial tuber necrosis (potato tuber necrotic ringspot disease) and may also cause necrotic foliar symptoms. The distribution of PVY strains is global; however, some virus strains are restricted to certain continents. The most common strain PVY^{O} occurs worldwide, while PVY^{N} occurs in Europe, parts of Africa, and South America; PVY^{C} strain has been reported from Australia, India, and Europe. PVY^{NTN} is reported in Europe and lately been intercepted in North America. PVY^{Z} and PVY^{E} are the least significant strains in the potato production system (Singh et al. 2008). Infected seed tubers are the major sources of PVY inoculum and disease spread. The primarily aphid-borne virus is transmitted in a non-persistent manner but can also be transmitted by sap inoculation and grafting. Aphids feeding on infected plants acquire PVY within a few seconds as the virus is stylet borne and also transmit the virus to healthy plants within seconds.

The virus has flexuous, 740 × 11 nm long filaments and single-stranded RNA, usually occurring in low titer in potato leaves.

12.3.2 *Potato Leafroll Virus* (PLRV)

Potato leafroll virus (PLRV) belonging to the genus *Polerovirus* within the family *Luteoviridae* is the second most important virus of potato globally. Although the virus is confined to phloem cells and detected in low titer in plants, it causes severe yield loss (up to 90%) and quality reduction due to internal damage to tubers (net necrosis). Some reports have estimated that the virus causes 20 million tons of potato production loss globally (Taliansky et al. 2003; Abbas et al. 2016). PLRV is the only aphid transmitted RNA virus infecting potato which is transmitted in a persistent circulative non-propagative manner. The virus persists in the aphid body throughout its life cycle. The disease becomes more aggravating in developing crops since all instars (stages) of the aphid can transmit the virus. However, the nymph stage is more efficient in virus acquisition and transmission than the adult. The long-distance spread of the virus occurs through winged aphids. The virus is also tuber-borne (Peiman and Xie 2006). The virion particles are isometric virions of 24 nm diameter and single-stranded RNA. The primarily infected plants show erect growing habits with chlorosis of young leaves. Other secondary symptoms include shoots stunting, old leaver becoming chlorotic and rolling upward, leathery and brittle young leaves, etc. Lower leaflets rolling inward and extended ultimately to the upper leaves is peculiar for PLRV infections in potato. The characteristic rustling noise appears from the dry and brittle leaves of the plants. A purple discoloration is also observed in the infected plant leaves. Already established infection leads to the development of normal-shaped but small tubers.

12.3.3 *Potato Virus X* (PVX)

Potato virus X (PVX) of genus *Potexvirus*, family *Flexiviridae*, is distributed worldwide. Despite its low yield reduction rate, the virus is important due to its high incidence in the potato crop (usually 15–20%) globally (Jeevalatha et al. 2016; Ozkaynak 2020). Moreover, severe yield losses are documented in the combined occurrence of PVX, PVA, and PVY. Mechanical transmission is the most common way for PVX spread. The highly contagious nature of the virus makes it much more notorious in the potato field. It can spread through the clothing of the workers and remain infective for many hours if the conditions are suitably wet (Kumar et al. 2019). Consequently, the virus is transmitted easily from one field to another. Since the virus is tuber-borne and its accumulation occurs in seed tubers, the cutting of seed tubers using implements may also spread the virus across the fields. The virus particles are filamentous virions of 515 × 13 nm size and single-stranded RNA. The viral symptoms are mild in most potato varieties, but coinfection with PVY causes severe mosaic symptoms in potatoes. The synergy of PVX and PVY gives symptoms of a faint or fleeting mottle to a severe necrotic streak (Agindotan et al. 2007). The symptom is more apparent in low-light conditions. The rare symptoms are leaf distortion, rugosity, and necrotic spotting which occur only in severe infections.

12.3.4 *Potato Virus S* (PVS)

Potato virus S (PVS) belongs to the genus *Carlavirus* within the family *Flexiviridae* and significantly shows worldwide loss in potatoes. Globally, it is known to cause up to 20% tuber yield reduction (Duan et al. 2018; Santillan et al. 2018). PVS alone is less devastating as compared to its coinfection with PVX where they can cause greater yield loss in potatoes. Interestingly, the virus is reported to be involved in breaking late blight resistance in some potato cultivars. Based on the non-systemic or systemic type of infection in the *Chenopodium* species, two strains of PVS have been identified as PVS^{O} (Ordinary) and PVS^{A} (Andean) (Wang et al. 2016). Several minor strains have also been identified apart from PVS^{O} and PVS^{A}. As the names indicate, PVS^{O} has worldwide occurrence, but PVS^{A} is restricted to the Andean region only. The virus PVS is highly contagious, and it can spread through seed cutting as well as plant contact. An aphid vector *Myzus persicae* is also reported to spread PVS in a non-persistent manner. The virion particle is slightly flexuous filamentous of size 660 x 12 nm and having single-stranded RNA (Gutiérrez et al. 2013). Although the latent infections are most common, the infected plant varieties look almost healthy, but occasional/transient leaf symptoms of faint rugosity, vein deepening, and leaf bronzing can be seen. The electron micrographic studies revealed that infected plant cells accumulate aggregated virus particles, observed as paracrystalline inclusions or banded bodies. These inclusions comprise of virus particles, endoplasmic reticulum, and proliferated ribosomes.

12.3.5 *Potato Virus M* (PVM)

Potato virus M (PVM), a member of the genus *Carlavirus* in the family *Flexiviridae* has global significance wherever potatoes are grown (Brunt 2001). The virus has a polyadenylated single-stranded, positive-sense genomic RNA of an estimated size of 8.5 kb in length. It normally causes a reduction in yield up to 15–45% as observed in eastern Europe and Russia, but it may go up to 100 percent in some cultivars. The host range of this virus is mainly confined to the family *Solanaceae*. The virus is primarily sap transmissible and also spreads through infected tubers and aphids in a non-persistent manner (He et al. 2019). Serologically, PVM and PVS are related, and both cause latent infections in potato plants. The visible symptoms of PVM include mottle, mosaic, rolling and crinkling of leaves, and stunting of shoots. Symptoms of PVM infection are similar to the symptoms caused by PVY, PVX, and PVS (Kowalska and Waś 1976). Symptom severity may vary depending upon the PVM isolate and potato cultivars.

12.3.6 *Potato Virus A* (PVA)

Like other viruses described earlier, *Potato virus A* (PVA) also has a worldwide occurrence. The virus is serologically related to PVY and belongs to the genus *Potyvirus*, family *Potyviridae*, with single-stranded positive-sense RNA (Fuentes et al. 2021). PVA alone causes yield reduction in tubers up to 30–40 percent, but its coinfection with PVX and/or PVY may cause more losses in potato. The virion particles are flexuous 730 × 11 nm long filaments, and its genome comprise of a messenger-polarity ssRNA of 9.5 kb including a virus-encoded protein (VPg) covalently attached to the 5′-end and a 3′-poly(A) tail. PVA-infected plants bear shiny leaves showing vein mottling and have a more open growth habit. PVA is transmitted by many aphids (*Myzus persicae*, *Macrosiphum euphorbiae*, *Aphis frangulae*, and *A. nasturtii*) species in a non-persistent manner (Puurand et al. 1994). The acquisition access period of *M. persicae* is less than 1 min, and virus transmission occurs equally rapidly with no latent period. The virus may be retained by the vector for 20 min to 2 h before it loses its infectivity. The host range of PVA is narrow and confined to *Solanaceae* such as tobacco or tomato.

12.3.7 *Groundnut Bud Necrosis Virus* (GBNV)

Groundnut bud necrosis virus (GBNV) belongs to the genus *Tospovirus* in the family *Bunyaviridae* and causes stem necrosis disease in potatoes (PSND). The host range of this virus includes crops like tomato (*Lycopersicon esculentum*), tobacco (*Nicotiana* spp.), peanut (*Arachis hypogaea*), soybean (*Glycine max*), and cotton (*Gossypium* spp.). The virus is reported from potato-growing areas in India, Argentina, Australia, and Brazil (Jain et al. 2004). The virus is reported to cause incidence up to 90% in Madhya Pradesh and Rajasthan states of India and up to 50% in Uttarakhand (Pundhir et al. 2012). The major vector for GBNV transmission is thrips species belonging to genera *Thrips* and *Frankliniella*. The virus is acquired by thrips in the nymphal stage only, while both nymph and adult can transmit the virus. In localized areas where the thrip vector incidence is high in potatoes, the chance of GBNV emergence is also very high (Verma and Vashisth 1985). The latent period after virus acquisition is around 4–9 days after which the virus is transmitted persistently. The early planting is more vulnerable to thrips attack and PSND incidence as compared to late planting. The virion particles are spherical and enveloped ranging from 70 to 110 nm in diameter and having single-stranded RNA (Ansar et al. 2015). Infected plants show typical necrotic lesions bearing concentric rings or spots on leaves and stems. The vegetative plant part shows a rosette appearance and stunted growth if not killed. The infected GBNV plant produces fewer tubers having smaller size, but the virus does not invade the developing tubers.

12.3.8 *Tomato Leaf Curl New Delhi Virus-Potato* (ToLCNDV)

Under the influence of sub-tropicalization of potato, early planting, and year-round potato cultivation, a new viral disease called potato apical leaf curl disease (PALCD) is emerging as a major threat in India (Jeevalatha et al. 2018). The disease is caused by a strain of *Tomato leaf curl New Delhi virus-potato* (ToLCNDV) belonging to the genus *Begomovirus* within the family *Geminiviridae* (Zaidi et al. 2017). The virus ToLCNDV itself is emerging as a major threat in the horticulture production system as it infects several crops. ToLCNDV is a bipartite begomovirus having dual genomic components referred to as DNA-A and DNA-B. The percent sequence identity revealed that DNA-A components of the ToLCNDV isolates shared more than 90.0% similarity to ToLCNDV isolates tomato and okra, 89.0–90.0% to papaya isolates, and 70.4–74.0% to other ToLCNDVs (Usharani et al. 2004). In the vast Indo-Gangetic plains, about 40–100 percent sporadic infection of this virus has been reported from potato crops. In susceptible potato varieties, its infection leads to severe seed degeneration. Symptoms of this viral disease appear as a conspicuous mosaic with curling/crinkling of apical leaves, while secondary infection leads to stunting of the potato plants (Kumar et al. 2019). The whiteflies *Bemisia tabaci* transmit this virus in a persistent manner, and the high incidence is correlated with a high whitefly population in early potato planting (Jeevalatha et al. 2014).

12.3.9 *Potato Spindle Tuber Viroid* (PSTVd)

Potato spindle tuber viroid (PSTVd) is a member of the family *Pospiviroidae* (type species: *Potato spindle tuber viroid*; PSTVd). The viroid has five structural/function domains such as the Terminal Left (TL), the Central (C), the Pathogenicity (P), the Variable (V), and the Terminal Right (TR) domains. These viroids replicate through an asymmetric rolling circle mechanism in the host's nucleus. The PSTVd is commonly 359 nucleotides in length (Matoušek et al. 2014). The occurrence of mild and severe strain in the ratio of 1:10 is reported in potatoes. Mild strains with latent symptoms cause tubers to yield a loss of 15–25 percent, while severe strains with peculiar symptoms may result in 65% yield loss in potatoes (Katsarou et al. 2016). The viroid is highly contagious and may spread through contaminated cutting tools. The pollens and true potato seeds can also transmit these viroids. The symptom is highly expressed in warmer conditions compared with cooler ones. The infected plants show erect growth habits with a slender stem and blossom pedicels. Leaflets are curved inward and overlap other leaflets.

12.3.10 *Tomato Yellow Vein Streak Virus* (ToYVSV) and *Tomato Severe Rugose Virus* (ToRSV)

The overlapping cropping seasons of potato and tomato and favorable weather conditions has caused the shift of tomato viruses onto the potato crop. Two species of *Begomovirus* (family *Geminiviridae*), *Tomato yellow vein streak virus* (ToYVSV) and *Tomato severe rugose virus* (ToRSV)), have been described intercepted in potato. Both of these viruses are reported to cause deforming mosaic symptoms in potato and tomato crops. The vector whitefly *Bemisia tabaci* has been reported to transmit both these viruses (Kreuze et al. 2020). The increasing whitefly populations in the potato growing areas during the early potato season are alarming and demand rigorous monitoring of begomoviruses.

12.3.11 *Potato Mop-Top Virus* (PMTV)

PMTV is classified in genus *Pomovirus*, and it is regarded as a *Furovirus* (fungus-transmitted, rod-shaped virus). The virus was first reported in Britain in 1966 and later on discovered in cooler areas in Europe, Asia, and the Andes of South America. PMTV has also been reported in Canada in 1991–1992 and the United States in 2002. It was later detected in Poland (Santala et al. 2010). The mode of transmission in the resting spores of *Streptomyces subterranean* (fungus causing powdery scab in potato) made it a unique virus infecting potatoes. However, other means of PMTV transmission include contaminated soils, equipment or vehicles, traded materials, etc. The virus transmission through seed tubers is variable, and in the absence of *S. subterranea*, "self-curing" of PMTV-infected tubers takes place after three generations (Xu et al. 2004; Santala et al. 2010). That implies, in the absence of the vector, that PMTV-infected plants may become free of the virus after three generations. The virus requires movement factor proteins for cell-to-cell movement, and it moves within-host through the xylem. The localization of PMTV has been reported in the cytoplasm. Limited work has been done on this widespread potato virus, and an exhaustive study is needed to establish the factors responsible for the occurrence and spread of this disease in potatoes (Gil et al. 2016).

12.3.12 *Alfalfa Mosaic Virus* (= *Calico*) (AMV)

Leaves of infected plants appear bright yellow with a smooth shiny surface (calico symptoms), very noticeable, and plants stand out in a field. Tubers may be malformed and may crack due to viral infection (Nie et al. 2020). The *Alfalfa mosaic virus* (AMV) is carried by many aphids affecting alfalfa and clover and sometimes around wheat. Spread is from nearby alfalfa and clover fields often just after the time of cutting or harvest as the aphids fly away from these fields into potato fields where water is present. Aphids pick up the virus and deposit all in only a few plant visits (non-persistent). Although the problem is not usually a major concern, sections of fields can be severely affected. Planting over a mile from an alfalfa or wheat field will eliminate any problem. If planted near these fields, an aphid control program may be needed (Xu and Nie 2006; Nie et al. 2020). Controlling volunteer alfalfa is helpful. The problem appears worst when a circle is planted partially in potato and partially in alfalfa.

12.4 Management of Potato Viruses

The consistent increase in intercontinental trade of potato, coupled with the process of tropicalization, poses a major challenge of emergence and spread of viral disease in potato in nontraditional areas. The shortening of the potato growing cycle, development of early maturing cultivars, and overlapping potato and other vegetable cropping systems influence the virus vector dynamics in the potato cropping system. A unified management strategy involves the strengthened seed certification system, good phytosanitary measures, use of host plant resistance, need-based pesticide application for vector control, and managing the tubers during harvest and storage. Integrated management of the viral disease is also a suitable ecosystem-based management approach for crop production and protection with a focus on environmental sustainability and economic feasibility. With the rapid advancement in molecular biology and computational technology coupled with increasing awareness of information technology, several advancements have been observed in the management of viral diseases in potatoes.

12.4.1 Production of Virus-Free Planting Material

Meristem tip culture The initial virus-free planting material released by breeders may be free from viral diseases; however, it becomes infected once grown nearby virus reservoir source and vector-prone areas. So, the consistent need for virus-free planting material is essential. To supply the virus-free planting materials, certification schemes have been developed. Special stocks are built up by propagating from single virus-free plants that form the basis of certification schemes. Serological studies, indexing, indicator hosts, and molecular detection are the major tools to identify virus-free stocks. Sometimes, a clone showing good agronomical trait is rejected due to the presence of viral infection (Kreuze et al. 2020). "Meristem tip culture" is a highly sophisticated method to develop virus-free planting material. A small piece of meristematic tissue (0.2–0.5 mm) is excised and cultured on the nutrient media, which leads to the development of a virus-free plantlet. The apical dome of leaf primordia of the subapical region is utilized for meristem culture (Chauhan et al. 2019). The apical meristem without leaf primordia has the highest probability to produce virus-free planting material but the lowest probability to survive in culture media. Virus elimination through meristem tip culture occurs possibly due to:

- Absence of virus in the non-differentiated meristematic region of the plant.
- The pace of virus replication is slower in the actively dividing meristem cells which restricts virus replication.
- A high concentration of endogenous auxin level may also inhibit the virus replication.

The meristem tip culture has been usefully applied and utilized for the generation of virus-free potato plants (Gong et al. 2019). It is also reported that virus-free plants, regenerated from meristem tips, are genetically stable and yield true-to-type plants. The potato viruses in order of increasing difficulty of their eradication are PLRV, PVA, PVY, PAMV, PVX, PVM, PVS, and PSTVd. PVS and PSTVd are notoriously difficult to eradicate through meristem culture, whereas PVA and PVY are easily eliminated by meristem culture alone (Cassells and Long 1982). The meristem culture-mediated eradication leads to the elimination of PVA and PVY from 85% to 90% of the meristem cultures, while PVX and PVS, being stable, were eliminated up to 10%.

Thermotherapy Heat therapy is highly effective in inactivating the viruses in the planting materials and explants. Different potato viruses have a variable response in their sensitivity to heat. Potato viruses like PVY and PVA are easily eliminated at 36/39 °C, while other viruses need higher temperatures (Ali et al. 2013). The elimination of virus from the infected plant as well as plant survival depends on the age of the plant, duration of therapy, initial virus loads, etc. In general, well-established hardy plants should be treated with heat. Heat treatment causes temporary abnormalities in the color and shape of foliage which normalize after few weeks. Using thermotherapy before meristem culture is effective in virus elimination. Thermotherapy is successfully used for many potato viruses. For eradication of PVS and PVX, a combination of 36 °C for 16 h and 29 °C for 8 h over 20–24 weeks is ideal (Lozoya-Saldaña and Merlin-Lara 1984). A temperate of 37 °C is bearable for most potato cultivars for up to a few weeks. Heat treatments of tubers sprouts are also very effective to obtain PVS-free plantlets. Ali et al. (2013) obtained 43.79% of PVX-free plants through meristem tip culture and thermotherapy at 35 ± 1 °C.

Chemotherapy The application of antiviral chemicals to the virus-infected plant or incorporation to the culture media to restrict multiplication and spread of the virus is known as chemotherapy. An effective antiviral chemical must possess the abilities to inhibit virus multiplication, spread, or symptom induction, be nontoxic to the host, and have a broad-spectrum effect against many viral diseases and systemic movement in the host. The antiviral antibiotic ribavirin (Virazole) has shown efficacy in inhibiting the virus multiplication of CMV, PVY, PVS, and PVM in potatoes. 2-Thiouracil was proved to be effective against PVY (Faccioli and Colombarini 1996). The amalgamation of riboside in the medium showed efficacy in the eradication of PVX, PVY, PVM, and PLRV. The sequential or simultaneous treatment of heat therapy and chemotherapy is also effective in managing potato viruses. Kumar et al. (2020) synergistically used a combination of thermotherapy (33 °C) and chemotherapy (20 ppm ribavirin) and eradicated PVS from nodal shoots of potato. A combined treatment of thermotherapy (37 °C) with ribavirin (RBV)/5-azacytidine (AZA)/3-deazauridine (DZD) caused the elimination of PVY up to 83.3%, 70%t, and 50%, respectively. Chemotherapy containing 20 mg/ml ribavirin coupled with thermotherapy at 37 ± 1 °C for 2 weeks eliminated PVX and PVS (Gopal and Garg 2011).

Electrotherapy Electrotherapy is a simple and inexpensive method of virus eradication that uses the application of electric current to plant tissues to disrupt or degrade viral nucleoprotein. Electrotherapy is found useful in the elimination of PVX from potato by using the current of 15 mA for 5 min which eliminated 60–100% of the virus (Lozoya-Saldaña et al. 1996). Similarly, elimination of PVY, PVA, PVS, and PLRV is also feasible through electrotherapy. The electric current of 15 mA for 10 minutes produced the highest degree of virus elimination for PLRV, PVY, and PVS (26–100%). Meylbodi et al. (2011) found that an electric current of 35 mA for 20 min was the most effective electrotherapy treatment for eliminating PVY and PVA with regeneration of 54–70%.

Cryotherapy In cryotherapy, low-temperature treatment and liquid nitrogen (−196 °C) exposure in the main component to eradicate the viruses from the planting materials. The technique results in a high frequency of virus-free regeneration. A combination of thermotherapy followed by cryotherapy of shoot tips can be used to enhance virus eradication. Cryotherapy of shoot tips is easy and conveniently allows treatment of large numbers of samples to produce pathogen-free regenerants (Wang et al. 2006; Gong et al. 2019). Some potato viruses like CMV, PLRV, and PVY have been eradicated using cryotherapy. Wang et al. (2006) utilized cryogenic protocols, i.e., encapsulation-dehydration, encapsulation-vitrification, and droplet, to obtain PLRV- and PVY-free plants at 83–95 percent frequencies, higher than those by meristem culture and thermotherapy. The elimination of notorious viroid in potato, PSTVd, can also be performed using cryotherapy.

12.4.2 Strengthening Seed Testing in Potato

The developed countries are mainly managing potato viruses through a sound seed tuber testing and certification program and by deploying host plant resistance in common cultivars. On the other hand, many developing countries are still struggling to develop a formal seed testing and certification channel, and most of the farmers in these countries obtain their seeds from previous crops that lead to the planting of virus-infected low-quality planting materials. With the lack of appropriate investment and infrastructure and insufficient return on investment for smallholders, the utilization of certified seed is negligible in developing countries (Thomas-Sharma et al. 2016). The innovative "seed plot technique" in India has revolutionized the quality seed tuber production in potatoes by utilizing the climatic conditions of hills and plains. India is the largest potato producer in Southwest Asia, and potatoes are grown in diverse agroecologies. Potato is grown both in the hills and in plains in summer and autumn/winters, respectively. Nearly 82% of the potato area is in the subtropical plains where it is cultivated in the winter under short-day conditions. Nearly 10% of the potato area lies in the plateau region of south peninsular India and the remaining 8% in the mountains. Potatoes in plateau and mountains are grown under long-day conditions during summer and spring seasons, respectively. To effectively use this natural phenomenon, indexing for virus freedom is done in hills for the crop to be raised in the plains and vice versa. In the preliminary surveys, it was observed that areas above 2000 m in the northern temperate hills in a few pockets were the most suitable for the production of quality seed potatoes (Naik and Buckseth 2018). During summer months in these areas when the potato crops are generally grown, low temperature, high-velocity winds, and frequent rains are unfavorable for the build-up of aphid populations that are vectors for potato viral diseases. Because of this observation, procedures were developed at the Central Potato Research Institute (CPRI), Simla, for the production of disease-free stocks in the high hills. CPRI is the backbone of potato seed production in India. Nearly 2500 tons of breeder seed are supplied every year to different organizations for further multiplication for foundation and certified seeds. After studying the epidemiology of the insect vectors in the subtropical plains, the "seed plot technique" was developed (Kumar et al. 2019; Pradel et al. 2019). Many countries are currently effectively utilising various rapid multiplication strategies in potato seed multiplication, such as tissue culture alone/micropropagation or in combination with various means of generating cuttings. In most of the countries, the in vitro material was being used to produce basic nucleus disease-free material, and further in vitro methods of multiplication were employed to multiply the basic disease-free material. However, by following some of the techniques discussed here, the majority of the potato growing countries have become self-sufficient in seed potato production for their country or depending on some other country. The rapid multiplication techniques are fast becoming important in developing self-sufficiency in seed production. To obtain cheaper and cleaner propagules, a compromise of the technology, i.e., minituber technology, was developed in the United States and later adopted in many other countries; in vitro plantlets are planted in beds in a screen house (screens for aphids with sterilized soil); the plantlets produce 3–6 minitubers.

12.4.3 Tuber Indexing and Clonal Multiplication

Indexing of seed tubers for viruses is done to discard the infected ones. Detection of viruses is affected by many factors like temperature, physiological stage of tubers (dormant or sprouted), and the technique employed (Naik and Buckseth 2018). ELISA detection of PVX and PVS was higher in physiologically advanced or non-dormant material when grown and tested 12 weeks after harvest at 6.0–29.5 °C than the dormant stocks of the same age at 12.5–28.5 °C (Singh and Somerville 1983). The clonal selection from stage I instead of stage III resulted in an overall improvement in the disease freedom of seed stocks being multiplied. Thorough roguing of seed crop helps an effective check of the natural spread of viruses, etc. ELISA is now routinely employed for tuber indexing. It has replaced biological, chloroplast agglutination testing that required more space/time and labor as well as allowed some latent infections of PVS/M, PVY, etc. going unchecked.

12.4.4 Avoidance and Elimination of Virus Vectors

Several insect vectors (aphids, whiteflies, thrips,) are responsible for the transmission of potato viruses. Most of the damaging viruses are transmitted primarily by aphid species in a non-persistent manner. The two major principles of vector control include avoidance of vectors and chemical-based management. In avoidance of vector, the seed plot technique plays a key role in managing aphid-borne viral diseases in potatoes more specifically in India. Several potato viruses (viz., PVY, PLRV, PVA, PVM) spread mainly through one or more aphid vectors (Tsedaley 2014). The seed crop in northwestern Indian plains is raised only in aphid-free periods or locations in the designated seed-producing areas through the seed plot technique. The seed crop is being grown during aphid-free period (October–December/January) including use of the healthy seed, application of systemic insecticides, field inspection for roguing all infected/off-type plants, and dehaulming the crop as soon as the aphids cross the critical limit of 20 aphids/100 compound leaves. Therefore, the seed crop in NW Indian plains is raised only in aphid-free periods or locations in the designated seed-producing areas through the seed plot technique (Pushkarnath 1967). Verma and Vashisth (1985) found that the incidence of the viral diseases was maintained within the permissible limit of 1 percent for several years if the haulms were cut as soon as the aphid build-up started reaching the critical limit. PSND can be effectively managed by manipulating planting dates and the use of systemic insecticides for managing thrips. It was shown that by planting alternate rows of tomatoes and cucumbers (cucumbers planted 30 days before transplanting tomatoes), the spread of *Tomato yellow leaf curl virus* (TYLCV) in the tomatoes was significantly delayed during the first 2 months (Al-Musa 1982).

Among the chemical-based management, systemic soil insecticides are applied at planting, while foliar sprays of certain contact cum systemic insecticides help check the aphid population on seed crops but only for a short period until dehaulming. Insecticides are often effective against the spread of persistently aphid-transmitted viruses like PLRV but not against the spread of non-persistently aphid-transmitted viruses. The pyrethroid application gave maximum control of the aphid vectors *Macrosiphum euphorbiae* and *M. persicae* and halved the incidence of PVY. Mowry (2005) reported that chloronicotinyl, imidacloprid, pyridine azomethine, and pymetrozine were highly effective in reducing transmission of PLRV from infected to healthy potato plants by *M. persicae*. The synthetic pyrethroid esfenvalerate was effective in reducing inoculation of PLRV by virus-infected aphids into potatoes due to its repellent effect, but not virus acquisition by aphids from infected plants. Insecticides are less effective in controlling PVY infection, because the PVY is non-persistent and borne on the aphid's stylet and may be transmitted before the aphid is killed.

12.4.5 Role of Botanicals and Oils in Potato Virus Management

Efforts have been made in different countries to test the efficacy of certain plant products (botanicals) to induce systemic resistance for managing the incidence of plant viral diseases. The botanical insecticides composed of essential oils may be an alternative to the more persistent synthetic pesticides for the management of vectors responsible for disease spread. A proprietary emulsifiable concentrate containing 25 percent essential oil extract of *Chenopodium ambrosioides* (EOCA) as the active ingredient at 0.5 percent caused mortality (43.6 percent) and insecticide soap (55.2 percent) and was effective against *Myzus persicae*. The extract of *C. ambrosioides* at the concentration of 0.5 percent gave excellent control of thrips *Frankliniella schultzei* (95.7 percent) than insecticidal soap (83.6 percent), neem oil (17.7 percent), and water (10.8 percent) (Cloyd and Chiasson 2007). EOCA proved to be more effective than commercial products in controlling major virus vectors such as *M. persicae* and *F. schultzei* than neem extract and insecticidal soap endosulfan and abamectin. Intensive researches on viral disease management have also led to the discovery of oils for the control of viral diseases. The motivation for using these nontoxic materials at recommended concentrations is great because they are less likely to cause environmental pollution than chemical pesticides, have excellent spreading and sticking properties, are not subject to resistance development, and are economical to use. The transmission of PVY by *M. persicae* was impeded by coating either the source plants or the test plants with mineral oil (liquid paraffin). However, the use of mineral oils can be effective only if seed potato production is located under low infection pressure conditions.

12.4.6 Utilizing Resistance Mechanism

The use of cultivar which is resistant to the various pathogen and viruses is an inexpensive method which has no adverse effect on the environment. Moreover, the cost involved in the additional purchase of equipment, chemicals, and decision-making shall not be borne by the farmer. There are many potato cultivars which are developed for resistance against insect and nematodes which are released and beneficial for farmers in terms of cost and inputs (Solomon-Blackburn and Barker 2001). Potato varieties which are specifically bred for fungal, bacterial, viral, and viroids are resistant to these pathogens and ultimately aimed for the reduction of pesticides. Various approaches such as morphological, biochemical, physiological, and molecular are taken into consideration for the mechanism of resistance against insect and pest in potato. The strategies against resistance to viruses include the absence of symptoms, reduction of viruses in the plant by restriction of multiplication, the spread of viruses in the field, and various effective measures and methods for management of viral diseases. The use of the cultural and chemical method for the management of viral diseases may not often be practicable in the field condition. However, the use of modern cultivars which are resistant or tolerant against virus-mediated diseases was found to be effective and efficient (Thomas-Sharma et al.

2016). There are various stable resistant sources available against viral diseases. The development of virus-resistant cultivar should be encouraged in areas where disease pressure is high, whereas cultural methods should be promoted in areas where disease pressure is moderate or low in potato-growing regions (Lin et al. 2014).

12.4.6.1 Coat Protein (CP)-Mediated Resistance

In virus, CP acts as the protective function which insulates nucleic acid (RNA or DNA) from degradation due to various hydrolyzing enzymes and inactivation by ultraviolet rays (Kumar et al. 2020). The CP is also reported to be involved in the host recognition in the early phases of infection. The first attempt to develop a potato cultivar against the virus was made by Hemenway et al. (1988), where they transformed the potato with the coat protein (CP) of *Potato virus X* (PVX). The higher level of resistance against viruses was associated with a low level of expression of CP in the plant. This could be attained using antisense CP or non-translatable gene constructs. The "homology bases gene silencing" is a transgenic approach where the RNA-mediated resistance is used with an inducible cellular RNA surveillance mechanism against viruses containing sequences homologous to the transgene (Beachy et al. 1990). The transgene which is developed includes movement protein gene, CP, and nuclear inclusion protein gene (NIb). In potato, multiple genes which are derived from various viruses have been effectively introduced into potato concurrently for multi-virus resistance. The dsRNA derived from the 3′ terminal part of the CP of *Potato virus Y* (PVY) is highly conserved among the different strains of PVY which include PVY^{N}, PVY^{O}, and PVY^{NTN}. The use of the multigene transgenic approach is also reported to be effective against a set of viral diseases in potato plants. Three partial gene sequences derived from ORF2 gene of PVX, CP gene of PVLRV, and the helper component protease gene of PVY were designed in a chimeric vector to develop a broad-spectrum transgenic potato cultivar. The potato cultivar was found to be resistant and effective against all three viruses (Arif et al. 2011).

Another approach in providing resistance against viral diseases is the introduction of CP gene either in full length or in truncated/shortened constructs. These CP genes then express in transgenic plants against viral infection. The aforementioned strategy of providing resistance against viral infection is based on the concept of pathogen-derived resistance where the introduction of viral sequence in the plant could interfere with the life cycle of the virus (during the infection stage) which later provides resistance against a particular or group of virus attacking plant (Kumar et al. 2020).

12.4.6.2 Movement Protein-Mediated Resistance

Movement protein-mediated viral resistant is another method for combating viruses in the infected plant. This method is effective against the reduction of virus movement in the plant thus providing a tolerance mechanism against the plant (Chauhan et al. 2019). The pathogenicity of the virus (virulence) is determined by the efficiency of movement of the virus from cell to cell and in another case the host range of plant virus (Waigmann et al. 2004). The advantage of this strategy over other strategies is that this approach offers a striking possibility to deliberate the broad-spectrum resistance against other kinds of viruses infecting plants. The plant may acquire resistance to viral infection by reducing the efficiency of transport function and the rate of virus movement. Synthesis of nonfunctional, degraded, and partially active movement proteins in transgenic plant is reported to confer resistance to viral infection. The mechanism behind acquiring resistance against the virus is that the modified movement protein competes with wild-type virus-coded movement protein (Kumar et al. 2019).

Plasmodesmata in plants act as the entry point for the virus movement between the cells which is known as a symplastic movement. Movement proteins were shown to accumulate in the plasmodesma junction and thus interact with virus movement protein, helping in the progression of viral infection. Thus, the study of plasmodesmata and their composition and protein targeting in plant cells is important for providing resistance against viral infection. Cooper et al. (1995) reported that the mutant movement protein of TMV also provides resistance against potex, cucumo, and tobraviruses. More interestingly, the transgenic potato plant of PLRV expressing movement protein also reported the resistance against PVX and PVY (Tacke et al. 1996).

12.4.6.3 Using Antisense RNA Technology

Another approach is the use of antisense RNA technology to provide resistance against viral infection. Along with this method, catalytic RNAs like the ribozyme method of providing resistance against viral infection are also an effective technique. However, the in vivo method of ribozyme-mediated virus degradation is not as effective as the in vitro method, and it requires refinement of field application. The classical example of the use of antisense constructs targeted against the viral coat protein and replicase enzyme has been successfully used against PVX, PVS, and PLRV (Kawchuk et al. 1991). These aforementioned methods induce resistance against viruses almost at par with that of CP-mediated virus resistance.

The use of RNAi-mediated virus resistance was first demonstrated by Waterhouse et al. (1998) in transgenic tobacco plants against PVY. The advantage of RNA-mediated resistance over other methods is that it provides the transgenic plant more durability than protein-mediated resistance.

12.5 Integrated Management of Potato Viruses

No single approach as outlined above will yield a desirable result. A combination of them or most of them will be the only lasting solution. Schedule of integrated control of potato viruses includes inspection of seed production areas and rejection of fields with a mosaic incidence higher than the prescribed level, killing of vines of seed crop at the recommended date or earlier and not allowing re-growth of vines, destroying volunteer potato plants and weeds in and around the seed crop, monitoring the population of vectors and application of insecticides to keep the aphid vectors below the critical level, use of properly disinfected tools, maintaining proper

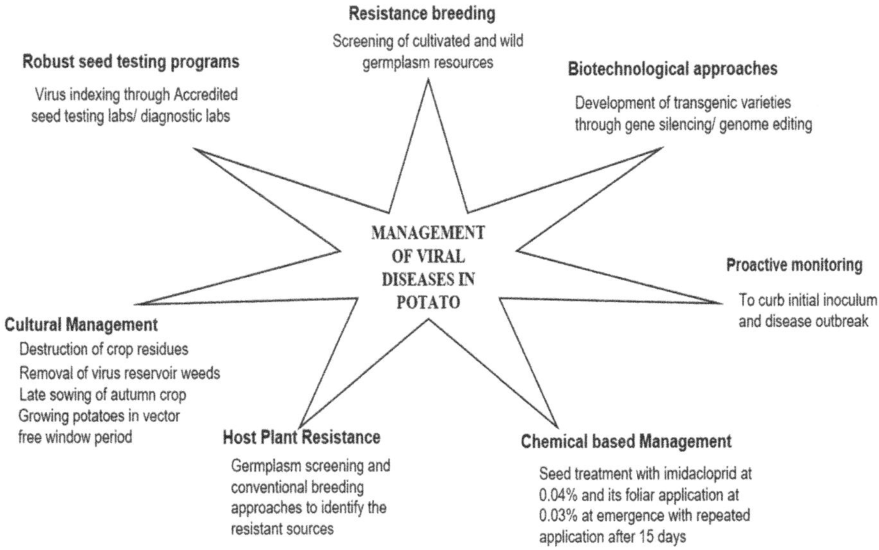

Fig. 12.1 Possible management strategies for viruses and viroid of potato

isolation of the seed crop from virus sources, use of the best-quality certified seed tubers for planting, avoiding the use of cut tubers as seed for seed crop, planting seed crop at a specified period to avoid exposure of the crop to the vectors, minimizing chances of virus spread through farm machinery, and stopping irrigation 10–15 days before harvest to allow skin curing (Fig. 12.1). Due to the differences in mode of spread and perpetuation of viruses, different indirect and direct control measures have to be adopted. It is normally done through an integrated package. Indirect measures of control of viruses and viroid are more important, while direct control measures depend mainly upon the cultivation of virus-resistant varieties. The best policy for management is to prevent viral infection of seed crops/stocks. Reliable detection methods have a great significance in the production of high-quality (virus-free) seed potatoes. To achieve this goal, detailed information on various aspects about nature of virus; the mode of transmission; health standards of the planting materials, which may act as an internal source of the virus spread in the crop; weed hosts which may act as external sources of the viruses; and factors affecting the build-up of the vector and viral diseases must be available.

12.6 Conclusion and Future Outlook

The importance of the virus as a serious threat in global potato production has increased consistently in developed as well as developing nations. Trade globalization, emergence of new strains, changing vector dynamics, and implementation of cropping system are some of the major factors which aggravate viral diseases in potatoes. The most common and effective method of pathogen management, viz., chemical control, is also ineffective against viruses due to the acellular nature of viruses and its lack of physiology. Indirect methods of managing viral diseases have to involve cultural practices like crop rotation, plant population, date of planting, etc. These management tools are most promising in minimizing virus spread under field conditions. If these cultural practices are well planned, they are for the most part low-cost management tactics aimed at minimizing vector populations and subsequent virus spread. These practices should be factored into management decisions that are made before crop production season. Plant resistance in crop plants having transmission through vegetative plant material has a great advantage for component that it usually enhances the effectiveness of other virus management measures. Plant resistance can reduce viral infection and disease development, and a large number of disease resistance genes were identified throughout the world. It is well documented that host plant and vector resistance are the most effective control measures against certain seed-borne diseases. The usefulness and success of the resistance strategy depend on our knowledge of the mechanism(s) of resistance and its effects on the virus-vector-host interactions. More complex and probably more durable resistance can be more difficult to establish and certainly more difficult to achieve by conventional plant breeding. This does not preclude their existence or possible future utilization. Paradoxically, application of knowledge of the genetics of major gene resistance in breeding programs may have militated against breeding for polygenic resistance by more empirical approaches. Novel methods for revealing or creating variation and for transferring it between genotypes by nontraditional methods are already available and are increasingly being applied to resistance genes. Biotechnologies should be seen not only as a means of solving problems when traditional methods have failed but also as a way of generating a better understanding of crop plants and the genes of plant pathogens through the cooperation of scientists from different disciplines, who for the first time are basing the model for their studies on plants.

References

Abbas A, Arif M, Ali MA (2016) Review paper on *Potato leaf roll virus* (PLRV) of potato in Pakistan. Asian J Agric Biol 4:77–86

Agindotan BO, Shiel PJ, Berger PH (2007) Simultaneous detection of potato viruses, PLRV, PVA, PVX and PVY from dormant potato tubers by TaqMan® real-time RT-PCR. J Virol Methods 142:1–9

Ali MA, Nasiruddin KM, Haque MS, Faisal SM (2013) Virus elimination in potato through meristem culture followed by thermotherapy. SAARC J Agric 11(1):71–80

Al-Musa A (1982) Incidence, economic importance, and control of tomato yellow leaf curl in Jordan. Plant Dis 66(7):561–563

Ansar M, Akram M, Singh R, Pundhir V (2015) Epidemiological studies of stem necrosis disease in potato caused by Groundnut bud necrosis virus. Indian Phytopathol 68(3):321–325

Arif M, Azhar U, Arshad M, Zafar Y, Mansoor S, Asad S (2011) Engineering broad- spectrum resistance against RNA viruses in potato. Transgenic Res 21:303–311

Baldo NH, Elhassan SM, Elballa MM (2010) Occurrence of viruses affecting potato crops in Khartoum State-Sudan. Potato Res 53(1):61–67

Beachy RN, Loesch-Fries S, Tumer NE (1990) Coat Protein-mediated resistance against virus infection. Annu Rev Phytopathol 28:451–472

Brunt AA (2001) Potato Virus M (PVM; Genus Carlavirus). In: Virus and virus-like diseases of potatoes and production of seed-potatoes. Springer, Cham, pp 101–107

Cassells AC, Long RD (1982) The elimination of potato viruses X, Y, S and M in meristem and explant cultures of potato in the presence of virazole. Potato Res 25:165–173

Chauhan P, Singla K, Rajbhar M, Singh A, Das N, Kumar K (2019) A systematic review of conventional and advanced approaches for the control of plant viruses. J Appl Biol Biotech 7(04):89–98. https://doi.org/10.7324/JABB.2019.70414

Chikh Ali M, Maoka T, Natsuaki KT (2008) The occurrence of potato viruses in Syria and the molecular detection and characterization of Syrian Potato virus S isolates. Potato Res 51:151–161

Chiunga E, Valkonen JPT (2013) First report of five viruses infecting potatoes in Tanzania. Plant Dis 97:1260

Cloyd RA, Chiasson H (2007) Activity of an essential oil derived from *Chenopodium ambrosioides* on greenhouse insect pests. J Econ Entomol 100(2):459–466. https://doi.org/10.1603/0022-0493(2007)100[459:AOAEOD]2.0.CO;2

Cooper B, Lapidot M, Heick M, Dodds JA, Beachy RN (1995) A defective movement protein of TMV in transgenic plants confers resistance to multiple viruses whereas the functional analog increases susceptibility. Virol 206:307–313

Dahal K, Li X-Q, Tai H, Creelman A, Bizimungu B (2019) Improving potato stress tolerance and tuber yield under a climate change scenario – a current overview. Front Plant Sci 10:563. https://doi.org/10.3389/fpls.2019.00563

Devaux A, Goffart JP, Petsakos A, Kromann P, Gatto M, Okello J, Suarez V, Hareau G (2020) Global food security, contributions from sustainable potato agri-food systems. In: Potato Crop. Springer, Cham, pp 3–35

Duan G, Zhan F, Du Z, Ho SYW, Gao F (2018) Europe was a hub for the global spread of potato virus S in the 19th century. Virology 525:200–204

Faccioli G, Colombarini A (1996) Correlation of potato virus S and virus M contents of potato meristem tips with the percentage of virus-free plantlets produced in vitro. Potato Res 39:129–140

Fuentes S, Gibbs AJ, Adams IP, Wilson C, Botermans M, Fox A, Kreuze J, Boonham N, Kehoe MA, Jones RAC (2021) Potato Virus A Isolates from Three continents: their biological properties, phylogenetics, and prehistory. Phytopathology 111(1):217–226. https://doi.org/10.1094/PHYTO-08-20-0354-FI

Gil JF, Adams I, Boonham N, Nielsen SL, Nicolaisen M (2016) Molecular and biological characterisation of *Potato mop-top virus* (PMTV, Pomovirus) isolates from potato growing regions in Colombia. Plant Pathol 65:1210–1220

Gong H, Igiraneza C, Dusengemungu L (2019) Major in vitro techniques for potato virus elimination and post eradication detection methods. A review. Am J Potato Res 96:379–389. https://doi.org/10.1007/s12230-019-09720-z

Gopal J, Garg ID (2011) An efficient protocol of chemo cum thermotherapy for elimination of potato (*Solanum tuberosum*) viruses by meristem-tip culture. Indian J Agric Sci 81(6):544–549

Green KJ, Quintero-Ferrer A, Chikh-Ali M, Jones RAC, Karasev AV (2020) Genetic diversity of nine non-recombinant potato virus y isolates from three biological strain groups: Historical and geographical insights. Plant Dis 104:2317–2323

Gutiérrez PA, Alzate JF, Marín-Montoya MA (2013) Complete genome sequence of a novel potato virus S strain infecting Solanum phureja in Colombia. Arch Virol 158:2205–2208

Harahagazwe D, Condor B, Barreda C et al (2018) How big is the potato (*Solanum tuberosum* L.) yield gap in Sub-Saharan Africa and why? A participatory approach. Open Agric 3(1):180–189

He Z, Gan H, Liang X (2019) Analysis of Synonymous codon usage bias in potato virus m and its adaption to hosts. Viruses 11(8):752. https://doi.org/10.3390/v11080752

Hemenway C, Fang RX, Kaniewski WK, Chua NH, Tumer NE (1988) Analysis of the mechanism of protection in transgenic plants expressing the potato virus X coat protein or its antisense RNA. EMBO J 7:1273–1280

Jain RK, Khurana SMP, Bhat AI, Chaudhary V (2004) Nucleocapsid protein gene sequence studies confirm that potato stem necrosis is caused by a strain of groundnut bud necrosis virus. Indian Phytopathol 57:169–173

Jeevalatha A, Kaundal P, Kumar R, Raigond B, Gupta M, Kumar A, Sharma S, Sagar V, Nagesh M, Singh BP (2016) Analysis of the coat protein gene of Indian Potato virus X isolates for identification of strain groups and determination of the complete genome sequence of two isolates. Eur J Plant Pathol 145:447–458

Jeevalatha A, Kaundal P, Kumar R, Raigond B, Kumar R, Sharma S, Chakrabarti SK (2018) Optimized loop-mediated isothermal amplification assay for Tomato leaf curl New Delhi virus-[potato] detection in potato leaves and tubers. Eur J Plant Pathol 150(3):565–573

Jeevalatha A, Singh BP, Kaundal P, Kumar R, Raigond B (2014) RCA-PCR: a robust technique for the detection of Tomato leaf curl New Delhi virus-[potato] at ultra-low virus titre. Potato J 41: 76–80

Katsarou K, Wu Y, Zhang R, Bonar N, Morris J, Hedley PE et al (2016) Insight on genes affecting tuber development in potato upon *Potato spindle tuber viroid* (PSTVd) infection. PLoS One 11(3):e0150711. https://doi.org/10.1371/journal.pone.0150711

Kawchuk LM, Martin RR, McPherson J (1991) Sense and antisense RNA-mediated resistance to potato leaf roll virus in Russet Burbank potato plants. Mol Plant-Microbe Interact 4:247–253

Kehoe MA, Jones RAC (2016) Improving Potato virus Y strain nomenclature: lessons from comparing isolates obtained over a 73-year period. Plant Pathol 65:322–333

Khurana SMP, Singh MN (1988) Yield loss potential of potato viruses X and Y in Indian potatoes. J Indian Potato Assoc 15:27–29

Kowalska A, Waś M (1976) Detection of potato virus M and potato virus S on test plants. Potato Res 19:131–139. https://doi.org/10.1007/BF02360415

Kreuze JF, Souza-Dias JAC, Jeevalatha A, Figueira AR, Valkonen JPT, Jones RAC (2020) Viral diseases in potato. In: Campos H, Ortiz O (eds) The potato crop. Springer, Cham. https://doi.org/10.1007/978-3-030-28683-5_11

Kumar R, Jeevalatha A, Baswaraj R, Tiwari RK (2020) Viral and viroid diseases of potato and their management. In: Singh AK, Chakrabarti SK, Singh B, Sharma J, Dua VK (eds) Potato science & technology for sub tropics, 1st edn. New India Publishing Agency, New Delhi, India, pp 267–292

Kumar R, Tiwari RK, Jeevalatha A, Kaundal P, Sharma S, Chakrabarti SK (2019) Potato viruses and their diagnostic techniques-an overview. J Pharm Phytochem 8:1932–1944

Lin YH, Abad JA, Maroon-Lango CJ, Perry KL, Pappu HR (2014) Molecular characterization of domestic and exotic potato virus S isolates and a global analysis of genomic sequences. Arch Virol 159:2115–2122

Lozoya-Saldaña H, Abelló JF, De La García RG (1996) Electrotherapy and shoot tip culture eliminate potato virus X in potatoes. Am Potato J 73:149–154

Lozoya-Saldaña H, Merlin-Lara O (1984) Thermotherapy and tissue culture for elimination of potato virus X (PVX) in Mexican potato cultivars resistant to late blight. Am Potato J 61:735–739

Matousek J, Piernikarczyk RJJ, Dedic P, Mertelik J, Uhlirova K, Duraisamy GS, Orctova L, Kloudova K, Ptacek J, Steger G (2014) Characterization of Potato spindle tuber viroid (PSTVd) incidence and new variants from ornamentals. Eur J Plant Pathol 138:93–101

Meylbodi DE, Mozafari J, Babaeiyan N, Rahimian H (2011) Application of electrotherapy for the elimination of potato *Potyvirus*. J Agric Sci Technol 13:921–927

Mowry TM (2005) Insecticidal reduction of Potato leafroll virus transmission by *Myzus persicae*. Ann Appl Biol 146:81–88

Naik PS, Buckseth T (2018) Recent advances in virus elimination and tissue culture for quality potato seed production. Biotechnol Crop Improv 1:131–158

Nie X, Dickison V, Singh M, de Koeyer D, Xu H, Bai Y, Hawkins G (2020) Potato tuber necrosis induced by alfalfa mosaic virus depends on potato cultivar rather than on virus strain. Plant Dis 104:340–347

Ozkaynak E (2020) Development of PVX resistant potato breeding lines using marker-assisted selection dergipark.org.tr 25: 66–73

Peiman M, Xie C (2006) Sensitive detection of potato viruses, PVX, PLRV and PVS, by RT-PCR in potato leaf and tuber. Australas Plant Dis Notes 1:41

Pradel W, Gatto M, Hareau G, Pandey SK, Bhardway V (2019) Adoption of potato varieties and their role for climate change adaptation in India. Clim Risk Manag 23:114–123

Pundhir VS, Akram M, Ansar M, Rajshekhara H (2012) Occurrence of stem necrosis disease in potato caused by *Groundnut bud necrosis virus* in Uttarakhand. Potato J 39(1):81–83

Pushkarnath (1967) Seed potato production in the sub-tropical plains of India. American Potato J 44:429–441

Puurand U, Makinen K, Paulin L, Saarma M (1994) The nucleotide sequence of potato virus A genomic RNA and its sequence similarities with other potyviruses. J Gen Virol 75:457–461

Santala J, Samuilova O, Hannukkala A, Latvala S, Kortemaa H, Beuch U, Kvarnheden A, Persson P, Topp K, Ørstad K, Spetz C, Nielsen S, Kirk H, Budziszewska M, Wieczorek P, Obr A, Pospieszny H, Valkonen J, Klein Flottbek B et al (2010) Detection, distribution and control of Potato mop-top virus, a soil-borne virus, in northern Europe. PMTV Intensive Ann Appl Biol 19:163–178

Santillan FW, Fribourg CE, Adams IP, Gibbs AJ, Boonham N, Kehoe MA, Maina S, Jones RAC (2018) The biology and phylogenetics of potato virus s isolates from the Andean region of South America. Plant Dis 102:869–885

Singh RP, Somerville TH (1983) Effect of storage temperatures on potato virus infectivity levels and serological detection by ELISA. Plant Dis 67:1133–1136

Singh RP, Valkonen JPT, Gray SM, Boonham N, Jones RAC, Kerlan C, Schubert J (2008) Discussion paper: The naming of *Potato virus Y* strains infecting potato. Arc Virol 153:1–13

Solomon-Blackburn R, Barker H (2001) Breeding virus resistant potatoes (*Solanum tuberosum*): a review of traditional and molecular approaches. Heredity 86:17–35. https://doi.org/10.1046/j.1365-2540.2001.00799.x

Tacke E, Salamini F, Rohde W (1996) Genetic manipulation of potato for broad spectrum protection against virus infection. Nat Biotechnol 14:1597–1601

Taliansky M, Mayo MA, Barker H (2003) Potato leafroll virus: a classic pathogen shows some new tricks. Mol Plant Pathol 4(2):81–89

Thomas-Sharma S, Abdurahman A, Ali S, Andrade-Piedra JL, Bao S, Charkowski AO, Crook D, Kadian M, Kromann P, Struik PC, Torrance L, Garrett KA, Forbes GA (2016) Seed degeneration in potato: the need for an integrated seed health strategy to mitigate the problem in developing countries. Wiley Online Libr 65:3–16

Tsedaley B (2014) Review on early blight (*Alternaria* spp.) of potato disease and its management options. J Biol Agric Healthc 4(27):191–199

Usharani KS, Surendranath B, Paul-Khurana SM, Garg ID, Malathi VG (2004) Potato leaf curl - a new disease of potato in northern India caused by a strain of Tomato leaf curl New Delhi virus. Plant Pathol 53:235–235

Verma AK, Vashisth KS (1985) Manipulation of planting and haulms cutting dates for better health standards in seed potato crop. Indian J Virol 1:54–60

Waigmann E, Ueki S, Trutnyeva K, Citovsky V (2004) The Ins and outs of nondestructive cell-to-cell and systemic movement of plant viruses. CRC Crit Rev Plant Sci 23:195–250

Wang J, Meng F, Chen R, Liu J, Nie X, Nie B (2016) RT-PCR differentiation, molecular and pathological characterization of Andean and ordinary strains of potato virus S in potatoes in China. Plant Dis 100:1580–1585

Wang Q, Liu Y, Xie Y, You M, Wang Q, Xie ÁY, You ÁM, Liu Y (2006) Cryotherapy of potato shoot tips for efficient elimination of *Potato leafroll virus* (PLRV) and *Potato virus Y* (PVY). Potato Res 49:119–129

Waterhouse PM, Graham MW, Wang MB (1998) Virus resistance and gene silencing in plants can be induced by simultaneous expression of sense and antisense RNA. Proc Natl Acad Sci U S A 1095(23):13959–13964. https://doi.org/10.1073/pnas.95.23.13959

Xu H, DeHaan TL, De Boer SH (2004) Detection and confirmation of potato mop-top virus in potatoes produced in the United States and Canada. Plant Dis 88:363–367

Xu H, Nie J (2006) Identification, characterization, and molecular detection of *Alfalfa mosaic virus* in potato. Am Phytopath Soc 96:1237

Zaidi SSEA, Martin DP, Amin I, Farooq M, Mansoor S (2017) Tomato leaf curl New Delhi virus: a widespread bipartite begomovirus in the territory of monopartite begomoviruses. Mol Plant Pathol 18:901–911

Serological and Molecular Diagnosis of Potato Viruses: An Overview

13

Baswaraj Raigond, Gaurav Verma, Ravinder Kumar, and Rahul Kumar Tiwari

Abstract

Potatoes are generally propagated by vegetative mode where the viral pathogens tend to accumulate over successive generations, which leads to degeneration of planting material, limiting the productivity and quality of potato. The development of rapid and precise diagnostic assays for the detection of potato viruses is absolutely necessary. It will assist in the production of virus-free produce and also ensure the restricted entry of quarantine viruses through germplasm and other propagative materials across the world. In early days, the viruses/diseases were identified based on symptoms expressed by potato plants. The limitations of visual diagnosis were overcome by the application of serological assays (ELISA) for the detection of viruses. Later, with the advancement in molecular biology, more sensitive techniques like PCR, quantitative PCR (qPCR), and NGS have evolved as promising methods for detecting and discovering new viruses. However these techniques remained limited to well-equipped laboratories. Hence more recently, novel isothermal assays, i.e., loop-mediated isothermal amplification (LAMP) and recombinase polymerase amplification, have been developed and applied for the detection of potato viruses in low-resource laboratories and also for near-site detection. In this chapter, an overview of potato virus research and evolution of diagnostics techniques, followed by a brief narrative on serological and molecular methods that are commonly used for the detection of potato viruses, is provided.

Keywords

Potato · Virus · Antibody · Diagnostics · ELISA · Lateral flow · PCR · NGS · Loop · Isothermal · Recombinase

---

B. Raigond (✉) · G. Verma · R. Kumar · R. K. Tiwari
ICAR-Central Potato Research Institute, Shimla, Himachal Pradesh, India

S. Kumar Chakrabarti et al. (eds.), *Sustainable Management of Potato Pests and Diseases*, https://doi.org/10.1007/978-981-16-7695-6_13

13.1 Introduction

Potato (*Solanum tuberosum*) is currently the third most important food crop globally after rice and wheat. The production of potato is vulnerable to several biotic and abiotic factors. The diseases caused by biotic factors, such as fungal, bacterial, and viral pathogens, are a major threat, which can cause a considerable yield loss and reduce the quality of the tubers (Khurana and Singh 1988). Among the pathogens, viral diseases are a major concern in healthy seed potato production system as they can cause a decline in yield and tuber quality or can lead to degeneration of seed potatoes. Globally, several viruses were reported to infect potato crop, of which few are responsible for serious yield and economic losses (Horvâth 1967; Martyn 1968). The rest of the viruses have either regional relevance or minor importance across the world. *Potato virus Y* (PVY) and *potato leaf roll virus* (PLRV) that infect potato are the most damaging viruses spread across the world. In the last two to three decades, PVY has evolved into several biological strains (PVY^{N}, PVY^{N}, PVY^{C}, PVY^{Z}, and PVY^{NTN}) and caused significant economical losses in Europe and North America, as well as Asia and South America (Singh et al. 2008; Verbeek et al. 2009; Scholthof et al. 2011; Kehoe and Jones 2016; Fuentes et al. 2019). As such, PVY has gained more importance over PLRV over time. In addition to viruses, potato spindle tuber viroid (PSTVd) is also an important quarantine plant pathogen in most countries. In the Indian scenario, the viruses like PVY, PLRV, *tomato leaf curl New Delhi virus* (ToLCNDV) and *groundnut bud necrosis virus* (GBNV) are a major threat in potato production (Jain et al. 2004; Pundhir et al. 2012; Jeevalatha et al. 2017; Raigond et al. 2017). The other major viruses present in India are *potato virus X* (PVX), *potato virus S* (PVS), *potato virus A* (PVA), and *potato virus M* (PVM).

The diagnosis of viruses is greatly important to ensure the safe movement of germplasms, seeds, and other propagative materials across the world, as well as germplasms across the borders through, national quarantine services, which can guarantee quality potato seed production system. Further, to achieve an effective management and to prevent further spread of the viral diseases of potato, early, reliable, and accurate detection and diagnosis of the virus and virus-like agents are absolutely necessary. In the last decade, much effort has been exerted on the development of rapid and sensitive detection methods for specific potato viruses and their strains. In this chapter, an overview of potato virus research and evolution of diagnostic techniques, followed by a brief narrative on serological and molecular methods that are commonly used for the detection of potato viruses, is provided.

13.2 Significance/Need of Developing Diagnostic Assays for Potato Viruses

Potatoes are propagated via both the sexual and asexual (vegetative) modes, where the vegetative mode of propagation, i.e., via tuber, is mostly followed. Among the pathogens, viral pathogens are the obligate parasites that are viable only in living cells and pass more easily from one generation to another through seed tubers or other vegetative propagules like tissue culture based planting material. In case of sexually propagated crops, the viral pathogens are filtered out in the process of sexual seed production. Contrarily, in vegetatively propagated crops like potato, the viral pathogens tend to accumulate over successive generations. The successive accumulation of viral pathogens in vegetative-propagated crops (potato) leads to the degeneration of planting materials. This is one of the major problems that limit the productivity and quality of potato. In addition, the worldwide movement of potato germplasm in the form of seed or other propagative materials and the threat posed by the new viruses have increased. Moreover, the viral diseases cannot be managed directly by chemical pesticides. Therefore, pathogen detection and exclusion constitute a major step for the production of healthy (virus-free) produce and the avoidance of introducing new viruses through germplasm exchange and international trade. This will indirectly manage the incidence of viral diseases of potato. Therefore, it has always been a priority to develop robust and sensitive diagnostic assays for the detection of potato viruses.

13.3 Evolution of Potato Virus Diagnostics: A Brief Historical Journey

In the early days, the plant viruses were generally identified through visual inspection of disease symptoms, which constituted a major step for virus detection. But the major limitation was that the symptoms can be variable, and also a similar symptom could be observed due to certain nutrient deficiencies or other abiotic/biotic factors. In addition, virus-infected potato plants may not exhibit distinguishable symptoms when infected with PVX, PVA, PVA, and PVM in certain varieties. To a certain extent, this was overcome by using indicator hosts. Under standardized controlled conditions, the diagnostic host plants when infected/inoculated with a particular virus will express consistent and characteristic disease symptoms. Several herbaceous plants that can express local or systemic symptoms are used for the detection and diagnosis of potato viruses. This is considered to be an accurate technique for diagnosing a disease that is still used for diagnosing also some viruses and viroids. But this technique is not suitable for handling a large number of samples in a short period of time. In addition, it demands more time, labor, and space. Therefore, during the 1960s through the 1970s, serological techniques and histochemical tests were standardized for the detection of viruses and phytoplasma diseases. Among the serological tests, chloroplast agglutination, microprecipitation, and gel diffusion were introduced. But, to increase the sensitivity, enzyme-linked immunosorbent assay (ELISA) emerged as the most suitable serological approach and became a standard laboratory-based virus testing method. The technique was first applied for the detection of plant virus by Clark and Adams in 1977. Since the technique is simple and sensitive, it is widely used for the detection of plant virus at a large scale. However, it has some drawbacks, such as being labor-intensive, chances of cross-contamination across the wells (numerous washing steps), requirement of a large sample volume, and sensitivity being affected by high background readings.

During the same period, the introduction of transmission electron microscopy (TEM) in life sciences, it played a crucial role in detection and its morphological characterization of virus particle (Derrick 1973; Pares and Whitecross 1982; Lin and Langenberg 1983; Lin 1984; Garg et al. 1989; Singh et al. 1990; Milne 1992). But this technique could be difficult to use as all the laboratories cannot afford to have a costly equipment, i.e., TEM.

Polymerase chain reaction (PCR), which was developed in 1983 by Kary Mullis, was utilized for the detection of plant viruses in 1990 (Vunsh et al. 1990). Being a highly sensitive technique, the technology gained importance in plant virus detection and diagnosis. Its sensitivity depends on the design of oligonucleotide primers targeting the specific region of the viral DNA. To date, PCR/RT-PCR has been used to detect almost all the viruses that infect potato. Even though PCR is sensitive, it demands a well-quipped laboratory and well-qualified manpower. Since it is laborious and time-consuming, a limited number of samples can be processed. Meanwhile, with the introduction of nanotechnology in the field of diagnostics and with the increasing demand for rapid and onsite (in-field) detection of plant viruses, lateral flow immunoassay (LFIA) was developed. The assay can confirm the presence or absence of a particular virus from suspected plant tissues in only 15–20 min (Byzova et al. 2009; Panferov et al. 2018).

With the advancement of science, the next-generation sequencing (NGS) technology has evolved as a promising technology to discover and diagnose viruses. This technology involves nucleic acid extraction (RNA) and high-throughput sequencing approaches, followed by analysis of the sequenced data by bioinformatics. But the viruses with low titer may not be detected as a huge amount of RNA sequence corresponding to the host gets wasted (Mlotshwa et al. 2008; Adams et al. 2009; Kreuze et al. 2009; Rwahnih et al. 2009; Wylie and Jones 2011; Li et al. 2012; Sela et al. 2012; Kreuze et al. 2013; Song et al. 2013). One of the limitations could be that the technique demands a well-equipped laboratory to carry out NGS. However, this can be eliminated by outsourcing the same work. The NGS data for virus discovery and diagnosis in the form of bioinformatics could be a major limitation before bringing the NGS technology in regular or routine viral disease detection and diagnostics.

Despite all the beauty of PCR and NGS-based detection technologies, they are restricted within the walls of sophisticated laboratories. Hence, these techniques are not applied in low-resource laboratories. Therefore, looking at the need or significance of rapid, onsite/near-site, and sensitive (molecular) detection techniques that can be applied in low-resource laboratories is essential. To address this issue, new isothermal amplification technologies like loop-mediated isothermal amplification (LAMP) (Notomi et al. 2000) and recombinase polymerase amplification (Piepenburg et al. 2006) have been developed, which can break through the boundary of traditional laboratory. These assays can provide nucleic acid replication at a single constant temperature in a short period of time, i.e., within 30–50 min. In the following sections, all the aforementioned techniques for virus diagnosis are discussed.

In the Indian perspective, Galloway (1936) was the first to record the occurrence of a potato virus. The authentic records on the occurrence of potato viruses X, Y, and A were made by pioneers like B.P. Pal, R.S. Vasudeva, S.B. Lal, Ramamoorthy, and R.N Azad during the 1940s to 1950s. The ICAR-Central Potato Research Institute, Shimla, has started a regular intensive research on potato viruses during the 1960s, together with Dr. B.B. Nagaich as the virologist. Later on, Dr. S.M. Paul Khurana, I.D. Garg, and colleagues developed and implemented the virus/viroid diagnostic techniques like ELISA and immunosorbent electron microscopy (ISEM) (Khurana 1999). This was followed by the development of molecular assays, i.e., PCR, RT-PCR, qPCR, and qRT-PCR. Soon after, RT-LAMP and RPA for rapid and sensitive detection of viruses and viroid infection in potatoes were also developed. The developed diagnostic techniques were successfully used to clean all the mother stocks of important popular potato varieties and many advanced hybrids. These techniques were also used for the post-entry quarantine testing of imported germplasm. The diagnostic techniques that developed/evolved over the years have improved the health status of breeder seed, which in turn helped India expand its potato production.

13.4 Techniques for the Detection and Identification of Potato Viruses

Potato suffers from infection with several viral pathogens. To get rid of these viral pathogens, techniques for the detection and identification of such pathogens were developed; early and accurate diagnosis of viral diseases is important for their effective management. Now, we present an overview of serological and molecular diagnostic techniques, which are used for the detection and identification of viruses that infect potato.

13.4.1 Serological Techniques

Plant viruses are generally strong immunogens, and they stimulate the production of virus-specific antibodies when injected to experimental animals. The antibodies produced can be utilized in various serological tests. The antigenicity of virus particle and the amount of highly purified virus particle immunized determines the titer of the antisera. The antibodies produced against host proteins (impurities) can be removed from the virus-specific antiserum via cross-adsorption with the host proteins. Polyclonal and monoclonal antibodies were successfully produced for the majority of viruses infecting potato. However, the production of monoclonal antibodies is labor-intensive. Among the serological tests, chloroplast agglutination, microprecipitation tests, and gel diffusion test were used in the early days. These techniques require a large volume of antisera; in addition, they have low sensitivity and cannot be used to detect viruses with low titer. Hence, these tests are no longer used and therefore will not be discussed here. To increase the sensitivity of the tests, a solid phase to adsorb an antibody/antigen followed by the attachment of an antigen/antibody with an enzyme conjugate to detect the presence of a specific antigen/virus has emerged as an important test for the detection of plant viruses. Important variants of these tests are described as follows.

13.4.1.1 Enzyme-Linked Immunosorbent Assay (ELISA)

ELISA was first used to detect plant viruses by Clark and Adams in 1977. Henceforth it has been increasingly used and became the most widely adopted method for the detection of potato viruses as well. There are variants available in ELISA, the most popular of which are double-antibody sandwich (DAS-ELISA) and direct antigen coating (DAC-ELISA); DAS-ELISA is widely adopted.

DAS-ELISA involves trapping of virus particles on specific antibody adsorbed on a solid surface, followed by binding of virus particles with enzyme-antibody conjugate. In case the sample is infected with the virus, the enzyme-antibody conjugate will bind the trapped particles, which are confirmed by adding the enzyme substrate; a positive sample will produce a color in the well, as in the case of the alkaline phosphatase-antibody conjugate. The reaction is measured either visually or using a spectrophotometer (reader). The advantage of DAS-ELISA is that, the adsorbed antibodies specifically trap the virus of interest, and the other virus particles are removed during washing. Therefore, it is the most commonly used technique for detecting viruses, especially in a complex mixture (Clark and Adams 1977; Goodwin and Banttari 1984; Torrance and Robinson 1989).

Commercial ELISA kits are available for almost all viruses infecting potato. ICAR-CPRI introduced the use of ELISA in 1984 and started producing its own ELISA kits for potato viruses, i.e., PVY, PVA, PVM, PVS, and PVX, in 1990. Pure cultures of the important potato viruses are being maintained at the "virus pure culture" facility in the Division of Plant Protection, CPRI, Shimla. These kits were used for large-scale screening of samples under seed production and certification. To minimize human interference and human errors and to make it more robust, an "automated ELISA washer and reader" facility is generally used.

13.4.1.2 Dot ELISA

The assay works according to the principle of ELISA, in which nitrocellulose or nylon membrane is used to directly spot/trap the purified antigens (virus) or crude sap from infected plant tissues and air-dried. This is followed by saturation of the membrane with bovine serum albumin (BSA) and addition of viral specific antibody. Later a second antibody enzyme conjugate is added, followed by a substrate. The enzyme reacts with the soluble substrate to produce an insoluble color at the reaction site. This technique can be applied for the detection of virus under field conditions. It was also applied for the detection of PVY, PVX, and PVS (Banttari and Goodwin 1985; Kumar and Khurana 1989), PLRV (Smith and Bantarri 1987), and PVM, PVS, PVX, PVY, and PVA (Hans 1988; Kumar and Singh 1999). The crucial element of the assay is a highly specific antigen-antibody interaction, and the concentration of the antigen is indicated by the intensity of the color of the dot.

13.4.1.3 Tissue Blotting and Tissue Squashes

In this case, freshly cut tissues are allowed to blot gently but firmly onto the nitrocellulose membrane. The antigen/virus from the tissues is detected on blots with the help of enzyme-labeled probes. It was applied for the detection of PVX and PVY from potato tubers (Brvo-Almonacid et al. 1992). It is also helpful in detecting the nucleic acid of plant viruses via hybridization of infected plant tissues squashed on a nylon membrane using a specific radioactive probe (Navot et al. 1989). This technique is a simple tool for a specific and rapid detection of plant viruses.

13.4.1.4 Immunosorbent Electron Microscopy (ISEM)

ISEM is a highly sensitive technique for the detection of plant viruses as it involves the combination of electron microscopy and serology. In addition, it can determine the shape and size of the virus particle. It was introduced by Derrick (1973). In this study, the electron microscope grid is coated with virus-specific antibodies, and after incubation it is floated onto the purified/crude sap of the infected plant. This step promotes the binding of homologous virus particles present in the sap and inhibits nonspecific binding of other proteins, which is termed as "trapping." An improved detection of the virus can be achieved by adding a second layer of virus-specific antibodies to the step of "trapping," which is termed as "decoration" (Milne 1992). Finally, the trapped and decorated virus particles can be observed in the TEM, which indirectly confirms the identity of the virus. It was reported to be 10 times more sensitive than conventional leaf-dip electron microscopy in the detection of PLRV. Moreover, it was standardized for the detection of PLRV in vector aphids and tuber sprouts (Garg et al. 1989; Singh et al. 1990).

13.4.1.5 Gold-Labeled Antibody Decoration (GLAD)

GLAD was coined by Pares and Whitecross (1982). In the same year, Lin and Langenberg (1983) used colloidal gold (CG)-labeled IgG for the localization of barley stripe mosaic virus (BSMV) in ultrathin sections of wheat cells. Here, the antigen-antibody reaction can be visualized using CG-labeled antibodies. This technique is more sensitive compared with direct leaf-dip electron microscopy and IEM. It has an additional advantage of being able to detect the viruses under low concentration in the infected tissues.

In 1984, Lin applied and described the use of gold-labeled IgG complexes for the rapid and specific detection of different viruses in different hosts, i.e., tobacco mosaic virus (TMV) in tobacco, BSMV in wheat, cowpea mosaic virus (CPMV) in cowpea, wheat streak mosaic virus (WSMV) in wheat, PLRV in potato, Brome mosaic virus (BMV) in barley, and barley yellow dwarf virus (BYDV) in oats using the simple leaf-dip method. Similarly, Raigond et al. (Raigond et al. 2013a) developed the immunogold electron microscopy technique for the clear detection of *potato viruses A and M* in suspected potato leaf samples. However, the application of TEM is limited to a small number of samples and could be costly.

13.4.1.6 Lateral Flow Immune Assay (LFIA)/Dip Sticks/Onsite Detection

The LFIA is a platform where an interaction between the target analyte (virus) and its antibodies along with their conjugates (colored colloidal particles) is applied on a membrane carrier (lateral flow test strips). When the strip is placed or dipped into the leaf sample crushed in an appropriated buffer, the liquid flows through the membrane with attached molecules, which interact with the analyte, generating a signal that is visually detected in test and reference lines. This assay can be performed within 10–15 min under field conditions. The strips were developed by Byzova et al. (2009) for the detection of viruses with different shapes and sizes, i.e., spherical carnation mottle virus, rod-shaped TMV, bean mild mosaic virus, and filamentous potato viruses X and Y. The test strips are commercially available in few private firms. ICAR-CPRI has also developed LFIA for the detection of PVA, PVX, PVM, PVS, and PVY individually or in combinations, viz., PVA and PVS, PVA and PVX, and PVY and PVM, in a single strip.

The assay is economical, portable, and very user-friendly. Efforts are being exerted on the identification of new signal amplification strategies to enhance the sensitivity of lateral flow assay (LFA), like Panferov et al. (2018) reported the silver-enhance LFIA for a highly sensitive detection of PLRV.

13.4.2 Molecular Techniques

13.4.2.1 Nucleic Acid Hybridization

The technique is based on the pairing of viral nucleic acid with a specific nucleic acid probe immobilized on filter papers, i.e., nitrocellulose, or on nylon. For plant virus diagnosis, the viral nucleic acid is first denatured and immobilized on nitrocellulose or nylon paper. This immobilized paper will be immersed in the solution of labeled probe and kept under the conditions that favor hybridization. Annealing of labeled nucleic acid sequences can occur if the sample contains complementary viral nucleic acid immobilized on filter paper. The endpoint detection would be by using autoradiography, which shows a positive reaction. Nucleic acid hybridization was reviewed by Hull (1993) for the detection of plant viruses. Digoxigenin-labeled dUTP, a nonradioactive tag for labeling probes, was most widely used for the detection of plant viruses. In potatoes, the technique was used successfully for the detection of PLRV in nonsprouting tubers (Loebenstein et al. 1997).

13.4.2.2 Polymerase Chain Reaction (PCR)

Development of PCR in 1983 by Kary Mullis, the first report of adopting it for the detection of plant virus cane in the year 1990 (Vunsh et al. 1990). Thereafter, its application was expanded to the diagnosis of plant viruses and viroids. It enables the detection of a specific virus through the amplification (several million folds) of the specific targeted part of the genome. This method involves three steps: (i) denaturation at temperatures above 90 °C, (ii) annealing of primers at 50–75 °C, and (iii) extension at 72–78 °C. It is performed on a programmable thermal cycler. The most commonly used enzyme is Taq DNA polymerase, which has 5′–3′ nuclease activities. Apart from DNA polymerase, the Mg^{2+} and dNTP concentrations need to be precisely used because Mg^{2+} affects the enzyme activity, and imbalanced dNTP mixtures can lead to reduced polymerase fidelity.

Majority of the plant viruses possess RNA genome and cannot serve as a template for PCR. Therefore, RNA is first converted to complementary DNA (cDNA), which involves the use of *Avian myeloblastosis virus* reverse transcriptase (AMV-RT) and *Moloney murine leukaemia virus* reverse transcriptase (MMLV-RT). The synthesized cDNA will be used as a template in PCR-based amplification. This process is called reverse transcription-polymerase chain reaction (RT-PCR). The amplified product of PCR can be loaded in agarose-based gel electrophoresis and visualized using ultraviolet light. The PCR and RT-PCR protocols for the detection of majority of the viruses infecting potato crops have been developed in addition to their detection in insect vector systems (Singh and Singh 1996; Singh et al. 1996; Nie and Singh 2001; Du et al. 2006; Raigond et al. 2013b; Raigond et al. 2014; Kumar et al. 2017). Over the period of time, several variants of PCR such as immunocapture PCR (IC-PCR), nested PCR, and multiplex RT-PCR have been developed for an improved detection of potato viruses or for differentiating strains of a particular virus.

13.4.2.3 Fluorescent Probe-Based Quantitative PCR (qPCR)

High-end sensitivity and specificity of a diagnostic assay are prerequisites for a precise detection of a pathogen. In line with the different variants of PCR assays, qPCR is generally the gold standard method for the detection of pathogens. Real-time PCR is a method for detecting the presence of a specific genetic material in any pathogens, including a virus. This testing method combines PCR chemistry with fluorescent probe detection of amplified product in the same reaction vessel. The assay not only detects the pathogen at very low concentrations (10 viral copies) (Wang et al. 2015) but also simultaneously determines its presence in the sample, which ultimately discourages endpoint detection as in RT-PCR, which is generally based on agarose gel electrophoresis. Additionally, accelerated PCR thermocycling and detection of amplified product allow the provision of a test result much sooner for real-time PCR than for conventional PCR. The qPCR-based protocols have been developed for the detection of major RNA and DNA viruses infecting potato, including PSTVd, aiming at eliminating the pathogen in sensitive materials (Kogovsek et al. 2008; Sheila et al. 2009; Jeevalatha et al. 2016; Jeevalatha et al. 2015; Raigond et al. 2019; Verma et al. 2020).

Multiplex detection of different viruses is possible with real-time PCR using novel fluorescent probes. The real-time fluorescent probes commonly referred to as TaqMan probes are short oligonucleotides that contain 5′-fluorescent dye and 3-′-quenching dye. Generally, the probe must bind to a complementary strand of DNA, and at a certain temperature, Taq polymerase, the same enzyme used for the PCR, must cleave the 5′-end of the TaqMan probe, separating the fluorescent dye from the quenching dye. With the application of a specific probe in a PCR reaction, the specificity of the assay is increased significantly. Different fluorescent-labeled probes can be incorporated in a single reaction to target more than one virus in order to multiplex the detection assay.

13.4.2.4 Rolling Circle Amplification (RCA)

Rolling circle amplification (RCA) is an isothermal enzymatic activity in which a short DNA primer is amplified to form a long strand of presumably a small single-stranded circular DNA template using a special DNA polymerase. The technique can amplify any circular DNA without the initial knowledge of the sequence (Haible et al. 2006). Due to its simplicity and versatility to amplify the DNA template, it has been adopted for the development of diagnostic methods for a variety of targets. In plant viruses, it was successfully adopted for the detection of geminiviruses. Apart from detection, the RCA-amplified (viral DNAs) products can be used for the cloning and characterization of the geminiviruses. RCA assay in combination with PCR assay has increased the sensitivity and specificity of geminivirus detection. The RCA-PCR assay has been successfully used for the detection of a DNA virus that infects potatoes, i.e., ToLCNDV-potato with a very low titer (Jeevalatha et al. 2014). Even though RCA-PCR is time-consuming and costly compared with PCR, it is of great help in the virus indexing of mother stocks of seed potato.

13.4.3 Next-Generation Sequencing (NGS) in Viral Diagnostics

The NGS technology has revolutionized the field of virus diagnosis and discovery. This technology involves a huge amount of sequencing, followed by bioinformatics analysis of the sequenced viral genome. Several NGS approaches have been published since 2009 to identify plant viruses (Adams et al. 2009; Kreuze et al. 2009). One of the successful approaches reported was sequencing of total mRNA (Rwahnih et al. 2009; Wylie and Jones 2011). But, a huge amount of total RNA of host gets sequenced while the viruses with low titer become difficult to identify or analyze, making it as a drawback. In addition, this approach may not capture the viruses lacking poly A sequences that are used to enrich mRNA. Aside from the total mRNA sequencing approach, several other approaches were reported, of which small RNA sequencing and assembly have been widely adopted. This approach generates a huge amount of relatively short sequences and is used successfully for the identification of many plant viruses, including viruses of potato (Mlotshwa et al. 2008, Li et al. 2012; Sela et al. 2012; Kreuze et al. 2013; Song et al. 2013). Therefore, the NGS technologies can be used as a diagnostic tool to identify a potato virus in an unbiased fashion when no prior knowledge of the etiology of the virus is available.

13.4.4 Loop-Mediated Isothermal Amplification (LAMP) Assay

Recently, an isothermal LAMP assay was reported by Notomi et al. (2000), which was successfully used for the detection of many plant viruses. It is simple, rapid, specific, and highly sensitive and has the potential to replace PCR. It can amplify the target gene at a constant temperature and at a short incubation period of 15–50 min in different apparatus like water bath/dry bath/thermal cycler. This LAMP assay employs four to six primers that can anneal at forward, backward as well as internal sites of the targeted region of the DNA. Gene amplification or LAMP-amplified product can be visualized by the naked eye either as turbidity or in the form of a color change when SYBR Green, a fluorescent dye, is added. The assays have been developed for the detection of major viruses and viroid infecting potatoes, i.e., PVY, PVX, PLRV, PVA, PVS, ToLCNDV-potato, and PSTVd (Ju 2011; Przewodowska et al. 2015; Jeong et al. 2015; Jeevalatha et al. 2018; Raigond et al. 2019; Kumar et al. 2020; Verma et al. 2020; Raigond et al. 2020).

13.4.5 Recombinase Polymerase Assay (RPA)

RPA assay is an ideal method for the detection of plant viruses owing to its simplicity, rapidity, and sensitivity. The assay was first introduced by Piepenburg et al. (2006) in 2006; it is suitable for diagnostic laboratory with limited resources. Most importantly, the assay operates at a low and constant temperature ranging from 37 to 40 °C. It amplifies the target gene in a short period of 20–30 min with a single pair of primers (contrary to LAMP) and without initial denaturation. For endpoint detection, post-reaction or post-amplification treatments are critical for the successful visualization of the amplicons. This is because, protein mixtures in the reaction tube interfere by forming lumps of smears that hinder gel electrophoresis. This is a major limitation in using the gel-based endpoint detection (Babu et al. 2017; Glais and Jacquot 2015). The limitation was overcome by denaturation of protein at 65 °C or 95 °C for 10 min or treatment SDS or enzymatic digestion with proteinase K or through high-speed centrifugation or even by purification using commercially available DNA clean-up kit. With respect to primers, it was reported that even PCR primers can be used in the assay (Mayboroda et al. 2015; Yamanaka et al. 2017).

This technique has been successfully used for the detection of numerous human, animal, and plant pathogens. It has also been successfully used for the detection of several *Begomoviruses* like ToLCNDV-*potato*, *bean golden yellow mosaic virus*, *tomato mottle virus*, and *tomato yellow leaf curl virus* (Londono et al. 2016; Raigond et al. 2021). In addition, it was successfully used to detect RNA viruses like *little cherry virus* 2, *plum pox virus*, and *rose rosette virus* as well as two *Potyviruses*, i.e., *yam mosaic virus* and *yam mild mosaic virus* (Mekuria et al. 2014; Zhang et al. 2014; Babu et al. 2017; Silva et al. 2018).

In addition to detection, RT-RPA has been applied for robust and rapid real-time detection of PVY O and N types from the crude extracts of potato leaves and tubers (Babujee et al. 2019). The RT-RPA assay has also been combined with nucleic acid LFA for the detection of RNA virus, i.e., PVX (Vanov et al. 2020). With respect to sensitivity, the assay was 260-fold more sensitive compared with conventional antibody-based LFA assay. It is equally sensitive when compared with PCR-based detection.

13.5 Conclusion and Future Outlook

Rapid and accurate onsite detection of viral pathogens of potato is essential to take up timely management practices. Development of diagnostic techniques will assist in the production of virus-free produce, thereby reducing the risk of further spread of viral diseases under field conditions and more importantly minimizing yield losses. Globally, serological and molecular assays are being used in virus testing laboratories; however these assays were limited within the boundaries of well-equipped laboratories. Hence, isothermal assays like LAMP and RPA have been developed and applied for the rapid and accurate detection of potato viruses in low-resource laboratories and also for near-site detection.

Despite the significant success in the development of plant virus diagnostics tools, there is a great scope of developing much simplified tools that can have a wider applicability and that are economical, rapid, sensitive, and suitable for onsite detection. In addition, multiple/mixed infections under field conditions pose a big hurdle in the diagnosis of viral diseases of potato, which deserve attention. Insect vectors like aphids, whiteflies, and thrips play a crucial role in the spread of viral inoculum during the copping season. Therefore, rapid and accurate diagnostic assays need to be developed to determine the viruliferous nature of vectors so as to take timely management practices. The NGS technology should be used so as to identify new viral pathogens and variants entering into the crop. This will ensure the preparedness for developing suitable management strategies and in turn reduce yield loss due to viral pathogens.

References

Adams IP, Glover RH, Monger WA, Mumford R, Jackeviciene E, Navalinskiene M, Samuitiene M, Boonham N (2009) Next-generation sequencing and metagenomic analysis: a universal diagnostic tool in plant virology. Mol Pl Path 10:537–545

Babu B, Washburn BK, Miller SH, Poduch K, Sarigul T, Knoxa GW, Ochoa-Corona FM, Paret ML (2017) A rapid assay for detection of rose rosette virus using reverse transcription-recombinase polymerase amplification using multiple gene targets. J Virol Meth 240:78–84

Babujee L, Witherell RA, Mikami K, Aiuchi D, Charkowski AO, Rakotondrafara AM (2019) Optimization of an isothermal recombinase polymerase amplification method for real-time detection of potato virus Y O and N types in potato. J Virol Methods 267:16–21. https://doi.org/10.1016/j.jviromet.2019.02.006

Banttari EF, Goodwin PH (1985) Detection of potato viruses S, X, Y by enzyme-linked immunosorbent assay on nitrocellulose membranes (dELISA). Plant Dis 69:202–205

Brvo-Almonacid F, Hain L, Mentaberry A (1992) Rapid immunological detection of potato viruses in plant tissue squashes. Plant Dis 76:574–578

Byzova NA, Safenkova IV, Chirkov SN, Zherdev AV, Blintsov AN, Dzantiev BB, Atabekov IG (2009) Development of Immunochromatographic test systems for express detection of plant viruses. App Biochem Microbiol 45(2):204–209

Clark MF, Adams AN (1977) Characteristics of the microplate method of enzyme linked immunosorbent assay for the detection of plant viruses. J Gen Virol 34:475–483

Derrick KS (1973) Quantitative assay for plant viruses using serologically specific electron microscopy. Virol 56:652–658

Du Z, Chen J, Hiruki C (2006) Optimization and application of a multiplex RT-PCR system for simultaneous detection of five potato viruses using 18S rRNA as an internal control. Plant Dis 90:185–189

Fuentes S, Jones RAC, Matsuoka H, Ohshima K, Kreuze JF, Gibbs A (2019) Potato virus Y; the Andean connection. Vir Evol 5(2):vez037. https://doi.org/10.1093/ve/vez037

Galloway LD (1936) Virus diseases of potato in India. Scientific Report of Imperial Agricultural Research Institute, Pusa 1934–1935:120–130

Garg ID, Khurana SMP, Singh MN (1989) Immuno electron microscopy for the detection of potato leaf roll virus in *Myzus persicae*. J Aphidology 3:196–200

Glais L, Jacquot E (2015) Detection and characterization of viral species/subspecies using isothermal recombinase polymerase amplification (RPA) assays. Methods Mol Biol 130:207–225

Goodwin PH, Banttari EE (1984) Increased sensitivity of ELISA for potato viruses S,X and Y by polystyrene pretreatments, additives and a modified assay procedure. Pl Dis 68:944–947

Haible D, Kober S, Jeske H (2006) Rolling circle amplification revolutionizes diagnosis and genomics of geminiviruses. J Virol Methods 135:9–16

Hans LW (1988) Rapid detection of potato viruses by dot-ELISA. Potato R 31:485–492

Horvâth J (1967) Separation and determination of viruses pathogenic to potatoes with special regard to potato virus Y. Acta Phytopath Acad Sei Hung 2:319–360

Hull R (1993) Nucleic acid hybridization procedures. In: Mathews REF (ed) Diagnosis of plant virus diseases. CRC Press, Boca Raton, FL, pp 253–272

Jain RK, Khurana SMP, Bhat AI, Chaudhary V (2004) Nucleocapsid protein gene sequence studies confirm that potato stem necrosis is caused by a strain of groundnut bud necrosis virus. Ind Phytopath 57:169–173

Jeevalatha A, Chakrabarti SK, Sharma S, Sagar V, Malik K, Raigond B, Singh BP (2017) Diversity analysis of *tomato leaf curl* New Delhi virus–[potato], causing apical leaf curl disease of potato in India. Phytoparasitica 45:33–43

Jeevalatha A, Kaundal P, Kumar A, Guleria A, Sundaresha S, Pant RP, Sridhar J, Venkateswarlu V, Singh B (2016) SYBR green based duplex RT-qPCR detection of a begomovirus, tomato leaf curl New Delhi virus-[potato] along with potato virus X and potato leafroll virus in potato. Potato J 43:125–137

Jeevalatha A, Kumar R, Raigond B, Sundaresha S, Sharma S, Singh BP (2015) Duplex realtime RT-PCR assay for the detection of potato spindle tuber viroid (PSTVd) along with ef 1-α gene of potato. Phytoparasitica 43:317–325

Jeevalatha A, Priyanka K, Kumar R, Raigond B, Kumar R, Sharma S, Chakrabarti SK (2018) Optimized loop-mediated isothermal amplification assay for *tomato leaf curl New Delhi virus-[potato]* detection in potato leaves and tubers. Eur J Plant Pathol 150:565–573

Jeevalatha A, Singh BP, Kaundal K, Kumar R, Raigond B (2014) RCA-PCR: a robust technique for the detection of *tomato leaf curl New Delhi virus*-potato at ultra-low virus titre. Potato J 41(1): 76–80

Jeong J, Cho SY, Lee WH, Lee KJ, Ju HJ (2015) Development of a rapid detection method for potato virus X by reverse transcription loop-mediated isothermal amplification. Plant Pathol J 31:219–225

Ju H (2011) Simple and rapid detection of *potato leaf roll virus* by reverse transcription loop-mediated isothermal amplification. Pl Pathol J 27(4):385–389

Kehoe MA, Jones RAC (2016) Improving potato virus Y strain nomenclature: lessons from comparing isolates obtained over a 73-year period. Pl Pathol 65:322–333

Khurana, SMP (1999) *Potato Viruses and Viral Diseases*. Technical Bulletin No. 35 (Revised). Central Potato Research Institute, Shimla, India. 94 p

Khurana SMP, Singh MN (1988) Yield loss potential of potato viruses X and Y in Indian potatoes. J Indian Potato Assoc 16:165

Kogovsek P, Gow L, Pompe M, Novak K, Gruden GD, Fpster N, Boonham RM (2008) Single-step RT real-time PCR for sensitive detection and discrimination of potato virus Y isolates. J Virol Methods 149:1–11

Kreuze J, Koenig R, De Souza J, Vetten HJ, Muller G, Flores B, Ziebell H, Cuellar W (2013) The complete genome sequences of a Peruvian and a Colombian isolate of Andean potato latent virus and partial sequences of further isolates suggest the existence of two distinct potato-infecting tymovirus species. Virus Res 173:431–435

Kreuze JF, Pérez A, Untiveros M, Quispe D, Fuentes S, Barker I, Simon R (2009) Complete viral genome sequence and discovery of novel viruses by deep sequencing of small RNAs: a generic method for diagnosis, discovery and sequencing of viruses. Virol 388:1–7. https://doi.org/10.1016/j.virol.2009.03.024

Kumar R, Jeevalatha A, Baswaraj R, Kumar R, Sharma S, Nagesh M (2017) A multiplex RT-PCR assay for simultaneous detection of five viruses in potato. J Plant Pathol 99(1):37–45

Kumar R, Kaundal P, Arjunan J, Sharma S, Chakrabarti SK (2020) Development of a visual detection method for potato virus S by reverse transcription loop-mediated isothermal amplification. 3 Biotech 10:213. https://doi.org/10.1007/s13205-020-02214-4

Kumar S, Khurana SMP (1989) Dot-ELISA or DIBA for potato viruses X, S, Y and leafroll. Proc. 5th Ann. IVS convention, CPRI, Shimla, Oct. 17-19, 1989. SP-VIII/8, p. 95

Kumar S, Singh S (1999) Sensitivity of polyvalent dot-ELISA in comparison to dot and DAS-ELISA for detection of potato viruses. J Indian Potato Assoc 26:39–42

Li R, Gao S, Hernandez AG, Wechter WP, Fei Z, Ling KS (2012) Deep sequencing of small RNAs in tomato for virus and viroid identification and strain differentiation. PLoS One 7:e37127. https://doi.org/10.1371/journal.pone.0037127

Lin NS (1984) Gold-IgG complexes improve the detection and identification of viruses in leaf dip preparations. J Virol Methods 8:181–190

Lin NS, Langenberg WG (1983) Immunohistochemical localization of barley stripe mosaic virus (BSMV) in infected wheat cells. J Ultrastruct Res 84:16–23

Loebenstein G, Akad F, Filatov V, Sadvakasova G, Manadilova A, Bakelman H, Teverovsky E, Lachman O, David A (1997) Improved detection of potato leaf roll luteovirus in leaves and tubers with a digoxigenin-labelled cRNA probe. Plant Dis 81:489–449

Londono MA, Harmon CL, Polston JE (2016) Evaluation of recombinase polymerase amplification for detection of begomoviruses by plant diagnostic clinics. Virol J 22:13–48

Martyn EB (1968) Plant virus names. An annotated list of names and synonyms of plant viruses and diseases. Commonwealth Mycological Institute, Surrey, UK, p 204

Mayboroda O, Benito AG, Del Rio JS, Svobodova M, Julich S, Tomaso H, O'Sullivan CK, Katakis I (2015) Isothermal solid-phase amplification system for detection of *Yersinia pestis*. Anal Bioanal Chem 408:671–676

Mekuria TA, Zhang S, Eastwella KC (2014) Rapid and sensitive detection of little cherry virus 2 using isothermal reverse transcription-recombinase polymerase amplification Tefera. J Virol Meth 205:24–30

Milne RG (1992) Immunoelectron microscopy of plant viruses and mycoplasma. Adv Dis Vector Res 9:283–297

Mlotshwa S, Pruss GJ, Vance V (2008) Small RNAs in viral infection and host defense. Trends Plant Sci 13:375–382

Navot N, Ber R, Czosnek H (1989) Rapid detection of tomato yellow leaf curl virus is squashes of plants and insect vectors. Phytopathology 79:562–568

Nie X, Singh RP (2001) A novel usage of random primers for multiplex RT-PCR detection of virus and viroid in aphids, leaves, and tubers. J Virol Methods 91:37–49

Notomi T, Okayama H, Masubuchi H, Yonekawa T, Watanabe K, Amino N, Hase T (2000) Loop-mediated isothermal amplification of DNA. Nucleic Acids Res 2(12):E63

Panferov VG, Safenkova IV, Byzova NA, Varitsev YA, Zherdev AV, Dzantiev BB (2018) Silver-enhanced lateral flow immunoassay for highly-sensitive detection of potato leafroll virus. Food Agric Immunol 29(1):445–457. https://doi.org/10.1080/09540105.2017.1401044

Pares RD, Whitecross MI (1982) Gold-labelling antibody decoration (GLAD) in the diagnosis of plant viruses by immuno-electron microscopy. J Immunol Methods 51:23–28
Piepenburg O, Williams CH, Stemple DL, Armes NA (2006) DNA detection using recombinase proteins. PLoS Biol 4:1115–1121
Przewodowska A, Zacharewska JC, Treder K (2015) A one-step, real-time reverse transcription loop mediated isothermal amplification assay to detect *potato virus Y*. Am J Potato Res 92:303–311
Pundhir VS, Akram M, Ansar M, Rajshekhara H (2012) Occurrence of stem necrosis disease in potato caused by groundnut bud necrosis virus in Uttarakhand. Potato J 39:81–83
Raigond B, Ambika V, Shruti P, Sridhar J, Tarvinder K, Chakrabarti SK (2020) Development of a reverse transcription loop-mediated isothermal amplification for detection of *potato virus a* in potato and in insect vector aphids. Crop Prot 137:105296
Raigond B, Chauhan Y, Verma G, Jeevalatha A, Kochhar T, Sharma S, Singh BP, Chakrabarti SK (2019) Standardization and validation of quantitative RT-PCR technique for specific detection of *potato virus a*. Potato J 46(2):91–100
Raigond B, Jevalatha A, Singh BP, Kochhar T (2013a) Electron microscopic detection of potato virus a using gold nano particles. Potato J 40(2):104–108
Raigond B, Pathania S, Verma A, Verma G, Kochhar T, Chakrabarti SK (2021) Recombinase polymerase amplification assay for rapid detection of a geminivirus associated with potato apical leaf curl disease. J Plant Dis Prot 128:1061–1071. https://doi.org/10.1007/s41348-021-00466-7
Raigond B, Sharma M, Chauhan Y, Jeevalatha A, Singh BP, Sharma S (2013b) Optimization of duplex RT-PCR for simultaneous detection of *potato virus Y* and *S*. Potato J 40(1):22–28
Raigond B, Sharma P, Kochhar T, Roach S, Verma A, Jeevalatha A, Verma G, Sharma S, Chakrabarti SK (2017) Occurrence of groundnut bud necrosis virus on potato in north western hills of India. Indian Phytopathol 70(4):478–482. https://doi.org/10.24838/ip.2017.v70.i4.76993
Raigond B, Vallepu V, Sridhar J, Jeevalatha A, Sharma S, Singh BP (2014) RT-PCR detection of *potato leaf roll virus* (PLRV) in aphids from northern and north-eastern India. Ind J Pl Prot 42(4):430–436
Rwahnih AM, Daubert S, Golino D, Rowhani A (2009) Deep sequencing analysis of RNAs from a grapevine showing Syrah decline symptoms reveals a multiple virus infection that includes a novel virus. Virology 387:395–401
Scholthof KBG, Adkins S, Czosnek H, Palukaitis P, Jacquot E, Hohn T, Barbara H, Keith S, Thierry C, Paul A, Cynthia H, Gary DF (2011) Top 10 plant viruses in molecular plant pathology. Mol Plant Pathol 12:938–954. https://doi.org/10.1111/j.1364-3703.2011.00752.x
Sela N, Luria N, Dombrovsky A (2012) Genome assembly of bell pepper endornavirus from small RNA. J Virol 86:7721–7721
Sheila M, Mortimer J, Michael GKJ, Roger ACJ, Gordon T, Geoffrey I (2009) A single tube, quantitative real-time RT-PCR assay that detects four potato viruses simultaneously. Virol Methods 161:289–296
Silva G, Oyekanmi J, Nkere CK, Bömer M, Kumar PL, Seal SE (2018) Rapid detection of potyviruses from crude plant extracts. Anal Biochem 546:17–22
Singh M, Singh RP (1996) Factors affecting detection of PVY in dormant tubers by reverse transcription polymerase chain reaction and nucleic acid spot hybridization. J Virol Methods 60:47–57
Singh MN, Khurana SMP, Garg ID (1990) Antiserum production and immune electron microscopy of potato leafroll virus. Indian Phytopathol 43:13–19
Singh R, Valkonen J, Gray S, Boonham N, Jones RAC, Kerlan C, Schubert J (2008) Discussion paper: the naming of *potato virus Y* strains infecting potato. Arch Virol 153:1–13. https://doi.org/10.1007/s00705-007-1059-1
Singh RP, Jolanta K, Boiteau G (1996) Detection of stylet-borne and circulative potato viruses in aphids by duplex reverse transcription polymerase chain reaction. J Virol Methods 59:189–196

Smith TD, Bantarri EE (1987) Dot-ELISA on nitrocellulose membranes for detection of potato leaf roll virus. Plant Dis 71:795–799

Song JJ, Meng JR, Zou CW, Li P, Wang ZQ, Chen BS (2013) Identification of viruses from potato planted in winter in Guangxi by small RNA deep sequencing. Scientia Agric Sinic 46(19): 4075–4081

Torrance L, Robinson D (1989) Modern methods of detecting and identifying potato viruses. AgBiotech News Inf 1:891–896

Vanov AV, Safenkova IV, Zherdev AV, Dzantiev BB (2020) Nucleic acid lateral flow assay with recombinase polymerase amplification: solutions for highly sensitive detection of RNA virus. Talanta 210(2020):120616

Verbeek M, Piron P, Cuperus C, Dullemans A, Wiegers G, de Bruin A, Raaij HMG, Vlugt RAA, van den Bovenkamp G, de Haan E, Miedema G, Stolt T (2009). Non-persistent virus transmission by aphids, Potato virus Y in potato. http://edepot.wur.nl/185257

Verma G, Raigond B, Pathania S, Kochhar T, Naga K (2020) Development and comparison of reverse transcription-loop-mediated isothermal amplification assay (RT-LAMP), RT-PCR and real time PCR for detection of potato spindle tuber viroid in potato. Eur J Plant Pathol 158:951–964

Vunsh R, Rosner A, Stein A (1990) The use of polymerase chain reaction (PCR) for the detection of bean yellow mosaic virus in gladiolus. Ann Appl Biol 117:561–569

Wang Y, Liu Y, Lian H, Li N, Zhang L, Hu R (2015) A PCR method for detection of mink circivirus. Chin J Vet Sci 35:1774–1776. https://doi.org/10.16303/j.cnki.1005-4545.2015.11.09

Wylie SJ, Jones MGK (2011) The complete genome sequence of a passion fruit woodiness virus isolate from Australia determined using deep sequencing, and its relationship to other potyviruses. Arch Virol 156:479–482

Yamanaka ES, Tortajada-Genaro LA, Maquieira A (2017) Low-cost genotyping method based on allele-specific recombinase polymerase amplification and colorimetric microarray detection. Microchim Acta 184:1453–1462

Zhang S, Ravelonandrom M, Russell P, McOwen N, Briard P, Bohannon S, Vrient A (2014) Rapid diagnostic detection of plum pox virus in Prunus plants by isothermal AmplifyRP® using reverse transcription-recombinase polymerase amplification. J Virol Meth 207:114–120

Pesticide Residues and International Regulations

14

J. K. Dubey and Ajay Sharma

Abstract

The presence of residues in different food commodities is a very serious issue worldwide. It not only has a direct effect on human health but also affects human beings by entering the environment and getting incorporated in the food chain. People throughout the world are concerned about these residues, and different legislations have been enacted to manage this menace since long. Efforts are being made continuously by imposing stringent limits on the levels of residues on food commodities. Since the first action of ban on dichloro-diphenyl-trichloroethane (DDT) was taken in respect of the agricultural chemicals, many chemicals have been either restricted or put out of use by many countries. The Food and Agriculture Organization (FAO) of the United Nations and the World Health Organization (WHO) recognized the importance of developing international standards in this context as early as 1960. The Codex Alimentarius Commission has worked since 1963 to create harmonized international food standards to protect the health of consumers and ensure fair trade practices. The *US* EPA regulates and enforces *pesticide* actions in the USA under FIFRA and FFDCA. The European Food Safety Authority and CIB&RC in India are responsible for monitoring these chemicals in their respective countries.

Keywords

Pesticide · Residues · Carcinogens · Restrictions · Regulations · Illegal · Provisions

J. K. Dubey (✉) · A. Sharma
Department of Entomology, Dr. Y. S. Parmar University of Horticulture and Forestry, Nauni Solan, Himachal Pradesh, India

S. Kumar Chakrabarti et al. (eds.), *Sustainable Management of Potato Pests and Diseases*, https://doi.org/10.1007/978-981-16-7695-6_14

14.1 Introduction

The world population is growing at an alarming speed, and as per an estimate, it will be approximately 8.5 billion in 2030 (Crist et al. 2017). The United Nations Population division has expected the world population to reach 9.7 billion by 2050, which is almost 30% more than that in 2017. The major contribution in this increase is attributed to the developing countries. In order to feed this ever-increasing population, we have to keep pace with the food production. There is very less scope of expansion of land, and majority of food has to come from the same piece of land, thus taking a greater number of crops in a year or putting high-yielding varieties in place. In addition to this, the prevention of food commodities from pests will also play a pivotal role. Pesticides play an important role in the prevention of food commodities from different pests like insects, fungi, weeds, bacteria, etc. As per an estimate, the Indian agriculture suffers an annual loss of about $36 billion due to insects only (Dhaliwal et al. 2015). In addition, different pesticides are also used in public health programs in order to prevent the increase in the population of many insects, which act as vectors of various diseases. Pesticide refers to any substance purposely released into the environment for preventing, destroying, repelling, attracting, or controlling any pest, including unwanted species of plant or animals (FAO 1997; Yamada 2017). As these chemicals have potential to kill the living organisms, it will not be wrong to term these chemicals as biocides.

The unstoppable race of pesticides is supposed to have started with the discovery of insecticidal properties of dichloro-diphenyl-trichloroethane (DDT) by Paul Muller in 1939–1940 for which he was awarded Nobel Prize in Medicine in 1948. Prior to that, the oldest known pesticides were of plant origin. Paris green is the first known inorganic synthetic chemical to be used in the control of Colorado potato beetle in the USA, followed by the use of lead arsenate for the control of gypsy moth in 1892. In the 1930s, some synthetic organic compounds like alkyl thiocyanates were used as insecticides. The unprecedented success of DDT in pest control put a full stop on all these compounds in pest control. But this honeymoon did not last long, and the adverse effects of the indiscriminate usage of DDT were evident as soon as in the year 1962, when Rachel Carson in her book *Silent Springs* highlighted the terrifying ill effects of DDT on the environment and human health. She suggested that DDT causes cancer and that its agricultural use is a threat to diversity. It resulted in a huge outrage among the people in the USA and subsequently resulted in the ban on the use of DDT in the USA in 1972. Subsequently the Environmental Protection Agency cancelled order of DDT due to its adverse environmental effects like those to wild life and potential risk to human health. The use of DDT in agriculture however was globally banned with the UN's Stockholm Convention on POPs listing it in Annexure B (meant for its restrictive and conditional use); India debarred its use in agriculture way back 1989 owing to credible scientific evidences of its ill impact on ecology and life (Betne and Rajankar 2011). The production of pesticides had started in India in 1952 with the establishment of a plant for the production of BHC near Calcutta, and India was the second largest manufacturer of pesticides in Asia during the late 1990s after China and ranked twelfth globally (Mathur 1999).

Ban on DDT is evidently the first instance where restriction was imposed on the use of chemicals in agriculture. But the success achieved by DDT in pest control had triggered the production of many other chemicals on war footing. Metcalf (1980) has rightly called the period after 1940 as the age of pesticides and had divided the period into three phases, viz., era of optimism (1946–1962), era of doubt (1962–1967), and then era of IPM. A large number of pesticides were discovered after the success of DDT. Moreover, a large number of synthetic insecticides like organophosphates (OP) gained popularity in the 1960s, followed by carbamates in the 1970s and synthetic pyrethroids in the 1980s, which was also followed by the advent of the use of herbicides and fungicides in the 1970s. The different pesticides developed have been classified on the basis of their mode of action, chemical structure, active ingredients, and toxicity (Botitsi et al. 2017). Approximately two million tons of pesticides were utilized annually worldwide, which has been increasing rapidly, and by the year 2020, the global use of pesticide had been estimated to increase up to 3.5 million tons (Sharma et al. 2019). There is no doubt that the use of pesticides has increased the crop productivity manifolds; in addition, the control of vector-borne diseases for which insects play the role of a vector is also a significant achievement to the credit of these chemicals. However, these chemicals are also very deleterious in their impacts on human health and have many adverse effects on the environment. In the year 2009, two pesticides, arsenical insecticides and TCOD (a dioxin), had been designated by the International Agency for Research on Cancer (IARC), a component of the World Health Organization (WHO), as known human carcinogens, but many others that are widely used are suspected human carcinogens, and few pesticides on the market today are directly genotoxic (Alavanja 2009). It is not only that these chemicals have the bad effects on human health; the indiscriminate and injudicious usage of pesticides had also surpassed their beneficial effects in relation to related environment and ecosystem. Pesticides have drastic effects on nontarget species and affect animal and plant biodiversity, aquatic environment, and terrestrial food webs and ecosystems. The problems like pest resurgence, pesticide resistance, and pesticide residues are also major drawbacks in the use of the pesticides. A striking and alarming example of the presence of pesticide residues on food commodities is the presence of some pesticides in mother's milk. The single or multiple pesticide contamination of *p,p′*-DDE, *p,p′*-DDT, and chlorpyrifos was revealed in 27.45% of mothers' milk samples from Himachal Pradesh, India. Among these, *p,p′*-DDE was the major contaminant found in 26.79% of the samples, followed by *p,p′*-DDT (1.31%) and chlorpyrifos (0.65%). The regional difference in xenobiotic levels of breast milk varied with the demographic characteristics of mothers and altitudinal variations (Sharma et al. 2017).

Pesticides may enter the human body either through inhalation or ingestion or through the skin; however, people are affected by the effects of pesticides mainly because of consumption of food contaminated with pesticides. Taking into account the importance of pesticides in agricultural and public health sector as well as the adverse effect being induced by the large-scale application of pesticides, the developed and developing countries have proposed some laws or regulations to regulate the production, sale, and usage of pesticides. Pesticide legislation varies greatly worldwide, because countries have different requirements, guidelines, and legal limits for plant protection. They include limits for pesticide residues on food, product registration requirements, and pesticide use restrictions. Developed nations have more stringent regulations than developing countries, with the latter lacking the resources and expertise to adequately implement and enforce legislation (Handford et al. 2015).

14.2 Pesticide Residues: The Concern

Pesticides are the chemicals used to control the growth of weeds and presence of insects, fungi, and other pests in plants. The application of pesticides to crops and animals may leave residues on food. Pesticide residues on food commodities and their entry into the food chain have become a major concern worldwide. For any given compound, a residue may exist as the unaltered parent compound or as one or more degradation products, toxic or nontoxic. Bioaccumulation of chlorinated hydrocarbons and their derivatives in the human body and the environment can reach harmful levels. Some persistent chemicals have tendency of biomagnifications in the food chain and can be detected at very high levels in the aquatic fauna, poultry, vegetable oils, nuts, fruits, etc. (Crinnion 2009). Pesticide residues reach consumers through the following: (a) use of pesticides on farms, (b) application of pesticides to harvested produce, (c) presence of pesticides in imported foods, and (d) discharge of banned substances into the environment (Ballestros and Martos 2010). These residues should not be excessively used as they may pose risks to human health. The concepts of maximum residue limits (MRLs), acceptable daily intake (ADI), and theoretical maximum daily intake (TMDI) for pesticides have been devised to keep track of the pesticide residues in the food chain and keep them within safe limits.

MRLs are the maximum residues of pesticides, which may be expected in a treated produce, considering that Good Agricultural Practices (GAPs) have been followed. ADI is the maximum intake of pesticide that can be tolerated by all dietary sources in a day without posing any chronic health risk. TMDI is an estimate of the maximum intake of the pesticide with the existing MRLs for a person following a particular dietary practice.

Numerous types of chemicals kill insects by interfering with their nervous system; thus, these chemicals are also harmful to humans as humans also have nervous system. The specific effects of pesticides include damage to the central and peripheral nervous systems, cancer, allergies and hypersensitivities, reproductive disorders, and disruption of the immune system (Mishra et al. 2014). Moreover, the presence of pesticide residues on food commodities affect the wealth of people and interrupts with the import and export of such commodities.

India is among the World trade organisation (WTO) members for export of food commodities with its key export markets in the countries like the USA and European Union. Products like basmati rice, grapes, mangoes, etc. have been rejected and even banned in the markets of the USA, Vietnam, EU, Saudi Arabia, etc. as they were found to contain pesticide residues. In the short run, these rejections and bans had resulted in huge monetary losses, while in the long run, the farmers had lost markets for their produce (Sushil 2016). India's basmati export to the EU had declined by around 60% from corresponding period of last year to 1.62 lakh tons in the period of April–December 2018. The European Commission (EC) had reduced the MRL for tricyclazole to 0.01 parts per million (ppm) from 1 ppm for all crops effective January 1, 2018. Import of any agricultural product with a higher reading would not be permitted in the EU. In 2016, the Agricultural and Processed Food Products Export Development Authority (APEDA) had warned that exports of mangoes from India are under severe threat due to higher pesticide residues than the prescribed limit by the global standard. This issue reached the Indian authorities when the UAE Ministry of Climate Change and Environment (MCCE) issued a warning to Indian exporters, informing them of the high level of pesticide residues (exceeding the allowed limits) in Indian mangoes. Aside from mangoes, the UAE MCCE also observed the same in chilli, pepper, and cucumber consignments. The UAE market receives over 70% of India's overall annual mango exports. What is worrying is that if Indian exporters do not adhere to the global guidelines of the Codex MRLs, then the UAE may permanently ban mango import from India, which would be a big blow to the growers.

14.3 Pesticide Regulations

14.3.1 International Food Standard (FAO/WHO Codex Alimentarius)

In the early 1960s, the Food and Agriculture Organization (FAO) of the United Nations and the WHO recognized the importance of developing international standards to protect public health and minimize the disruption of international food trade. Thus, the Joint FAO/WHO Food Standards Program was established, and the Codex Alimentarius Commission (CAC) was designated to administer the program. The CAC, an international food standards body, was established jointly by the FAO and the WHO in May 1963, with the objective of protecting consumer's health and ensuring fair practices in food trade. The Codex Alimentarius, or "the Food Code," is a collection of standards, guidelines, and codes of practice adopted by the CAC. The Commission is a joint intergovernmental body of the FAO of the United Nations and WHO with 188 Member Countries and 1 Member Organization (the European Union). Codex has been working since 1963 to create harmonized international food standards in order to protect the health of consumers and ensure fair trade practices. India became a member of Codex Alimentarius in 1964 (FSSAI n.d.)

The Codex Alimentarius covers all foods, whether processed, semiprocessed, or raw. In addition to standards for specific foods, the Codex Alimentarius contains general standards covering matters such as food labeling, food hygiene, food additives, and pesticide residues, as well as procedures for assessing the safety of foods derived from modern biotechnology. It also contains guidelines for the management of official import and export inspection and certification systems for foods. To protect the health of people from the adverse effects of pesticide residues on food commodities, most countries have maximum legal limits for pesticide residues on foods. Trade difficulties arise when limits differ between countries.

The Codex Committee on Pesticide Residues (CCPR) is responsible for establishing Codex MRLs for pesticide residues in specific food items or in groups of food or feed that move in international trade. MRL is the highest level of pesticide residue that can be tolerated on food or feed when pesticides are correctly applied in accordance with GAPs. Before a Codex MRL can be established, human health risk assessments must be conducted to ensure the food supply is safe. It is the responsibility of the Joint FAO/WHO Meeting on Pesticide Residues (JMPR) to review the appropriate toxicology and data obtained mainly from supervised trials that reflect approved pesticide use in accordance with GAPs. JMPR conducts dietary risk assessments and recommends specific MRLs to the CCPR.

The Commission maintains a Codex Pesticide Residues on Food Online Database. This database contains Codex MRLs for pesticides and extraneous MRLs adopted by the CAC up to and including its 42nd session held in July 2019 (FAO n.d.).

The Organisation for Economic Co-operation and Development (OECD) is yet another international organization that works to establish better policies to achieve better lives. Its headquarter is in Paris. The OECD helps governments cooperate in assessing and reducing the harms of agricultural pesticides. The OECD encourages the governments to share the work of pesticide registration and develops tools to monitor and minimize the risks of pesticides to human health and the environment.

14.3.2 Regulations in the USA

In the *USA*, *pesticide regulation* is largely overseen by the *US* EPA (Environmental Protection Agency), which regulates and enforces *pesticide* actions under the *Federal Insecticide*, Fungicide, and Rodenticide Act (FIFRA) and the *Federal* Food, Drug, and Cosmetic Act (FFDCA).

14.3.2.1 Federal Insecticide, Fungicide, and Rodenticide Act (FIFRA)

The Federal Insecticide Act of 1910 was replaced by FIFRA in 1947, and since then, this law has undergone many changes (EPA n.d.-a). A significant revision in 1972 by the Federal Environmental Pesticide Control Act (FEPCA) and several others have expanded EPA's present authority to oversee the sales and use of pesticides, with emphasis on the preservation of human health and protection of the environment as follows:

- Strengthening the registration process by shifting the burden of proof to the chemical manufacturer.
- Enforcing compliance against banned and unregistered products.
- Promulgating the regulatory framework missing from the original law.

All pesticides distributed or sold in the USA must be registered (licensed) by EPA. Under FIFRA, EPA has registered approximately 50,000 pesticide products. Before EPA can register a pesticide under FIFRA, the applicant must demonstrate, among other things, that using the pesticide according to the specifications "will not generally cause unreasonable adverse effects on the environment," which is defined by FIFRA as follows:

- Any unreasonable risk to man or the environment, taking into account the economic, social, and environmental costs and benefits of the use of any pesticide.
- A human dietary risk from residues that result from a use of a pesticide in any food inconsistent with the standard under Sect. 408 of the Federal Food, Drug, and Cosmetic Act.

Aside from registering and approving the product label of pesticide being put into use in the USA and providing for the penalties for inconsistent labeling, FIFRA has also many other responsibilities. FIFRA governs the pesticides to be of general use or of restricted use. It establishes tolerances for residues that may remain on raw agricultural products or in processed food. It also governs the storage and disposal of the pesticides and makes provisions for penalties for illegal handling of containers.

FIFRA gives EPA the authority to develop regulations so as to provide standards for worker protection and provide reentry standards for treated areas. It defines *restricted-entry intervals* as the time immediately following the application of a pesticide when unprotected workers may not enter the treated area. These laws are enacted to protect the unprotected persons from being harmed by the adverse effects of pesticides. It clearly says that there shall be no unprotected person in the area during the time of pesticide application. It also defines the time of reentry of person in the treated area.

The Agricultural Worker Protection Standards (WPS) were issued by EPA in 1992, which covers both workers in areas treated with pesticides and employees who handle pesticides for use in these areas. The revised regulations (Title 40 CFR Part 170) govern the protection of employees on farms, forests, nurseries, and greenhouses from occupational exposures to agricultural pesticides. These regulations define agricultural workers as persons who perform tasks related to the cultivation and harvesting of plants on farms or in greenhouses, nurseries, or forests. They also define pesticide handlers as those who handle agricultural pesticides (mix, load, apply, clean, or repair equipment, act as flaggers, etc.). These regulations are enacted with the objective of reducing the pesticide hazards to the agricultural workers and pesticide handlers (EPA n.d.-b).

14.3.2.2 Federal Food, Drug, and Cosmetic Act (FFDCA)

This act is administered by the Food and Drug Administration (FDA) of the Department of Health and Human Services, and it gives the EPA authority to set limits on the amount of pesticide residues allowed on food or animal feed. These limits are called tolerances. Since its inception in 1938, this law has been amended several times (FDA n.d.).

Under the FFDCA, EPA has the responsibility of setting tolerances, or maximum legal limits, for pesticide residues on food commodities marketed in the USA. In setting tolerances, EPA must ensure that the tolerance is "safe." Safe means that there is a "reasonable certainty that no harm will result from aggregate exposure to the pesticide residue." To make the safety finding, EPA considers, among other things, the following:

- The toxicity of the pesticide and its breakdown products.
- Aggregate exposure to the pesticide on foods and from other sources of exposure.
- Any special risks posed to infants and children.

Some pesticides are exempted from the requirement to have tolerance level. EPA may grant exemption in cases where the pesticide residues do not pose a dietary risk under reasonably foreseeable circumstances. The purpose of the tolerance program is to ensure that the consumers in the USA are not exposed to unsafe food-pesticide residue levels. The FDA is responsible for enforcing the tolerance levels set by EPA. This law:

- Mandates the monitoring of food crops for pesticide residues and enforces tolerances.
- Mandates the monitoring and enforcement of food additive tolerances and prosecutes violators.
- Works jointly with EPA to register pesticides used on animals.
- Mandates the monitoring of pesticide residues in animals by the Meat Inspection Division of the US Department of Agriculture.

In addition to these two laws, the Food Quality Protection Act of 1996 (FQPA) amended FIFRA and FFDCA by increasing the safety standards for new pesticides used on foods. The FQPA also required older pesticides and previously established tolerances to be periodically reassessed using the new, tougher standards. Another act, Pesticide Registration Improvement Act (PRIA), establishes the fees and timelines associated with pesticide registration actions. The Endangered Species Act (ESA) of 1973 requires the EPA to assess the risk of pesticides to threatened or endangered species and their habitats.

14.3.3 Regulations in the European Union

In the EU, the European Food Safety Authority (EFSA) is responsible for managing the food risks. It includes advice, laws, and policymaking to protect people from risks in the food chain. It covers food and food safety as well as plant protection in addition to some other works. EFSA was established in 2002. Prior to that, individual EU member countries and the EC have a long history of controlling pesticide use through a myriad of country-specific programs (Skevas et al. 2013). In 1979, pesticide policies were introduced for the first time in the EU. The regulations impacting pesticide use were first introduced through the waste framework directive (2006/12/EC) and the directive on hazardous waste (91/689/EEC). These two directives established provisions for safe collection/disposal of empty pesticide packages and unused or expired pesticides. The MRLs of pesticide residues are addressed by the regulation on MRL 396/2005, which also addresses the residues of active substances in plant protection products. As per EFSA, plant protection products are chemical compounds used to protect crops by killing or controlling pests or weeds, whereas active substances – such as chemicals or microorganisms – are the essential ingredients in the products that enable them to do their job. Regulation 1107/2009 covers the placement of plant protection products in the market and directive 2009/128/EC on sustainable use of pesticides has replaced the earlier directive 91/414/EC. All approved active substances are listed in Implementing Regulation (EU) No. 540/2011 and included in the EU Pesticides Database.

Every active substance is evaluated for safety before it can be placed on the market and used in a plant protection product. At least one safe use of the substances in plant protection products must be demonstrated for people's and animals' health, including their residues on food, and must not have any negative effects on the environment before they can be approved. The initial approval of an active substance is valid for a limited period and needs to be reviewed periodically. A renewal of approval is only granted after the substance is reevaluated for its safety. The details of the renewal procedure are set out in the Commission Implementing Regulation (EU) No. 2020/1740, which will come into force from 27 March 2021 and replaces the previous procedure under Implementing Regulation (EU) No. 844/2012 (EC). According to the timelines set out in Regulation (EC) No. 1107/2009, completing the evaluation of applications for the first approval of an active substance should take between 2.5 and 3.5 years from the date of admissibility of the application to the publication of a regulation on the approval or nonapproval of the active substance. For renewals of approval, applications must be submitted at the latest 3 years before the expiry of the current approval of the active substance.

Since 2003, the EFSA has investigated the EU peer review of active substances used in plant protection products. This task is carried out by the EFSA's Pesticides Unit, supported by a network of experts from Member States, following procedures set out in the legislation and internal EFSA decisions and applying the methodology endorsed by risk managers for regulatory assessments. Experts from EFSA's Scientific Panel on Plant Production Products are not regularly involved in the peer review process, although the Panel has been requested to endorse some scientifically complex conclusions in the past. EFSA is composed of four bodies: (a) management board, (b) executive director, (c) advisory forum, and (d) scientific committee and scientific panels. The management board has 15 members. The board members act in the public interest. They do not represent any government, organization, or industry sector. The board sets EFSA's budget and approves its annual work program. EFSA's executive director is responsible for operational and staffing matters. He also draws up the annual work program together with the Commission, the European Parliament, and the EU countries. The Advisory Forum advises the executive director. In particular, it advises the executive director in drafting the proposal for the work program. The forum is made up of representatives of national bodies responsible for risk assessment in the EU countries. There are also observers from Norway, Iceland, Switzerland and the Commission.

Depending on their characteristics, some active substances can be approved as so-called low-risk substances or as candidates for substitution. Active substances with certain properties defined in Regulation (EC) No. 1107/2009 are considered as candidates for substitution. Once active substances are approved, companies can submit in the Member States applications for authorization for placing on the market and use of plant protection products containing them. Basic substances are active substances, not predominantly used as plant protection products but which may be of value for plant protection and for which the economic interest in applying for approval may be limited. The criteria for their approval are laid down, and specific provisions are set to ensure that such active substances, as long as they do not have an immediate or delayed harmful effect on human and animal health nor an unacceptable effect on the environment, can be legally used in the EU after having been approved as "basic" under Regulation 1107/2009. A specific procedure is set out for the approval of so-called basic substances.

14.3.4 Regulations in India

Pesticide regulations in India are governed by two different bodies: the Central Insecticides Board and Registration Committee (CIB&RC) and the Food Safety and Standards Authority of India (FSSAI). The use of pesticides in India was triggered by the introduction of DDT in 1948 to control malaria and BHC for locust management. The production of pesticides in India started with the establishment of manufacturing unit for DDT and benzene hexachloride (BHC) (HCH) in the year 1952, and by 1958, India was producing over 5000 metric tons of pesticides. The rampant use of these chemicals has given rise to several short- and long-term adverse effects. Kerala witnessed the first direct effect of poisoning by any pesticides in India in 1958 during which over 100 people died after consuming wheat flour contaminated with parathion. As a result, the government of India appointed a commission of enquiry to suggest remedial measures. The expert committee of ICAR was headed by Prof. M.S. Thacker. Based on the recommendations of the committee, the Insecticide Act was passed in the year 1968 so as to regulate the manufacture, import, registration, sale, transport, distribution, and use of pesticides in India.

14.3.4.1 Insecticide Act, 1968

This act was enforced throughout India from 1 August 1971, and the rules were framed and brought into force on 30 October 1971.There are nine chapters in the

Insecticide Rule, 1971 relating to the functions of CIB&RC, Central Insecticides Laboratory (CIL), grant of licenses, packing, labeling, first aid, antidote, protective clothing, etc. After the introduction of this act, many difficulties were faced in its implementation and enforcement. As a result, the act was amended in 1972 and 1977. In order to overcome the administrative and technical difficulties, a bill was introduced in the parliament of India, and some amendments were made to the Insecticide Act on 7 August 2000.

The salient features of the Insecticide Act (1968) are as follows:

- Compulsory registration of the product at the central level and licenses for manufacture, formulation, and sale at the state level.
- Interdepartmental/ministerial/organizational coordination is achieved by a high-level advisory board "Central Insecticides Board" with 24 members (to be raised to 29 by an amendment) drawn from various fields having expert knowledge of the subject.
- "Registration Committee" to look after the registration aspects of all insecticides.
- Establishment of enforcement machinery like Insecticide Analysts and Insecticide Inspectors by the Central or State Government.
- Establishment of central laboratory.
- Power to prohibit the import, manufacture, and sale of pesticides and also confiscate the stocks. The offences are punishable, and other penalties are prescribed.
- Both the Central and State Governments are empowered to make rules and prescribe forms and fees.

The Central Insecticides Board (CIB) The Central Insecticides Board advises on matters relating to the risk to human beings or animals involved in the use of insecticides and the safety measures necessary to prevent such risk. The manufacture, sale, storage, transport, and distribution of insecticides with a view to ensure the safety of human beings and animals are under the preview of the board. The board also fixes tolerance limits for insecticide residues, safety period, and shelf life based on data provided by manufacturers on the basis of research conducted by scientists in the country. The board is headed by the Director General of Health Services as its chairman.

Registration Committee An insecticide can be sold in India after it has been approved and registered by this committee. In order to register a chemical, the manufacturer has to apply to the secretary of the committee and CIB for consideration and acceptance of the result and pay registration fees. The committee, after satisfying itself with respect to the effectiveness of the chemical and its safety for the human beings and animals, issues a registration number and certificate of registration for the chemical. New insecticides introduced in the country for the first time are registered provisionally for a period of 2 years. Complete data is required during this time for regular registration of the chemical. Once the insecticide is found to be effective and safe, a regular registration is granted for its import or manufacture.

The registration committee consists of a chairman and not more than five members from CIB, including the drug controller and plant protection advisor. The secretary of the committee is appointed by the Central Government and is assisted by seven technical officers, an entomologist, plant pathologist, agronomist, medical toxicologist, chemist, packaging engineer, and law officer, along with some ministerial staff.

The Insecticides Act (1968) was to be replaced by the proposed Pesticide Management Bill (2017), with more focus on protecting farmers and promoting the safe use of pesticides. The data requirements and guidelines for registration under this bill were almost same as those in the Insecticides Act but have major changes for punishment and fines for misbranded products and are being governed more by state governments in dealing with such issues. The bill also proposed extension of data protection to 5 years with application not to be reused by another applicant for 3 years. A draft copy of the bill was released in February 2018 for comments from representatives of industry, farmers, retail sector, environmental groups, and center and states.

The Pesticide Management Bill (2020) is a long overdue law on critical segment of agriculture, in the making since 2008, to replace the obsolete Insecticides Act, 1968. Taking into account advances in modern pest management science and the ill effects of synthetic pesticides, the Pesticide Management Bill should bring India's pesticide sector in line with global norms, to some of which India has signed up. The food safety law already has limits on pesticide residue. The present law addresses the manufacture, sale, import, transport, use, and distribution of insecticides. The bill will cover the life cycle of pesticides from manufacture to disposal and will include regulation of export, packaging, labeling, pricing, storage, and advertisement. Penalties on manufacturers for noncompliance with rules and regulations would be stiffer.

14.3.4.2 The Prevention of Food Adulteration Act, 1954

This act was enacted on 29 September 1954 for the prevention of adulteration of food. It also specified tolerance limits of pesticides on different raw agricultural products. The Central Committee for Food Standards (CCFS) under the Ministry of Health recommended the quality of food commodities under the PFA Act. One of its subcommittees, the pesticide residue subcommittee, advised CCFS on the tolerance limits of pesticides in different food articles based on use pattern, dietary habits, and nutritional status as well as restrictions on the selling of insecticides to persons selling, storing, or manufacturing food.

14.3.4.3 Food Safety and Standards Act, 2006

Food safety issues are also gaining attention in India mainly because of customers' increasing demand for better-quality food. The FSSAI came into existence under the Food Safety and Standards Act, 2006. FSSAI has been created to develop scientific standards for food products and to regulate food manufacturers, warehousing, distribution channels, domestic sales, exports, and imports to ensure the availability of safe and food fit for human consumption. The acts like Vegetable Oil Products

(Control) Order 1947, Prevention of Food Adulteration Act 1954, Fruit Products Order 1955, Meat Food Products Order 1973, Edible Oils Packaging (Regulation) Order 1988, Milk and Milk Products Order, 1992, are replaced by the Food Safety and Standards Act, 2006 (Mahajan and Garg 2014). This act is enacted to consolidate the laws relating to food and to establish the FSSAI for laying down science-based standards for articles of food and to regulate their manufacture, storage, distribution, sale, and import in order to ensure the availability of safe and wholesome food for human consumption and for matters connected therewith or incidental thereto.

The FSSAI is an autonomous body established under the Ministry of Health and Family Welfare, Government of India. The FSSAI is responsible for protecting and promoting public health through the regulation and supervision of food safety. The FSSAI is headed by a nonexecutive chairperson, appointed by the Central Government, either holding or has held the position of not below the rank of secretary to the Government of India. It has set certain guidelines for food safety research. The FSSAI has been mandated to perform various functions related to the quality and standards of food. Standards framed by FSSAI are prescribed under Food Safety and Standards (Food Product Standards and Food Additives) Regulation, 2011; Food Safety and Standards (Packaging and Labeling) Regulation, 2011; and Food Safety and Standards (Contaminants, Toxins, and Residues) Regulations, 2011.

In India, pesticide use is regulated by the Central Insecticides Board and Registration Committee (CIBRC) and the FSSAI. The CIBRC registers pesticides for crops, while the FSSAI sets the MRLs of pesticides for the crops it has been registered for.

14.4 Conclusion and Future Outlook

Food safety is now a days most important aspect not only in respect of human health but it also plays a crucial role in the world trade related with food commodities. Pesticide residues on food commodities are one of the biggest hurdles in the import and export of food commodities across the borders. In order to control the use of pesticides as well as their manufacture, sale purchase, storage, etc., different courtiers have enacted laws as per their needs and utilities. The tolerance limits are also imposed by different countries that other nations need to abide by before transporting their food commodities to that particular country. The tolerance limits set by the different agencies need to be strictly followed so as to prevent adverse effects of the pesticides on human beings.

Acknowledgments The material incorporated in the text is either rules or laws framed by different countries or unions or committees so as to regulate the pesticide use in the area of their domain. All the material is published by the respective agencies of different countries either on the websites or by some authors in their publications. The present authors have taken the material from all these sources and compiled the literature so as to give an overview about the international regulations to the readers. The material does not belong to authors but is just collection of material from different sources. All the sources have been duly cited in the text and references. The present authors acknowledge the labor put in by the different authors from whom the material has been collected to compose the present manuscript.

References

Alavanja MCR (2009) Pesticides use and exposure extensive worldwide. Rev Environ Health 24(4):303–309

Ballesteros E, Martos NR (2010) Residues of pesticides and polycyclic aromatic hydrocarbons in olive and olive-pomace oils by gas chromatography/tandem mass spectrometry. In: Preedy VR, Watson RR (eds) Olives and olive oil in health and disease prevention. Academic Press, San Diego, CA, pp 425–436

Betne R, Rajankar P (2011) DDT losing grounds in India: alternatives imminent Toxic Link Factsheet No 38. http://toxicslink.org/docs/Factsheet_38_DDT.pdf. Accessed 10 January 2021

Botitsi H, Tsipi D, Economou A (2017) Current legislation on pesticides. In: Romers-Gonzalez R, Frenich AG (eds) Applications in high resolution mass spectrometry: food safety and pesticide residue analysis. Elsevier, Amsterdam, pp 83–130

Crinnion WJ (2009) Chlorinated pesticides: threats to health and importance of detection. Altern Med Rev 14(4):347–359

Crist E, Mora C, Engelman R (2017) The interaction of human population, food production, and biodiversity protection. Science 356(6335):260–264

Dhaliwal GS, Jindal V, Mohindru B (2015) Crop losses due to insect pests: global and Indian scenario. Indian J Entomol 77:165–168

EPA (n.d.-a) Agricultural Worker Protection Standard (WPS). https://www.epa.gov/pesticide-worker-safety/agricultural-worker-protection-standard-wps. Accessed 10 January 2021

EPA (n.d.-b) Federal Insecticide, Fungicide, and Rodenticide Act (FIFRA). https://www.epa.gov/enforcement/federal-insecticide-fungicide-and-rodenticide-act-fifra-and-federal-facilities#Summary. Accessed 10 January 2021

European Commission (n.d.) Approval of active substances. https://ec.europa.eu/food/plants/pesticides/approval-active-substances_en. Accessed 10 January 2021

FAO (2017) NSP - The Joint FAO/WHO Meeting on Pesticide Residues (JMPR). http://www.fao.org/agriculture/crops/thematic-sitemap/theme/pests/jmpr/en/. Accessed 10 January 2021

FAO (n.d.) Codex pesticides residues in food online database. http://www.fao.org/fao-who-codexalimentarius/codex-texts/dbs/pestres/en/. Accessed 10 January 2021

FDA (n.d.) Federal Food, Drug, and Cosmetic Act (FD&C Act). https://www.fda.gov/regulatory-information/laws-enforced-fda/federal-food-drug-and-cosmetic-act-fdc-act. Accessed 10 January 2021

FSSAI (n.d.) Codex Alimentarius Commission. https://www.fssai.gov.in/cms/codex.php. Accessed January 10, 2021

Handford CE, Elliott CT, Campbell K (2015) A review of the global pesticide legislation and the scale of challenge in reaching the global harmonization of food safety standards. Integr Environ Assess Manag 11(4):525–536

Mahajan R, Garg S (2014) Global food safety: determinants are Codex standards and WTO's SPS food safety regulations. J Adv Manag Res 11(2):176–191

Mathur SC (1999) Future of Indian pesticides industry in next millennium. Pestic Info 24(4):9–23

Metcalf RL (1980) Changing role of insecticides in crop protection. Annu Rev Entomol 25:219–256

Mishra P, Sharma A, Sharma D (2014) A study on harmful effects of pesticide residue in vegetables. Int J Recent Res Rev 7(1):45–48

Sharma A, Kumar V, Shahzad B, Tanveer M, Sidhu GPS, Handa N, Kohli SK, Yadav P, Bali AS, Parihar RD, Dar OI, Singh K, Jasrotia S, Bakshi P, Rama Krishnan M, Umar S, Bhardwaj R, Thukral AK (2019) Worldwide pesticide usage and its impacts on ecosystem. SN Appl Sci 1:1446

Sharma ND, Sharma ID, Chandel RS, Wise JC (2017) Presence of pesticides in breast milk and infants' formulae in Himachal Pradesh. India Int J Environ Anal Chem 96(3):225–236
Skevas T, Lansink AO, Stefanou SE (2013) Designing the emerging EU pesticide policy: a literature review. NJAS-Wageningen J Life Sci 64:95–103
Sushil SN (2016) Emerging issues of plant protection in India. Natural Resource Management: Ecological Perspectives. International Conference, SKUAST, Jammu
Yamada Y (2017) Importance of codex maximum residue limits for pesticides for the health of consumers and international trade. In: Ambrus A, Hamilton D (eds) Food safety assessment of pesticide residues. World Scientific Europe Publishing, London, pp 269–282

RNA Interference: A Versatile Tool to Augment Plant Protection Strategies in Potato

15

S. Sundaresha, A. Jeevalatha, Ravinder Kumar, Salej Sood, Sanjeev Sharma, Vinay Bhardwaj, Birpal Singh, and S. K. Chakrabarti

Abstract

Potato is one of the most important nongrain food crops grown worldwide. The potato genome sequencing consortium has not yet identified and marked its functionality for a large number of hypothetical genes. In order to systematically assign functions to all predicted genes in its genome and their specific application in potato improvement, the RNAi tool played a major and functional role over the years to generate functional mutants. Assigning of the functional role of genes has wide applicability for the development of biotic, abiotic, and quality improvement program aiding not only in elucidating the function of genes but also in developing healthy potato varieties and future feed. In this book chapter, we have summarized the use of the RNAi tool in assigning and understanding the functional role of the genes for developing potato varieties for biotic and abiotic stress tolerance. Further, the advantages and limitations of these techniques suitable for mining the genomic data have been discussed in the context of potato functional genomics. This detailed analysis will lay the foundation for genetic improvement of potato for food and nutritional security.

Keywords

RNAi · PTGS · dsRNA · Potato · ToLCNDV · *Phytophthora infestans*

S. Sundaresha (✉) · A. Jeevalatha · R. Kumar · S. Sood · S. Sharma · V. Bhardwaj · B. Singh · S. K. Chakrabarti
Division of Plant Protection, ICAR-Central Potato Research Institute, Shimla, Himachal Pradesh, India

S. Kumar Chakrabarti et al. (eds.), *Sustainable Management of Potato Pests and Diseases*, https://doi.org/10.1007/978-981-16-7695-6_15

15.1 Introduction

Ribonucleic acid (RNA) is a class of macromolecules that is found in every living cell. For many years, it was believed that RNA was simply a "messenger" carrying genetic information from the DNA to the cell's protein-manufacturing machinery. But in recent years, it has become clear that RNA has multiple other roles in the cell, and perhaps most significantly, it is now known that RNA is directly involved in the control of gene expression. This discovery has revolutionized our understanding of gene regulation, and it holds great promise for significant advances in both basic science and biotechnology and medicine. Based on research during the 1960s, the Nobel Laureate Francis Crick formulated the "Central Dogma" of molecular biology stating that genetic information flows in one direction: from DNA to RNA to protein (Morangae 2008).

Recent years have witnessed the discovery of RNAi from the dsRNA and gain of unprecedented knowledge from ribonucleic acid (RNA) research. This caused a paradigm shift from the 40 year old central dogma theory, and also could be the RNA is a key molecule that led to the origins of life on earth the so called RNA World hypothesis (Gilbert 1986; Robertson and Joyce 2012). High expectations and great hope arose from the discovery of RNAi, a mechanism widely employed by eukaryotic cells to inhibit protein production at a posttranscriptional level, which allows gene silencing in experimental settings and has enormous therapeutic potential. RNAi established itself very quickly as a useful molecular biology tool, making large-scale functional genomic screens and high-throughput drug target screening model in medicines and plant protection strategy in agricultural research possible. There is an increasing hope that RNA-based approaches would bring significant advances to the diagnosis, management, and functional studies of various signaling pathways in plant biology.

15.2 RNA Interference (RNAi): History and Mechanism

RNAi is an intriguing phenomenon in which short, double-stranded RNA (dsDNA) can prevent the expression of specific genes. First discovered in plants, RNAi is now recognized as a widespread, if not ubiquitous, phenomenon, and it is causing great excitement as an experimental technique for selectively blocking gene expression. The mechanisms of RNA silencing have been intensively studied. One important step is the formation of single-stranded RNA pieces (called siRNAs) from the double-stranded triggers. In lower organisms, including plants, protozoa, fungi, and nematode worms, it also involves an enzyme called RNA-dependent RNA polymerase, which can generate a strand of RNA using existing RNA as a template. This means that it can create double-stranded RNA (dsRNA) from single-stranded pieces of RNA. By doing so, it generates more triggers and so amplifies the effect of RNA silencing.

15.2.1 History of RNAi

In 1990, researchers noticed for the first time that RNA could potentially suppress gene expression in plants. The underlying mechanism of this erratic and reversible gene suppression was not clear. It was not until almost a decade later when Andrew Fire and Craig Mello showed in the worm *Caenorhabditis elegans* that dsRNA, but not single-stranded RNA (ssRNA), was involved in gene silencing. Moreover, their report established that RNAi occurs in a sequence-specific manner (Fire et al. 1998). This finding triggered a series of studies unraveling the detailed mechanism of RNAi over the years. In 2006, Fire and Mello were awarded the Nobel Prize for Physiology or Medicine for their seminal work.

Over the last several years, much progress has been made in unraveling the mechanism of RNA silencing, a process leading to the degradation of homologous mRNAs, which is also termed RNAi in animals, posttranscriptional gene silencing (PTGS) in plants, and quelling in fungi (Kooter et al. 1999; Matzke et al. 2001; Vaucheret et al. 2001; Waterhouse et al. 2001; Hannon 2002). Although the phenomenon of RNA inhibition was first described in petunia as "co-suppression" (Napoli et al. 1990), more extensive studies have been carried out on the functional analysis of genes in *Caenorhabditis elegans* (Fire et al. 1998). The occurrence of apparently similar underlying mechanisms for this phenomenon in different species indicates a conserved biological function of RNA silencing during the evolution of organisms.

The phenomenon involves the use of dsRNA to silence gene expression through binding, cleaving, and degrading complimentary endogenous mRNA. RNAi is used as a method to study the function of genes and provide a basis of discovering new drugs capable of silencing viral or human genes that cause diseases (e.g., HIV infection).

Synonyms Used for RNAi

- PTGS and co-suppression – plant biologists
- RNAi in *C. elegans* and drosophila – animal biologists
- Quelling in fungi – fungi scientist

Given below is a timeline of the research and development with respect to RNAi:

- 1987 – RNA technology in plants – inhibition of nopaline synthase gene in tobacco
- 1990 – Co-suppression – first report of gene silencing, overexpression of chalcone synthase gene (CHS) (Napoli et al. 1990)
- 1994 – Quelling – (bleached fungi) *Neurospora crassa* inhibition of gene involved in carotenoid production
- 1995 – RNAi in *Caenorhabditis elegans* – by the introduction of asRNA to silence gene that regulates embryo symmetry
- 1998 – PTGS in plants – with the class of small RNAs as triggering signal for gene silencing.

15.2.2 RNAi Machinery and Mechanism

The main components of RNAi machinery are *DICER* (a protein with RNAse activity that detects dsRNAs into siRNAs) and *RISC* (RNA-induced silencing complex) protein (a large protein complex assembled and activated in the presence of siRNA).

The DICER family of RNase III enzymes recognizes and processes dsRNA into siRNA; each dicer enzyme has an amino terminal helicase domain, separates two strands into single-strand, 2 RNase III catalytic domains, dsRNA-binding domain, and PAZ domain. The cleavage of dsRNA is performed by RNase III catalytic domains where they cleave dsRNA into 21–23 bp fragments.

RISC consists of 2 RNA-binding proteins, RNA/DNA helicase, translation initiation factor, RNA-dependent RNA polymerase (RdRP), and transmembrane protein.

Current Mechanism of PTGS in Plants Characterization of the PTGS mechanism is still in its infancy. Much effort is being made to elucidate the genes and biochemistry involved in the defensive response since it affects gene silencing. The mechanisms seem to be conserved throughout evolution since homologous genes involved in the process have been found in various species of fungi and plants, as well as in animals.

The present model for gene silencing (Fig. 15.1) includes three phases:

- **Initiation phase:** dsRNA synthesis or formation and production of small interfering RNA (siRNAs) fragments.
- **Maintenance phase:** association of siRNA complex protein (RISC) to guide nuclease activity and degradation of target mRNA.
- **Signal amplification and spreading phase:** siRNA acts as promoter for dsRNA polymerization which moves through cell to cell.

In initiation phase, dsRNA is the triggering factor for gene silencing, and it involves its synthesis or formation, its recognition, and the production of siRNAs fragments (Hamilton and Baulcombe 1999). dsRNAs can be generated by the RNA virus replication mechanism, which includes the formation of dsRNA by an RNA-dependent RNA polymerase (RdRP); by hpRNA, which originated from a bidirectionally cloned transgene; or by an asRNA cloning strategy.

Maintenance phase involves mRNA targeting and degradation by the enzyme *Dicer* (Bernstein et al. 2001), which recognizes and leaves dsRNAs from both ends into siRNAs of 21 to 23 nt (Zamore et al. 2000). These siRNAs, alternatively referred to as guide RNAs, are identification sequences for the RNA-induced silencing complex (RISC), a protein-RNA effector nuclease formation of about 500 kDa, in the maintenance phase (Hammond et al. 2000). RISC has exo- and endonuclease activities, an RNA homology-searching activity, and a helicase to unwind the dsRNA. This complex can be activated by unwinding siRNAs in order to use their single-stranded siRNA sequences for identification and degradation of complementary transcripts (Nykänen et al. 2001).

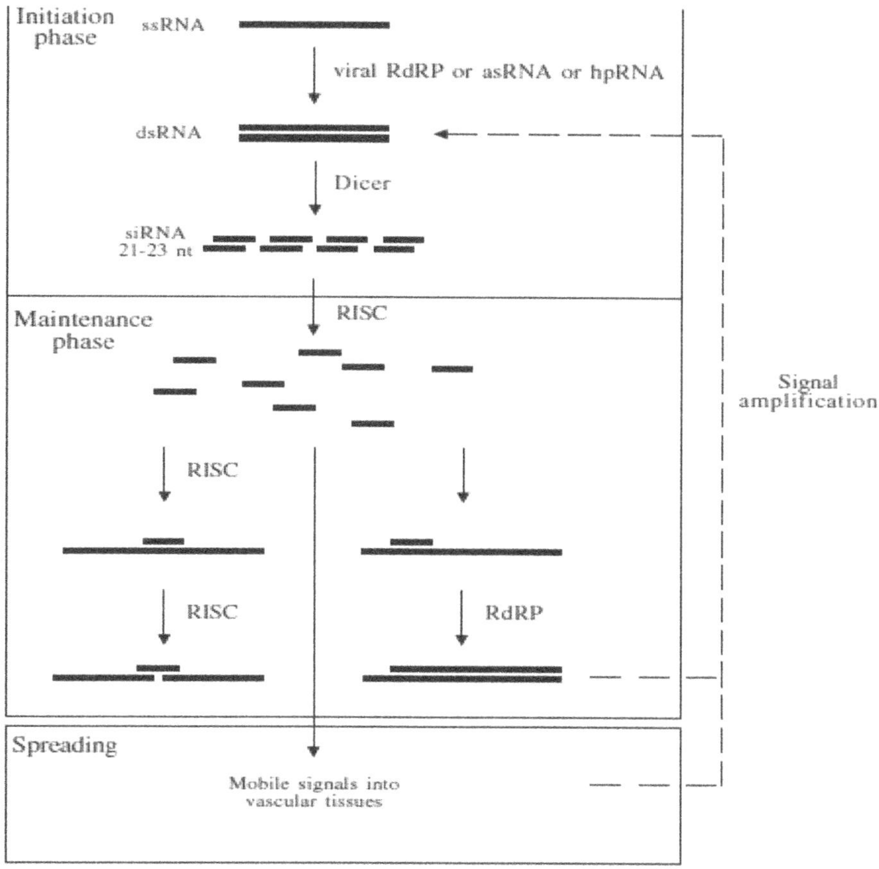

Fig. 15.1 Current model of posttranscriptional gene silencing

The initiation phase can be triggered by a viral RNA-directed RNA polymerase (RdRP) that makes a complementary RNA strand, using the single stranded as a template, by a transgenic antisense RNA (asRNA) or by a hairpin RNA (hpRNA). Double-stranded RNA (dsRNA) is recognized by the Dicer enzyme and is chopped into pieces of 21–23 nucleotides called siRNAs. In the maintenance phase, RNA-induced silencing complex (RISC) unwinds the siRNAs, triggers the surveillance mechanism to find siRNA complementary RNA in the cell, and inactivates them with its RNase activity. Alternatively, siRNAs can serve as primers for an RdRP to make double-stranded RNA during the silencing signal amplification in a feedback fashion.

Signal amplification and spreading phase concerns the amplification of a silencing signal and its dissemination throughout the target transcript. This signal is being identified as the siRNAs that originated in the preceding phases, and it is alternatively referred to as transitive RNAi. Guided by an RdRP, in order to amplify and potentiate the silencing response, siRNAs act as activators (primers) for dsRNA polymerization with the use of ssRNA as a template. In plants, siRNA can induce RNA polymerization in both $3' \rightarrow 5'$ and $5' \rightarrow 3'$ directions, whereas in animals, RdRP travels in only one direction ($3' \rightarrow 5'$), along a certain mRNA, to amplify the silencing signal (Vaistij et al. 2002). Production of dsRNA feeds the initiation phase for the production of more siRNAs, and the process continues in a reiterated fashion. Moreover, siRNAs are believed to constitute (at least partially) mobile signals that spread the gene silencing mechanism to other parts of the plant through vascular tissues.

Involvement of PTGS in virus protection was first evident in transgenic plants using potyviral CP cDNA sequence (Lindbo and Dougherty 1992; Van der Vlugt et al. 1992). Lindbo et al. (1993) first proposed PTGS as an antiviral state in plants. This is best achieved when plants are transformed with constructs that express a self-complementary RNA, containing sequences homologous to the target plant virus. Transgene constructs encoding intron-spliced RNA with hairpin structure provided stable silencing to nearly 100% efficiency against homologous plant viruses (Smith et al. 2000). Hairpin constructs can be made using generic vectors such as pHANNIBAL and pHELLSGATE (Wesley et al. 2001; Helliwell and Waterhouse 2003), and 98 to 853 nt sense/antisense arms in hairpin constructs were efficient in silencing 90 to 100% of independent transgenic plants. In addition to transgene expression, transient expression of double-stranded RNA corresponding to viral sequences, either by mechanical inoculation or by *Agrobacterium*-mediated leaf infiltration, can also impart resistance to plant viruses and has been reviewed by Tenllado et al. (2004).

miRNA-Mediated Resistance miRNAs, a class of noncoding (untranslated) RNAs of 20–24 nucleotides, are another type of small RNA products processed from dsRNA hairpin precursors by Dicers. So far, more than 200 miRNA genes have been identified in animals and plants, which are mainly derived from the regions between protein coding genes (Lagos-Quintana et al. 2001, Lau et al. 2001). The loci that encode miRNAs, the MIR genes, can occur in clusters in the genome and may even be transcribed polycistronically and processed sequentially into pre-miRNA and miRNA (Lee et al. 2002).

The involvement of the microRNA (miRNA) pathway in RNA silencing is a notable feature in plants. In *Arabidopsis*, endogenous developmental signals may trigger the formation of some imperfect dsRNAs, which are subsequently diced by DCL1 and/or other DCLs into double-stranded miRNAs. These miRNAs participate in a variety of regulatory processes: Some serve as siRNA molecules in the RNA silencing pathway with perfect or near perfect base complementarity to their mRNA target; some might be recruited into the microRNA ribonucleoprotein complex (miRNP) that further regulates other PTGS processes, such as translational inhibition, with imperfect base-pairing interaction with their targets. The interaction between DCL and ARGONAUTE protein (AGO) may mediate the identification and processing of different dsRNA precursors, which produces different types of small RNAs that are required for either plant defense or development.

15.3 RNAi: Potential Tool in Plant Disease Management

RNAi is a PTGS process which downregulates the gene expression of target gene in a precise manner without affecting the expression of other genes. In this process, a DNA construct is introduced into a cell that produces dsRNA complementary to the gene of interest, which is cleaved into siRNAs by a ribonuclease called DICER or Dicer-like enzyme. The artificial microRNAs (miRNAs) have also been commonly used to activate the RNAi pathway in plants (Gilchrist and Haughn 2010). These miRNA and siRNA in association with RNA-induced silencing complex (RISC), Argonaute, and other effector proteins activate the RNA silencing pathway leading to sequence-specific degradation of target mRNA (Saurabh et al. 2014). The siRNA and miRNA have similar mode of action; however, siRNAs are specific and degrade the expression of only one gene in homology-dependent manner, whereas the miRNAs nonspecifically target the expression of numerous genes (Lam et al. 2015; Moin et al. 2017). RNAi is frequently used to generate mutant lines lacking the expression of some genes, which can be used to identify the function of the gene knockdowns by examining them for variant phenotypes. This technique simplifies the phenotypic assays required in a functional genomics effort, which will otherwise require presence of specific markers and several compelling generations of crosses to detect a specific mutant allele for a genotype (McGinnis 2010). Host gene silencing hairpin RNAi (HGS-hpRNAi) has also been reported as stable gene silencing method employed to increase disease resistance in wide range of host plant species through genetic engineering (Pattanayak et al. 2005; Senthil-Kumar and Mysore 2010).

15.3.1 RNAi for Fungal and Viral Disease Resistance

In potato, RNAi approach has been used extensively for imparting resistance and identifying genes responsible for resistance against pathogens, insects, and viruses that cause significant economic losses. Bhaskar et al. (2009) demonstrated double agro-infiltration of RNAi-based silencing construct and a late blight pathogen effector which can be used for screening candidate genes involved in late blight resistance pathway mediated by the corresponding resistance gene. In another study, Eschen-Lippold et al. (2012) enhanced defense status of potato against *Phytophthora infestans* by downregulating the expression of *syntaxin* gene. They generated transgenic plants expressing RNAi constructs targeted against plasma membrane-localized syntaxin-related 1 (StSYR1) which reduced the growth of *P. infestans* in potato. Further, Sanju et al. (2015) studied host-mediated gene silencing of an RXLR effector Avr3a gene, which is responsible for *P. infestans* virulence, the causal agent of late blight in potato. They observed that siRNA targeted against single-effector Avr3a gene conferred partial resistance to *P. infestans* and indicated the need of targeting cumulative effect of effector genes to achieve complete resistance in potato (Fig. 15.2). Similarly, Thakur et al. (2015) used artificial microRNA for silencing of *P. infestans* single-effector Avr3a gene causing pathogen death or loss of virulence which imparts resistance against late blight in potato. Jahan et al. (2015) also designed and introduced an hpRNA construct containing *GFP* marker gene in potato. They found *hp-PiGPB1* targeting the G protein β-subunit (*PiGPB1*) important for pathogenicity resulting in most restricted disease progress.

Recently, Hameed et al. (2017) have designed the expression cassette to generate dsRNAs having a hairpin loop configuration and developed transgenic potato lines expressing fused viral coat protein coding sequences from potato virus X (PVX), potato virus Y (PVY), and potato virus S (PVS). They have obtained nearly 100% resistance against three RNA viruses PVX, PVY, and PVS infection in transgenic lines compared to untransformed controls exhibiting severe viral disease symptoms. Lately, Tomar et al. (2018) targeted replication-associated protein gene (*AC1*) of ToLCNDV-potato virus by PTGS using hairpin loop construct to confer resistance against apical leaf curl disease in potato (Fig. 15.3).

15.3.2 RNAi for Bacterial Disease Resistance

Bacterial wilt caused by *Ralstonia solanacearum* is one of the most serious diseases in potatoes, with a wide host range of over 200 plant species. Worldwide loss in potato yield due to wilt ranges from 33 to 90%, while it is as high as 70% in India (Sagar et al. 2014). Few available resistant germplasm (wild diploid sources) in potato have failed to provide durable resistance against Indian *R. solanacearum* races since the 2n endosperm balance number (2EBN) of diploid potato is difficult to introgress with the commercial cultivated tetraploid potatoes having 4EBN and which restricts its use in resistance breeding programs. Thus, eradication of this pathogen is a major challenge for potato production.

Recent studies have identified PAP2 as a crucial protein that controls pathogenesis in various plant species. In response to *R. solanacearum* infection, PAP2 acts as a negative regulator and makes it unavailable for ROS burst (Nakano et al. 2013, 2015). Therefore, PAP2 inhibits the accumulation of PA which further interferes with HR response in the host. In some cases, the PAP2 act as a negative regulator by being temporarily inactive under pathogen infection to stimulate plant adaptive defense. Therefore, PAP2 could be silenced for regulating the defense action against *R. solanacearum* in potato. RNAi lines were developed which showed an immune response to wilt symptoms, and few lines showed delayed infection compared to control plants. The 80 Kufri Jyoti RNAi lines were developed and subjected to efficient, quick in vitro hydroponic method of bioassay to avoid the escape of the pathogen (Kajal et al. 2020) as well as whole plant bioassay using the root inoculation method. These lines were screened up to the third clonal generation. The integration, expression, and inheritance of three promising lines displayed enhanced resistance to wilt disease. Avirulent colony assay and bacterial ooze test were used to

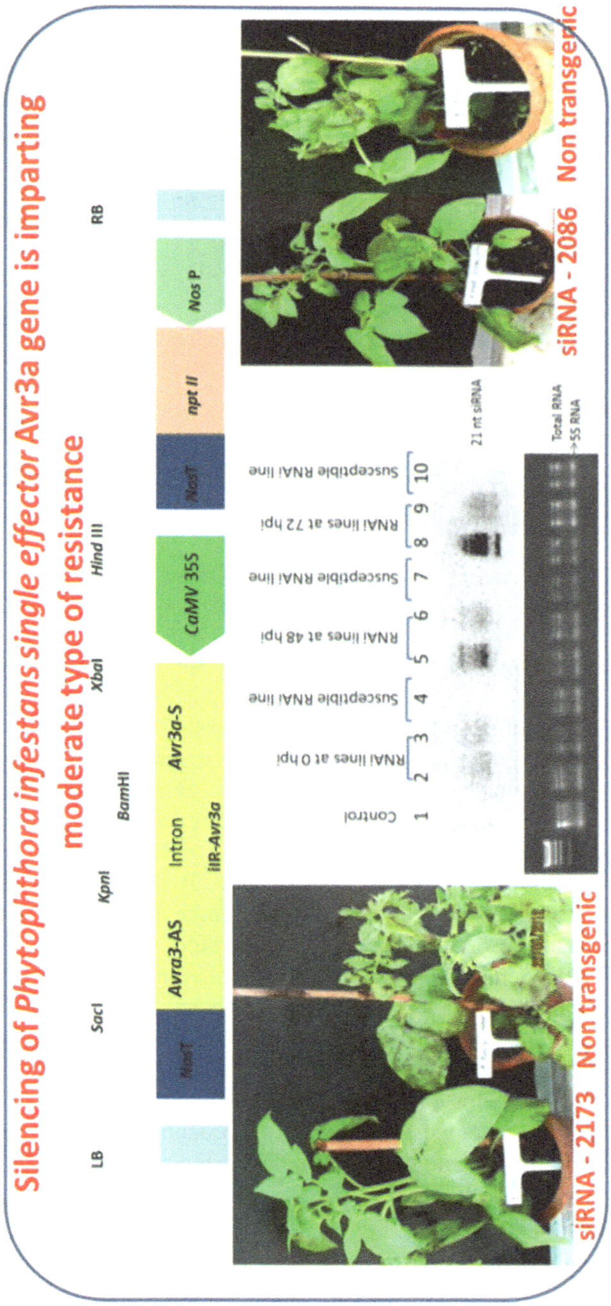

Fig. 15.2 RNAi potato cultivars expressing Avr3a siRNA and moderate response against late blight disease

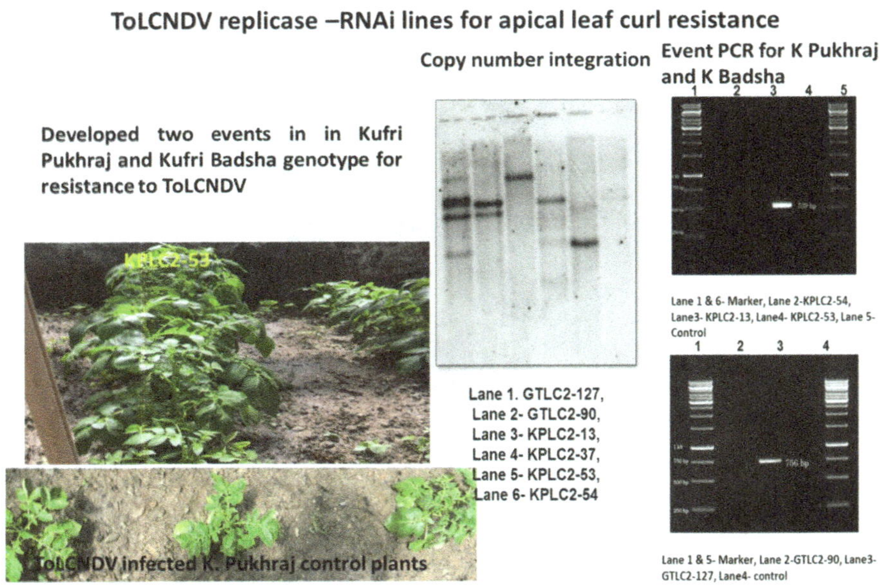

Fig. 15.3 RNAi Kufri Pukhraj resistance response to ToLCNDV and copy number analysis of replicase and event characterization of K. Pukhraj and K. Badsha events

authenticate the silencing of PAP2 in inhibiting the colonization and bacterial load (Fig. 15.4).

It is important to identify and mine the PAP2-regulated PA intracellular signaling molecule to understand and assign its downstream defense pathway in *R. solanacearum* and potato host interaction. This system can be manipulated at a molecular level and can be used as an economic and eco-friendly disease control method. Moreover, it is difficult for the pathogens to overcome PRR recognition as the PAMPs are conserved molecules (Lacombe et al. 2010; Lu et al. 2018). This is the first evidence that provides the resistance source in potato for developing bacterial wilt-resistant varieties. These sources would serve as future breeding material for the management of bacterial wilt disease in potatoes. This approach provides plant-mediated silencing of susceptible gene PAP2 for creating novel genetic resources in potato for the management of bacterial wilt disease. This invention provides the tool for management of bacterial wilt disease in other solanaceous crops and helps in discovery of R gene in potato. This approach provides a first genetic resources in potato in the country for the management of bacterial wilt since none of the potato germplasms have resistance source against *R. solanacearum* (Kajal et al. 2020).

(a) Symptoms of plants

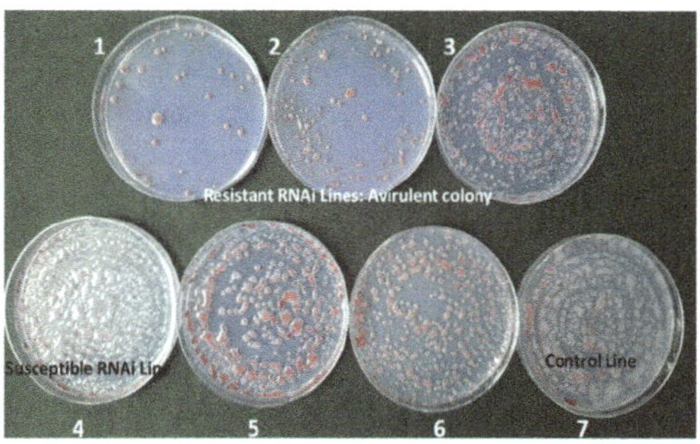

(b) TZC Assay for virulence

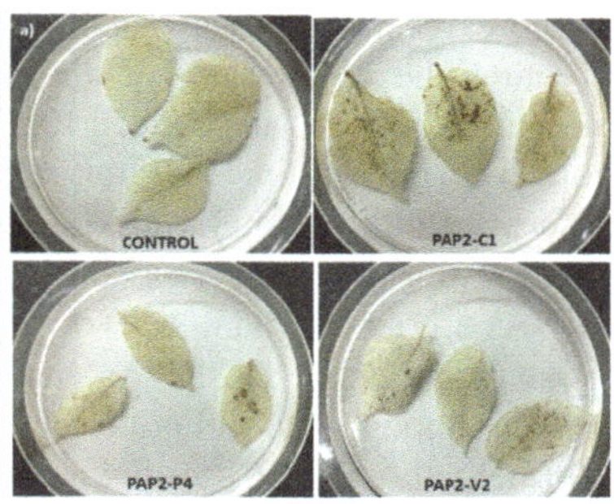

(c) DAB Staining for HR

Fig. 15.4 Resistance response of RNAi lines against wilt disease: (**a**) symptoms on plants, (**b**) TZC test for virulence assay, and (**c**) DAB staining for HR response of RNAi and susceptible lines (Adapted from Kajal et al. 2020)

15.4 Topical Application of dsRNA: New Innovation Area of RNAi

The importance of RNAi in sustainable agriculture research on a range of potential applications of RNAi in crop protection is increasing, and it is becoming apparent that RNAi-based approaches could make a major contribution toward integrated pest management and sustainable agriculture.

Mitter et al. (2017a) explored the use of dsRNA complexed with layered double hydroxide (LDH) nanosheets, termed BioClay, as a spray application (Mitter et al. 2017b). Employing BioClay allowed the window of protection from viral pathogens to be expanded to 20 or more days. Importantly, LDH itself is biocompatible and used in human therapeutics (Del Hoyo 2007; Kuthati et al. 2015). LDH also safely degrades in the presence of mildly acidic conditions, thus minimizing the risk of excessive persistence of the dsRNA in the environment. Abating risk while maintaining effectiveness will require similarly novel solutions during the conception of many RNAi-based products, indicating the benefits of risk identification at the earliest stages of development.

Given the devastation caused by fungal pathogens to crop yield worldwide, the successful topical application of dsRNA to control a fungal infection is significant. Koch et al. (2016) showed that *Fusarium graminearum* growth could be inhibited by direct application on detached barley leaves of a dsRNA targeting three CYP450 genes (Koch et al. 2016). By targeting two Dicer-like genes in *Botrytis cinerea*, Wang and coworkers effectively controlled the pathogen on fruit, vegetable, and flower surfaces, demonstrating that RNAi could play a role in the postharvest protection of agricultural produce in addition to preharvest protection (Wang et al. 2016). McLoughlin et al. (2018) were also able to decrease fungal infection and reduce symptoms in *B. cinerea*, as well as *Sclerotinia sclerotiorum*, via foliar application of dsRNA on Arabidopsis and Brassica napus leaves.

More recently, it was reported that spraying of in vitro transcribed dsRNA targeting the myosine 5 of *Fusarium asiaticum* in wheat (*Triticum aestivum*) coleoptiles resulted in reduced fungal virulence (Song et al. 2018). Recent studies from the Hailing Jin laboratory (Wang et al. 2017) have clearly established the concept of bidirectional transkingdom transport of siRNAs and mechanistic details of RNA transport between plants and fungi. *B. cinerea* was able to not only deliver siRNAs into host plant cells to suppress host immunity genes but also uptake exogenously applied dsRNAs and siRNAs that inhibit its growth (Weiberg et al. 2013). Although it is still not clearly defined how the in vitro-applied dsRNA travels into the pathogen cells to downregulate the expression of the target genes, studies indicate that the small RNAs can be transferred between tissues and cells, inside an organism, by either symplast (direct internal connection) or apoplast (externally). Based on these research hypotheses, Sundaresha et al. proposed the hypothetical model (Fig. 15.5) for the dsRNA interaction between potato and *P. infestans* assuming that sprayed dsRNA molecule enters the host as well as processed siRNA into fungal cells via nutrient uptake.

The use of transient gene silencing was first reported in *P. infestans* by direct in vitro delivering of synthesized dsRNA in its protoplast (Whisson et al. 2005). The study proved that the secretion of extracellular membrane vesicles called exosomes assists the delivery and transport of small RNAs between Arabidopsis and pathogenic fungi *Botrytis cinerea*. These exosomes deliver the host small RNAs to the fungi and silence their infection and virulence genes (Cai et al. 2018). *P. infestans* produces haustorium, which penetrates the host cells serving as a portal for the secretion of proteins, signal molecules, virulence effectors, and nutrient uptake (Wang et al. 2018).

Numerous studies have aimed to reveal that the gene silencing of the pathogen (fungi) cannot be monitored until the formation of haustoria. Effective silencing was

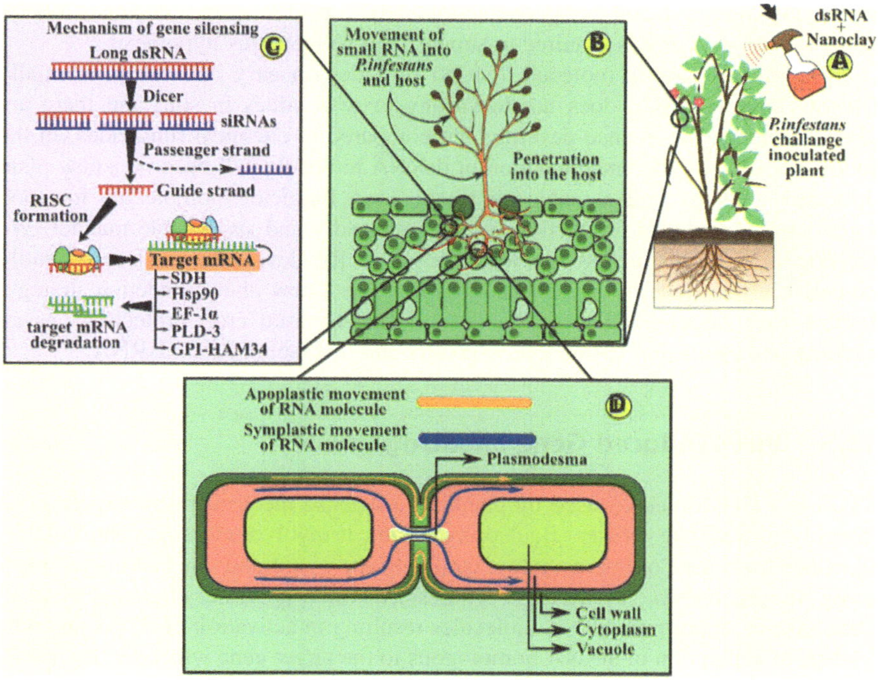

Fig. 15.5 Illustration of mechanism of *P. infestans* gene silencing by spray-induced gene silencing. (**a**) Spraying of dsRNA-nanoclay formulation on *P. infestans* challenge-inoculated plants. (**b**) Penetration of *P. infestans* into the host plant. (**c**) Mechanism of gene silencing by RNAi. dsRNA is acted upon by RNase III-type endonucleases called Dicer specific to dsRNA and cleaves it into smaller 21–23 nucleotide long dsRNA intermediates. These dsRNA intermediates are then acted upon by RNA helicase and other accessory proteins to form a single-stranded siRNA-containing RNA-induced silencing complex (RISC). siRNA having sequence complementarity to the target RNA serves as guide RNA, while the other strand called passenger RNA is degraded. RISC then guides the sequence-specific degradation of near-complementary or complementary target mRNAs of the target genes whose information was present in the dsRNA. (**d**) Movement of small RNA into the host via apoplastic and symplastic movement. (Adapted from Sundaresha et al. 2021, Unpublished)

observed after the haustorial structure formation, and silencing was more efficient against the genes that were greatly expressed in the haustoria than that expressed in other parts of the plants (Yin et al. 2010; Panwar et al. 2013). These findings clearly indicate that the siRNA moves into the pathogen from the host through haustoria or similar structures and also that membrane-bound extracellular vesicles of plants are most probable determinants for RNA delivery into the pathogen (Cai et al. 2018; Micali et al. 2011). In the pathosystem of *P. infestans,* the HIGS signals (RNAs) could have possibly traveled from host to the parasite (*P. infestans*) by using the haustorial interphase and using vesicle-mediated transport using exosomes though experimental evidence is required to support and validate this hypothesis.

Spraying of dsRNA molecule (SIGS) provides an easy and environmentally friendly approach as it does not leave any toxic residues in soil, and there are fewer chances of resistance development compared to chemical fungicides. In the future, the utility and versatile action of dsRNA molecule will become a new plant protection strategy and a viable next-generation fungicide/biomolecule for food safety and agricultural production in an eco-friendly and sustainable manner. We are hopeful that this technology would result in the development of eco-friendly potential biomolecule in the field of agriculture as a new plant protection strategy. Further, our strategy relies on the use of "RNAi-based crop protection" as an exciting and promising option due to greater and diverse utility of dsRNA.

15.5 Virus-Induced Gene Silencing (VIGS)

VIGS is a PTGS method used by plants as a defense mechanism by targeting the integrity of invading viruses (Baulcombe 1996). It involves cloning a short cDNA sequence from gene of interest into a viral delivery vector and transfecting the plant using *Agrobacterium*. A dsRNA is synthesized which is further degraded by plant Dicer-like enzymes into siRNA molecules resulting in activation of PTGS and thus leading to generation of siRNA homologous to the target gene which finally results into silencing of the endogenous plant gene (Senthil-Kumar and Mysore 2014). Several vectors such as tobacco mosaic virus (TMV), potato virus X (PVX), tobacco rattle virus (TRV), tomato golden mosaic virus (TGMV), apple latent spherical virus (ASLV), cabbage leaf curl virus (CbLCV), and barley stripe mosaic virus (BSMV) have been developed for VIGS usage in various plant species (Burch-Smith et al. 2004; Senthil-Kumar and Mysore 2010). However, in potato, PVX and TRV vectors have been found suitable for VIGS-based silencing. Faivre-Rampant et al. (2004) found PVX-based VIGS vector efficiently silencing the phytoene desaturase (PDS) gene in leaves and tubers of both wild diploid and cultivated tetraploid *Solanum* species. They have also reported that VIGS could be triggered and sustained in in vitro micropropagated tetraploid potato and on in vitro-generated microtubers. Brigneti et al. (2004) used TRV vector for VIGS and silenced PDS gene in the diploid wild species *S. bulbocastanum* and *S. okadae*, in the cultivated tetraploid *S. tuberosum*, and in the distant hexaploid relative *S. nigrum*. They have also silenced known resistance genes R1 and Rx in *S. tuberosum* and RB in *S. bulbocastanum* and obtained susceptible phenotypes in detached leaf tests. Recently, Jeevalatha et al. (2017) used VIGS system for functional analysis of genes in *Tomato leaf curl New Delhi virus* (ToLCNDV)-susceptible potato cultivar Kufri Pukhraj by silencing three genes, viz., TMV-induced protein 1–2 gene, peripheral-type benzodiazephine receptor, and conserved gene of unknown function. So VIGS has been proved as a valuable tool in identification of plant genes involved in infection and in resistance to begomoviruses. Neha and Sundaresha's group demonstrated the use of VIGS tool to confirm the role of CDF allele in earliness in

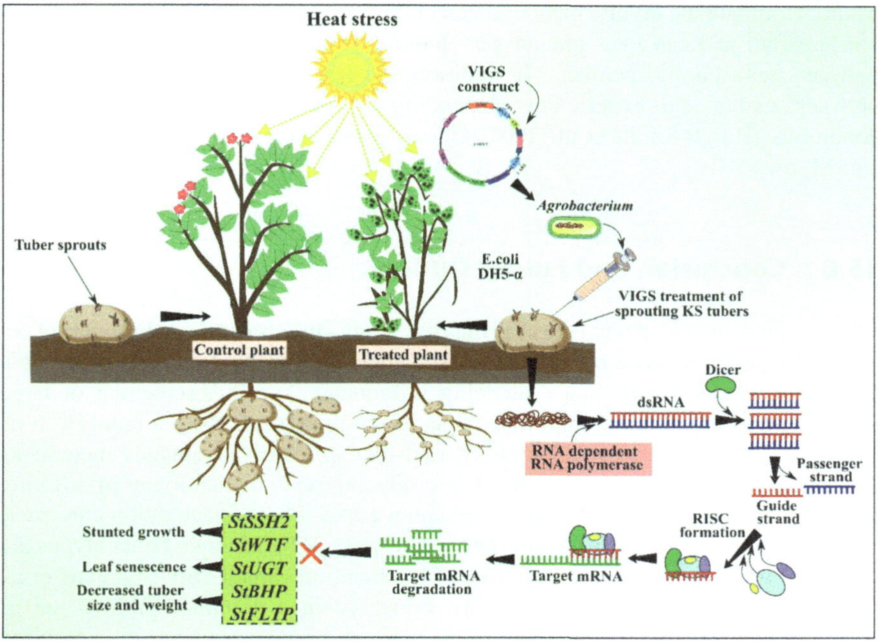

Fig. 15.6 Schematic illustration of the mechanism of gene silencing by VIGS and the phenotypic effect of StSSH2, StWTF, StBHP, StFLTP, and StUGT silencing on tuberization and plant growth under heat stress. (Adapted from Tomar et al. 2021)

Indian potato cultivar (Salaria et al. 2020). Similarly, Tomar et al. (2021) proved the molecular response of tuberization in heat-tolerant Kufri Surya cultivars using VIGS technique and proposed the model for VIGS in functional assignment of tuberization genes (Fig. 15.6). Functional validation through silencing tools could help to better understand the complexity in the tuberization pathway. This could help in authenticating the role of each gene in tuberization induction and tuber development under elevated temperatures. This study of silencing tuberization networking genes could open new research avenues for mining the possibilities of metabolic pathways and their impact on potato physiology and tuber development in relation to an elevated temperature for the development of potato cultivars for tropical regions of the country.

The main advantages of VIGS include its low cost and rapid performance by identifying a loss of function phenotype for a particular gene within a single generation. Since, its expression is transient in nature; therefore, it does not require the laborious transformation procedures for the development of transgenic plants (Burch-Smith et al. 2004; Singh et al. 2018). Therefore, it is extensively used as powerful tool for decoding the functional relevance of the genes (Becker and Lange 2010). However, this technique has certain limitations as the phenotypes obtained are not heritable; hence, it cannot be used for genetic engineering. Also, VIGS cannot eliminate the involvement of a gene for a particular function if a phenotype is not apparent and can miss phenotypes that are masked by functional redundancy between gene family members. In addition, the levels of silencing can also vary between plants and experiments depending on the construct and the growth conditions (Burch-Smith et al. 2004; Gilchrist and Haughn 2010; Senthil-Kumar and Mysore 2010).

15.6 Conclusion and Future Outlook

RNAi offers many advantages, primarily being its usage for discovering or validating gene functions along with genetic engineering studies due to its heritable expression. Additionally, since silencing is sequence specific, screening of large populations is not required, and transcripts of multiple genes from a family can be silenced by a single construct in polyploid plants. Other advantages include its partial loss of function characteristic, thus producing several phenotypes of differing severity which can aid in analysis of essential genes whose inactivation can cause lethality or extremely severe pleiotropic phenotypes. RNAi-induced phenotypes are dominant which can be observed in the T1 generation (Small 2007; Eamens et al. 2008; McGinnis 2010). Genes that are expressed in a dominant fashion are of particular interest since backcrossing to achieve homozygosity is not required especially in potato. However, gene silencing approaches allow for reduction in the expression of specific genes, resulting in a dominant negative effect (Pandey et al. 2015), eliminating the need to achieve homozygosity for traits that are normally inherited recessively, such as resistance to some diseases.

Meanwhile, the RNAi technology might improve vastly with better designed virus-based vectors for delivery of siRNAs as well as CRISPR gRNA to the appropriate tissues at the appropriate time. Such technology is bound to give a new shape to theory of RNAi silencing as well. The science and technology of RNAi has given us a cultural ocean of virtually bottomless depth as RNA- and DNA-based vectors apply as a versatile tool in understanding the gene function and genome editing technology to boost crop improvement program.

References

Baulcombe DC (1996) Mechanisms of pathogen-derived resistance to viruses in transgenic plants. Plant Cell 8:1833–1844

Becker A, Lange M (2010) VIGS genomics goes functional. Trends Plant Sci 15:1–4

Bernstein E, Caudy AA, Hammond SM, Hannon CJ (2001) Role for a bidentate ribonuclease in the initiation step of RNA interference. Nature 409:363–366

Bhaskar PB, Venkateshwaran M, Wu L, Ane JM, Jiang J (2009) Agrobacterium-mediated transient gene expression and silencing: a rapid tool for functional gene assay in potato. PLoS One 4(6): e5812. https://doi.org/10.1371/journal.pone.0005812

Brigneti G, Martín-Hernández AM, Jin H, Chen J, Baulcombe DC, Baker B, Jones JD (2004) Virus-induced gene silencing in *solanum* species. Plant J 39(2):264–272

Burch-Smith TM, Anderson JC, Martin GB, Dinesh-Kumar SP (2004) Applications and advantages of virus-induced gene silencing for gene function studies in plants. Plant J 39(5):734–746

Cai Q, Qiao L, Wang M, He B, Lin FM, Palmquist J, Da HS, Jin H (2018) Plants send small RNAs in extracellular vesicles to fungal pathogen to silence virulence genes. Science 360:1126–1129

Del Hoyo C (2007) Layered double hydroxides and human health: An overview. Appl Clay Sci 36: 103–121. https://doi.org/10.1016/j.clay.2006.06.010

Eamens A, Wang MB, Smith NA, Waterhouse PM (2008) RNA silencing in plants: yesterday, today, and tomorrow. Plant Physiol 147:456–468

Eschen-Lippold L, Landgraf R, Smolka U, Schulze S, Heilmann M, Heilmann I, Hause G, Rosahl S (2012) Activation of defense against *Phytophthora infestans* in potato by down-regulation of syntaxin gene expression. New Phytol 193(4):985–996

Faivre-Rampant O, Gilroy EM, Hrubikova K, Hein I, Millam S, Loake GJ, Birch P, Taylor M, Lacomme C (2004) Potato virus X-induced gene silencing in leaves and tubers of potato. Plant Physiol 134:1308–1316

Fire A, Xu SQ, Montgomery MK, Kostas SA, Driver SE, Mello CC (1998) Potent and specific genetic interference by double-stranded RNA in *Caenorhabditis elegans*. Nature 391:806–811

Gilbert W (1986) Origin of life: the RNA world. Nature 319:618. https://doi.org/10.1038/319618a0

Gilchrist E, Haughn G (2010) Reverse genetics techniques: engineering loss and gain of gene function in plants. Brief Func Genom 9(2):103–110

Hameed A, Tahir MN, Asad S, Bilal R, Eck JV, Jander G, Mansoor S (2017) RNAi-mediated simultaneous resistance against three RNA viruses in potato. Mol Biotechnol 59:73–83

Hamilton AJ, Baulcombe DC (1999) A species of small antisense RNA in post-transcriptional gene silencing in plants. Science 286:950–952

Hammond SM, Bernstein E, Beach D, Hannon CJ (2000) An RNA-directed nuclease mediates post-transcriptional gene silencing in *drosophila* cells. Nature 404:293–296

Hannon GJ (2002) RNA interference. Nature 418:244–251

Helliwell C, Waterhouse P (2003) Constructs and methods for high-throughput gene silencing in plants. Methods 30:289–295

Jahan SN, Asman AKM, Corcoran P, Fogelqvist J, Vetukuri RR, Dixelius C (2015) Plant-mediated gene silencing restricts growth of the potato late blight pathogen Phytophthora infestans. J Exp Bot 66(9):2785–2794

Jeevalatha A, Siddappa S, Kumar A, Kaundal P, Guleria A, Sharma S, Nagesh M, Singh BP (2017) An insight into differentially regulated genes in resistant and susceptible genotypes of potato in response to tomato leaf curl New Delhi virus-[potato] infection. Virus Res 232:22–33

Kajal T, Sagar V, Sundaresha S, Salaria N, Goutam U, Sharma S (2020) An efficient and quick in-vitro method of evaluation of potato genotypes against *Ralstonia solanacearum*. Plant cell Biotechnol Mol Biol 21:46–52

Koch A, Biedenkopf D, Furch A, Weber L, Rossbach O, Abdellatef E et al (2016) An RNAi-based control of *fusarium graminearum* infections through spraying of long ds RNAs involves a plant passage and is controlled by the fungal silencing machinery. PLoS Pathog 12(10):e1005901. https://doi.org/10.1371/journal.ppat.100590

Kooter JM, Matzke MA, Meyer P (1999) Listening to the silent genes: transgene silencing, gene regulation and pathogen control. Trends Plant Sci 4:340–347

Kuthati Y, Kankala RK, Lee CH (2015) Layered double hydroxide nanoparticles for biomedical applications: current status and recent prospects. Appl Clay Sci 112:100–116. https://doi.org/10.1016/j.clay.2015.04.018

Lacombe S, Rougon-Cardoso A, Sherwood E, Peeters N, Dahlbeck D, Peter van Esse P, Smoker M, Rallapllli G, Thomma BPH, Staskawicz B, Jones JDG, Zipfel C (2010) Interfamily transfer of a plant pattern-recognition receptor confers broad-spectrum bacterial resistance. Nat Biotech 28: 365–369

Lagos-Quintana M, Rauhut R, Lendeckel W, Tuschl T (2001) Identification of novel genes coding for small expressed RNAs. Science 294:853–858

Lam JK, Chow MY, Zhang Y, Leung SW (2015) siRNA versus miRNA as therapeutics for gene silencing. Mol Therapy-Nucleic Acids 4:e252. https://doi.org/10.1038/mtna.2015.23

Lau NC, Lim LP, Weinstein EG, Bartel DP (2001) An abundant class of tiny RNAs with probable regulatory roles in *Caenorhabditis elegans*. Science 294:858–862

Lee Y, Jeon K, Lee JT, Kim S, Kim VN (2002) MicroRNA maturation: stepwise processing and subcellular localization. EMBO J 21:4663–4670

Lindbo J, Silva-Rosales L, Proebsting W, Dougherty W (1993) Induction of a highly specific antiviral state in transgenic plants: implications for regulation of gene expression and virus resistance. Plant Cell 5:1749–1759

Lindbo JA, Dougherty WG (1992) Pathogen-derived resistance to a potyvirus: immune and resistant phenotypes in transgenic tobacco expressing altered forms of a potyvirus coat protein nucleotide sequence. Mol Plant Microbe Interact 5:144–153

Lu H, Lema AS, Planas-Marques M, Alonso-Diaz A, Valls M, Col NS (2018) Type III secretion-dependent and -independent phenotypes caused by *Ralstonia solanacearum* in *Arabidopsis* roots. MPMI 31:175–184

Matzke MA, Matzke AJ, Pruss GJ, Vance VB (2001) RNA-based silencing strategies in plants. Curr Opin Genet Dev 11:221–227

McGinnis KM (2010) RNAi for functional genomics in plants. Brief Funct Genom 9(2)1:111–117

McLoughlin AG, Wytinck N, Walker PL, Girard IJ, Rashid KY, De Kievit T, Fernando WGD, Whyard S, Belmonte MF (2018) Identification and application of exogenous dsRNA confers plant protection against *Sclerotinia sclerotiorum* and *Botrytis cinerea*. Sci Rep 8:7320. https://doi.org/10.1038/s41598-018-25434-4

Micali CO, Neumann U, Grunewald D, Panstruga R, O'Connell R (2011) Biogenesis of a specialized plant-fungal interface during host cell internalization of *Golovinomyces orontii* haustoria. Cell Microbiol 13:210–226

Mitter N, Worrall EA, Robinson KE, Li P, Jain RG, Taochy C et al (2017b) Clay nanosheets for topical delivery of RNAi for sustained protection against plant viruses. Nat Plants 3:16207. https://doi.org/10.1038/nplants.2016.207

Mitter N, Worrall EA, Robinson KE, Xu ZP, Carroll BJ (2017a) Induction of virus resistance by exogenous application of double-stranded RNA. Curr Opin Virol 26:49–55. https://doi.org/10.1016/j.coviro.2017.07.009

Moin M, Bakshi A, Maheswari M, Kirti PB (2017) Small interfering RNA-mediated regulation of gene expression and its role as a plant reverse genetic tool. Ind J Plant Physiol 22(4):549–557

Morange M (2008) What history tells us. XIII. Fifty years of the central dogma. J Biosci 33:171–175

Nakano M, Nishihara M, Oshioka HY, Takahashi H, Sawasaki T, Ohnishi K, Hikichi Y, Kiba A (2013) Suppression of DS1 phosphatidic acid phosphatase confirms resistance to *Ralstonia solanacearum* in *Nicotiana benthamiana*. PLoS One 8:e75124. https://doi.org/10.1371/journal.pone.0075124

Nakano M, Yoshioka H, Ohnishic K, Hikichia Y, Kiba A (2015) Cell death-inducing stresses are required for defense activation in DS1-phosphatidic acid phosphatase-silenced *Nicotiana benthamiana*. J Plant Physiol 184:15–19

Napoli C, Lemieux C, Jorgensen R (1990) Introduction of chimeric chalcone synthase gene in petunia results in reversible co-suppression of homologous genes in trans. Plant Cell 2:279–289

Nykänen A, Haley B, Zamore PD (2001) ATP requirements and small interfering RNA structure in the RNA interference pathway. Cell 107:309–321

Pandey P, Senthil-Kumar M, Mysore K (2015) Advances in plant gene silencing methods. In: Mysore KS, Senthil-Kumar M (eds) Plant gene silencing. Springer, New York, pp 3–23

Panwar V, McCallum B, Bakkeren G (2013) Host-induced gene silencing of wheat leaf rust fungus *Puccinia triticina* pathogenicity genes mediated by the barley stripe mosaic virus. Plant Mol Biol 81:595–608

Pattanayak D, Agarwal S, Sumathi S, Chakrabarti SK, Naik PS, Khurana SMP (2005) Small but mighty RNA-mediated inference in plants. Indian J Exp Biol 43:7–24

Robertson MP, Joyce GF (2012) The origins of the RNA world. Cold Spring Harbour Perspect Biol 4(5):a003608. https://doi.org/10.1101/cshperspect.a003608

Sagar V, Gurjar MS, Arjunan J, Bakade RR, Chakrabarti SK, Arora RK, Sharma S (2014) Phylotype analysis of Ralstonia solanacearum strains causing potato bacterial wilt in Karnataka in India. African J Microbiol Res 8:1277–1281

Salaria N, Sidappa S, Thakur K, Tomar M, Goutam U, Sharma N, Sood S, Bhardwaj V, Singh B (2020) Solanum tuberosum (CYCLING DOF FACTOR) CDF1.2 allele: A candidate gene for developing earliness in potato. S Afr J Bot 132:242–248

Sanju S, Siddappa S, Thakur A, Shukla PK, Srivastava N, Pattanayak D, Sharma S, Singh BP (2015) Host-mediated gene silencing of a single effector gene from the potato pathogen Phytophthora infestans imparts partial resistance to late blight disease. Funct Integr Genomics 15:697–706

Saurabh S, Vidyarthi AS, Prasad D (2014) RNA interference: concept to reality in crop improvement. Planta 239:543–564

Senthil-Kumar M, Mysore KS (2010) RNAi in plants: recent developments and applications in agriculture. In: Catalano AJ (ed) Gene silencing: theory, techniques and applications. Nova Scientific Publishers Inc., New York, pp 183–199

Senthil-Kumar M, Mysore KS (2014) Tobacco rattle virus-based virus-induced gene silencing in *Nicotiana benthamiana*. Nat Protoc 9(7):1549–1562

Singh B, Kukreja S, Goutam U (2018) Milestones achieved in response to drought stress through reverse genetic approaches. F1000Research 7:1311. https://doi.org/10.12688/f1000research.15606.1

Small I (2007) RNAi for revealing and engineering plant gene functions. Curr Opin Biotech 18: 148–153

Smith NA, Singh SP, Wang MB, Stoutjesdijk PA, Green AG, Waterhouse PM (2000) Gene expression: total silencing by intron-spliced hairpin RNAs. Nature 407:319–320

Song XS, Gu KX, Duan XX, Xiao XM, Hou YP, Duan Y, Wang JX, Na Y, Zhou MG (2018) Secondary amplification of siRNA machinery limits the application of spray-induced gene silencing. Mol Pl Pathol 19:2543–2560

Sundaresha S, Sharma S, Bairwa A, Tomar M, Kumar R, Bhardwaj V, Jeevalatha A, Bakade R, Salaria N, Thakur K, Singh BP, Chakrabarti SK (2021) Spraying of dsRNA molecules derived from Phytophthora infestans, as a plant protection strategies for the management of potato late blight. Preprints 2021020280. https://doi.org/10.20944/preprints202102.0280.v1

Tenllado F, Llave C, Diaz-Ruiz JR (2004) RNA interference as a new biotechnological tool for the control of virus diseases in plants. Virus Res 102:85–96

Thakur A, Sanju S, Siddappa S, Srivastava N, Shukla PK, Pattanayak D, Sharma S, Singh BP (2015) Artificial microRNA mediated gene silencing of *Phytophthora infestans* single effector Avr3a gene imparts moderate type of late blight resistance in potato. Plant Pathol J 14(1):1–12

Tomar G, Chakrabarti SK, Sharma NN, Jeevalatha A, Sundaresha S, Vyas K, Azmi W (2018) RNAi based transgene conferred extreme resistance to the geminivirus causing apical leaf curl disease in potato. Plant Biotechnol Rep 12:195–205

Tomar M, Sundaresha S, Singh B, Bhardwaj V, Sood S, Singh B, Salaria N, Thakur K, Kumar A, Sharma N, Goutam U (2021) Validation of molecular response of tuberization in response to elevated temperature by using a transient virus induced gene silencing (VIGS) in potato. Funct Integr Genomics 21:215–229

Vaistij FE, Jones L, Baulcombe DC (2002) Spreading of RNA targeting and DNA methylation in RNA silencing requires transcription of the target gene and a putative RNA dependent RNA polymerase. Plant Cell 14:857–867

Van der Vlugt RA, Ruiter RK, Goldbach R (1992) Evidence for sense RNA-mediated protection to *PVYN* in tobacco plants transformed with the viral coat protein cistron. Plant Mol Biol 20:631–639

Vaucheret H, Beclin C, Fagard M (2001) Post-transcriptional gene silencing in plants. Cell Sci 114: 3083–3091

Wang M, Thomas N, Jin H (2017) Cross-kingdom RNA trafficking and environmental RNAi for powerful innovative pre- and post-harvest plant protection. Curr Opin Pl Biol 38:133–141

Wang M, Weiberg A, Lin FM, Thomma BPHJ, Da HH, Jin H (2016) Bidirectional cross-kingdom RNAi and fungal uptake of external RNAs confer plant protection. Nature Pl 2:16151. https://doi.org/10.1038/nplants.2016.151

Wang S, Welsh L, Thorpe P, Whisson SC, Boevink PC, Birch PRJ (2018) The *Phytophthora infestans* haustorium is a site for secretion of diverse classes of infection-associated proteins. mBio 9:e01216–e01218. https://doi.org/10.1128/mBio.01216-18

Waterhouse PM, Wang MB, Lough T (2001) Gene silencing as an adaptive defence against viruses. Nature 411:834–842

Weiberg A, Wang M, Lin FM, Zhao H, Zhang Z, Kaloshian I, Da HH, Jin H (2013) Fungal small RNAs suppress plant immunity by hijacking host RNA interference pathways. Science 342: 118–123

Wesley SV, Helliwell CA, Smith NA (2001) Construct design for efficient, effective and high-throughput gene silencing in plants. Plant J 27:581–590

Whisson SC, Avrova AO, Van West P, Jones JT (2005) A method for double-stranded RNA-mediated transient gene silencing in *Phytophthora infestans*. Mol Pl Pathol 6:153–163

Yin GH, Sun ZN, Song YZ, An HL, Zhu CX, Wen FJ (2010) Bacterially expressed double-stranded RNAs against hot-spot sequences of tobacco mosaic virus or potato virus y genome have different ability to protect tobacco from viral infection. App Biochem Biotech 162:1901–1914

Zamore PD, Tuschl T, Sharp PA, Bartel DP (2000) RNAi: double-stranded RNA directs the ATP-dependent cleavage of mRNA at 21 to 23 nucleotide intervals. Cell 101:25–33

New Chemistry Pesticides for Management of Potato Pests 16

Thomas P. Kuhar and Chris McCullough

Abstract

Pesticides have been the most widely used tactic to protect potatoes from various insect pests and disease-causing agents. For over a century, a litany of chemicals has been applied to the crop, leading to problems with environmental and human safety, resistance development in certain key pests, and destruction of natural enemies that otherwise would contribute to a sound IPM program. Pesticides will likely remain the base of pest management for the foreseeable future, but a switch to more selective and less environmentally toxic tools has and will continue to lead to more sustainable potato production. In this chapter we briefly review the history and pitfalls of pesticide use in potato production and discuss strategies and novel insecticide chemistries that are available to growers today.

Keywords

Insecticide resistance · Potato pests · Organochlorines · IRAC · Neonicotinoids · IGRs · Spinosyns · Botanicals · Microbial pesticides

16.1 Introduction

Chemical pesticides have been one of the most widely used tools for management of insects and diseases in potato production for over a century. While their use has led to significant increases in potato yields, the indiscriminate and excessive use of certain insecticides, nematicides, and fungicides has resulted in environmental and

T. P. Kuhar (✉) · C. McCullough
Department of Entomology, Virginia Tech, Blacksburg, VA, USA
e-mail: tkuhar@vt.edu

S. Kumar Chakrabarti et al. (eds.), *Sustainable Management of Potato Pests and Diseases*, https://doi.org/10.1007/978-981-16-7695-6_16

human health concerns, destruction of nontarget beneficial organisms, as well as pesticide resistance development in key pests (Zehnder and Warthen 1988). Fortunately, in the past couple of decades, more selective pesticides have been registered for use on potatoes with minimal impact on human health, lower toxicity to nontarget organisms, low potential for environmental contamination, and low use rates. Judicial use of these chemicals within the framework of an integrated pest management strategy is a key component toward more sustainable pest management in potatoes.

16.1.1 Lessons Learned from Early Chemical Control in Potatoes

One of the first groups of pesticides used on potatoes was the arsenicals. In the late 1800s to mid-1900s in North America, arsenic-containing compounds such as acetoarsenite of copper (called Paris green) as well as lead arsenate and calcium arsenate were used to control Colorado potato beetle, *Leptinotarsa decemlineata* (Say) (Coleoptera: Chrysomelidae) as well as other pests (Gauthier et al. 1981). Arsenicals were also widely used as fungicides. However, human health concerns related to the accumulation of arsenic in soils and ground water, as well as inherent challenges with mixing, application, and phytotoxicity, resulted in an end to their use on potatoes by the 1980s (Gauthier et al. 1981). Other heavy metal-containing poisons such as bichloride of mercury and Bordeaux powder, a mixture of copper sulfate and lime as well as nicotine sulfate, derived from tobacco, were also commonly used on agricultural crops in the early 1900s. However, as was the case with the arsenicals, human health and environmental concerns eventually brought an end to the use of these compounds on food crops such as potatoes.

By the middle of the twentieth century, the discovery of the insecticidal properties of dichlorodiphenyltrichloroethane, commonly known as DDT, offered a powerful long-lasting insecticide that killed a broad spectrum of pests with reduced acute mammalian toxicity (Ware and Whitacre 2004; Conis 2017). Beginning in the 1950s, DDT as well as several other chlorinated hydrocarbon insecticides including dieldrin, heptachlor, methoxychlor, endosulfan, and aldrin were used on potatoes to provide effective control of most aboveground insect pests (Gauthier et al. 1981). In addition, soil-applied DDT and aldrin became the standard treatment for subterranean pests such as wireworms (Coleoptera: Elateridae) in many parts of the world (Merrill 1952; Gunning and Forrester 1984; Parker and Howard 2001). However, in the 1950s, resistance to DDT occurred in Colorado potato beetle populations in the United States (Hofmaster 1956; Cutkomp et al. 1958), and cross-resistance to several of the other chlorinated hydrocarbons would soon follow (Gauthier et al. 1981; Alyokhin et al. 2008a). In addition to pesticide resistance development, DDT and other chlorinated hydrocarbons had serious environmental and health concerns including persistence in the environment, nontarget effects, and bioaccumulation that impacted vertebrate animals (Kelce et al. 1995; Holm et al. 2006, Cohn et al. 2007; Mrema et al. 2013). The United States banned the use of DDT in 1972, as well as the agricultural uses of most of the other chlorinated hydrocarbons by 1980. As chlorinated hydrocarbons were retired, different classes of neuroactive insecticides, including organophosphates, carbamates, and pyrethroids, were relied upon to provide control of potato pests (Gerhardt and Turley 1961; Harding 1962; Gerhardt 1966; Cranshaw 1997; Kuhar et al. 2012).

Organophosphates and carbamates are cholinesterase-inhibiting neurotoxins that have broad-spectrum activity against most insect pests attacking potatoes (Ware and Whitacre 2004). In addition, many of the organophosphates like disulfoton, fensulfothion, and phorate, as well as the carbamates aldicarb, oxamyl, and carbofuran, offered systemic insecticide options that could be applied in the soil for control of aboveground insect pests. Organophosphates and carbamates also became the top choice for control of soil pests such as wireworms (Hancock et al. 1986; Toba 1987; Noetzel and Ricard 1988; Jansson et al. 1988; Parker et al. 1990; Sorensen and Kidd 1991; Pavlista 1997; Shamiyeh et al. 1999; Nault and Speese 2000; Kuhar et al. 2003) and nematodes (Haydock et al. 2006; Deliopoulos et al. 2010). However, most of these insecticides like aldicarb, bendiocarb, carbofuran, chlorpyrifos, diazinon, disulfoton, fensulfothion, and fonofos, among others, are quite toxic to humans and other nontarget organisms (Morris et al. 2014; DiBartolomeis et al. 2019), and subsequently, most are no longer registered for use on potatoes in the United States or other countries.

The first synthetic pyrethroid insecticides were registered on potatoes in the 1970s. Pyrethroids are modeled after the natural *Chrysanthemum*-derived pyrethrins, which modulate voltage-gated sodium channels on neuronal membranes (Ware and Whitacre 2004). Pyrethroids are generally less toxic to mammals than most organophosphates or carbamates and are efficacious at much lower use rates. This group of insecticides became a popular choice of potato growers in the 1980s and 1990s for control of many pests (Kuhar et al. 2012).

However, after multiple years of use on potatoes, Colorado potato beetle populations developed resistance to virtually all carbamates, organophosphates, and pyrethroids rendering them ineffective against this pest in the United States by the late 1980s (Harris and Turnball 1986; Casagrande 1987; Roush et al. 1990; Tisler and Zehnder 1990; French et al. 1992; Alyokhin et al. 2008a). Multiple types of resistance mechanisms were determined including target site insensitivity, enhanced metabolic enzyme activity, reduced insecticide penetration, and increased excretion (Rose and Brindle 1985; Ioannidis et al. 1991; Argentine et al. 1994; Wierenga and Hollingworth 1994; Alyokhin et al. 2008a).

In addition to resistance problems, the frequent applications of organophosphates, carbamates, and pyrethroids often destroyed arthropod natural enemy populations in fields (Metcalf 1980). Continued use of the aforementioned broad-spectrum insecticides in potatoes as foliar sprays was not sustainable, and their use has declined worldwide. However, soil applications of organophosphates such as ethoprop and phorate as well as the pyrethroid bifenthrin remain some of the most efficacious insecticides for control of wireworms and other soil-dwelling pests of potatoes (Shamiyeh et al. 1999; Nault and Speese 2000; Kuhar et al. 2003; Kuhar and Alvarez 2008).

In recent decades, there has been a shift in insecticide development to safer and more targeted (or narrow spectrum) insecticides that are often less toxic to nontarget species compared with carbamates, organophosphates, and pyrethroids. Today, there is a wide diversity of insecticide mode of actions and products that can aid in a more sustainable approach to pest management for potatoes (Table 16.1). Herein we discuss insecticide options that are considered more selective and sustainable than the previously discussed broad-spectrum nerve poisons. They are grouped by their modes of action following the Insecticide Resistance, and Mode of Action Classification (IRAC) numbering system summarized by Nauen et al. (2012).

16.2 Selective Neuroactive Insecticides

A wide range of insecticides target the nervous system but are selective to certain organisms or insect or mite groups. Inherently, these narrower-spectrum insecticides are typically safer for the end user and often compatible to various degrees with beneficial organisms.

16.2.1 Neonicotinoids and Related Pesticides

Similar to nicotine, the synthetic neonicotinoids (IRAC Group 4) are neuroactive chemicals that act on selective acetylcholine nicotinic receptors (Matsuda et al. 2020). The first neonicotinoid, imidacloprid, was introduced in the 1990s and, by the 2000s, became the most commonly used insecticide on potatoes because of its low mammalian toxicity and ability to be applied to seed pieces at planting to provide long-term systemic protection from aboveground foliar-feeding pests like Colorado potato beetle and sucking pests such as leafhoppers, aphids, and psyllids (Boiteau et al. 1997; Pavlista 2002; Kuhar et al. 2003; Kuhar and Speese 2005b, 2005c; Kuhar et al. 2007; Kuhar and Doughty 2018). Other neonicotinoids registered for use on potatoes include thiamethoxam, clothianidin, acetamiprid, thiacloprid, and dinotefuran. Neonicotinoids quickly became the most popular insecticide class used on potatoes by the late 1990s. Soil applications of neonicotinoids also were shown to significantly reduce wireworm damage to potato tubers (Kuhar et al. 2003; Kuhar and Alvarez 2008).

Although neonicotinoids have relatively low mammalian toxicity, their impact on other nontarget organisms has been a major concern in recent years. Arthropod predators not only can be affected by direct contact of foliar-applied neonicotinoids but also may be adversely affected by systemic applications when they feed on contaminated pollen, nectar, or other plant fluids or when they feed on prey that have consumed leaves contaminated with the active ingredient (Cloyd and Bethke 2011). Toxicity bioassays and field experiments with neonicotinoids on natural enemy species have shown conflicting results; however, it is generally believed that neonicotinoids pose at least a moderate risk to arthropod predators and parasitoids in agroecosystems (Frank 2012; Roubos et al. 2014; Yeary et al. 2015; Cheng et al.

Table 16.1 Insecticides and miticides currently registered for use on potatoes in the United States as of 2020 (all products are not registered for use in all states)

Insecticide Group (IRAC Classification Number*)	Mode of action	Insecticide(s)	Reduced risk**
Carbamate (1A)	Acetylcholine esterase inhibitor (reversible)	Carbaryl	N
		Methomyl	N
		Oxamyl (nematicide)	N
Organophosphate (1B)	Acetylcholine esterase inhibitor (irreversible)	Dimethoate	N
		Ethoprop (nematicide)	N
		Malathion	N
		Phorate	N
		Phosmet	N
Phenylpyrazoles (2B)	GABA-gated chloride channel blocker	Fipronil	Y
Pyrethroids (3A)	Sodium channel modulator	Beta-cyfluthrin	N
		Bifenthrin	N
		Cyfluthrin	N
		Esfenvalerate	N
		Lambda-cyhalothrin	N
		Permethrin	N
		Zeta-cypermethrin	N
Neonicotinoid (4A)	Nicotinic acetylcholine receptor competitive modulator	Acetamiprid	Y
		Clothianidin	Y
		Dinotefuran	Y
		Imidacloprid	Y
		Thiamethoxam	Y
Sulfoximines (4C)		Sulfoxaflor	Y
Butenolides (4D)		Flupyradifurone	Y
Spinosyns (5)	Nicotinic acetylcholine receptor allosteric modulator - site I	Spinetoram	Y
		Spinosad	Y
Avermectins (6)	Glutamate-gated chloride channel allosteric activator	Abamectin	N
JH mimics (7C)	Juvenile hormone mimic	Pyriproxyfen	Y
Fluorides (8C)	Miscellaneous nonspecific (multisite) inhibitor	Cryolite	Y
Pyridine azomethine derivatives (9B)	Chordotonal organ TRPV channel modulator	Pymetrozine	Y
		Pyrifluquinazon	Y
Pyropenes (9D)		Afidopyropen	Y
–	Mite growth inhibitor affecting CHS1	Hexythiazox	Y
Bt (11A)	Microbial disruptors of insect midgut membranes	*Bacillus thuringiensis* var. *tenebrionensis*	Y

(continued)

Table 16.1 (continued)

Insecticide Group (IRAC Classification Number*)	Mode of action	Insecticide(s)	Reduced risk**
		Bacillus thuringiensis var. *kurstaki*	Y
(12C)	Inhibitors of mitochondrial ATP synthase	Propargite	Y
Benzoylureas (15)	Inhibitors of chitin biosynthesis affecting CHS1	Novaluron	Y
Dipteran IGR (17)	Molting disruptor, dipteran	Cyromazine	Y
METI complex III (20D)	Mitochondrial complex III electron transport inhibitor	Bifenazate	Y
METI complex I (21A)	Mitochondrial complex I electron transport inhibitor	Tolfenpyrad	Y
22A	Voltage-dependent sodium channel blocker	Indoxacarb	Y
Tetronic and tetramic acid derivatives (23)	Inhibitors of acetyl CoA carboxylase	Spiromesifen	Y
		Spirotetramat	Y
Diamides (28)	Ryanodine receptor modulator	Chlorantraniliprole	Y
		Cyantraniliprole	Y
		Cyclaniliprole	Y
(29)	Chordotonal organ modulator - undefined target site	Flonicamid	Y
Unknown	Botanical essence - unknown MoA	*Chenopodium* extract	Y
Unknown	Unknown or uncertain MoA	Azadirachtin	N

*Insecticide Resistance Action Committee (IRAC) mode of action classification is the definitive global authority on the target site of insecticides (Nauen et al. 2012)
**The from reduced risk designation comes the United States Environmental Protection Agency list, https://www.epa.gov/pesticide-registration/reduced-risk-and-organophosphate-alternative-decisions-conventional

2018; Esquivel et al. 2020; Jones et al. 2020). Nitro-containing neonicotinoids such as imidacloprid, thiamethoxam, and clothianidin are generally more toxic than cyano-containing neonicotinoids such as acetamiprid and thiacloprid (Lundin et al. 2015).

Much of the ecotoxicology research on neonicotinoids has focused on pollinators, especially honey bees, *Apis mellifera* L. (Hymenoptera: Apidae) (Blacquiere et al. 2012; Lundin et al. 2015). In addition to direct mortality at higher concentrations, sublethal concentrations of neonicotinoids have caused reduced foraging activity, impaired navigation, and decreased viable sperm in drone bees (Pisa et al. 2015; Brandt et al. 2016; Tison et al. 2016; Straub et al. 2016; Cressey 2017; Wood and Goulson 2017). These concerns have resulted in a ban on outdoor use of clothianidin, thiamethoxam, and imidacloprid in the European Union (European Commission 2020). Additional regulatory restrictions on neonicotinoids around the world appear inevitable.

Resistance development to neonicotinoids in Colorado potato beetle populations has also become a serious problem, which has made the reliance on this class of insecticides in potatoes even more problematic (Alyokhin et al. 2008a). Either through governmental regulatory action, commercial industry bans, or insecticide resistance management plans (Huseth et al. 2014), many potato growers have either been required or encouraged to rotate to non-neonicotinoid alternatives for insecticides in potato production.

More recently insecticides such as sulfoxaflor (IRAC Group 4C) and the butenolide flupyradifurone (IRAC Group 4D), which have similar mode of actions as neonicotinoids, but which act upon different nicotinic acetylcholine receptors (Sparks et al. 2013; Jeschke et al. 2015; Nauen et al. 2015), purportedly provide a more IPM and pollinator-compatible option for potato growers. Although sulfoxaflor and flupyradifurone have been listed by the US EPA as reduced risk, research on the nontarget effects of sulfoxaflor and flupyradifurone have only been undertaken in the last 5 years. In laboratory assays, sulfoxaflor did not impact the survival of the ladybeetle *Hippodamia convergens* Guérin-Méneville (Coleoptera: Coccinellidae) or green lacewing *Chrysoperla rufilabris* (Burmeister) (Neuroptera: Chrysopidae) but caused similar mortality of minute pirate bug, *Orius insidiosus* (Say) (Hemiptera: Anthocoridae), as a pyrethroid insecticide (Tran et al. 2016; Aita et al. 2020). In field experiments, Bordini et al. (2021) reported that cotton fields sprayed sulfoxaflor or flupyradifurone had similar arthropod predator abundance as untreated control plots. In addition, the sublethal effects of these insecticides on beneficial arthropods need further investigation (Siviter and Muth 2020).

16.2.2 Spinosyns

Spinosyns (IRAC Group 5) are macrocyclic lactones derived from the fermentation of the soil actinomycete *Saccharopolyspora spinosa* (Thompson et al. 2000). There are multiple types of spinosyns that act by disrupting the binding of the neurotransmitter acetylcholine in nicotinic receptors at the postsynaptic nerve cell (Salgado 1998). The insecticide spinosad is a natural mixture of spinosyns A and D, and the insecticide spinetoram is a derived from spinosyns J and L, which have been chemically modified to produce a semisynthetic insecticide (Sparks et al. 2008). Both insecticides provide excellent control of lepidopteran, thysanopteran, and coleopteran pests (Thompson et al. 2000), including Colorado potato beetle (Byrne et al. 2006; Kuhar and Doughty 2009, 2016; Groves et al. 2017). However, some populations of Colorado potato beetle populations that are resistant to neonicotinoids have demonstrated reduced susceptibility to spinosad (Mota-Sanchez et al. 2006).

Spinosyns are generally considered to be IPM compatible with low to moderately low impact on natural enemy populations (Elzen 2001; Chapman et al. 2009; Roubos et al. 2014; Mills et al. 2016; D'Ávila et al. 2018; Dale and Borden 2018; Sarkar et al. 2020). Although some formulations of spinosad are permitted for use in certified organic systems, spinetoram is not because it is synthetically produced.

Among the options for organic growers, spinosad is arguably the most effective insecticide for controlling potato beetles, flea beetles, thrips, and lepidopteran larvae (Dively et al. 2020).

16.2.3 Avermectins

Avermectins (IRAC Group 6) are biologically derived macrocyclic lactones from the fermentation of the soil actinomycete *Streptomyces avermitilis* (Campbell 1989). Abamectin is a mixture of macrocyclic lactones 80% avermectin B1a and less than 20% avermectin B1b. The insecticide stimulates the release and binding of gamma-aminobutyric acid (GABA) at nerve endings, which causes an influx of chloride ions into the cells leading to hyperpolarization and paralysis of the neuromuscular systems (Bloomquist 1996). Abamectin controls a wide range of mites and insects including Colorado potato beetle (Nault and Speese 1999a; Kuhar et al. 2006a; Marčić´ et al. 2009; Sewell and Alokhin 2010b; Kuhar and Doughty 2016; Groves et al. 2017).

Although avermectins are considered valuable tools for Colorado potato beetle insecticide resistance management programs (Huseth et al. 2014), they are not considered reduced risk insecticides because of their high toxicity to mammals and other nontarget organisms.

16.2.4 Phenylpyrazole (Fipronil)

Fipronil is a phenylpyrazole insecticide that was registered for use on potatoes in the United States in the mid-2000s. Fipronil blocks the gamma-aminobutyric acid (GABA)-regulated chloride channel in neurons similar to the action of the cyclodienes (IRAC Group 2). Fipronil is highly effective as a foliar insecticide on Colorado potato beetle (Moffat 1993; Noetzel and Holder 1996; Nault and Speese 1999b) and as a systemic material for control of European corn borer (Nault and Speese 1999a; Kuhar et al. 2010), but the primary target for this insecticide in potatoes and other crops is wireworms (van Herk et al. 2015). Kuhar and Alvarez (2008) and Vernon et al. (2013) both showed that in-furrow applications of fipronil significantly reduced wireworm damage to potato tubers similar to organophosphate standards.

Fipronil does not meet the criteria for reduced risk status due to toxicity to nontarget organisms, particularly *A. mellifiera* (Pisa et al. 2015). Additionally, lethal and sublethal effects have been observed with *O. insidiosus* and *Geocoris punctipes* Fallén (Hemiptera: Geocoridae) that are comparable to levels seen with organophosphates (Elzen 2001). However, it is much safer to mammals than the organophosphate or carbamate insecticides that otherwise would be used for wireworm control.

16.2.5 Oxadiazines

Oxadiazines (IRAC Group 22A) are voltage-dependent sodium channel blockers on nerve axons (Wing et al. 2000). Two insecticides within this group include indoxacarb and metaflumizone. Indoxacarb controls most lepidopteran larvae, as well as Colorado potato beetle and potato leafhopper (Davis et al. 2003; Linduska et al. 2002; Kuhar and Speese 2005a). The addition of the synergist piperonyl butoxide enhances the efficacy of indoxacarb (Linduska et al. 2002; Sewell and Alyokhin 2003).

When exposed to residues of indoxacarb, survival of adult *O. insidiosus* was similar to that of the control group, but survival for nymphs of *O. insidiosus* was lower than the control group (Roubos et al. 2014; Andorno et al. 2019). Additionally, sublethal effects of indoxacarb were manifested as females produced fewer eggs than the control group (Andorno et al. 2019). Similar to indoxacarb, metaflumizone is a semicarbazone oxadiazine insecticide that is currently not registered for use in the United States. The insecticide provides excellent control of Colorado potato beetle (Sewell and Alyokhin 2009; Sewell and Alokhin 2010b; Hitchner et al. 2012).

16.3 Anthranilic Diamides

Introduced in the 2000s, the anthranilic diamides (IRAC Group 20), which include chlorantraniliprole, cyantraniliprole, cyclaniliprole, flubendiamide (no longer registered in the United States), and tetraniliprole, activate the insect ryanodine receptors affecting calcium release during muscle contraction (Cordova et al. 2006). These selective insecticides provide strong efficacy against lepidopteran and coleopteran pests, including Colorado potato beetle, in potatoes (Kuhar and Doughty 2009; Sewell and Alokhin 2009; Kuhar and Doughty 2010, 2016; Groves et al. 2017). Chlorantraniliprole and cyantraniliprole are systemic and provide long-lasting insect control from seed-piece or in-furrow applications (Sewell and Alokhin 2009; Sewell and Alokhin 2011; Kuhar and Doughty 2010; Groves et al. 2011a, b).

Diamides are considered reduced risk and IPM compatible (Roubos et al. 2014; Mills et al. 2016; Whalen et al. 2016; Dale and Borden 2018; Machado et al. 2019; Bordini et al. 2021) and are excellent options for resistance management rotations with neonicotinoids (Huseth et al. 2014).

16.4 Chemicals Affecting the Chordotonal Organ in Insects

In recent years several highly selective insecticides have been developed that affect the chordotonal organ interfering with the regulatory mechanism of food intake particularly in hemipteran insects such as aphids, whiteflies, psyllids, and leafhoppers (Kristinsson 1995). Pymetrozine is a pyridine-azomethine derivative (IRAC Group 9B) that was one of the first insecticides registered that modulate the

TRPV channel of the chordotonal organ (Kristinsson 1995). Pymetrozine is highly effective against aphids (Sewell and Alokhin 2010a; Bradford et al. 2019) but only moderately effective against potato psyllid in field trials (Russell et al. 2001; Liu and Trumble 2005). Pyrifluquinazon and afidopyropen are relatively new insecticides with similar mode of actions (Group 9) and pest spectrums as pymetrozine.

Flonicamid is a pyridinecarboxamide insecticide (IRAC Group 29) that also modulates the chordotonal organ causing the normally rigid stylet of piercing sucking pests to become flaccid and unable to pierce leaf tissue causing the insect to starve to death (Morita et al. 2007). Flonicamid is a systemic compound with activity on aphids (Bradford et al. 2019), whiteflies, and thrips.

Insecticides in either IRAC Group 9 or 29 are considered quite compatible with IPM programs causing no to very little mortality to predatory arthropods (Barbosa et al. 2018; Kim and Kim 2019; Machado et al. 2019; Koch et al. 2020). The effects of these chemicals on hymenopterans, however, warrant further investigation; afidopyropen was more toxic to *Aphelinus certus* Yashnosh (Hymenoptera: Aphelinidae) than the control but less toxic than the pyrethroid lambda-cyhalothrin (Koch et al. 2020).

16.5 Mitochondrial (Respiratory) Poisons

Tolfenpyrad is a pyrazole (IRAC group 21A) insecticide that inhibits mitochondrial electron transport at the NADH-CoQ reductase site, leading to the disruption of adenosine triphosphate (ATP) formation (Ware and Whitacre 2004). The insecticide provides excellent control of Colorado potato beetle larvae and adults (Sewell and Alokhin 2011; Wimer et al. 2015; Buzza and Alyokhin 2017; Bradford et al. 2020) as well as other pests such as leafhoppers and lepidopteran larvae. Although tolfenpyrad is considered to be a reduced risk insecticide, it is not the most compatible option with regard to biological control. Tolfenpyrad was shown to be moderately toxic to the predatory insects *O. insidiosus* and *Dalotia coriaria* (Kraatz) (Coleoptera: Staphylinidae) (Cloyd and Herrick 2018) and highly toxic to the parasitoid wasp *Trichogramma pretiosum* Riley (Hymenoptera: Trichogrammatidae) (Khan et al. 2015).

Bifenazate is a broad-spectrum miticide that inhibits the mitochondrial electron transport system (IRAC Group 20D). It has been shown to be a very good miticide for tetranychid mites, with little to no impact on the predatory mite *Phytoseiulus persimilis* Athias-Henriot (Acari: Phytoseiidae) (Bergeron and Schmidt-Jeffris 2020) but some moderate toxicity to *Neoseiulus californicus* (McGregor) (Acari: Phytoseiidae) predatory mites.

Propargite is an acaricide that inhibits mitochondrial ATP synthase (IRAC Group 12C). It has demonstrated to be safe for the predatory mite *Neoseiulus cucumerus* (Oudermans) (Acari: Phytoseiidae) (Stara et al. 2011) but to cause some mortality in minute pirate bugs, *O. insidiosus* (Ashley et al. 2006).

16.6 Insect Growth Regulators and Disruptors of Metabolic Processes

A number of reduced risk insecticides and miticides have been developed that impact arthropod development or molting in different ways. Collectively, these are referred to as insect growth regulators.

16.6.1 Juvenile Hormone Mimics

Pyriproxyfen is an insect juvenile hormone mimic (IRAC Group 7C) that primarily targets whiteflies (Ishaaya and Horowitz 1995). Juvenile hormone mimics have favorable safety profiles for mammals, including humans, and are much less toxic to nontarget organisms than many insecticides. In a greenhouse study, survival of *O. insidiosus* and *C. externa* was not different between control and pyriproxyfen treatments (Machado et al. 2019). However, some sublethal effects were detected on *Tenuisvalvae notata* (Mulsant) (Coleoptera: Coccinellidae). While female beetles did not lay fewer eggs compared to control treatments, reduced egg hatch was observed in beetles that were exposed to pyriproxyfen (Barbosa et al. 2018).

16.6.2 Chitin Biosynthesis Inhibitors

Novaluron is an insect growth regulator that belongs to the benzoylphenyl urea (or benzoylurea) class of chemicals (IRAC Group 15). These insecticides target and disrupt chitin biosynthesis on the larval stages of many insects (Ishaaya et al. 2003; Ware and Whitacre 2004). Novaluron is very effective at controlling the larval stage of Colorado potato beetle (Cutler et al. 2007) but can also cause egg mortality (Alyokhin et al. 2008b) and a decrease in reproductive viability of adult females when ingested (Alyokhin et al. 2010). Two foliar applications of novaluron will provide effective control of Colorado potato beetle (Kuhar et al. 2006b; Kuhar and Doughty 2009; Sewell and Alokhin 2009; Sewell and Alokhin 2010b; Groves et al. 2017). Novaluron also controls European corn borer (Kuhar et al. 2006b).

In bioassays, novaluron did not affect the survival of the predators *C. rufilabris*, *O. insidiosus*, or *H. convergens* (Roubos et al. 2014). However, sublethal effects of novaluron were detected for *N. californicus*, *H. convergens*, or *Chrysoperla carnea* (Stephens) (Neuroptera: Chrysopidae), in reduced fecundity compared to control treatments (Mills et al. 2016; Kim and Kim 2019).

Cyromazine, a triazine (IRAC Group 17), is a chitin synthesis inhibitor (Ware and Whitacre 2004). It is selective toward dipterous insects and is used for the control of leafminers and root maggots (Thetford 1993). The insecticide has also been shown to provide effective control of Colorado potato beetle larvae (Sirota and Grafius 1994; Linduska et al. 1996).

Cyromazine did not impact the survival of *Coleomegilla maculata* Timberlake (Coleoptera: Coccinellidae), larvae compared to the control treatment when topically applied or fed treated Colorado potato beetle eggs (Lucas et al. 2004). Cryomazine had the highest LD_{50} value for *C. carnea* when compared to chlorpyrifos and cypermethrin, indicating it is a good candidate for pest control use while conserving natural enemy populations (Mansoor and Shad 2020).

16.6.3 Mite Growth Inhibitors

Hexythiazox is a mite growth inhibitor (IRAC Group 10A) that provides effective control of spider mites while minimizing effects on predatory mites such as *N. californicus*, *N. fallacis* Garman, and *Phytoseiulus persimilis* Athias-Henriot (Acari: Phytoseiidae) (Bergeron and Schmidt-Jeffris 2020), as well as parasitoids like *Aphytis melinus* DeBach (Hymenoptera: Aphelinidae) that were exposed to hexythiazox were not higher than the control treatments (Vanaclocha et al. 2013).

16.6.4 Lipid Biosynthesis Inhibitor

Spirotetramat is a lipid biosynthesis inhibitor (IRAC Group 23) derived from spirocyclic tetramic acid and has been shown to inhibit ecdysis in immature insects and reduce fecundity and fertility in adult insects (Nauen et al. 2008). Spirotetramat penetrates the leaf surface and is hydrolyzed to an active form that can enter both phloem and xylem in plants, resulting in two-way systemicity (Brücka et al. 2009). Spirotetramat has activity on a number of pests such as aphids, psyllids, mealy bugs, and whiteflies (Bretschneider et al. 2007; Nauen et al. 2008; Sewell and Alokhin 2010a; Bradford et al. 2019). It has demonstrated excellent long-lasting efficacy against potato psyllid, *Bactericerca cockerelli* (Sulc) (Hemiptera: Triozidae), which transmits the alphaproteobacteria *Candidatus Liberibacter solanacearum* that causes Zebra chip syndrome (Munyaneza et al. 2007; Gao et al. 2009). Spirotetramat also controls two spotted spider mites although it is not registered as a miticide (Popov and Alyokhin 2019). In addition, spirotetramat has been shown to suppress wireworm damage; because of its two-way systemicity, foliar applications of spirotetramat can travel from the point of application on the foliage into the roots where it suppresses root and tuber feeding organisms such as phytoparasitic nematodes and wireworms (Bayer CropScience 2019; Shirley et al. 2019).

Spirotetramat is also IPM compatible having little to no impact on predatory insects such as *G. puncitpes*, *O. insidiosus*, *H. convergens*, *C. rufilabris* (Prabhaker et al. 2017), or *Cryptolaemus montrouzieri* Mulsant (Coleoptera: Coccinelidae) (Planes et al. 2013). Although it has been shown to negatively affect reproduction in the predatory mite species *N. fallacis*, *N. californicus*, and *P. persimilis*, it is much less toxic than standard broad-spectrum insecticides (Kim and Kim 2019).

Spiromesifen is another tetramic acid derivative (IRAC Group 23) insecticide/miticide that provides effective control of mites and whiteflies, while minimizing impacts on natural enemies (Prabhaker et al. 2017, Bergeron and Schmidt-Jeffris 2020). It is considered to be an IPM compatible miticide but has demonstrated negative effects on reproduction in predatory mites (Bergeron and Schmidt-Jeffris 2020).

16.6.5 Fluorides

Cryolite is an inorganic fluoride-based insecticide (IRAC Group 8C) that inhibits the activity of enzymes that contain iron, calcium, or magnesium such as phosphatases and phosphorylases (Ware and Whitacre 2004). Cryolite has selective activity providing effective control of lepidopteran and coleopteran pests like Colorado potato beetle (Noetzel and Holder 1996) while having minimal impact on important arthropod predators such as *C. maculata*) (Lucas et al. 2004).

16.7 Insecticide Options for Organic Potatoes

Although organic growers rely heavily upon nonchemical methods such as biological control, promoting natural enemies, and cultural control tactics to prevent insect damage to crops, chemical control may be needed when pest pressures exceed acceptable or economic thresholds (Zehnder et al. 2007). A wide range of naturally derived insecticidal compounds are available. The soil microbe-derived insecticide spinosad (discussed previously) is arguably the most efficacious insecticide option in organic potato production for the control of coleopteran and lepidopteran pests (Groves et al. 2017; Nault and Seaman 2019; Dively et al. 2020). However, the extensive use of spinosad has already resulted in resistance development in some Colorado potato beetle populations in the Untied States (Schnaars-Uvino and Baker 2021). Therefore, overreliance on this active ingredient should be avoided. Several other insecticides that are derived from various living organisms are permitted for use in organically certified potatoes, and these are discussed below.

Pyrethrins are obtained from the flowers of *Chrysanthemum cinerariifolium* (Casida 1980) and, like their synthetic counterpart pyrethroids, are nerve poisons that modulate the sodium channel on axon neuronal membranes (IRAC Group 3). They are fast-acting broad-spectrum contact poisons but do not have long residual efficacy, and because of the resistance to DDT and pyrethroids in key pest populations, these insecticides are generally ineffective against pests such as Colorado potato beetle, beet armyworm, and green peach aphids.

Veratrine alkaloids, which are extracted from the seeds of the tropical lily plant Sabadilla (*Schoenocaulon officinale*), have a mode of action similar to pyrethrins (Bloomquist 1996). Sabadilla seeds are aged, heated, or alkali-treated to activate the insecticidal alkaloids.

Azadirachtin is a tetranortriterpenoid derived from the seeds of the neem tree (*Azadirachta indica*). This compound has been shown to be an antifeedant and disrupt insect growth by blocking the release of peptide hormones (Mordue and Blackwell 1993; Seymour et al. 1995; Abudulai et al. 2003). It has been shown to be effective on a wide range of insects including lepidopteran and coleopteran larvae

(Zehnder and Warthen 1988; Marčic et al. 2009). Azadirachtin is most effective as a growth regulator on eggs and small larvae, and therefore, application timing is important for effective control, particularly when targeting Colorado potato beetle (Trisyono and Whalon 1999; Kowalska 2007).

Bacillus thuringiensis is a bacterium that produces delta-endotoxins that are toxic to the midgut of insect pests. If the endotoxins are ingested, they form an ion channel that causes shrinking or swelling in the epithelium cells, leading to cell lysis and eventual death of the insect (Slaney et al. 1992). *Bt* subsp. *kurstaki* or *Bt aizawai* are very effective insecticides for control of lepidopteran larvae such as armyworms, whereas *Bt* subsp. *tenebrionis* applications are effective against Colorado potato beetle larvae (Wantuch et al. 2016; Nault and Seaman 2019). *Bt* products are most effective against small larvae, and thus, as with azadirachtin, early application timing is critical for effective control in the field (Ghidiu and Zehnder 1993). Even with proper application timing, the efficacy of this insecticide against Colorado potato beetle has been moderate at best (Sewell and Alokhin 2009), and resistance to *Bt* subsp. *tenebrionis* has been reported in isolated populations of Colorado potato beetle in the United States (Whalon et al. 1993).

Bt is one of the most IPM-compatible insecticides available with no detectable negative effects on predatory arthropods when exposed to *Bt tenebrionis* topically, through consumption of treated eggs or larvae, or on leaf residues (Zwahlen et al. 2000; Lucas et al. 2004; Vasileiadis et al. 2017). It is worth noting that parasitoids can experience sublethal effects such as reduced fertility and fecundity when attacking hosts that feed on *Bt kurstaki* (da Rolim et al. 2020). However, these effects are more likely due to the poor host quality than the *Btk* itself (Lundgren et al. 2009).

Betaproteobacteria that produce insecticidal compounds have recently been formulated into organic insecticides. *Chromobacterium subtsugae* produces insecticidal compounds that are active against a variety of insect pests including Colorado potato beetle (Martin et al. 2007). Another relatively new biological insecticide is derived from heat-killed cells and fermentation solids of the bacteria *Burkholderia* spp. The insecticide interferes with molting and disrupts insect exoskeletons (Asolkar et al. 2013).

Nault and Seaman (2019) evaluated the organic insecticides, pyrethrin, spinosad, azadirachtin, sabadilla alkaloids, *Bacillus thuringiensis*, and *Chromobacterium subtsugae*, on potatoes over 2 years in the United States. In that study, spinosad consistently had the highest level of efficacy against Colorado potato beetle followed by *Bt tenebrionis* and azadirachtins. Very little control was provided by any of the other organic insecticides. Groves et al. (2017) also achieved effective control of Colorado potato beetle with spinosad and virtually no control with either of the betaproteobacteria insecticides.

16.8 Final Thoughts

Chemical control remains one of the most widely used strategies for eliminating potato damage by insect pests (Alyokhin 2009; Kuhar et al. 2012). However, chemical control should be used in a more sustainable and responsible manner, one that avoids past mistakes of indiscriminate applications of broad-spectrum poisons and uses insecticides efficiently, with a better understanding of the pest's biology, and as part of an integrated pest management program (Huseth et al. 2014). A number of reduced risk insecticides have been registered on potatoes that provide effective control of the major potato pests, and many more are in development that will undoubtedly be safer for the user, have less of an impact on nontarget insects, and fit better into potato IPM programs (Kuhar et al. 2012). Rotation of insecticide classes is also recommended to minimize insecticide resistance development in Colorado potato beetle and other pests.

References

Abudulai M, Shepard BM, Mitchell PL (2003) Antifeedant and toxic effects of a neem (*Azadirachta indica* a. Juss) based formulation Neemix against *Nezara viridula* (L.) (Hemiptera: Pentatomidae). J Entomol Sci 38:398–408

Aita RC, Tran AK, Koch RL (2020) Susceptibility of first instar Hippodamia convergens (Coleoptera: Coccinellidae) and *Chrysoperla rufilabris* (Neuroptera: Chrysopidae) to the insecticide Sulfoxaflor. Florida Entomol 103:191–196

Alyokhin A (2009) Colorado potato beetle management on potatoes: current challenges and future prospects. Fruit Veg Cereal Sci Biotech 3:10–19

Alyokhin A, Baker M, Mota-Sanchez D, Dively G, Grafius E (2008a) Colorado potato beetle resistance to insecticides. Am J Potato Res 85:395–413

Alyokhin A, Guillemette R, Choban R (2010) Stimulatory and suppressive effects of novaluron on the Colorado potato beetle reproduction. J Econ Entomol 102:2078–2083

Alyokhin A, Sewell G, Choban R (2008b) Reduced viability of Colorado potato beetle, *Leptinotarsa decemlineata*, eggs exposed to novaluron. Pest Manag Sci 64:94–99

Andorno AV, Carrizo PI, López SN (2019) Toxicity of four pesticides *on Orius insidiosus* under laboratory conditions. Bull Insectol 72:69–75

Argentine JA, Zhu KY, Lee SH, Clark JM (1994) Biochemical mechanisms of azinphosmethyl resistance in isogenic strains of Colorado potato beetle. Pestic Biochem Physiol 48:63–78

Ashley JL, Herbert DA, Lewis EE, Brewster CC, Huckaba R (2006) Toxicity of three acaricides to *Tetranychus urticae* (Tetranychidae: Acari) and *Orius insidiosus* (Anthocoridae: Hemiptera). J Econ Entomol 99:54–59

Asolkar R, Koivunen M, Marrone P (2013) Isolated bacterial strain of the genus Burkholderia and pesticidal metabolites there from formulations and uses. US Patent No WO-2013032693-A2

Barbosa PRR, Oliveira MD, Barros EM, Michaud JP, Torres JB (2018) Differential impacts of six insecticides on a mealybug and its coccinellid predator. Ecotox Environ Saf 147:963–971

Bayer Crop Science (2019) Movento insecticide label. US61384489C 170628Cv2 01/19. http://legacy.picol.cahnrs.wsu.edu/~picol/pdf/WA/51483.pdf

Bergeron PE, Schmidt-Jeffris RA (2020) Not all predators are equal: miticide non-target effects and differential selectivity. Pest Manag Sci 76:2170–2179

Blacquiere T, Smagghe G, van Gestel CA, Mommaerts V (2012) Neonicotinoids in bees: a review on concentrations, side-effects and risk assessment. Ecotoxicology 21:973–992

Bloomquist JR (1996) Ion channels as targets for insecticides. Annu Rev Entomol 41:163–190

Boiteau G, Osborn WPL, Drew ME (1997) Residual activity of imidacloprid controlling Colorado potato beetle (Coleoptera: Chrysomelidae) and three species of potato colonizing aphids (Homoptera: Aphidae). J Econ Entomol 90:309–319

Bordini I, Ellsworth PC, Naranjo SE, Fournier A (2021) Novel insecticides and generalist predators support conservation biological control in cotton. Biol Control 154:104502. https://doi.org/10.1016/j.biocontrol.2020.104502

Bradford BZ, Chapman SA, Crubaugh LK, Groves RL (2019) Evaluation of registered and experimental foliar insecticides for the control of potato aphid in potato, 2018. Arthropod Manag Tests 44:tsz045. https://doi.org/10.1093/amt/tsz045

Bradford BZ, Chapman SA, Groves RL (2020) Evaluation of Torac, warrior, and Coragen for first-Ggeneration Colorado potato beetle management in Wisconsin, 2019. Arthropod Manag Tests 45:tsaa018. https://doi.org/10.1093/amt/tsaa018

Brandt A, Gorenflo A, Siede R, Meixner M, Büchler R (2016) The neonicotinoids thiacloprid, imidacloprid, and clothianidin affect the immunocompetence of honey bees (*Apis mellifera* L.). J Insect Physiol 86:40–47

Bretschneider T, Fischer R, Nauen R (2007) Inhibitors of lipid synthesis (acetyl-CoA-carboxylase inhibitors). In: Kramer W, Schirmer U (eds) Modern crop protection compounds. Wiley, Weinheim, Germany, pp 909–925

Brücka E, Elberta A, Fischer R et al (2009) Movento®, an innovative ambimobile insecticide for sucking insect pest control in agriculture: biological profile and field performance. Crop Prot 28: 838–844

Buzza AM, Alyokhin A (2017) Control of Colorado potato beetle on potato with tolfenpyrad, 2017. Arthropod Manage Tests 43:tsy078. https://doi.org/10.1093/amt/tsy078

Byrne AM, Grafius E, Pett W (2006) Colorado potato beetle control, 2005. Arthropod Manage Tests 31:E54

Campbell WC (1989) Ivermectin and Abamectin. Springer, New York

Casagrande RA (1987) The Colorado potato beetle: 125 years of mismanagement. Bull Entomol Soc Am 33:142–150

Casida JE (1980) Pyrethrum flowers and pyrethroid insecticides. Environ Health Perspect 34:189–202

Chapman A, Kuhar T, Schultz P, Leslie T, Fleischer S, Dively G, Whalen J (2009) Integrating chemical and biological control of European corn borer in bell pepper. J Econ Entomol 102: 287–295

Cheng S, Lin R, Zhang N, Yuan S, Zhou X, Huang J, Ren X, Wang S, Jiang H, Yu C (2018) Toxicity of six insecticides to predatory mite *Amblyseius cucumeris* (Oudemans) (Acari: Phytoseiidae) in- and off-field. Ecotoxicol Environ Saf 161:715–720

Cloyd RA, Bethke JA (2011) Impact of neonicotinoid insecticides on natural enemies in greenhouse and interiorscape environments. Pest Manag Sci 67:3–9

Cloyd RA, Herrick NJ (2018) Effects of pesticides on the survival of rove beetle (Coleoptera: Staphylinidae) and insidious flower bug (Hemiptera: Anthocoridae) adults. J Econ Entomol 111: 78–88

Cohn BA, Wolff MS, Cirillo PM, Sholtz RI (2007) DDT and breast cancer in young women: new data on the significance of age at exposure. Environ Health Perspect 115:1406–1414

Conis E (2017) Beyond silent spring: an alternate history of DDT. Distillations 2(4):16–23

Cordova D, Benner EA, Sacher MD, Rauh JJ, Sopa JS, Lahm GP (2006) Anthranilic diamides: a new class of insecticides with a novel mode of action, ryanodine receptor activation. Pestic Biochem Physiol 84:196–214

Cranshaw WS (1997) The potato (tomato) psyllid, *Paratrioza cackerelli* (Sulc) as a pest of potatoes. In, Zehnder GW, Powelson ML, Jansson RK, Raman KV Advances in potato pest management. American Phytopathological Society, St. Paul, MN, 83–95

Cressey D (2017) The bitter battle over the world's most popular insecticides. Nature 551:156–158

Cutkomp LK, Peterson AG, Hunter PE (1958) DDT-resistance of the Colorado potato beetle. J Econ Entomol 51:828–831

Cutler GC, Scott-Dupree CD, Tolman JH, Harris CR (2007) Field efficacy of novaluron for control of Colorado potato beetle (Coleoptera: Chrysomelidae) on potato. Crop Prot 26:760–767

D'Ávila VA, Barbosa WF, Guedes RNC, Cutler GC (2018) Effects of Spinosad, Imidacloprid, and lambda-cyhalothrin on survival, parasitism, and reproduction of the aphid parasitoid *Aphidius colemani*. J Econ Entomol 111:1096–1103

da Rolim GS, Plata-Rueda A, Martínez LC et al (2020) Side effects of bacillus thuringiensis on the parasitoid *Palmistichus elaeisis* (hymenoptera: Eulophidae). Ecotoxicol Environ Safety 189: 109978. https://doi.org/10.1016/j.ecoenv.2019.109978

Dale AG, Borden MA (2018) Evaluation of reduced-risk insecticides to control chilli thrips (Thysanoptera: Thripidae) and conserve natural enemies on ornamental plants. Fla Entomol 101:237–243

Davis JA, Radcliff EB, Ragsdale DW (2003) Control of potato leafhopper on potatoes using foliar insecticides, 2002. Arthropod Manage Tests 28:E52

Deliopoulos T, Minnis ST, Jones PW, Haydock PPJ (2010) Enhancement of the efficacy of a carbamate nematicide against the potato cyst nematode, *Globodera pallida*, through mycorrhization in commercial potato fields. J Nematol 42:22–32

DiBartolomeis M, Kegley S, Mineau P, Radford R, Klein K (2019) An assessment of acute insecticide toxicity loading (AITL) of chemical pesticides used on agricultural land in the United States. PLoS One 14(8):e0220029. https://doi.org/10.1371/journal.pone.0220029

Dively GP, Patton T, Barranco L, Kulhanek K (2020) Comparative efficacy of common active ingredients in organic insecticides against difficult to control insect pests. Insects 11:614. https://doi.org/10.3390/insects11090614

Elzen GW (2001) Lethal and sublethal effects of insecticide residues on *Orius insidiosus* (Hemiptera: Anthocoridae) and *Geocoris punctipes* (Hemiptera: Lygaeidae). J Econ Entomol 94:55–59

Esquivel CJ, Martinez EJ, Baxter R et al (2020) Thiamethoxam differentially impacts the survival of the generalist predators, *Orius insidiosus* (Hemiptera: Anthocoridae) and Hippodamia convergens (Coleoptera: Coccinellidae), when exposed via the food chain. J Insect Sci 20:1–10

European Commission (2020) Neonicotinoids. https://ec.europa.eu/food/plant/pesticides/approval_active_substances/approval_renewal/neonicotinoids_en. Accessed 15 July 2021

Frank SD (2012) Reduced risk insecticides to control scale insects and protect natural enemies in the production and maintenance of urban landscape plants. Environ Entomol 41:377–386

French NM II, Heim DC, Kennedy GG (1992) Insecticide resistance patterns among Colorado potato beetle, *Leptinotarsa decemlineata* (say) (Coleoptera: Chrysomelidae), populations in North Carolina. Pestic Sci 36:1091

Gao F, Jifon J, Yang X, Liu T (2009) Zebra chip disease incidence on potato is influenced by timing of potato psyllid infestation, but not by the host plants on which they were reared. Insect Sci 16: 399–408

Gauthier NL, Hofmaster RN, Semel M (1981) History of Colorado potato beetle control. In: Lashomb JH, Casagrande RA (eds) Advances in potato pest management. Hutchinson Ross, Stroudsburg, PA, pp 13–23

Gerhardt PD (1966) Potato psyllid and green peach aphid control on Kennebec potatoes with Temik and other insecticides. J Econ Entomol 59:9–11

Gerhardt PD, Turley DL (1961) Control of certain potato insects in Arizona with soil applications of granulated phorate. J Econ Entomol 54:1217–1221

Ghidiu GM, Zehnder GW (1993) Timing of the initial spray application of *bacillus thuringiensis* for control of the Colorado potato beetle (Coleoptera: Chrysomelidae) in potatoes. Biol Control 3: 348–352

Groves RL, Chapman S, Crubaugh LK, Duerr E, Bradford B, Clements J (2017) Registered and experimental foliar insecticides for control of Colorado potato beetle and potato leafhopper in potato, 2016. Arthropod Manage Tests 42:tsx057. https://doi.org/10.1093/amt/tsx057

Groves RL, Chapman S, Schramm S (2011a) Registered and experimental foliar insecticides to control Colorado potato beetle and potato leafhopper in potato, 2010. Arthropod Manage Tests 36:E60

Groves RL, Chapman S, Schramm S (2011b) Evaluations of systemic insecticides for the control of Colorado potato beetle, potato leafhopper and aphids in potato, 2010. Arthropod Manage Tests 36:E61

Gunning RV, Forrester NW (1984) Cyclodiene lindane resistance in *Agrypnus variabilis* (Candeze) (Coleoptera: Elateridae) in northern New South Wales. J Aust Entomol Soc 23:247–248

Hancock M, Green D, Lane A, Mathias PL, Port CM, Tones SJ (1986) Evaluation of insecticides to replace aldrin for the control of wireworms on potatoes. Ann Appl Biol 108:28–29

Harding JA (1962) Tests with systemic insecticides for control of insects and certain diseases on potatoes. J Econ Entomol 55:62–64

Harris CR, Turnbull SA (1986) Contact toxicity of some pyrethroid insecticides, alone and in combination with piperonyl butoxide, to insecticide-susceptibile and pyrethroid-resistant strains of the Colorado potato beetle (Coleoptera: Chrysomelidae). Can Entomol 118:1173–1176

Haydock PPJ, Woods SR, Grove IG, Hare MC (2006) Chemical control of nematodes. In: Perry RN, Moens M (eds) Plant nematology. CABI Publishing, Wllingford, UK, pp 392–410

Hitchner EM, Kuhar TP, Dively GP, Youngman RR, Philips CR, Anderson TD (2012) Baseline toxicity and field efficacy of metaflumizone on Colorado potato beetle (Coleoptera: Chrysomelidae). J Econ Entomol 105:207–213

Hofmaster RN (1956) Resistance of the Colorado potato beetle to DDT. Veg Growers News 10:3–4

Holm L, Blomqvist A, Brandt I, Brunström B, Ridderstråle Y, Berg C (2006) Embryonic exposure to o,p'-DDT causes eggshell thinning and altered shell gland carbonic anhydrase expression in the domestic hen. Environ Toxicol Chem 25(10):2787–2793

Huseth AS, Groves RL, Chapman SA, Alyokhin A, Kuhar TP, Macrae IV, Szendrei Z, Nault BA (2014) Managing Colorado potato beetle insecticide resistance: new tools and strategies for the next decade of pest control in potato. J Int Pest Manag 5:A1–A8

Ioannidis PM, Grafius E, Whalon ME (1991) Patterns of insecticide resistance to azinphosmethyl, carbofuran, and permethrin in the Colorado potato beetle (Coleoptera: Chrysomelidae). J Econ Entomol 84:1417–1423

Ishaaya I, Horowitz AR (1995) Pyriproxyfen, a novel insect growth regulator for controlling whiteflies: mechanisms and resistance management. Pestic Sci 43:227–232

Ishaaya I, Kontsedalov S, Horowitz AR (2003) Novaluron (Rimon), a novel IGR: potency and cross-resistance. Arch Insect Biochem Physiol 54:157–164

Jansson RK, Lecrone SH, Tyson R, Daigle CF Jr (1988) Wireworm control on potato, 1987. Insecticide Acaricide Tests 13:145

Jeschke P, Nauen R, Gutbrod O, Beck ME, Matthiesen S, Haas M, Velten R (2015) Flupyradifurone (Sivanto) and its novel butenolide pharmacophore: structural considerations. Pestic Biochem Physiol 121:31–38

Jones AG, Hoover K, Pearsons K, Tooker JF, Felton GW (2020) Potential impacts of translocation of neonicotinoid insecticides to cotton (*Gossypium hirsutum* (Malvales: Malvaceae)) extrafloral nectar on parasitoids. Environ Entomol 49:159–168

Kelce WR, Stone CR, Laws SC, Gray LE, Kemppainen JA, Wilson EM (1995) Persistent DDT metabolite p,p'–DDE is a potent androgen receptor antagonist. Nature 375(6532):581–585

Khan MA, Khan H, Ruberson JR (2015) Lethal and behavioral effects of selected novel pesticides on adults of *Trichogramma pretiosum* (hymenoptera: Trichogrammatidae): pesticide effects on survival and behavior of adult *Trichogramma pretiosum*. Pest Manag Sci 71:1640–1648

Kim DS, Kim SS (2019) Susceptibility of the predatory mite, *Neoseiulus californicus* (Acari: Phytoseiidae), to seven insecticides. Entomol Res 49:131–135

Koch RL, da Silva QO, Aita RC, Hodgson EW, Potter BD (2020) Efficacy of afidopyropen against soybean aphid (Hemiptera: Aphididae) and toxicity to natural enemies. Pest Manag Sci 76:375–383

Kowalska J (2007) Azadirachtin as a product for control of Colorado beetles. J Res App Agric Engin 52:78–81

Kristinsson H (1995) Pyridine-azomethines - a novel class of insecticides with a new mode of action. Opportunities for pymetrozine. Agro Food Ind Hi Tech 6:21–26

Kuhar TP, Alvarez JM (2008) Timing of injury and efficacy of soil-applied insecticides against wireworms on potato in Virginia. Crop Prot 27:792–798

Kuhar TP, Doughty HB (2009) Evaluation of foliar insecticides for the control of Colorado potato beetle in potato, 2008. Arthropod Manage Tests 34:E71

Kuhar TP, Doughty HB, Hitchner EM, Chapman AV (2007) Evaluation of seed-piece insecticides on potatoes, 2006. Arthropod Manage Tests 32:E37

Kuhar TP, Doughty HB (2010) Evaluation of foliar insecticides for the control of Colorado potato beetle and potato leafhopper in potatoes in Virginia, 2009. Arthropod Manage Tests 35:E17

Kuhar TP, Doughty HB (2016) Evaluation of foliar insecticides for the control of Colorado potato beetle in potatoes, 2015. Arthropod Manage Tests 2016:E1–E1. https://doi.org/10.1093/amt/tsw014

Kuhar TP, Doughty HB (2018) Evaluation of soil and foliar insecticides for control of Colorado potato beetle on potatoes, 2017. 43. https://doi.org/10.1093/amt/tsy080

Kuhar TP, Ghidiu G, Doughty HB (2010) Decline of European corn borer as a pest of potatoes. Plant Health Prog 11:1535

Kuhar TP, Hitchner EM, Chapman AV (2006a) Evaluation of foliar insecticides on potatoes, 2005. Arthropod Manage Tests 31:E56

Kuhar TP, Hitchner EM, Chapman AV (2006b) Evaluation of Rimon 0.83EC for control of insect pests on potatoes, 2005. Arthropod Manage Tests 31:E57

Kuhar TP, Kamminga K, Philips CR, Wallingford A, Wimer A (2012) Chemical control of potato pests. In: Alyokhin A, Giordanengo P, Vincent C (eds) Insect pests of potato: biology and management. Elsevier, pp 375–397

Kuhar TP, Speese J (2005a) Evaluation of foliar insecticides in potatoes, 2004. Arthropod Manage Tests 30:E63

Kuhar TP, Speese J (2005b) Evaluation of seed-piece treatment insecticides in potatoes, 2004. Arthropod Manage Tests 30:E64

Kuhar TP, Speese J (2005c) Evaluation of soil-applied insecticides in potatoes, 2004. Arthropod Manage Tests 30:E65

Kuhar TP, Speese J, Whalen J, Alvarez JM, Alyokhin A, Ghidiu G, Spellman MR (2003) Current status of insecticidal control of wireworms in potatoes. Pestic Outlook 14:265–267

Linduska JJ, Ross M, Abbott B, Steele S, Ross E, Eastman R (2002) Colorado potato beetle control on potatoes, 2001. Arthropod Manage Tests 27:E69

Linduska JJ, Ross M, Mulford K, Baumann D (1996) Colorado potato beetle control on potatoes with foliar insecticide sprays, 1995. Arthropod Manage Tests 21:E83

Liu D, Trumble JT (2005) Interactions of plant resistance and insecticides on the development and survival of *Bactericerca cockerelli* [Sulc] (Homoptera: Psyllidae). Crop Prot 24:111–117

Lucas É, Giroux S, Demougeot S, Duchesne R-M, Coderre D (2004) Compatibility of a natural enemy, *Coleomegilla maculata* Lengi (Col., Coccinellidae) and four insecticides used against the Colorado potato beetle (Col., Chrysomelidae): compatibility of *C. maculata* and insecticides. J Appl Entomol 128:233–239

Lundgren JG, Gassmann AJ, Bernal J, Duan JJ, Ruberson J (2009) Ecological compatibility of GM crops and biological control. Crop Prot 28:1017–1030

Lundin O, Rundlof M, Smith HG, Fries I, Bommarco R (2015) Neonicotinoid insecticides and their impacts on bees: a systematic review of research approaches and identification of knowledge gaps. PLoS One 10:e0136928. https://doi.org/10.1371/journal.pone.0136928

Machado AVA, Potin DM, Torres JB, Silva Torres CSA (2019) Selective insecticides secure natural enemies action in cotton pest management. Ecotoxicol Environ Safety 184:109669. https://doi.org/10.1016/j.ecoenv.2019.109669

Mansoor MM, Shad SA (2020) Biochemical mechanism, inheritance and cross-resistance to cyromazine in a non-target *Chrysoperla carnea*: a potential predator of whiteflies and aphids. Chemosphere 260:127620. https://doi.org/10.1016/j.chemosphere.2020.127620
Marcic D, Peric P, Krasteva L, Panayotov N (2009) Field evaluation of natural and synthetic insecticides against *Leptinotarsa decemlineata* say. Acta Hort 830:391–396
Martin PA, Gundersen-Rindal DE, Blackburn MB, Buyer J (2007) Chromobacterium subtsugae sp. nov., a betaproteobacterium toxic to Colorado potato beetle and other insect pests. Int J Syst Evol Microbiol 57:993–999
Matsuda K, Ihara M, Sattelle DB (2020) Neonicotinoid insecticides: molecular targets, resistance, and toxicity. Annu Rev Pharmacol Toxicol 60:241–255
Merrill LG (1952) Reduction of wireworm damage to potatoes. J Econ Entomol 45:548–549
Metcalf RL (1980) Changing role of insecticides in crop protection. Annu Rev Entomol 25:219–256
Mills NJ, Beers EH, Shearer PW, Unruh TR, Amarasekare KG (2016) Comparative analysis of pesticide effects on natural enemies in western orchards: a synthesis of laboratory bioassay data. Biol Control 102:17–25
Moffat AS (1993) New chemicals seek to outwit insect pests. Science 5121:550–551
Mordue AJ, Blackwell A (1993) Azadirachtin: An Update. J Insect Physiol 39:903–924
Morita M, Ueda T, Yoneda T, Koyanagi T, Haga T (2007) Flonicamid, a novel insecticide with a rapid inhibitory effect on aphid feeding. Pest Manag Sci 10:969–973
Morris C, Savy C, Judge S, Blain P (2014) Acute toxicity of organophosphorus compounds. In: Balali-Mood M, Abdollahi M (eds) Basic and clinical toxicology of organophosphorus compounds. Springer, London. https://doi.org/10.1007/978-1-4471-5625-3_3
Mota-Sanchez D, Hollingworth RM, Grafius EJ, Moyer DD (2006) Resistance and cross-resistance to neonicotinoid insecticides and spinosad in the Colorado potato beetle, *Leptinotarsa decemlineata* (say) (Coleoptera: Chrysomelidae). Pest Manag Sci 62:30–37
Mrema EJ, Rubino FM, Brambilla G, Moretto A, Tsatsakis AM, Colosi C (2013) Persistent organochlorinated pesticides and mechanisms of their toxicity. Toxicology 307:74–88
Munyaneza JE, Crosslin JM, Upton JE (2007) Association of *Bactericera cockerelli* (Homoptera: Psyllidae) with "zebra chip", a new potato disease in southwestern United States and Mexico. J Econ Entomol 100:656–663
Nauen R, Elbert A, McCaffery A, Slater R, Sparks TC (2012) IRAC: insecticide resistance, and mode of action classification of insecticides. In: Jeschke P, Witschel M, Krämer W, Schirmer U (eds) Modern crop protection compounds, 3rd edn. Wiley, Weinheim, Germany, pp 935–955
Nauen R, Jeschke P, Velten R, Beck ME, Ebbinghaus-Kintscher U, Thielert W, Wolfel K, Haas M, Kunz K, Raupach G (2015) Flupyradifurone: a brief profile of a new butenolide insecticide. Pest Manag Sci 71:850–862
Nauen R, Reckmann U, Thomzik J, Thielert W (2008) Biological profile of spirotetramat (Movento®)–a new two-way systemic (ambimobile) insecticide against sucking pest species. Bayer Crop Sci J 61(2):245–278
Nault BA, Seaman A (2019) Colorado potato beetle control with insecticides allowed for organic production, 2017 and 2018. Arthropod Manage Tests 44:tsz081. https://doi.org/10.1093/amt/tsz081
Nault BA, Speese J (1999a) Evaluation of Agri-mek for CPB control in potatoes, 1998. Arthropod Manage Tests 24:E74
Nault BA, Speese J (1999b) Evaluation of agenda for control of ECB in potatoes, 1998. Arthropod Manage Tests 24:E75
Nault BA, Speese J (2000) Evaluation of insecticides to control soil insect pests on potatoes, 1999. Arthropod Manage Tests 25:148
Noetzel D, Ricard M (1988) Wireworm and white grub control in potato, 1987. Arthropod Manage Tests 13:160–161
Noetzel DM, Holder B (1996) Control of resistant Colorado potato beetles Andover, MN 1994. Arthropod Manage Tests 21:E87

Parker WE, Clarke A, Ellis SA, Oakley JN (1990) Evaluation of insecticides for control of wireworms (*Agriotes* spp.) on potato. Ann Appl Biol 116:28–29

Parker WE, Howard JJ (2001) The biology and management of wireworms (*Agriotes* spp.) on potato with particular reference to the U.K. Agric Forestry Entomol 3:85–98

Pavlista AD (1997) Mocap and Aztec on reducing wireworm damage to potato tubers, 1995. Arthropod Manag Tests 22:158

Pavlista AD (2002) Comparison of systemic insecticides applied in-furrow or as a seed dust for potato psyllid and leafhopper control on potato, 2001. Arthropod Manag Tests 27:E75

Pisa LW, Amaral-Rogers V, Belzunces LP et al (2015) Effects of neonicotinoids and fipronil on non-target invertebrates. Environ Sci Pollut Res 22:68–102

Planes L, Catalán J, Tena A et al (2013) Lethal and sublethal effects of spirotetramat on the mealybug destroyer, *Cryptolaemus montrouzieri*. J Pest Sci 86:321–327

Popov SY, Alyokhin A (2019) Gender-specific acaricidal properties and sexual transmission of spirotetramat in two-spotted spider mite (Tetranychidae: Acariformes). J Econ Entomol 112: 2186–2192

Prabhaker N, Naranjo S, Perring T, Castle S (2017) Comparative toxicities of newer and conventional insecticides: against four generalist predator species. J Econ Entomol 110:2630–2636

Rose RL, Brindley WA (1985) An evaluation of the role of oxidative enzymes in Colorado potato beetle resistance to carbamate insecticides. Pestic Biochem Physiol 23:74–84

Roubos CR, Rodriguez-Saona C, Holdcraft R, Mason K, Isaacs R (2014) Relative toxicity and residual activity of insecticides used in blueberry pest management: mortality of natural enemies. J Econ Entomol 107:277–285

Roush RT, Hoy CW, Ferro DN, Tingey WM (1990) Insecticide resistance in the Colorado potato beetle (Coleoptera: Chrysomelidae): influence of crop rotation and insecticide use. J Econ Entomol 83:315–319

Russell JS, Grichar J, Besler B, Brewer K (2001) Evaluation of selected insecticides against potato psyllids on potatoes, 2000. Arthropod Manage Tests 26:E62

Salgado VL (1998) Studies on the mode of action of spinosad: insect symptoms and physiological correlates. Pestic Biochem Physiol 2:91–102

Sarkar N, Rhodes EM, Spies J et al (2020) Evaluation of non-target effects of OMRI - listed insecticides for management of *Drosophila suzukii* Matsumura in berry crops. J Appl Entomol 144:12–25

Schnaars-Uvino K, Baker MB (2021) High-level field-evolved resistance to spinosad in Colorado potato beetle, *Leptinotarsa decemlineata*, in organically managed fields. Pest Manag Sci 77: 4393–4399. https://doi.org/10.1002/ps.6473

Sewell GH, Alokhin A (2009) Control of Colorado potato beetle on potato, 2008. Arthropod Manage Tests 34:E52

Sewell GH, Alokhin A (2010a) Control of aphids on Irish potato, 2009. Arthropod Manag Tests 35: E18

Sewell GH, Alokhin A (2010b) Control of Colorado potato beetle on potato, 2009. Arthropod Manag Tests 35:E19

Sewell GH, Alokhin A (2011) Control of Colorado potato beetle on potato, 2010. Arthropod Manag Tests 36:E63

Sewell GH, Alyokhin A (2003) Control of Colorado potato beetle on potato, 2002. Arthropod Manag Tests 28:E63

Seymour J, Bowman G, Crouch M (1995) Effects of neem seed extract on feeding frequency of *Nezara viridula* L. (Hemiptera, Pentatomidae) on pecan nuts. J Aust Entomol Soc 34:221–223

Shamiyeh NB, Pereira R, Straw RA, Follum RA (1999) Control of wireworms in potatoes, 1998. Arthropod Manag Tests 24:164–165

Shirley AM, Noe JP, Nyczepir AP, Brannen PM, Shirley BJ, Jagdale GB (2019) Effect of spirotetramat and fluensulfone on population densities of *Mesocriconema xenoplax* and *Meloidogyne incognita* on peach. J Nematol 51:34. https://doi.org/10.21307/jofnem-2019-012

Sirota JM, Grafius E (1994) Effects of cyromazine on larval survival, pupation, and adult emergence of Colorado potato beetle (Coleoptera: Chrysomelidae). J Econ Entomol 87:577–582

Siviter H, Muth F (2020) Do novel insecticides pose a threat to beneficial insects? Proc R Soc B 287:1265. https://doi.org/10.1098/rspb.2020.1265

Slaney AC, Robbins HL, English L (1992) Mode of action of *bacillus thuringiensis* toxin CryIIIA: an analysis of toxicity in *Leptinotarsa decemlineata* (say) and *Diabrotica undecimpunctata* Howardi barber. Insect Biochem Mol Biol 22:9–18

Sorensen KA, Kidd KA (1991) Wireworm control, Currituck County, 1989. Arthropod Manag Tests 16:E106

Sparks TC, Crouse GD, Dripps JE, Anzeveno P, Martynow J, DeAmicis CV, Gifford J (2008) Neural network-based QSAR and insecticide discovery: spinetoram. J Comput Aided Mol Des 6:393–401

Sparks TC, Watson GB, Loso MR, Geng C, Babcock JM, Thomas JD (2013) Sulfoxaflor and the sulfoximine insecticides: chemistry, mode of action and basis for efficacy on resistant insects. Pestic Biochem Physiol 107:1–7

Stara J, Ourednickova J, Kocourek F (2011) Laboratory evaluation of the side effects of insecticides on *Aphidius colemani* (hymenoptera: Aphidiidae), *Aphidoletes aphidimyza* (Diptera: Cecidomyiidae), and *Neoseiulus cucumeris* (Acari: Phytoseidae). J Pest Sci 84:25–31

Straub L, Villamar-Bouza L, Bruckner S, Chantawannakul P et al (2016) Neonicotinoid insecticides can serve as inadvertent insect contraceptives. Proc Biol Sci 283:0506. https://doi.org/10.1098/rspb.2016.0506

Thetford LC (1993) Trigard a new tool for resistance management of the Colorado potato beetle. Resist Pest Manag 5:18–19

Thompson GD, Dutton R, Sparks TC (2000) Spinosad - a case study: an example from a natural products discovery programme. Pest Manag Sci 56:696–702

Tisler AM, Zehnder GW (1990) Insecticide resistance in the Colorado potato beetle (Coleoptera: Chrysomelidae) on the eastern shore of Virginia. J Econ Entomol 83:666–671

Tison L, Hahn M-L, Holtz S et al (2016) Honey bees' behavior is impaired by chronic exposure to the neonicotinoid thiacloprid in the field. Environ Sci Technol 50:7218–7227. https://doi.org/10.1021/acs.est.6b02658

Toba HH (1987) Treatment regimens for insecticidal control of wireworms on potato. J Agric Entomol 4:207–212

Tran AK, Alves TM, Koch RL (2016) Potential for Sulfoxaflor to improve conservation biological control of Aphis glycines (Hemiptera: Aphididae) in soybean. J Econ Entomol 109:2105–2114

Trisyono A, Whalon ME (1999) Toxicity of neem applied alone and in combinations with *bacillus thuringiensis* to Colorado potato beetle (Coleoptera: Chrysomelidae). J Econ Entomol 92:1281–1288

van Herk WG, Vernon RS, Vojtko B, Snow S, Fortier J, Fortin C (2015) Contact behaviour and mortality of wireworms exposed to six classes of insecticide applied to wheat seed. J Pest Sci 88: 717–739. https://doi.org/10.1007/s10340-015-0697-4

Vanaclocha P, Vidal-Quist C, Oheix S et al (2013) Acute toxicity in laboratory tests of fresh and aged residues of pesticides used in citrus on the parasitoid *Aphytis melinus*. J Pest Sci 86:329–336

Vasileiadis VP, Veres A, Loddo D, Masin R, Sattin M, Furlan L (2017) Careful choice of insecticides in integrated pest management strategies against *Ostrinia nubilalis* (Hübner) in maize conserves *Orius* spp. in the field. Crop Prot 97:45–51

Vernon RS, van Herk WG, Clodiu M, Harding C (2013) Further studies on wireworm management in Canada: damage protection versus wireworm mortality in potatoes. J Econ Entomol 106:786–799

Wantuch HA, Kuhar T, Hessler A, Doughty H (2016) Evaluation of organic insecticides for the control of Colorado potato beetles in potatoes in Virginia, 2014. Arthropod Manag Tests 41: tsv128. https://doi.org/10.1093/amt/tsv128

Ware GW, Whitacre DM (eds) (2004) The pesticide book, 6th edn. Meister Media Worldwide, Willoughby, OH, p 496

Whalen RA, Herbert DA, Malone S, Kuhar TP, Brewster CC, Reisig DD (2016) Effects of diamide insecticides on predators in soybean. J Econ Entomol 109:2014–2019

Whalon ME, Miller DL, Hollingworth RM, Grafius EJ, Miller JR (1993) Selection of a Colorado potato beetle (Coleoptera: Chrysomelidae) strain resistant to *bacillus thuringiensis*. J Econ Entomol 86:226–233

Wierenga JM, Hollingworth RM (1994) The role of metabolic enzymes in insecticide- resistant Colorado potato beetles. Pestic Sci 40:259–264

Wimer A, Philips C, Kuhar T, Adams J, Szendrei Z (2015) A new tool for resistance management: baseline toxicity, ovicidal activity, and field efficacy of the novel insecticide tolfenpyrad on Colorado potato beetle, *Leptinotarsa decemlineata*. Adv Entomol 3:139–147. https://doi.org/10.4236/ae.2015.34017

Wing KD, Sacher M, Kagaya Y, Tsurubuchi Y, Mulderig L, Connair M, Schnee M (2000) Bioactivation and mode of action of the oxadiazine indoxacarb in insects. Crop Prot 19:537–545

Wood TJ, Goulson D (2017) The environmental risks of neonicotinoid pesticides: a review of the evidence post 2013. Environ Sci Pollut Res 24:17285–17325

Yeary W, Fulcher A, Klingeman W, Grant J, Xiaocun S (2015) Responses of three natural enemy species to contact and systemic insecticide exposures in confined assays. J Enotoml Sci 50:35–46

Zehnder G, Warthen D (1988) Feeding inhibition and mortality effects of neem-seed extract on the Colorado potato beetle (Coleoptera: Chrysomelidae). J Econ Entomo 81:1040–1044

Zehnder G, Gurr G, Kühne S, Wade M, Wratten S, Wyss E (2007) Arthropod pest management in organic crops. Annu Rev Entomol 52:57–80

Zwahlen C, Nentwig W, Bigler F, Hilbeck A (2000) Tritrophic interactions of transgenic *bacillus thuringiensis* corn, *Anaphothrips obscurus* (Thysanoptera: Thripidae), and the predator *Orius majusculus* (Heteroptera: Anthocoridae). Environ Entomol 29:846–850

Genome Editing Prospects to Develop Disease/Pest-Resistant Potato Varieties

17

Salej Sood, Arun Kumar, S. Sundaresha, and Vinay Bhardwaj

Abstract

Potato is a wholesome food crop and has tremendous significance for its acceptance across the globe. Potato production has kept pace with the growing population to meet the ever-increasing demand; however, the crop suffers substantial yield losses due to incidence of several diseases and pests. Limited success has been achieved to control major diseases and pests through traditional breeding approaches. Moreover, the heterozygosity, autopolyploidy, and clonal propagation enhance the difficulties to transfer resistance genes for better crop varieties. The recent genetic engineering approaches particularly CRISPR/Cas have shown great potential through precise genetic modifications of the crop species including potato genome for several traits of economic importance. A number of studies successfully demonstrated CRISPR-Cas engineered resistance to diseases and pests in crop plants. Although studies on disease/pest resistance are very limited in potatoes, a combination of base editing and DNA-free genome editing could significantly improve its application, product development, end use, and commercialization.

Keywords

Gene editing · CRISPR-Cas · Late blight · Viruses · Potatoes · Resistant variety

S. Sood (✉) · S. Sundaresha · V. Bhardwaj
ICAR-Central Potato Research Institute, Shimla, Himachal Pradesh, India
e-mail: salej.sood@icar.gov.in

A. Kumar
CSIR-Institute of Himalayan Bioresource Technology, Palampur, Himachal Pradesh, India

Academy of Scientific and Innovative Research, Ghaziabad, Uttar Pradesh, India

S. Kumar Chakrabarti et al. (eds.), *Sustainable Management of Potato Pests and Diseases*, https://doi.org/10.1007/978-981-16-7695-6_17

17.1 Introduction

Potato (*Solanum tuberosum* L.) is the third important food crop in the world in terms of human consumption and thus contributes significantly to food and nutritional security across the globe. The global estimate of potato production was 370 million tons (FAOSTAT 2019). It is consumed by >1 billion people daily and is grown in more than 178 countries. India produces around 50 MT of potatoes from a 2.17 Mha area (FAOSTAT 2019). Now, India stands second after China in terms of total production with annual productivity of 22.3 tons per hectare.

Potato is vulnerable to many pests and diseases which include viruses, fungi, bacteria, and nematodes, which cause substantial yield losses every year. Late blight caused by *Phytophthora infestans* is a major production constraint in all potato-producing countries. Other important fungal diseases are early blight, potato wart, and powdery scab. Among viruses, *Potato virus Y* (PVY), *Potato virus X* (PVX), *Potato leaf roll virus* (PLRV), and *Potato apical leaf curl virus* (PALCV) are important and cause severe damage in subtropical potato production areas, including India. Bacterial wilt, black leg, and common scab are serious bacterial diseases in warm areas and cause huge losses directly and indirectly (Charkowski et al. 2020). Similarly, many insect pests damage potato crops globally, major among them are potato tuber moth, potato weevils, aphids, whitefly, etc. (Kroschel et al. 2020). Up to 30% of potato production is lost to insect pests globally (Oerke et al. 1994). The use of chemicals for the management of diseases and pests is uneconomical, environmentally unsafe, and sometimes ineffective. Therefore, conferring resistance to biotic stresses is of prime importance to develop resistant cultivars. Host resistance is the major strategy for the management of diseases and pests. This requires screening of germplasm accessions against the diseases and pests and identification of resistance sources for use in breeding programs. The resistant genes for late blight, viruses, and potato cyst nematodes (PCN) have been discovered from potato wild species and introduced in cultivated potato backgrounds. Potato breeding strategies for resistance to pests and diseases have been discussed in detail by various workers (Bradshaw and Mackay 1994; Jansky 2000; 2009). However, the progress is slow due to the genetic complexity of the crop, i.e., heterozygosity and autotetraploidy (Andersson et al. 2018). It is difficult to introgress a resistant gene into a cultivated background in potato as the original genetic constitution of the variety can never be brought back through breeding. Moreover, it takes more than 12–15 years to develop a new variety in potatoes following a conventional breeding program.

Like field crops, host resistance is considered to be an economical and ecologically safe approach to combat and control major diseases and pests in potatoes (Yoshida et al. 2013). The resistance to biotic stresses is either qualitative (major genes controlled) or quantitative (minor genes controlled). Qualitative or major gene resistance is due to one or few *R* genes, which have been well documented and have been extensively used in potato breeding programs across the globe (Watanabe 2015). These *R* genes confer vertical resistance and are often defeated by the pathogen producing a new effector as is happening in the case of late blight of potatoes. On the other hand, quantitative or horizontal resistance is most durable, but developing crop varieties with horizontal resistance is difficult and challenging due to polygenic inheritance. Besides, the information on molecular and biochemical mechanisms governing horizontal resistance is scanty and inadequate (Kou and Wang 2010).

Latest developments in the next-generation sequencing technologies integrated with omics tools like genomics, transcriptomics, proteomics, and metabolomics have facilitated the identification of functional genes and their allelic variants (Karre et al. 2019; Karre et al. 2016; Kumar et al. 2016; Yogendra et al. 2015). The advancements in functional genomics have helped in exploring the mechanisms involved in resistance. Several genes for diseases/pests resistance have been identified, and their resistance functions have been validated. Several techniques are being explored to carry out gene replacement of polymorphic genes with a functional copy of resistance genes. Transcription activator-like effector nucleases (TALEN) and clustered regularly interspersed short palindromic repeats (CRISPR)-Cas9 systems are two promising genome editing technologies (Rinaldo and Ayliffe 2015). Genome editing, through gene knockout or mutation and knock-in, replacement, or targeting of crop plants, is evolving as a new era of opportunities, especially in molecular breeding. However, advancements in precise gene insertion and replacement are yet to come. In this chapter, we will discuss the possibilities and achievements of gene editing in potato with an emphasis on biotic stress resistance breeding.

17.2 Genome Editing for Crop Improvement

Genome editing technology relies on engineered sequence-specific nucleases (SSNs) that cleave DNA in a sequence-specific manner because of the presence of a sequence-specific DNA-binding domain or RNA sequence. Through recognition of the specific DNA sequence, these nucleases can efficiently and precisely cleave the targeted genes. The double-strand breaks (DSBs) of DNA consequently result in cellular DNA repair mechanisms, including homology-directed repair (HDR) and error-prone nonhomologous end-joining breaks (NHEJ), leading to gene modification at the target sites. Thus, SSNs, delivery methods, and DNA repair are three important components of genome editing technology.

17.2.1 Sequence-Specific Nucleases

Genome editing uses engineered SSNs to remove, add, or substitute a DNA sequence. Engineered endonucleases, zinc finger nucleases (ZFNs), TALENs, and type II CRISPR/CRISPR-Cas9 approaches showed the path of single nucleotide editing mechanism for the improvement of crops (Arora and Narula 2017). The genome editing process modifies the genes precisely and accurately at specific target sites with in a genome. The targeted alterations of genes in different cells and organisms are done with the use of nucleases. Although genome editing looks similar to traditional breeding technique, mutation breeding, the preciseness, high efficiency, and target accuracy make it more alluring and useful (Butler et al. 2018). Currently genome editing can be used to make several alternations in DNA sequences, which include loss, change, or gain of novel gene function. The genome editing products are mostly similar to ones developed through spontaneous or induced mutation. But the major difference in genome editing is specified DNA sequence changes in the target region than genome-wide variations to be identified in spontaneous or induced mutants.

Genome editing is possible with the use of SSNs, which bind the target DNA in a sequence-specific manner, cut, and follow DNA repair mechanisms. The first class of SSNs was restriction endonucleases, called homing endonucleases. The second class of SSNs was designer nucleases, which includes ZFNs and TALENs. The paradigm shift in gene editing technology came after the discovery of a completely new class of nucleases, based on the naturally occurring editing system in bacterial against invading viruses, known as clustered regularly interspaced short palindromic repeats/CRISPR-associated system. It is a new highly efficient technology which directs a nuclease using a guide RNA to bind and cleave specific DNA sequences (Butler et al. 2018). CRISPR/Cas system has the major advantage of using single nuclease, Cas9, to target any DNA sequence with protospacer adjacent motif (PAM) in the genome. There are several other Cas9-like proteins identified for use in gene editing. These proteins/endonucleases act on the target DNA by forming a complex with a single-guide RNA (sgRNA). The sgRNA is designed within the target DNA sequence upstream of PAM and consists of 17–20 bp. First the sgRNA and Cas9 form a complex and sgRNA pairs with the target DNA sequence, which follows cleavage of the double-strand target DNA using RuvC and HNH nuclease domains of Cas9. Multiple targets or traits can be targeted using CRISPR-Cas system by designing multiple sgRNAs for target genes by using single nuclease, known as multiplexing. In the case of earlier designer nucleases, the nucleases were required for each target separately and expressed individually, while single CRISPR-Cas nuclease acts simultaneously on all target sequences (Cong et al. 2013). The CRISPR-Cas technology has been demonstrated and found application in virtually every biological kingdom, which shows the robustness of the technology (Sternberg and Doudna 2015).

CRISPR-Cas9 system, apart from its use as a genome editing tool, has been exploited as a sequence-specific, non-mutagenic gene regulation tool to understand the gene function using programmable gene regulation studies and for engineering novel genetic regulatory circuits intended for synthetic biology applications (Fig. 17.1) (Bortesi and Fischer 2015; Rath et al. 2015). This reorientation of function was achieved by introducing mutations in the HNH and RuvC nuclease domains of *Streptococcus pyogenes* Cas9, which made it nuclease-deficient Cas9 mutant. The disabled nuclease also known as dead Cas9, dCas9, interferes with RNA polymerase binding or elongation and can be used for regulating gene expression (Larson et al. 2013). This was first proved in *E. coli* as CRISPR interference (CRISPRi) mechanism, where dCas9 was paired with a sequence-specific sgRNA that resulted in either interference with transcription elongation by blocking RNA

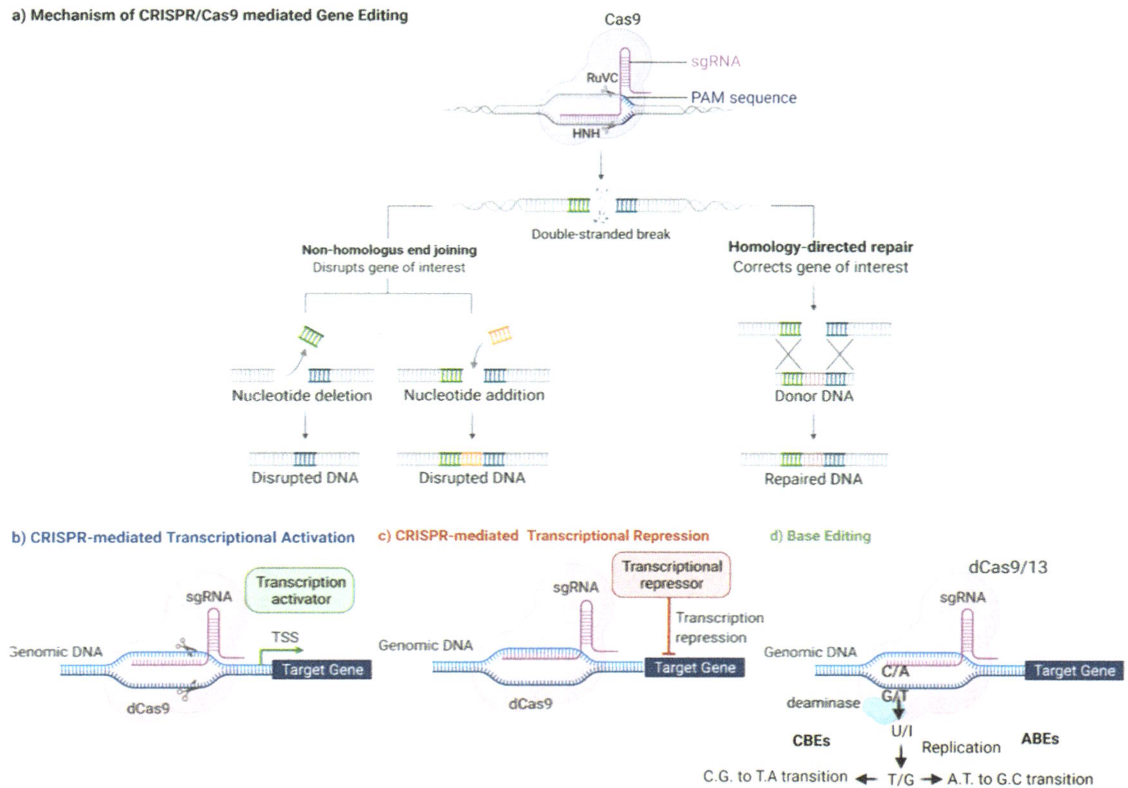

Fig. 17.1 Overview of (**a**) mechanism of CRISPR-Cas9-mediated genome editing and its applications in regulation of gene expression (**b**) transcriptional-activation, (**c**) repression, and (**d**) precise modifications using base editors

polymerase or by impeding transcription initiation by disrupting transcription factor binding (Bikard et al. 2013; Qi et al. 2013; Dominguez et al. 2015). However, the CRISPRi is less effective in eukaryotic system, and its repressive function can be enhanced by tethering dCas9 to transcriptional repressor domains to promote epigenetic silencing (Gilbert et al. 2013; Konermann et al. 2014). Similarly, CRISPR-mediated transcriptional activation, known as CRISPRa, depends upon fusion of dCAS9 with transcriptional activators. Fusion of dCas9 with transactivation or transrepression domain of a transcription factor (TF) can lead to precise and reversible control of target genes. CRISPR/Cas9 system also has potential in studying the chromosome structure and dynamics (Chen et al. 2013; Anton et al. 2014), identifying target proteins involved in histone DNA methylation, allowing the selective interrogation of the epigenome, and editing it to unravel the critical links of epigenetic mechanisms in gene regulation (Day 2014). Although progress has been made in achieving HDR-mediated gene replacement in plants, it still remains challenging owing to low frequency of HDR and limited number of donor repair templates. Recently, base editing has emerged as an alternative tool to HDR-mediated replacement in which either cytidine base editors (CBEs) or adenine base editors (ABEs) are used to make single substitutions in a programmable manner (Fig. 17.1). It is important to note that the base editing does not require a repair donor template and also not does involve double-stranded break in the genome (Bharat et al. 2020). Base editing using CBEs is achieved by fusing Cas9 nickase or dCas9 with a cytidine deaminase. The latter converts cytidine (C) to uracil (U) which is detected as thymine (T) during the replication process resulting in C.G to T.A transition. Base editing using ABEs utilizes hypothetical deoxyadenosine deaminase domain with dCa9 to catalyze the conversion of A to inosine (I) transition which is read as guanine (G) leading to A.T to G.C transition. Since single nucleotide polymorphisms (SNPs) are associated with the agronomic and disease resistance traits of crops, the use of BEs has started gaining momentum during last few years, and success has been achieved in plants such as rice, tomato, maize, etc. (Mishra et al. 2020).

17.2.2 Plant Transformation

The major limitation of any genetic engineering strategy including genome editing is DNA, RNA, or protein delivery in the plant cells and regeneration of new plants. Agrobacterium-mediated transformation, biolistics, and protoplast transfection are the three main genetic transformation methods in plants. Among these methods, the agrobacterium-mediated transformation is by far the commonly used method for introducing editing cassette in potato (Nadakuduti et al. 2018). The vast majority of cultivated potato is amenable to genetic transformation by *Agrobacterium tumefaciens*; however, certain wild relatives of potato used in plant breeding programs remain recalcitrant for the same. Recently, efforts were made to standardize a protocol for hairy root transformation using different strains of *Agrobacterium rhizogenes*. One of the strains, namely, *A. rhizogenes* MSU440, was found efficient in delivering the CRISPR reagents, and plants with high regeneration and target efficiency were generated for the *phytoene desaturase* (*PDS*) gene in a diploid clone of potato (Butler et al. 2020). Besides *Agrobacterium*, plant viral systems have been developed and being used to deliver editing cassettes (Baltes et al. 2014; Butler et al. 2018). Considerable efforts have been made to develop transient plant transformation methods, i.e., protoplasts and particle biolistics, which can rapidly transform crop plants and assess the editing efficiency. Although these methods have been widely used, the procedures used for transformation and regeneration are labor-intensive and time-consuming. Efforts are being made to overcome these challenges by using plant viral systems which are capable of complementing an *Agrobacterium*-mediated transformation system by enhancing the nuclease expression and DNA repair template copy number (Ali et al. 2015). Tobacco rattle virus (TRV) and bean yellow dwarf virus (BeYDV) are the two plant viruses which have been adapted as delivery vehicles for transformation. The ability of plant RNA viruses like TRV to deliver sgRNAs has proven to be a powerful approach to NHEJ-based modifications. On the other hand, the DNA virus BeYDV and other geminiviruses replicate to a high copy number within the nucleus, creating a geminivirus replicon (GVR), and provide a unique opportunity for synthesis-dependent strand annealing (SDSA) repair.

17.2.3 DNA Repair Mechanisms

DNA repair is an integral part of genome editing process. The SSNs cut the DNA at target sites, forcing the cell machinery to repair the DNA to survive or perish at all. However, it is important to make DNA breaks in the target DNA sequence for desired modifications. The SSNs cleave the DNA and create the breaks by minimizing the opportunity for excessive off-targeting which is critical for the modified cells growth and regeneration. The DNA breaks by the SSNs are largely repaired through nonhomologous end-joining (NHEJ) and homologous recombination (HR). The brief details on both these repair mechanisms have been given below.

In most plant cells, NHEJ is the preferred DNA repair pathway which results in error-prone repair. Nonavailability of homologous template DNA results in various sequence changes, i.e., insertions or deletions at the DNA break site, which may range up to 1 kilobase (kb) in size (Manova and Gruszka 2015). NHEJ is the most common DNA repair mechanism in genome editing as most site-specific nucleases induce double-stranded DNA breaks invoking NHEJ response. The NHEJ-mediated modifications are mostly deleterious resulting in loss of function, but it can be used for gain of function mutations too (Budhagatapalli et al. 2015).

HR is an error-free, high-fidelity repair mechanism, less frequently used in most plant cells. The main difference in HR compared to NHEJ relies in the use of DNA repair template which has sequence similarity with the break site. The homologous recombination between break sites and nonallelic, repetitive sequence regions of the genome can result in rearrangements in plants with large genomes. This explains why NHEJ is the preferred DNA repair pathway in most plant cells. Yet the choice for HR or NHEJ cannot be overly simplified, since both pathways repair a single break. In genome editing, the HR-mediated modifications are less commonly used due to the strong predilection for NHEJ repair. Moreover, the difficulty of supplying a readily available repair template favors NHEJ repair.

17.3 Genome Editing Approaches to Design Disease-Resistant Plants

Five different approaches to generate plants with disease resistance traits are discussed in the section, and a generalized workflow of genome editing is depicted in Fig. 17.2.

17.3.1 Gene Disruption in Coding Region

Insertion or deletion of one or more nucleotides at the sgRNA-guided site results in gene disruption in the coding region. This happens due to the error-prone NHEJ DNA-repair mechanism in the cell. This results in a frameshift mutation which disrupts the gene function by abnormal protein production. The technique has been successfully used to disrupt plant susceptibility (*S*) genes, which affects the host-pathogen communication, resulting in reduced pathogen fitness on the plant.

17.3.2 Gene Disruption in Promoter Region

The following modifications in the promoter region can be done for gene disruption:

(a) Disrupt the promoter sequence.
(b) Block the gene expression entirely.
(c) Disrupt an effector-binding site.
(d) Prevent a pathogen effector binding to the promoter.

17.3.3 Gene Deletion

For large deletions of DNA fragments, multiple sgRNAs designed from different sites introduce double-stranded breaks at target locations in the genome. For example, two sgRNAs binding each before the start and after the stop codon of the gene will produce double-stranded breaks at the respective locations. The DSBs result in the removal of the DNA fragment between the two breaks in the gene of interest, before the NHEJ repair mechanism repairs the DSBs. This approach could be used to delete large DNA fragments with in the genome.

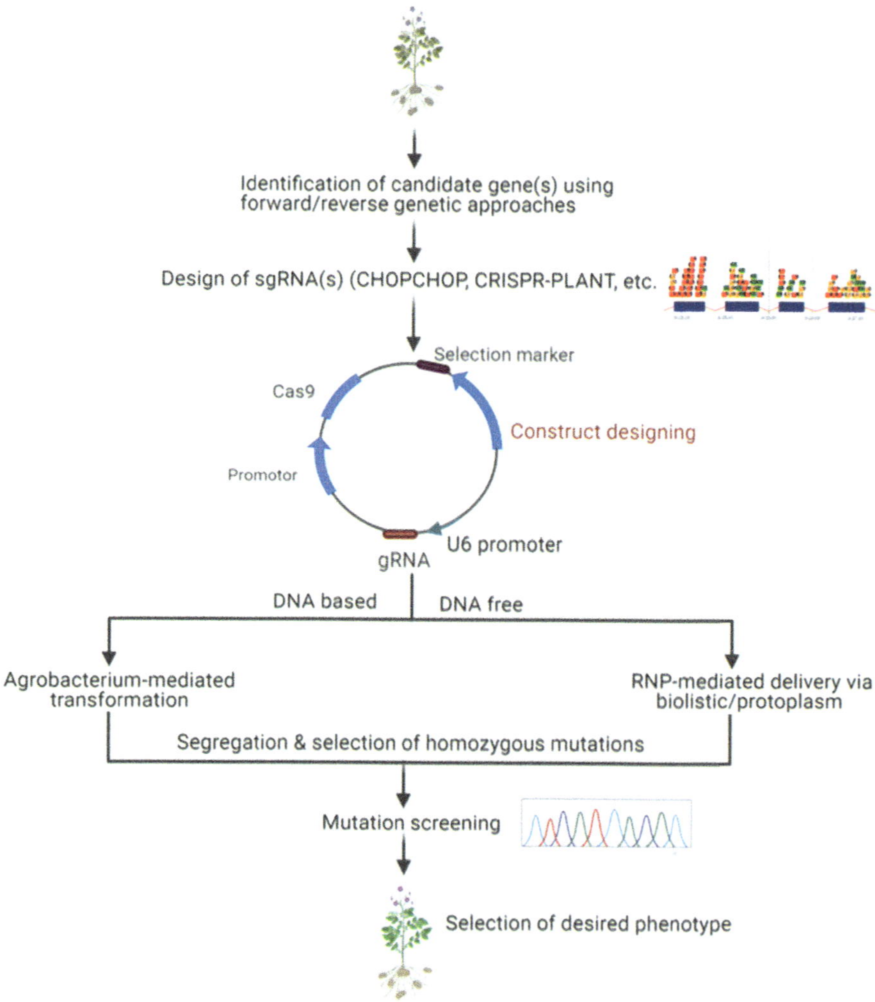

Fig. 17.2 Workflow of CRISPR-Cas9-mediated genome editing in plants. Short guide RNAs (sgRNAs) are prepared from the gene of interest, followed by construct design and delivery in the host. Desired mutants are screened for homozygous mutations

17.3.4 Gene Insertion

All the abovementioned CRISPR technologies could be used for disease resistance through alteration of *S* gene(s). But disrupting the product of *S* genes might affect plant health and productivity as most plant proteins are important and multifunctional including *S* gene proteins. Therefore, CRISPR-mediated gene insertion is an appropriate alternative. The gene insertion operates via an alternative route that works after Cas9 has created DSB guided by sgRNA. The route utilizes the cellular homology-directed repair (HDR), rather than NHEJ repair mechanism. The delivery DNA sequence containing the *R* gene surrounded by a sequence homologous to the DSB ends is supplemented with Cas9 and the sgRNAs. This guides the insertion of the *R* gene through HDR between the two DSBs. The HDR efficiency in crop plants is very low and needs improvement.

17.3.5 Gene Replacement

CRISPR technique can also be used to replace a defective or poorly acting *R* gene in a crop variety with the functional *R* gene variant from a disease-resistant variety through multiplexed HDR methodology. It is also known as biomimicking, which refers to the introduction of functional variant through CRISPR-mediated mutations, i.e., conversion of sequence of target gene in disease-susceptible variety similar to that of disease-resistant variety. In this process, the specific mutations associated with disease resistance are introduced in the target site rather than replacing the whole gene. This is based on the assumption that nucleotide differences in the gene of interest in cultivated and wild varieties are not otherwise significant to plant viability and production.

17.4 Genome Editing for Disease Resistance in Potato: Strategies and Achievements

Late blight (*Phytophthora infestans*) is the gravest problem and menace to potato production globally (Fisher et al. 2012). The approaches to combat the disease are developing disease resistance varieties and use of fungicide sprays. Recognized disease resistance genes, i.e., *R* genes, belong to the nucleotide-binding, leucine-rich repeat (NLR) class of intracellular immune receptor proteins which recognize pathogen effectors to initiate defense responses in the potato plants (Jones et al. 2016). The late blight pathogen has high rate of evolution of effector proteins, limiting the durability of resistance (Dong et al. 2014). Therefore, genome editing by base editors could be applied to genetically modify the potato varieties for resistance to late blight through targeting the codons encoding specific amino acids in *R* genes essential for effector recognition. Loss of susceptibility is considered as an alternative breeding strategy for durable broad-spectrum resistance and could be potentially applied in potatoes (Pavan et al. 2009). RNAi have been used to silence multiple susceptibility genes (*S* genes) resulting in increased late blight resistance in potato (Sun et al. 2016). But RNAi does not always result in a complete knockout, thereby genome editing could potentially be used to simultaneously knock out genes belonging to the *S*-locus for late blight susceptibility in potatoes. Susceptibility genes (*S* genes) are important for pathogenesis, and loss of *S* gene function confers increased resistance in several plants, such as rice, wheat, citrus, and tomatoes. Van den Hoogen and Govers (2018) targeted *Avr1*, *PiTubA2*, and

PiAP5 using CRISPR/Cas9 editing system for resistance to *Phytophthora infestans* as previously demonstrated for *P. sojae* by Ma et al. (2017b); however, they could not get transformants displaying mutagenized target genes. An alternate approach of base substitution for conversion of late blight susceptible cultivar to resistant was demonstrated in Russet Burbank cultivar (Hegde et al. 2021). The gene encoding caffeoyl-CoA O-methyltransferase (*StCCoAOM*T), which methylates caffeoyl-CoA to feruloyl-CoA and 5-hydroxyferuloyl-CoA to sinapoyl-CoA, was selected as the candidate for gene editing. They carried out a precise single nucleotide polymorphism (SNP) mutation correction of the *StCCoAOMT* gene in Russet Burbank potato using *Geminivirus* replicon-based CRISPR-Cas9-mediated homology-directed repair (HDR). Recently, tetra-allelic deletion mutants were generated to knock out the function of susceptibility genes, *StDND1*, *StCHL1*, and DMG400000582 (*StDMR6-1*), using a CRISPR/Cas9 system, which resulted in increased resistance to late blight (Kieu et al. 2021).

The management of plant viruses involve use of improved agricultural practices and virus-resistant crop varieties. Many genome editing studies for biotic stresses resistance in crop plants have involved resistance to viruses. All these studies demonstrate two main ways to engineer virus resistance: (1) targeting the virus genome directly and (2) targeting plant *S* genes crucial for the development of the viral disease.

In the first case, important viral genome sites are targeted inside the plant cells to protect the plants from viral diseases. The important gene targets in this approach are coat protein and replication genes. Targeting the coat protein and replication gene sites introduces indels in the virus genome subsequently resulting in the lower virus titer and significantly reduced disease symptoms. While in the second approach, instead of targeting the virus genome, plant S genes are targeted through sgRNA targets. The most widely targeted *S* genes in CRISPR-Cas editing of virus resistance are the *eIF4E* genes that encode cap-binding proteins essential for the cellular infection cycle of various RNA potyviruses. Notably, the most economically important plant virus diseases are caused by geminiviruses, and all studies to date of CRISPR-mediated geminivirus resistance have used direct virus DNA targeting. This approach however has its limitations, owing to the possibility of virus escape and generation of resistance-blocking strains. The most probable solution is use of host susceptibility factors involved specifically in the plant-geminivirus interaction. But precise *S* genes are not known and available for editing for geminiviruses.

Attempts to control viral diseases with conventional approaches such as breeding and RNA interference have met with limited success (Chaudhary 2018). Researchers have high hopes on CRISPR/Cas-mediated genome editing to control viral diseases. In a study to control PVY, Zhan et al. (2019) used class 2 type VI CRISPR/Cas effector Cas13a to protect potato plants. Transgenic potato lines expressing Cas13a/sgRNA (small guide RNA) constructs showed suppressed PVY accumulation and disease symptoms. The levels of viral resistance correlated with the expression levels of the Cas13a/sgRNA construct in the plants. It was suggested that appropriately designed sgRNAs can specifically interfere with multiple PVY strains. Potato plants are exposed to a variety of different viruses. Thus, it would be worthwhile to investigate whether the CRISPR/Cas13a system can be used to express multiple sgRNAs against several RNA viruses to produce multiresistant plants. Multiplexing is also a suitable strategy to prevent the development of resistance of the virus to cleavage by the CRISPR/Cas13a system.

17.5 Gene Editing Attempts for Other Traits in Potato

The first successful report of gene editing in potato appeared in 2014, where the use of TALEN was demonstrated by knocking out all the four alleles of *Sterol side chain reductase 2* (*StSSR2*) involved in anti-nutritional sterol glycoalkaloid (SGA) synthesis (Sawai et al. 2014). Later CRISPR/Cas9 was used in numerous studies (Table 17.1). CRISPR/Cas9 was employed using *Agrobacterium*-mediated transformation to exert an efficient site-specific mutation (up to 83%) in the host gene, *Auxin/indole-3-acetic acid* (*StIAA2*) (Wang et al. 2015). The knockout of the *StIAA2* gene resulted in engineered potato lines having an altered Aux/IAA protein expression and paved the way for the efficient CRISPR-mediated targeted mutagenesis in tetraploid cultivated potato (Hameed et al. 2020). The endogenous *Acetolactate synthase1* (*StALS1*) gene was modified to incorporate mutations using a donor repair template leading to herbicide tolerance, and mutations were shown to be heritable (Butler et al. 2015, 2016). *StALS1* was also targeted by TALENs via protoplast transfection, and successful regeneration of *StALS1* knockout lines from transformed protoplasts was demonstrated in tetraploid potato (Nicolia et al. 2015). Similarly, improvement in tuber cold storage quality of a commercial tetraploid cultivar, Ranger Russet, was achieved by targeting *Vacuolar invertase* (*StVInv*) using TALENs via protoplast transformation and regeneration (Clasen et al. 2016). The tubers of *StVInv* knockout lines had very low levels of reducing sugars, low acrylamide, and produced light colored chips along with no foreign DNA in their genome (Clasen et al. 2016). Nakayasu et al. (2018) targeted steroidal glycoalkaloids biosynthesis gene *St16DOX* encoding a steroid 16α-hydroxylase to knockout its function and causes the complete abolition of the SGA accumulation in potato and generated two lines free of SGA. Similarly, editing of sgRNAs targets from different regions of phytoene desaturase (PDS) from the carotenoid biosynthesis pathway showed visual phenotype as depigmentation (Khromov et al. 2018; Bánfalvi et al. 2020). The host transcription factor gene (*StMYB44*), involved in phosphate mobilization, was knocked out using CRISPR/Cas9 technology, and stable transformants were generated through *Agrobacterium*-mediated transformation (Zhou et al. 2017). In another study, a translational enhancer (dMac3) was used in the CRISPR/Cas9 system to enhance the targeted mutation frequency in the tetraploid genome of potato. The enhanced expression of Cas9 resulted in 25% of targeted mutagenesis in the four alleles of the potato *granule-bound starch synthase I* (*GBSSI*) gene that generated the potato tubers having a lower amylose starch (Kusano et al. 2018). Andersson et al. (2017, 2018) produced waxy potato with altered tuber starch quality by knocking out all four alleles of *Granule-bound starch synthase* (*GBSS*) in a tetraploid potato cultivar via CRISPR/Cas9. Andersson et al. (2018) used the

Table 17.1 Genome editing studies in potato

Target gene/ sequences	Delivery method/main strategy	Molecular function	Application perspective	References
Sterol side chain reductase 2	*Agrobacterium*-mediated transformation/ TALEN	Involved in anti-nutritional sterol glycoalkaloid synthesis	Low glycoalkaloid potatoes	Sawai et al. (2014)
Acetolactate synthase1 (*StALS1*)	Transient expression of TALENs in potato protoplasts for targeted mutagenesis and regeneration	Involved in the acetolactate synthase biosynthesis (amino acid biosynthesis)	Herbicide Resistance	Nicolia et al. (2015)
Auxin/ indole-3-acetic acid	*Agrobacterium*-mediated transformation for delivery of CRISPR–Cas9 system	Targeted mutagenesis using native StU6 promoter driving the sgRNA involved in indole 3-acetic acid synthesis	Petiole hyponasty and shoot morphogenesis	Wang et al. (2015)
Acetolactate synthase 1 (*ALS1*)	*Agrobacterium*-mediated transformation for GVR-mediated delivery of CRISPR-Cas9 system and donor template// gene knockout and replacement	Involved in the acetolactate synthase biosynthesis (amino acid biosynthesis)	Herbicide resistance	Butler et al. (2015); Butler et al. (2016)
Vacuolar invertase (*StVInv*)	Transient expression of TALENs in potato protoplasts for targeted mutagenesis and regeneration	Cold-induced sweetening, acrylamide content in tubers	Tuber improvement for cold storage, minimizing reducing sugars	Clasen et al. (2016)
StALS1	*Agrobacterium*-mediated transformation/ TALEN	Involved in the acetolactate synthase biosynthesis (amino acid biosynthesis)	Herbicide resistance	Forsyth et al. (2016)

(continued)

Table 17.1 (continued)

Target gene/ sequences	Delivery method/main strategy	Molecular function	Application perspective	References
StGBSS	*Agrobacterium*-mediated transformation/ TALEN	Development of a gateway system for rapid assembly of TALENs in a binary vector	Tuber starch quality	Kusano et al. (2016)
Three different regions of the gene encoding *Granule-bound starch synthase* (*GBSS*)	PEG-mediated protoplast transfection with CRISPR-Cas9 expression plasmid constructs//gene knockout with Cas9/gRNA	Enzyme responsible for the synthesis of amylose (encoded by a single locus)	Starch quality (amylopectin potato starch)	Andersson et al. (2016)
1,4-alpha-glucan Branching enzyme Gene (*SBE1*), *StVInv*	Agroinfiltration/ TALEN	Degree of starch branching, cold-induced sweetening	Tuber improvement for starch quality and minimizing reducing sugars in cold storage	Ma et al. (2017a)
Transcription factor Gene *StMYB44*	*Agrobacterium*-mediated transformation/ CRISPR-Cas9	Phosphate transport via roots	Understand the molecular basis of phosphate stress responses in potato	Zhou et al. (2017)
Granule-bound Starch synthase (*StGBSS*)	Transient expression of CRISPR/Cas9 in potato protoplasts for targeted mutagenesis and regeneration	Degree of starch branching	Potato tuber with altered starch content	Andersson et al. (2017), Andersson et al. (2018)
Phytoene desaturase and *coilin*	In vitro study without delivery in the plant	Phytoene desaturase gene involved in carotenoid biosynthesis was targeted	Visible albino phenotype of PDS null mutants	Khromov et al. (2018)
Steroidal glycoalkaloids	*Agrobacterium rhizogenes* strain ATCC15834	Knockout of *St16DOX* encoding a	Reduced glycoalkaloid potatoes	Nakayasu et al. (2018)

(continued)

Table 17.1 (continued)

Target gene/ sequences	Delivery method/main strategy	Molecular function	Application perspective	References
biosynthesis gene *St16DOX*	using electroporation	steroid 16α-hydroxylase in GA biosynthesis caused the complete abolition of the SGA accumulation in potato		
Granule-bound starch synthase I (*GBSSI*) gene	*A. tumefaciens* EHA105-mediated transformation	Degree of starch branching	Potato tuber with altered starch content	Kusano et al. (2018)
Stylar ribonuclease Gene (*S-Rnase*)	*Agrobacterium*-mediated transformation for delivery of CRISPR-Cas9 system	Knockout of self-incompatibility gene S-RNase in diploid potato line resulted in self compatibility	Self-incompatibility	Ye et al. (2018)
P3, CI, NIb and CP regions	Agrobacterium-mediated transformation for delivery of CRISPR-Cas13a system	Targeted mutagenesis of P3, CI, Nlb, and CP. P3 protein is the potyviral membrane protein involved in virus replication, systemic infection, pathogenicity and movement; the CI protein forms the laminate cytoplasmic inclusion bodies involved in virus movement and infection; the NIb protein is the RNA-dependent RNA polymerase that participates in the replication of the viral RNA; the coat protein	PVY resistance (PVY^{O}, PVY^{N} and the recombinant $PVY^{N:O}$)	Zhan et al. (2019)

(continued)

Table 17.1 (continued)

Target gene/ sequences	Delivery method/main strategy	Molecular function	Application perspective	References
		(CP) is required for virion assembly, cell-to-cell and systemic movement, and aphid transmission		
Phytoene desaturase (*PDS*) gene	Transient *agrobacterium*-mediated CRISPR/Cas9 system	Phytoene desaturase gene involved in carotenoid biosynthesis was targeted	Visible albino phenotype of PDS null mutants	Bánfalvi et al. (2020)
StPPO2 polyphenol oxidase gene	PEG-mediated protoplast transfection with CRISPR-Cas9 expression plasmid Constructs	Polyphenol oxidases (PPOs) catalyze the conversion of phenolic substrates to quinones, leading to the formation of dark-colored precipitates. Targeted editing of *StPPO2* gene lower the activity	Reduced enzymatic browning	Gonzalez et al. (2020)
Caffeoyl-CoA O-methyltransferase gene (*StCCoAOMT*)	*Agrobacterium*-mediated transformation/ geminivirus replicon-based CRISPR-Cas9 system	SNP mutation correction of *StCCoAOMT* gene enhanced late blight resistance through production of functional protein	Late blight resistance	Hegde et al. (2021)
S genes, namely, *StDND1*, *StCHL1*, And DMG400000582 (*StDMR6-1*)	*Agrobacterium*-mediated transformation for delivery of CRISPR-Cas9 system	*DND1* encodes a cyclic nucleotide-gated ion channel protein which has a role in conducting Ca^{2+} into plant cells; *DMR6* encodes a salicylic acid 5-hydroxylase that fine-tunes	Late blight resistance	Kieu et al. (2021)

(continued)

Table 17.1 (continued)

Target gene/ sequences	Delivery method/main strategy	Molecular function	Application perspective	References
		salicylic acid homeostasis; *CHL1* encodes a transcription factor, involved in brassinosteroid (BR) hormone signaling, which interacts with the RXLR effector AVR2		
Sbe1 and *Sbe2*	PEG-mediated protoplast transfection with CRISPR-Cas9 expression plasmid constructs	Targeted mutagenesis of *Sbe 1* and *Sbe 2* to increase amylose content in starch	Starch quality	Zhao et al. (2021)

RNPs-mediated delivery of CRISPR/Cas9 constructs in potato using the protoplast-mediated transfection to knockout an endogenous *GBSS* gene. The results showed full-knockout with a mutation frequency of up to 9%, and all the regenerated knockout lines transfected with synthetically produced RNPs were transgene-free, contrary to their previous study (Andersson et al. 2017). The RNPs mediated delivery could be potentially adopted and further optimized for producing commercial potato lines having transgene-free status. Another study reported the use of CBEs coupled with CRISPR/Cas9 technology to produce transgene-free potato plants resistant to herbicide "chlorsulfuron." They targeted the host *acetolactate synthase* (*ALS*) gene in potato using *Agrobacterium*-mediated stable transformation of CRISPR/Cas9 components and produced 10% of transgene-free regenerated potato plants lacking any stable-integration of T-DNA, however, confirmed the base edited (C-to-T base) mutation at the targeted site (Veillet et al. 2019). Replacement of Arabidopsis U6 promotor with endogenous potato U6 promotor increased editing efficiencies for the target *GBSS* gene in potato.

17.6 Future Prospects

The pace and scale of genome editing have triggered a major revolution in plant biology and are poised to impact plant breeding also. Genome editing technologies can accelerate crop improvement, as they produce precise genetic modifications in a variety of species and can yield the desired trait in a relatively short time compared to traditional breeding. How genome-edited crops will be regulated is still under evaluation, but may not be as stringent as is the case for earlier genetically engineered technologies. The genome editing technology has shown inordinate potential in agriculture, but it is still limited by the low efficacy, off-target effects, limiting protospacer adjacent motif (PAM) sequences, and many other issues. To address these limitations, novel innovations are being added continually (Zhang et al. 2018). The CRISPR-Cas technology is being progressively used to introduce resistance to various diseases and pests in numerous economically important crop plants including potatoes. Several independent studies have demonstrated successful introduction of CRISPR-mediated major gene resistance and, in some cases, broad-spectrum resistance against multiple pathogens. Conventional genome editing involves the delivery and integration into the host genome of DNA cassettes encoding editing components. Integration occurs at random and therefore can generate undesirable genetic changes. The DNA-free genome editing is a ground-breaking technology, producing genetically edited crops with a reduced risk of undesirable off-target mutations, and meeting current and future agriculture demands from both a scientific and regulatory standpoint. A combination of base editing and DNA-free genome editing has been recently described in potato, which could greatly facilitate both the application of base editing to plant breeding and the commercialization of edited plants.

Acknowledgments Arun Kumar is thankful to DST-SERB, Govt. of India, for funding his research through Ramanujan fellowship (SB/S2/RJN-022/2017).

References

Ali Z, Abulfaraj A, Idris A, Ali S, Tashkandi M, Mahfouz MM (2015) CRISPR/Cas9-mediated viral interference in plants. Genome Biol 16:238–249. https://doi.org/10.1186/s13059-015-0799-6

Andersson M, Turesson H, Nicolia A, Fält AS, Samuelsson M, Hofvander P (2016) Efficient targeted multiallelic mutagenesis in tetraploid potato (*Solanum tuberosum*) by transient CRISPR-Cas9 expression in protoplasts. Plant Cell Rep 36:117–128. https://doi.org/10.1007/s00299-016-2062-3

Andersson M, Turesson H, Nicolia A et al (2017) Efficient targeted multiallelic mutagenesis in tetraploid potato (Solanum tuberosum) by transient CRISPR-Cas9 expression in protoplasts. Plant Cell Rep 36:117–128. https://doi.org/10.1007/s00299-016-2062-3

Andersson M, Turesson H, Olsson N et al (2018) Genome editing in potato via CRISPR-Cas9 ribonucleoprotein delivery. Physiol Plant 164:378–384. https://doi.org/10.1111/ppl.12731

Anton T, Bultmann S, Leonhardt H, Markaki Y (2014) Visualization of specific DNA sequences in living mouse embryonic stem cells with a programmable fluorescent CRISPR/Cas system. Nucleus 5:163–172

Arora L, Narula A (2017) Gene editing and crop improvement using CRISPR-Cas9 system. Front Plant Sci 8:1932. https://doi.org/10.3389/fpls.2017.01932

Baltes NJ, Gil-Humanes J, Cermak T, Atkins PA, Voytas DF (2014) DNA replicons for plant genome engineering. Plant Cell 26:151–163

Bánfalvi Z, Csákvári E, Villányi V, Kondrák M (2020) Generation of transgene-free PDS mutants in potato by agrobacterium-mediated transformation. BMC Biotechnol 20:25. https://doi.org/10.1186/s12896-020-00621-2

Bharat SS, Li S, Li J, Yan L, Xia L (2020) Base editing in plants: current status and challenges. Crop J 8:384–395

Bikard D, Jiang W, Samai P, Hochschild A, Zhang F, Marraffini LA (2013) Programmable repression and activation of bacterial gene expression using an engineered CRISPR-Cas system. Nucleic Acids Res 41:7429–7437

Bortesi L, Fischer R (2015) The CRISPR/Cas9 system for plant genome editing and beyond. Biotechnol Adv 33:41–52

Bradshaw JE, Mackay GR (eds) (1994) Potato genetics. CAB International, Wallingford, UK

Budhagatapalli N, Rutten T, Gurushidze M, Kumlehn J, Hensel G (2015) Targeted modification of gene function exploiting homology-directed repair of TALEN-mediated double strand breaks in barley. G3 5:1857–1863

Butler NM, Atkins PA, Voytas DF, Douches DS (2015) Generation and inheritance of targeted mutations in potato *Solanum tuberosum* L. using the CRISPR/Cas system. PLoS One 10: e0144591

Butler NM, Baltes NJ, Voytas DF, Douches DS (2016) Geminivirus-mediated genome editing in potato (Solanum tuberosum L.) using sequence-specific nucleases. Front. Plant Sci 7:1045. https://doi.org/10.3389/fpls.2016.01045

Butler NM, Jansky SH, Jiang J (2020) First-generation genome editing in potato using hairy root transformation. Plant Biotechnol J 18(11):2201–2209

Butler NM, Jiang J, Stupar RM (2018) Crop improvement using genome editing. Plant Breed Rev 41:55–101

Charkowski A, Sharma K, Parker ML, Secor GA, Elphinstone J (2020) Bacterial diseases of potato. In: Campos H, Ortiz O (eds) The potato crop: its agricultural, nutritional and social contribution to humankind. Springer, Cham, pp 351–388

Chaudhary K (2018) CRISPR/Cas13a targeting of RNA virus in plants. Plant Cell Rep 37:1707–1712

Chen B, Gilbert LA, Cimini BA, Schnitzbauer J, Zhang W, Li GW et al (2013) Dynamic imaging of genomic loci in living human cells by an optimized CRISPR/Cas system. Cell 155:1479–1491

Clasen BM, Stoddard TJ, Luo S, Demorest ZL, Li J, Cedrone F et al (2016) Improving cold storage and processing traits in potato through targeted gene knockout. Plant Biotechnol J 14:169–176. https://doi.org/10.1111/pbi.12370

Cong L, Ran FA, Cox D, Lin S, Barretto R et al (2013) Multiplex genome engineering using CRISPR/Cas systems. Science 339:819–823

Day JJ (2014) New approaches to manipulating the epigenome. Dialogues Clin Neurosci 16:345

Dominguez AA, Lim WA, Qi LS (2015) Beyond editing: repurposing CRISPR-Cas9 for precision genome regulation and interrogation. Nat Rev Mol Cell Biol 17:5

Dong S, Stam R, Cano LM, Song J, Sklenar J, Yoshida K et al (2014) Effector specialization in a lineage of the Irish potato famine pathogen. Science 343:552–555. https://doi.org/10.1126/science.1246300

FAOSTAT (2019). http://faostat.fao.org. Accessed 25 March 2021

Fisher MC, Henk DA, Briggs CJ, Brownstein JS, Madoff LC, McCraw SL et al (2012) Emerging fungal threats to animal, plant and ecosystem health. Nature 484:186–194. https://doi.org/10.1038/nature10947

Forsyth A, Weeks T, Richael C, Duan H (2016) Transcription activator-like effector nucleases (TALEN)-mediated targeted DNA insertion in potato plants. Front Plant Sci 7:1572. https://doi.org/10.3389/fpls.2016.01572

Gilbert LA, Larson MH, Morsut L, Liu Z, Brar GA, Torres SE et al (2013) CRISPR-mediated modular RNA-guided regulation of transcription in eukaryotes. Cell 154:442–451

Gonzalez MN, Massa GA, Andersson M, Turesson H, Olsson N, Fält AS, Storani L, Décima Oneto CA, Hofvander P, Feingold SE (2020) Reduced enzymatic Browning in potato tubers by specific editing of a polyphenol oxidase gene via ribonucleoprotein complexes delivery of the CRISPR/Cas9 system. Front Plant Sci 10:1649. https://doi.org/10.3389/fpls.2019.01649

Hameed A, Mehmood MA, Shahid M, Fatma S, Khan A, Ali S (2020) Prospects for potato genome editing to engineer resistance against viruses and cold-induced sweetening. GM Crops Food 11: 185–205

Hegde N, Joshi S, Soni N et al (2021) The caffeoyl-CoA *O*-methyltransferase gene SNP replacement in russet Burbank potato variety enhances late blight resistance through cell wall reinforcement. Plant Cell Rep 40:237–254. https://doi.org/10.1007/s00299-020-02629-6

Jones JD, Vance RE, Dangl JL (2016) Intracellular innate immune surveillance devices in plants and animals. Science 354:aaf6395. https://doi.org/10.1126/science.aaf6395

Karre S, Kumar A, Dhokane D, Kushalappa AC (2016) Metabolo-transcriptome profiling of barley reveals induction of a novel chitin elicitor receptor kinase gene (*HvCERK1*) conferring resistance against *fusarium graminearum*. Plant Mol Biol 93:247–267

Karre S, Kumar A, Yogendra KN, Kage U, Kushalappa AC, Charron JB (2019) HvWRKY23 regulates flavonoid glycoside and hydroxycinnamic acid amide biosynthetic genes in barley to combat fusarium head blight. Plant Mol Biol 100:591–605

Khromov AV, Gushchin VA, Timerbaev VI, Kalinina NO, Taliansky ME, Makarov VV (2018) Guide RNA design for CRISPR/Cas9-mediated potato genome editing. Dokl Biochem Biophys 479:90–94

Kieu NP, Lenman M, Wang ES, Petersen BL, Andreasson E (2021) Mutations introduced in susceptibility genes through CRISPR/Cas9 genome editing confer increased late blight resistance in potatoes. Sci Rep 11:4487. https://doi.org/10.1038/s41598-021-83972-w

Konermann S, Brigham MD, Trevino AE, Joung J, Abudayyeh OO, Barcena C et al (2014) Genome-scale transcriptional activation by an engineered CRISPR-Cas9 complex. Nature 517:583–588

Kou Y, Wang S (2010) Broad-spectrum and durability: understanding of quantitative disease resistance. Curr Opin Plant Biol 13:181–185

Kroschel J, Mujica N, Okonya J, Alyokhin A (2020) Insect pests affecting potatoes in tropical, subtropical, and temperate regions. In: Campos H, Ortiz O (eds) The potato crop: its agricultural, nutritional and social contribution to humankind. Springer, Cham, pp 251–306

Kumar A, Yogendra KN, Karre S, Kushalappa AC, Dion Y, Choo TM (2016) WAX INDUCER1 (HvWIN1) transcription factor regulates free fatty acid biosynthetic genes to reinforce cuticle to resist fusarium head blight in barley spikelets. J Exp Bot 67:4127–4139

Kusano H, Ohnuma M, Mutsuro-Aoki H, Asahi T, Ichinosawa D, Onodera H et al (2018) Establishment of a modified CRISPR/Cas9 system with increased mutagenesis frequency using the translational enhancer dMac3 and multiple guide RNAs in potato. Sci Rep 8:13753

Kusano H, Onodera H, Kihira M, Aoki H, Matsuzaki H, Shimada H (2016) A simple gateway-assisted construction system of TALEN genes for plant genome editing. Sci Rep 6:30234. https://doi.org/10.1038/srep30234

Larson MH, Gilbert LA, Wang X, Lim WA, Weissman JS, Qi LS (2013) CRISPR interference (CRISPRi) for sequence-specific control of gene expression. Nat Protoc 8:2180–2196

Ma J, Xiang H, Donnelly DJ, Meng FR, Xu H, Durnford D et al (2017a) Genome editing in potato plants by agrobacterium-mediated transient expression of transcription activator-like effector nucleases. Plant Biotechnol Rep 11:249–258. https://doi.org/10.1007/s11816-017-0448-5

Ma Z, Zhu L, Song T, Wang Y, Zhang Q, Xia Y et al (2017b) A paralogous decoy protects *Phytophthora sojae* apoplastic effector *PsXEG1* from a host inhibitor. Science 355:710–714. https://doi.org/10.1126/science.aai7919

Mishra R, Joshi RK, Zhao K (2020) Base editing in crops: current advances, limitations and future implications. Plant Biotechnol J 18:20–31

Nadakuduti SS, Buell CR, Voytas DF, Starker CG, Douches DS (2018) Genome editing for crop improvement–applications in clonally propagated polyploids with a focus on potato (Solanum tuberosum L.). Front Plant Sci 9:1607. https://doi.org/10.3389/fpls.2018.01607

Nakayasu M, Akiyama R, Lee HJ, Osakabe K, Osakabe Y, Watanabe B, Sugimoto Y, Umemoto N, Saito K, Muranaka T, Mizutani M (2018) Generation of α-solanine-free hairy roots of potato by CRISPR/Cas9 mediated genome editing of the St16DOX gene. Plant Physiol Biochem 131:70–77. https://doi.org/10.1016/j.plaphy.2018.04.026

Nicolia A, Proux-Wéra E, Åhman I, Onkokesung N, Andersson M, Andreasson E et al (2015) Targeted gene mutation in tetraploid potato through transient TALEN expression in protoplasts. J Biotechnol 204:17–24. https://doi.org/10.1016/j.jbiotec.2015.03.021

Oerke EC, Dehne HW, Schönbeck F, Weber A (eds) (1994) Crop production and crop protection. Elsevier, Amsterdam

Pavan S, Jacobsen E, Visser RGF, Bai Y (2009) Loss of susceptibility as a novel breeding strategy for durable and broad-spectrum resistance. Mol Breed 25:1. https://doi.org/10.1007/s11032-009-9323-6

Qi LS, Larson MH, Gilbert LA, Doudna JA, Weissman JS, Arkin AP et al (2013) Repurposing CRISPR as an RNA-guided platform for sequence-specific control of gene expression. Cell 152: 1173–1183

Rath D, Amlinger L, Hoekzema M, Devulapally PR, Lundgren M (2015) Efficient programmable gene silencing by Cascade. Nucleic Acids Res 43:237–246

Rinaldo AR, Ayliffe M (2015) Gene targeting and editing in crop plants: a new era of precision opportunities. Mol Breed 35:1–15

Sawai S, Ohyama K, Yasumoto S, Seki H, Sakuma T, Yamamoto T et al (2014) Sterol side chain reductase 2 is a key enzyme in the biosynthesis of cholesterol, the common precursor of toxic steroidal glycoalkaloids in potato. Plant Cell 26:3763–3774. https://doi.org/10.1105/tpc.114.130096

Sternberg SH, Doudna JA (2015) Expanding the biologist's toolkit with CRISPRCas9. Mol Cell 58: 568–574

Sun N, Zhang X, Wu C, He Y et al (2016) Engineering herbicide-resistant rice plants through CRISPR/Cas9-mediated homologous recombination of acetolactate synthase. Mol Plant 9:628–631

van den Hoogen J, Govers F (2018) Attempts to implement CRISPR/Cas9 for genome editing in the oomycete *Phytophthora infestans*. bioRxiv. https://doi.org/10.1101/274829

Veillet F, Perrot L, Chauvin L, Kermarrec MP, Guyon-Debast A, Chauvin JE, Nogué F, Mazier M (2019) Transgene-free genome editing in tomato and potato plants using agrobacterium-mediated delivery of a CRISPR/Cas9 Cytidine Base editor. Int J Mol Sci 20:402. https://doi.org/10.3390/ijms20020402

Wang S, Zhang S, Wang W, Xiong X, Meng F, Cui X (2015) Efficient targeted mutagenesis in potato by the CRISPR/Cas9 system. Plant Cell Rep 34:1473–1476. https://doi.org/10.1007/s00299-015-1816-7

Watanabe K (2015) Potato genetics, genomics, and applications. Breeding Sci 65:53–68. https://doi.org/10.1270/jsbbs.65.53

Ye M, Peng Z, Tang D, Yang Z, Li D, Xu Y et al (2018) Generation of self-compatible diploid potato by knockout of S-RNase. Nat Plants 4:651–654. https://doi.org/10.1038/s41477-018-0218-6

Yogendra KN, Kumar A, Sarkar K, Li Y, Pushpa D, Mosa K, Duggavathi R, Kushalappa AC (2015) Transcription factor *StWRKY1* regulates phenylpropanoid metabolites in potato to resist late blight. J Exp Bot 66:7377–7389

Yoshida K, Schuenemann VJ, Cano LM (2013) The rise and fall of the Phytophthora infestans lineage that triggered the Irish potato famine. eLife 2:e00731. https://doi.org/10.7554/eLife.00731

Zhan X, Zhang F, Zhong Z, Chen R, Wang Y, Chang L, Bock R, Nie B, Zhang J (2019) Generation of virus-resistant potato plants by RNA genome targeting. Plant Biotechnol J 17:1814–1822
Zhang Y, Massel K, Godwin ID, Gao C (2018) Applications and potential of genome editing in crop improvement. Gen Biol 19:210. https://doi.org/10.1186/s13059-018-1586-y
Zhao X, Jayarathna S, Turesson H, Fält AS, Nestor G, González MN, Olsson N, Beganovic M, Hofvander P, Andersson R, Andersson M (2021) Amylose starch with no detectable branching developed through DNA-free CRISPR-Cas9 mediated mutagenesis of two starch branching enzymes in potato. Sci Rep 11:4311. https://doi.org/10.1038/s41598-021-83462-z
Zhou X, Zha M, Huang J, Li L, Imran M, Zhang C (2017) StMYB44 negatively regulates phosphate transport by suppressing expression of PHOSPHATE1 in potato. J Exp Bot 68:1265–1281. https://doi.org/10.1093/jxb/erx026

18 Biological Suppression of Insect Pests of Potato

M. Nagesh, J. Sridhar, Mohd Abas Shah, V. Venkateswarulu, and Anuj Bhatnagar

Abstract

Global and Indian agri-horti-dairy markets are transforming agriculture to smart agriculture with commodities driven by quality and safety. Natural biological control agents and biological processes are gaining prominence in input market for organic farming and plant and soil heath management. There has been paradigm shift in the consumer preference toward organic and residue-free healthy agricultural products for better well-being and longevity. Biological control of insects and diseases is one of promising pest management technologies in organic farming and conservation agriculture. Potato being one of the most important food crops in India after rice and wheat is slowly shifting toward organic farming to harness health benefits. Potato production is challenged by a number of biotic stresses such as aphids, whiteflies, leaf hoppers, thrips, white grubs, cutworms, potato tuber moth, flea beetle, and mites. Though many promising pest control strategies are available, biological control offers great potential in healthy potato production. Current global biopesticide consumption

M. Nagesh (✉)
ICAR-National Bureau of Agricultural Insect Resources, Bengaluru, Karnataka, India

J. Sridhar
ICAR-National Institute of Biotic Stress Management, Raipur, Chhattisgarh, India

M. A. Shah
ICAR-Central Potato Research Institute-Regional Station, Jalandhar, Punjab, India

V. Venkateswarulu
ICAR-National Institute of Biotic Stress Management, Raipur, Chhattisgarh, India

ICAR-Central Tobacco Research Institute, Rajahmundry, Andhra Pradesh, India

A. Bhatnagar
ICAR-Central Potato Research Institute-Regional Station, Modipuram, Uttar Pradesh, India

S. Kumar Chakrabarti et al. (eds.), *Sustainable Management of Potato Pests and Diseases*, https://doi.org/10.1007/978-981-16-7695-6_18

is 4.5% of the total pesticide market. The USA, Canada, Mexico, and Europe consume 65% of global biopesticides, while India's share is 3–4% for its 140 mha of cropped area, with a production growth rate of 2.5–3.0%. Biological control agents such parasitoids, predators, entomopathogenic fungi (*Metarhizium anisopliae*, *Beauveria bassiana*, *Verticillium lecanii*, etc.), bacteria (*Bacillus thuringiensis*, *Bacillus cereus*, etc.), and viruses (granulosis viruses) are some of the promising alternatives to chemical control. In this chapter, we discuss the potential biological control avenues for healthy potato production in India.

Keywords

Natural farming · Residue-free · Parasitoids · Predators · Microbials · *Bacillus thuringiensis*

18.1 Introduction

Potato is the third most important food crop in the world after rice and wheat with a record global production of 365 million tons (MT). India is the second largest producer after China with a production and productivity of 53.11 MT and 23.0 t/ha, respectively. Although potato is largely consumed as fresh vegetable, its potential in processing sector is expanding enormously at present (7.5%), and potato would play key role in food and nutritional security of the country in near future. However, it is projected that India would need to produce 55 and 122 MT by 2025 and 2050, respectively, to meet the demand of the enormously growing population. Potato crop would certainly meet out all these requirements worldwide due to its highly diverse distribution pattern, and its current cultivation and demand, particularly in developing countries fighting with poverty, hunger, and malnutrition. Challenges of balancing sustainable potato cropping include increased productivity in developing countries while conserving the biodiversity and species richness, better resource management, and optimization. But often the potential yield of potato is limited by a number of biotic and abiotic factors in hills, plains, and plateau regions in India. Potato pest management technologies have controlled insect pests to a greater extent. Insecticide application has been found to be one of the effective strategies to control biotic stresses in potato. However, they are also accountable for degrading soil quality. Significant ill effects on nontarget organisms such as pollinator insects, birds, fishes, beneficial microorganisms, etc., are quite noticeable because it disrupts the entire natural ecosystems and contaminates food chain (Saha et al. 2020; Giordanengo et al. 2012). Therefore, environmentally safe alternative pest management strategies such as biological control are the need of the hour globally.

Biological control including biopesticides are viable alternatives to chemical pesticides, typically derived from living organisms, microorganisms, and other natural sources, pose less risk to people and the environment, and hence gain worldwide attention as a new tool to kill insects. Biopesticides are being widely used to manage biotic stresses as a component of IPM under protected cultivation (Sabir et al. 2011; Ramasamy and Ravishankar 2018). On considering the international market of export commodities and health-conscious, microbial biopesticides in pest management gaining more importance globally. Therefore, awareness among the farmers and consumers on health and environmental benefits offered by microbial biopesticides ultimately enhance the marketability of microbial biopesticides in India. Significantly, the challenges for uptake and adoption levels of biological control agents and biopesticides are timely availability, scale of production, quality, effective strains, pricing, "one-size-fits-all" label claims, and number of products (currently only 14) that are eligible for marketing under national and international regulatory regime. Paradoxically, to bridge the growing gap between supply and demand of biopesticides, the market is thriving with spurious, contaminated, and unregulated products. Biopesticides are yet to take off in India in commercial scale despite their enormous market potential. Smart agriculture essentially requires agri start-ups for linking farm produce to market chain, production to demand-supply, investment to employment generation, and profitability. We intend to discuss the opportunities that entail these challenges from the entrepreneurial point of view that ensures investment, employment, quality input supply, value addition to the agri products, and enhanced profitability. India, with the increasing agritech start-up entrepreneurs, is primed to tap potential opportunities with a variety of smart agricultural technologies in biological control per se. In this chapter, we discuss about the potential of biological control strategies for efficient management of insect pests in potato in India.

18.2 Significance of Biological Control in Potato

Biological control is the use of living organisms to suppress pest populations, making them less damaging than they would otherwise be. Biological control can be used against all types of pests, including vertebrates, plant pathogens, and weeds as well as insects, but the methods and agents used are different each type of pest. Biological control includes predators and parasitoids which actively seek out the pest and have an enormous potential to suppress potato insect pests in the context of a truly integrated pest management approach, locally adapted to include essential cultural controls, pest thresholds, and a variety of compatible intervention tactics such as biopesticides, pheromone-based technologies, and trap cropping. Microbial control agents, including viruses, bacteria, fungi, and entomopathogenic nematodes are natural products such as botanical and semiochemical preparations. There are three categories of biological control agents of insect pests: predators, parasitoids, and entomopathogens. Under favorable conditions, entomopathogens can cause zoonotics in insect pests and give excellent control as good as insecticidal application (Boiteau 2010).

A well complied, exhaustive, and informative global review is available on biological control of potato pests (Weber 2012). Before discussing the status of biological control in management of potato pests, a brief relook at the pests of potato would be relevant.

18.3 Biological Control of Major Insect Pests

18.3.1 White Grubs (Coleoptea: Scarabaeidae)

Twenty species of white grub have been reported on potato from India. Of these, *Brahmina coriacea* (Hope), *Holotrichia seticollis* Moser, *H. longipennis* (Blanchard), *Anomala dimidiata* Hope, and *Melolontha indica* Hope are the most destructive in northwestern hilly region, and others are found in different parts of the country (Chandel et al. 2005).

White grubs are polyphagous and cosmopolitan in nature, and these are worldwide in distribution, viz., Europe, Asia, Africa, and American countries. In India, white grub is the most destructive insect threatening potato production in the hilly regions (Fig. 18.1). In plains, white grub has long been associated with sugarcane crop, but now it is causing damage to potato crop as well (Chandla et al. 1988).

White grubs have become serious pests of agricultural crops, many horticultural crops, grasses, lawns, and forest trees such as *Robinia*, *Polygonum*, *Kathie*, neem (*Azadirachta indica*), wild ber (*Ziziphus* sp.), babul (*Acacia arabica*), and khair (*Acacia catechu*), which serve as host to white grubs. In India, adult beetles emerge from the soil during April to June in response to pre-monsoon showers. The second fortnight of June is the peak period of emergence of June beetles, and the emergence continues until the first fortnight of August. Larvae of white grub characteristically curl up in a C shape when disturbed. Usually only one generation is produced per year, but in some parts of the country a second generation may occur. In summer, the overwintering May beetles emerge from the ground at dusk, feed on the leaves of trees, and mate during the night. At dawn, they return to the ground, where the females lay 15–20 eggs in earthen cells several centimeters below the surface. Most

Fig. 18.1 White grubs damaging potato

May beetles lay eggs in grassy surface. Eggs hatch 3–4 weeks later. The young grubs feed on plant roots throughout the summer during the monsoon period, and they burrow down to a depth of 1.5 m and hibernate. The usual length of time for one complete generation (adult to adult) is 2–4 years depending upon latitude. The larvae live and pupate in soil, after which the emerging adults may move to new feeding sites. Adults are present year-round in low numbers, but peaks are found in October–December and then again in March–April. It is during these months that damage to potatoes can be expected.

Biological management Natural enemies that control white grubs include parasitic wasps in the genera *Tiphia*, *Myzinum* (Hymenoptera: Tiphiidae), and *Pelecinus polyturator* Drury (Hymenoptera: Pelecinidae) and the fly, *Pyrgota undata* Wiedemann (Diptera: Pyrgotidae). Botanicals or extracts from the different parts of plants like *Azadirachta indica*, *Nicotiana tabacum*, *Sapindus mukorossi*, *Jatropha curcas*, *Melia*, *Urtica dioica*, and *Nerium oleander* are effective in management of white grubs. *Metarhizium anisopliae* and *Beauveria bassiana* can be used to reduce population of white grub. Entomopathogenic nematodes (EPN) like *Steinernema carpocapsae* and *Heterorhabditis indica* have been reported to be effective in management of white grubs in mid and higher hills of Himachal Pradesh, Uttarakhand, and NE states (Misra and Chandel 2003). Microbial control can be combined with soil application of phorate 10G at 15.0 kg/ha at the time of planting or at the time of earthing up to reduce the white grub infestation (Yadav et al. 1977). Spores of the pathogens *Bacillus popilliae*, *B. lentimorbus*, and *Metarhizium anisopliae* can be used to inoculate the soil. Nematodes species such as *Steinernema* sp. and *Heterorhabditis* sp. can also be effectively used against white grubs (Sanchez and Vergara 2002).

In Assam, Bhattacharya and Pujari (2014) evaluated *B. brongniartii* and *M. anisopliae* alone and in combination with insecticides against white grubs in green gram, and it was found that both entomopathogens in combination with imidacloprid 200 SL were effective in reducing plant mortality caused by white grubs resulting in a significant increase in grain yield (Chandel et al. 2019). Entomopathogenic nematode, *H. indica*, reduced grub population by 66–80%, while 83% reduction in grub population was reported with *S. carpocapsae* with 60% reduction in tuber damage (Sharma et al. 2009). Entomopathogenic bacteria also play crucial role in suppressing the grubs under field conditions. Sharma et al. (2013) reported that ten bacteria belonging to genera *Bacillus*, *Psychrobacter*, *Paracoccus*, *Paenibacillus*, *Mycobacterium*, *Staphylococcus*, and *Novosphingobium* from infected grubs of *B. coriacea*. Bioassay studies revealed 100% mortality with *B. cereus* and 88.89% with *P. pulmonis* after 30 days of treatment. In addition, wettable powder formulations of *H. indica* and *H. bacteriophora* were developed by NBAIR, Bengaluru, for the control of white grubs and other soil-borne pests (Kumar et al. 2019). ICAR-NBAIR has developed green technologies, namely, aarmour, nema power, grubcure, and soldier (novel WP formulations with a shelf-life of 10–12 months), which ensured very effective control of white grubs in areca nut, banana, cashew, sugarcane, maize, groundnut, and redgram covering more than

30,000 ha, and the same have been sold to >16 companies and are readily available to farmers in the market (Kumar et al. 2019; NBAIR 2013).

18.3.2 Cutworm (Lepidoptera: Noctuidae)

Cutworms are polyphagous, cosmopolitan, and most destructive insect pests present throughout the world. In India, cutworms are more serious in northern region. *Agrotis segetum* is commonly found in hills, and *A. ipsilon* is common in plains. Peak activity of cutworms occurs during May–June in Shimla Hills, in August in peninsular India, and in March–April in Bihar and Punjab. In Bihar the tuber up to 12.7% and in Himachal Pradesh 9.0–16.0% has been recorded to be damaged by cutworm. *Agrotis spinifera* has been observed in Punjab, Bihar, Andhra Pradesh, and Karnataka (Fig. 18.2).

Cutworms feeds on variety of crops, viz., pulses, vegetables, cereals, oilseeds, and many agricultural and horticultural crops. Crop is damaged by the caterpillar (larva) stage only. The young larvae cause damage by feeding on leaves, cutting stem of the plant just near the ground level by mature larvae and making irregular holes in the tubers. They feed at night on young shoots or underground tubers and

Fig. 18.2 Cutworm damaging potato

hide themselves in the soil near to the stem during daytime. In early stages of the crop, the caterpillars cut the seedlings at ground level and feed on shoots and leaves. After tuber formation, they start feeding on tubers and roots, resulting in a variety of holes, ranging from small and superficial to very large deep ones resulting in reduction in tuber yield and its market value. The eggs are ribbed, globular, and small measuring about 0.5 mm in diameter. The newly laid eggs are cream colored which turn reddish yellow to blackish before hatching. They are laid singly or in clusters up to 2000 eggs but generally between 600 and 800 eggs. Eggs are laid on vegetation, on moist ground around plants, or in cracks in the soil. Eggs hatch in 10–28 days. In the plains, the first peak of population of cutworm *Agrotis ipsilon* (Hufn.) is attained in mid of December, while the second peak is observed during third to fourth week of March. The nature of the soil has a large influence on the rate of infestation. Cutworms tend to be more frequent in soil with plenty of decaying organic material or where organic manure has been applied. Damage is worse where cutworms are present in large numbers before planting. Cutworms often reoccur in the same field, coming with crop residues and dense stands of weeds.

Biological management Exposing the larvae to bird predators is the best way to manage the cutworm in a natural way. Cutworms are hosts for numerous parasitoid wasps and flies, including species of Braconidae *Cotesia ruficrus* (Haliday), *Snellenius manila* (Ashmead), Ichneumonidae *Tenichneumon panzer* (Wesmael), Tachinidae *Bonnetia comta* (Fallen), and *Euplectrus plathypenae*, and a good control of 60% of cutworm larvae has been reported by an entomopathogenic nematode, *Hexamermis arvalis*, in central USA. Entomopathogenic nematodes that have been used in the control of *A. ipsilon* larvae include *Steinernema glaseri*, *S. riobrave*, and *Steinernema carpocapsae* (Korschel et al. 2020). Black cutworm larvae, *Agrotis ipsilon*, was effectively managed by using pellets containing *Steinernema feltiae* strains Mexican and Kapow, *S. bibionis*, and *Heterorhabditis heliothidis*.

18.3.3 Potato Tuber Moth (Lepidoptera: Gelechiidae)

The potato tuber moth (PTM), *Phthorimaea operculella* (Zeller), is the most significant insect pest of the potato. The PTM is a cosmopolitan already reported in more than 90 countries worldwide. The PTM moth occurs in all tropical and subtropical potato-producing countries in Africa, Asia, and North, Central, and South America. Potato tuber moth was introduced into India in 1906 through seed potato imported from Italy, and since then this insect has been causing damage both in potato stores and in the field. The damage has been reported from Maharashtra, Bihar, Madhya Pradesh, Kangra Valley of Himachal Pradesh, Tamil Nadu, North Eastern hill states, plateau region, and Karnataka. The PTM is principally a storage pest damaging potatoes though it causes damage in field also. The range of infestation could be 30–70% in stored potato (Chandel et al. 2005).

PTM alternate host plants include Solanaceous plants such as brinjal (*Solanum melongena* L.), tomato (*Solanum lycopersicum* L.), black nightshade (*S. nigrum* L.), bell pepper (*Capsicum frutescens* L.), tobacco (*Nicotiana tabacum* L.), *Physalis peruviana* L., and *Datura stramonium* L. The larvae enter into the tubers and feed on them causing mines. The activity of larvae in tubers, placed in heaps, results in production of heat which promotes significant rotting of the produce. In country stores, 18–83% tuber damage due to PTM has been reported in the NE hills. The male and female moths are brownish gray in color, and wings are folded to form a rooflike structure. Maximum population growth of PTM occurs at 20–25 °C. Life cycle of PTM is completed in 21–30 days at 27–35 °C. Upper and lower threshold limit of temperatures for PTM are 40 °C and 5 °C. In addition to temperature, precipitation also influences development and abundance of *P. operculella*. The damage is severe under low rainfall and high temperature conditions (Raj 1991; Rondon 2010).

Biological management Use of water traps can catch the good number of adult moths per trap as compared to the cylinder-shaped traps and funnel traps. Among all, the delta trap was found as the most effective. Sex pheromones have been used for mass trapping of PTM male adults which are available for most of the PTM species. An attracticide consisting of pure pheromone formulated with plant oils and ultraviolet screens was found successful both in laboratory and under field conditions with installation of PTM sex pheromone traps at 20 traps/ha.

Habitat manipulation through use of either inundative/inoculative releases or by conservation of natural enemies can help effective management of PTM. Covering potato heaps with 2.5-cm-thick layer of chopped dried leaves of lantana or eucalyptus can prevent tuber infestation of PTM. *Eucalyptus globosus*, *Lantana camara*, and *Minthostachys* both in dried and powdered forms were found effective in controlling *P. operculella.* Parasitoids such as *Apanteles subandinus* and *Copidosoma koehleri* are being widely used in classical biological control of *P. operculella* in different parts of the world. The parasitic wasps identified are *Diadegma pulchripes* (Kokujev) and Ichneumonidae, *Bracon gelechiae* Ashmead (Braconidae) which can parasitise PTM. A total of 20 parasitoids species from Braconidae (9 species), Encyrtidae (2 species), and Ichneumonidae family (9 species) have been reported parasitizing on *P. operculella. Diadegma molliplum* has been found parasitizing *P. operculella* with very high degree of parasitism (>80%) in Yemen. *Copidosoma koehleri* (Blanchard), *Apanteles subandinus* (Blanchard), and *Orgilus lepidus* (Muesebeck) have been widely and successfully used. Predators such as *Coccinella septempunctata* Linnaeus (Coccinellidae), *Chrysoperla carnea* Stephens (Chrysopidae), and *Orius albidipennis* (Reuter) (Anthocoridae) and use of granulosis virus (GV) are extremely effective in reducing PTM damage.

Spray *Bacillus thuringiensis* (10^9 cfu/ml) at 0.05% or GV at 4 LE/lit of water, and store the healthy potatoes covered with thick chopped dried leaves of *Lantana camara*, soapnut, *Neem*, and *Eucalyptus* spp. Dusting of the seed tubers with malathion gives good control of PTM in seed potatoes but treated potatotes should not be consumed. Pheromone trap at 4 traps/100 m^3 of stored godowns are recommended. Microbial biopesticides for *P. operculella* field control have been tested based on *Bacillus thuringiensis* subsp. *kurstaki* (Btk) and *P. operculella* specific granulovirus (PhopGV, *Baculoviridae*). Btk was effective but required repeated applications because it is quickly degraded by UV light. Likewise, PhopGV has shown mixed results which can be protected from UV inactivation using a variety of adjuvants (e.g., dyes, optical brighteners). Applications of PhopGV doses sufficient to cause >95% mortality are considered not being economical, hence low-dose treatments are proposed for a relatively inexpensive partial suppression of the field population (Lacey and Kroschel 2009).

18.3.4 Flea Beetle (Coleoptera: Elateridae)

Flea beetle (*Psyllodes plana* Maulik) feeds on a variety of crops, including eggplant, tomatoes, peppers, and potatoes. Weeds around the fields serve as their protected homes; when crop is not there, adult feeds on leaves and larvae on roots of weed plants. They feed on a variety of herbaceous plants until potatoes emerge. Adults feed by making small holes on leaves by chewing which can be easily identified. The damage starts as soon as plant comes out of the soil. Adults could be identified as they jump immediately if disturbed.

Eggs are deposited on the soil at the base of plants, about 100 eggs per female. Larvae emerge after about 10 days and burrow into the soil, where they feed on roots, sprouts, and tubers, weakening plants and sometimes killing seedlings. They pupate in the soil. Adult beetles emerge and burrow to the surface. They climb onto plants and chew small holes in the foliage which can facilitate the entry of plant pathogens. Beetles may also spread diseases from plant to plant as they feed. The life cycle takes 4–6 weeks to complete.

Removing of weeds and crop debris in and around planting sites will deprive flea beetle larvae of food sources and overwintering places and may help to lessen the flea beetle population. Trap crops may be helpful in managing the flea beetles (e.g., radish, Chinese cabbage, turnip, mustard, etc.). Radish is a highly favored crop, so plant radish before the main crop, in an effort to attract flea beetles away from the main crop. Adult flea beetles will be attracted to the tallest and earliest crops. Once beetles are actively feeding in the trap crop, they can be sprayed with recommended insecticide. Most flea beetle treatments are applied as foliar sprays to protect the foliage against the feeding of the adult beetle. *Microctonus vittatae* is a native braconid that kills the adult flea beetle and can effectively manage the flea beetle population.

18.3.5 Aphids

Aphids are key insect vectors of potato viruses globally including India. They cause more economic damage indirectly to potato crop by transmitting wide variety of viruses such as PVY, PLRV, PVA, PVM, etc. To date, we have a record of 14 distinct aphid species: green peach aphid, *Myzus persicae* (Sulzer); cotton aphid, *Aphis gossypii* Glover; potato aphid, *Macrosiphum euphorbiae* (Thomas); cabbage aphid, *Brevicoryne brassicae* Linnaeus; foxglove aphid, *Aulacorthum solani* (Kaltenbach); mustard aphid, *Lipaphis erysimi*; bean aphid, *Aphis fabae* Scopoli; coriander aphid, *Hyadaphis coriandri* (Das); rice root aphid, *Rhopalosiphum rufiabdominalis* (Sasaki); spiraea aphid, *Aphis spiraecola*; oleander aphid, *Aphis nerii* Boyer de Fonscolombe; groundnut aphid, *Aphis craccivora* Koch; bird cherry-oat aphid, *Rhopalosiphum padi* (L.); and maize aphid, *Rhopalosiphum maidis* (Fitch) on potato in India (Bhatnagar et al. 2017a).

Important parasitoids of aphids in potato are *Lysiphlebus* sp., *Diaeretiella* sp., *Aphelinus* sp., and *Aphidius colemani* followed by predators such as ladybird beetle, green lacewing, spiders, hover fly, etc., which keep the aphid populations under control in natural ecosystem. However, insecticidal treatments to control these virus vector species adversely affect the abundance of natural enemies under field conditions. As of today, inundative and inoculative release of parasitoids and predators for the management of aphid vectors in potato are not in practice. However, there are several more-or-less specific aphidicides which minimally impact aphid predators and parasitoids, avoiding pest resurgence. Ghosh (2015) reported that plant extract of *Polygonum* flower at 5% and tobacco leaf extract at 10% suppressed aphid populations by 70% and 65%, respectively. The botanical insecticide azadirachtin was also found very effective against aphid, achieving more than 60% suppression. In another study Kahar et al. (2016) reported that soil treatment with Bt followed by two foliar sprays of azadirachtin and one spray of Bt suppressed the populations of *Myzus persicae* and *Aphis gossypii* by 76–78% in potato, indicating the significant and potential of biological control agents. Biological control will be effective when pest population has not exceeded economic injury levels (EIL). *Beauveria*, *Metarhizium*, *Isaria* spp., and some *Lecanicillium* spp. (including Vertalec) are common, naturally occurring pathogens of aphids and generally exhibit greater virulence against *A. gossypii* and *A. splani* in Europe and the USA, while their potential use as biological control agents is yet to be realized in India.

18.3.6 Whiteflies

Whiteflies are small milky white insects belonging to sap-sucking group causing huge economic losses to crops by transmitting begomoviruses. The population of whitefly is highly diverse, and many biotypes have been identified world over during the past few years. *Bemisia tabaci* (Gennadius) is a complex of 11 well-defined high-level groups containing at least 44 morphologically indistinguishable species. So far, three distinct genetic groups, namely, Asia I, Asia II-1, Asia II-5, were reported on potato in India. *Bemisia tabaci* have been widely distributed on potato in India and became a serious vector of ToLCNDV-potato since 2000. Whiteflies have a characteristic life cycle of six stages: the egg, four immature stages (nymphal instars), and the adult stage. Management of whitefly vector plays a crucial role in begomoviral incidence in potato.

Although a great number of predators, parasitoids, and fungal pathogens are known to attack *B. tabaci*, no biotic agents are known to provide adequate suppression alone. Under field conditions, parasitism does not usually build to high levels (Capinera 2004). Insecticides often interfere with parasitoids, of course, and effective use of parasitoids like *Encarsia formosa* and *Eretmocerus* sp. will probably be limited to cropping systems where insecticide use is minimized and other cultural techniques and biorational pesticides are used which favor action of predators, parasitoids, and entomopathogens (Capinera 2001).

Mycoinsecticide products based on *Verticillium lecanii*, *Paecilomyces* sp., and *Beauveria bassiana* have the capacity to provide good control of whiteflies even under field conditions (Arno et al. 2010) to greater extent. The negative side is slow action and poor adulticidal activity. Important parasitoids are *Encarsia formosa* and *Eretmocerus* sp. having considerable potential for whitefly control (Zaki et al. 1999). *Prospaltella flava* is another important nymphal parasitoid of *B. tabaci* (Arno et al. 2010). Important predators such as *Chrysoperla carnea*, *C. scelestes*, *Geocoris bicolor*, and *Mallada boninensis* feed on eggs and nymphs of *B. tabaci*. Some of predators such as *Brumoides suturalis*, *Menochilus sexmaculatus*, and *Euseius hibisci* could able to predate only on nymphs, while other predators like *Scymnus syriacus*, *Euseius hibisci*, and *Serangium* sp. are important predators of both nymphs and adults of *B. tabaci* (Dhawan et al. 2007). Further studies on conservation, augmentation, inundation, and commercial exploitation of these bioagents need to explored. *Beauveria bassiana* (Balsamo-Crivelli) Vuillemin is one of the most commonly encountered entomopathogens and has been commercially developed as a microbial insecticide to control *B. tabaci*. The combination of *B. bassiana* with *Bacillus thuringiensis* for the biological control of *B. tabaci* was shown to have an antagonistic effect, causing mortality greater than 50% observed over a period of 7 days (Somoza-Vargas et al. 2018). The application of *Aschersonia aleyrodis* to control *B. tabaci* is a promising entomopathogenic fungi candidate, which has been proven effective in parasitizing whiteflies (Sani et al. 2020). The mechanism of action of entomopathogenic fungi has been found to be attachment of spores to the cuticle, germination of hyphae over the surface of insects, penetration of hyphae through the integument, growth of fungus in the hemocoel, and ultimately death of the whitefly (Sani et al. 2020).

18.3.7 Thrips

So far, four species of thrips, namely, *T. hawaiensis*, *M. distalis*, *T. palmi*, and *Scirtothrips* sp., have been reported on potato by Khurana et al. (2001). Recently *Haplothrips tenuipennis* have been reported to occur for the first time on potato in India (Sridhar et al. 2020). Thrips population moves from the preceding crop to the other cultivated host plants and weeds in the vicinity. Adults tend to feed on young growth and prefer to hide in complex plant parts and so are found on new leaves. Potato stem necrosis disease causes 15–30% yield loss in potato in northern Gujarat, parts of Madhya Pradesh, and Rajasthan. Thrips are able to multiply during all seasons that crops are cultivated but are favored by warm weather. When crops mature, their suitability for thrips declines, so thrips growth rate diminishes even in the presence of warm weather. Thrips is a polyphagous species but is best known as a pest of Cucurbitaceae and Solanaceae plants. Thrips not only spread disease in potato crop but also suck the sap from tender parts of potato plant. The high temperature (30–35 °C) and dry weather during September/October are favorable for thrips activity and higher disease incidence (Bhatnagar et al. 2017b). Thrips transmitted GBNV is an emerging problem on potato in India.

There is a strong positive correlation between early planting and thrips dynamics in potato as warmer temperatures prevail early in the crop season. Therefore, early planting (September/October) must be avoided whenever possible in thrips-prone areas. Certain varieties are resistant to thrips injury such as the Kufri Sutlej, Kufri Badshah, and Kufri Jawahar. Most IPM programs are based on the utilization of insect parasitoids. Less attention has been paid to predators, and this group of natural enemies can be very effective in warmer areas where they are spontaneous. Biocontrol agents are a major factor in preventing outbreaks in populations of many pest species including sucking pests like thrips (Loomans et al. 1995). Farm yard manure (FYM) enriched with *Trichoderma harzianum* (4 g/kg) is used to control thrips, mites, and soil-borne diseases. Very few plant-based products such as Ryania and Sabadilia have been registered outside India for the control of thrips (Chauhan et al. 2018). However, investigations are needed in exploiting biological control of thrips in India.

18.3.8 Leafhoppers

Potato leafhopper is a polyphagous pest and distributed worldwide. It is most damaging insect causing major loss to potato. In India, the leafhoppers are distributed in all potato-growing regions of Indo-Gangetic plains. There are several species of leafhoppers: *Amrasca biguttula biguttula* (Ishida), *Alebroides nigroscutulatus* Distant, *Seriana equata* Singh, *Empoasca solanifolia* Pruthi, *Empoasca kerri motti* Pruthi, *E. fabae* Harris, and *E. punjabensis* Pruthi which damage potato crop. Both the nymphs and adults of the leafhoppers suck the sap from lower side of the leaves causing extensive damage by direct feeding of the plants.

18.4 Constraints that Offer a Wide Opportunity for Entrepreneurship and Start-Ups

Production and timely availability Benefits of use of biological control agents are now well documented and validated in different crops against several pests, including microbials, macrobials, plant products, pheromones, etc. World over, several bioagents are in regular use at farmers levels with production storage supply chain supported by the industry and private entrepreneurs. There are several products that are now available on board and through e-supply chain elsewhere in the world (Table 18.1). However, in developing countries like India, the production supply

Table 18.1 List of commercially available natural enemies, their hosts, and mode of action for biological control of potato insect and mite pests

Pest	Type of natural enemy	Name of natural enemy	Mode of action
Thrips and mites	Predatory mite	*Amblyseius andersoni*	These mites predate their prey by piercing and suing with their sucking mouthparts and suck out the contents
Thrips species, whitefly (*Trialeurodes vaporariorum* and *Bemisia tabaci*) and tarsonemid mites (*Polyphagotarsonemus latus*)	Predatory mite	*Amblyseius swirskii*	Thrips species, eggs and larvae of whitefly, and tarsonemid mites. Adult predatory mites search for their prey or wait for it to pass by and subsequently feed on their prey. Adults and nymphs pierce thrips larvae and adults with their sucking mouthparts and suck out the contents
Thrips and whitefly	Predatory beetle (ladybird)	*Adalia bipunctata*	Thrips larvae as well as eggs and crawler stages of whitefly
Thrips and mites	Predatory mite	*Amblydromalus limonicus*, *Typhlodromalus limonicus*	Larvae of various species of thrips (first and second larval stages). Eggs and larvae of greenhouse and cotton whitefly
Thrips	Predatory bug	*Orius laevigatus*	For control of various species of thrips (larvae and adults). When thrips is absent, *Orius* can also survive on aphids, spider mite, butterfly eggs, and pollen
Aphids	Aphid midge	*Aphidoletes aphidimyza*	Aphid colonies disperse a smell of honeydew that attracts adult gall-midges. They deposit their eggs in these colonies, providing an immediate food source for the larvae. Upon hatching, the larvae paralyze and then consume the aphids
Aphid: Potato aphid *Macrosiphum euphorbiae* and the glasshouse potato aphid *Aulacorthum solani*	Parasitic wasp	*Aphelinus abdominalis*	Female adult parasitic wasps parasitize the aphid. Host feeding also takes place. Parasitized aphids convert into a leathery black-colored mummy. The adult parasitic wasp emerges through a hole

(continued)

Table 18.1 (continued)

Pest	Type of natural enemy	Name of natural enemy	Mode of action
			with a jagged edge at the rear of the mummy. The first mummies are noticed in the crop around 2 weeks after the first introduction. It also works against *Myzus persicae*
Aphids	Parasitic wasp	*Aphidius matricariae*	Adult female parasitic wasps lay their eggs parasitically in the aphids, causing them to swell and harden into leathery, grey/brown mummies. The first adult parasitic wasps emerge through a round hole at the rear of the mummies approximately 10–14 days after introduction
Cotton aphid (*Aphis gossypii*), tobacco aphid (*Myzus persicae* spp. *nicotianae*), peach-potato aphid (*Myzus persicae* spp. *persicae*)	Parasitic wasp	*Aphidius colemani*	Adult female parasitic wasps lay their eggs parasitically in the aphids, causing them to swell and harden into leathery, grey/brown mummies. The first adult parasitic wasps emerge through a round hole at the rear of the mummy approximately 2 weeks after introduction
Aphids: Potato aphid (*Macrosiphum euphorbiae*) and greenhouse potato aphid (*Aulacorthum solani*)	Parasitic wasp	*Aphidius ervi*	–
Aphids, *Echinothrips*, and mealybugs	Lacewing	*Chrysoperla carnea*	After emerging from the eggs, larvae of the lacewing attack prey and suck their body fluids. The remainder of the dead pest insect shrivels up completely and is difficult to find. Larvae are mainly active and feeding during nighttime and hiding during the day
Mealybug species	Predatory beetle	*Cryptolaemus montrouzieri*	Predatory beetles and larvae eat the mealybugs completely

(continued)

Table 18.1 (continued)

Pest	Type of natural enemy	Name of natural enemy	Mode of action
Greenhouse whitefly, tobacco whitefly	Parasitic wasp	*Encarsia Formosa*	Greenhouse whiteflies (*Trialeurodes vaporariorum*) and tobacco whiteflies (*Bemisia tabaci*) in the third and fourth larval stage. Female adult parasitic wasps parasitize the third and fourth larval stage of the whitefly. Additionally, host feeding also occurs
	Parasitic wasp	*Eretmocerus eremicus*	Greenhouse whiteflies (*Trialeurodes vaporariorum*) and tobacco whiteflies (*Bemisia tabaci*) in the third and fourth larval stage. Female adult parasitic wasps parasitize the third and fourth larval stage of the whitefly. Additionally, host feeding also occurs
	Parasitic wasp	*Aphidius ervi*	Adult female parasitic wasps parasitize the aphids. Parasitized aphids swell and turn into leathery, gray- or brown-colored mummies. The first adult parasites emerge through a round hole at the rear of the mummy approximately 2 weeks after introduction
Leaf miner larvae	Parasitic wasp	*Dacnusa sibirica*	Adult female parasitic wasps of *Dacnusa sibirica* lay their eggs inside the leaf miner larva. The parasitic wasps develop inside the leaf miner pupae. Adult wasps will emerge from the pupae
Greenhouse whitefly; tobacco whitefly; two-spotted spider mite; thrips, eggs of butterflies and tomato leaf miner moth; aphids; leaf miner larvae	Predatory bug	*Macrolophus pygmaeus*	–

(continued)

Table 18.1 (continued)

Pest	Type of natural enemy	Name of natural enemy	Mode of action
Two-spotted spider mite (*Tetranychus urticae*), fruit tree red spider mite, also known as European red mite (*Panonychus ulmi*), citrus red mite (*Panonychus citri*), broad mite (*Polyphagotarsonemus latus*), cyclamen mite (*Tarsonemus pallidus*)	Predatory mite	*Neoseiulus californicus*	Adult predatory mites, nymphs, and larvae actively search and consume their prey

chain to meet the demand is in infancy in biological control input chain. With the advantage of short production cycles, amenability to in vitro production, storage at NTP, long shelf-life, formulations, etc., microbials have been reaching the market and farmers' fields more consistently. However, macrobial biocontrol agents, viz., predators and parasitoids, need altogether different approaches to address their production, storage, supply, and application methods. The demand-supply for several of the natural enemies recorded is wide or not been addressed so far. Primary need for a successful operation of biological control in field is the timely availability of quality bioagents, which now offers an opportunity for entrepreneurship and support from industry.

Handling and application technology Natural enemies have a very short life and therefore need to be introduced into the crop as soon as possible after receipt. Technologies and applicators in field need to be evolved for rapid and safe delivery in niche/cryptic crop-pest situations. Failure to do so can have a negative impact on their quality. This is another opportunity for AI and drone technologies.

Coordination of biological control with pest-crop situations Prediction models on onset of pests in potato in different production zones across the country can be evolved or, if evolved, can be utilized for timely release of the natural enemies.

18.5 Conclusion

Biological control offers immense potential for its regular use in IPM and niche programs to suppress the pest populations in potato and becomes most viable and tangible technology in organic potato production.

References

Arnó J, Gabarra R, Liu TX, Simmons AM, Gerling D (2010) Natural enemies of *Bemisia tabaci*: predators and parasitoids. In: Stansly PA, Naranjo SE (eds) Bemisia: bionomics and management of a global pest. Springer, Dordrecht, pp 385–421

Bhatnagar A, Jandrajupalli S, Venkateswarlu V, Malik K, Shah MA, Singh BP (2017a) Mapping of aphid species associated with potato in India using morphological and molecular taxonomic approaches. Potato J 44(2):126–134

Bhatnagar A, Singh SP, Sridhar J, Dua VK, Ahmad I (2017b) Effect of planting dates on thrips population and transmission of groundnut bud necrosis virus in early potato. Potato J 44(2): 117–121

Bhattacharyya B, Pujari D (2014) Field evaluation of *Beauveria brongniartii* and *Metarhizium anisopliae* against white grubs damaging green gram in Assam. Int J Plant Prot 7(1):67–70

Boiteau G (2010) Insect pest control on potato: harmonization of alternative and conventional control methods. Am J Pot Res 87:412–419

Capinera JL (2001) Handbook of vegetable pests. Academic Publishers, San Diego, CA

Capinera JL (2004) Encyclopedia of entomology, vol I. Kluwer Academic Publishers, Dordrecht

Chandel RS, Chandla VK, Singh BP (2005) Potato tuber moth *Phthorimaea operculella* (Zeller). Technical bulletin no. 65. CPRI, Shimla

Chandel RS, Soni S, Vashisth S, Pathania M, Mehta PK, Rana A, Bhatnagar A, Agrawal VK (2019) The potential of entomopathogens in biological control of white grubs. Int J Pest Manag 65(4): 348–362

Chandla VK, Misra SS, Bist BS, Bhalla OP, Thakur JR (1988) White grubs, *Lachnosterna (Holotrichia) coracea* Hope infesting potatoes in Shimla hills. Seeds Farms 14(2):12–13

Chauhan A, Ranjan A, Jindal T (2018) Biological control agents for sustainable agriculture, safe water and soil health. In: Jindal T (ed) Paradigms in pollution prevention. Springer briefs in environmental science. Springer, Cham, pp 71–83. https://doi.org/10.1007/978-3-319-58415-7_6

Dhawan AK, Butter NS, Narula AM (2007) The cotton whitefly *Bemisia tabaci* (Gennadius). Technical Bulletin, Department of Entomology, Punjab Agricultural University, Ludhiana

Ghosh SK (2015) Integrated field management of aphids (*Myzus persicae* Sulz. And *Aphis gossypii* Glov. Together) on potato (*Solanum tuberosum* L.) using bio-pesticides. Int J Sci Env Technol 4(3):682–689

Giordanengo P, Vincent C, Alyokhin A (2012) Insect pests of potato: global perspectives on biology and management. Academic Press, Cambridge, MA, pp 1–616

Kahar B, Mondal P, Konar A (2016) Comparative field efficacy of different treatment schedules against aphid spp. *Myzus persicae* (Sulzer) and *Aphis gossypii* Glover on Potato. Inter J Hortic 13(6):1–6

Khurana SMP, Bhale U and Garg ID (2001). Stem necrosis disease of potato. Central Potato Research Institute, Shimla, India, Technical bulletin-54: 5

Kroschel J, Mujica N, Okonya J, Alyokhin A (2020) Insect pests affecting potatoes in tropical, subtropical, and temperate regions. In: Campos H, Ortiz O (eds) The potato crop. Springer, Cham, pp 251–306

Kumar KK, Sridhar J, Murali-Baskaran RK, Senthil-Nathan S, Kaushal P, Dara SK, Arthurs S (2019) Microbial biopesticides for insect pest management in India: current status and future prospects. J Inverteb Pathol 165:74–81

Lacey L, Kroschel J (2009) Microbial control of the potato tuber moth (Lepidoptera: Gelechiidae). Fruit Veg Cereal Sci Biotech 3:46–54

Loomans AJM, Van Lenteren JC, Tommasini MG, Maini S, Riudavets J (1995) Biological control of thrips pests. Wageningen Agricultural University Papers, p 202

Misra SS, Chandel RS (2003) Potato white grubs in India. Technical bulletin no. 60. ICAR-CPRI, Shimla

NBAIR (2013) Annual Progress report 2003-04. ICAR-National Bureau of Agricultural Insect Resources, Bengaluru, Karnataka

Raj BT (1991) Potato tuber moth with special reference to India. Tech. Bull no. 16. CPRI, Shimla, p 16

Ramasamy S, Ravishankar M (2018) Integrated pest management strategies for tomato under protected structures. In: Wakil W, Brust GE, Perring TM (eds) Sustainable Management of Arthropod Pests of tomato. Academic Press, pp 313–322

Rondon SI (2010) The potato tuber worm: a literature review of its biology, ecology, and control. Am J Pot Res 87:149–166

Sabir N, Deka S, Singh B, Sumitha R, Hasan M, Kumar M, Tanwar RK, Bambawale OM (2011) Integrated pest management for greenhouse cucumber: a validation under north Indian plains. Indian J Hortic 68(3):357–363

Saha AL, Kishor V, Bauddh K (2020) Impacts of synthetic pesticides on soil health and non-targeted flora and fauna. In: Bauddh K, Kumar S, Singh R, Korstad J (eds) Ecological and practical applications for sustainable agriculture. Springer, Singapore, pp 65–88. https://doi.org/10.1007/978-981-15-3372-3_4

Sanchez G, Vergara C (2002) Plagas del cultivo de papa. UNALM, Lima, Peru

Sani I, Ismail SI, Abdullah S, Jalinas J, Jamian S, Saad N (2020) A review of the biology and control of whitefly, *Bemisia tabaci* (Hemiptera: Aleyrodidae), with special reference to biological control using entomopathogenic fungi. Insects 11(9):619

Sharma A, Thakur DR, Chandla VK (2009) Use of Steinernema and Heterorhabditis nematodes for control of white grubs, *Brahmina coriacea* Hope (Coleoptera: Scarabaeidae) in potato crop. Potato J 36(3–4):160–165

Sharma A, Thakur DR, Kanwar S, Chandla VK (2013) Diversity of entomopathogenic bacteria associated with the white grub, *Brahmina coriacea*. J Pest Sci 86(2):261–273

Somoza-Vargas CE, Hernández-Velázquez VM, Pena-Chora G, Torres-Garcia G, Huerta-De La Pena A, Ortega-Martínez LD, Salazar-Magallón JA (2018) Interaction of *Beauveria bassiana* strain HPI-019/14 and *bacillus thuringiensis* strain GP139 for the biological control of *Bemisia tabaci* in strawberry. Bull Insectol 71(2):201–209

Sridhar J, Kumari N, Raigond B, Venkateswarlu V, Bhatnagar A, Malik K, Sharma S, Nagesh M, Chakrabarti SK (2020) First report of Haplothrips tenuipennis bagnall (Thysanoptera, Phlaeothripidae) on potato in north western hills of India. Potato J 47(1):92–96

Weber DC (2012) Biological control of potato insect pests. In: Giordanengo P, Vincent C, Alyokhin A (eds) Insect pests of potato: global perspectives on biology and management. Academic Press, pp 399–437

Yadav CPS, Saxena RC, Misra RC, Dadheech LD (1977) Evaluation of some granular insecticides for the control of grubs of *Holotrichia consanguinea* blanch. Indian J Agric Sci 47:139–142

Zaki FN, Shaarawy MF, Farag NA (1999) Release of two predators and two parasitoids to control aphids and whiteflies. Anzeiger fur Schadlingskunde Pflanzenschutz Umweltschutz 72:19–20

Bio-Intensive Management of Fungal Diseases of Potatoes 19

Mehi Lal, Sorabh Chaudhary, Sanjeev Sharma, S. Subhash, and Manoj Kumar

Abstract

Potato is an important food crop in the world including India. Potato crop is affected by various phytopathogens, viz., fungi, bacteria, viruses, and nematodes. Among these, fungal pathogens may cause significant economic yield losses, if proper plant protection measures are not applied. Among the fungal pathogens, *Phytophthora infestans*, *Alternaria* spp., *Rhizoctonia solani*, *Fusarium* spp. are the major pathogens, while *Sclerotinia sclerotiorum*, *Sclerotium rolfsii*, *Synchytrium endobioticum*, *Helminthosporium solani*, and *Spongospora subterranea* f. sp. *subterranea* are considered as minor pathogens. For effective management of these fungal pathogens various methods, i.e., chemical control, biological control, planting resistant varieties, cultural control, and physical control are applied. Chemical management is highly effective to manage the diseases in short span; however, due to continuous and irrational use of the chemicals, pathogens may develop resistance against certain classes of fungicides. Moreover, these chemicals can lead to environmental pollution and toxicity in the crop produce. Bio-intensive management is an integrated approach, which involved biological control, cultural practices/agronomical practices and resistant varieties, etc. These approaches not only aid in managing the diseases but also increased the crop yield with sustainable approaches. In the present chapter major fungal diseases of potato, their causal organism, symptoms, losses, epidemiology, and bio-intensive approaches for management are discussed.

M. Lal (✉) · S. Chaudhary · S. Subhash · M. Kumar
ICAR-Central Potato Research Institute-Regional Station, Modipuram, Meerut, Uttar Pradesh, India

S. Sharma
Division of Plant Protection, ICAR-Central Potato Research Institute, Shimla, Himachal Pradesh, India

S. Kumar Chakrabarti et al. (eds.), *Sustainable Management of Potato Pests and Diseases*, https://doi.org/10.1007/978-981-16-7695-6_19

Keywords

Bio-intensive · Potato · Diseases · Management · Biological control · Cultural control · Resistant varieties · Crop rotation

19.1 Introduction

Potato originated in the hills of Andes and Bolivia in South America. It was introduced into Europe by Spaniards in the second half of the sixteenth century. From there it spread throughout Europe and the rest of the world in the mid-seventeenth to mid of eighteenth century. In India, it was introduced by Portuguese in the seventeenth century. Potato is the third most important food crop in the world in terms of human consumption. It is affected by various diseases and pests. Diseases are the major cause of concern for reducing the economic yield and affecting economy of the potato growers. Among the fungal diseases, late blight, early blight, black scurf and stem canker, *Fusarium* wilt and dry rot, *Sclerotinia* rot, *Sclerotium* rot, silver scurf, powdery scab, wart of potato, etc., are cause of concern. These diseases may cause losses up to 90%, depending upon varieties grown and adopted plant protection measures. These diseases can be managed by various methods, viz., chemical control, cultural control, biological control, and physical and resistant varieties. Chemical control is used extensively for managing the diseases because of quick response and managing the disease effectively. However, due to extensive use of chemicals with non-judicious application for longer periods to manage the diseases, the pathogens have developed resistance against certain chemicals. Moreover, awareness among the environmentalist and consumers about the toxic effect of these chemicals in the nature as well as in the plant produces is increasing. Therefore, it requires adopting strategy like bio-intensive management to avoid development of resistance in pathogens and toxicity in the environment. Use of bioagents/biological control is the best option. In simple way, biological control can be defined as the partial or total inhibition or destruction of pathogen population by other microorganisms. Baker and Cook (1974) defined it as the reduction of inoculum density or disease-producing activities of a pathogen or parasite in its active or dormant state, by one or more organisms, accomplished naturally or through manipulation of environment, host, or antagonist or by mass introduction of one or more antagonist. Cultural practices including nutrients management, crop rotation, and biofumigation are used in bio-intensive management. Besides, host resistance is also widely used in bio-intensive management. Symptoms, causal organism, losses, epidemiology, and management of the fungal diseases are discussed in the following heads.

19.2 Early Blight

19.2.1 Symptoms

For the first time, Ellis and Martin (1882) observed the symptoms on dying potato leaves. The name came from the fact that early blight infects early maturing cultivars more severely than medium or late maturing cultivars (van der Waals et al. 2001). Foliar infection generally becomes visible with the onset of tuber formation (Runno-Paurson et al. 2015). Typical foliage symptoms of early blight infection are characterized as dark brown to black necrosis. The first foliar symptoms usually appear on the lowermost leaves and then progress on the upper leaves just a few weeks after infection. Initially, the infected leaves show dark brown dot-like blotches which may be angular, circular, or oval with a few millimeters in diameter. The spots may enlarge and coalesce to form large necrotic area (Fig. 19.1). The necrotic area gradually expands, and the leaf symptoms grow to take up the whole of the green leaf tissue and to a lesser extent on stems at a late stage of the plant growth. As the lesion enlarges, a series of dark concentric rings are visible as a result of irregular growth patterns of the pathogen. This characteristic "target-spot" or "bull's eye" pattern is typical of early blight symptoms. Subsequently, the necrotic leaf tissue is often surrounded by a chlorotic border caused by fungal mycotoxin (i.e., alternaric acid), which turn the leaf tissue yellow. The chlorosis can extend to the whole infected leaf resulting in dried up leaf which hangs along the stem.

Conidia of *Alternaria* spp. are washed from the leaves and enter in the soil which can also infect potato tubers. The affected tubers show dark brown, slightly sunken lesions on the tubers. Diseased tuber tissue underneath lesion is dark brown, firm, and 10–12 mm deep. The dry or hard rot of tubers causes storage losses, decreases potatoes quality, and reduces emergence capacity of seed tubers.

Fig. 19.1 Symptoms of early blight of potato foliage

19.2.2 Causal Organism

The main causal organism of early blight on potato crop is *Alternaria solani* Sorauer (Ell. and Mart.). However, many other large-spored *Alternaria* spp., which infect potato plants, have also been reported. Rodrigues et al. (2010) observed that *A. grandis* Simmons was the causal organism which infects potato plants in various regions in Brazil. In an artificially inoculated field study, Duarte et al. (2014) found that *A. grandis* can cause infection on potato crops. In Algeria, Ayad et al. (2017) detected *A. protenta* as the causal agent of early blight and together with *A. grandis* and *A. solani* found to be part of the complex of *Alternaria* spp. detected in potato fields in Belgium (Landschoot et al. 2017). Hauslanden and Bassler (2004) reported that in Germany the occurrence of *A. alternata* and *A. solani* in potato crop was almost equal, whereas in Poland, the frequency of *A. alternata* was higher than that of *A. solani* (Kapsa 2007).

19.2.3 Epidemiology

Alternaria spp. overwinter as mycelium, chlamydospores, or conidia in the soil and on crop residues (Wale et al. 2008). The infection occurs through primary inoculum (conidia) carried to the older leaves by rain water. *Alternaria* is able to penetrate the leaf tissue directly through the intact epidermis or through natural openings and wounds. Initially, the lower leaves closer to the ground are infested. The fungus is restricted to the lower leaf level for several days. Formation of conidia starts on the necrotic leaf tissue at temperatures between 5 °C and 30 °C (optimum 20 °C). The secondary inoculum is dispersed through wind and causes infections on the nearby plant leaves and stems. The latent period is about 3–7 days. When a condition becomes favorable for infection, and at a certain age of the plant, *A. solani* colonizes the middle and upper leaves very rapidly. In fields, a cascade-like progression of the pathogen from the lower, via the middle, to the upper leaves is visible. Heavily infected leaves fall off and serve as inoculum source on and in the soil.

Weather conditions, plant growth stage and their health, cultivar maturity, susceptibility of the cultivar, and inoculum level play an important role in the progression of the disease. Temperature above 22 °C and alternating high relative humidity are the favorable weather conditions for *A. solani* infection. Besides potato, early blight can also occur on other crops. It has been observed on many solanaceous host plants such as tomato (*Solanum lycopersicum* L.), eggplants (*S. melongena* L.), hairy nightshade (*S. sarrachoides* Sendt), black nightshade (*S. nigrum* L), horse nettle (*S. carolinense* L.), pepper (*Capsicum* spp.), and non-solanaceous weeds (Jones et al. 1993; Hausladen and Aselmeyer 2017).

19.2.4 Economic Impact

Nowadays, under climatic change scenario early blight is considered to be one of the most important fungal diseases of potato after late blight. It is found in almost all countries where potatoes are cultivated (Woudenberg et al. 2014). However, *A. solani* is described as an important fungal pathogen especially in warmer regions because it requires high temperature for growth and disease development. Depending on the cultivar susceptibility and geographical regions, *A. solani* can cause considerable yield losses up to 2–58% (Shtienberg et al. 1996; van der Waals et al. 2001; Campo Arana et al. 2007; Horsfield et al. 2010). In India, yield loss has been estimated up to 79% due to early blight damage in severe condition.

19.2.5 Management

Crop rotation *A. solani* survives in the form of mycelium or conidia on the crop residue or soil in the field from one growing season to the next. Therefore, crop rotation with nonhost crop and control of the host weed plants like black shadow reduces the inoculum level of the pathogen. Additionally, removal and burning of infected plants also reduce the pathogen inoculum level.

Biofumigation It is an alternative option to reduce the primary pathogen inoculum in the soil. Biofumigation is a process to suppress the pathogen inoculums by isothiocyanates (ITCs), which derive from hydrolyzation of glucosinolates by myrosinase in disrupted plant cells. Bio-fumigant plants such as white mustard, leaf radish, etc., can reduce the early blight incidence in the crop (Volz et al. 2013).

Use of disease-free seed Diseased and virus-infected potato plants are more susceptible to early blight infection than normal healthy plants; therefore planting the diseased or virus-free seed tubers can reduce the pathogen attack.

Abiotic stresses Potato plants stressed by biotic or abiotic factors are more susceptible to early blight disease compared to non-stressed plants. Various abiotic stresses such as drought, frost, high temperature, and over-irrigation affected potato plants during the cropping season. Salt stress enhanced the symptoms of early blight disease. Additionally, prolonged leaves wetness period due to overhead irrigation allows successful fungal infection.

Nutrition management For optimum potato plant growth and tuber yield, a balanced nutrition is required during the growing season. Specially, N-fertilizer should be applied properly; otherwise susceptibility of plant against early blight will be higher. Better soil fertility and plant nutrition can decrease the severity of early blight (Lambert et al. 2005; MacDonald et al. 2007). Under drought condition, when plants are unable to take enough nutrients from the soil through the roots, foliar spraying of fertilizer can decrease the nutrient deficiency that reduces plant susceptibility to the disease. The fertilizer form can also influence the disease progression of *A. solani*. Application of calcium cyanamide results in a delay of early blight disease, as the fungicidal side effects of degradation products of calcium cyanamide can reduce the initial inoculum in the soil (Volz et al. 2013).

Varietal resistant Genetic resistance offers the most effective means to control early blight; however, no completely resistant genotypes have been reported so far. Most of the cultivated potato varieties are much more susceptible to early blight than wild species. Generally, early maturing cultivars are more susceptible to *A. solani* than those of late maturing cultivars. Screening of wild diploid relatives, breeding clones, and some tetraploid cultivars for resistance to early blight have been reported (Xue et al. 2019). Few clones of *Solanum tarijence*, *S. neorossii*, and *S. commersonii* showed high degree of resistance (Jansky et al. 2008), while moderate resistance was observed in *S. chacoense*. Some potato cultivars such as "Kufri Jeevan," "Kufri Pukhraj," "Kufri Badshah," "Kufri Sherpa," and "Kufri Sindhuri" show moderate resistance to early blight.

Biological control Biocontrol is the application of microorganism (bioagents) to reduce the plant pathogen population and is considered to be an eco-friendly alternative for disease management. Several potential antagonists have been evaluated; among them PGPR (*Pseudomonas* spp., *Bacillus* spp.) and fungi (*Trichoderma polysporum*, *T. harzianum*, *T. viride*, *Chaetomium globosum*) are common. In a field study, *T. viride* (0.5%) was found effective against early blight for reducing disease intensity (Yadav and Pathak 2011). A combination of *T. harzianum* and *P. fluorescens* was applied as seed treatment and foliar spraying for reducing the disease intensity under field conditions (Mane et al. 2014). *Trichoderma longibrachiatum* inhibited mycelial growth of *A. solani* by up to 87.6% under in vitro conditions (Prabhakaran et al. 2015). Volatile organic compounds (VOCs) produced by *B. subtilis* ZD01 can inhibit the conidia germination and reduce the lesion areas in vivo (Zhang et al. 2020). Recently, Gorai et al. (2021) evaluated the biocontrol efficacy of endophytic *B. velezensis* SEB1 and concluded that cell-free extract at 1000 ug/ml was effective to inhibit the conidial germination and reduces the radial growth up to 82.34% in vitro and decrease disease severity up to 52.5% under field conditions.

19.3 Late Blight

19.3.1 Symptoms

The aerially dispersed asexual sporangia are responsible for epidemics on potato crops. When the flying sporangia arrive on the plant surface, it can germinate directly or release zoospores, which encyst, germinate, and penetrate the host tissues (Fry et al. 2015). This infection stage is not seen by naked eye, but inside the cells, complex mechanism of molecular interactions takes place. After entering, formation of haustoria begins inside the plant cells, from where many effector proteins are secreted (Whisson et al. 2016; Wang et al. 2017). At this stage *P. infestans* follow biotrophic mechanism to obtain nutrients.

The visible symptoms started to appear within 2–3 days when the pathogen achieves the necrotrophic stage. Symptoms appear at first as water-soaked irregular pale green lesions, usually at edges of lower leaves. These lesions grow rapidly and

Fig. 19.2 Symptoms of late blight of potato foliage

turn brown to purplish black within 1–2 days. During morning, a white mildews growth develops around the lesion on the underside of leaves (Fig. 19.2), which consists of sporangiophores and sporangia, which emerge through the stomata (Nowicki et al. 2012) and are the typical characteristics of potato late blight. On stems or petiole dark brown lesions develop which elongate and encircle the stems. Underground tubers may be infected by sporangia which are washed off the diseased foliage and enter the soil. Infected tubers show irregular, slightly depressed areas with brown coloration which extend deep in to the tubers.

19.3.2 Causal Organism

Phytophthora infestans (Mont.) de Bary is the main causal organism of late blight disease of potato. Previously, it was described as a fungus due to the superficial resemblance to filamentous fungi but is now classified as oomycete in the kingdom of stramenopiles (Kamoun et al. 2014). The vegetative stage of *P. infestans* is diploid, whereas it is haploid in true. Recent research has shown that in the present-day lineages the progenies from sexual *P. infestans* populations are diploid, while the clonal lineages responsible for most important pandemic are triploid (Li et al. 2017).

Phytophthora infestans populations are constantly evolving and novel, and usually highly pathogenic races appear periodically dominating the previously existing races. Divergence, recombination, and migration are the main reasons responsible for the emergence of new genotypes (Knaus et al. 2016). *Phytophthora infestans* reproduce mainly through asexual reproduction, and diverse numbers of clonal lineages occur in different countries and locations. Many studies have found that emergence of new races can often be credited to migration (Fry et al. 2015; Knaus et al. 2016; Saville et al. 2016). Previously, the mating type A1 was dominating worldwide, except its presumed center of origin, Central Mexico, where both mating types (A1 and A2) exist in equal frequencies (Goodwin et al. 1992). This situation has changed dramatically, and migration of A2 mating type to various countries of the world during late 1980s has resulted in increased emergence and severity of late blight disease (Goodwin 1997; Zhu et al. 2015; Chowdappa et al. 2015; Montes et al. 2016; Rojas and Kirk 2016; Rekad et al. 2017). Existence of both A1 and A2 genotypes at the same location has opened up the possibility of development of thick-walled oospores which could survive either extreme winter (Medina and Platt 1999) or summers conditions. Recent investigations have shown that these self-fertile isolates are found more frequently, constituting a new threat to potato crops because of their increased genotypic variability, better fitness, and greater aggressiveness (Zhu et al. 2016; Casa-Coila et al. 2017).

19.3.3 Epidemiology

Phytophthora infestans overwinters as mycelium in infected seed tubers, refuse piles, and host plant. Infected seed tubers serve as a primary source of inoculum. When A1 and A2 mating types are present, formation of oospore takes place which has potential to initiate the disease (Stevenson et al. 2001). Under favorable environmental conditions, the pathogen may sporulate and discharge zoospores in the soil which move upward and infect the plant at ground level. Older leaves touching soil level get infected first. Severe infection takes place under low temperature and high relative humidity with heavy dews or alternate raining. Sporangia are produced rapidly at 18–20 °C and high relative humidity (>90%). Sporangia are sensitive to desiccation, and, after dispersal by wind or splashing water, they require free water to germinate. The sporangia may germinate by two ways: indirect or direct. The optimal temperature for indirect germination via zoospores is 10 °C, whereas that for direct germination of sporangia via germ tubes is 24 °C. In the presence of water, zoospore enters to the host tissue through germ tubes and appresoria within few hours at 8 °C and 25 °C. After entering in the plant, subsequent development of the diseases is most rapid at 21 °C, and lesions with new sporangia appear within few days.

19.3.4 Economic Impact

The potential economic and social impact of potato late blight disease is best illustrated by the well-publicized role it played in the Irish Famine in the middle of the nineteenth century when it completely destroyed potato crop, either by killing foliage prior to the harvest or by causing massive tuber rot in storage condition. As a result of the famine, millions of Irish people died or emigrated (Bourke 1993).

Haverkort et al. (2009) recorded the global costs and losses due to late blight that take 16% of all global potato production. The yield loss due to late blight ranged from 20% to 70%, and it can destroy the whole crop under epidemic conditions (Haq et al. 2008; Lal et al. 2015; Lal et al. 2019).

19.3.5 Management

Late blight disease can be controlled by a combination of integrated disease management approaches. Various management measures include elimination or reduction of initial inoculum sources such as infected seed, cull piles, infected neighboring fields, and host plants, spraying fungicide before the appearance of disease, and use of resistant cultivars to reduce the rate of disease development. Planting early-maturing cultivars to reduce the crop duration or planting the crop in seasons or locations where the environment is not favorable for the disease development may also be helpful.

Cultural practices Cultural practices include all the activities carried out during cropping season for agronomic management which change the microclimate, host condition, and pathogen behavior to reduce phytopathogen activity, viz., their survival, dispersal, and reproduction (Garrett and Dendy 2001). Control of inoculum sources such as host weed plants and cull piles and plant debris, disease in neighboring fields can help in management of the disease (Turkensteen and Mulder 1999). Use of disease-free certified seed, growing resistant varieties, well-drained aerated fields, adequate space between rows and plants, rotation with nonhost, adequate hilling, timely mechanical weeding, harvesting in dry conditions, and when the tubers are mature could minimize late blight (Garrett and Dendy 2001; Perez and Forbes 2010). Scouting all stored potatoes frequently and removing diseased tubers from storage are desirable to prevent disease spread. Increased use of nitrogen fertilizers can lead to increase in disease severity resulting in yield reduction; therefore, moderate nitrogen fertilization is often recommended as cultural practices to delay the development of late blight. However, higher use of phosphorus and potassium fertilizers gives a positive response to yield in a late blight year (Roy et al. 2001).

Varietal resistance The use of resistant cultivars is among the most effective and eco-friendly means of controlling the late blight disease particularly in tropical conditions. Cultivars having high degree of resistance can allow them to be grown without fungicide application or less fungicide either by lowering the fungicide dose or using longer application intervals (Liljeroth et al. 2016; Haverkort et al. 2016). Ideally, a late blight resistance variety should have high level of resistance to both foliage and tuber blight. Binyam et al. (2014) observed that appearance of the potato late blight disease was delayed almost by 20 days on the moderately resistant varieties as compared to the moderately susceptible and susceptible varieties. Advanced hybrid "Kufri Garima" derived from cross PH/F-1045 X MS/82-638 has been released for commercial cultivation. "Kufri Mohan," "Kufri Fryom," "Kufri Sangam," and "Kufri Karan" are new varieties with field resistance to late blight. However, the race-specific oligogenic resistance in the existing released potato varieties can be rapidly broken down by compatible races of *P. infestans* rendering the varieties to be susceptible to the disease within a short period (Shtienberg et al. 1994). Potato breeders are therefore working to develop late-resistant genotypes to improve tolerance in genes of indigenous species that have been hit hard by non-native invasive plant pathogens.

Organic amendments Application of compost in crop production not only improves the physicochemical properties and soil fertility but also controls various soilborne diseases and increase crop yields (Adebayo and Ekpo 2001; Remade 2006; Yadessa et al. 2010). Different organic materials such as seashells, vegetable waste, farmyard manure, and other waste products are used to promote plant growth. The most common soil organic amendments are compost and animal manure. The efficiency of compost in controlling plant diseases is attributed to its content in antagonistic microorganisms such as bacteria and actinomycetes (Yadessa et al. 2010). Various benefits derived from the application of compost as fertilizer include increase in organic carbon content and microbial activity (Scotti et al. 2015), a greater concentration of plant macro- and micronutrients, i.e., N, P, K, and Mg, and root reinforcement (Donn et al. 2014). Organic compost has capability to influence soil microflora by suppressing various soilborne pathogens diseases such as *Pythium*, *Phytophthora*, and *Fusarium* spp. (Szczech and Smolińska 2001; Borrero et al. 2004).

Biological control Biological control consists of minimizing plant diseases by the interaction of one or more live microorganisms with the pathogen or use of extract of plants. Some findings report the use of *Trichoderma* isolates (Yao et al. 2016), *Chaetomium globosum* (Shanthiyaa et al. 2013), *T. viride*, and *Penicillium viridicatum* (Gupta 2016) and bacteria from the genera *Bacillus*, *Pseudomonas*, *Rahnella*, and *Serratia* (Daayf et al. 2003) as biocontrol agents in the management of late blight disease in potato. In Ethiopia, Zegeye et al. (2011) evaluate the antagonistic activity of *T. viride* and *P. fluorescens* against *P. infestans* under in vitro and greenhouse conditions. The result revealed that both the antagonists have the potential to inhibit the mycelium growth of *P. infestans* in vitro; however, foliar spray of the *T. viride* suspensions was found to be more efficient than *P. fluorescens* and mixed culture. Integrated approaches using fungal and bacterial bioagents has been adopted for managing late blight disease (Lal et al. 2017). Use of biosurfactant from *P. aeruginosa* was found effective in minimizing late blight disease (Tomar et al. 2019a). Recently, in a field study of Lal et al. (2021), neem-based products were found effective for controlling the late blight as well as increase tuber yields. *Trichoderma viride* and *P. fluorescens* were also found effective. *Allium sativum* (garlic) has been suggested as a potential intercropping plant for the management of potato late blight disease under Ethiopian condition (Kassa and Sommartya 2006). Still, few biological control measures are used by nonorganic growers due to low efficacy and farmers' lack of knowledge about these options and access to the most efficient products. Leaf extracts of onions, garlic, *Malus toringo*, *Reynoutria japonica*, and *Rheum coreanum* inhibited mycelial growth of

P. infestans in vitro. Further, extracts of *Malus toringo* were found effective in controlling late blight under greenhouse experiments (Paik 1989).

19.4 Black Scurf and Stem Canker

19.4.1 Symptoms

Black scurf on potato tubers and stem canker are two distinct phases of the same disease. Black scurf, characterized by the presence of varying size of sclerotia on the surface of tuber, is the best-known symptom of *Rhizoctonia* disease in potato (Fig. 19.3). In addition, symptoms due to severe infection of the stolons and tubers include atypical cracks, corky lesion, malformation, pitting, and desquamation, and elephant hide may also be observed (Campion et al. 2003; Muzhinji et al. 2014). After planting, the fungus may attack young sprouts through the epidermis and produce dark brown lesions, thereby killing underground sprouts much before the plant emergence resulting germination reduction. On the newly developing sprouts reddish brown to gray sunken lesions can be observed. These lesions can girdle the young sprout completely causing the part above the lesion to die. As these lesions mature, they become cankers that are rough and brown and have craters, cracks, or both (Baker 1970; Banville 1978). Infection of the stem causes stunting and rosetting of plant tops resulting in curling the upper leaves which sometime turn red or yellow (Wharton et al. 2007). In a recent study, Ito et al. (2017) observed that leaf curling is not a direct symptom of *Rhizoctonia*, but prior infection of *Potato leafroll virus* enhanced the severity of *Rhizoctonia* diseases. Aerial tubers could be formed in the leaf axils of stems due to interference of carbohydrate movement (Beukema and van der Zang 1990).

Fig. 19.3 Black scurf on potato tubers and stem cankers

19.4.2 Causal Organism

The causal organism of black scurf and stem/stolon/root canker of potato is *Rhizoctonia solani* Kühn AG-3 (anamorph) and *Thanatephorus cucumeris* (Frank) Donk (teleomorph) (Virgen-Calleros et al. 2000). Stevens et al. (1993) differentiated AG-3 isolates from potato and tobacco on the basis of culture appearance, fatty acid profile, and pathogenicity. *Rhizoctonia solani* does not produce asexual spores and exists as mycelia (hyphal growth form), sclerotia (dense asexual hyphal resting structures), or basidiospores (sexual spores) (Keijer et al. 1996). Anamorphic classification of *Rhizoctonia* spp. is based on a characterization of the cell nuclear condition (multi,- bi-, or uninucleate) and the ability of hyphae to anastomose with tester isolates of designated anastomosis groups (AGs) (Sneh et al. 1991).

Occurrence of *R. solani* anastomosis group (AGs) in potato Among the AGs, AG-3 is the most prevalent AG infecting potato (Woodhall et al. 2007; Lehtonen et al. 2008). However, a range of other AGs at lower frequency have been found in potato fields around the world. AG2-1 has been found in potato fields in Alaska (Carling et al. 1986), France (Campion et al. 2003), Turkey (Yanar et al. 2005), the Great Britain (Woodhall et al. 2007), and Finland (Lehtonen et al. 2008). Black scurf caused by AG4 has been observed under warm conditions from Peru (Anquiz and Martin 1989), Australia (Balali et al. 1995), Canada (Bains and Bisht 1995), and Mexico (Virgen-Calleros et al. 2000). Isolates of AG4 HG-I and AG4 HG-III (Muzhinji et al. 2014, 2015) and AG4 HG-II (Woodhall et al. 2012) cause stem canker symptoms on potato plants, but sclerotia formation and blemishes were not observed on the progeny tubers. In Maine, USA, AG-5 was widespread in soil but infrequently found on the stem, stolon, and root of potato plants and not on the tubers (Bandy et al. 1988). In Canada, isolates of AG-5 were not restricted to any particular region (Bains and Bisht 1995), but in France it was found in geographically distinct locations. AG-5 and some AG-3 and AG2-1 isolates were recovered from superficial tuber alterations, such as deformations, or corky or scabby lesions (Campion et al. 2003). Isolates of AG-8 have been recovered from Australian potato field soil (Balali et al. 1995), and symptoms of canker on stems, stolons, and roots and decreased numbers of feeder roots were reported in glasshouse experiment but not sclerotia on tubers. *Rhizoctonia solani* AG-9 has been isolated from Alaskan (Carling et al. 1986) and Turkish (Yanar et al. 2005) potato fields. It causes slight to moderate tuber damage on susceptible cultivars in the field and in greenhouse experiments. Also, binucleate *Rhizoctonia* (BNR) isolates were obtained from potato plants (Carling et al. 1986). Farrokhi-Nejad et al. (2007) reported 12 BNR isolates (out of 58), and Lehtonen et al. (2008) found a single BNR isolate (out of 119) that causes mild symptoms on potato sprouts. However, BNR AG A and AG R causing stem canker, black scurf, and tuber defects on potato were reported from South Africa (Muzhinji et al. 2015; Zimudzi et al. 2017).

19.4.3 Epidemiology

Rhizoctonia solani overwinters as sclerotia on seed tubers or as mycelium in plant debris in soil or on alternate hosts. The pathogen has a wide host range including many solanaceous and non-solanaceous plants. But the main sources of inoculums are infested seed tubers and infected soil. At the end of growing season, sclerotia remaining in soil serve as primary inoculum for infection of plants in the next growing season (Keijer et al. 1996). Soil temperature plays critical role in the initiation of *Rhizoctonia* disease in potato, with severity of the disease being positively correlated with the temperature. Low temperature with high soil moisture, organic matter, and a neutral to acidic soil (pH 7 or less) are suitable conditions for the development of stem canker. Sclerotia start forming on daughter tubers late in the crop growing season, mainly after harms cutting, but sclerotia can also be seen at mid of the cropping season.

19.4.4 Economic Impact

In potato production, *Rhizoctonia* diseases are responsible for both quantitative and qualitative yield losses (Fiers et al. 2011; Das et al. 2014). Quantitative yield losses occur due to infection of the stems, stolon, and roots, which affect tuber size and numbers (Carling et al. 1989), whereas qualitative losses occur mainly by the production of misshapen tubers and the development of sclerotia on the tuber surface (James and McKenzie 1972). It is reported that *Rhizoctonia* disease was responsible for 10–25% yield loss in India (Sharma 2015), up to 30% in Canada, and up to 50% in other countries, thereby affecting potato production severely (Woodhall et al. 2008). The marketable yield losses caused by *Rhizoctonia* spp. on potato have been estimated to reach up to 30% (Platt et al. 1993; Tsror 2010). *Rhizoctonia* disease in potato is hard to control due to the wider host range of the pathogen and long survivability in the form of dormant sclerotia under unfavorable environmental conditions. Further, the pathogen evolves with time allowing the pathogen to overcome the resistance level that may have been serious problem of the potato producers and breeders.

19.4.5 Management

Cultural practices Agronomic practices such as disease-free planting material, soil disinfection, crop rotation, haulm destruction, harvest timing, soil management, irrigation, and plant residues all have an influence on the *Rhizoctonia* disease development and crop quality and quantity.

Disease-free planting material Since black surf is tuber- and soilborne disease, infested seed tubers play an important role in disease development. Disease can be managed to a large extent by the use of certified seed free from sclerotia or any type of *R. solani* inoculums; thus quarantine of potato seed tubers should be done before planting.

Disease free soil *Rhizoctonia solani* inoculum density level in soil can be used as criteria in a risk-prediction system to decide control measure of the disease. Solarized soils are frequently more suppressive and less conducive to certain soil-borne pathogens than nonsolarized soils (Greenberger et al. 1987). Soil solarization also improves soil structure and increases the availability of essential plant nutrients for rapid growth and development of plants (Elmore et al. 1997). Soil solarization with transparent polyethylene mulching during hot summer months in Indian sub-tropical plains was found effective against black scurf (Arora et al. 1997).

Crop rotation Besides the advantages like maintenance of soil fertility, soil organic matter, reduction in soil erosion, etc., crop rotation specifically decreases the incidence of plant diseases caused by soilborne pathogens (Pedersen and Hughes 1992). Monocropping systems generally led to the increase of soil density of specific pathogens resulting in the decline of crop yield and quality (Honeycutt et al. 1996). An increased number of potato cropping cycles enhanced the incidence and severity of stem canker due to the increase in soilborne inoculum level (Scholte 1992; Honeycutt et al. 1996). Although 2-year rotations are found effective to reduce disease levels compared with continuous potato cultivation (Little et al. 2004; Manici and Caputo 2009), longer rotation lengths of 3 or 4 years between potato crops are known to be more effective in controlling soilborne diseases (Buyer et al. 1999; Little et al. 2004; Larkin et al. 2010). Rotations of 3–5 years are often recommended for effectively reducing the black scurf severity. The use of crops with known disease-suppressive capabilities, such as *Brassica* spp., cereals, millets, sunhemp, and non-solanaceous crops, may provide additional resources for reducing disease through improved cropping systems. Various other plant species (including weeds) have been shown to sustain *R. solani* (Jager et al. 1982; Carling et al. 1986) and should be considered in crop rotation and weed control. In three cropping sequences, viz., potato-wheat-paddy, potato-onion-maize, and potato-green gram-groundnut, highest incidence of black scurf was recorded in potato-onion-maize cropping sequence (Anonymous 2019).

Haulm destruction and harvest timing Potato crop may be harvested as soon as it is possible. The harvesting methods applied for potato production can affect the level of black scurf (Dijst et al. 1986). The incidence of infested tubers increased with the length of interval between haulm destruction and harvest. When the temperature and moisture conditions are favorable, the sclerotia keep on appearing and developing on the tubers in the soil. Sclerotial production was stimulated similarly with individual practices of cutting off shoots, chemical haulm destruction, and cutting off roots (Dijst 1985). Green-crop harvesting (harvesting the immature crop mechanically and replacing the tubers to the soil for a curing and final harvesting 2–4 weeks later) and immature-crop harvesting often result in a low level of black scurf (Mulder et al. 1992; Lootsma and Scholte 1996). Green-crop harvesting has the advantage of involving the application of fungicides or antagonistic organisms with the first lifting of the tubers, resulting in effective control of black scurf (Mulder et al. 1992).

Organic matter amendment Organic compost matter, such as cattle manure, is an essential component of organic crop management as it improves soil structure, water holding capacity, and cation exchange capacity and promotes plant growth. The organic amendments provide an effective measure for soilborne black scurf disease management, and it represents a substitute to reliance on fungicides. Tsror et al. (2001) reported that in a field experiment application of *Trichoderma harzianum*, nonpathogenic *Rhizoctonia* and cattle manure compost in furrow could reduce black scurf incidence. Kumar and Kumar (2018) found that the soil amendment with vermicompost reduced disease severity up to 50%, followed by neem cake and mustard cake. The highest reduction in disease severity was observed when farm yard manure was applied in combination with white mustard or when oats were grown as a green manure crop (Scholte and Lootsma 1998), whereas least reduction was reported when farmyard manure was applied alone (Kumar and Kumar 2018). *Brassica* spp. and barley reduced inoculums level of *R. solani* by 20–56% in greenhouse tests (Larkin and Griffin 2007). Green manuring of *Brassica* crops by biofumigation at flowering stage was found effective to minimize disease incidences of black scurf of potato (Anonymous 2017).

Plant extract Plant extract or phytobiocides may be an effective alternative to control *Rhizoctonia* diseases due to their rapid degradation, narrow range of activity, and nonhazardous effects. Earlier reports have shown antifungal potential of *Azadirachta indica*, *Eucalyptus camaldulensis*, *Allium cepa*, *Allium sativum*, *Lantana camara*, *Capparis decidua*, *Dodonaea viscosa*, and *Peganum harmala* extracts against *R. solani* (Atiq et al. 2014; Khan et al. 2016). The bulb extract of *Allium sativum* and rhizome extract of *Zingiber officinale* were found effective in suppressing the mycelial growth of *R. solani* in vitro (Kumar et al. 2017). Recently, Rafiq et al. (2021) reported the in vitro antifungal activity of methanolic leaf extract of *Carthamus oxyacantha* against *R. solani.*

Biological control PGPR strains that were found effective against *R. solani* included *Pseudomonas* spp., *Bacillus* spp., and *Enterobacter* spp. (Tabassum et al. 2017). Two strains of *Pseudomonas* spp. (StT2 and StS3) were found effective against potato black scurf which reduced disease severity up to 65.1% and 73.8%, respectively (Tariq et al. 2010). In a greenhouse experiment, interaction of potato seeds with *Bacillus* spp. showed 30–41.4% disease reduction of black scurf and 28.5–40.2% of stem canker (Kumar et al. 2012). In an in vitro study, *B. subtilis* (V26) strain was found effective against *R. solani* and reduced disease incidence up to 63% and 81% of root canker and black scurf, respectively, as well as enhanced plant growth in planta (Khedher et al. 2015). *Pseudomonas* sp. strain (S8.Fb11) reduced the proportion of infected tubers by *R. solani* to 40% for cv Spunta and to 74% for cv Nicola (Mrabet et al. 2013).

Trichoderma spp. and *Gliocladium* spp. reduce *R. solani* growth by competition for nutrients and space, antibiosis, and by mycoparasitism involving antifungal secondary metabolites (Harman 2007). Tsror et al. (2001) reported that application of *T. harzianum* to the soil surface had relatively small effect compared to the in-furrow treatments. Wilson et al. (2008) reported that application of *T. harzianum*, either in-furrow or in combination with flutolanil applied to seed tubers, increased marketable tuber yield (from 35% to 60%) and reduced black scurf incidence on progeny tubers from 31% to 11%, which could not be achieved using flutolanil alone. In another study, Hicks et al. (2014) reported that isolates of *Trichoderma* spp. (*T. virens*, *T. atroviride*, and *T. barbatum*) reduced percentage of diseased stolon by 41–46% *in planta*. Rahman et al. (2014) evaluated *Trichoderma* spp. against *R. solani* on potato and suggested that integrated or combination approaches could be effective for the management of black scurf. A combination of *B. subtilis* and *T. virens* demonstrated a better control of stem canker than each organism alone (Brewer and Larkin 2005). Arora (2008) reported the treatment of *T. viride* after seed dressing with boric acid (1.5%) significantly minimized the black scurf disease on potato tubers. In a field study, tuber treatment with 2% boric acid along with *T. viride* at 10 g/kg seed recorded the lowest disease incidence (15.33%) and index (0.38) with highest yield (324.68 q/ha) (Patel and Singh 2020). Less control percent ability of *P. aeruginosa* and its metabolites was found to manage black scurf of potato (Tomar et al. 2019b). Recently, Chaudhary et al. (2020a) reported antagonistic activity of native *T. harzianum* against *R. solani* in vitro and greenhouse experiments. Despite the promising results with bioagents, the introduction of new biocontrol agents involves various considerations such as the tedious work of selection and screening, optimization of mode of application to achieve best results (Tabassum et al. 2017), shelf life of the bioagents, efficacy in the field experiments, eco-friendly measures, and registration to be used as a PGPR (Etesami and Maheshari 2018).

19.5 *Sclerotinia* Stem Rot

19.5.1 Symptoms

The first visible symptom of stem rot appears as water-soaked spots usually at stem and branch axils or on branches or stems in contact with the soil. A cottony white mycelium growth develops around the lesion, and the infected tissue becomes soft and watery. Lesions often expand in size rapidly and may girdle the stem which causes foliage wilting. Lesions become dry and will turn beige, tan, or bleached white in color and show papery appearance (Fig. 19.4). Hard, irregularly shaped sclerotia develop in and on decaying plant tissues. Generally, sclerotia are few millimeters in diameter and up to 25 mm in length. Initially, white to cream in color but become black after maturity and are frequently found in the hollowed-out center of infected stems.

19.5.2 Causal Organism

Sclerotinia stem rot, white mold, or watery soft rot of potato is caused by the necrotrophic fungus *Sclerotinia sclerotiorum* (Lib.) de Bary (Ojaghian et al. 2016; Chaudhary et al. 2020b). Generally, *S. sclerotiorum* is more important pathogen of

Fig. 19.4 Symptoms of *Sclerotinia* rot of potato and formation of sclerotia on stems

vegetables in the field during transit and in store. The fungus is both soil- and airborne and geographically widespread in nature, but the disease occurs in relatively cool moist conditions areas.

19.5.3 Epidemiology

Sclerotinia sclerotiorum overwinters in the soil for long time periods under dry and high temperature conditions in the form of dormant structures called sclerotia. The sclerotia may germinate myceliogenically to produce hyphae that infect stems of host plants directly or germinate carpogenically to produce apothecia depending on environmental conditions (Bardin and Huang 2001). The apothecia release millions of airborne ascospores thereby initiating plant infection. Extensive foliage growth which increases humidity and extends leaf wetness within crop canopies promotes development and spread of disease. Increased disease incidence is associated with overhead sprinkler irrigation, a non-upright cultivar architecture, higher crop density, close row width, continuous wetness, and excess nitrogen fertilization in potato and other crops (Grogan and Abawi 1975; Grau and Radke 1984; Gutierrez and Shew 1998). Epidemics of potato stem rot are initiated when airborne ascospores land on open potato blossoms attached to the canopy (Atallah and Johnson 2004). Apothecia present in the potato field, in neighboring potato fields, or in fields of other crops in rotation with potatoes or crops susceptible to *S. sclerotiorum* are likely sources of ascospore inoculum. Ascospores originating external to a potato field appear to be an important and abundant source of inoculum (Johnson and Atallah 2014). Over the last decade, a wide adaptation of monocropping cultural practices and cultivation of susceptible varieties under irrigated conditions has increased *S. sclerotiorum* inoculum in the soil that has made stem rot a serious threat for potato production.

19.5.4 Economic Impact

The economic impact of *S. sclerotiorum* is more limited and varies among the host plant species. In potato crop, it is capable of reducing crop yields up to 60% in a large number of potato fields in India (Dutta et al. 2009). In Germany, *S. sclerotiorum* causes yield reduction up to 30% in potato crop in some areas of Niedersachsen (Quentin 2004). Recently, Alam et al. (2021) observed that about 23% potato plants were wilted and died before harvest in affected fields in Pakistan. The *Sclerotinia* stem rot is reemerging in Western Uttar Pradesh due to change in climatic condition.

19.5.5 Management

The control of stem rot diseases is difficult due to the pathogen's wider host range, long-term persistence of sclerotia in the soil, and the production of airborne ascospores. Management practices to control *S. sclerotiorum* can be developed at several growth stages of the potato crops. Effective disease management strategies usually require implementation and integration of multiple methods.

Cultural practices Traditional agricultural practices such as use of disease-free clean seed tubers, early planting, soil tillage, and adjustment of row width and density of plant population contribute to a reduction of stem rot severity, but the effectiveness of these measures can be very limited (Steadman 1979; Mueller et al. 2002). Irrigation practices that promote leaf wetness or develop high relative humidity within the crop canopy should be avoided. Irrigation should be restricted during rainy weather and on cool, cloudy days, whenever possible.

Crop rotation *Sclerotinia sclerotiorum* survives in soil as sclerotia for long time under adverse environmental conditions. When conditions become congenial for its growth, dormant sclerotia germinate and develop inoculum-laden apothecia (Bolton et al. 2006). The most effective way to reduce the number of sclerotia in the fields is crop rotation. By rotating potato with nonhost crops, the annual life cycle of pathogen can be disrupted, resulting in decreased annual number of sclerotia in the fields. For effective implication of crop rotation, it must be coupled with an efficient weed control program that minimizes the chances of establishing and allowing *S. sclerotiorum* to persist in fields (Derbyshire and Denton-Giles 2016).

Varietal resistance Disease-resistant varieties remain the most economical and long-term approach for controlling the potato stem rot disease. However, no potato cultivars are available with resistance to infection of *S. sclerotiorum*. Further, the expression of the field resistance may be influenced by inoculum potential and other environmental conditions (Mueller et al. 2002). Higher disease incidence was found in "Kufri Garima" and "Kufri Chipsona-1," and less incidence was in "Kufri Pushkar" and "Kufri Pukhraj" under Indian conditions.

Organic amendments Organic matters are rich sources of nutrients for soil microorganism causing quantitative and qualitative changes in bacterial and fungal communities (Emmerling et al. 2002) which improves soil properties, plant health, and yield. In a study, Huang et al. (2002) tested 87 organic residues for their potential of controlling carpogenic germination of sclerotia. Among them, 46 effectively inhibited the development of the fungus when the materials were applied to the soil at a dose of 3% w/w. However, only three kinds of residues were effective at 0.5% w/w. The most effective in preventing ascospore production were materials with elevated levels of nitrogen, e.g., fish meal. They concluded that the loss of viability of sclerotia in the soil was connected with the production of ammonia and ammonia-related compounds. The most promising method to decrease inoculum level of *Sclerotinia* from infested field soil and pathogen multiplication is the use of organic matters combined with bioagents. Huang et al. (2002) reported that soil amendment with organic residues infested with *Coniothyrium minitans* and *T. virens* decreased carpogenic germination of sclerotia by killing the sclerotia. Similarly, Smolinska et al. (2016) found that the application of some selected *Trichoderma* species multiplied on the organic carriers prepared from agro-industrial wastes allowed the complete eradication of sclerotia of *S. sclerotiorum*. After analysis of about 2432 experiments, Bonanomi et al. (2007) concluded that compost was the most suppressive material and showing more than 50% disease control. The conducive conditions for *Sclerotinia* and addition of plant residues to the soil infested with sclerotia significantly decreased the yield of lettuce plants (Smolinska et al. 2016).

Biological control Several bioagents have been studied and identified for controlling stem rot disease in different crops. *Trichoderma harzianum* parasitizes both the sclerotial and hyphal growth stages of *S. sclerotiorum* (Abdullah et al. 2008; Troian et al. 2014). The mycoparasitic properties of *Trichoderma* species play a crucial role in the antagonistic activity against *S. sclerotiorum*. Hydrolytic enzymes, viz., chitinases, glucanases, proteases, and cellulases, are secreted by *Trichoderma* that disintegrate the cell wall of the pathogens (Chet et al. 1998; Kaur et al. 2005; Lopez-Mondejar et al. 2011; Chaudhary et al. 2020c).

In a field experiment, Geraldine et al. (2013) observed reduction in *S. sclerotiorum* apothecia number and disease severity after application of *T. asperellum* spore suspension with common bean. Under field conditions, *T. hamatum* reduced *Sclerotinia* disease by 31–57%, showing that *T. hamatum*-colonized sclerotia had reduced apothecial production and a lower carpogenic infection of cabbage (Jones et al. 2015). The white mold of cucumber fruit and stems was reduced by 64 and 30–35%, respectively, after *T. harzianum* T39 application under commercial greenhouse conditions (Elad 2000). *Trichoderma harzianum* isolate T-22 was found effective against *S. sclerotiorum* and decreased the disease severity index (DSI) by 38.5% in a field-grown soya bean crop (Zeng et al. 2012a). *Coniothyrium minitans* is another parasitic fungus that has been used for the biocontrol of *S. sclerotiorum*. Like *Trichoderma* spp., *C. minitans* parasitizes the sclerotia and mycelia of *S. sclerotiorum* (McQuilken et al. 1995; McLaren et al. 1996). During the seedling stage of canola, active spreading of *C. minitans* can reduce the amount of carpogenic germination of *S. sclerotiorum* later in the growing season (Yang et al. 2009). Studies showed that parasitization of *S. sclerotiorum* by *C. minitans* probably involves the degradation of oxalic acid, a pathogenicity factor of *S. sclerotiorum* (Cessna et al. 2000).

Additionally, many diverse bacterial genera have been studied and found effective against stem rot pathogen, *S. sclerotiorum*. *Bacillus* species were most commonly used as biocontrol agents (Hou et al. 2006; Hu et al. 2011, 2013; Gao et al. 2014; Wu et al. 2014); other BCAs including *Streptomyces platensis* (Wan et al. 2008), *S. lydicus* (Zeng et al. 2012b), *P. fluorescens* (Aeron et al. 2011), *P. chlororaphis* (Fernando et al. 2007; Selin et al. 2010), and *Serratia plymuthica* (Thaning et al. 2001) were also found effective against *S. sclerotiorum*.

19.6 *Sclerotium* Wilt

19.6.1 Symptoms

The pathogen first attacks the collar region, and a grayish brown, slightly sunken lesion appears on the stem just below the soil surface. Stem lesions expand upward the stem and downwards to cover the entire underground part of the plant leading to yellowing and wilting of the foliage (Mullen 2001). The wilting plants show a white weft of course fungal threads which girdle the basal part of the stem with selerotial bodies resembling mustard seeds on the collar region and roots (Fig. 19.5). The pathogen also infects tubers which showed small sunken, tan-colored spots with brownish margin. The affected tissues are tough and become soft and watery due to secondary rot-causing organisms.

The internal tissue decays and collapses, and the skin becomes broken exposing sunken cavities in the flesh. The white mycelium of the pathogen grows rapidly over the tuber surface in a fan-shaped outline. Sclerotia are formed in abundance on the hyphae.

19.6.2 Causal Organism

Sclerotium rolfsii (teleomorph: *Athelia rolfsii*) is the causal organism of stem rot or southern blight of many plant species in warm temperate, subtropical, and tropical regions (Punja 1985). It is a soilborne phytopathogen, distributed worldwide, and infects a wide range of plant species. *Sclerotium rolfsii* is a polyphagous plant pathogen which infects more than 500 species of monocotyledonous and dicotyledonous plants but especially severe on legumes, solanaceous crops, cucurbits, and other vegetable crops.

19.6.3 Epidemiology

Sclerotium rolfsii overwinters as sclerotia on seed tubers and in the soil or as mycelium on plant debris or on alternate hosts. In dry conditions the sclerotia remain viable for more than 2 years. The mycelial strands from an affected plant grow over the soil and cause infection of the adjoining plants. In crop fields, the wilted plants

Fig. 19.5 Symptoms of *Sclerotium* rot of potato and formation of sclerotia on stem

may be seen in patches indicating the center of infection. On potatoes, it attains major importance only occasionally and in certain locations. In the United States, *S. rolfsii* is an important pathogen in the tropics and subtropics and in areas of the southern and southeastern regions where temperatures are sufficiently high to permit the growth and survival of the fungus (Punja 1985), therefore known as southern blight, southern wilt, and southern *Sclerotium* wilt.

19.6.4 Economic Impact

Sclerotium wilt or rot is a disease of the warmer regions and attacks on a wide range of vegetable and field crops causing considerable yield losses. During the early 1960s, the disease was of annual occurrence at Pune, Maharashtra, especially during *kharif* season, and yield loss of 1–3% was recorded. However, severely affected crops recorded more than 50% crop loss. Postharvest losses of potato to the extent of 15% have been recorded in West Bengal state of India (Dasgupta and Mandal 1989). In Karnataka (India), the wilt incidence up to 30% and tuber rot up to 43% were recorded by Baswaraj (2005). In Bangladesh, it is responsible for the potato tuber yield reduction up to 60% (Rubayet et al. 2017).

19.6.5 Management

Cultural practices Cultural practices such as use of healthy seed tubers, excluding the pathogen from an area, soil removal and replacement, and rouging of infected plants and weed plants might help decrease disease incidence. Deep plowing is another effective method for removal of primary inoculum sources, i.e., sclerotia and infested plant debris, and prevents from contacting with plant tissues (Mullen 2001). Irrigating the fields at regular intervals to avoid too much dry helps in reducing the disease incidence.

Crop rotation Planting rotational crops that are non-susceptible such as corn, sorghum, cotton, or switchgrass was reported to reduce *S. rolfsii* disease incidence (Rodriguez-Kabana et al. 1994).

Organic amendments Incorporating organic amendments such as compost, oat or corn straw, and cotton-gin trash reduced the incidence of southern blight and also enhanced populations of beneficial soil microbes (Bulluck and Ristaino 2002). Neem oil and pine bark extracts or pine bark powder also reduce the growth of *S. rolfsii* (Kokalis-Burelle and Rodriquez-Kabana 1994). Organic matters such as neem cake with and without oil were found effective in reducing the potato *Sclerotium* rot incidence under field conditions (Gurjar et al. 2004; Baswaraj 2005).

Soil treatments In temperate and humid regions, soil solarization has been found effective in control of *S. rolfsii* (Hagan 2004). Other cultural practices that suppress *S. rolfsii* growth include adjusting the soil pH to about 6.5 by adding lime (Bulluck and Ristaino 2002) and aerification of the soil (Mullen 2001). A combined application of soil solarization with biofumigation was found most effective method for the management of *Sclerotium* rot disease in potato (Rubayet et al. 2017).

Varietal resistance Planting the resistant varieties or cultivars is a potentially preferable management method of stem rot disease (Mullen 2001). Potato cultivars show variations in their reaction to *S. rolfsii*; however, to date no cultivars have been reported to show complete resistance. "Kufri Chandramukhi," "Kufri Sindhuri," and some hybrid varieties showed moderate resistance to *Sclerotium* rot. An early maturing cultivar "Kufri Jawahar" recorded least disease incidence against *S. rolfsii* (Baswaraj 2005).

Biological control Various studies have reported the inhibition of mycelial growth and sclerotial production of *S. rolfsii* by using PGPRs, actinomycetes, mycorrhizal fungus, and *Trichoderma* species (Punja 1985). However, most of the studies were conducted under controlled in vitro conditions, and only few reports have demonstrated the biocontrol efficacy of these bioagents for control of *S. rolfsii* in the field (Cattalan et al. 1999; Tsahouridou and Thanassoulopoulos 2002). Many

Trichoderma spp. have been reported to control seed, root, and stem rots in many crops including potato. Under field conditions, isolates of *T. harzianum* and *T. longibrachiatum* have reported about 35–50% reduction in *Sclerotium* rot (Sreenivasaprasad and Manibhusanrao 1990; Asghari and Mayee 1991). In a study, Anahosu (2001) recorded least wilting (10%) with *T. harzianum* followed by *T. viride* (14%) in reducing potato wilt caused by *S. rolfsii*. Isolates of *T. harzianum* and *T. viride* were also reported the best bioagents in reducing the disease incidence in potato *Sclerotium* wilt (Baswaraj 2005). A combination of *T. harzianum* and mycorrhizal fungus *Glomus clarum* was found effective in suppression of *Sclerotium* rot (Sennoi et al. 2013). In a field study, Meena et al. (2018) found that soil treatment with *T. harzianum* (Th-BKN) at 10 kg/ha was the most effective treatment against *Sclerotium* rot.

19.7 *Fusarium* Wilt and Dry Rot

A study on the problems caused by *Fusarium* began with an investigation on the rotting of potatoes by Martius in 1840–1841, who found the causal organism to be a fungus which he called *Fusisporium solani* that was later transferred to *Fusarium* as *Fusarium solani* (Mart.) Sacc (Saccardo 1882). In India, Ajrekar and Kamat (1923) reported that *Fusarium coeruleum* affects potato. Padwick (1943) isolated *Fusarium solani* from rotting tubers at Shimla. Afterwards, several species of *Fusarium* are known to cause dry rot of potato, nine of which were reported from different parts of India (Singh et al. 1987). In India, the first report of dry rot caused by *F. sambucinum* was documented by Sagar et al. (2011). *Fusarium* wilt and dry rot has been reported in China, Tunisia, Egypt, the Great Britain, South Africa, Canada, Australia, the USA, Iran, and Poland.

19.7.1 Symptoms

In dry rot, the skin of infected tubers first becomes brown, then turns darker, and develops wrinkles. These wrinkles are often irregular concentric circles. In later stage, a hole may be observed in the center of ring with whitish or pinkish growth with one or more cavities (Fig. 19.6). At wilting stage, lower leaves turn yellow and affected plant dried off of fungal mycelium. After cutting the affected tubers, whitish to brownish colored tissues are visible. Sometimes partial stem infection is also observed where leaf symptoms may appear only on one side of the infected plants. Both stems and tubers at stolon end show vascular browning. Moreover, internal flecking of stem extending to upper leaves is also observed. Sometimes, damping off seedling type symptoms are also observed when temperature is high at early planting stage. Other symptoms like stem rot, damping off of seedlings, spots and necrosis on tubers, and seed pieces decay are also reported due to different *Fusarium*.

Fig. 19.6 Symptoms of dry rot in potato tubers

19.7.2 Epidemiology

Fusarium spp. are considered as both seed- and soilborne phytopathogen. Infected tubers and field soil are the primary source of inoculum. In general, the fungus remains viable in soil for 9–12 months. However, its resting structure (chlamydospores) can survive in soil for several years. *Fusarium* spp. have good saprophytic ability to survive in soil. The fungus grows well between 15 °C and 28 °C, and high humidity favors infection of tubers, and also congenial for secondary organisms such as *Erwinia* spp. can invade the infected tubers and cause soft rot. Infection of tubers occurs through wounds produced during harvesting operations, and dry rot develops slowly in storage. Temperature > 10 °C favors *Fusarium* growth, while temperature < 5 °C inhibits fungal growth. The pathogenicity of

Fusarium species varies significantly (Peters et al. 2008) with the potato cultivar and temperature during inoculation (Esfahani 2005). *Fusarium* wilt of potato is mainly affected by soil temperature and relative humidity. High wilt incidence in early planted crop is mainly associated with high temperature (Singh et al. 1990). The production of fusaric acid is also correlated with virulence of different *Fusarium oxysporum* (Yenter Sonja and Steyn 1998). Positive correlation was reported among thumb nail injury, wet rot, and *Fusarium* dry rot (Kumar et al. 2021).

19.7.3 Economics

Fusarium wilt and dry rot are caused by *Fusarium* spp. The wilt is caused under field condition and dry rot mainly at postharvest stages. Dry rot of seed tubers can reduce crop establishment by affecting the development of potato sprouts, decaying seed pieces, and causing crop losses up to 25% and occasionally losses up to 60% during long-term storage (Desjardins 2006; Wharton and Kirk 2007). In Tunisia, *Fusarium* wilt was reported to cause losses estimated at 30–50% of potato yield and decreased tuber quality (Kerkeni et al. 2013). Dry rot mainly occurs in storage condition, which causes 5–23% storage loss in plains (Sharma and Lal 2015), whereas wilt disease causes up to 19% losses under field condition in Western India; however, it causes 25–35% yield loss in highly infested field (Singh 2002). Recently, wide variation (0–90%) of *Fusarium* rot was recorded in seed lots of potato in Punjab (Kumar et al. 2016).

19.7.4 Management

Sanitation Use only clean and healthy seed tubers for planting and storage. The tuber damage and injury must be avoided during harvest, grading, transport, storage, etc. Adhering of soil on tubers must be avoided during harvesting. Washing of tubers to remove contaminated soil which adhere to the surface, besides, dry in shade can reduce the risk of infection. Curing of the seed tubers for 7–12 days at warm condition with dry atmosphere is suitable for wound healing. As far as possible, avoid cut tubers for planting because such tubers may get infected under infested soil.

Shallow planting and adjustment date of planting Deep planting should be avoided because it may cause more damage of the seed tubers. By adjusting 1 month delaying date of planting, *Fusarium* wilt can be reduced up to 36% disease incidence (Singh et al. 1990).

Crop rotation and soil solarization It will be better to follow longer (more than 3 years) crop rotation. Because 1or 2 years rotation had not shown significant results in managing *Fusarium* diseases. Crop rotations with Italian ryegrass, red clover, barley, or Italian red clover did not show significant reduction of dry rot diseases (Carter et al. 2003). In another study also, 3-year rotations with red clover, barley, and potato did not reduce significantly the severity of dry rot in 2 of the 3 years observed (Peters et al. 2004).

Soil solarization can be utilized to reduce the inoculum of soilborne pathogens. This process harnesses solar energy to increase soil temperature of moistened soil by covering soil with plastic films. Soil solarization minimizes the inoculum level of *Fusarium* spp. after 6 weeks of treatment (Saremi et al. 2011).

Soil amendments and botanicals Immature crop plant amendments, viz., pearl millet, sesbania, sunhemp, maize, and eucalyptus leaves, are used against *Fusarium* wilt of potato. Among these, eucalyptus leaves and maize show maximum suppression, and least was for sesbania. The groundnut cake was most effective than mustard cake and cotton seed cake for reducing the buildup of *Fusarium* wilt (Singh et al. 1988). Methanolic extracts of different plant species (eucalyptus, datura, thyme, lavender) revealed higher efficacy against *F. solani*, whereas aqueous extracts of these plant species showed less efficacy under lab and storage condition (Zaker 2014). Garlic and clove extracts (10%) were highly effective against *F. solani* under laboratory conditions (Awad et al. 2020).

Hot water treatment Artificially, wounded potato tubers may be dipped into hot water at 45 °C for 10 min. It was observed that hot water dipping was effective for wound healing of these tubers, thereby reducing weight loss and minimizing dry rot losses (Yanga et al. 2020).

Biological control The combined effect of antagonists *(Trichoderma* and *Pseudomonas)* with modified montmorillonite particles (Mod- MMT) against *Fusarium oxysporum* f. sp. *tuberose* showed less disease incidence and also enhanced plant height, fresh and dry weight, number of tubers/plant, and weight of tubers (Abeer and Makhlouf 2015). Application of *T. koningii* and *B. megaterium* alone or in combination 7 days earlier than soil infestation with *F. oxysporum* and/or the mixed population of *Meloidogyne* spp. significantly reduced *Fusarium* wilt disease incidence and nematode infection on potato and improved plant growth components under greenhouse condition. Generally, the mixture of the two biocontrol agents was more effective in controlling the plant disease and improving plant growth components than either of the two organisms used singly (El-Shennawy et al. 2012). The fungi *Aspergillus*, *Penicillium*, *Trichoderma*, and *Colletotrichum* showed positive response under in vitro against *F. sambucinum* and *F. solani.* These bioagents were isolated from roots, stems, and tubers of healthy plants (Trabelsi et al. 2016).

Varietal resistance Varieties like Baraka, Asterix, Alaska, Safrane, and Timate have some resistance against *Fusarium oxysporum* f. sp. *tuberose* in Tunisia (Ayd et al. 2006). The cultivar "Owyhee Russet" showed significantly higher resistance to dry rot than "Russet Burbank." The cultivar Saturna and Frontier Russet and clone B-7200-33 are also reported as resistant and immune against *Fusarium* spp., respectively (Angelique et al. 2013).

19.8 Potato Wart

Potato wart is caused by *Synchytrium endobioticum* (Schilb) Perc. Potato wart was first reported in Trentschen, Slovakia, in Czechoslovakia in 1895 (Schilberszky 1896). Then it was reported to other countries. In India, wart disease of potato was first reported by Ganguly and Paul (1952) from Darjeeling hills, and it continues to be endemic to that area. It is a quarantine disease. The disease is known to many common names as per appearance of the symptoms, black wart, black scab, canker and cancer, cauliflower disease, etc. It was reported in Africa, Asia, Europe, South and North Asia, and New Zealand.

19.8.1 Symptoms

The disease shows cauliflower-like warty growths on tubers, stolons, and stem bases but not roots. The warts on tuber initially appear as small white granular swelling on the eyes. These warts on potato tubers may remain minute or may become as large as tuber. It depends on level of infection, variety, and available soil moisture. Under wet conditions, it may be seen in the form of greenish-yellow excrescences on the stem and leaves at or near the soil level. It is not necessary that all tubers from a diseased plant show wartlike symptoms. Diseased tubers may show either one or more tumors but sometimes are completely transformed into warty mass. Size of warts on tuber may be minute at harvesting time, but it may enlarge in stores.

19.8.2 Epidemiology

Wart disease is seed- and soilborne in nature. The pathogen spreads from one locality to another through infected seed tubers, infested soil adhering tubers, machinery, and other carriers of contaminated soil. The wart is favored by periodic flooding followed by drainage and aeration since free water is required for germination of sporangia and dispersal of zoospores of the pathogen. The resting sporangia are thick walled and may remain viable in soil for almost three to four decades. The resting sporangia may germinate over a wide range of temperature, the optimum being between 14 °C and 24 °C. The optimum temperature for wart development is found to be from 16.7 °C to 17.8 °C (Dutt 1979).

19.8.3 Management

Resistant varieties Host resistance is the best option to manage this disease. Wart-immune varieties, viz., "Kufri Jyoti," "Kufri Bahar," "Kufri Sherpa," "Kufri Kanchan," "Pimpernel," and "Aeckersegen," should be grown.

Quarantine Introduction of the disease in a field or locality can be effectively checked by strict quarantine legislation. Many countries have made possible to confine the disease with strictly enforcing quarantine measures.

Crop rotation Disease is both soil- and tuber-borne in nature. Therefore, application of long-term crop rotation (5 years or more) with non-solanaceous crops preferably maize, radish, cabbage, and pea would be helpful in minimizing the disease.

Agronomics Diseased seed tubers should not be used for planting. Rogue out plant of susceptible varieties. Warted lumps and potato peelings should not be thrown in the field or in the manure pit but destroyed by burning.

Soil amendments Infested soil needs to be amended with crushed crab shell for minimizing wart severity.

19.9 Silver Scurf

Silver scurf disease was recorded in Europe in 1871 and in Ireland in 1903 (Mckay 1955). After that, it was reported in Denmark, England, Brazil, the USA, India, Canada and Australia. In India, it was reported since 1962 in Nilgiris and in Darjeeling district (Srivastva 1965).

19.9.1 Symptoms

This disease does not affect foliage of the potato plant except stolons and tubers. Lesions could be seen on stolons after tuber initiation. On tuber skin, blemishes appear which start as small, round, silvery patches on the skin. When moistened the tuber lesions often appear as very clear silvery patches. These patches expand and merge during storage. Silver scurf does not usually cause any yield losses at harvest, but it does increase the permeability of the tuber skin, which leads to water losses and shrinkage during storage leading to weight losses reaching up to 17% (Read and Hide 1984).

19.9.2 Causal Organism and Epidemiology

Silver scurf is caused by *Helminthosporium solani*. Both tubers and soil may serve as primary sources of inoculum. The disease is favored by 12–26 °C along with 95% humidity. Symptoms are not normally present at harvest, but the disease can develop rapidly in store under humid, warm (>3 °C) conditions. The infection can spread from diseased to healthy tubers under storage. The disease is more common in sandy soils and red color varieties.

19.9.3 Management

Biological Control The infection of *H. solani* was reduced by fungal bioagent, *Clonostachys rosea* (*Gliocladium rosea*). A combination of different mechanisms, i.e., mycoparasitism, biocontrol-activated stimulation of plant defense mechanisms, microbial competition for nutrients, space, and antibiosis, etc., is involved to minimize silver scurf disease (Lysøe et al. 2017). Phosphorus acid-based products are effective to manage silver scurf of potato when applied at low to medium level of infection at postharvest (Hamm et al. 2013).

Crop Rotation Three years rotation with nonhost crop will reduce inoculum in the soil; subsequently infection would be reduced (Hamm et al. 2013).

Sanitation Potato seeds should be free from infection. It is essential to maintain hygienic condition in storage and avoid condensation of tuber surface for longer period.

Agronomics After haulms cutting and maturing of the skin of the tubers, harvesting should be followed. Harvested tubers should be shade dried before storage to reduce chance of infection.

19.10 Powdery Scab

This disease is sometimes known as corky scab. It is found mainly in cool and wet climates. Powdery scab was first reported as a disease in Germany in 1841 (Harrison et al. 1997). In India, it is mainly found in the higher hills specially Kumaon, Himalayas, Darjeeling, and Nilgiri (Ootacamund). It is also reported in Australia, Africa, America, Columbia, Japan, New Zealand, Russia, the UK, Pakistan, and Korea.

19.10.1 Symptoms

This disease attacks only the underground parts of the potato plants and does not show any effect on the growth of the plant. The underground parts include roots, stolons, tubers, and newly emerged shoots. On roots and stolons, small gall formation takes place, which is confused sometimes with symptoms of root knot nematode. Pimple-like spots appears on the surface of young tubers. These spots are circular, smooth, and light brown which gradually increase in size and later turn to scab-like lesions. However, unlike common scab, the lesions of powdery scab are round, raised, filled with powdery mass of spores, and surrounded by ruptured remains of the epidermis. Under certain conditions, wartlike protuberances may develop. Sometimes canker-like symptoms are also observed; whenever the eyes of the tubers are infected by the zoospores, canker formation takes place.

19.10.2 Causal Organism and Epidemiology

Powdery scab caused by *Spongospora subterranea* f. sp. *subterranea.* It is soilborne and obligate biotrophic pathogen. *Spongospora subterranea* f. sp. *subterranea* has both diploid and haploid phase in life cycle. The spore balls of pathogen on the tubers as well as in the soil serve as a source of infection. It can also survive in soil up to 10 years. The temperature below 18 °C and high soil water content favor the development of the disease. The infected root/stolons galls, which contain sporosori, are released into the soil. The pathogen also acts as the vector of potato mop-top virus (Harrison et al. 1997). The disease is more severe in heavy soil than the light soils.

19.10.3 Management

Cultural management By manipulating soil temperature during tuber initiation using plant covering with nonwoven fabric minimizes powdery scab on potato tubers up to 93%. In this process an increased average minimum and maximum soil temperature of 1.8 °C and 4.2 °C was achieved during experimentation (Tsror et al. 2020). Generally, tuber initiation phase is considered as the susceptible phase.

Sanitary measures Farm implements and container should be avoided from disease-affected areas, because they are sources to spread of spore ball/contaminated soil. Moreover, rotted tubers should not be kept in the manure pit, and also manures from animal fed with affected tubers should be avoided.

Disease-free seed It is essential to use of healthy seeds for planting; otherwise after planting diseased tubers, the inoculum level in the soil will be increased.

Drainage of field The disease can be managed by proper drainage facility in the fields because high moisture is conducive for powdery scab disease.

Crop rotation It was reported that crop rotation with *Brassica* crops (Indian mustard and rye grass) has been effective for minimizing incidence and severity of powdery scab (O'Brien and Milroy 2017). Growing non-solanaceous hosts in longer crop rotation also minimizes the disease up to certain extent.

Biological management During 3 years experimentation in Hokkaido, a fungus (*Aspergillus versicolor* Im6-50) was found effective to suppress pathogen *Spongospora subterranea* f. sp. *subterranea* with a protection value of 54–70%, when mycelia were applied directly on seed tubers, compared with a protection value of 77–93% by fluazinam (Nakayama 2021).

19.11 Conclusion and Future Outlook

The potato crop is an important vegetable crop in India and the world. It is directly utilized in vegetables and other processing products. Therefore, it would be better if we can use minimum chemical-based management strategies for managing fungal diseases of potato crop. In above chapter a comprehensive bio-intensive integrated management strategy for potato fungal diseases has been discussed. Bio-intensive management strategies enable management of diseases, besides maintaining soil health and ecological balance of microbes, which is helpful in sustaining the better crop yields with nutritious food.

References

Abdullah MT, Ali NY, Suleman P (2008) Biological control of *Sclerotinia sclerotiorum* (lib.) de Bary with *Trichoderma harzianum* and *bacillus amyloliquefaciens*. Crop Prot 27:1354–1359

Abeer H, Makhlouf RA (2015) Biological and nanocomposite control of *fusarium* wilt of potato caused by *fusarium oxysporum* f. sp. *tuberose*. Global J Biol Agric Health Sci 4:151–163

Adebayo OSI, Ekpo EJA (2001) Effects of organic amendments on tomato diseases caused by Ralstonia solanacearum and fusarium oxysporum f. *sp. lycopersicum*. In: Tanywa JS, Nampala P, Tusiime G, Osiru M African crop science conference proceeding, p 305–307

Aeron A, Dubey RC, Maheshwari DK et al (2011) Multifarious activity of bioformulated *Pseudomonas fluorescens* PS1 and biocontrol of *Sclerotinia sclerotiorum* in Indian rapeseed (*Brassica campestris* L.). Eur J Plant Pathol 131:81–93

Ajrekar SL, Kamat MN (1923) The relationship of the species of fusarium causing wilt and dry rot of potatoes in Western India. Agric J India 18:515–520

Alam MW, Rehman A, Malik A et al (2021) First repost of white mould of potato caused by *Sclerotinia sclerotiorum* in Pakistan. J Plant Pathol 103:669

Anahosu KH (2001) Integrated management of potato sclerotium wilt caused by *Sclerotium rolfsii*. Indian Phytopathol 54:158–166

Angelique B, Avisa TJ, Sophie P, Russell JT (2013) Management of potato dry rot. Postharvest Biol Technol 84:99–109

Anonymous (2017) Annual Scientific Report. ICAR-Central Potato Research Institute, Shimla, pp 60–61

Anonymous (2019) Annual Scientific Report. Central Potato Research Institute, Shimla, India, p 68

Anquiz R, Martin C (1989) Anastomosis groups, pathogenecity, and other characteristics of *Rhizoctonia solani* isolated from potatoes in Peru. Plant Dis 77:199–201

Arora RK (2008) Management of black scurf of potato with the integrated use of *Trichoderma viride* and boric acid. Potato J 35:130–133

Arora RK, Trehan SP, Sharma J, Khanna RN (1997) Soil solarization for improving potato health and production. In: Golden Jubilee International Conference on Integrated Plant Disease Management for Sustainable Agriculture Proceedings Indian Phytopathological Society, vol. II, pp 1190–1191

Asghari MA, Mayee CD (1991) Comparative efficiency of management practices on stem and pod rots of groundnut. Indian Phytopathol 44:328–332

Atallah Z, Johnson DA (2004) Development of *Sclerotinia* stem rot in potato fields in south-Central Washington. Plant Dis 88:419–423

Atiq M, Karamat A, Khan NA et al (2014) Antifungal potential of plant extracts and chemicals for the management of black scurf disease of potato. Pakistan J Phytopathol 26:161–167

Awad MA, Amer GA, Farag AH (2020) Control of potato tuber dry rot disease during storage. Menoufia J Plant Prot 5:169–183

Ayad D, Leclerc S, Hamon B et al (2017) First report of early blight caused by Alternaria protenta on potato in Algeria. Plant Dis 101:836

Ayd F, Dammi-Remadi M, Jabnoun-Khiareddine H, El Mahjoube M (2006) Effect of potato cultivars on incidence of *fusarium oxysporum f.sp tuberosi* and its transmission to progeny tubers. J Agron 5:430–434

Bains PS, Bisht VS (1995) Anastomosis group identity and virulence of *Rhizoctonia solani* isolates collected from potato plants in Alberta, Canada. Plant Dis 79:241–242

Baker KF (1970) Types of *Rhizoctonia* disease and their occurrence. In: Parmeter JR (ed) *Rhizoctonia solani*: biology and pathology. University of California Press, Berkeley, CA, pp 125–148

Baker KF, Cook RJ (1974) Biological control of plant pathogens. American Phytopathological Society, St. Paul, MN, pp 35–50

Balali GR, Neate SM, Scott ES et al (1995) Anastomosis group and pathogenecity of isolates of *Rhizoctonia solani* from potato crops in south Austraila. Plant Pathol 44:1050–1057

Bandy BP, Leach SS, Tavantzis SM (1988) Anastomosis group 3 is the major cause of *Rhizoctonia* disease of potato in Maine. Plant Dis 72:596–598

Banville GJ (1978) Studies on the *Rhizoctonia* disease of potatoes. Am Potato J 55:56

Bardin SD, Huang HC (2001) Research on biology and control of *Sclerotinia* diseases in Canada. Can J Palnt Pathol 23:88–98

Baswaraj R (2005) Studies on potato wilt caused by *Sclerotium rolfsii* Sacc., m.Sc. (Agri.) thesis, University of Agricultural Sciences, Dharwad

Beukema HP, van der Zang DE (1990) Introduction to potato production, Centre for Agriculture Publishing and Documentation, The Netherlands

Binyam T, Temam H, Tekalign T (2014) Efficacy of reduced dose of fungicide sprays in the management of late blight (*Phytophthora infestans*) disease on selected potato (*Solanum tuberosum* L.) varieties Haramaya, eastern Ethiopia. J Biol Agric Healthcare 4:46–52

Bolton MD, Thomma BPHJ, Nelson BD (2006) Sclerotinia sclerotiorum (lib.) de Bary: biology and molecular traits of a cosmopolitan pathogen. Mol Plant Pathol 7:1–16

Bonanomi G, Antignani V, Pane C, Scala F (2007) Suppression of soilborne fungal diseases with organic amendments. J Plant Pathol 89:311–324

Borrero C, Trillas MI, Ordovás J, Tello JC, Avilés M (2004) Predictive factors for the suppression of *fusarium* wilt of tomato in plant growth media. Phytopathology 94:1094–1101

Bourke A (1993) The visitation of god? The potato and the great Irish famine. Lilliput Press, Dublin, Ireland

Brewer MT, Larkin RP (2005) Efficacy of several potential biocontrol organisms against *Rhizoctonia solani* on potato. Crop Protect 24:939–950

Bulluck LR, Ristaino JB (2002) Effect of synthetic and organic soil fertility amendments on southern blight, soil microbial communities, and yield of processing tomatoes. Phytopathology 92:181–189

Buyer JS, Roberts DP, Russeck-Cohen E (1999) Microbial community structure and function in the spermosphere as affected by soil and seed type. Can J Microbiol 45:138–144

Campion C, Chatot C, Perraton B, Andrivon D (2003) Anastomosis groups, pathogenicity and sensitivity to fungicides of *Rhizoctonia solani* isolates collected on potato crops in France. Eur J Plant Pathol 109:983–992

Campo Arana RO, Zambolim L, Costa LC (2007) Potato early blight epidemics and comparison of methods to determine its initial symptoms in a potato field. Rev Fac Nac Agron Medellin 60: 3877–3890

Carling DE, Kebler KM, Leiner RH (1986) Interactions between *Rhizoctonia solani* AG-3 and 27 plant species. Plant Dis 70:577–578

Carling DE, Leiner RH, Westphale PC (1989) Symptoms, signs and yield reduction associated with *Rhizoctonia* disease of potato induced by tuber-borne inoculum of *Rhizoctonia solani* AG-3. Am Potato J 66:693–701

Carter MR, Kunelius HT, Sanderson JB et al (2003) Productivity parameters and soil health dynamics under long-term 2-year potato rotations in Atlantic Canada. Soil Till Res 72:153–168

Casa-Coila VH, Lehner MDS, Hora Júnior BT et al (2017) First report of *Phytophthora infestans* self-fertile genotypes in southern Brazil. Plant Dis 101:1682–1682

Cattalan AJ, Hartel PG, Fuhrmann JJ (1999) Screening for plant growth-promoting rhizobacteria to promote early soybean growth. Soil Sci Soc Amer J 63:1670–1680

Cessna SG, Sears VE, Dickman MB, Low PS (2000) Oxalic acid, a pathogenicity factor for *Sclerotinia sclerotiorum,* suppresses the oxidative burst of the host plant. Plant Cell 12:2191–2199

Chaudhary S, Lal M, Sagar S et al (2020b) Genetic diversity studies based on morpho-pathological and molecular variability of the *Sclerotinia sclerotiorum* population infecting potato (*Solanum tuberosum* L.). World J Microbiol Biotechnol 36:1–15
Chaudhary S, Sagar S, Kumar M et al (2020c) Molecular cloning, characterization and semi-quantitative expression of endochitinase gene from the mycoparasitic isolates of *Trichoderma harzianum*. Res J Biotchnol 15:40–45
Chaudhary S, Sagar S, Lal M et al (2020a) Biocontrol and growth enhancement potential of *Trichoderma* spp. against *Rhizoctonia solani* causing sheath blight disease in rice. J Environ Biol 41:1034–1045
Chet I, Benhamou N, Haran S (1998) Mycoparasitism and lytic enzymes. In: Harman GE, Kubicek CP (eds) *Trichoderma* and *Gliocladium*, vol 2. Taylor and Francis Ltd., London, pp 153–169
Chowdappa P, Nirmal Kumar BJ, Madhura S et al (2015) Severe outbreaks of late blight on potato and tomato in South India caused by recent changes in the *Phytophthora infestans* population. Plant Pathol 64:191–199
Daayf F, Adam L, Fernando WGD (2003) Comparative screening of bacteria for biological control of potato late blight (strain US-8), using *in-vitro*, detached-leaves, and whole-plant testing systems. Can J Plant Pathol 25:276–284
Das S, Shah FA, Butler RC et al (2014) Genetic variability and pathogenicity of *Rhizoctonia solani* associated with black scurf of potato in New Zealand. Plant Pathol 63:651–666
Dasgupta MK, Mandal NC (1989) Postharvest pathology of perishables. Oxford and IBH Publisher, New Delhi, p 623
Derbyshire MC, Denton-Giles M (2016) The control of sclerotinia stem rot on oilseed rape (*Brassica napus*): current practices and future opportunities. Plant Pathol 65:859–877
Desjardins AE (2006) *Fusarium* mycotoxins, chemistry, genetics, and biology. American Phytopathological Society, St. Paul, MN
Dijst G (1985) Investigations on the effect of haulm destruction and additional root cutting on black scurf on potatoes. Neth J Plant Pathol 91:153–162
Dijst G, Bouman A, Mulder A, Roosjen J (1986) Effect of haulm destruction supplemented by cutting of roots on the incidence of black scurf and skin damage, flexibility of harvest period and yield of seed potatoes in field experiments. Neth J Plant Pathol 92:287–303
Donn S, Wheatley RE, McKenzie BM et al (2014) Improved soil fertility from compost amendment increases root growth and reinforcement of surface soil on slope. Ecol Eng 71:458–465
Duarte HSS, Zambolim L, Rodrigues FA et al (2014) Field resistance of potato cultivars to foliar early blight and its relationship with foliage maturity and tuber skin types. Trop Plant Pathol 39: 294–306
Dutt BL (1979) Bacterial and fungal diseases of potato. ICAR, New Delhi
Dutta S, Ghosh PP, Kuiry SP (2009) Stem rot, a new disease of potato in West Bengal, India. Aust Plant Dis Notes 4:80–81
Elad Y (2000) Biological control of foliar pathogens by means of *Trichoderma harzianum* and potential modes of action. Crop Prot 19:709–714
Ellis JB, Martin GB (1882) Macrosporium solani E&M. Am Nat 16:1003
Elmore CL, Stapelton JJ, Bell CE, Devay JE (1997) Soil solarization: a non-pesticidal method for controlling diseases, nematodes and weeds, UC DANR pub. 21377 Oakland, pp 10–14
El-Shennawy MZ, Khalifa MM, Ammar EM, Mousa Hafez SL (2012) Biological control of the disease complex on potato caused by root-knot nematode and *fusarium* wilt fungus. Nematol Medit 40:169–172
Emmerling C, Schloter M, Hartmann A, Kandeler E (2002) Functional diversity of soil organisms-a review of recent research activities in Germany. J Plant Nutr Soil Sci 165:408–420
Esfahani MN (2005) Present status of *fusarium* dry rot of potato tubers in Isfahan (Iran). Indian Phytopath 59:142–147
Etesami H, Maheshari DK (2018) Use of plant growth promoting rhizobacteria (PGPRs) with multiple plant growth promoting traits in stress agriculture: action mechanisms and future prospects. Ecotoxicol Environ Saf 156:225–246

Farrokhi-Nejad R, Cromey MG, Moosawi-Jorf SA (2007) Determonation of the anastomosis grouping and virulence of *Rhizoctonia* spp. associated with potato tubers grown in Lincoln, New Zealand. Pak J Biol Sci 10:3786–3793

Fernando WGD, Nakkeeran S, Zhang Y, Savchuk S (2007) Biological control of *Sclerotinia sclerotiorum* (lib.) de Bary by *pseudomonas* and *bacillus* species on canola petals. Crop Prot 26:100–107

Fiers M, Edel-Hermann V, Heraud C et al (2011) Genetic diversity of *Rhizoctonia solani* associated with potato tubers in France. Mycologia 103:1230–1244

Fry WE, Birch PRJ, Judelson HS et al (2015) Five reasons to consider *Phytophthora infestans* a reemerging pathogen. Phytopathology 105:966–981

Ganguly A, Paul DN (1952) Wart disease of potatoes in India. Sci Cult 18:605–606

Gao XN, Han QM, Chen YF et al (2014) Biological control of oilseed rape *Sclerotinia* stem rot *bacillus* subtillis strain Em7. Biocontrol Sci Tech 24:39–52

Garrett KA, Dendy SP (2001) Cultural practices in potato late blight management. In: Complementing resistance to late blight in the Andes, 13-16 February 2001. International Potato Center, Cochabamba, Bolivia

Geraldine AM, Lopes FAC, Carvalho DDC et al (2013) Cell wall-degrading enzymes and parasitism of sclerotia are key factors on field biocontrol of white mold by *Trichoderma* spp. Biol Control 67:308–316

Goodwin SB (1997) The population genetics of *Phytophthora*. Phytopathology 87:462–473

Goodwin SB, Spielman LJ, Matuszak JM et al (1992) Clonal diversity and genetic differentiation of *Phytophthora infestans* populations in northern and Central Mexico. Phytopathology 82:955–961

Gorai PS, Ghosh R, Konra S, Mandal NC (2021) Biological control of early blight disease of potato caused by *Alternaria alternata* EBP3 by an endophytic bacterial strain *bacillus velezensis* SEB1. Biol Control 156:104551

Grau CR, Radke VL (1984) Effects of cultivars and cultural-practices on *Sclerotinia* stem rot of soybean. Plant Dis 68:56–58

Greenberger A, Yogev A, Katan J (1987) Induced suppressiveness in solarized soils. Phytopathology 77:1663–1667

Grogan RG, Abawi GS (1975) Influence of water potential on growth and survival of *Whetzelinia sclerotiorum*. Phytopathology 65:122–128

Gupta J (2016) Efficacy of biocontrol agents against *Phytophthora infestans* on potato. Int J Eng Sci Comput 6:2249–2251

Gurjar RBS, Bansal RK, Gupta RBL (2004) Viability of Sclerotia of *Sclerotium rolfsii* at different depth and duration in soil of Northwest India. J Mycol Plant Pathol 34:558–559

Gutierrez WA, Shew HD (1998) Identification and quantification of ascospores as the primary inoculum for collar rot of greenhouse-produced tobacco seedlings. Plant Dis 82:485–490

Hagan A (2004) Southern blight on flowers, shrubs, and trees. Alabama Cooperative Extension System Publication, ANR-1157

Hamm PB, Johnson DA, Miller JS et al (2013) Silver scurf management in potato. A Pacific northwest extension publication, PNW-596, Revised April 2013, p 1-7

Haq I, Rashid A, Kahn SA (2008) Relative efficacy of various fungicides, chemicals and biochemicals against late blight of potato. J Phytopathol 21:129–133

Harman GE (2007) Overview of mechanisms and uses of *Trichoderma* spp. Phytopathology 96: 190–194

Harrison JG, Searle RJ, Williams NA (1997) Powdery scab disease of potato - a review. Plant Pathol 46:1–25

Hausladen H, Aselmeyer A (2017) Studies about infection of different *Alternaria solani* isolates on *Solanum tuberosum*, *Lycopersicon esculentum* and *Solanum nigrum*. PPO-Special Report no 18: 201

Hauslanden H, Bassler E (2004) Early blight disease in potatoes. What are the causes? Kartaffelbau 6:210–212

Haverkort AJ, Boonekamp PM, Hutten R et al (2016) Durable late blight resistance in potato through dynamic varieties obtained by cisgenesis: scientific and societal advances in the DuRPh project. Potato Res 59:35–66

Haverkort AJ, Struik PC, Visser RGF, Jacobsen E (2009) Applied biotechnology to combat late blight in potato caused by *Phytophthora infestans*. Potato Res 52:249–264

Hicks E, Bienkowski D, Braithwaite M et al (2014) Trichoderma strains suppress Rhizoctonia diseases and promote growth of potato. Phytopathol Mediterr 53:502–514

Honeycutt CW, Clapham WM, Leach SS (1996) Crop rotation and N fertilization effects on growth, yield, and disease incidence in potato. Am Potato J 73:45–61

Horsfield A, Wicks T, Davies K, Wilson D, Paton S (2010) Effect of fungicide use strategies on the control of early blight (*Alternaria solani*) and potato yield. Australas Plant Pathol 39:368–375

Hou XW, Boyetchko SM, Brkic M et al (2006) Characterization of the antifungal activity of bacillus spp. associated with sclerotia from *Sclerotinia sclerotiorum*. Appl Microbiol Biotechnol 72:644–653

Hu X, Roberts DP, Maul JE et al (2011) Formulation of the endophytic bacterium *Bacillus subtilis* Tu-100 suppresses *Sclerotinia sclerotiorum* on oilseed rape and improves plant vigor in field trials conducted at separate locations. Can J Microbiol 57:539–546

Hu X, Roberts DP, Xie L et al (2013) Bacillus megaterium A6 suppresses Sclerotinia sclerotiorum on oil-seed rape in the field and promotes oilseed rape growth. Crop Prot 52:151–158

Huang HC, Erickson RS, Chang C et al (2002) Organic soil amendments for control of apothecial production of *Sclerotinia sclerotiorum*. Plant Pathol Bull 11:207–214

Ito M, Meguro-Maoka A, Maoka T, Akino S, Masuta C (2017) Increased susceptibility of potato to *Rhizoctonia* diseases in potato leafroll virus-infected plants. J Gen Plant Pathol 83:169–172

Jager G, Hekman W, Deenen A (1982) The occurrence of *Rhizoctonia solani* on subterranean parts of wild plants in potato fields. Neth J Plant Pathol 88:155–161

James WC, McKenzie AR (1972) The effect of tuberborne sclerotia of *Rhizoctonia solani* Kühn on the potato crop. Am Potato J 49:296–301

Jansky SHS, Simon R, Spooner DM (2008) A test of taxonomic predictivity: resistance to early blight in wild relatives of cultivated potato. Phytopathology 98:680–687

Johnson DA, Atallah ZK (2014) Disease cycle, development and management of *Sclerotinia* stem rot of potato. Am J Plant Sci 5:3717–3726

Jones EE, Rabeendran N, Stewart A (2015) Biocontrol of *Sclerotinia sclerotiorum* infection of cabbage by *Coniothyrium minitans* and *Trichoderma* spp. Biocontrol Sci Tech 24:1363–1382

Jones JB, Jones JP, Stall RE, Zitter TA (1993) Compendium of tomato diseases. American Phytopathological Society, St. Paul, MN

Kamoun S, Furzer O, Jones JD et al (2014) The top 10 oomycete pathogens in molecular plant pathology. Mol Plant Pathol 16:413–434

Kapsa J (2007) Application of the Burkard spore trap to determine a composition of the genus *Alternaria* in potato crops. Biuletyn-Instytutu Hodowli-i- Aklimatyazacji-Roslin 244:223–229

Kassa B, Sommartya T (2006) Effect of intercropping on potato late blight, *Phytophthora infestans* (Mont.) de Bary development and potato tuber yield in Ethiopia. Kasetsart J (Nat Sci) 40:914–924

Kaur J, Munshi GD, Singh RS, Koch E (2005) Effect of carbon source on production of lytic enzymes by the sclerotial parasites *Trichoderma atroviride* and *Coniothyrium minitans*. J Phytopathol 153:274–279

Keijer J, Houterman PM, Dullemans AM, Korsman MG (1996) Heterogeneity in electrophoretic karyotype within and between anastomosis groups of *Rhizoctonia solani*. Mycol Res 100:789–797

Kerkeni A, Daami-Remadi M, Khedher MB (2013) In vivo evaluation of compost extracts for the control of the potato fusarium wilt caused by fusarium oxysporum f. sp. tuberosi. Afr J Plant Sci Biotechnol 7:36–41

Khan I, Alam S, Hussain H et al (2016) Study on the management of potato black scurf disease by using biocontrol agents and phytobiocides. J Entomol Zool Stud 4:471–475

Khedher SB, Kilani-Feki O, Dammak M et al (2015) Efficacy of *Bacillus subtilis* V26 as a biological control agent against *Rhizoctonia solani* on potato. C R Biol 338:784–792
Knaus BJ, Tabima JF, Davis CE et al (2016) Genomic analyses of dominant US clonal lineages of *Phytophthora infestans* reveals a shared common ancestry for clonal lineages US11 and US18 and a lack of recently shared ancestry among all other US lineages. Phytopathology 106:1393–1403
Kokalis-Burelle N, Rodriquez-Kabana R (1994) Effects of pine bark extracts and pine bark powder on fungal pathogens, soil enzymatic activity, and microbial populations. Biol Control 4:269–276
Kumar M, Kumar A (2018) Evaluation of efficacy of different organic amendments against *Rhizoctonia solani* under the screen house condition. J Pharma Phytochem 7:191–194
Kumar S, Sekhon PS, Kaur J (2016) Status and etiology of potato dry rot in Punjab under cold store conditions. Potato J 43:182–192
Kumar S, Singh SP, Singh A (2021) Incidence of potato diseases in cold storage and performance of seed lots under field conditions. J Pharma Phytochem 10:477–482
Kumar SS, Rao MRK, Kumar RD et al (2012) Biocontrol by plant growth promoting rhizobacteria against black scurf and stem canker disease of potato caused by *Rhizoctonia solani*. Arch Phytopathol Plant Protect 46:487–502
Kumar V, Chaudhary VP, Kumar D et al (2017) Efficacy of botanicals and fungicides against *Rhizoctonia solani* inciting sheath blight disease on Rice (*Oryza sativa* L.). J Appl Nat Sci 9: 1916–1920
Lal M, Chaudhary S, Rawal S et al (2021) Evaluation of bio-agents and neem based products against late blight disease (*Phytopthora infestans*) of potato. Indian Phytopathol 74:181–187
Lal M, Chaudhary S, Yadav S et al (2019) Development of spray schedules for management of late blight of potato using new chemicals. J Mycol Plant Pathol 49:405–412
Lal M, Yadav S, Chand S et al (2015) Evaluation of fungicides against late blight (*Phytophthora infestans*) on susceptible and moderately resistant potato cultivars. Indian Phytopathol 68:345–347
Lal M, Yadav S, Sharma S et al (2017) Integrated management of late blight of potato. J Appl Nat Sci 9:1821–1824
Lambert DH, Powelson ML, Stevenson WR (2005) Nutritional interactions influencing diseases of potato. Am J Potato Res 82:309–319
Landschoot S, Carrette J, Vandecasteele M et al (2017) Identification of *a. arborescens, A. grandis,* and *A. protenta* as new members of the European *Alternaria* population on potato. Fungal Biol 121:172–188
Larkin RP, Griffin TS (2007) Control of soil-borne potato diseases using *brassica* green manures. Crop Protect 26:1067–1077
Larkin RP, Griffin TS, Honeycutt CW (2010) Rotation and cover crop effects on soilborne potato diseases, tuber yield, and soil microbial communities. Plant Dis 94:1491–1502
Lehtonen MJ, Ahvenniemi P, Wilson PS et al (2008) Biological diversity of *Rhizoctonia solani* (AG-3) in a northern potato-cultivation environment in Finland. Plant Pathol 57:141–151
Li Y, Shen H, Zhou Q, Qian K et al (2017) Changing ploidy as a strategy: the Irish potato famine pathogen shifts ploidy in relation to its sexuality. Mol Plant-Microbe Interact 30:45–52
Liljeroth E, Lankinen Å, Wiik L et al (2016) Potassium phosphite combined with reduced doses of fungicides provides efficient protection against potato late blight in large-scale field trials. Crop Prot 86:42–55
Little SA, Hocking PJ, Greene RSB (2004) A preliminary study of the role of cover crops in improving soil fertility and yield for potato production. Commun Soil Sci Plant Anal 35:471–494
Lootsma M, Scholte K (1996) Effects of soil disinfection and potato harvesting methods on stem infection by *Rhizoctonia solani* Kühn in the following year. Potato Res 39:15–22
Lopez-Mondejar R, Ros M, Pascual JA (2011) Mycoparasitism-related genes expression of *Trichoderma harzianum* isolates to evaluate their efficiency as biological control agents. Biol Control 56:59–66

Lysøe E, Dees MW, May BB (2017) A three-way transcriptomic interaction study of a biocontrol agent (*Clonostachys rosea*), a fungal pathogen (*Helminthosporium solani*), and a potato host (*Solanum tuberosum*). Mol Plant-Microbe Interact 30:646–655

Mac Donald W, Peters R, Coffin R, Lacroix C (2007) Effect of strobilurin fungicides on control of early blight (*Alternaria solani*) and yield of potatoes grown under two N fertility regimes. Phytoprotection 88:9–15

Mane MM, Lal AA, Zghair QN, Simon S (2014) Efficacy of certain bio agents and fungicides against early blight of potato (*Solanum tuberosum* L.). Int J Plant Protect 7:433–436

Manici LM, Caputo F (2009) Fungal community diversity and soil health in intensive potato cropping systems of the East Po valley, northern Italy. Ann Appl Biol 155:245–258

Mckay R (1955) Potato diseases. At the sign of the three candles, Fleet Street, Dublin

McLaren DL, Huang HC, Rimmer SR (1996) Control of apothecial production of *Sclerotinia sclerotiorum* by *Coniothyrium minitans* and *Talaromyces flavus*. Plant Dis 80:1373–1378

McQuilken MP, Mitchell SJ, Budge SP et al (1995) Effect of *Coniothyrium minitans* on sclerotial survival and apothecial production of *Sclerotinia sclerotiorum* in field grown oilseed rape. Plant Pathol 44:883–896

Medina MV, Platt HW (1999) Viability of oospores of *Phytophthora infestans* under field conditions in north eastern North America. Can J Plant Pathol 21:137–143

Meena MC, Meena AK, Meena PN, Meena RR (2018) Management of stem rot of groundnut incited by *S. rolfsii* through important bioagents. Chem Sci Rev Lett 7:1012–1017

Montes MS, Nielsen BJ, Schmidt SG et al (2016) Population genetics of *Phytophthora infestans* in Denmark reveals dominantly clonal populations and specific alleles linked to metalaxyl-M resistance. Plant Pathol 65:744–753

Mrabet M, Djebali N, Elkahouri S et al (2013) Efficacy of selected *pseudomonas* strains for biocontrol of *Rhizoctonia solani* in potato. Phytopathol Medeterr 52:449–456

Mueller DS, Dorrance AE, Derksen RC et al (2002) Efficacy of fungicides on *Sclerotinia sclerotiorum* and their potential for control of sclerotinia stem rot on soybean. Plant Dis 86: 26–31

Mulder A, Turkensteen LJ, Bouman A (1992) Perspectives of green-crop-harvesting to control soil-borne and storage diseases of seed potatoes. Eur J Plant Pathol 98:103–114

Mullen J (2001) Southern blight, southern stem blight, white mold the plant health instructor DOI: https://doi.org/10.1094/PHI-I-2001-0104-01

Muzhinji N, Truter M, Woodhall JW, van der Waals JE (2015) Anastomosis groups and pathogenecity of *Rhizoctonia solani* and Binucleate *Rhizoctonia* from potato in South Africa. Plant Dis 99:1790–1802

Muzhinji N, Woodhall JW, Truter M, van der Waals JE (2014) Elephant hide and growth cracking on potato tubers caused by *Rhizoctonia solani* AG 3-PT in South Africa. Plant Dis 98:570

Nakayama T (2021) Biocontrol of powdery scab of potato by seed tuber application of an antagonistic fungus, *aspergillus versicolor*, isolated from potato roots. J Gen Plant Pathol 83: 253–263

Nowicki M, Foolad MR, Nowakowska M, Kozik EU (2012) Potato and tomato late blight caused by *Phytophthora infestans:* an overview of pathology and resistance breeding. Plant Dis 96:4–17

O'Brien PA, Milroy SP (2017) Towards biological control of *Spongospora subterranea* f. sp. *subterranea*, the causal agent of powdery scab in potato. Australas Plant Pathol 46:1–10

Ojaghian MR, Zhang J, Zhang F et al (2016) Early detection of white mold caused by *Sclerotinia sclerotiorum* in potato fields using real-time PCR. Mycol Prog 15:959–965

Padwick GW (1943) Notes on Indian fungi. III Mycol Pap Mycop Inst 12:15

Paik SB (1989) Screening for antagonistic plants for control of *Phytophthora* spp. in soil. Korean J Mycol 17:39–47

Patel VM, Singh N (2020) Management of black scurf (*Rhizoctonia solani*) of potato through organic approaches. Indian J Agri Res 55:157–162

Pedersen EA, Hughes GR (1992) The effect of crop rotation on development of the septoria disease complex on spring wheat in Saskatchewan. Can J Plant Pathol 14:152–158
Perez W, Forbes G (2010) Potato late blight: technical manual. International Potato Center (CIP), Lima. http://www.cipotato.org/publications/pdf/005446.pdf
Peters RD, Platt HW, Drake KA et al (2008) First report of fludioxonil-resistant isolates of *fusarium* spp. causing potato seed-piece decay. Plant Dis 92:172
Peters RD, Sturz AV, Carter MR, Sanderson JB (2004) Influence of crop rotation and conservation tillage practices on the severity of soil-borne potato diseases in temperate humid agriculture. Can J Soil Sci 84:397–402
Platt HW, Canale F, Gimenez G (1993) Effect of tuber-borne inoculum of *Rhizoctonia solani* and fungicidal seed potato treatment of plant growth and *Rhizoctonia* disease in Canada and Uruguay. Am Potato J 70:553–559
Prabhakaran N, Prameeladevi T, Sathiyabama M, Kamil D (2015) Screening of different *Trichoderma* species against agriculturally important foliar pathogens. J Environ Biol 36: 191–198
Punja ZK (1985) The biology, ecology, and control of *Sclerotium rolfsii*. Annu Rev Phytopathol 23: 97–127
Quentin U (2004) Sclerotinia sclerotiorum, occurrence and control. Kartoffelbau 8:318–319
Rafiq M, Javaid A, Shoaib A (2021) Antifungal activity of methanolic leaf extract of *Carthamus oxycantha* against *Rhizoctonia solani*. Pak J Bot 53:1133–1139
Rahman M, Ali MA, Dey TP et al (2014) Evolution of disease and potential biocontrol activity of *Trichoderma* spp. against *Rhizoctonia solani* on potato. Biosci J 30:1108–1117
Read PJ, Hide GA (1984) Effects of silver scurf (*Helminthosporium solani*) on seed potatoes. Potato Res 27:145–154
Rekad FZ, Cooke DEL, Puglisi I et al (2017) Characterization of *Phytophthora infestans* populations in northwestern Algeria during 2008-2014. Fungal Biol 121:467–477
Remade K (2006) Compost and disease suppression, Organics Factsheet, Scotland, UK
Rodrigues TTMS, Berbee ML, Simmons EG et al (2010) First report of *Alternaria tomatophila* and *A. grandis* causing early blight on tomato and potato in Brazil. New Dis Rep 22:28–28
Rodriguez-Kabana R, Kokalis-Burelle N, Robertson DG et al (1994) Rotations with coastal Bermuda grass, cotton, and bahiagrass for management of *Meloidogyne arenaria* and southern blight in peanut. J Nematol 26:665–668
Rojas A, Kirk WW (2016) Phenotypic and genotypic variation in Michigan populations of *Phytophthora infestans* from 2008 to 2010. Plant Pathol 65:1022–1033
Roy SK, Sharma RC, Trehan SP (2001) Integrated nutrient management by using farmyard manure and fertilizers in potato-sunflower-paddy rice rotation in the Punjab. J Agric Sci 137:271–278
Rubayet MT, Bhuiyan MKA, Hossain MM (2017) Effect of soil solarization and biofumigation of stem rot disease of potato caused by *Sclerotium rolfsii*. Ann Bangladesh Agric 21:49–59
Runno-Paurson E, Loit K et al (2015) Erly blight destroys potato foliage in the northern Baltic region. Acta Agric Scand 65:422–432
Saccardo PA (1882) Sylloge fungorum omnium hucusque congitorum, vol 1. Edwards Brothers, Ann Arbor, MI
Sagar V, Sharma S, Jeevalatha A et al (2011) First report of *fusarium sambucinum* causing dry rot of potato in India. New Dis Rep 24:5
Saremi H, Okhovvat SM, Ashrafi SJ (2011) Fusarium diseases as the main soil borne fungal pathogen on plants and their control management with soil solarization in Iran. Afr J Biotechnol 10:18391–18398
Saville AC, Martin MD, Ristaino JB (2016) Historic late blight outbreaks caused by a widespread dominant lineage of Phytophthora infestans (Mont.) de Bary. PLoS One 11:e0168381
Schiberszky K (1896) Ein neuer Schorfparasit der Kartoffelknollen. Ber Deut Bot Ges 14:36–37
Scholte K (1992) Effect of crop rotation on the incidence of soil-borne fungal diseases of potato. Neth J Plant Pathol 98:93–101

Scholte K, Lootsma M (1998) Effect of farmyard manure and green manure crops on populations of mycophagous soil fauna and *Rhizoctonia* stem canker of potato. Pedobiologia 42:223–231
Scotti R, D'Ascoli R, Caceres MG et al (2015) Combined use of compost and wood scraps to increase carbon stock and improve soil quality in intensive farming systems. Eur J Soil Sci 66: 463–475
Selin C, Habibian R, Poritsanos N et al (2010) Phenasines are not essential for *pseudomonas chlororaphis* PA23 biocontrol of *Sclerotinia sclerotiorum*, but do play a role in biofilm formation. FEMS Microbiol Ecol 71:73–83
Sennoi R, Singkham N, Jogloy S et al (2013) Biological control of southern stem rot caused by *Sclerotium rolfsii* using *Trichoderma harzianum* and arbuscular mycorrhizal fungi on Jerusalem artichoke (*Helianthus tuberosus* L.). Crop Protect 54:148–153
Shanthiyaa V, Saravanakumar D, Rajendran L et al (2013) Use of *Chaetomium globosum* for biocontrol of potato late blight disease. Crop Prot 52:33–38
Sharma S (2015) Black scurf. In: Singh BP, Nagesh M, Sharma S et al (eds) A manual on diseases and pest of potato, Tech Bull No. 101. ICAR-Central Potato Research Institute, Shimla, India, pp 11–13
Sharma S, Lal M (2015) Dry rot. In: Singh BP, Nagesh M, Sharma S et al (eds) A manual on diseases and pest of potato-Technical Bulletin No. 101. ICAR-Central Potato Research Institute, Shimla, India, pp 17–19
Shtienberg D, Blachinsky D, Ben-Hador G, Dinoor A (1996) Effects of growing season and fungicide type on the development of *Alternaria solani* and on potato yield. Plant Dis 80: 994–998
Shtienberg D, Raposo R, Bergerson SN et al (1994) Inoculation of cultivar resistance reduced spray strategy to suppress early and late blight on potato. Plant Dis 78:23–26
Singh BP, Bhattacharya SK, Saxena SK, Nagaich BB (1990) Managing *fusarium* wilt of potato by adjusting date of planting. J Indian Potato Assoc 17:75–77
Singh BP, Nagaich BB, Saxena SK (1987) Fungi associated with dry rot of potatoes, their frequency and distribution. Indian J Plant Pathol 5:142–145
Singh BP, Nagaich BB, Saxena SK (1988) Studies on the effect of organic amendments on *fusarium* wilt of potato. J Indian potato Assoc 15:60–67
Singh PH (2002) Training course on research methodology in potato: In: Identification and management of fungal diseases. p 146
Smolinska U, Kowalska B, Kowalczyk W et al (2016) Eradication of *Sclerotinia sclerotiorum* sclerotia from soil using organic waste materials as *Trichoderma* fungi carriers. J Horticultural Res 24:101–110
Sneh B, Burpee L, Ogoshi A (1991) Identification of *Rhizoctonia* species. The American Phytopathological Society, St Paul, MN
Sreenivasaprasad S, Manibhusanrao K (1990) Biocontrol potential of the fungus antagonist *Gliocladium wrens* and *Trichoderma longibrachiatum*. Zeitschrift fur Plazenkrankheiten und Plazenschutz 97:570–579
Srivastva SNS (1965) The occurrence of silver scurf of potato in India and its control. Sci Cult 31: 537–538
Steadman JR (1979) Control of plant diseases caused by *Sclerotinia* species. Phytopathology 69: 904–907
Stevens JJ, Jones RK, Shew HD, Carling DE (1993) Characterization of populations of *Rhizoctonia solani* AG-3 from potato and tobacco. Phytopathology 83:854–858
Stevenson WR, Loria R, Franc GD, Weingartner DP (2001) Compendium of potato diseases, 2nd edn. The American Phytopathological Society, St. Paul, MN
Szczech M, Smolińska U (2001) Comparison of suppressiveness of vermicomposts produced from animal manures and sewage sludge against *Phytophthora nicotianae* Breda de Haan var. *nicotianae*. J Phytopathol 149:77–82
Tabassum B, Khan A, Tariq M et al (2017) Bottlenecks in commercialisation and future prospects of PGPR. Appl Soil Ecol 121:102–117

Tariq M, Yasmin S, Hafeez (2010) Biological control of potato black scurf by rhizosphere associated bacteria. Braz J Microbiol 41:439–451

Thaning C, Welch CJ, Borowicz JJ et al (2001) Suppression of *Sclerotinia sclerotiorum* apothecial formation by the soil bacterium Serratia plymuthica: identification of a chlorinated macrolide as one of the causal agents. Soil Biol Biochem 33:1817–1826

Tomar S, Khan MA, Lal M, Singh BP (2019a) Efficacy of biosurfactant producing bacteria (*Pseudomonas aeruginosa*) against black scurf (*Rhizoctonia solani*) of potato. Pesticides Res J 31:126–128

Tomar S, Lal M, Khan MA, Singh BP, Sharma S (2019b) Characterization of glycolipid biosurfactant from *Pseudomonas aeruginosa* PA 1 and its efficacy against *P. infestans*. J Envin Biol 40:725–730

Trabelsi BM, Abdallah RAB, Kthiri Z et al (2016) Assessment of the antifungal activity of non-pathogenic potato-associated fungi toward *fusarium* species causing tuber dry rot disease. J Plant Pathol Microbiol 7:343

Troian RF, Steindorff AS, Ramada MHS et al (2014) Mycoparasitism studies of *Trichoderma harzianum* against *Sclerotinia sclerotiorum:* evaluation of antagonism and expression of cell wall-degrading enzymes genes. Biotechnol Lett 36:2095–2101

Tsahouridou PC, Thanassoulopoulos CC (2002) Proliferation of *Trichoderma koningii* in the tomato rhizosphere and the suppression of damping off by *Sclerotium rolfsii*. Soil Biol Biochem 34:767–776

Tsror L (2010) Biology, epidemiology and management of *Rhizoctonia solani* on potato. J Phytopathol 158:649–658

Tsror L, Barak R, Sneh B (2001) Biological control of black scurf on potato under organic management. Crop Prot 20:145–150

Tsror L, Lebiush S, Hazanovsky M, Erlich O (2020) Control of potato powdery scab caused by *Spongospora subterranea* by foliage cover and soil application of chemicals under field conditions with naturally infested soil. Plant Pathol 69:1070–1082

Turkensteen LJ, Mulder A (1999) The potato disease *Phytophthora infestans*. Gewasbescherming 30:106–112

Van der Waals JE, Korsten L, Aveling TAS (2001) A review of early blight of potato. African Plant Protect 7:91–102

Virgen-Calleros G, Olalde-Portugal V, Carling DE (2000) Anastomosis groups of *Rhizoctonia solani* on potato in Central Mexico and potential for biological and chemical control. Am J Potato Res 77:219–224

Volz A, Tongle H, Hausladen H (2013) An integrated concept for early blight control in potatoes. PPO special report 14:12–15

Wale S, Platt HW, Cattlin N (2008) Diseases, pests and disorders of potatoes-a color handbook. Manson Publishing, London

Wan MG, Li GQ, Zhang JB et al (2008) Effect of volatile substances of *Streptomyces platensis* F-1 on control of plant fungal diseases. Biol Control 46:552–559

Wang H, Ren Y, Zhou J et al (2017) The cell death triggered by the nuclear localized RxLR effector PITG_22798 from *Phytophthora infestans* is suppressed by the effector AVR3b. Int J Mol Sci 18(2):409

Wharton P, Kirk W (2007) *Fusarium* dry rot. www.potatodiseases.org/dryrot.html

Wharton P, Kirk W, Berry D, Snapp S (2007) *Rhizoctonia* stem canker and black scurf of potato. Michigan potato diseases series, MSU extension bulletin E-2994, Michigan State University, Lansing, MI

Whisson SC, Boevink PC, Wang S, Birch PRJ (2016) The cell biology of late blight disease. Curr Opin Microbiol 34:127–135

Wilson PS, Ketola EO, Ahvenniemi PM et al (2008) Dynamics of soilborne *Rhizoctonia solani* in the presence of *Trichoderma harzianum*: effects on stem canker, black scurf and progeny tubers of potato. Plant Pathol 57:152–161

Woodhall JW, Belcher AR, Peters JC et al (2012) First report of *Rhizoctonia solani* AG2-2IIIB infecting potato stem and stolon in the united sates. Plant Dis 96:460

Woodhall JW, Lees AK, Edwards SG, Jenkinson P (2007) Characterization of *Rhizoctonia solani* from potato in Great Britain. Plant Pathol 56:286–295

Woodhall JW, Lees AK, Edwards SG, Jenkinson P (2008) Infection of potato by *Rhizoctonia solani:* effect of anastomosis group. Plant Pathol 57:697–905

Woudenberg JHC, Truter M, Groenewald JZ, Crous PW (2014) Large-spored *Alternaria* pathogens in section Porri disentangled. Stud Mycol 79:1–47

Wu YC, Yuan J, Raza W et al (2014) Biocontrol traits and antagonistic potential of *bacillus amyloliquefaciens* strain NJZJSB3 against *Sclerotinia sclerotiorum,* a causal agent of canola stem rot. J Microbiol Biotechnol 24:1327–1336

Xue W, Haynes KG, Qu X (2019) Characterization of early blight resistance in potato cultivars. Plant Dis 104:629–637

Yadav R, Pathak SP (2011) Management of early blight of potato through fungicides and botanical and bioagents. Plant Arch 11:1143–1145

Yadessa GB, van Bruggen A, Ocho FL (2010) Effects of different soil amendments on bacterial wilt caused by *Ralstonia solanacearum* and on the yield of tomato. J Plant Pathol 92:439–450

Yanar Y, Yilmaz G, Cesmeli I, Coskum S (2005) Characterization of *Rhizoctonia solani* isolates from potatoes in Turkey and screening potato cultivars for resistance to AG-3 isolates. Phytoparasitica 33:370–376

Yang L, Li GQ, Jiang DH, Huang HC (2009) Water assists dissemination of conidia of the mycoparasite *Coniothyrium minitans* in soil. Biocontrol Sci Tech 19:779–796

Yanga R, Hana Y, Hana Z et al (2020) Hot water dipping stimulated wound healing of potato tubers. Postharvest Biol Tech 167:111245

Yao Y, Li Y, Chen Z et al (2016) Biological control of potato late blight using isolates of *Trichoderma.* Am J Potato Res 93:33–42

Yenter Sonja L, Steyn PJ (1998) Correlation between fusaric acid production and virulence of isolates of *fusarium oxysporum* that causes potato dry rot in South Africa. Potato Res 41:289–294

Zaker M (2014) Antifungal evaluation of some plant extracts in controlling *fusarium solani*, the causal agent of potato dry rot *in vitro* and *in vivo*. Int J Agric Biosci 3:190–195

Zegeye ED, Santhanam A, Gorfu D et al (2011) Biocontrol activity of *Trichoderma viride* and *Pseudomonas fluorescens* against *Phytophthora infestans* under greenhouse conditions. J Agric Technol 7:1589–1602

Zeng W, Kirk W, Hao J (2012a) Field management of *Sclerotinia* stem rot of soybean using biological control agents. Biol Control 60:141–147

Zeng WT, Wang DC, Kirk W, Hao JJ (2012b) Use of *Coniothyrium minitans* and other microorganisms for reducing *Sclerotinia sclerotiorum*. Biol Control 60:225–232

Zhang D, Yu S, Yang Y et al (2020) Antifungal effects of volatiles produced by *Bacillus subtilis* against *Alternaria solani* in potato. Front Microbiol 11:1196

Zhu W, Shen L, Fang Z et al (2016) Increased frequency of self-fertile isolates in *Phytophthora infestans* may attribute to their higher fitness relative to the A1 isolates. Sci Rep 6:29428

Zhu W, Yang LN, Wu EJ et al (2015) Limited sexual reproduction and quick turnover in the population genetic structure of *Phytophthora infestans* in Fujian, China. Sci Rep 5:10094

Zimudzi J, Coutinho TA, van der Waals JE (2017) Pathogenecity of fungi isolated from atypical skin blemishes on potatoes in South Africa and Zimbabwe. Potato Res 60:119–144

Use of Green Chemicals in Pest and Disease Management

20

S. Subhash, K. V. Raghavendra, Rekha Balodi, Deepika, and N. K. Dubey

Abstract

To alleviate the biotic stress in crops, farmers rely on the use of synthetic pesticides. The indiscriminate use of synthetic chemical pesticides has posed adverse effects on the beneficial organisms, human beings, and other nontargeted organisms. The use of plant-derived green chemicals is believed to bring some relief to this situation. Besides being safer, green chemicals offer varied modes of action due to the variation in their chemical composition, and unlike synthetic ones, the green chemicals due to their biodegradable nature do not persist in the environment for over longer period of time. Historically, green chemicals such as rotenone, pyrethrum, azadirachtin, veratrines, ryanodine, and nicotine have been used for the management of various insect pests. Green chemicals exhibit a myriad of modes of actions against insects including rapid or slow kill, feeding inhibition, repellents, oviposition deterrent, and growth regulatory effects. Besides insect pests, various green chemicals have been demonstrated to possess antibiotic, antifungal, nematicidal, and herbicidal activities. Although green chemicals have many advantages over the synthetic pesticides, green chemicals suffer from very short residual life due to UV-induced degradation. Many aspects are being explored to increase the usage of green chemicals in IPM in a sustainable manner.

S. Subhash (✉)
ICAR-Central Potato Research Institute-Regional Station, Modipuram, Meerut, Uttar Pradesh, India

K. V. Raghavendra · R. Balodi
ICAR-National Research Centre for Integrated Pest Management, Pusa Campus, New Delhi, India

Deepika · N. K. Dubey
Laboratory of Herbal Pesticides, Department of Botany, Institute of Science, Banaras Hindu University, Varanasi, India

S. Kumar Chakrabarti et al. (eds.), *Sustainable Management of Potato Pests and Diseases*, https://doi.org/10.1007/978-981-16-7695-6_20

Keywords

Biotic stress · Pests · Diseases · Synthetic pesticides · Environmental contamination · Nontarget effects · Rotenone · Pyrethrins · Essential oils

20.1 Introduction

Agriculture plays an important role in developing countries since it is the driving force for broad-based economic growth. Tropical and subtropical regions have a greater potential for food production as they grow multiple crops annually in order to cater the food, fodder, and fiber requirements of the populace. However, pests (insects, nematodes, weeds, and pathogenic microbes) are major biotic constraint to crop production and to ensuring food security. Losses due to pests and diseases including weeds range between 10% and 30% depending upon the genetic makeup of crop and the prevailing environment. In some cases, losses are much higher, producing disastrous results for those who rely on agriculture for their livelihood. To protect the crops from pest attack, farmers generally rely mainly on synthetic pesticides as instant pest management option (Nkechi et al. 2018). The indiscriminate use of synthetic chemical pesticides has posed adverse effects on the beneficial organisms, human beings, and other nontargeted organisms. To overcome these ill effects, the green pesticides or botanical pesticides or plant-based pesticides are identified as safe alternatives over the synthetic chemical pesticides of today (Packiam 2018).

20.2 Impact of Synthetic Pesticides

Synthetic pesticides as well as analogues of natural products are continuously used for the management of pests in agriculture due to lack of environmentally benign, effective, and safe alternatives of biological origin. They have minimized the threat from pest manifestation by rapid knockdown effect on them. Broadly, it has been estimated that hardly 0.1% of the agrochemical used in crop protection reaches the target pest leaving the remaining 99.9% to enter the environment thus causing hazards to nontarget organisms. Many of these effective synthetic pesticides and analogues are responsible for adverse effects on environment, humans, and plant health and crop resistance, which led the ban and or restricted production, sale, and use of various synthetic chemical pesticides.

The indiscriminate use of chemical pesticides over a period of time has not only proven to be harmful to the environment but also contributed to various side effects such as development of resistance to pesticides by pests, resurgence and outbreak of new pests, persistence of pesticide residues on seeds, vegetables, fruits above maximum residue limit (MRL), toxicity to nontargeted organisms, and border alteration in dynamics of pest species population, cumulatively causing hazardous effects on environment endangering the sustainability of ecosystem (Sande et al.

2011). Repeated frequency of application and higher doses of chemical pesticides have also caused approximately about one million people to suffer every year from pesticide poisoning and other chronic diseases. These dreadful facts demand eco-friendly and environmentally safer alternate methods for crop protection.

20.3 Green Chemicals: A Safe and Eco-Friendly Alternative to Synthetic Pesticides

The hazardous effect of synthetic pesticides on environment and other nontargeted organisms has led to a resurgence of interest in pesticides of biological origin due to their less or no effect on environment and living being. Biological origin pesticides, especially extracts and natural substances originating from plant, microorganisms, algae, and animals, are called as *green pesticides*, also called ecological pesticides, which are considered environmentally friendly and are causing less harm to human and animal health and to habitats and the ecosystem are gaining a lot of interest for the integrated management of crop pests and diseases.

Green chemicals, especially essential oils (EOs), have attracted a great deal of attention among consumers because of their biological origin. These eco-friendly chemicals would play a crucial role in minimizing the adverse effects of chemical pesticides and maintaining environmental balance. The varied modes of action of green chemicals are due to the variation in their chemical composition. Unlike synthetic ones, the green chemicals due to biodegradable nature do not persist in the environment for longer period of time. Further, EOs have the potential for being used as natural preservatives because of their remarkable antimicrobial and antioxidant properties. The advantages of green chemicals over gray chemicals are depicted in Fig. 20.1.

Among biological origin pesticides, botanical pesticides rich in flavonoids, alkaloids, glycosides, esters, and fatty acids as well as essential oils and compounds having pest and disease control activities are being utilized for the development of an alternative to chemical compounds. Several plant species have been reported for their pest control activity. About 2300 plant species are reported to possess pest control properties, and additional 1000 plants may have pest control properties because of the poisonous nature of some of their constituents or because of their use in management of pests and diseases. The species generally represent an assortment of plant types, i.e., from aquatic weeds to giant trees, from tropical evergreens to desert succulents, and from highly poisonous to completely edible.

Usually, the pest management properties of these plants are being utilized in two ways: one approach is using plant tissues or crude derivatives, such as an aqueous or organic extract directly, and the second approach is to isolate, identify, and process the active compound and then, if possible, to produce it or its analogues through industrial processes (Grainge and Ahmed 1988). These plant species are found to possess pest management properties due to presence of detrimental proteins and secondary metabolites, which has resulted in identifying some promising plant species as well as molecules/products.

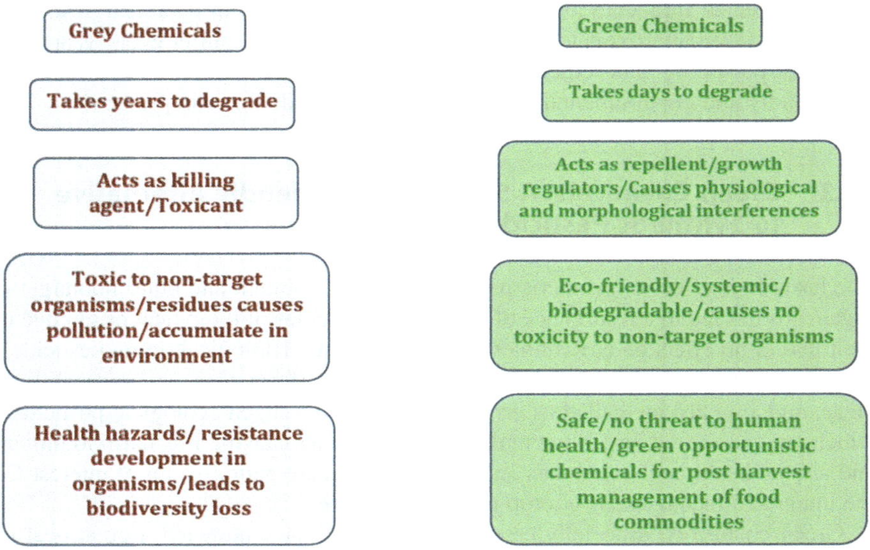

Fig. 20.1 Comparison of the characteristics of gray and green chemicals

20.4 Routinely Used Green Pesticide Compounds of Plant Origin

Plants have evolved a protection mechanism to defend themselves from insect attack in the form of repellents and even insecticidal effects. Plants with bioactive compounds have been used to manage different crop pests with notable success. Commonly used active compounds in insect pest management are described in this section. Prakash and Rao (1997) and Dubey (2011) have given detailed accounts on the subject.

20.4.1 Rotenone ($C_{23}H_{228}O_6$)

Rotenone is an insecticidal flavonoid (Fig. 20.2) extracted from the roots of two plants: *Derris* spp. (Fabaceae) in Asia and *Lonchocarpus* spp. (Fabaceae) in South America and several other related tropical legumes. The first one gives up to 13% of rotenone while the second only about 5%. *Derris* spp. are native to Eastern tropics, while *Lonchocarpus* spp. are native to the western hemisphere. Commercial rotenone was once produced from Malaysian *Derris*. At present, the main commercial source of rotenone is Peruvian *Lonchocarpus*, which often is referred to as cube root. Generally, the *Lonchocarpus* and *Derris* roots are dried, powdered, and mixed directly with an inert carrier to form an insecticidal dust. Rotenone is a contact and stomach poison, which acts as a repellent too. Its mode of action involves the

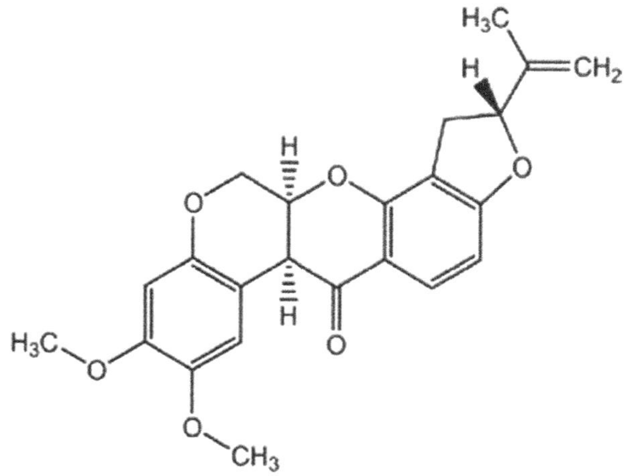

Fig. 20.2 Molecular structure of rotenone

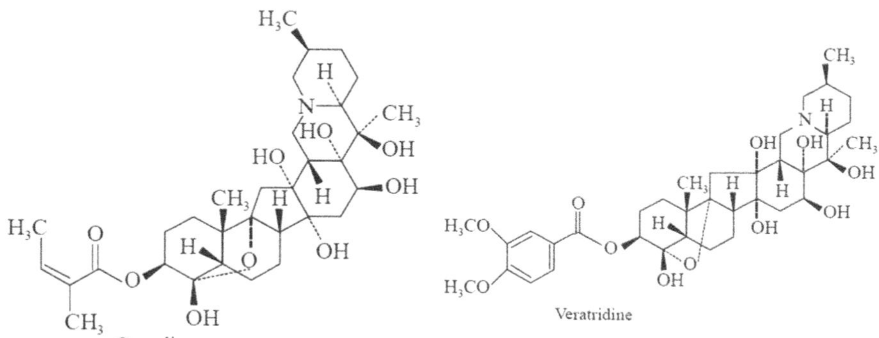

Fig. 20.3 Molecular structure of active ingredients in sabadilla

inhibition of the electron transport at the mitochondrial level, thus blocking phosphorylation of ADP to ATP and thereby inhibiting insect metabolism.

20.4.2 Sabadilla (Veratrine Alkaloids; Cevadine, $C_{32}H_{49}NO_9$, and Veratridine, $C_{36}H_{51}NO_{11}$)

Sabadilla is derived from the ripe seeds of *Schoenocaulon officinale*, a tropical lily plant which grows in Central and South America. Seeds of this plant have been shown to have high concentrations of alkaloids which impart its toxic properties. Sabadilla is also sometimes known as cevadilla or caustic barley. The alkaloids (cevadine and veratridine, Fig. 20.3) in sabadilla are known collectively as veratrine or as the veratrine alkaloids which are most active insecticidal compounds. The mode of action of sabadilla is disruption of neuron cell membranes causing reduction of nerve activity, paralysis, and death. Sabadilla kills insects of some species immediately, while others may survive in a state of paralysis for several days before dying. The ground seeds are one of the plant insecticides with the lowest mammal toxicity, but that is not the case with their isolated alkaloids which are both highly toxic and skin irritants.

20.4.3 Nicotine ($C_{10}H_{16}N_2O_4S$)

Nicotine is an alkaloid (Fig. 20.4) derived from tobacco, *Nicotiana tabacum*, and other *Nicotiana* species belonging to Solanaceae family. The insecticidal property of nicotine was earlier recognized during the sixteenth century. Nicotine is predominantly a nonpersistent contact insecticide, wherein the nicotine is found in the form of maleates and citrates in the tobacco plants. Insecticidal formulations generally contain nicotine in the form of 40% nicotine sulfate and are currently exported in small quantities from India. Nicotine activity causes the production of new nerve impulses which cause convulsions and death. Its mode of action consists in mimicking acetylcholine when it binds with its receptor in the postsynaptic membrane of the muscular union. The acetylcholinic receptor is a site of action of the postsynaptic membrane which reacts with acetylcholine and alters the membrane permeability.

20.4.4 Ryania (Ryanodine, $C_{25}H_{35}NO_9$)

Ryania is obtained from the roots and stems of *Ryania speciosa* (Flacourtiaceae), a plant native to South America. Powdered *Ryania* stem wood is combined with carriers to produce a dust or is extracted to produce a liquid concentrate. From *Ryania speciosa*, a series of alkaloids are obtained, of which the most important active compound is ryanodine (Fig. 20.5). Ryania is a slow-acting stomach poison and effective as both contact and stomach poison. Even though it does not produce rapid knockdown effect, it directly prevents muscles from contraction, causing paralysis of the insect and makes insects to stop feeding soon after ingesting it.

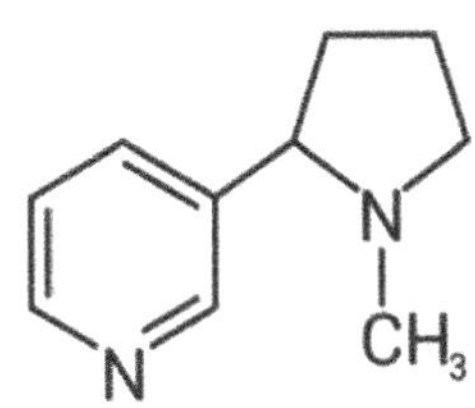

Fig. 20.4 Molecular structure of nicotine

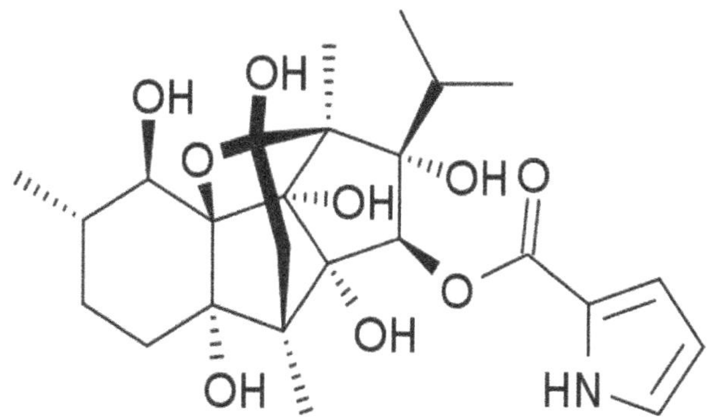

Fig. 20.5 Molecular structure of ryanodine

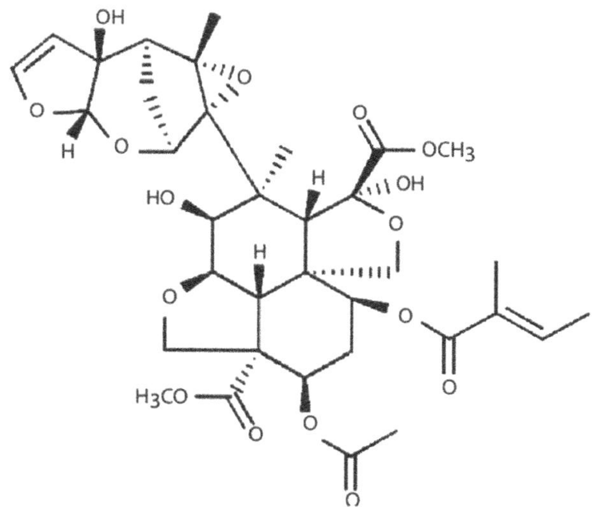

Fig. 20.6 Molecular structure of azadiractin

20.4.5 Azadirachtin ($C_{35}H_{44}O_{16}$)

Azadirachtin (Fig. 20.6) is the principal active compound in neem tree *Azadirachta indica*, which is grown in arid tropical and subtropical regions on several continents, native to India. Azadirachtin is a tetraterpenoid, found in bark, leaves, and fruits of the tree, but seeds have the highest concentration. Azadiractin has not yet been synthesized in the laboratory and in addition to azadiractin, neem tree is also rich in other 17 limonoid compounds, among which azadiractin, salanine, and meliantrol are most prominent, the earlier being in the highest concentration. Azadirachtin is a bitter, complex chemical that is a repellent, feeding deterrent, and a growth regulator. It also exhibits oviposition inhibition and is also a sterilizing compound. As a repellent, neem prevents insects from initiating feeding. As a feeding deterrent, it causes insects to stop feeding. As a growth regulator, neem is thought to disrupt normal development interfering with chitin synthesis. Susceptibility to the various effects of neem differs by species. Today, commercial formulations of neem may be found with names like Neem Gold, Neemazal, Econeem, Neemark, Neemcure, and Azatin, among others, in many countries including the United States, India, Germany, and several Latin American countries.

20.4.6 Pyrethrum and Pyrethrins (Pyrethrin I, $C_{21}H_{28}O_3$; Pyrethrin II, $C_{22}H_{28}O_5$; Cinerin I, $C_{20}H_{28}O_3$; Cinerin II, $C_{21}H_{28}O_5$; Cinerin III, $C_{21}H_{30}O_3$; Jasmolin I, $C_{21}H_{30}O_3$; Jasmolin II, $C_{22}H_{30}O_5$)

Pyrethrum is the powdered, dried seeds of *Chrysanthemum cinerariaefolium*. Most of the world's pyrethrum crop is grown in Kenya. The term "pyrethrum" is the name for the crude flower dust itself, and the term "pyrethrins" refers to the six related insecticidal compounds (esters) that occur naturally in the crude material of chrysanthemum flowers formed by the combination of the acids chrysanthemic and pyrethric acid and the alcohols pyrethrolone, cinerolone, and jasmolone (Fig. 20.7). These compounds act both on the central nervous system and in the peripheral nervous system causing repetitive discharges, followed by convulsions. Pyrethrins exert their toxic effects by disrupting the sodium and potassium ion exchange process in insect nerve fibers and interrupting the normal transmission of nerve impulses. Pyrethrin insecticides are extremely fast acting and cause an immediate "knockdown" paralysis in insects. Despite their rapid toxic action, however, many insects are able to metabolize pyrethrins quickly.

20.4.7 Citrus Oil Extracts (Limonene, $C_{10}H_{16}$, and Linalool, $C_{10}H_{18}O$)

Crude citrus oils and the refined compounds d-limonene and linalool (Fig. 20.8) are extracted from fruit peels belonging to the citrus family. Limonene is a terpene, and linalool is a terpene alcohol predominantly found in citrus peel and in over 200 other herbs, flowers, fruits, and woods. The modes of action of limonene and linalool in insects are not fully understood. Limonene is believed to cause an increased spontaneous activity of sensory nerves in insects resulting in lack of coordination, twitching, and convulsions. Massive overstimulation of motor nerves leads to rapid knockdown paralysis. The central nervous system may also be affected, resulting in additional stimulation of motor nerves. Little has been studied regarding the mode of action of linalool in insects.

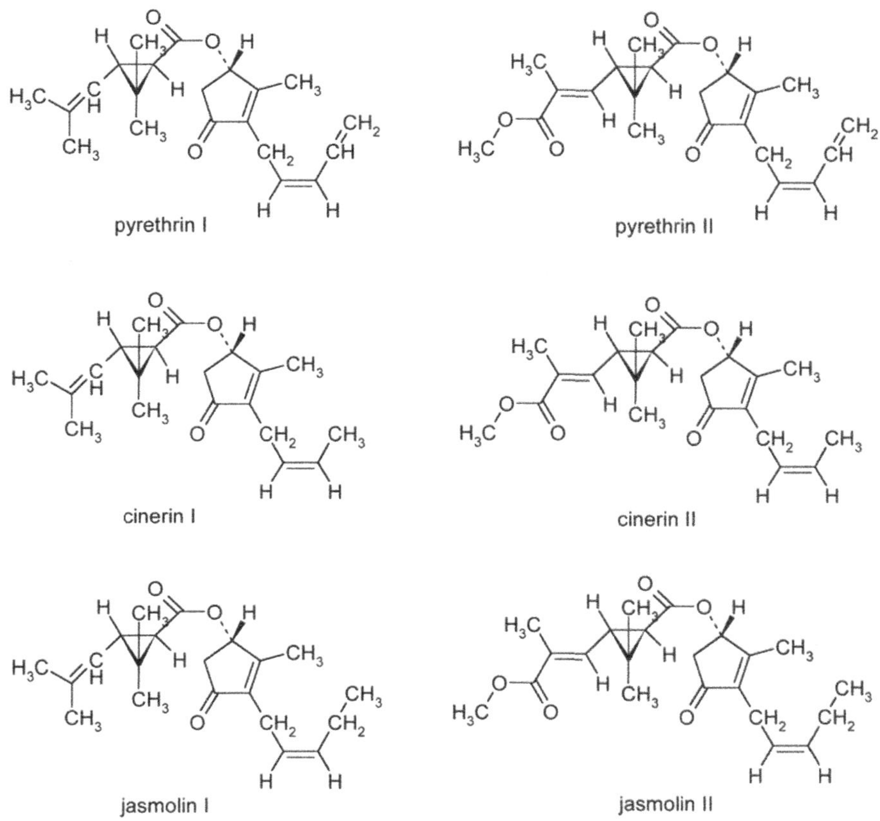

Fig. 20.7 Molecular structure of pyrethrum and pyrethrins

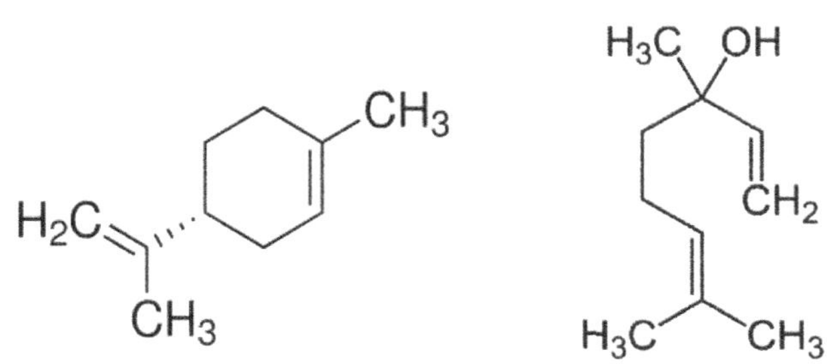

Fig. 20.8 Molecular structure of limonene and linalool

20.5 Potential Application of Green Chemicals

20.5.1 Insecticidal Activity

There are different types of plant-based products such as essential oil, flavonoids, alkaloids, terpenoids, saponins, fatty oil, crude extracts, etc., found in the different parts of the plants and are reported to possess pest and disease control activities. Plants and insects have co-evolved over millions of years; plants have accumulated specific secondary metabolites to counteract insect damage. These bioactive secondary metabolites act as insecticides, antifeedants, insect growth regulators, juvenile hormones, ecdysones, repellents, attractants, arrestants, etc. *Nicotiana tabacum* (Tobacco) is an oldest known pesticidal plant and rich source of nicotine that possesses promising insecticidal activity. Rotenone, a pest control agent identified from the species of the genera *Derris*, *Lonchocarpus*, *Millettia*, and *Tephrosia*, is another well-known insect control agent worldwide. *Chrysanthemum cinerariifolium* (Pyrethrum) flowers are rich source of pyrethrins, which have quick knockdown effect on flying insects. Similarly, azadirachtin from *Azadirachta indica* (neem) is a very good insect repellent and growth regulator. The bioactive extracts/pest control agents of these plant species are now in commercial use in crop fields (Koul 2008; Walia et al. 2014; Unsworth 2020; Rana et al. 2020).

The botanical pesticides (essential oils) have been recognized as safe by US-FDA than synthetic pesticides (Regnault-Roger et al. 2012). Essential oils are obtained from the aromatic plants and used in medicinal, perfumery, and flavoring purposes. In addition, essential oils are also being used as insecticide because of their repellent, insecticidal, antifeedant, growth inhibitor, oviposition inhibitor, ovicidal, and growth-reducing effects on a variety of insects. For example, *Mentha piperita* oil repels ants, flies, lice, and moths and is effective against *Callosobruchus maculatus* and *Tribolium castaneum*. Nepetalactone, the active constituent in catnip (*Nepeta cataria*) essential oil, is highly effective for repelling mosquitoes, bees, and other flying insects. It repels mosquitoes more than DEET. It is particularly effective against *A. aegypti* mosquito, a vector for yellow fever virus (Koul 2008; Unsworth 2020; Rana et al. 2020). Eugenol, a key compound of basil oil and cloves, has a strong mosquito repellency effect. Linalool from basil oil also shows a toxic effect on Bruchid (*Zabrotes subfasciatus*) and other storage insect pests. Essential oils from *Eucalyptus globulus* are toxic to *A. aegypti* worms. *Eucalyptus citriodora* oil is used as a mosquito repellent in Africa. Lemon eucalyptus oil (containing p-menthane-3,8-diol, as an active ingredient) is used for protection against mosquitoes (Walia et al. 2014; Koul 2008; Rana et al. 2020).

20.5.2 Antifungal and Antibacterial Activity

Phytopathogenic fungi are responsible for nearly 30% of all crop diseases (Jain et al. 2019), affecting them during cultivation or postharvest, during storage. These fungi cause high yield losses by damaging host plant, and some of them such as

Aspergillus sp. and *Fusarium* sp. are known to produce mycotoxins. Among all the phytopathogenic fungi, *Alternaria*, *Botrytis*, *Fusarium*, *Penicillium*, and *Rhizoctonia* are the most studied ones. Recent studies showed that *Aspergillus* spp. were found to be susceptible to lemongrass, clove, oregano, and thyme oil but not susceptible to cinnamon and ginger oil (Fig. 20.9) (Božik et al. 2017), while *Penicillium digitatum* was highly affected by thyme and summer savory essential oil and less by fennel and sweet basil ones (Ortiz de Elguea-Culebras et al. 2016). *Mentha piperita* oil showed promising activity against *Rhizoctonia solani* and *Macrophomina phaseolina* but was less effective against *Fusarium oxysporum* (Yangui et al. 2017) and *Penicillium verrucosum*. Lemongrass (*Cymbopogon citratus*) was found efficient against *Colletotrichum gloeosporioides* (Sharma et al. 2017) and *Aspergillus* spp. (Božik et al. 2017). Zerumbone showed promising antifungal activity against three phytopathogenic fungi, namely, *Rhizoctonia solani* (EC_{50} 39.6 ppm), *Sclerotium rolfsii* (EC_{50} 59.3 ppm), and *Macrophomina phaseolina* (EC_{50} 147.4 ppm) compared with hexaconazole (EC_{50} 18.3, 13.4, and 4.5 ppm), respectively (Rana et al. 2017). Antifungal activity of neem nano-emulsion (NNE10) and citronella nano-emulsion (CNE10) was carried out against *Rhizoctonia solani* and *Sclerotium rolfsii*, and results showed that neem nano-emulsion and citronella nano-emulsion were most active against *R. solani* (ED_{50} 13.67 mg/L and 25.64 mg/L) and *S. rolfsii* (ED_{50} 14.71 mg/L and 20.88 mg/L) in in vitro study (Ali et al. 2017a, b).

20.5.3 Nematicidal Activity

Nematicidal compounds identified from the various plant species have been previously reviewed (Chitwood 2002; Ntalli and Caboni 2012). Neem (*Azadirachta indica*), neem seed cake, and leaves and whole plant of *Crotalaria* as raw materials are being used in the management of *Meloidogyne* species. Various bioactive compounds with nematicidal activity have been identified. *Lantana camara* Linn. (family: Verbenaceae), an aromatic plant, is known to contain a number of nematicidal compounds which are reported to be active against *M. incognita*, root knot nematode. Off these, pomolic acid, lantanolic acid, and lantoic acid (Fig. 20.9) showed 100% mortality at 1 mg/mL concentration after 24 h. Its other four compounds, namely, camarin, lantacin, camarinin and ursolic acid, were also found to exhibit 100% mortality at 1 mg/mL concentration after 48 h and were comparable to Furadan (100% mortality at 1 mg/mL concentration after 24 h), a conventional nematicide (Begum et al. 2008). Lantanilic acid, camaric acid, and oleanolic acid (Fig. 20.9) also isolated from *Lantana camara* showed 98%, 95%, and 70% mortality of *M. incognita* at 0.5% concentration compared to Furadan which showed 100% mortality at this concentration (Qamar et al. 2005).

Mucuna pruriens (family: Leguminosae), a medicinal climber, is known to be natural source of 3,4-dihydroxyphenylalanine, commonly known as L-Dopa, used in the treatment of Parkinson's disease. L-DOPA is mainly found in the seeds (6–9%) of *Mucuna* species. *Mucuna aterrima* is reported to have nematicidal activity against *M. incognita* (LC_{50}, 21 μg/mL) and *H. glycines* (LC_{50}, 0.17 μg/mL), respectively

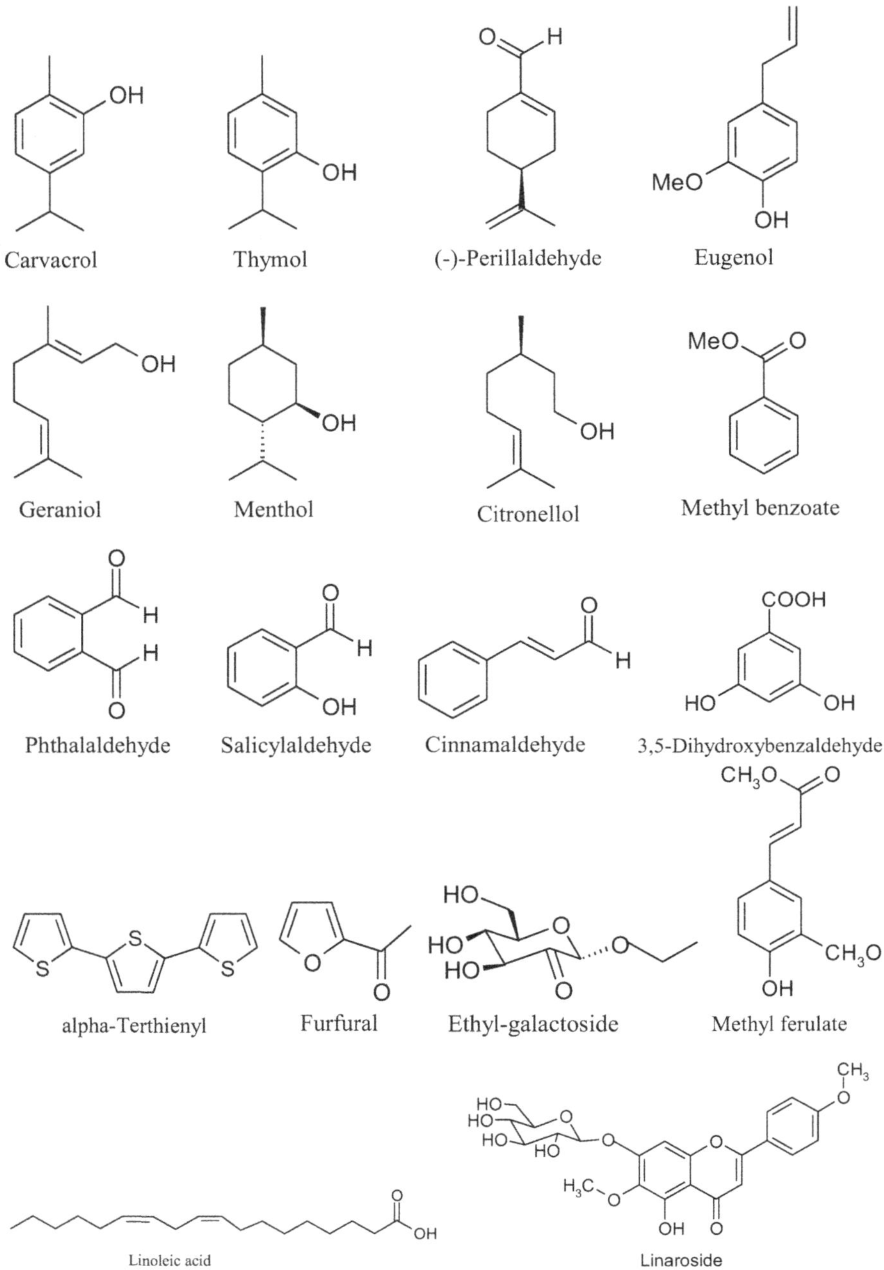

Fig. 20.9 Molecular structures of active ingredients identified from plants known to exhibit biocidal or biostatic effects against insects, fungal pathogens,and nematodes

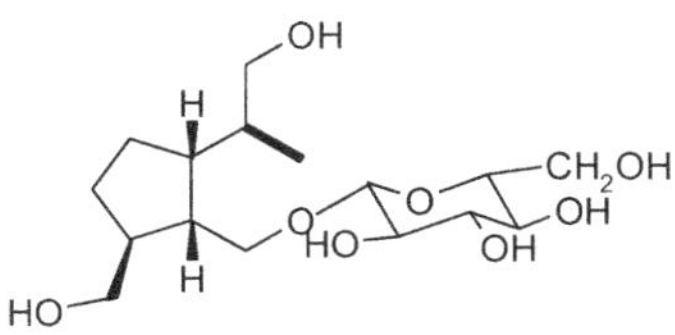

Lantanoside

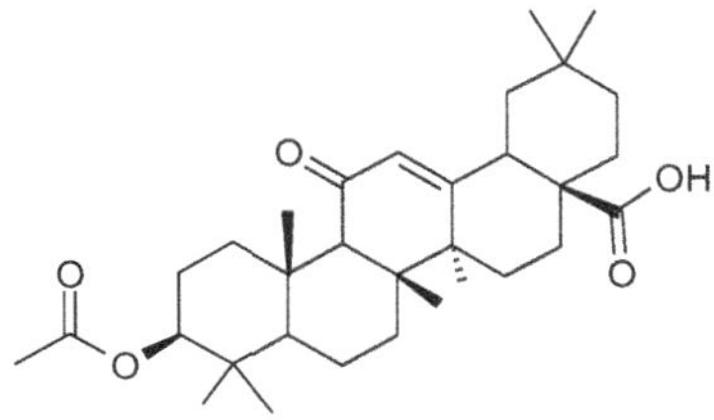

Lantanone

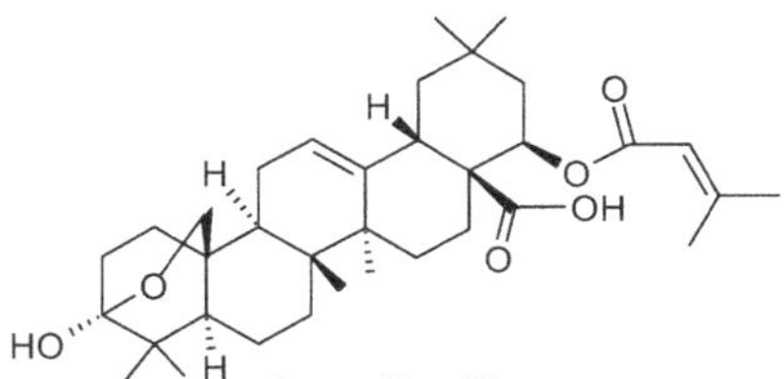

Lantanilic acid

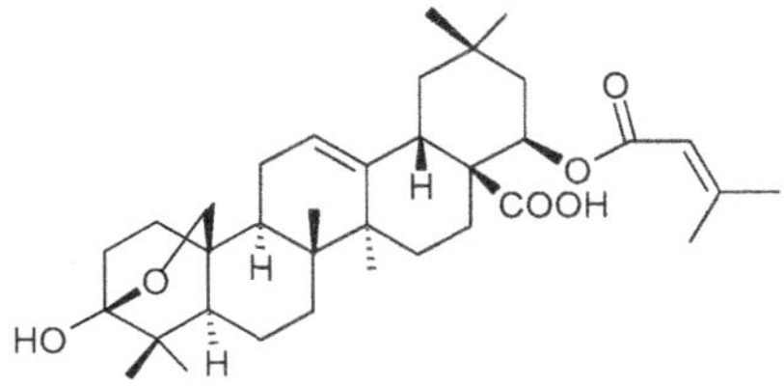

Camaric acid

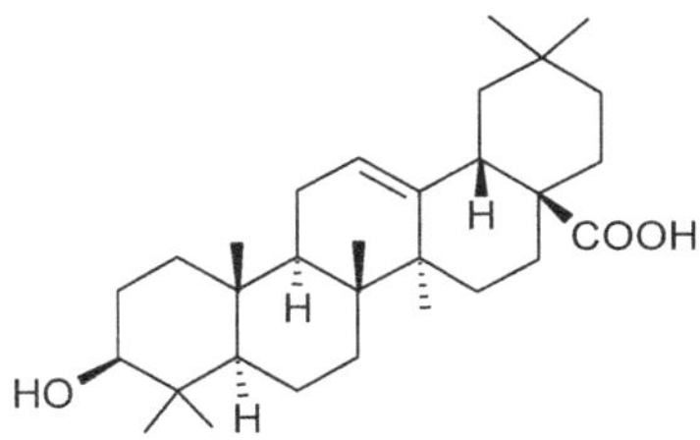

Oleanolic acid

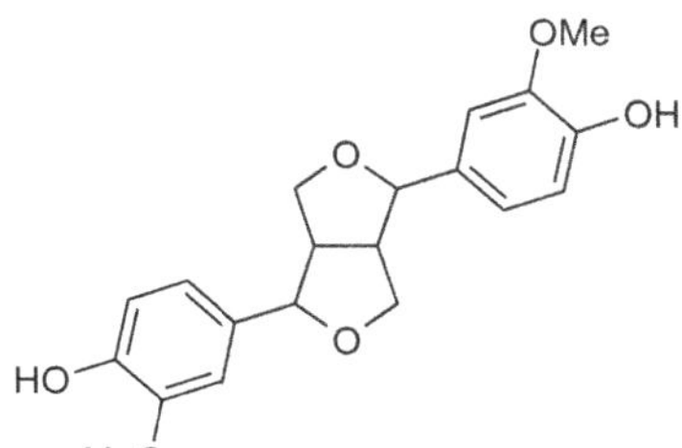

Pinoresinol

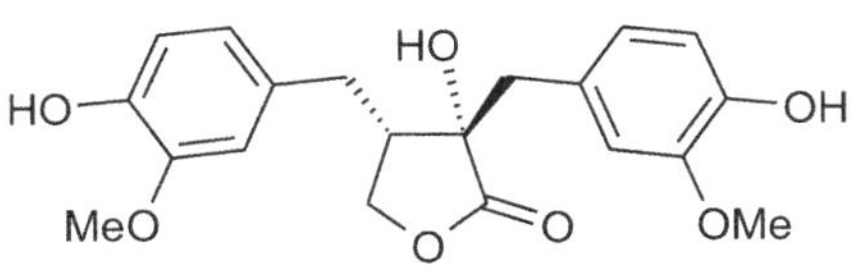

(-)-Nortrachelogenin

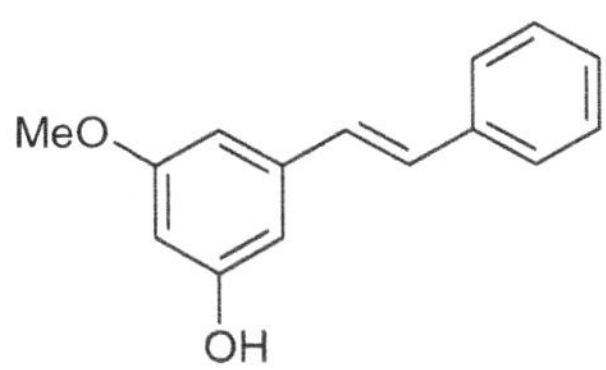

Pinosylvin monomethyl ether

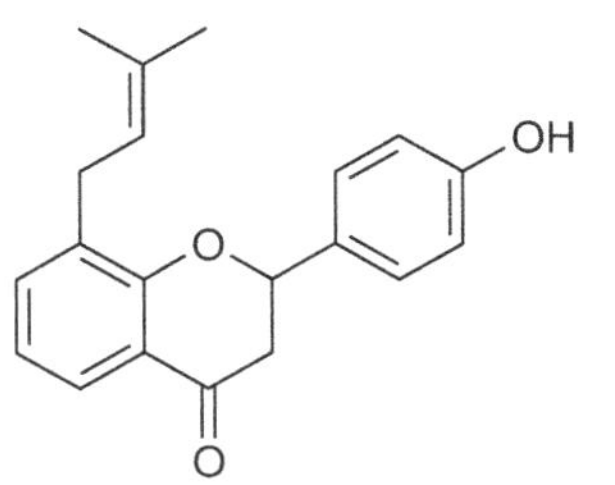

2-(4-hydroxyphenyl)-8-(3-methyl-but-2-enyl)-chroman-4-one

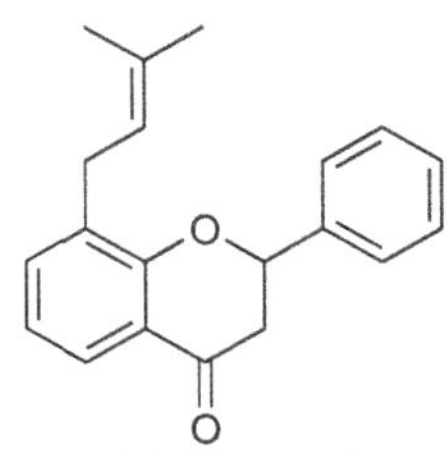

8-(3-methyl-but-2-enyl)-2-phenyl-chroman-4-one

Fig. 20.9 (continued)

(Barbarosa et al. 1999). α-Terthienyl identified in *Tagetes* species and gallic acid and linoleic acid in many plant species were found to show 100% mortality at concentrations of 0.125% after 24 h. Further, it has been reported that in the structure-activity relationships, an increase in the number of hydroxyl groups in phenolic acids increased the nematicidal activity, while the bioactivity fatty acids depended on chain length and the number and position of double bonds (Faizi et al. 2011). The crude extracts from leaves and seed cake from Indian neem tree (*Azadirachta indica* L.) also showed nematicidal activity against *Meloidogyne javanica* (Javed et al., 2007).

The heartwood of *Pinus massoniana* contains two nematicidal substances, pinosylvin monomethylether (PSM) and (−)-nortrachelogenin for the nematode. The bark and heartwood of *Pinus massoniana*, *P. strobus*, and *P. palustris* were also found to contain two nematicidal compounds methyl ferulate and (+)-pinoresinol, which showed the highest nematicidal (LD_{50} 4 ppm) activity (Suga et al. 1993). 2,3-Dihydro-2-hydroxy-3-methylene-6-methylbenzofuran, a compound with a strong nematicidal action, was isolated from the roots of the *Helenium* hybrid Moerheim Beauty (Gommers 1971). 2-(4-hydroxyphenyl)-8-(3-methyl-but-2-enyl)-chroman-4-one and 8-(3-methyl-but-2-enyl)-2-phenyl-chroman-4-one exhibited nematicidal activity against *Meloidogyne incognita* (LC_{50}, 14.5 and 70.9 ppm) and *Rotylenchulus reniformis* (LC_{50}, 3.3 and 102.9 ppm). 2-(4-hydroxyphenyl)-8-(3-methyl-but-2-enyl)-chroman-4-one showed activity at par with the standard carbofuran (LC_{50}, 3.1 ppm, respectively) against *Rotylenchulus reniformis* nematode (Shakil et al. 2008). Acetogenins from the seeds of *Annona squamosa* were found to possess promising nematicidal activity against *Bursaphelenchus xylophilus* and *Meloidogyne incognita* (Dang et al. 2011), while fatty acids and caprylic and capric acid showed about 50% mortality after a 24 h exposure (Zhang et al. 2012). 3,5-dihydroxy benzoic acid (100%), gallic acid (94%), and ethyl galactoside (100%) isolated from *R. niveus* and methyl benzoate from *Buddleja crispa* (92%) showed nematicidal activity against freshly hatched second stage juveniles of *Meloidogyne incognita* (root-knot nematode) after 48 h at 0.5% concentration and were more potent than the nematicide, *Azadirachta indica*, at the same concentration (Sultana et al. 2010a, b). A recent study showed that the lemongrass, clove, and palmarosa oils and their major compounds, citral, eugenol, and geraniol showed promising nematicidal activity and could be alternative to synthetic nematicides (Ajith et al. 2020).

The search for nematicidal agents in green chemicals led to the identification of many potent nematode controlling molecules such as geraniol, thymol, camphor, carvacrol, anethole, (+)-carvone, linalool, (−)-perillaldehyde, citronellol, undecanone, borneol, carveol, citral, α-terpineol, furfural, benzaldehyde, p-anisaldehyde, *trans*-cinnamaldehyde, (R)-(+)-pulegone, (*E*,*E*)-2,4-decadienal, oleic acid, and (*E*)-2-decenal which hold some hope for future use in integrated nematode management (Oka et al. 2000; Oka 2001; Echeverrigaray et al. 2010; Ntalli et al. 2011; Ibrahim et al. 2006; Rodriges-Kabana et al. 1993; Ntalli et al. 2010; Caboni et al. 2012; Tsao and Yu 2000).

20.5.4 Herbicidal Activity

Botanicals have been investigated for their effect on seed germination, shoot growth, and development (Alipour et al. 2019; Fagodia et al. 2017; Kaur et al. 2010). The essential oils from *Origanum acutidens* showed phytotoxic effect against *Amaranthus retroflexus*, *Chenopodium album*, and *Rumex crispus* (Kordali et al. 2008), while the oils from *Thymus vulgaris*, *Verbena officinalis*, and *Melissa officinalis* were effective against *Raphanus sativus*, *Lactuca sativa*, and *Lepidium sativum* (De Almeida et al. 2010). Similarly, oils from *Achillea gypsicola* and *Achillea biebersteinii* were found effective against *A. retroflexus*, *Cirsium arvense*, and *Lactuca serriola* (Kordali et al. 2009). Zerumbone has also shown concentration-dependent effect on seedling growth of *Phalaris minor* Retz. and strongly suppressed the root and shoot growth of *P. minor* seedling at 1000 ppm compared with control. It exhibited no or less effect on the germination of seeds of *T. aestivum* (Rana et al. 2017).

20.6 Use of Green Chemicals in Potato Pest and Disease Management

A large number of green chemicals (botanicals and essential oils), crude extracts, or refined and formulated products have been evaluated against various insect pests such as the potato tuber moth, Colorado potato beetle, soil arthropods, potato aphids, and disease like late blight, bacterial wilt, nematodes, etc. Even some compounds have been reported to have some degree of antiviral effects. The pest and disease-wise summary of studies have been provided in other chapters of this volume (see Bio-Intensive Management of Potato Diseases by Lal et al. (2021), Biological Suppression of Potato Pests by Nagesh et al. (2021), and New chemistry pesticides for the management of potato pests by Kuhar and McCullough (2021)).

Recent summaries on the potential of botanicals in potato IPM are given in Sharaby and Fallatah (2019), Natikar and Balikai (2019), Mulugeta et al. (2020), and Middya et al. (2021), to mention a few. Although some botanicals are not comparable with conventional pesticides (Chandler et al. 1994), these can be integrated in to common fungicide spray schedules. The green chemicals may help to reduce the use of conventional fungicides with synergistic effect and maintained efficacy (Liljeroth et al. 2016). Forrer et al. (2017) evaluated three botanicals against late blight and found that the bark of buckthorn (*Frangula alnus*) was as effective as copper in the multiyear field experiment. Therefore, advances in discovering botanicals benefit organic potato farmers both in developed and developing countries. Though most of the studies were conducted under in vitro conditions, the potential of botanicals against major potato pests and diseases has also been demonstrated under field condition (Table 20.1). Until now, *Allium sativum*, *Azadirachta indica*, *Cymbopogon citratus*, *Datura stramonium*, *Lantana camara*, and *Ocimum gratissimum* are most commonly studied green chemicals both in field and lab conditions.

Table 20.1 Summary of studies on the use of botanicals and their effects on potato pathogens/pests in laboratory and field conditions

Botanicals	Plant parts used for extraction	Solvent used for extraction	Pathogens/pests	Study condition	References
Achillea wilhelmsii	Whole plant	Ethanol	Potato tuber moth (PTM)	In vitro	Erdogan and Yilmaz (2018)
Ageratum conyzoides	Whole plant	Oil, hot and cold water, ethanol	*Phytophthora infestans*	In vitro	Hubert et al. (2013)
Allium cepa	Seeds	Aqueous	*P. infestans*, *Alternaria solani*	In vitro, in vivo, field	Abd-El-Khair and Haggag (2007)
Allium sativum	Leaves	Acetone, ethyl acetate, and water	*P. infestans*	In vitro, in vivo	Ngadze (2014)
		Powder	PTM	Potato store	Ibrahim and Sisay (2011)
	Bulb	Aqueous	*Pectobacterium* spp., *Erwinia* spp.	In vitro	Simeon and Abubakar (2014)
		Aqueous	*P. infestans, A. solani*	In vitro, in vivo, field	Abd-El-Khair and Haggag (2007)
Aloe vera	Leaves	Aqueous	*Pectobacterium* spp., *Erwinia* spp.	In vitro	Simeon and Abubakar (2014)
Anthemis deserti	Whole plant	Ethanol	PTM	In vitro	Sharaby et al. (2020)
Azadirachta indica	Seeds	Aqueous	*Pectobacterium* spp., *Erwinia* spp.	In vitro	Simeon and Abubakar (2014)
	Leaves	Acetone, ethyl acetate, and water	*P. infestans*	In vitro, in vivo	Ngadze (2014)
		Aqueous, ethanol	*Rhizopus oryzae*	In vitro, in vivo	Amadioha (2001)
		Powder	PTM	Potato store	Ibrahim and Sisay (2011)
	Oil	a	PTM	In vitro	Tefera (2017)
Callistemon citrinus	Leaves	Oil, hot and cold water, ethanol	*P. infestans*	In vitro	Hubert et al. (2013)

Capsicum annum	Fruits	Methanol, ethanol, aqueous	*P. infestans*	In vitro	Abayhne and Chauhan (2016)
Capsicum frutescens	Fruits	Aqueous	*P. infestans*, *A. solani*	In vitro, in vivo, field	Abd-El-Khair and Haggag (2007)
Carica papaya	Leaves	Aqueous	*P. infestans*	In vitro	Amienyo and Onunze (2015)
Cassia sieberiana	–	–	*P. infestans*	In vivo, field	Deshi et al. (2015)
Cassia senna	Leaves	Ethanol	PTM	In vitro	Sharaby et al. (2020)
Citrus Limon	Fruits	Methanol, ethanol, aqueous	*P. infestans*	In vitro	Abayhne and Chauhan (2016)
Citrullus spp.	Fruits	Ethanol	PTM	In vitro	Sharaby et al. (2020)
Chrysanthemum cinerariaefolium	Oil	–	PTM	In vitro	Tefera (2017)
Coriandrum sativum	Leaves	Ethanol	PTM	In vitro	Sharaby et al. (2020)
Cuminum cyminum	Oil	–	PTM	In vitro	Eftkhar et al. (2011)
Cymbopogon citratus	Seeds	Methanol, ethanol, aqueous	*P. infestans*	In vitro	Abayhne and Chauhan (2016)
	Leaves	Aqueous, ethanol	*Rhizopus oryzae*	In vitro, in vivo	Amadioha (2001)
		Aqueous	*P. infestans*, *A. solani*	In vitro, in vivo, field	Abd-El-Khair and Haggag (2007)
		Aqueous	*Pectobacterium* spp., *Erwinia* spp.	In vitro	Simeon and Abubakar (2014)
	Oil vapor	Oil vapor	PTM	In vitro	Sharaby et al. (2014)
Datura stramonium	Leaves	Methanol, ethanol, aqueous	*P. infestans*	In vitro	Abayhne and Chauhan (2016)
Distemonanthus benthamianus	Roots	Methanol	*Fusarium oxysporum*	In vitro, in vivo	Ogunsola and Aduramigba (2014)
Dolichoskilimand scharicus	Roots	Methanol	*F. Oxysporum*	In vitro	Tegegne and Pretorius (2007)

(continued)

Table 20.1 (continued)

Botanicals	Plant parts used for extraction	Solvent used for extraction	Pathogens/pests	Study condition	References
Eucalyptus citriodora	Leaves	Aqueous	*Pectobacterium* spp., *Erwinia* spp.	In vitro	Simeon and Abubakar (2014)
Eucalyptus globulus	Leaves	Aqueous	*P. infestans, A. solani*	In vitro, in vivo, field	Abd-El-Khair and Haggag (2007)
		Aqueous	*Ralstonia solanacearum*	In vivo, field	Hassan et al. (2009)
		Powder	PTM	Potato store	Ibrahim and Sisay (2011)
Foeniculum vulgare	Oil	–	PTM	In vitro	Eftkhar et al. (2011)
Hibiscus sabdariffa	Leaves	Aqueous	*R. Solanacearum*	In vivo, field	Hassan et al. (2009)
Jasminum grandiflorum	Leaves, flowers	Ethanol	PTM	In vitro	Sharaby et al. (2020)
Lantana camara	Leaves	Aqueous	*P. infestans*	In vitro	Amienyo and Onunze (2015)
	Leaves, flowers	Aqueous	*P. infestans, A. solani*	In vitro, in vivo, field	Abd-El-Khair and Haggag (2007)
Lepidium sativum	Seeds	Methanol, ethanol, aqueous	*P. infestans*	In vitro	Abayhne and Chauhan (2016)
Leptospermum petersonii	Whole plant	Ethanol	PTM	In vitro	Erdogan and Yilmaz (2018)
Majorana hortensis	Leaves	Aqueous	*P. infestans, A. solani*	In vitro, in vivo, field	Abd-El-Khair and Haggag (2007)
Maytenus senegalensis	Leaf, stem, bark, root	–	*P. infestans*	In vivo, field	Deshi et al. (2015)
Mentha piperita	Leaves	Aqueous	*P. infestans, A. solani*	In vitro, in vivo, field	Abd-El-Khair and Haggag (2007)
	Oil	–	PTM	In vitro	Eftkhar et al. (2011)
Mentha viridis	Leaves	Ethanol	PTM	In vitro	Sharaby et al. (2020)
Millettia ferruginea	Oil		PTM	In vitro	Tefera (2017)

Moringa stenopetala	Leaves	Methanol, ethanol, aqueous	*P. infestans*	In vitro	Abayhne and Chauhan (2016)
Nauclea latifolia	Leaf, stem, root	–	*P. infestans*	In vivo, field	Deshi et al. (2015)
Nicotiana tobaccum	Leaves	Methanol, ethanol, aqueous	*P. infestans*	In vitro	Abayhne and Chauhan (2016)
Nerium indicum	Leaves	Powder	PTM	In vitro	Thakur and Chandla (2013)
Ocimum basilicum	Leaves	Aqueous	*P. infestans, A. solani*	In vitro, in vivo, field	Abd-El-Khair and Haggag (2007)
	Leaves, flowers	Ethanol	PTM	In vitro	Sharaby et al. (2020)
Ocimum gratissimum	Leaves	Oil, methanol	*R. Solanacearum*	In vitro	Wagura et al. (2011)
		Oil, hot and cold water, ethanol	*P. infestans*	In vitro	Hubert et al. (2013)
		Aqueous, ethanol	*R. oryzae*	In vitro, in vivo	Amadioha (2001)
Ocimum lamiifolium	Leaves	Methanol, ethanol, aqueous	*P. infestans*	In vitro	Abayhne and Chauhan (2016)
Plantago albicans	Leaves	Ethanol	PTM	In vitro	Sharaby et al. (2020)
Piliostigma thonningii	Flower, fruit, leaf, bark	–	*P. infestans*	In vivo, field	Deshi et al. (2015)
Pelargonium graveolens	Leaves	Ethanol	PTM	In vitro	Sharaby et al. (2020)
Punica granatum	Leaves	Aqueous	*R. Solanacearum*	In vivo, field	Hassan et al. (2009)
Rhamnus prinoides	Leaves	Methanol, ethanol, aqueous	*P. infestans*	In vitro	Abayhne and Chauhan (2016)
Rhazya stricta	Leaves	Ethanol	PTM	In vitro	Sharaby et al. (2020)
Ruta chalepensis	Leaves	Methanol, ethanol, aqueous	*P. infestans*	In vitro	Abayhne and Chauhan (2016)
Ruta graveolens	Leaves	Powder	PTM	In vitro	Thakur and Chandla (2013)

(continued)

Table 20.1 (continued)

Botanicals	Plant parts used for extraction	Solvent used for extraction	Pathogens/pests	Study condition	References
Sapindus spp.	Fruit	Powder	PTM	In vitro	Thakur and Chandla (2013)
Solanum villosum	Leaves	Ethanol	PTM	In vitro	Sharaby et al. (2020)
Tanacetum parthenium	Whole plant	Ethanol	PTM	In vitro	Erdogan and Yilmaz (2018
Tagetes erecta	Leaves	Powder	PTM	In vitro	Thakur and Chandla (2013)
Vernonia amygdalina	Leaf, stem, root, seed	–	*P. infestans*	In vivo, field	Deshi et al. (2015)
Zanthoxylum zanthoxyloides	Stem	Methanol	*F. Oxysporum*	In vitro, in vivo	Ogunsola and Aduramigba (2014)
Zingiber officinale	Bulb	Aqueous	*Pectobacterium* spp., *Erwinia* spp.	In vitro	Simeon and Abubakar (2014)
	Oil vapor	Oil vapor	PTM	In vitro	Sharaby et al. (2014)
Zygophyllum coccineum	Leaves	Ethanol	PTM	In vitro	Sharaby et al. (2020)

[a]Plant part and extraction solvent not indicated.

Allium sativum (garlic), is herb with potential antimicrobial and insecticidal activities. It contains an active fraction of organosulfur compounds that are responsible for broad-spectrum insecticidal/antimicrobial activities. For example, in vitro assay showed that aqueous extract of *A. sativum* reduced mycelial growth (about 67%) and spore germination (about 64%) of *A. solani* (Abd-El-Khair and Haggag 2007). Similarly, field evaluation revealed *A. sativum* to reduce severity of early blight by 81% along with increase in potato yield (Abd-El-Khair and Haggag 2007). Significant reduction in the potato late blight severity was observed in the field with *A. sativum* extract (Ngadze 2014), but the efficiency was concentration dependent. Under in vitro conditions, colony growth of *P. infestans* was significantly reduced by acetone and water extracted *A. sativum* and performed well as the fungicide metalaxyl in a greenhouse trial (Ngadze 2014). In addition, it has shown potent antibacterial activity against *R. solanacearum* causing bacterial wilt in potato (Abo-Elyousr and Asran 2009). Likewise, studies have proved its efficacy against potato insect pests, for instance, two seasons of field trials conducted in Kenya revealed efficacy of *A. sativum* against whitefly, *B. tabaci*, causing the nymph populations to reduce to half (Lengai et al. 2017).

Azadirachta indica (neem) is reported to possess antimicrobial, antiviral, insecticidal, and nematicidal properties. Neem extracts contain several phytochemicals with antioxidant, antimicrobial, and anti-inflammatory properties (Susmitha et al. 2013; Itelima et al. 2016; Dash et al. 2017). Several neem products have been used in potato crop protection (Table 20.1; Atawodi and Atawodi 2009; Alzohairy 2016; Galeane et al. 2017). A study by Ibrahim and Sisay (2011) showed that neem leaf powder reduced the infestation of PTM and its damage in potato stores in Ethiopia.

Cymbopogon citratus (lemongrass), known for its medicinal use in different parts of the world (Negrelle and Gomes 2007), also exhibits pesticidal properties. Methanol, ethanol, and aqueous seed extracts of lemongrass are found to have direct antifungal effect on *P. infestans* mycelial growth and zoosporangia germination (Abayhne and Chauhan 2016). Aqueous leaf extract of lemongrass effectively reduced the mycelial growth, spore germination, and severity of *P. infestans* and *A. solani* both in laboratory and field conditions (Abd-El-Khair and Haggag 2007). Likewise, leaf extract significantly reduced bacterial populations of *Pectobacterium* spp. along with several other species that cause soft rot in potato (Simeon and Abubakar 2014). In vitro test conducted by Hubert et al. (2013) showed the effectiveness of essential oil of *C. citratus*, wherein it completely inhibited mycelial growth of *P. infestans*. The response was greatly dependent on the concentration of oil used; 0.03–0.5% gave cent percent mycelial growth inhibition. The major active fractions of essential oil were citral, myrcene monoterpene, and geranial (Negrelle and Gomes 2007). A study conducted by Sharaby et al. (2014) showed that life spans of male and female PTM adults reduced to 1 and 2 days, respectively, when exposed to *C. citratus* oil vapor.

Datura stramonium (jimson weed) is a well-known widespread solanaceous medicinal herb. It contains alkaloids such as hyoscyamine, scopolamine, aposcopolamine, and apoatropine (Soni et al. 2012). An in vitro assay confirmed efficacy of ethanol, methanol, and aqueous extracts of *D. stramonium* against

P. infestans in potato. It inhibits mycelial growth with reduction in spore germination by 75% (Abayhne and Chauhan 2016). In a greenhouse trial, soil treatment with hot and cold-water extracts of *D. stramonium* before and after artificial bacterial inoculations significantly reduced the bacterial wilt (Abo-Elyousr and Asran 2009). *Lantana camara* (lantana), an evergreen shrub and worst invasive weed in the world, works well in crop protection against pests and pathogens. For instance, it reduced half of the mycelial growth of potato late blight pathogen, *P. infestans*, in vitro (Amienyo and Onunze 2015). Leaf and flower extracts of lantana are reported to inhibit mycelial growth and sporangia germination of *P. infestans* and *A. solani* by more than 50% under in vitro (Abd-El-Khair and Haggag 2007). An experiment conducted in locally made potato stores revealed that *L. camara* leaf powder applied on potato tubers at 2-month intervals at a rate of 50 g per bed (2 m × 3 m) reduced potato infestation and damage by PTM by more than six times in comparison to the controls (Ibrahim and Sisay 2011). Furthermore, an efficacy trial was conducted under storage condition with extracts of fruits of soapnut, *Sapindus* spp.; leaves of rue, *Ruta graveolens*; yellow sage, *Lantana camara*; marigold, *Tagetes erecta*; kaner, *Nerium indicum*; and guldaudi, *Chrysanthemum cinerariifolium* against PTM (Thakur and Chandla 2013). The results showed that *L. camara* (1.08% infestation) was found to be effective in reducing PTM infestation on tubers followed by *Sapindus* (3.41%) and *R. graveolens* (4.38%) after 45 days of treatment (Thakur and Chandla 2013).

A laboratory study conducted by El Ghanam (2016) on efficacy of four plant oils, orange oil, colocynth oil, marjoram oil, and chili oil, and four plant powder, ginger, cinnamon, thyme, and rosemary, against *P. operculella* (PTM) larval penetration, pupation, and adult emergence during storage revealed that marjoram oil at concentration of 10 ml/L recorded the highest efficiency against larval penetration (7.7%), pupation (2.3%), and moth emergence (1.6%). Plant powders of ginger and cinnamon restrained the moth emergence from pupa at concentration of 3%. Erdogan and Yilmaz (2018) studied the efficacy of extracts from *Leptospermum petersonii* Bailey (Myrtaceae), *Achillea wilhelmsii* C. Koch (Asteraceae), and *Tanacetum parthenium* L. (Asteraceae) on PTM using two different methods (tuber dipping and larvae dipping). Bioassays were carried out to determine the effect of varied concentrations of extracts (for *L. petersonii* 0.05%, 0.1%, 0.3%, and 0.4%; for *A. wilhelmsii* and *T. parthenium* 1%, 3%, 6%, and 12%). In tuber dipping method, the highest mortality (100%) occurred at concentration of 0.4%, while the lowest mortality was at 0.05% when the extracts of *L. petersonii* were used. It was determined that when the extract of *A. wilhelmsii* and *T. parthenium* were used at highest concentration, mortality of 85% and 90% was reported, respectively. In larva dipping method, the extracts of *L. petersonii*, *A. wilhelmsii*, and *T. parthenium* caused 100%, 82%, and 87% mortality at their highest concentrations, respectively.

Sharaby et al. (2020) conducted an experiment to evaluate botanical extracts against PTM under storage conditions. Approximately 80% ethanolic extracts of 12 plants were tested on PTM during storage condition (30 ± 2 °C and 70 ± 5% RH). Biological parameters of the pest, evidence of the potato tubers damage, and continuation of protection to the tubers were recorded. Extracts of mint, zygophyllum, coriander, arnoglosse, harmel, and solanum indicated a total inhibition of egg deposition at 2.5% concentration; also, they provided high protections to the potato tubers from the PTM infestation for about 3 months and without affecting tuber germination. Senna, colocynth, and basil reduced the number of deposited eggs per female. Jasmine, geranium, and chamomile recorded a low potential on egg deposition. Basil showed the highest potency in decreasing development of larvae that hatched from eggs and therefore reduced the number of next adult offspring and followed by jasmine and geranium. Furthermore, a field trial was conducted to study the effectiveness of four essential oils in attracting aphids in potato crop. The oils used were basil oil, lavender oil, geranium oil, and tea tree oil. Two series of experiments were carried out with yellow sticky traps and colorless sticky traps along with control without any oil. The results revealed that in both the series basil oil is found to be more effective attractant, and 85% increase in the trap catches was observed in series 1 with yellow sticky traps with basil oil and 98% increase in the trap catches was observed in series 2 with colorless sticky trap with basil oil (unpublished data). Basil oil was found to be most effective attractant to potato aphids (various species) and having synergistic effect when used along with sticky traps in both the series of experiments. It can be used in organic potato production for monitoring aphid vectors.

20.7 Advantages of Green Pesticides over Chemical Pesticides

- Green pesticides are economically viable and ecologically feasible, and they are highly compatible as one of the major components in integrated pest management programs.
- In contrary to chemical pesticides, most of the plant and animal origin green pesticides have more than one biochemical compound, which possess the biological activity. These chemicals may exhibit single biological effect or may express diverse biological effects. Therefore, there are unlikely chances of developing quick resistance by the pests to green pesticides.
- Poor and marginal farmers who suffer from increasing costs and hazards of synthetic pesticides can grow their own pesticide-yielding plants for using the same for plant protection purpose and thus economically viable.
- Since the green pesticides are easily biodegradable in nature, the chances of pesticide persistence and residue effects are meagre and thus environmentally safer.
- Plants with pesticidal properties are known by the farmer because most of the time they grow in the same general area and therefore can be easily used for plant protection at local level.
- Often pesticidal plants also have other uses like household insect repellents or are plants with medicinal applications.
- The rapid degradation of the active ingredients is convenient as it reduces the risk of residues on food commodities.

- Some of the green pesticides may be used shortly before harvesting against late-stage pests.
- Many of the products act very quickly inhibiting insect feeding even though in long term they do not cause insect death.
- Since most of the green pesticide products have a stomach action and are rapidly decomposed, they may be more selective to insect pests and less aggressive with natural enemies.
- Most of the biological origin pesticides are non-phytotoxic.
- Resistance to these compounds is not developed as quickly as with synthetic insecticides.

20.8 Limitations of Green Chemicals

- Green pesticides are slow in action against pests. Mostly green chemicals are not associated with immediate knockdown/suppressant effect.
- There is a lack of residual action for most of the green pesticides as they are rapidly degraded by UV light in open.
- In case of green insecticides, most of them are not truly insecticides since many are merely insect deterrents and their effect is slow.
- Not all plant insecticides are less toxic to other animals than the synthetic ones.
- They are not necessarily available season long.
- Most of them have no established residue tolerances.
- There are no legal registrations establishing their use.
- Not all recommendations followed by growers have been scientifically verified.

20.9 Conclusion and Future Outlook

Tens of thousands of secondary products of plants have been identified, and there are estimates that hundreds of thousands of such compounds exist. These secondary compounds represent a large reservoir of chemical structures with biological activity. Therefore, higher plants can be exploited for the discovery of new bioactive products that could serve as lead compounds in pesticide development because of their novel modes of action (Philogene et al. 2005). In nature, essential oils play an important role in the protection of the plants as antibacterials, antivirals, antifungals, insecticides, and also against herbivores by reducing their appetite for such plants. They also may attract some insects to favor the dispersion of pollens and seeds or repel undesirable others. Some essential oils have been recognized as an important natural source of pesticides. Aromatic plants produce many compounds that are insect repellents or act to alter insect feeding behavior, growth and development, ecdysis (molting), and behavior during mating and oviposition. Recently researchers have demonstrated such compounds showing larvicidal and antifeedant activity (Larocque et al. 1999), capacity to delay development, adult emergence and fertility (Marimuthu et al. 1997), deterrent effects on oviposition (Naumann and Isman

1995), and arrestant and repellent action (Landolt et al. 1999). Plants with strong smells, such as French marigold and coriander, act as repellents and can protect the crops nearby.

The current consumer's demand for healthier and natural food products would result in rapid use of green chemicals in the food sectors. Innovative approach to enhance the global food security can be achieved through implementation of sustainably safe and environmentally sound green chemicals. In this context, the plant-based chemicals are gaining momentum that can fulfil the need of the hour (McClements et al. 2017). Some of these chemicals (especially EOs) are also recognized as safe and hence placed under GRAS category by US FDA depicting its use without further approval (Chaudhari et al. 2020a). However, some limitations associated with the EOs could be solved by encapsulating these green chemicals (EOs) into polymeric matrices through nano- or microencapsulation technology. The nano-encapsulation has several advantages such as increased protection from degradation, more stability, better bioavailability, masked aroma of bioactive components, and potential to be applied on large scale (Chaudhari et al. 2019; Chaudhari et al. 2020b). Some of the challenges associated with formulation of nano-encapsulated green chemicals are suitable polymer selection for encapsulation, selection of effective bioactive compounds, assessment of toxicological effects on human consumption, and impact on environment of prepared nanostructures (Pandey 2018). There is need of further studies to overcome the problems associated with nano-encapsulation technology and improvement in existing technology to fulfil the increasing demand of nano-encapsulated green chemicals (Shishir et al. 2018).

Biotechnology tools offer some exciting opportunities to manipulate the production of green chemicals. The production of secondary metabolites depends mainly on pathway engineering/metabolic engineering, rather than engineering the gene responsible for the final product. Various compounds of plant and animal origin have been successfully produced in tailor-made microbes under controlled conditions. Efforts are underway to upscale the systems to make them economically feasible. On the other hand, development of tissue and cell cultures of desired species of plants for the purpose of induction/increasing the quantity of secondary metabolites is also being explored.

References

Abayhne MA, Chauhan NM (2016) Antifungal activity of various medicinal plants against late blight of potato from Ethiopia. J Sci Res 12:1–9

Abd-El-Khair H, Haggag M (2007) Application of some Egyptian medicinal plant extracts against potato late and early blights. Res J Agric Biol Sci 3:166–175

Abo-Elyousr K, Asran MR (2009) Antibacterial activity of certain plant extracts against bacterial wilt of tomato. Arch Phytopathol Plant Prot 42:573–578

Ajith M, Pankaj SNA, Kaushik P, Rana VS (2020) Chemical composition and nematicidal activity of essential oils and their major compounds against *Meloidogyne graminicola* (Rice root-knot nematode). J Essential Oil Res 32(6):526–535

Ali EOM, Shakil NA, Rana VS, Sarkar DJ, Majumder S, Kaushik P, Singh BB, Kumar J (2017a) Antifungal activity of nano emulsions of neem and citronella oils against phytopathogenic fungi, *Rhizoctonia solani* and *Sclerotium rolfsii*. Ind Crop Prod 108:379–387

Ali EOM, Shakil NA, Rana VS, Srivastava C, Sarkar DJ, Kaushik P (2017b) Bioefficacy of nano emulsion mixture of citronella and neem oils against *Spodoptra litura* (fab). Ann Plant Prot Sci 25(2):231–238

Alzohairy MA (2016) Therapeutics role of *Azadirachta indica* (neem) and their active constituents in diseases prevention and treatment. Evid Based Complementary Altern Med 2016:7382506. https://doi.org/10.1155/2016/7382506

Amadioha AC (2001) Fungi toxic effects of some leaf extracts against *Rhizopus oryzae* causing tuber rot of potato. Arch Phytopathol Plant Prot 33:499–507

Amienyo C, Onuze E (2015) In vitro evaluation of *Carica papaya* L. and *Lantana camara* L. extracts in the control of late blight of potato caused by *Phytophthora infestans* (Mont.) de Bary. Res J Agric Sci 5:187–194

Atawodi SE, Atawodi JC (2009) Azadirachta indica (neem): a plant of multiple biological and pharmacological activities. Phytochem Rev 8:601–620

Barbarosa LCA, Barcelos FF, Demuner AJ, Santos MA (1999) Chemical constituents from *Mucuna aterrima* with activity against *Meloidogyne incognita* and *Heterodera glycines*. Nematropica 29:81–88

Begum S, Zehra SQ, Siddiqui BS, Fayyaz S, Ramzan M (2008) Pentacyclic triterpenoids from the aerial parts of *Lantana camara* and their nematicidal activity. Chem Biodivers 5:1856–1866

Božik M, Císarová M, Tancinová D, Kourimská L, Hleba L, Kloucek P (2017) Selected essential oil vapours inhibit growth of *aspergillus* spp. in oats with improved consumer acceptability. Ind Crop Prod 98:146–152

Caboni P, Ntalli NG, Aissani N, Cavoski I, Angioni A (2012) Nematicidal activity of (E,E)-2,4-decadienal and (E)-2-decenal from *Ailanthus altissima* against *Meloidogyne javanica*. J Agric Food Chem 60:1146–1151

Chandler D, Bailey AS, Tatchell GM, Davidson G, Greaves JJ, Grant WP (1994) The development, regulation and use of biopesticides for integrated pest management. Philos Trans R Soc B 366: 1987–1998

Chaudhari AK, Dwivedy AK, Singh VK, Das S, Singh A, Dubey NK (2019) Essential oils and their bioactive compounds as green preservatives against fungal and mycotoxin contamination of food commodities with special reference to their nanoencapsulation. Env Sci Pollution Res 26(25):25414–25431

Chaudhari AK, Singh VK, Das S, Prasad J, Dwivedy AK, Dubey NK (2020a) Improvement of in vitro and in situ antifungal, AFB1 inhibitory and antioxidant activity of *Origanum majorana* L. essential oil through nanoemulsion and recommending as novel food preservative. Food Chem Toxicol 143:111536

Chaudhari AK, Singh VK, Das S, Singh BK, Dubey NK (2020b) Antimicrobial, aflatoxin B1 inhibitory and lipid oxidation suppressing potential of anethole-based chitosan nanoemulsion as novel preservative for protection of stored maize. Food Bioprocess Technol 13:1462–1477

Chitwood DJ (2002) Phytochemical based strategies for nematode control. Annu Rev Phytopathol 40:221–249

Dang QL, Kim WK, Nguyen CM, Choi YH, Choi GJ, Jang KS, Park MS, Lim CH, Luu NH, Kim JC (2011) Nematicidal and antifungal activities of *Annonaceous acetogenins* from *Annona squamosa* against various plant pathogens. J Agric Food Chem 59:11160–11167

Dash SP, Dixit S, Sahoo S (2017) Phytochemical and biochemical characterizations from leaf extracts from *Azadirachta Indica*: an important medicinal plant. Anal Biochem 6:1–4

Deshi S, Wonang D, Imoni E, Gowok HG (2015) Efficacy of naturally occurring antifungal agents from five plants for the control of late blight (*Phytophthora infestans*) on Irish potato. J Sci Eng Res 6:1990–2004

Dubey NK (ed) (2011) Natural products in plant pest management. CABI, Oxfordshire, UK

Echeverrigaray S, Zacaria J, Beltrão R (2010) Nematicidal activity of monoterpenoids against the root-knot nematode *Meloidogyne incognita*. Phytopathology 100:199–203

Eftkhar AHE, Arafat NF, Riad NF (2011) Efficacy of some plant extracts and the commercial biocide, dolphin on potato tuber worm, *Phthorimaea operculella* (Zeller) (Lepidoptera: Gelechiidae) under field and storage conditions. J Plant Prot Pathol 2:1123–1131

El Ghanam MS (2016) Efficacy of some plant powders and essential oils against potato tuber moth, *Phthorimaea operculella* (Zeller) (Lepidoptera: Gelechiidae) in potato store. Egypt J Agric Res 95(2):2017

Erdoğan P, Yilmaz BS (2018) Insecticidal effect of three different plant extracts on potato tuber moth [*Phthorimaea operculella* Zeller (Lep.: Gelechiidae)]. J Food Sci Eng 8:215–221

Faizi S, Fayyaz S, Bano S, Yawar IE, Lubna L, Siddiqi H, Naz A (2011) Isolation of nematicidal compounds from *Tagetes patula* L. yellow flowers: structure-activity relationship studies against cyst nematode, *Heterodera zeae* infective stage larvae. J Agric Food Chem 59:9080–9093

Forrer HR, Vogelgsang S, Musa T (2017) Botanicals and phosphonate show potential to replace copper for control of potato late blight. J Fungi 3:65

Galeane MC, Martins CHG, Massuco J, Bauab TM, Sacramento LVS (2017) Phytochemical screening of *Azadirachta indica* a. Juss for antimicrobial activity. Afr J Microbiol Res 11: 117–122

Gommers FJ (1971) A nematicidal principle from the roots of a *Helenium* hybrid. Phytochemistry 10:1945–1946

Grainge M, Ahmed S (1988) Handbook of plants with pest-control properties. Wiley, Hoboken, NJ

Hassan M, Bereika M, Abo-Elnaga H, Sallam M (2009) Direct antimicrobial activity and induction of systemic resistance in potato plants against bacterial wilt disease by plant extracts. Plant Pathol J 25:352–360

Hubert JY, Nguefack J, Charles D, Daniel F, Sandrine S, Romain F, Pza H (2013) Antifungal potential and phytochemical analysis of extracts from seven Cameroonian plants against late blight. Int J Curr Microbiol Appl Sci 2:140–154

Ibrahim A, Sisay A (2011) Evaluation of some botanicals to control potato tuber moth *Phthorimaea operculella* (Lepidoptera: Gelechiidae) at Bako, west Shoa, Ethiopia. East Afr J Sci 5:37–40

Ibrahim SK, Traboulsi AF, El-Haj S (2006) Effect of essential oils and plant extracts on hatching, migration and mortality of *Meloidogyne incognita*. Phytopathology 45:238–246

Itelima JU, Nwokedi VC, Ogbonna AI, Nyam MA (2016) Phytochemical screening and antimicrobial activity evaluation of aqueous and ethanolic extracts of the leaf of *Azadirachta indica* Juss (neem) on some microorganisms. World J Microbiol 3:56–60

Javed N, Gowen SR, Inam-ul-Haq M, Abdullah K, Shahina F (2007) Systemic and persistent effect of neem (*Azadirachta indica*) formulations against root-knot nematodes, *Meloidogyne javanica* and their storage life. Crop Prot 26:911–916

Koul O (2008) Essential oils as green pesticides: potential and constraints. Biopestic Int 4(1):63–84

Landolt PJ, Hofstetter RW, Biddick LL (1999) Plant essential oils as arrestants and repellents for neonate larvae of the codling moth (Lepidoptera: Totricidae). Environ Entomol 28:954–960

Larocque N, Vincent C, Belanger A, Bourassa JP (1999) Effects of tansy essential oil from Tanacetum vulgare on biology of oblique-banded leafroller, *Choristoneura rosaceana*. J Chem Ecol 25:1319–1330

Lengai GMW, Muthomi JW, Narla RD (2017) Efficacy of plant extracts and antagonistic fungi in managing tomato pests and diseases under field conditions. J Agri Life Sci 4:20–27

Liljeroth E, Lankinen Å, Wiik L, Burra D, Alexandersson E, Andreasson E (2016) Potassium phosphite combined with reduced doses of fungicides provides efficient protection against potato late blight in large-scale field trials. Crop Protect 86:42–55

Marimuthu S, Gurusubramanian G, Krishna SS (1997) Effect of exposure of eggs to vapours from essential oils on egg mortality, development and adult emergence in *Earias vittella* (F.) (Lepidoptera: Noctuidae). Biol Agric Hortic 14:303–307

McClements DJ, Bai L, Chung C (2017) Recent advances in the utilization of natural emulsifiers to form and stabilize emulsions. Annu Rev Food Sci Technol 8:205–236

Middya R, Islam S, Mondal B (2021) Field efficacy of fungicides and bio-botanicals against late blight disease of potato. J Plant Physiol Pathol 9(7):254

Mulugeta T, Muhinyuza JB, Gouws-Meyer R, Matsaunyane L, Andreasson E, Alexandersson E (2020) Botanicals and plant strengtheners for potato and tomato cultivation in Africa. J Integ Agric 19(2):406–427

Natikar PK, Balikai RA (2019) Evaluation of botanicals and bio-pesticides against major insect pests of potato during kharif season. J Eco-friendly Agric 14(1):72–77

Naumann K, Isman MB (1995) Evaluation of neem Azadirachta indica seed extracts and oils as oviposition deterrents to noctuid moths. Entomol Exp Appl 76:115–120

Negrelle RRB, Gomes EC (2007) Cymbopogon citratus (dc) Stapf: chemical composition and biological activities. Revolution of Brasilian Plant Medical Botucatu 9:80–92

Ngadze E (2014) In vitro and greenhouse evaluation of botanical extracts for antifungal activity against *Phythopthora infestans*. J Biopest 7:199–204

Nkechi EF, Ejike OG, Ihuoma NJ, Maria-goretti OC, Francis U, Godwin N, Njokuocha R (2018) Effects of aqueous and oil leaf extracts of *Pterocarpus santalinoides* on the maize weevil, *Sitophilus zeamais* pest of stored maize grains. Afr J Agric Res 13(13):617–626

Ntalli NG, Caboni P (2012) Botanical Nematicides: a review. J Agric Food Chem 60:9929–9940

Ntalli NG, Ferrari F, Giannakou IO, Menkissoglu-Spiroudi U (2010) Phytochemistry and nematicidal activity of the essential oils from 8 Greek Lamiaceae aromatic plants and 13 terpene components. J Agric Food Chem 58:7856–7863

Ntalli NG, Ferrari F, Giannakou IO, Menkissoglu-Spiroudi U (2011) Synergistic and antagonistic interactions of terpenes against *Meloidogyne incognita* and nematicidal activity of essential oils from 7 plants indigenous in Greece. Pest Manag Sci 67:341–351

Ogunsola JF, Aduramigba-Modupe A (2014) Evaluation of four plant extracts in the control of post-harvest fungi tuber rot of Irish potato (*Solanum tuberosum*). Afr J Root Tuber Crops 11:1–9

Oka Y (2001) Nematicidal activity of essential oil components against the root-knot nematode *Meloidogyne javanica*. Nematol 3:159–164

Oka Y, Nacar S, Putievsky E, Ravid U, Yaniv Z, Spiegel Y (2000) Nematicidal activity of essential oils and their components against the root-knot nematode. Phytopathology 90:710–715

Ortiz de Elguea-Culebras G, Sánchez-Vioque R, Santana-Méridas O, Herraiz-Peñalver D, Carmona M, Berruga MI (2016) In vitro antifungal activity of residues from essential oil industry against *Penicillium verrucosum*, a common contaminant of ripening cheeses. LWT Food Sci Technol 73:226–232

Packiam S (2018) Green pesticides: eco-friendly technology for integrated pest management. Acta Sci Agric 2(11):1

Pandey G (2018) Challenges and future prospects of Agri-nanotechnology for sustainable agriculture in India. Environ Technol Innov 11:299–307

Philogene BJR, Regnault-Roger C, Vincent C (2005) Botanicals: yesterday's and today's promises. In: Regnault-Roger C, Philogene BJR, Vincent C (eds) Biopesticides of plant origin. Intercept, Andover, UK, pp 1–15

Prakash A, Rao J (1997) Botanical pesticides in agriculture. CRC Press, Boca Raton, FL. https://doi.org/10.1201/9781315138572

Qamar F, Begum S, Raza SM, Wahab A, Siddiqui BS (2005) Nematicidal natural products from the aerial parts of *Lantana camara* Linn. Nat Prod Res 19:609–613

Rana VS, Ahluwalia V, Shakil NA, Prasad L (2017) Essential oil composition, antifungal and seedling growth inhibitory effects of zerumbone from *Zingiber zerumbet* smith. J Essent Oil Res 29(4):320–329

Rana VS, Shakil NA, Kumar J (2020) Potential of some of the botanical pesticides as insect pest control agents. In, souvenir 4th IUPAC Internation conference on agrochemical protecting crops, health and natural environment: discovery and development of synthetic and natural products for health and pest management IARI, PUSA. New Delhi 2020:46–53

Regnault-Roger C, Vincent C, Arnason JT (2012) Essential oils in insect control: low-risk products in a high-stakes world. Annu Rev Entomol 57:405–424

Rodriges-Kabana R, Kloepper JW, Weaver CF, Robertson DG (1993) Control of plant parasitic nematodes with furfural-a naturally occurring fumigant. Nematropica 23:63–73

Sande D, Mullen J, Wetzstein M, Houston J (2011) Environmental impacts from pesticide use: a case study of soil fumigation in Florida tomato production. Int J Environ Res Public Health 8(12):4649–4661

Shakil NA, Pankaj KJ, Pandey RK, Saxena DB (2008) Nematicidal prenylated flavanones from *Phyllanthus niruri*. Phytochemistry 69:759–764

Sharaby A, Rahman HA, Abdel-Aziz SS, Moawad SS (2014) Natural plant oils and terpenes as protector for the potato tubers against *Phthorimaea operculella* infestation by different application methods. Ecol Balkanica 6:45–59

Sharaby AMF, Fallatah SB (2019) Protection of stored potatoes from infestation with the potato tuber moth, *Phthorimaea operculella* (Zeller)(Lepidoptera: Gelechiidae) using plant powders. Bull Natl Res Cent 43:79. https://doi.org/10.1186/s42269-019-0119-5

Sharaby AMF, Gesraha MA, Fallatah SAB (2020) Botanical extracts against the potato tuber moth, *Phthorimaea operculella* (Zeller 1873) (Lepidoptera: Gelechiidae), during storage conditions. Egypt J Biol Pest Control 30:93

Sharma A, Rajendran S, Srivastava A, Sharma S, Kundu B (2017) Antifungal activities of selected essential oils against *fusarium oxysporum* f. sp. *lycopersici* 1322, with emphasis on *Syzygium aromaticum* essential oil. J Biosci Bioeng 123:308–313

Shishir MRI, Xie L, Sun C, Zheng X, Chen W (2018) Advances in micro and nano-encapsulation of bioactive compounds using biopolymer and lipid-based transporters. Trend Food Sci Technol 78:34–60

Simeon AU, Abubakar A (2014) Evaluation of some plant extracts for the control of bacterial soft rot of tubers. Am J Exp Agric 4:1869–1876

Soni P, Siddiqui AA, Dwivedi J, Soni V (2012) Pharmacological properties of *Datura stramonium* L. as a potential medicinal tree: an overview. Asian Pac J Trop Biomed 2:1002–1008

Suga T, Ohta S, Munesada K, Ide N, Kurokawa M, Shimizu M, Ohta E (1993) Endogenous pine wood nematicidal substances in pines, *Pinus massoniana*, *P. strobus* and *P. palustris*. Phytochemistry 33:1395–1401

Sultana N, Akhter M, Khan RA, Afza N, Tareen RB, Malik A (2010b) Nematicidal natural products from the aerial parts of *Buddleja crispa*. Nat Prod Res 24:783–788

Sultana N, Akhter M, Khatoon Z (2010a) Nematicidal natural products from the aerial parts of *Rubus niveus*. Nat Prod Res 24:407–415

Susmitha S, Vidyamol KK, Ranganayaki P, Vijayaragavan R (2013) Phytochemical extraction and antimicrobial properties of *Azadirachta indica* (neem). Global J Pharmacol 7:316–320

Tefera YK (2017) Evaluation of botanical insecticides for the management of potato tuber moth, *Phthorimaea operculella* (Zeller) (Lepidoptera: Gelechiidae) under laboratory condition. Academic Res J Agri Sci Res 5:291–308

Tegegne G, Pretorius J (2007) In vitro and in vivo antifungal activity of crude extracts and powdered dry material from Ethiopian wild plants against economically important plant pathogens. Biol Control 52:877–888

Thakur M, Chandla VK (2013) Evaluation of bio-pesticides for potato tuber moth control, *Phthorimaea opercullela* (Zeller) under polyhouse and rustic storage conditions. Potato J 40(2):135–141

Tsao R, Yu Q (2000) Nematicidal activity of monoterpenoid compounds against economically important nematodes in agriculture. J Essent Oil Res 12:350–354

Unsworth J (2020) Chemistry and pest control. In souvenir 4th IUPAC Internation conference on agrochemical protecting crops, health and natural environment: discovery and development of synthetic and natural products for health and pest management IARI, PUSA, New Delhi 2020, p 1–25

Wagura A, Kimenju J, Gichimu B (2011) Comparative antibacterial effects of raw extracts and essential oils of *Ocimum gratissimum* L. against *Ralstonia solanacearum* (smith). Int J Plant Pathol 2:144–152

Walia S, Saha S, Rana VS (2014) Phytochemical pesticides. In: Singh D (ed) Advances in biopesticides. Springer, New Delhi, India, pp 295–322

Yangui I, Zouaoui Boutiti M, Boussaid M, Messaoud C (2017) Essential oils of Myrtaceae species growing wild in Tunisia: chemical variability and antifungal activity against *Biscogniauxia mediterranea*, the causative agent of charcoal canker. Chem Biodivers 14:e1700058

Zhang WP, Ruan WB, Deng YY, Gao YB (2012) Potential antagonistic effects of nine natural fatty acids against *Meloidogyne incognita*. J Agric Food Chem 60(46):11631–11637

Management of Major Fungal and Fungal-Like Soilborne Diseases of Potato

21

R. S. Tegg and C. R. Wilson

Abstract

This brief review focuses on the three most widely researched potato soilborne fungal or fungal-like diseases in recent years. A scan of published literature over the past 10 years identified *Verticillium dahliae*, *Rhizoctonia solani*, and *Spongospora subterranea* that cause verticillium wilt, black scurf, and powdery scab, respectively, as those for where considerable recent research had been undertaken. Their similarities, the production of long-lived resting structures within the soil, combined with some unique differences including pathogen life cycle and control options, provide the opportunity to discuss recent relevant research that provides insight into their management and identify research gaps that would aid understanding of these recalcitrant pathogens and subsequent diseases.

Keywords

Soilborne diseases · Sclerotia · Black scurf · Verticillium wilt · Powdery scab · Alternate hosts · Soil health · Disease control

21.1 Introduction

Soilborne diseases are of critical importance to potato production. This reflects both the impact on root and vascular function on plant growth and the direct impact on tuber numbers, size, and quality.

R. S. Tegg (✉) · C. R. Wilson
Tasmanian Institute of Agriculture, University of Tasmania, Hobart, TAS, Australia
e-mail: robert.tegg@utas.edu.au

S. Kumar Chakrabarti et al. (eds.), *Sustainable Management of Potato Pests and Diseases*, https://doi.org/10.1007/978-981-16-7695-6_21

Table 21.1 The major fungal and fungal-like soilborne pathogens and diseases of potato and the number of articles that studied these pathogens/diseases in the decade 2011–2021

Potato pathogen species	Disease	Number of WoS articles (2011–2021)[a]
Colletotrichum coccodes	Black dot	53
Fusarium spp.	Fusarium dry rots	85
Helminthosporium solani	Silver scurf	41
Phytophthora erythroseptica	Pink rot	26
Pythium ultimum var. *ultimum*	Leak	22
Rhizoctonia solani	Black scurf/stem canker	117
Sclerotium rolfsii	Stem rot	33
Spongospora subterranea	Powdery scab (PMTV vector)	98
Synchytrium endobioticum	Wart	35
Verticillium dahliae and *V. albo-atrum*	Verticillium wilt	184

[a]Number of articles was determined using Web of Science search for disease and/or pathogen in title and/or abstract from 2011 to 2021

In their widely cited review of soilborne pathogens, Fiers and colleagues (Fiers et al. 2012) described 17 major fungal or fungal-like soilborne potato pathogens, of which 10 have broad global distribution. These coincide with key fungal species identified elsewhere (Locke 2002). In this review we examined results from Web of Science database searches over the last 10 years to identify the three pathogens and their diseases that have received the greatest research attention, and these are the focus of this chapter. The pathogens identified were *Verticillium dahliae* and *V. albo-atrum* (causing verticillium wilt), *Rhizoctonia solani* (causing stem canker and black scurf), and *Spongospora subterranea* f. sp. *subterranea* (causing powdery scab) which were the focus of 184, 117 and 98 published articles (as determined by the presence of the pathogen name or disease in the article title or abstract) over the past 10 years, respectively (Table 21.1). That these pathogens and their diseases are at the forefront of current research initiatives can be attributed to their importance in terms of impact, cost of management, and/or the lack of effective control options.

These three represent typical soilborne diseases affecting the potato crop that can be divided into two groups depending on symptoms those damaging tubers and those damaging other parts of the plant (Fiers et al. 2012; Gudmestad et al. 2007). Those that impact stems or roots can produce vascular wilts, stem, stolon or root lesions or root galls as demonstrated by *V. dahliae* (root and stem vascular wilt), *R. solani* (root, stem and stolon cankers), and *S. subterranea* (root galls). All these non-tuber symptoms can impact the performance of the plant leading to a reduction in yield (Fiers et al. 2012) although quantification of these specific losses is not easy and generally underestimated (Wilson 2016). Typical tuber-based symptoms can include galls, blemishes, and the deeper penetrating rots demonstrated by *V. dahliae* (vascular ring discoloration of flesh), *R. solani* (sclerotia on tuber surface known as black scurf), and *S. subterranea* (powdery scab lesions).

These more visually obvious tuber symptoms can result in losses in many market areas, from failure of seed certification in the seed market, the requirement for extra peeling and tuber wastage in the processing sector, and consumer rejection and losses in the fresh market (Wilson 2016). This review chapter presents a succinct summary of each of these three soilborne pathogens and future challenges for managing them within potato production systems.

21.2 Verticillium Wilt (*Verticillium dahliae* and *V. albo-atrum*)

Verticillium wilt is an economically important disease and occurs wherever potatoes are commercially grown (Borza et al. 2018; Johnson and Dung 2010; Li et al. 2019; Nair et al. 2019; Rowe and Powelson 2002). The disease leads to the premature wilting (decline) of the crop (Fig. 21.1) and is commonly referred to as potato early dying. Within Australia, yield losses of up to 40% have been reported in severe cases. In North America yields can be reduced by 10–15% in moderately infested fields and 30–50% in heavily infested fields (Omer et al. 2008; Rowe and Powelson 2002). Quantification of losses can be difficult as infestation and yield loss are quite often associated with interaction with other soilborne pathogens such as root-infesting nematodes (Johnson and Dung 2010; Nair et al. 2019).

Verticillium dahliae Klebahn and *V. albo-atrum* Reinke & Berthold are the main causal agents of the disease (Rowe and Powelson 2002) with other minor pathogenic *Verticillium* species occasionally present. *Verticillium dahliae*, the most prevalent pathogen species, has a wide host range of more than 200 dicotyledon plant species, including annual herbs, perennial, and woody plants (Johnson and Dung 2010; Powelson and Rowe 1993; Steere and Kirk 2015). Key to the soilborne persistence

Fig. 21.1 Symptoms of *Verticillium dahliae* infection of potato; vascular browning of tubers (left) and chlorosis, necrosis, and leaf wilting of plants (right) (Images courtesy of A-M Donoghue and K Goulding)

of *V. dahliae* is the production of melanized dormant structures called microsclerotia, which form in the decaying tissues of infected host crops (Omer et al. 2008) and can survive for over a decade in infested soils (Steere and Kirk 2015). Potatoes and related solanaceous weeds such as blackberry and hairy nightshade have a greater potential than other hosts to increase or sustain inoculum levels in the soil. In comparison to *V. dahliae*, *V. albo-atrum* prefers cooler soil temperatures, it has a more limited host range, and its survival structures (melanized hyphae) persist in the soil for a shorter time period (Johnson and Dung 2010; Nair et al. 2019).

The microsclerotia of *V. dahliae* germinate, and hyphae penetrate plant roots to colonize the root and stem vascular tissues. The invasion of the xylem elements can disrupt water transport in plants resulting in vascular wilt (Johnson and Dung 2010). Although infection can occur early in the growing season, visible wilting symptoms are not usually seen until later in the season, generally at rapid tuber bulking, appearing as chlorosis, necrosis, and leaf wilting with entire stems senescing prematurely (Johnson and Dung 2010). Vascular browning (Fig. 21.1) may develop on tubers of susceptible cultivars (Johnson and Dung 2010; Nair et al. 2019). Early dying or senescence of the crop can occur 4–6 weeks earlier than normal crop senescence, with the severity of symptoms more pronounced during times of heat stress and under high rates of evapotranspiration. Infection and yield loss can be heightened by interaction with root-infesting nematodes, especially root lesion nematodes (*Pratylenchus* spp.) (Omer et al. 2008).

21.2.1 Disease Management

Management of verticillium wilt, and in particular *V. dahliae*, can be difficult due to pathogen prevalence, the diverse alternative hosts that propagate the pathogen, and the persistence of resting structures within the soil. Prior to planting potatoes, testing of fields for pathogen inoculum levels can provide a valuable estimate of disease risk. It has been shown that soil inoculum levels are directly related to subsequent disease levels and yield impacts (Omer et al. 2008). Determining preplant levels allows growers to select the most appropriate fields for planting or determine which fields require disease preventative measures.

Reducing initial inoculum in the soil prior to growing the potato crop is important. Tactics that may reduce initial soil inoculum include crop rotation, green manures and biofumigant crops, soil solarization, and fumigation (Gudmestad et al. 2007; Johnson and Dung 2010; Steere and Kirk 2015). Long crop rotations may aid inoculum depletion through natural decay processes, but as microsclorotia may persist for over 10 years in soils, standard rotation practices will be unlikely to fully eliminate the threat of disease. Further, due to the wide host range of *V. dahliae*, care needs to be taken in selecting appropriate rotation crops and managing alternative weed hosts (Johnson and Dung 2010; Steere and Kirk 2015).

Green manures and biofumigants can provide organic matter to the soil building structure and resilience, producing stronger plants that may better cope with pathogen attack. The addition of organic matter is also associated with greater levels of microbial activity that can include pathogen antagonists. Biofumigants such as the brassica mustards can provide additional suppression through production of glucosinolates (a class of organic molecules composed of a glucose and an amino acid and containing sulfur and nitrogen) within their tissues that break down following incorporation in the soil by microbial degradation into isothiocyanates which are toxic to a broad variety of soil organisms, including *Verticillium* sp. (Kirkegaard and Sarwar 1998; Neubauer et al. 2014).

In many parts of the world, fumigants, such as metam sodium, are applied to infested fields prior to planting to assist in management of verticillium wilt. These materials if correctly applied can reduce both the pathogen and its nematode synergist populations within the treated soil zones providing control (Taylor et al. 2005). In a similar manner, soil solarization, where the soil surface layers are heated to elevated temperatures by the sun following covering of the soil by black plastic, has been used to reduce pathogen populations in warmer climates (Johnson and Dung 2010).

Ensuring crops are planted with certified seed is always best practice; however visible infections of seed tubers can be difficult to detect, and certified seed may thus carry inoculum (Johnson and Dung 2010). Choice of variety to plant can also assist in reducing disease. Varietal resistance or tolerance may restrict pathogen infection and colonization of plants (Li et al. 2019) or reduce the impact of infection on yield. There are few resistant commercial varieties; however, varieties such as Ranger Russet have tolerance to *V. dahliae*, meaning that despite succumbing to infection they seldom express significant yield loss (Steere and Kirk 2015).

For best management practices, integration of several management tactics (Rowe and Powelson 2002; Steere and Kirk 2015) and a better understanding of soil health and how this can be enhanced and maintained will be needed for lasting and economic management of the disease (Johnson and Dung 2010).

21.3 Black Scurf (*Rhizoctonia solani*)

Black scurf of potato tubers and stem or stolon cankers (Fig. 21.2) are economically important diseases of potato of worldwide significance caused by the necrotrophic fungal pathogen *Rhizoctonia solani* Kühn (Brierley et al. 2016; Patil et al. 2018; Tsror 2010; Woodhall et al. 2013; Zrenner et al. 2020). The diseases result in both qualitative and quantitative impacts with losses of up to 50% reported in severe cases with black scurf disease (Woodhall et al. 2013). Losses associated with tuber symptoms may be through a reduction of marketable tubers and/or rejection of tubers for the premium seed market. The *Rhizoctonia* disease complex that causes sprout, stolon, and root infections can account for significant marketable tuber yield losses of up to 30% (Woodhall et al. 2013; Zrenner et al. 2020). In the Australian processing potato sector alone, there are conservative estimates of annual losses of A $5.4 M, just under 2% of the gross production value (Wilson 2016).

Fig. 21.2 Symptoms of *Rhizoctonia solani* infection of potato, (**a**) black scurf of tubers, and (**b**) canker of stems

Rhizoctonia solani is classified into 14 reproductively incompatible anastomosis groups (AGs) with AG3 considered the most important associated with potato disease. However, other AGs including 2-1, 2-2, 4, 5, and 8 have been associated with *Rhizoctonia* disease symptoms on potato (Brierley et al. 2016; Patil et al. 2018; Tsror 2010; Woodhall et al. 2013; Zrenner et al. 2020). The extent and significance of these other AG groups in promoting disease either alone or in combination with other AG groups are an area of evolving interest that requires further study.

There is a temporal pattern of symptom expression related to the two distinct disease phases observed. The *Rhizoctonia* disease complex associated with root/stem/stolon infection can occur throughout the growing season. Early infection is associated with hyphae growing from soil inoculum invading susceptible plant structures such as potato roots, sprouts, stems, stolons, and developing tubers (Tsror 2010; Woodhall et al. 2013; Zhang et al. 2015). The colonized hyphae form infection structures (infection cushion or appressoria) on the plant surface, beneath which host penetration and resulting lesions occur (Zhang et al. 2015). These necrotic infections on sprout tips can inhibit or delay emergence. Brown, dry, and sunken lesions may also develop on stems, stolons, and roots. Lesions which reach the vascular bundles develop into canker, typically brown and formed on stem bases, which can girdle the stems causing stunting (Tsror 2010). Aboveground symptoms can include production of aerial tubers, upward leaf roll, chlorosis, purple leaf pigmentation, and stunting or rosetting of plant shoots and leaves (Tsror 2010; Woodhall et al. 2013; Zhang et al. 2015).

The most conspicuous sign of *Rhizoctonia* disease occurs later in the season as tubers mature and is characterized by the formation of black, irregular sclerotia of various sizes on tubers (Tsror 2010). Other less well-characterized tuber symptoms also attributed to *R. solani* infection include misshapen tubers, growth cracks, netted scab or elephant hide, dry core, and tuber greening (Brierley et al. 2016; Tsror 2010).

Rhizoctonia solani AG3 has been described as having a relatively narrow host range by some (Fiers et al. 2012) but described by others as having a large spectrum of alternative hosts (Tsror 2010; Zhang et al. 2015). Both indicate the importance of solanaceous species with others also noting the importance of other host crop species that may be used in rotation with potato. Further clarification and testing of alternative hosts represents an ongoing area of interest, particularly elucidation of what alternate hosts support full life cycle development (sclerotial formation) of the pathogen and what AG groups are involved.

21.3.1 Disease Management

Effective disease management of *Rhizoctonia* disease requires implementation of an integrated disease management approach and knowledge of each of its stages. Cultural controls are the most important strategies for *Rhizoctonia* disease control, with seed and soil applied fungicides providing important additional control (Tsror 2010).

An initial determination of soilborne inoculum levels in the soil can aid future management decisions. High levels are generally correlated with a higher incidence of disease in the resultant crop, while lower levels favor a healthier crop (Brierley et al. 2016; Tsror 2010). It should be noted that quantification of *R. solani* at low inoculum levels in the soil, utilizing qPCR, is difficult (Brierley et al. 2016), so knowledge of the paddock, from a grower history perspective, is also useful for determining likelihood of *Rhizoctonia* diseases.

The persistence of resting structures (sclerotia) in the soil coupled with a moderate host range provides a challenge for reducing soilborne inoculum levels. Lengthening the rotation between potato crops is one strategy, combined with planting nonhost crops and effective management of potato volunteers (Powell et al. 2020; Tsror 2010). Other cultural management practices include careful management of irrigation scheduling and haulm destruction (Tsror 2010). Although there are differences in susceptibility, no resistant cultivars are available. While biological control options have been trialed and shown some promise, their commercial usage has not occurred (Tsror 2010).

Ensuring crops are planted with certified seed is recommended as the positive relationship between increasing seed pathogen levels and resultant disease has been clearly demonstrated in potato (Tegg et al. 2015). This is particularly relevant when planting in new ground or in areas where soilborne inoculum is low (Brierley et al. 2016). Where high soil inoculum exists, the treatment of seed and soil with chemicals should be considered.

Chemical control is widely practiced by industry with seed treatment and in-furrow applications of fungicides targeting *Rhizoctonia* a standard component of many crop protection programs. Various active ingredients (such as azoxystrobin) have been shown to reduce and provide disease control benefit, although chemical fungicides are not always fully effective (Gudmestad et al. 2007; Tsror 2010; Zhang et al. 2015). Fumigation strategies are becoming less popular and have variable efficacy against soil inoculum, although various fumigants are still utilized in some regions (Tsror 2010).

21.4 Powdery Scab (*Spongospora subterranea* f. sp. *subterranea*)

Powdery scab and associated root disease of potato (Fig. 21.3) are due to infections of roots and tubers with *Spongospora subterranea* f. sp. *subterranea*. The disease is common worldwide resulting in both cosmetic tuber damage and yield loss (Balendres et al. 2016b; Falloon 2008; Falloon et al. 2016). In the Australian processing potato sector alone, there are estimates of annual losses of A$13.4 M, approximately 4% of the gross production value (Wilson 2016). *Spongospora subterranea* is also the vector of *Potato mop top virus* (PMTV), which on infection can further diminish tuber quality and exacerbate economic loss. PMTV outbreaks within the Scottish seed industry have recorded significant yield reduction of up to 67% (Davey 2009).

Pathogen inoculum resides in soil and on infected seed tubers as dormant agglomerations of resting spores known as sporosori. The sporosori are highly robust, able to withstand environmental extremes, and have been known to persist within infested soils for several decades (Falloon 2008). Germination of resting

Fig. 21.3 Symptoms of *Spongospora subterranea* infection of potato, (**a**) powdery scab of tubers, (**b**) root galling, and (**c**) zoosporangia infection of root hairs (images b and c courtesy of MAO Balendres)

spores requires presence of soil water, a preference for temperatures within the range of 9–17 °C, and is stimulated by occurrence of certain root exudate compounds such as glutamine (Balendres et al. 2016a). Resting spores release short-lived motile zoospores that navigate through the soil water to the root surface. Zoospores utilize chemotaxis, following root exudate compound gradient signals, to locate host roots (Amponsah et al. 2021). The successful encystment and infection of roots lead to production of zoosporangia within root cells and subsequent production and release of secondary zoospores which increase root infections in a polycyclic manner (Balendres et al. 2016b). As the infection progresses and the crop matures, root galls may form from infected roots, and young developing tubers succumb to infection resulting in tuber lesions or powdery scab (Balendres et al. 2016a). Both root galls and tuber lesions contain new sporosori that can be released into the soil adding to the soil inoculum load.

21.4.1 Disease Management

There remains no single effective management strategy for *S. subterranea* (Falloon 2008). Anecdotal evidence has shown inoculum within pathogen-infested fields can remain infective for up to 50 years (Falloon 2008). Long crop rotations of at least 5 years are recommended, although it is unlikely that soil inoculum will be fully depleted by the periods between potato crops. Free-draining soils with good structure and organic matter content are less prone to disease. Maintaining healthy soils will promote plant heath, and robust soil microbial activity will promote pathogen antagonists. Disease-suppressive soils have been identified in some cropping regions. Where they have been studied, the soils appear to have possibly biological and/or chemical disease-suppressive mechanisms (Wright et al. 2021). Application of biological agents such as *Trichoderma* spp. has the ability to reduce disease, presumably through reduction in sporosori viability or zoospore activity and infectivity (Nielsen and Larsen 2004), but the commercial efficacy and cost benefit of such approaches remain to be proven.

Preplant testing of cropping soil can identify those fields with elevated inoculum levels, and this information can be used to assist in choosing fields to crop or those where greater attention to disease mitigation is required (Mallik et al. 2019; van de Graaf et al. 2003). It has been postulated that preplant treatment of soils with sporosori germination stimulants could be used to diminish soil inoculum as the released short-lived zoospores would perish in absence of a suitable host (Balendres et al. 2018). At planting, in-furrow fungicide treatments, where these are registered and available, have been shown to assist in slowing and suppressing disease (Tsror et al. 2020). The benefits of soil fumigation are less clear, with some data suggesting standard fumigants can exacerbate disease, presumably through diminishing antagonistic soil microflora (Bittara et al. 2017).

It is critical that disease-free certified seed tubers are planted, as infested seed can be a potent inoculum source for initiating infection cycles. Symptom-free seed tubers may still carry significant inoculum loads, and seed pathogen testing with possible use of seed treatments can be beneficial (Braithwaite et al. 1994; Tegg et al. 2016). Resistant varieties are an important management tool. While there are no varieties that are immune to infection, there exists significant variation in disease susceptibility among commercial potato varieties (Merz et al. 2012). Interestingly, however, while usually the case, relative resistance to root and tuber infection is not always linked with some important varieties (Falloon 2008).

As with the previous two disease examples, best disease management practices should utilize an integration of management strategies to achieve adequate control. There however remains a need for new more effective controls for powdery scab and root diseases due to *S. subterranea* infections.

21.5 Challenges and Research Gaps for these Soilborne Fungal Pathogens

21.5.1 Maintaining Soil Health

For many soilborne diseases including the three examples presented in this paper, many authors conclude that a better understanding of soil health is needed for lasting and economic management of these diseases (Johnson and Dung 2010; Larkin 2015). Soil health is a complex concept encompassing the physical, chemical, and biological properties of the soil and their role in ecosystem services and the growth of plants. Additionally, potato crops are demanding on the soil with significant heavy machinery traffic, intensive tillage operations, and high inputs of fertilizer, pesticides, and water. Maintaining or improving soil health can therefore be challenging for growers (Hills et al. 2020; Powell et al. 2020).

The effect of management practices designed to improve soil health on disease incidence has recently been reviewed (Larkin 2015). Although individual studies can produce both positive and negative results, Larkin concluded that management practices that improve soil health generally increase soil biota abundance, diversity, and activity which, over time, reduces disease incidence, even if these practices do not remove pathogens from the soil. The mechanisms involved in this resilience against disease are not entirely understood (Larkin 2015; Powell et al. 2020) and represent an area of evolving interest.

The relationship between maintaining soil health and ensuring plant health is critical as plants can succumb to disease significantly quicker when an additional plant stress is imposed (Johnson and Cummings 2015; Powell et al. 2020). Verticillium wilt symptoms develop more rapidly when plants are stressed from heat, under- or overirrigation, or nutrient deficiency. Field observations of greater disease in headland and high traffic areas, where soil compaction and poor soil aeration are more common, also support the link between reduced soil health and a greater propensity of plant and/or tuber disease. Where alleviation of physical soil stresses occurs, a resultant benefit to potato yield and quality can occur (Hills et al. 2020; Powell et al. 2020).

Soil fumigation is a strategy that can remove key pathogens from the soil, but its usage is detrimental to overall soil biological health, as it has a nonspecific impact. Many potato growing regions worldwide do not have access to these chemicals due to environmental and human safety concerns, and where they are still utilized, there is an increasing movement to source environmentally sustainable alternative options (Hills et al. 2020; Powell et al. 2020).

21.5.2 Pathogen Interactions

In a field situation, it is likely that more than one soilborne pathogen will be present. The interaction of these pathogens may negate the production of one of the diseases, have no significant impact, or have a synergistic effect promoting disease incidence/ severity. However, formal replicated studies in this area are sparse. Some observational work has hinted at the promotion of some diseases in the presence of others. For example, powdery scab tuber lesions may act as entry points for other opportunistic tuber-invading pathogens (Locke 2002; Johnson and Cummings 2015), increasing susceptibility to secondary storage diseases (Falloon 2008; Merz et al. 2012). Tubers with powdery scab lesions have been associated with increased incidences of late blight, black dot, and pink rot (Diriwachter and Parbery 1991; Locke 2002; Johnson and Cummings 2015). A well-documented interaction is that of potato early dying which can occur with *V. dahliae* alone, but earlier and more severe symptoms occur when both the fungus and the root lesion nematode (*Pratylenchus penetrans*) are present (Nair et al. 2019). Indeed, disease threshold levels of *V. dahliae* in soil for producing disease are significantly reduced (more than halved) in the presence of *P. penetrans*, indicating significant synergistic impacts (Powelson and Rowe, 1993; Omer et al., 2008).

Replicated pot trials showed no significant additive effect on disease expression and yield between the root lesion nematode (*Pratylenchus penetrans*) and the fungus (*R. solani* AG3, 2.1) (Edin et al. 2019; Viketoft et al. 2020). Further studies relating to potato soilborne pathogen interactions and quantification of yield and disease impacts are warranted.

21.5.3 Detection and Quantification of the Pathogen in Soil and Seed

A key management criterion for any pathogen is its detection, differentiation, and quantitation from both soil and in many cases seed tubers (Fiers et al. 2012; Nair et al. 2019; Tegg et al. 2015). Recent innovations in PCR-based technologies have enabled the development of commercial soil tests that quantify specific pathogens and provide a threshold risk analysis for planting potato tubers into these soils (Powell et al. 2020). Such soil pathogen tests exist in various regions and are well established and utilized for predicting a selection of potato pathogens including *S. subterranea*, *Colletotrichum coccodes* (causal agent of black dot), and root knot nematode (Stagnitti 2015; Tegg et al. 2015; Brierley et al. 2016; Mallik et al. 2019). Within Australia, the *S. subterranea* soil test is widely used by industry with a similar test also available for seed tubers (Tegg et al. 2015). The benefit of a qPCR test for seed is that it may detect significant pathogen from visually blemish-free tubers that may pose a risk to the grower (Tegg et al. 2016), which is unlikely to be detected using visual certification.

While the *S. subterranea* soil test is robust and validated with clear risk thresholds, PCR tests available for *R. solani* AG3 and *V. dahliae* are less able to detect field infection levels and do not provide accurate risk thresholds with further work required for commercial usage (Brierley et al. 2016; Stagnitti 2015). In the case of *R. solani* AG3, reasons attributed for the reduced quality and robustness of the soil test include sporadic distribution of the major propagules and the dynamic nature of inoculum in the soil, being able to survive as both hyphae on suitable host tissue and dormant sclerotia. For low levels of soil inoculum, improved sampling strategies need to be further explored (Brierley et al. 2016).

21.5.4 Quantifying Losses Attributable to Soilborne Diseases

It is well acknowledged that soilborne pathogens are responsible for considerable losses in potato crops. There is however often very poor data on what these losses are. Soilborne diseases may not necessarily be easily seen from crop scouting without excavation of roots and tubers or may be somewhat obscured resembling abiotic disorders.

Estimating disease incidence within a field can be difficult. Plants infected by soilborne pathogens commonly occur in aggregated clumps making subsampling problematic. There may also be insufficient healthy plants growing under identical conditions in commercial settings for meaningful comparisons of pathogen impact on plant growth and yields.

Yield and quality losses will have a direct impact on marketable yields. However, this often ignores additional costs incurred through requirements for investment in their control or avoidance. These often hidden indirect costs include the measures to avoid, prevent, or reduce infection and disease expression, and these can be substantial, perhaps greater than the direct losses observed.

These indirect costs can include:

- Investments on farm specifically for disease control, such as the need for specialized machinery, costs of pesticides, increased need for labor and fuel, etc. There will also be costs associated with restrictions of cultivar and crop choice as rotations are necessary to reduce pathogen buildup in soils to levels rendering fields unproductive.
- Increased costs with growing crops that are not harvested do not make retail grade. This would include all the inputs for cropping, fertilizer, water, fuel, labor, land etc.

- Costs associated with inspections, certification, and pathogen testing such as the costs of legislative controls via quarantine and inspection to exclude exotic pathogens, surveys and surveillance, seed certification, and diagnostic services.
- Investments in development of management tools and training, such as potato breeding programs, disease research, industry extension, and education programs.
- Issues around market failure where disease may result in loss of market access, disruption of continuity of supply, or noncompetitive pricing with increased costs due to disease.

There can also be significant social or environmental costs relating to areas like farm worker employment (e.g., if contracts are lost), consumer health, and environmental pesticide pollutants.

Plant pathologists are often questioned why there is such poor information on the costs of disease on our crops and, in particular, soilborne diseases such as verticillium wilt, stem canker, and powdery scab. This is partially due to the difficulty in assessing the extent of disease and comparing to what would have been the outcome in absence of disease. Where this has been estimated, this invariably is presented as yield or quality losses leading to direct financial penalties. However, it is also important to consider costs beyond those associated with direct yield or quality reductions and the obvious costs of control (e.g., fungicides). This would help better guide investment in disease mitigation activities on farm and research priorities.

21.6 Conclusions

Soilborne fungal and fungal-like pathogens are responsible for significant losses in potato production, exacerbated by the harvestable product (tubers) being produced within the soil environment. The three most commonly researched potato soilborne fungal diseases are verticillium wilt, rhizoctonia stem canker and black scurf, and powdery scab. While these represent diverse organisms, they share capacity for persistence within cropping soils through production of robust resting structures and for invasion of seed tubers which has enabled global dissemination and infestation of new cropping soils. Management of all three diseases can be difficult but relies on a good understanding of soil inoculum levels and strategies to reduce these prior to, or at, planting and variety resistance. There is a clear need for a better understanding of the general principles of maintaining soil health, particularly with a crop such as potato that necessitates soil disturbance during field preparation, planting, and harvest. Similarly, better methods for soil inoculum measurement and monitoring are required and studies to more clearly define mechanisms of soil pathogen interactions that exacerbate disease. Lastly, greater thought around how best to determine the full impacts of these diseases is required to better understand the investments needed for their control.

References

Amponsah J, Tegg RS, Thangavel T, Wilson CR (2021) Moments of weaknesses - exploiting vulnerabilities between germination and encystment in the Phytomyxea. Biol Rev 96:1603–1615

Balendres MA, Clark TJ, Tegg RS, Wilson CR (2018) Germinate to exterminate: chemical stimulation of Spongospora subterranea resting spore germination and its potential to diminish soil inoculum. Plant Pathol 67:902–908

Balendres MA, Nichols DS, Tegg RS, Wilson CR (2016a) Metabolomes of potato root exudates: compounds that stimulate resting spore germination of the soil-borne pathogen *Spongospora subterranea*. J Agric Food Chem 64:7466–7474

Balendres MA, Tegg RS, Wilson CR (2016b) Key events in pathogenesis of Spongospora diseases in potato: a review. Austral Plant Pathol 45:229–240

Bittara FG, Secor GA, Gudmestad NC (2017) Chloropicrin soil fumigation reduces *Spongospora subterranea* soil inoculum levels but does not control powdery scab disease on roots and tubers of potato. Am J Potato Res 94:129–147

Borza T, Beaton B, Govindarajan A, Gao XX, Liu YH, Ganga Z, Wang-Pruski G (2018) Incidence and abundance of *Verticillium dahliae* in soil from various agricultural fields in Prince Edward Island, Canada. Eur J Plant Pathol 151:825–830

Braithwaite M, Falloon RE, Genet RA, Wallace AR, Fletcher JD, Braam WF (1994) Control of powdery scab of potatoes with chemical seed tuber treatments. NZ J Crop Hortic Sci 22:121–128

Brierley JL, Hilton AJ, Wale SJ, Woodhall JW, Lees AK (2016) The relative importance of seed- and soil-borne inoculum of *Rhizoctonia solani* AG-3 in causing black scurf on potato. Potato Res 59:181–193

Davey T (2009) The importance of potato mop-top virus (PMTV) in Scottish seed potatoes. British Potato Council. http://hdl.handle.net/10399/2300. Accessed 15 Aug 2021

Diriwachter G, Parbery DG (1991) Infection of potato by *Spongospora subterranea*. Mycol Res 95: 762–764

Edin E, Gulsher M, Franko MA, Englund JE, Flohr KJ, Viketoft M (2019) Temporal interactions between root-lesion nematodes and the fungus *Rhizoctonia solani* lead to reduced potato yield. Agron 9(7):361. https://doi.org/10.3390/agronomy9070361

Falloon RE (2008) Control of powdery scab of potato: towards integrated disease management. Am J Potato Res 85:253–260

Falloon RE, Merz U, Butler RC, Curtin D, Lister RA, Thomas SM (2016) Root infection of potato by *Spongospora subterranea*: knowledge review and evidence for decreased plant productivity. Plant Pathol 65:422–434

Fiers M, Edel-Hermann V, Chatot C, Le Hingrat Y, Alabouvette C, Steinberg C (2012) Potato soil-borne diseases. A review. Agron Sust Develop 32:93–132

Gudmestad NC, Taylor RJ, Pasche JS (2007) Management of soilborne diseases of potato. Austral Plant Pathol 36:109–115

Hills K, Collins H, Yorgey G, McGuire A, Kruger C (2020) Improving soil health in Pacific northwest potato production: a review. Am J Potato Res 97:1–22

Johnson DA, Cummings TF (2015) Effect of powdery scab root galls on yield of potato. Plant Dis 99:1396–1403

Johnson DA, Dung JKS (2010) Verticillium wilt of potato - the pathogen, disease and management. Can J Plant Pathol 32:58–67

Kirkegaard JA, Sarwar M (1998) Biofumigation potential of brassicas - I. variation in glucosinolate profiles of diverse field-grown brassicas. Plant Soil 201:71–89

Larkin RP (2015) Soil health paradigms and implications for disease management. Annu Rev Phytopathol 53:199–221

Li HY, Wang ZP, Hu XP, Shang WJ, Shen RQ, Guo CJ, Guo QY, Subbarao KV (2019) Assessment of resistance in potato cultivars to Verticillium wilt caused by *Verticillium dahliae* and *Verticillium nonalfalfae*. Plant Dis 103:1357–1362

Locke T (2002) Compendium of potato diseases, 2nd edn. APS Press, St Paul, MN

Mallik I, Fulladolsa AC, Yellareddygari SKR, Bittara FG, Charkowski AO, Gudmestad NC (2019) Detection and quantification of *Spongospora subterranea* sporosori in soil by quantitative real-time PCR. Plant Dis 103:3189–3198

Merz U, Lees AK, Sullivan L, Schwarzel R, Hebeisen T, Kirk HG, Bouchek-Mechiche K, Hofferbert HR (2012) Powdery scab resistance in *Solanum tuberosum*: an assessment of cultivar x environment effect. Plant Pathol 61:29–36

Nair PVR, Wiechel TJ, Crump NS, Taylor PWJ (2019) Seed tuber incidence, identification and pathogenicity of Verticillium species infecting potatoes in south East Australia. Austral Plant Pathol 48:637–650

Neubauer C, Heitmann B, Müller C (2014) Biofumigation potential of Brassicaceae cultivars to *Verticillium dahliae*. Eur J Plant Pathol 140:341–352

Nielsen SL, Larsen J (2004) Two *Trichoderma harzianum*-based bio-control agents reduce tomato root infection with *Spongospora subterranea* (Wallr.) Lagerh., f. sp. *subterranea*, the vector of *potato mop-top virus*. J Plant Dis Prot 111:145–150

Omer MA, Johnson DA, Douhan LI, Hamm PB, Rowe RC (2008) Detection, quantification, and vegetative compatibility of *Verticillium dahliae* in potato and mint production soils in the Columbia Basin of Oregon and Washington. Plant Dis 92:1127–1131

Patil VU, Girimalla V, Sagar V, Bhardwaj V, Chakrabarti SK (2018) Draft genome sequencing of Rhizoctonia solani anastomosis group 3 (AG3-PT) causing stem canker and black scurf of potato. Am J Potato Res 95:87–91

Powell SM, McPhee JE, Dean G, Hinton S, Sparrow LA, Wilson CR, Tegg RS (2020) Managing soil health and crop productivity in potato: a challenging test system. Soil Res 58:697–712

Powelson ML, Rowe RC (1993) Biology and management of early dying of potatoes. Annu Rev Phytopathol 31:111–126

Rowe RC, Powelson ML (2002) Potato early dying: management challenges in a changing production environment. Plant Dis 86:1184–1193

Stagnitti F (2015) Parent project for APRP2 program: final report Hort innovation: PT09039. Horticulture Australia Limited, Sydney, Australia. https://www.horticulture.com.au/growers/help-your-business-grow/research-reports-publications-fact-sheets-and-more/pt09039/. Accessed 15 Aug 2021

Steere L, Kirk W (2015) Potato Early Die Complex (*Verticilium* wilt). Extension Bulletin E-3207, February 2015. Michigan State University, Michigan. https://www.canr.msu.edu/resources/potato_early_die_complex_e3207. Accessed 15 Aug 2021

Taylor RJ, Pasche JS, Gudmestad NC (2005) Influence of tillage and method of metam sodium application on distribution and survival of *Verticillium dahliae* in the soil and the development of verticillium wilt of potato. Am J Potato Res 82:451–461

Tegg RS, Corkrey R, Herdina H, McKay AC, Crump NS, de Boer RF, Wiechel TJ, Wilson CR (2015) Modeling pathogen DNA content and visual disease assessment in seed tubers to inform disease in potato progeny root, stolon, and tubers. Plant Dis 99:50–57

Tegg RS, Thangavel T, Balendres MA, Wilson CR (2016) Grading seed potato lots to remove tubers with powdery scab damage may not eliminate the pathogen threat. Am J Potato Res 93: 231–238

Tsror L (2010) Biology, epidemiology and management of Rhizoctonia solani on potato. J Phytopathol 158:649–658

Tsror L, Lebiush S, Hazanovsky M, Erlich O (2020) Control of potato powdery scab caused by *Spongospora subterranea* by foliage cover and soil application of chemicals under field conditions with naturally infested soil. Plant Pathol 69:1070–1082 van de Graaf P, Lees AK, Cullen DW, Duncan JM (2003) Detection and quantification of *Spongospora subterranea* in soil, water and plant tissue samples using real-time PCR. Eur J Plant Pathol 109:589–597

Viketoft M, Flohr A, Englund JE, Kardell J, Edin E (2020) Additive effect of the root-lesion nematode *Pratylenchus penetrans* and the fungus *Rhizoctonia solani* on potato yield and damage. J Plant Dis Prot 127:821–829

Wilson CR (2016) Plant pathogens - the great thieves of vegetable value. In: Birch CJ, Searle B, Heuvelink E (eds) xxix international horticultural congress on horticulture: sustaining lives, livelihoods and landscapes, vol 1123, pp 7–15

Woodhall JW, Adams IP, Peters JC, Harper G, Boonham N (2013) A new quantitative real-time PCR assay for *Rhizoctonia solani* AG3-PT and the detection of AGs of *Rhizoctonia solani* associated with potato in soil and tuber samples in Great Britain. Eur J Plant Pathol 136:273–280

Wright PJ, Falloon RE, Anderson C, Frampton RA, Curtin D, Hedderley D (2021) Factors influencing suppressiveness of soils to powdery scab of potato. Australas Plant Pathol 50 (6):715–728. https://doi.org/10.1007/s13313-021-00822-z

Zhang X, Huo H, Xi X, Liu L, Yu Z, Hao J (2015) Histological observation of potato in response to *Rhizoctonia solani* infection. Eur J Plant Pathol 145(2):289–303

Zrenner R, Genzel F, Verwaaijen B, Wibberg D, Grosch R (2020) Necrotrophic lifestyle of *Rhizoctonia solani* AG3-PT during interaction with its host plant potato as revealed by transcriptome analysis. Sci Rep 10:12574. https://doi.org/10.1038/s41598-020-68728-2

The manufacturer's authorised representative in the EU is Springer Nature Customer Service Centre GmbH, Europaplatz 3, 69115 Heidelberg, Germany. If you have any concerns regarding our products, please contact ProductSafety@springernature.com

Printed and bound by CPI Group (UK) Ltd, Croydon, CR0 4YY

30/09/2026

02225250-0019